W9-BPQ-654

CONVERSION FACTORS

Length

1 m = 39.37 in. = 3.281 ft
1 in. = 2.54 cm
1 km = 0.621 mi
1 mi = 5280 ft = 1.609 km
1 lightyear (ly) = 9.461×10^{15} m
1 angstrom (Å) = 10^{-10} m

Mass

1 kg = 10^3 g = 6.85×10^{-2} slug
1 slug = 14.59 kg
1 u = 1.66×10^{-27} kg

Time

1 min = 60 s
1 h = 3600 s
1 day = 8.64×10^4 s
1 yr = 365.242 days = 3.156×10^7 s

Volume

1 L = 1000 cm^3 = 3.531×10^{-2} ft^3
1 ft^3 = 2.832×10^{-2} m^3
1 gal = 3.786 L = 231 in.^{3}

Angle

180° = π rad
1 rad = 57.30°
1° = 60 min = 1.745×10^{-2} rad

Speed

1 km/h = 0.278 m/s = 0.621 mi/h
1 m/s = 2.237 mi/h = 3.281 ft/s
1 mi/h = 1.61 km/h = 0.447 m/s = 1.47 ft/s

Force

1 N = 0.2248 lb = 10^5 dynes
1 lb = 4.448 N
1 dyne = 10^{-5} N = 2.248×10^{-6} lb

Work and energy

1 J = 10^7 erg = 0.738 ft·lb = 0.239 cal
1 cal = 4.186 J
1 ft·lb = 1.356 J
1 Btu = 1.054×10^3 J = 252 cal
1 J = 6.24×10^{18} eV
1 eV = 1.602×10^{-19} J
1 kWh = 3.60×10^6 J

Pressure

1 atm = 1.013×10^5 N/m^2 (or Pa) = 14.70 lb/in.^{2}
1 Pa = 1 N/m^2 = 1.45×10^{-4} lb/in.^{2}
1 lb/in.^{2} = 6.895×10^3 N/m^2

Power

1 hp = 550 ft·lb/s = 0.746 kW
1 W = 1 J/s = 0.738 ft·lb/s
1 Btu/h = 0.293 W

COLLEGE PHYSICS

SIXTH EDITION

COLLEGE PHYSICS

SIXTH EDITION

RAYMOND A. SERWAY

JAMES MADISON UNIVERSITY

JERRY S. FAUGHN

EASTERN KENTUCKY UNIVERSITY

CONTRIBUTING AUTHOR:
CLEMENT J. MOSES

THOMSON
———
BROOKS/COLE

Australia · Canada · Mexico · Singapore · Spain
United Kingdom · United States

THOMSON

BROOKS/COLE

Editor in Chief: Michelle Julet
Senior Development Editor: Susan Dust Pashos
Assistant Editor: Alyssa White
Editorial Assistant: Jessica Howard
Technology Project Manager: Samuel Subity
Marketing Manager: Kelley McAllister
Marketing Assistant: Sandra Perin
Advertising Project Manager: Stacey Purviance
Senior Project Manager, Editorial Production: Charlene Catlett Squibb, Teri Hyde
Print/Media Buyer: Karen Hunt
Production Service: Progressive Publishing Alternatives

Text Designer: Lisa Adamitis
Art Director: Carol Bleistine
Photo Researcher: Dena Betz
Copy Editor: Progressive Publishing Alternatives
Illustrator: Rolin Graphics
Cover Designer: Larry Didona
Cover Image: Aurora Borealis over Trees, © Yutaka Iijima/ Photonica
Cover Printer: Transcontinental Printing
Compositor: Progressive Information Technologies
Printer: Transcontinental Printing

COPYRIGHT © 2003 by Raymond A. Serway

ALL RIGHTS RESERVED. No part of this work covered by the copyright hereon may be reproduced or used in any form or by any means—graphic, electronic, or mechanical, including but not limited to photocopying, recording, taping, Web distribution, information networks, or information storage and retrieval systems—without the written permission of the publisher.

Printed in Canada
3 4 5 6 7 06 05 04 03

For more information about our products, contact us at:
Thomson Learning Academic Resource Center
1-800-423-0563
For permission to use material from this text, contact us by:
Phone: 1-800-730-2214
Fax: 1-800-730-2215
Web: http://www.thomsonrights.com

ExamView® and ExamView Pro® are registered trademarks of FSCreations, Inc. Windows is a registered trademark of the Microsoft Corporation used herein under license. Macintosh and Power Macintosh are registered trademarks of Apple Computer, Inc. Used herein under license.

COPYRIGHT 2003 Thomson Learning, Inc. All Rights Reserved. Thomson Learning WebTutor is a trademark of Thomson Learning, Inc.

Brooks/Cole—Thomson Learning
511 Forest Lodge Road
Pacific Grove, CA 93950
USA

ISBN 0-030-35114-6
Text with passcard: ISBN 0-534-49258-4
International Student Edition: ISBN 0-534-40814-1
(Not for sale in the United States)

Asia
Thomson Learning
60 Albert Street, #15-01
Albert Complex
Singapore 189969

Australia
Nelson Thomson Learning
102 Dodds Street
South Melbourne, Victoria 3205
Australia

Canada
Nelson Thomson Learning
1120 Birchmount Road
Toronto, Ontario M1K 5G4
Canada

Europe/Middle East/Africa
Thomson Learning
Berkshire House
168-173 High Holborn
London WC1V 7AA
United Kingdom

Latin America
Thomson Learning
Seneca, 53
Colonia Polanco
11560 Mexico D.F.
Mexico

Spain
Paraninfo Thomson Learning
Calle/Magallanes, 25
28015 Madrid, Spain

Library of Congress Control Number: 2002113279

We dedicate this book

To our wives, Elizabeth Ann Serway and Mary Ann Faughn, to our children, Mark Serway, Michele Austin, David Serway, Jennifer Serway, David Faughn, and Laura Faughn

and

To our grandchildren, Nicole Austin, Christopher Austin, Brian Serway, Kaitlyn Serway, Nathan Serway, and Stone Evan Faughn.

We have been truly blessed by receiving so much love and support from these warm and thoughtful family members through all the years.

CONTENTS OVERVIEW

As this skateboarder moves gracefully through the air, the shape of the curved path he follows and the height he achieves depend on his velocity as he leaves the ground below. *(Allsport/Getty)*

1

1 Introduction

Chapter Outline

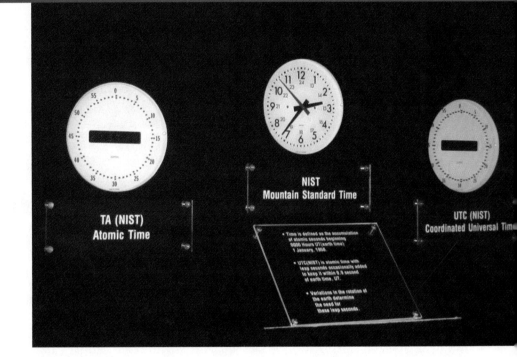

This hallway display of clocks indicates, from left to right, Atomic Time, Mountain Standard Time, and Universal Time. *(Courtesy of National Institute of Standards and Technology, Boulder Laboratories, U.S. Department of Commerce)*

When inserting numerical factors into equations, you must be sure that the dimensions of these factors are consistent throughout the equation. In many cases it will be necessary to use the table of conversion factors on the inside front cover to convert from one system of units to another.

Often it is useful to estimate an answer to a problem in which little information is given. Such estimates are called **order-of-magnitude calculations.**

The three most basic trigonometric functions for a right triangle are the sine, cosine, and tangent, defined as

$$\sin \theta = \frac{\text{side opposite } \theta}{\text{hypotenuse}}$$

$$\cos \theta = \frac{\text{side adjacent to } \theta}{\text{hypotenuse}} \qquad \textbf{[1.1]}$$

$$\tan \theta = \frac{\text{side opposite } \theta}{\text{side adjacent to } \theta}$$

The **Pythagorean theorem** is an important relationship between the lengths of the sides of a right triangle:

$$c^2 = a^2 + b^2 \qquad \textbf{[1.2]}$$

where c is the hypotenuse of the triangle and a and b are its other two sides.

CONCEPTUAL QUESTIONS

1. Estimate the order of magnitude of the length, in meters, of each of the following: (a) a mouse, (b) a pool cue, (c) a basketball court, (d) an elephant, (e) a city block.

2. Long ago, in England, the inch was defined as the length of three barley grains placed end to end. In what ways was this a good way to define a standard? In what ways was this a poor way to define a standard?

3. (a) Estimate the number of times your heart beats in a month. (b) Estimate the number of human heartbeats in an average lifetime.

4. An object with a mass of 1 kg weighs a little more than 2 lb. Use this information to estimate the mass of the following objects: (a) an apple seed; (b) the rest of the apple; (c) a football player; (d) a car.

5. Find the order of magnitude of your age in seconds.

6. Estimate the number of atoms in 1 cm³ of a solid. (Note that the diameter of an atom is about 10^{-10} m.)

7. The height of a horse is sometimes given in units of "hands." Why is this a poor standard of length?

8. How many of the lengths or time intervals given in Tables 1.2 and 1.3 could you verify using only equipment found in a typical dormitory room?

9. An ancient unit of length called the *cubit* was equal to six palms, where a palm was the width of the four fingers of an open hand. Noah's ark was 300 cubits long, 50 cubits wide, and 30 cubits high. Estimate the volume of the ark in cubic meters. Also, estimate the volume of a typical home in cubic meters and compare it to the volume of the ark.

10. If an equation is dimensionally correct, does this mean that the equation must be true?

11. Estimate the cost per pound of a typical automobile. How does your answer compare to the cost of a good steak?

12. Figure Q1.12 is a photograph showing unit conversions on the labels of some grocery-store items. Check the accuracy of these conversions. Are the manufacturers using significant figures correctly?

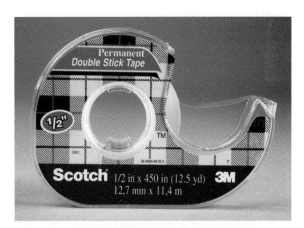

FIGURE Q1.12   *(Courtesy of George Semple)*

PROBLEMS

1, 2, 3 = straightforward, intermediate, challenging ☐ = full solution available in Student Solutions Manual/Study Guide

web = solution posted **http://info.brookscole.com/serway** 🐁 = biomedical application

Section 1.3 Dimensional Analysis

☐1. In a desperate attempt to come up with an equation to use during an examination, a student tries $v^2 = ax$. Use dimensional analysis to determine whether this equation might be valid.

2. The radius r of a circle inscribed in any triangle whose sides are a, b, and c is given by $r = [(s - a)(s - b)(s - c)/s]^{1/2}$, where s is an abbreviation for $(a + b + c)/2$. Check this formula for dimensional consistency.

3. The period of a simple pendulum, defined as the time for one complete oscillation, is measured in time units and is given by

$$T = 2\pi \sqrt{\frac{\ell}{g}}$$

where ℓ is the length of the pendulum and g is the acceleration due to gravity, in units of length divided by time squared. Show that this equation is dimensionally consistent. (You might want to check the formula using your keys at the end of a string and a stopwatch.)

4. Which of the equations below are dimensionally correct? (a) $v = v_0 + ax$ (b) $y = (2 \text{ m})\cos(kx)$, where $k = 2 \text{ m}^{-1}$.

5. Newton's law of universal gravitation is represented by

$$F = G\frac{Mm}{r^2}$$

Here F is the gravitational force, M and m are masses, and r is a length. Force has the SI units kg·m/s². What are the SI units of the proportionality constant G?

6. (a) One of the fundamental laws of motion states that the acceleration of an object is directly proportional to the resultant force on it and inversely proportional to its mass. If the proportionality constant is defined to have no dimensions, determine the dimensions of force. (b) The newton is the SI unit of force. According to the results for (a), how can you express a force having units of newtons using the fundamental units of mass, length, and time?

Section 1.4 Uncertainty in Measurements and Significant Figures

7. How many significant figures are there in (a) 78.9 ± 0.2, (b) 3.788×10^9, (c) 2.46×10^{-6}, (d) 0.0032?

8. Two I-beams are to be joined end to end. One has been measured to be 3.456 m long; the other, 4.3 m. When they are bolted snugly together, what value can you accurately give for the total length?

☐9. Carry out the following arithmetic operations: (a) the sum
web of the measured values 756, 37.2, 0.83, and 2.5; (b) the product 0.0032×356.3; (c) the product $5.620 \times \pi$.

10. The value of the speed of light is now defined to be $2.99 \ 7924 \ 58 \times 10^8$ m/s. Express the speed of light to (a) three significant figures, (b) five significant figures, and (c) seven significant figures.

11. A farmer measures the distance around a rectangular field. The length of each long side of the rectangle is found to be 38.44 m, and the length of each short side is found to be 19.5 m. What is the total distance around the field?

12. Calculate (a) the circumference of a circle of radius 3.5 cm and (b) the area of a circle of radius 4.65 cm.

13. A fisherman catches two striped bass. The smaller of the two has a measured length of 93.46 cm (two decimal places, four significant figures), and the larger fish has a measured length of 135.3 cm (one decimal place, four significant figures). What is the total length of fish caught for the day?

14. (a) Using your calculator, find, in scientific notation with appropriate rounding, the value of $(2.437 \times 10^4)(6.5211 \times 10^9)/(5.37 \times 10^4)$. (b) Find the value of $(3.14159 \times 10^2)(27.01 \times 10^4)/(1234 \times 10^6)$.

Section 1.5 Conversion of Units

15. A fathom is a method for measuring lengths, usually reserved for measuring the depth of water. A fathom is approximately 6 ft in length. Take the distance to the Moon to be 250 000 miles, and use the approximation above to find the distance in fathoms.

16. Find the height or length of these natural wonders in kilometers, meters, and centimeters. (a) The longest cave system in the world is the Mammoth Cave system in Central Kentucky. It has a mapped length of 348 miles. (b) In the United States, the waterfall with the greatest single drop is Ribbon Falls in California which drops 1 612 ft. (c) Mount McKinley in Alaska is America's highest mountain at 20 320 feet. (d) The deepest canyon in the United States is King's Canyon in California with a depth of 8 200 ft.

17. An aspirin tablet contains 325 mg of acetylsalicylic acid. Express this mass in grams.

18. Suppose your hair grows at the rate of 1/32 inch per day. Find the rate at which it grows in nanometers per second. Since the distance between atoms in a molecule is on the order of 0.1 nm, your answer suggests how rapidly atoms are assembled in this protein synthesis.

19. Find the distance to the nearest star, in feet, using the data in Table 1.1 and the appropriate conversion factors.

20. Find the age of Earth in years, using the data in Table 1.3 and the appropriate conversion factors.

☐21. The speed of light is about 3.00×10^8 m/s. Convert this to miles per hour.

22. A house is 50.0 ft long and 26 ft wide, and has 8.0-ft-high ceilings. What is the volume of the interior of the house in cubic meters and in cubic centimeters?

23. The amount of water in reservoirs is often measured in acre-ft. One acre-ft is a volume that covers an area of one acre to a depth of one foot. An acre is an area of 43 560 ft². Find the volume in SI units of a reservoir containing 25.0 acre-ft of water.

24. The base of a pyramid covers an area of 13.0 acres (1 acre = 43 560 ft²) and has a height of 481 ft (Fig. P1.24). If the volume of a pyramid is given by the expression

FIGURE 2.1 A jogger moving along the x axis from x_i to x_f undergoes a displacement of $\Delta x = x_f - x_i$.

$$\Delta x \equiv x_f - x_i \qquad\qquad [2.1]$$

From this definition, we see that Δx is positive if x_f is greater than x_i and negative if x_f is less than x_i. If the jogger moves from an initial position of $x_i = 3$ m to a final position of $x_f = 15$ m, his displacement is $\Delta x = 15$ m $- 3$ m $= 12$ m.

Displacement is an example of a vector quantity. Many physical quantities in this book, including displacement, velocity, and acceleration, are vector quantities. In general, **a vector quantity is a physical quantity that must be characterized by both a magnitude and a direction.** By contrast, **a scalar quantity is one that has magnitude but not direction.** Scalar quantities, such as mass and temperature, are completely specified by a numeric value with appropriate units; no direction is involved.

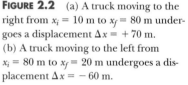

◀ A vector has both direction and magnitude

We shall usually designate vector quantities with boldface type. For example, **v** represents velocity and **a** denotes an acceleration, both vector quantities. In this chapter, which deals with one-dimensional motion, it will not be necessary to use this boldface notation. The reason is that in one-dimensional motion an object can only move in two directions, and these directions are easily specified by plus and minus signs. For example, if the truck in Figure 2.2a moves from an initial po-

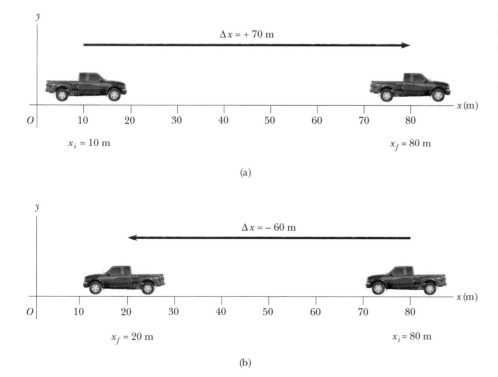

FIGURE 2.2 (a) A truck moving to the right from $x_i = 10$ m to $x_f = 80$ m undergoes a displacement $\Delta x = +70$ m. (b) A truck moving to the left from $x_i = 80$ m to $x_f = 20$ m undergoes a displacement $\Delta x = -60$ m.

sition of 10 m to a final position of 80 m, its displacement is

$$\Delta x = x_f - x_i = +80 \text{ m} - (+10 \text{ m}) = +70 \text{ m}$$

In this case, the displacement has a magnitude of 70 m and is directed in the positive *x* direction, as indicated by the plus sign of the result. (Sometimes the plus sign is omitted, but a result of 70 m for the displacement is understood to be the same as + 70 m.) Any vector quantity is usually represented by an arrow, as is the displacement in Figure 2.2a. The length of the arrow represents the magnitude of the displacement, and the head of the arrow indicates its direction.

If the truck moves to the left from an initial position of 80 m to a final position of 20 m, as in Figure 2.2b, the displacement is

$$\Delta x = x_f - x_i = +20 \text{ m} - (+80 \text{ m}) = -60 \text{ m}$$

The minus sign in this result indicates that the displacement is a vector quantity pointing in the negative *x* direction as in Figure 2.2b.

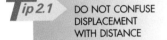

DO NOT CONFUSE DISPLACEMENT WITH DISTANCE

The displacement of an object is *not* the same as the distance it travels. For example, a model rocket fired straight upward and returning to the firing point travels some distance, but its displacement is zero.

2.2 AVERAGE VELOCITY

The motion of an object is completely known if its position is known at all times. Consider again the truck of Figure 2.2, moving along a highway (the *x* axis). Let the truck's position be x_i at some time t_i, and let its position be x_f at some time t_f. In the time interval $\Delta t = t_f - t_i$, the displacement of the truck is $\Delta x = x_f - x_i$.

▶ Definition of average velocity

The average velocity $\bar{v}$ during some time interval is defined as the displacement Δx divided by the time interval during which the displacement occurred:

$$\bar{v} = \frac{\Delta x}{\Delta t} = \frac{x_f - x_i}{t_f - t_i} \qquad \text{[2.2]}$$

The average velocity of an object can be either positive or negative, depending on the sign of the displacement. (The time interval Δt is always positive.) For example, in Figure 2.2b, the average velocity of the truck is negative, the minus sign indicating motion to the left along the *x* axis. Similarly, a positive average velocity for the truck in Figure 2.2a indicates that it moves to the right along the *x* axis.

In order to understand the vector nature of velocity, consider the following situation. Suppose a friend tells you that she will be taking a trip in her car and will travel at a constant rate of 55 mi/h in a straight line for 1 h. If she starts from her home, where will she be at the end of the trip? Obviously, you cannot answer this question because she did not specify the direction of her trip. All you can say is that she will be located 55 mi from her starting point. However, if she tells you that she will be driving at the rate of 55 mi/h directly northward, then her final location will be known exactly. The velocity of an object is known only if its direction of motion and its speed (a scalar) are specified.

As an example of the use of Equation 2.2, suppose the truck in Figure 2.2a moves 70 m to the right in 5.0 s. Substituting these values into Equation 2.2 gives the average velocity in this time interval as

$$\bar{v} = \frac{\Delta x}{\Delta t} = \frac{80 \text{ m} - 10 \text{ m}}{5.0 \text{ s}} = +14 \text{ m/s}$$

Note that the units of average velocity are units of length divided by units of time. These are meters per second (m/s) in SI. Other units for velocity might be feet per second (ft/s) in the U.S. customary system or centimeters per second (cm/s) in the cgs system.

FIGURE 2.3 A drag race viewed from a blimp. One car follows the red straight-line path from Ⓟ to Ⓠ, and a second car follows the blue curved path.

Let us assume we are watching a drag race from the Goodyear blimp. In one run we see a car follow the straight-line path from Ⓟ to Ⓠ shown in Figure 2.3 during the time interval Δt, and in a second run a car follows the curved path during the same time interval. If you examine the definition of average velocity (Eq. 2.2)

Example 2.2 A Fly Ball Is Caught

A baseball player moves in a straight-line path in order to catch a fly ball hit to the out-field. His velocity as a function of time is shown in Figure 2.11. Find his instantaneous acceleration at points Ⓐ, Ⓑ, and Ⓒ and translate the graph into a verbal description of the player's motion.

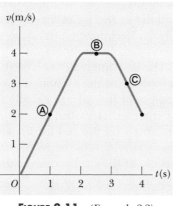

FIGURE 2.11 (Example 2.2)

Solution The instantaneous acceleration at any time is the slope of the velocity-time curve at that instant. The slope of the curve at point Ⓐ is

$$a = \frac{\Delta v}{\Delta t} = \frac{4 \text{ m/s}}{2 \text{ s}} = +2 \text{ m/s}^2$$

At point Ⓑ, the slope of the line is zero and hence the instantaneous acceleration at this time is also zero. Note that even though the instantaneous acceleration has dropped to zero at this instant, the velocity of the player is not zero. (In general, if $a = 0$, the velocity is constant in time but *not necessarily* zero.) Instead, the player is running at a constant velocity of $+4$ m/s. A value of zero for an instantaneous acceleration means that at a particular instant the velocity of the object is not changing.

Finally, at point Ⓒ we calculate the slope of the velocity-time curve and find that the instantaneous acceleration is

$$a = \frac{\Delta v}{\Delta t} = \frac{-2 \text{ m/s}}{1 \text{ s}} = -2 \text{ m/s}^2$$

For this situation, the minus sign and the fact that the velocities are positive indicate that the speed of the player is decreasing as he approaches the location where he will attempt to catch the baseball.

For the first 2 s the ball player is moving in the positive x direction (the velocity is positive) and steadily speeding up (the curve is rising) to a maximum speed of 4 m/s. He moves for 1 s at a steady speed of 4 m/s and then slows down in the last second (the v versus t curve is falling), still moving in the positive x direction (v is always positive).

Follow up: Do you see any way of using the v versus t curve to calculate how far he has gone in 4 s?

2.5 MOTION DIAGRAMS

The concepts of velocity and acceleration are often confused with each other, but in fact they are quite different quantities. It is instructive to use a motion diagram to describe the velocity and acceleration vectors as time progresses while an object is in motion. A **motion diagram** is a representation of a moving object at successive time intervals, with velocity and acceleration vectors sketched at each position. In order not to confuse these two vector quantities, we use red for velocity vectors and violet for acceleration vectors as in Figure 2.12. The time intervals between adjacent positions in the motion diagram are assumed to be equal.

A motion diagram is similar to the set of images resulting from a stroboscopic photograph of a moving object. Each image is made as the strobe light flashes. Figure 2.12 represents three sets of strobe photographs of cars moving along a straight roadway, from left to right. The time intervals between flashes of the stroboscope are equal in each diagram. Let us describe the motion of the car in each diagram.

In Figure 2.12a, the images of the car are equally spaced—the car moves the same distance in each time interval. Thus, the car moves with *constant positive velocity* and has *zero acceleration*. Note that the red arrows are all the same length, and there are no violet arrows.

In Figure 2.12b, the images of the car become farther apart as time progresses. In this case, the velocity vector increases in time because the car's displacement between adjacent positions increases as time progresses. Thus, the car is moving with a *positive velocity*, and a constant *positive acceleration*. The red arrows are successively longer in each image, and the violet arrows point to the right.

In Figure 2.12c, the car slows as it moves to the right because its displacement between adjacent positions decreases as time progresses. In this case, the car moves initially to the right with a constant negative acceleration. The velocity vector decreases in time (red arrows get shorter) and eventually reaches zero. (This type of motion is exhibited by a car that comes to a stop after applying its brakes.) In this diagram the acceleration and velocity vectors are *not* in the same direction. The car is moving with a *positive velocity* but with a *negative acceleration*.

You should be able to construct motion diagrams for a car that moves initially to the left with a constant positive or negative acceleration. You should also construct appropriate motion diagrams along with the mathematical solutions to kinematic problems, to see if your answers are consistent with the diagrams. In all of these exercises our convention is to choose the right-hand direction as positive.

FIGURE 2.12 (a) Motion diagram for a car moving at constant velocity (zero acceleration). (b) Motion diagram for a car whose constant acceleration is in the direction of its velocity. The velocity vector at each instant is indicated by a red arrow, and the constant acceleration vector by a violet arrow. (c) Motion diagram for a car whose constant acceleration is in the direction *opposite* the velocity at each instant.

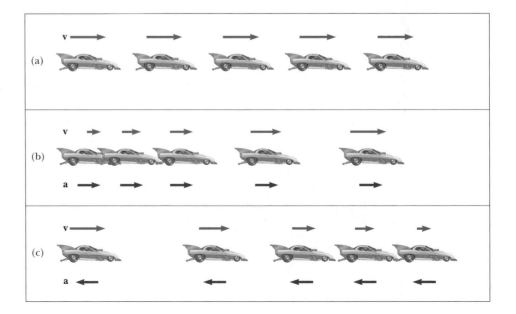

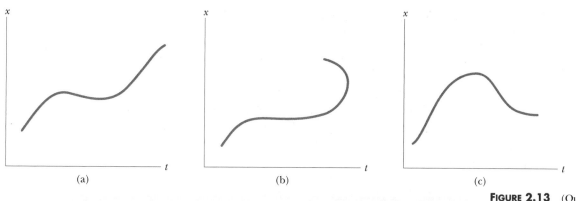

FIGURE 2.13 (Quick Quiz 2.4) Which position-time curve is impossible?

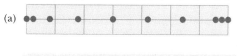

Quick Quiz 2.4 The three graphs in Figure 2.13 represent the position versus time for objects moving along the x axis. Which of these graphs is not physically possible?

Quick Quiz 2.5 Figure 2.14a is a diagram of a multiflash image of an airpuck moving to the right on a horizontal surface. The images sketched are separated by equal time intervals, and the first and last images show the puck at rest. (a) In Figure 2.14b, which graph best shows the puck's position as a function of time? (b) In Figure 2.14c, which graph best shows the puck's velocity as a function of time? (c) In Figure 2.14d, which graph best shows the puck's acceleration as a function of time?

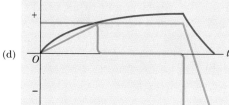

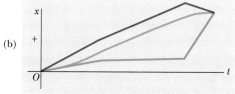

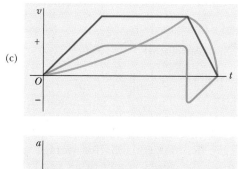

FIGURE 2.14 (Quick Quiz 2.5) Choose the correct graphs.

2.6 ONE-DIMENSIONAL MOTION WITH CONSTANT ACCELERATION

Many of the mechanics applications in this book will be concerned with objects moving with *constant acceleration*. This type of motion is important because it applies to many objects in nature. For example, an object in free fall near Earth's surface moves in the vertical direction with constant acceleration, assuming that air resistance can be neglected. When an object moves with constant acceleration, *the instantaneous acceleration at any point in a time interval is equal to the value of the average acceleration over the entire time interval*. Consequently, the velocity increases or decreases at the same rate throughout the motion, and a plot of v versus t gives a straight line with either positive or negative slope.

Because the average acceleration equals the instantaneous acceleration when a is constant, we can eliminate the bar used to denote average values from our defining equation for acceleration. That is, because $\bar{a} = a$, we can write Equation 2.4 as

$$a = \frac{v_f - v_i}{t_f - t_i}$$

For convenience, let $t_i = 0$ and t_f be any arbitrary time t. Also, let $v_i = v_0$ (the initial velocity at $t = 0$) and $v_f = v$ (the velocity at any arbitrary time t). With this notation, we can express the acceleration as

$$a = \frac{v - v_0}{t}$$

or

$$v = v_0 + at \qquad \text{(for constant } a\text{)} \qquad \text{[2.6]}$$

Equation 2.6 just states that the velocity after t seconds of constant acceleration is equal to what it was at the start plus the amount added after t seconds of acceleration at a. For example, if a car starts with a velocity of $+2.0$ m/s to the right and accelerates to the right with $a = +6.0$ m/s², it will have a velocity of $+14$ m/s after 2.0 s have elapsed:

$$v = v_0 + at = +2.0 \text{ m/s} + (6.0 \text{ m/s}^2)(2.0 \text{ s}) = +14 \text{ m/s}$$

One of the features of one-dimensional motion with constant acceleration is the manner in which the initial, final, and average velocities are related. Because the velocity is increasing or decreasing *uniformly* with time, we can express the average velocity in any time interval as the arithmetic average of the initial velocity v_0 and the final velocity v:

$$\bar{v} = \frac{v_0 + v}{2} \qquad \text{(for constant } a\text{)} \qquad \text{[2.7]}$$

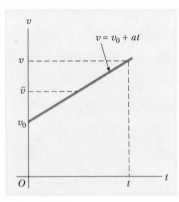

FIGURE 2.15 The velocity varies linearly with time for a particle moving with constant acceleration. The average velocity is the mean value of the initial and final velocities.

Note that this expression is valid only when the acceleration is constant—that is, when the velocity changes uniformly with time. The graphical interpretation of $\bar{v}$ is shown in Figure 2.15. As you can see, the velocity varies linearly with time according to Equation 2.6.

We can now use this result along with the defining equation for average velocity, Equation 2.2, to obtain an expression for the displacement of an object as a function of time. Again we choose $t_i = 0$ and $t_f = t$, and for convenience we write $\Delta x = x_f - x_i = x - x_0$. This gives

$$\Delta x = \bar{v}t = \left(\frac{v_0 + v}{2}\right)t$$

$$\Delta x = \tfrac{1}{2}(v_0 + v)t \qquad \text{[2.8]}$$

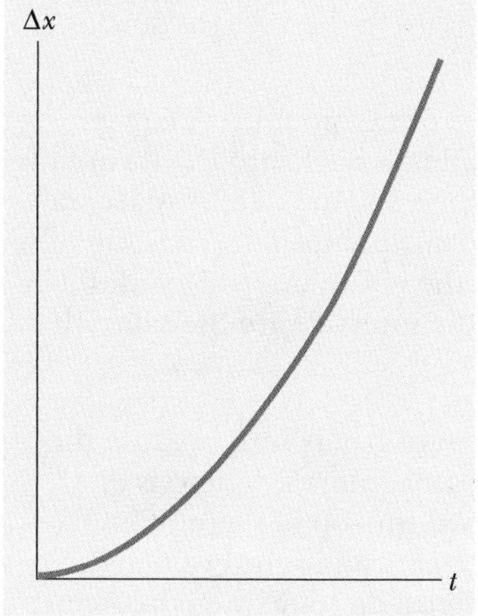

FIGURE 2.16 The displacement versus time for an object moving with constant acceleration. Note that the displacement varies as t^2. The constantly increasing slope indicates that the velocity increases with time.

We can obtain another useful expression for displacement by substituting the equation for v (Eq. 2.6) into Equation 2.8:

$$\Delta x = \tfrac{1}{2}(v_0 + v_0 + at)\,t$$

$$\Delta x = v_0 t + \tfrac{1}{2}at^2 \qquad \text{(for constant } a\text{)} \qquad\qquad \textbf{[2.9]}$$

Figure 2.16 shows a plot of Δx versus t for this equation. You should be able to show that the area under the curve in Figure 2.15 is equal to $v_0 t + \tfrac{1}{2}at^2$, which is equal to the displacement Δx. In fact, **the area under any curve of v versus t can be shown to be equal to the displacement of the object.**

Finally, we can obtain an expression that does not contain time by substituting the value of t from Equation 2.6 into Equation 2.8. This gives

$$\Delta x = \tfrac{1}{2}(v + v_0)\left(\frac{v - v_0}{a}\right) = \frac{v^2 - v_0{}^2}{2a}$$

$$v^2 = v_0{}^2 + 2a\,\Delta x \qquad \text{(for constant } a\text{)} \qquad\qquad \textbf{[2.10]}$$

Equations 2.6–2.10 may be used to solve any problem in one-dimensional motion with constant acceleration. For convenience, we list the four equations that are used most often in Table 2.3.

The best way to gain confidence in the use of these equations is to work a number of problems. Many times you will discover that there is more than one method for solving a given problem.

At appropriate places within each chapter we shall include interludes that we hope will motivate you to think critically about the material you have been reading. These interesting interludes, called "Applying Physics," are conceptual rather than mathematical and test your understanding of concepts. The answers serve as models for responding to the conceptual questions at the end of each chapter.

Tip 2.6   IS THE ANSWER REASONABLE?

When solving problems, you should think about your answer and determine if it seems reasonable. For example, if you are calculating the mass of a housefly and arrive at a value of 100 kg, this value is *unreasonable*—there is an error somewhere. If you are calculating the length of a spacecraft on a launch pad and end up with a value of 10 cm, this is *unreasonable*—look for a mistake.

TABLE 2.3	Equations for Motion in a Straight Line Under Constant Acceleration
Equation	**Information Given by Equation**
$v = v_0 + at$	Velocity as a function of time
$\Delta x = \tfrac{1}{2}(v_0 + v)\,t$	Displacement as a function of velocity and time
$\Delta x = v_0 t + \tfrac{1}{2}at^2$	Displacement as a function of time
$v^2 = v_0{}^2 + 2a\,\Delta x$	Velocity as a function of displacement

Note: Motion is along the x axis. At $t = 0$, the velocity of the particle is v_0.

APPLYING PHYSICS 2.1

Consider the following motions of an object in one dimension: (a) A ball is thrown directly upward, rises to a highest point and falls back into the thrower's hand. (b) A race car starts from rest and speeds up to 100 m/s. (c) A spacecraft drifts through space at constant velocity. Are there any points in the motion of these particles at which the average velocity (over the entire interval) and the instantaneous velocity (at an instant of time within the interval) are the same? If so, identify the point(s).

Explanation   (a) The average velocity for the thrown ball is zero because the ball returns to the starting point at the end of the time interval. There is one point at which the instantaneous velocity is zero—at the top of the motion. (b) The average velocity for the motion of the race car cannot be evaluated unambiguously with the information given, but it must be some value between 0 and 100 m/s. Because the car will have every instantaneous velocity between 0 and 100 m/s at some time during the interval, there must be some instant at which the instantaneous velocity is equal to the average velocity. (c) Because the instantaneous velocity of the spacecraft is constant, its instantaneous velocity over *any* time interval and its average velocity at *any* time are the same.

PROBLEM-SOLVING STRATEGY   Accelerated Motion

The following procedure is recommended for solving problems involving accelerated motion.

1. Make sure all the units in the problem are consistent. That is, if displacements are measured in meters, be sure that velocities have units of meters per second and accelerations have units of meters per second squared.
2. Choose a coordinate system. Choose instants to call the initial and final points. Make a labeled diagram of the problem including the direction of all displacements, velocities, and accelerations.
3. Make a list of all the quantities given in the problem and a separate list of those to be determined.
4. Think about the physical concepts involved, any restrictions that apply, and select from the list of kinematic equations the one or ones that will enable you to determine the unknowns.
5. Check to see if your answers are reasonable and consistent with the diagram.

Example 2.3   The Indianapolis 500

A race car starting from rest accelerates at a rate of 5.00 m/s^2. What is the velocity of the car after it has traveled 100 ft?

Reasoning and Solution   You should refer to the preceding Problem-Solving Strategy to see how we apply the steps indicated there to this example.

(Step 1) Be sure that the units you use are consistent. The units in this problem are *not* consistent. If we choose to leave the distance in feet, we must change the length dimension of the units of acceleration from meters to feet. Alternatively, we can leave the units of acceleration as they are and convert the distance traveled to meters. Let's do the latter. The table of conversion factors on the inside front cover gives 1 m = 3.281 ft, so we find that 1 ft = 0.305 m and 100 ft = 30.5 m.

(Step 2) You must choose a coordinate system. A convenient one for this problem is shown in Figure 2.17. The origin of the coordinate system is at the initial location of the car, and the positive direction is to the right. Using this convention, we require that velocities, accelerations, and displacements to the right are positive, and vice versa.

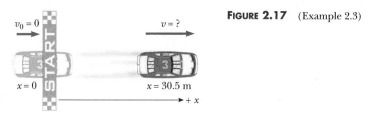

$v_0 = 0$

$v = ?$

$x = 0$

$x = 30.5$ m

$+ x$

FIGURE 2.17 (Example 2.3)

(Step 3) Make a list of the quantities given in the problem and a separate list of those to be determined.

Given	To Be Determined
$v_0 = 0$	v
$a = +5.00$ m/s^2	
$\Delta x = +30.5$ m	

(Step 4) Select from Table 2.3 those that will allow you to determine the unknowns. In the present case, the equation

$$v^2 = v_0{}^2 + 2a\,\Delta x$$

is our best choice because it will give us a value for v directly:

$$v^2 = (0)^2 + 2(5.00 \text{ m/s}^2)(30.5 \text{ m}) = 305 \text{ m}^2/\text{s}^2$$

$$v = \sqrt{305 \text{ m}^2/\text{s}^2} = \pm 17.5 \text{ m/s}$$

Because the car is moving to the right, we choose $+17.5$ m/s as the correct solution for v. Alternatively, the problem can be solved by using $\Delta x = v_0 t + \frac{1}{2}at^2$ to find t and then using the expression $v = v_0 + at$ to find v. Try it!

Example 2.4   Watch Out for the Speed Limit

A car traveling at a constant speed of 30.0 m/s (≈ 67 mi/h) passes a trooper hidden behind a billboard as in Figure 2.18. One second after the speeding car passes the billboard, the trooper sets off in chase with a constant acceleration of 3.00 m/s^2. (a) How long does it take the trooper to overtake the speeding car? (b) How fast is the trooper going when she overtakes the car?

$v_{\text{car}} = 30.0$ m/s

$a_{\text{car}} = 0$

$a_{\text{trooper}} = 3.00$ m/s^2

FIGURE 2.18 (Example 2.4) A speeding car passes a hidden trooper. When does the trooper catch up to the car?

$t = 0$

$t = ?$

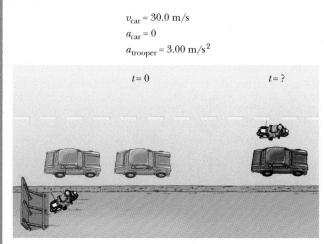

Reasoning and Solution   To solve this problem algebraically, we write an expression for the position of each vehicle as a function of time. It is convenient to choose the origin at the position of the billboard and take $t = 0$ as the time the trooper begins moving. At that instant, the speeding car has already traveled a distance of 30.0 m, because it travels at a constant speed of 30.0 m/s. Thus, the initial position of the speeding car is $x_0 = +30.0$ m. (Draw a diagram for your own benefit.)

A Because the car moves with constant speed, its acceleration is zero, and applying Equation 2.9 gives

$$\Delta x_{car} = x_{car} - x_0 = v_0 t + \tfrac{1}{2} a t^2$$

In this situation, $v_0 = v$, and $a = 0$, so

$$x_{car} = x_0 + vt = 30.0 \text{ m} + (30.0 \text{ m/s})t$$

For the trooper, who starts from the origin at $t = 0$, we have $x_0 = 0$, $v_0 = 0$, and $a = 3.00$ m/s^2. Hence, the position of the trooper as a function of time is

$$x_{trooper} = \tfrac{1}{2} a t^2 = \tfrac{1}{2}(3.00 \text{ m/s}^2) t^2$$

The trooper overtakes the car at the instant that $x_{trooper} = x_{car}$, or

$$\tfrac{1}{2}(3.00 \text{ m/s}^2) t^2 = 30.0 \text{ m} + (30.0 \text{ m/s})t$$

This gives the quadratic equation

$$(1.50 \text{ m/s}^2) t^2 - (30.0 \text{ m/s}) t - 30.0 \text{ m} = 0$$

Since the final point is after the original point, only the positive root is physically meaningful: $t = \boxed{21.0 \text{ s.}}$

B Note that in the time interval of 21.0 s, the trooper travels a distance of about 660 m:

$$\Delta x_{trooper} = v_0 t + \tfrac{1}{2} a t^2 = 0 + \tfrac{1}{2}(3.00 \text{ m/s}^2)(21.0 \text{ s})^2 = 6.59 \times 10^2 \text{ m}$$

Because the trooper starts from rest and accelerates at 3.00 m/s^2 for 21.0 s, when she overtakes the car she is traveling at a speed of

$$v_{trooper} = v_0 + at = 0 + (3.00 \text{ m/s}^2)(21.0 \text{ s}) = \boxed{62.9 \text{ m/s}}$$

Note that this is about twice the speed of the car.

EXERCISE   This problem also can be easily solved graphically. On the *same* graph, plot the position versus time for each vehicle, and from the intersection of the two curves determine the time at which the trooper overtakes the speeding car.

Example 2.5   Stop That 747!

A typical jetliner lands at a speed of 160 mi/h and brakes at the rate of 10 mi/h/s. If the jetliner travels at a constant speed of 160 mi/h for 1.0 s after landing before applying the brakes, what is the total displacement of the jetliner between touchdown on the runway and coming to rest?

Solution   First we convert the given quantities to SI units. Because 1.00 mi/h = 0.447 m/s, we have

$$v_0 = (160 \text{ mi/h})\left(\frac{0.447 \text{ m/s}}{1.00 \text{ mi/h}}\right) = 71.5 \text{ m/s}$$

$$a = (10.0 \text{ mi/h/s})\left(\frac{0.447 \text{ m/s}}{1.00 \text{ mi/h}}\right) = 4.47 \text{ m/s}^2$$

Next we choose our coordinate system and sketch the problem as shown in Figure 2.19.

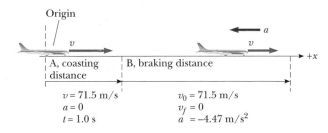

FIGURE 2.19   (Example 2.5) Coasting and braking distances for a landing jetliner.

From the sketch we see that the problem has two parts. In part A, the plane coasts with a constant velocity and so the acceleration is zero. Part B involves a braking distance over which the acceleration is constant at -4.47 m/s^2 and the velocity drops from 71.5 m/s to 0 m/s. Listing the given quantities and those to be determined, we find:

For the Coasting Displacement—Part A	
Given	**To Be Determined**
$v = 71.5 \text{ m/s}$	Δx
$a = 0$	
$t = 1.0 \text{ s}$	

Selecting $\Delta x = \overline{v}t$ from our list of equations for motion with constant acceleration ($a = 0$) gives the coasting displacement, Δx_A.

$$\Delta x_\text{A} = \overline{v}t = (71.5 \text{ m/s})(1.0 \text{ s}) = 71.5 \text{ m}$$

Note that using $\Delta x = v_0 t + \frac{1}{2}at^2$ would also give the same result, but $v^2 = v_0^2 + 2a\,\Delta x$ simply yields $v = v_0$.

For the Braking Displacement—Part B	
Given	**To Be Determined**
$v_0 = 71.5 \text{ m/s}$	Δx
$v = 0$	
$a = -4.47 \text{ m/s}^2$	

Because the acceleration is again steady, we use the equations in Table 2.3. The only equation which connects v, a, and Δx is $v^2 = v_0^2 + 2a\,\Delta x$. Solving for the displacement during braking, Δx_B,

$$\Delta x_\text{B} = \frac{v^2 - v_0^2}{2a} = \frac{0^2 - (71.5 \text{ m/s})^2}{2(-4.47 \text{ m/s}^2)} = 572 \text{ m}$$

The total displacement of the plane on the runway is

$$\Delta x_\text{A} + \Delta x_\text{B} = 72 \text{ m} + 572 \text{ m} = 644 \text{ m} = \boxed{6.4 \times 10^2 \text{ m}}$$

to the correct number of significant figures. Does the approximate answer of 640 m or 2100 ft seem reasonable?

Example 2.6 The Porsche of American Trains, the Acela

The sleek high-speed electric train known as the Acela (pronounced ahh-sell-ah) is currently in service on the Washington-New York-Boston run and is shown in Figure 2.20a. It consists of two power cars and six coaches and can carry 304 passengers at speeds up to 170 mi/h. In order to negotiate curves comfortably at high speed, the train carriages tilt as much as six degrees from the vertical to prevent passengers from being pushed to the side. A velocity-time graph for the Acela is shown in Figure 2.20b. (a) Describe the motion of the train in each interval. (b) Find the peak acceleration of Acela in miles per hour per second (mi/h/s) as it speeds up from 45 to 170 mi/h. (c) Find the train's displacement in miles between $t = 0$ and $t = 200$ s. (d) Find the average acceleration of Acela and its displacement in miles in the interval from 200 s to 300 s. (The train has regenerative braking, which means that it feeds energy back into the utility lines each time it stops!) (e) Find the total displacement in the interval from 0 to 400 s.

Solution **A** As we see in Figure 2.20b, from about -50 s to 50 s Acela is cruising at a constant positive velocity in the $+x$ direction. From 50 s to 200 s, Acela accelerates in the $+x$ direction reaching a top speed of about 170 mi/h. Around 200 s, the engineer applies the brakes, and the train, still traveling in the $+x$ direction, slows down and then stops at 350 s. Just after 350 s, Acela reverses direction (v becomes negative) and steadily gains speed in the $-x$ direction. **B** Because the acceleration is changing in this problem, we cannot use the equations for motion with constant a, but must resort to the most general graphical methods. The peak acceleration between 45 and 170 mi/h is given by the slope of the steepest tangent to the v versus t curve in this interval. From the tangent of the blue line shown in Figure 2.20c, we find

$$a = \text{slope} = \frac{\Delta v}{\Delta t} = \frac{(150 - 50)\ \text{mi/h}}{(100 - 50)\ \text{s}} = \boxed{2.0\ \text{mi/h/s}}$$

Note that if we work with *extreme caution* and treat units like algebraic quantities, we can manage mixed units of mi/h/s. There are two reasons to do this. One is that an acceleration of 2.0 mi/h/s is easy to visualize for most people, while 0.89 m/s² is not. The other is that it is easier to work in mixed units in this problem instead of converting all values to m/s and re-plotting the graph. At any rate, we urge extreme caution in using mixed units and point out they can have a hazardous effect on your physics grade.

C Let us use the fact that the area under the v versus t curve equals the displacement. The train's displacement between 0 and 200 s is equal to the area of the gray shaded region in Figure 2.20d, which we have approximated with a series of triangles and rectangles.

$$\Delta x_{0 \to 200\ \text{s}} = \text{area}_1 + \text{area}_2 + \text{area}_3 + \text{area}_4 + \text{area}_5$$
$$\approx (50\ \text{mi/h})(50\ \text{s}) + (50\ \text{mi/h})(50\ \text{s}) + (160\ \text{mi/h})(100\ \text{s})$$
$$+ \tfrac{1}{2}(50\ \text{s})(100\ \text{mi/h}) + \tfrac{1}{2}(100\ \text{s})(170\ \text{mi/h} - 160\ \text{mi/h}) = 24\ 000\ (\text{mi/h})(\text{s})$$

Now, at the end of our calculation, we can find the displacement in miles by converting hours to seconds. As 1 h = 3 600 s,

$$\Delta x_{0 \to 200\ \text{s}} \approx \left(\frac{24\ 000\ \text{mi}}{3\ 600\ \text{s}} \right)(\text{s}) = \boxed{6.7\ \text{mi}}$$

D The average acceleration is just the slope of the green line from 200 s to 300 s, shown in Figure 2.20c:

$$\bar{a} = \text{slope} = \frac{\Delta v}{\Delta t} = \frac{(10 - 170)\ \text{mi/h}}{100\ \text{s}} = \boxed{-1.6\ \text{mi/h/s}}$$

The negative sign means the train's velocity in the $+x$ direction is decreasing and it is going 1.6 mi/h slower each second. Acela's displacement is given by the area of the yellow shaded region of Figure 2.20d, composed of a large triangle and a thin rectangle.

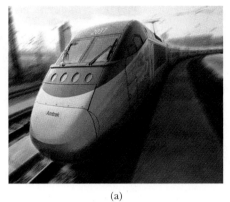

(a)

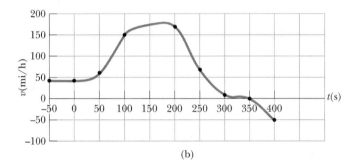

(b)

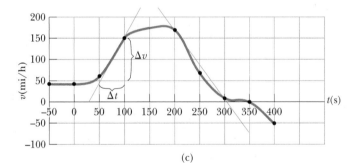

(c)

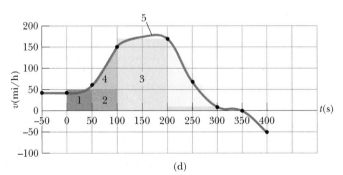

(d)

FIGURE 2.20 (Example 2.6) (a) The Acela—1 171 000 lb of cold steel thundering along at 150 mi/h. *(Courtesy Amtrak NEC Media Relations)* (b) Velocity versus time graph for the Acela. (c) The slope of the steepest tangent blue line gives the peak acceleration, while the slope of the green line is the average acceleration between 200 s and 300 s. (d) The area under the velocity-time graph in some time interval gives the displacement of Acela in that time interval.

$$\Delta x_{200 \to 300 \text{ s}} \approx \tfrac{1}{2}(100 \text{ s})(170 - 10) \text{ mi/h} + (10 \text{ mi/h})(100 \text{ s})$$

$$= 9\ 000 \ (\text{mi/h})(\text{s}) = \boxed{2.5 \text{ mi}}$$

E To get the total displacement, we simply add all the individual displacements. We need the displacements for the time intervals from 300 s to 350 s and from 350 s to 400 s:

$$\Delta x_{300 \to 350 \text{ s}} \approx \tfrac{1}{2}(50 \text{ s})(10 \text{ mi/h}) = 250 \ (\text{mi/h})(\text{s})$$

$$\Delta x_{350 \to 400 \text{ s}} \approx \tfrac{1}{2}(50 \text{ s})(-50 \text{ mi/h}) = -1\ 250 \ (\text{mi/h})(\text{s})$$

Finally we find

$$\Delta x_{0 \to 400 \text{ s}} \approx (24\ 000 + 9\ 000 + 250 - 1\ 250)(\text{mi/h})(\text{s}) = \boxed{8.9 \text{ mi}}$$

2.7   FREELY FALLING OBJECTS

It is now well known that, in the absence of air resistance, all objects dropped near Earth's surface fall toward Earth with the same constant acceleration. It was not until about 1600 that this conclusion was accepted. Prior to that time, the teachings of the great philosopher Aristotle (384–322 B.C.) had held that heavier objects fell faster than lighter ones.

It was Galileo who originated our present-day ideas concerning falling objects. There is a legend that he discovered the law of falling objects by observing that two different weights dropped simultaneously from the Leaning Tower of Pisa hit the ground at approximately the same time. Although it is doubtful that this particular experiment was carried out, we know that Galileo performed many systematic experiments on objects moving on inclined planes. In his experiments he rolled balls down a slight incline and measured the distances they covered in successive time intervals. The purpose of the incline was to reduce the acceleration and enable Galileo to make accurate measurements of the time intervals. (Some people refer to this experiment as "diluting gravity.") By gradually increasing the slope of the incline, he was finally able to draw mathematical conclusions about freely falling objects, because a falling ball is equivalent to a ball falling down a vertical incline. Galileo's achievements in the science of mechanics paved the way for Newton in his development of the laws of motion, which we shall study in Chapter 4.

You might want to try the following experiment. Simultaneously drop a coin and a crumpled piece of paper from the same height. If the effects of air friction are negligible, both objects will have the same motion and hit the floor at the same time. In the idealized case, where air resistance is absent, such motion is referred to as free fall. If this same experiment could be conducted in a vacuum, in which air friction is zero, the paper and coin would fall with the same acceleration, regardless of the shape of the paper. (See the photographs in Physics in Action.) On August 2, 1971, such a demonstration was conducted on the Moon by astronaut David Scott. He simultaneously released a hammer and a feather, and they fell with the same acceleration to the lunar surface. This demonstration surely would have pleased Galileo!

When we use the expression *freely falling object*, we do not necessarily mean an object dropped from rest. **A freely falling object is any object moving under the influence of gravity only, regardless of its initial motion. Objects thrown upward or downward and those released from rest are all falling freely once they are released. Any freely falling object experiences an acceleration directed *downward*, regardless of the direction of its motion at any instant.**

We shall denote the magnitude of the **free-fall acceleration** by the symbol *g*. The value of *g* decreases with increasing altitude. Furthermore, slight variations in *g* occur in latitude. At Earth's surface the value of *g* is approximately 9.80 m/s².

GALILEO GALILEI, ITALIAN PHYSICIST AND ASTRONOMER (1564–1642)

Galileo formulated the laws that govern the motion of objects in free fall. He also investigated the motion of an object on an inclined plane, established the concept of relative motion, invented the thermometer, and discovered that the motion of a swinging pendulum could be used to measure time intervals. After designing and constructing his own telescope, he discovered four of Jupiter's moons, found that the Moon's surface is rough, discovered sunspots and the phases of Venus, and showed that the Milky Way consists of an enormous number of stars. Galileo publicly defended Nicholaus Copernicus's assertion that the Sun is at the center of the Universe (the heliocentric system). He published *Dialogue Concerning Two New World Systems* to support the Copernican model, a view which the Church declared to be heretical. After being taken to Rome in 1633 on a charge of heresy, he was sentenced to life imprisonment, and later was confined to his villa at Arcetri, near Florence, where he died in 1642. *(North Wind)*

where *x* is measured in meters and *t* in seconds. (a) Plot a graph of position versus time. (b) Determine the instantaneous velocity at *t* = 4.0 s, using time intervals of 0.40 s, 0.20 s, and 0.10 s. (c) Compare the average velocity during the first 4.0 s with the results of (b).

17. Find the instantaneous velocities of the tennis player of Figure P2.7 at (a) 0.50 s, (b) 2.0 s, (c) 3.0 s, (d) 4.5 s.

Section 2.4 Acceleration

18. Secretariat ran the Kentucky Derby with times for the quarter mile of 25.2 s, 24.0 s, 23.8 s, and 23.0 s. (a) Find his average speed during each quarter-mile segment. (b) Assuming that Secretariat's instantaneous speed at the finish line was the same as the average speed during the final quarter mile, find his average acceleration for the entire race. (*Hint:* Recall that horses in the Derby start from rest.)

19. A particle is moving with a velocity of 60.0 m/s in the positive *x* direction at *t* = 0. Between *t* = 0 and *t* = 15.0 s the velocity decreases uniformly to zero. What was the acceleration during this 15.0-s interval? What is the significance of the sign of your answer?

20. A tennis ball with a speed of 10.0 m/s is thrown perpendicularly at a wall. After striking the wall, the ball rebounds in the opposite direction with a speed of 8.0 m/s. If the ball is in contact with the wall for 0.012 s, what is the average acceleration of the ball while it is in contact with the wall?

21. A certain car is capable of accelerating at a rate of + 0.60 m/s^2. How long does it take for this car to go from a speed of 55 mi/h to a speed of 60 mi/h?

22. The velocity-time graph for an object moving along a straight path is shown in Figure P2.22. (a) Find the average accelerations of this object during the time intervals 0 to 5.0 s, 5.0 s to 15 s, and 0 to 20 s. (b) Find the instantaneous accelerations at 2.0 s, 10 s, and 18 s.

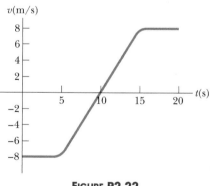

FIGURE P2.22

23. The engine of a model rocket accelerates the rocket vertically upward for 2.0 s as follows: At *t* = 0, its speed is zero; at *t* = 1.0 s, its speed is 5.0 m/s; at *t* = 2.0 s, its speed is 16 m/s. Plot a velocity-time graph for this motion, and from it determine (a) the average acceleration during the 2.0-s interval and (b) the instantaneous acceleration at *t* = 1.5 s.

Section 2.6 One-Dimensional Motion with Constant Acceleration

24. The French National Railroad holds the world's speed record for passenger trains in regular service. A TGV (*tres grand vitesse,* or very great speed) train traveling at a speed of 300 km/h requires 1.20 km to come to an emergency stop. Find the braking acceleration for this train, assuming constant acceleration.

25. Jules Verne in 1865 proposed sending men to the Moon by firing a space capsule from a 220-m-long cannon with final speed of 10.97 km/s. What would have been the unrealistically large acceleration experienced by the space travelers during launch? (A human can stand an acceleration of 15*g* for a short time.) Compare your answer with the free-fall acceleration, 9.80 m/s^2.

26. A truck covers 40.0 m in 8.50 s while smoothly slowing down to final speed 2.80 m/s. (a) Find its original speed. (b) Find its acceleration.

27. The minimum distance required to stop a car moving at 35.0 mi/h is 40.0 ft. What is the minimum stopping distance for the same car moving at 70.0 mi/h, assuming the same rate of acceleration?

28. A racing car reaches a speed of 40 m/s. At this instant, it begins a uniform negative acceleration, using a parachute and a braking system, and comes to rest 5.0 s later. (a) Determine the acceleration of the car. (b) How far does the car travel after the acceleration starts?

29. A Cessna aircraft has a lift-off speed of 120 km/h. (a) What minimum constant acceleration does this require if the aircraft is to be airborne after a takeoff run of 240 m? (b) How long does it take the aircraft to become airborne?

30. A truck on a straight road starts from rest and accelerates at 2.0 m/s^2 until it reaches a speed of 20 m/s. Then the truck travels for 20 s at constant speed until the brakes are applied, stopping the truck in a uniform manner in an additional 5.0 s. (a) How long is the truck in motion? (b) What is the average velocity of the truck for the motion described?

31. A drag racer starts her car from rest and accelerates at 10.0 m/s^2 for the entire distance of 400 m ($\frac{1}{4}$ mile). (a) How long did it take the race car to travel this distance? (b) What is the speed of the race car at the end of the run?

32. A jet plane lands with a speed of 100 m/s and can accelerate at a maximum rate of − 5.00 m/s^2 as it comes to rest. (a) From the instant the plane touches the runway, what is the minimum time needed before it can come to rest? (b) Can this plane land on a small tropical island airport where the runway is 0.800 km long?

33. A driver in a car traveling at a speed of 60 mi/h sees a deer 100 m away on the road. Calculate the minimum constant acceleration that is necessary for the car to stop without hitting the deer (assuming that the deer does not move in the meantime).

34. A record of travel along a straight path is as follows.

   1. Start from rest with constant acceleration of 2.77 m/s^2 for 15.0 s
   2. Constant velocity for the next 2.05 min
   3. Constant negative acceleration − 9.47 m/s^2 for 4.39 s

(a) What was the total displacement for the complete trip? (b) What were the average speeds for legs 1, 2, and 3 of the trip as well as for the complete trip?

35. A train is traveling down a straight track at 20 m/s when the engineer applies the brakes, resulting in an acceleration of -1.0 m/s^2 as long as the train is in motion. How far does the train move during a 40-s time interval starting at the instant the brakes are applied?

36. A car accelerates uniformly from rest to a speed of 40.0 mi/h in 12.0 s. (a) Find the distance the car travels during this time and (b) the constant acceleration of the car.

37. A car starts from rest and travels for 5.0 s with a uniform acceleration of $+1.5$ m/s^2. The driver then applies the brakes, causing a uniform acceleration of -2.0 m/s^2. If the brakes are applied for 3.0 s, (a) how fast is the car going at the end of the braking period, and (b) how far has it gone?

38. A train 400 m long is moving on a straight track with a speed of 82.4 km/h. The engineer applies the brakes at a crossing, and later the last car passes the crossing with a speed of 16.4 km/h. Assuming constant acceleration, determine how long the train blocked the crossing. Disregard the width of the crossing.

39. A hockey player is standing on his skates on a frozen pond when an opposing player, moving with a uniform speed of 12 m/s, skates by with the puck. After 3.0 s, the first player makes up his mind to chase his opponent. If he accelerates uniformly at 4.0 m/s^2, (a) how long does it take him to catch his opponent, and (b) how far has he traveled in this time? (Assume the player with the puck remains in motion at constant speed.)

40. A car, capable of a constant acceleration of 2.5 m/s^2, is stopped at a traffic light. When the light turns green, the car starts from rest with this acceleration. Also, as the light turns green, a truck traveling with a constant velocity of 40 km/h passes the car. Clearly, the car will eventually travel faster than the truck and will overtake it. Where will the car catch up with the truck?

41. A car has an initial velocity v_0 when the driver sees an obstacle in the road in front of him. His reaction time is t_r, and the braking acceleration of the car is a. Show that the total stopping distance is

$$s_{stop} = v_0 t_r - \frac{v_0{}^2}{2a}$$

Note that a is negative in this situation.

42. The yellow caution light on a traffic signal should stay on long enough to allow a driver to either pass through the intersection or safely stop before reaching the intersection. That is, if a car is more than the stopping distance of the previous problem away from the intersection, it can stop. If the car is less than this stopping distance from the intersection, the yellow light should stay on long enough to allow the car to pass entirely through the intersection. (a) Show that the yellow light should stay on for a time

$$t_{light} = t_r - \left(\frac{v_0}{2a}\right) + \left(\frac{s_i}{v_0}\right)$$

where t_r is the driver's reaction time, v_0 is the velocity of the car approaching the light at the speed limit, a is the braking acceleration, and s_i is the length of the intersection. (b) As city traffic planner, you expect cars to approach an intersection 16 m wide with a speed of 60 km/h. Be cautious and assume a reaction time of 1.1 s to allow for a driver's indecision. Find the length of time the yellow light should remain on. Use a braking acceleration of -2.0 m/s^2.

Section 2.7   Freely Falling Objects

43. A ball is thrown vertically upward with a speed of 25.0 m/s. (a) How high does it rise? (b) How long does it take to reach its highest point? (c) How long does it take to hit the ground after it reaches its highest point? (d) What is its velocity when it returns to the level from which it started?

44. In the mid-1960s, McGill University launched high-altitude weather sensors by firing them from a cannon made from two World War II Navy cannons bolted together for a total length of 18 m. It was proposed that this arrangement be used to launch a satellite. The orbital speed of a satellite is about 29 000 km/h. What would the average acceleration throughout the 18-m length of the cannon have to be in order to have a muzzle velocity of 29 000 km/h for this satellite?

45. Killer whales at Sea World routinely jump 7.5 m above the water. What is their speed as they leave the water to achieve this?

46. A peregrine falcon dives at a pigeon. The falcon starts downward from rest and falls with free-fall acceleration. If the pigeon is 76.0 m below the initial position of the falcon, how long does it take the falcon to reach the pigeon? Assume that the pigeon remains at rest.

47. A small mailbag is released from a helicopter that is descending steadily at 1.50 m/s. After 2.00 s, (a) what is the speed of the mailbag, and (b) how far is it below the helicopter? (c) What are your answers to parts (a) and (b) if the helicopter is rising steadily at 1.50 m/s?

48. A ball thrown vertically upward is caught by the thrower after 2.00 s. Find (a) the initial velocity of the ball and (b) the maximum height it reaches.

49. A model rocket is launched straight upward with an initial speed of 50.0 m/s. It accelerates with a constant upward acceleration of 2.00 m/s^2 until its engines stop at an altitude of 150 m. (a) What is the maximum height reached by the rocket? (b) How long after lift-off does the rocket reach its maximum height? (c) How long is the rocket in the air?

50. A parachutist with a camera, both descending at a speed of 10 m/s, releases that camera at an altitude of 50 m. (a) How long does it take the camera to reach the ground? (b) What is the velocity of the camera just before it hits the ground?

51. A student throws a set of keys vertically upward to her sorority sister, who is in a window 4.00 m above. The keys are caught 1.50 s later by the sister's outstretched hand. (a) With what initial velocity were the keys thrown? (b) What was the velocity of the keys just before they were caught?

52. A ball is thrown directly downward, with an initial speed of 8.00 m/s, from a height of 30.0 m. After what interval does the ball strike the ground?

ADDITIONAL PROBLEMS

53. A truck tractor pulls two trailers, one behind the other, at a constant speed of 100 km/h. It takes 0.600 s for the big rig to completely pass onto a bridge 400 m long. For what duration of time is all or part of the truck-trailer combination on the bridge?

54. A speedboat moving at 30.0 m/s approaches a no-wake buoy marker 100 m ahead. The pilot slows the boat with a constant acceleration of -3.50 m/s^2 by reducing the throttle. (a) How long does it take the boat to reach the buoy? (b) What is the velocity of the boat when it reaches the buoy?

55. A bullet is fired through a board 10.0 cm thick in such a way that the bullet's line of motion is perpendicular to the face of the board. If the initial speed of the bullet is 400 m/s and it emerges from the other side of the board with a speed of 300 m/s, find (a) the acceleration of the bullet as it passes through the board and (b) the total time the bullet is in contact with the board.

56. An indestructible bullet 2.00 cm long is fired straight through a board that is 10.0 cm thick. The bullet strikes the board with a speed of 420 m/s and emerges with a speed of 280 m/s. (a) What is the average acceleration of the bullet through the board? (b) What is the total time that the bullet is in contact with the board? (c) What thickness of boards (calculated to 0.1 cm) would it take to stop the bullet assuming the acceleration through all boards is the same?

57. A ball is thrown upward from the ground with an initial speed of 25 m/s; at the same instant, a ball is dropped from a building 15 m high. After how long will the balls be at the same height?

58. A ranger in a national park is driving at 35.0 mi/h when a deer jumps into the road 200 ft ahead of the vehicle. After a reaction time of t, the ranger applies the brakes to produce an acceleration of $a = -9.00$ ft/s^2. What is the maximum reaction time allowed if she is to avoid hitting the deer?

web

59. Two students are on a balcony 19.6 m above the street. One student throws a ball vertically downward at 14.7 m/s; at the same instant the other student throws a ball vertically upward at the same speed. The second ball just misses the balcony on the way down. (a) What is the difference in their time in air? (b) What is the velocity of each ball as it strikes the ground? (c) How far apart are the balls 0.800 s after they are thrown?

60. The driver of a truck slams on the brakes when he sees a tree blocking the road. The truck slows uniformly with acceleration -5.60 m/s^2 for 4.20 s, making skid marks 62.4 m long that end at the tree. With what speed does the truck then strike the tree?

61. A young woman named Kathy Kool buys a sports car that can accelerate at the rate of 4.90 m/s^2. She decides to test the car by drag racing with another speedster, Stan Speedy. Both start from rest, but experienced Stan leaves the starting line 1.00 s before Kathy. If Stan moves with a constant acceleration of 3.50 m/s^2 and Kathy maintains an acceleration of 4.90 m/s^2, find (a) the time it takes Kathy to overtake Stan, (b) the distance she travels before she catches him, and (c) the speeds of both cars at the instant she overtakes him.

62. A mountain climber stands at the top of a 50.0-m cliff that overhangs a calm pool of water. He throws two stones vertically downward 1.00 s apart and observes that they cause a single splash. The first stone has an initial velocity of -2.00 m/s. (a) How long after release of the first stone will the two stones hit the water? (b) What initial velocity must the second stone have if they are to hit simultaneously? (c) What will the velocity of each stone be at the instant they hit the water?

63. Using a rocket pack with full throttle, a lunar astronaut accelerates upward from the Moon's surface with a constant acceleration of 2.00 m/s^2. At a height of 5.00 m, a bolt comes loose. (The free-fall acceleration on the Moon's surface is about 1.67 m/s^2.) (a) How fast is the astronaut moving at that time? (b) When will the bolt hit the Moon's surface? (c) How fast will it be moving then? (d) How high will the astronaut be when the bolt hits? (e) How fast will the astronaut be traveling then?

64. In Bosnia, the ultimate test of a young man's courage once was to jump off a 400-year-old bridge (now destroyed) into the River Neretva, 23 m below the bridge. (a) How long did the jump last? (b) How fast was the diver traveling upon impact with the river? (c) If the speed of sound in air is 340 m/s, how long after the diver took off did a spectator on the bridge hear the splash?

65. A person sees a lightning bolt pass close to an airplane that is flying in the distance. The person hears thunder 5.0 s after seeing the bolt, and sees the airplane overhead 10 s after hearing the thunder. The speed of sound in air is 1 100 ft/s. (a) Find the distance of the airplane from the person at the instant of the bolt. (Neglect the time it takes the light to travel from the bolt to the eye.) (b) Assuming that the plane travels with a constant speed toward the person, find the velocity of the airplane. (c) Look up the speed of light in air, and defend the approximation used in (a).

66. Another scheme to catch the roadrunner has failed. A safe falls from rest from the top of a 25.0-m-high cliff toward Wile E. Coyote, who is standing at the base. Wile first notices the safe after it has fallen 15.0 m. How long does he have to get out of the way?

67. A stunt woman sitting on a tree limb wishes to drop vertically onto a horse galloping under the tree. The constant speed of the horse is 10.0 m/s, and the woman is initially 3.00 m above the level of the saddle. (a) What must be the horizontal distance between the saddle and limb when the woman makes her move? (b) How long is she in the air?

web

68. A hard rubber ball, released at chest height, falls to the pavement and bounces back to nearly the same height. When it is in contact with the pavement, the lower side of the ball is temporarily flattened. Before this dent in the ball pops out, suppose that its maximum depth is on the order of 1 cm. Compute an order-of-magnitude estimate for the maximum acceleration of the ball. State your assumptions, the quantities you estimate, and the values you estimate for them.

GROUP ACTIVITIES

Note: Following are a variety of group activities. Some are mini-experiments or demonstrations you can carry out with the assistance of a co-worker. Others are problems, frequently consisting of both conceptual and numerical parts, that stimulate group discussion. The first step in developing problem-solving skills and in developing a full understanding of physics is understanding the basic concepts. These activities are designed to provide assistance in that endeavor.

G.1 Estimate a few speeds in metric units, using a stopwatch or wristwatch chronometer. For example, roll a ball across a table and estimate the number of centimeters it moves each second to find its speed. Other speeds you might try are for someone walking across the room, a jogger running, a car moving through some distance, and so forth. How closely do your estimates match those of a co-worker? To see how well you did, make some actual measurements for those situations in which it is feasible to do so.

G.2 Use what you know about falling objects to measure your reaction time. Hold the index finger and thumb of your dominant hand about 2.5 cm apart, and then have your co-worker hold a ruler vertically in the space between your finger and thumb, as shown in Figure GA2.2. Note the position of the ruler relative to your index finger. Your co-worker must release the ruler and you must catch it (without moving your hand downward) as quickly as you can. The ruler (a freely falling object) falls through a distance $d = \frac{1}{2}gt^2$, where t is the reaction time and $g = 9.80 \text{ m/s}^2$. Repeat this measurement of d five times, average your results and calculate an average value of t. Now measure your co-worker's reaction time using the same procedure. Compare your results. For most people, the reaction time is at best about 0.2 s. As an extension to this experiment, replace the ruler with a crisp dollar bill. Hold the bill such that your thumb and index finger are just at the level of Washington's face. Unless you are anticipating the time of release, you will not be able to catch the bill when it is released because the time for the top to pass out of your hand is less than the typical 0.2-s reaction time.

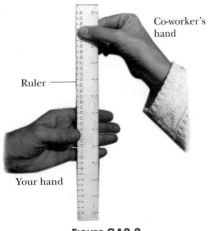

FIGURE GA2.2

Co-worker's hand

Ruler

Your hand

G.3 Galileo studied accelerated motion by allowing objects to roll down inclined planes so that their motion would be slow enough to make reasonable observations. Try a similar procedure. Make a mark at the top of an inclined plane as the starting point for the motion and use a metal barrier at the end as a sound cue for stopping a stopwatch. Measure the length of the plane. Record the average time for several trials of a ball rolling down the plane at a measured angle. From this information calculate the acceleration. Repeat the experiment for a larger angle of inclination. Do this for several trials until you can plot a graph of acceleration versus angle. From your graph can you guess what the acceleration would be if the inclined plane were vertical? Would the results of your experiment be different if you had used a significantly more massive ball? If you are unsure, repeat the experiment to see if there is a difference.

G.4 A ball is thrown straight upward and moves in free fall. Choose a coordinate system with its origin at the release point of the ball and the positive direction upward. What is the sign of the following quantities? (a) The velocity of the ball just before it reaches its maximum height, just after it reaches its maximum height, and at its maximum height. (b) The acceleration of the ball at just before it reaches its maximum height, just after its maximum height, and at its maximum height. (c) The displacement of the ball just before it reaches its maximum height, just after its maximum height, and at its maximum height. (d) If it takes time t_1 to reach its maximum height, how long will it take to return to ground level? (e) If it is thrown upward with a velocity of $+ v_0$, what will be its velocity upon returning to ground level? (f) Assume a ball is thrown upward with a velocity of 12 m/s. Find the various answers for the above questions to test your answers.

G.5 Figure GA2.5 is a graph of position versus time for the motion of two balls, A and B, moving on parallel tracks. Sketch this figure on a sheet of paper and answer the following questions.

(a) Mark with the symbol t_A along the t axis any instant or instants at which one ball is passing the other.

(b) Which ball is moving faster at clock reading t_B?

(c) Mark with the symbol t_C along the t axis any instant or instants at which the balls have the same velocity.

(d) Over the period of time shown in the diagram, which of the following is true of ball B? Explain your answer.

   (i) It is speeding up all the time.

   (ii) It is slowing down all the time.

   (iii) It is speeding up part of the time and slowing down part of the time.

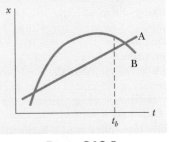

FIGURE GA2.5

placements $\mathbf{d}_{2m}$ and $\mathbf{d}_{2f}$ have magnitudes of 50.0 cm and 43.0 cm, respectively. (a) Find the vector sum of the displacements $\mathbf{d}_1$ and $\mathbf{d}_2$ in each case. (b) The male figure is 180 cm tall, the female 168 cm. Normalize the displacements of each figure to a common height of 200 cm, and re-form the vector sums as in part (a). Then find the vector difference between the two sums.

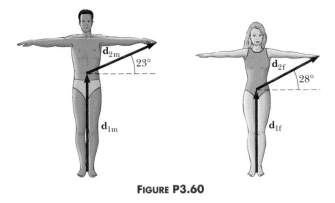

FIGURE P3.60

61. By throwing a ball at an angle of 45°, a boy can throw it a maximum horizontal distance of R on a level field. How far can he throw the same ball vertically upward? Assume that his muscles give the ball the same speed in each case. (Is this assumption valid?)

62. A projectile is fired with an initial speed of v_0 at an angle of θ_0 to the horizontal, as in Figure 3.13. When it reaches its peak, it has (x, y) coordinates given by $(R/2, h)$, and when it strikes the ground, its coordinates are $(R, 0)$, where R is called the horizontal range. (a) Show that it reaches a maximum height h given by

$$h = \frac{v_0{}^2 \sin^2 \theta_0}{2g}$$

(b) Show that its horizontal range is given by

$$R = \frac{v_0{}^2 \sin 2\theta_0}{g}$$

63. A hunter wishes to cross a river that is 1.5 km wide and flows with a speed of 5.0 km/h parallel to its banks. The hunter uses a small powerboat that moves at a maximum speed of 12 km/h with respect to the water. What is the minimum time necessary for crossing?

64. A water insect maintains a constant average position on the surface of a stream by darting upstream (against the current), then drifting downstream (with the current) to its original position. The current in the stream is 0.500 m/s relative to the shore, and the insect darts upstream 0.560 m (relative to a spot on shore) in 0.800 s during the first part of its motion. Take upstream as the positive direction. (a) Determine the velocity of the insect relative to the water (i) during its dash upstream and (ii) during its drift downstream. (b) How far upstream relative to the water does the insect move during one cycle of this motion? (c) What is the average velocity of the insect relative to the water?

65. A daredevil is shot out of a cannon at 45.0° to the horizontal with an initial speed of 25.0 m/s. A net is positioned a horizontal distance of 50.0 m from the cannon. At what height above the cannon should the net be placed in order to catch the daredevil?

66. A projectile is fired with an initial velocity of 15.0 m/s at 53.0° above the horizontal from the foot of a ramp inclined 20.0° above the horizontal, as in Figure P3.66. How far up the ramp does the projectile strike the ramp?

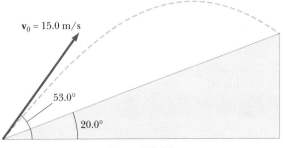

FIGURE P3.66

67. A student decides to measure the muzzle velocity of a pellet from his gun. He points the gun horizontally. He places a target on a vertical wall a distance x away from the gun. The pellet hits the target a vertical distance y below the gun. (a) Show that the position of the pellet when traveling through the air is given by $y = Ax^2$, where A is a constant. (b) Express the constant A in terms of the initial (muzzle) velocity and the free-fall acceleration. (c) If $x = 3.00$ m and $y = 0.210$ m, what is the initial speed of the pellet?

68. In ancient mythology, King Theseus of Athens is trapped in the Labyrinth (a maze) and finds his way out by following a thread given to him by Ariadne. He walks 10.0 m, makes a 90.0° right turn, walks 5.00 m, makes another 90.0° right turn, and walks 7.00 m. What is Theseus's displacement from his initial position?

69. Instructions for finding a buried treasure include the following: Go 75.0 paces at 240°, turn to 135° and walk 125 paces, then travel 100 paces at 160°. Determine the resultant displacement from the starting point.

70. When baseball outfielders throw the ball, they usually allow it to take one bounce on the theory that the ball arrives sooner this way. Suppose that after the bounce the ball rebounds at the same angle θ as it had when released (as in Fig. P3.70) but loses half its speed. (a) Assuming the ball is always thrown with the same initial speed, at what angle θ should the ball be thrown in order to go the same distance D with one bounce as one thrown upward at 45.0°

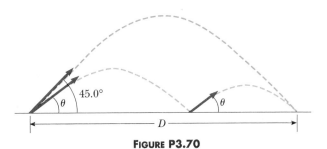

FIGURE P3.70

with no bounce? (b) Determine the ratio of the times for the one-bounce and no-bounce throws.

71. One strategy in a snowball fight is to throw a snowball at a high angle over level ground. While your opponent is watching the first one, you throw a second snowball at a low angle timed to arrive before or at the same time as the first one. Assume both snowballs are thrown with a speed of 25.0 m/s. The first one is thrown at an angle of 70.0° with respect to the horizontal. (a) At what angle should the second snowball be thrown to arrive at the same point as the first? (b) How many seconds later should the second snowball be thrown after the first to arrive at the same time?

GROUP ACTIVITIES

G.1 Take three steps, turn 90°, and then walk four steps. Now count the number of steps it takes to walk in a straight line back to your starting point. Verify your result mathematically.

G.2 For this investigation, you need to be outside with a small ball such as a tennis ball and a wristwatch with a second hand. Throw the ball vertically upward as hard as you can and determine the initial speed of your throw and the approximate maximum height of the ball using only your wristwatch. What happens when you throw the ball at some angle other than 90°? Does this change the time of flight? Can you still determine the maximum height and initial speed? Give careful explanations for your answers. Your co-worker can eyeball the maximum height by standing at a distance and noting the angle.

G.3 Using as projectiles the drops of water spraying from a garden hose at ground level, test the statement that the maximum range occurs when the angle of projection is 45 degrees. As an additional part to this experiment, hold the nozzle horizontally above the ground and have your co-worker position a marker at the location where the water strikes the ground. Increase the angle of inclination by about 10° and again record the strike position. Repeat until you have reached an angle of about 75°. Note the pattern produced. Are there two angles at which the range is the same? Explain the reason behind your observations.

G.4 Use a chessboard as a coordinate system with the intersection of the lines on the board as the positions of the coordinates. Select an origin for your coordinate system and on index cards write down several vector displacements that are at right angles to one another. For example, displacement (1) might be a movement of four units to the right, displacement (2) an upward movement of six units, and so forth. Continue this for a total of seven or eight movements until you end up at some particular location on the chessboard. Let your co-worker start at the origin and follow your vector directions to see if he/she arrives at the expected final location. Now, shuffle the cards and repeat the experiment. Does the order of the displacements make any difference as to where you eventually end up?

G.5 (a) A vector **A** is in the first quadrant of a Cartesian coordinate system. What is the sign of the x component of **A**? What is the sign of the y component? (b) A vector **B** is in the second quadrant of a Cartesian coordinate system. What is the sign of the x component of **B**? What is the sign of the y component? (c) The vector sum **A** + **B** _____. Choose the correct fill-in-the blank answer from: (i) must be in either the first or second quadrant. (ii) could be in any quadrant. (d) Let **A** = 30 m at an angle of 30° from the positive x axis, and **B** = 20 m at an angle of 40° from the negative x axis. Test your predictions of parts (a) through (c).

G.6 Roll a ball off a table. At the very instant the rolling ball leaves the table, drop a second ball from the same height above the floor. (This will require a sharp eye and good reflexes!) Do the two balls hit the floor at the same time? Try varying the speed at which you roll the ball off the table. Does this change affect the time at which the balls strike the floor? Finally, roll one of the balls down an incline and drop the other ball from the base of the incline at the instant the first ball leaves the slope. Which of these balls hits the floor first in this situation? Explain the reason behind your observations.

G.7 You can use any coordinate system you like in order to solve a projectile motion problem. To demonstrate this, let us consider a case in which we throw a ball off the top of a building with a velocity **v** at an angle θ with respect to the horizontal. We choose the origin of the coordinate system at the base of the building directly below the point where the ball is released. (a) When the ball reaches its highest point, which of the following are true? (i) The x component of the velocity is zero. (ii) The y component of the velocity is zero. (iii) Neither component of velocity is necessarily zero. (b) What is the displacement when the ball strikes the ground? (c) Let the building be 50.0 m tall, the initial horizontal velocity be 9.00 m/s, and the initial vertical velocity be 12.0 m/s. Find the maximum height and the time to reach the maximum height. Also, find the time to reach ground level. Use your choices in (a) and (b) to find your answers.

G.8 A boy and a girl are tossing an apple back and forth between them. Figure GA3.8 shows the path the apple follows when watched by an observer looking on from the side. The apple is moving from left to right. Five points are marked on the path. Ignore air resistance. (a) Make a copy of this figure. At each of the marked points, draw an arrow that indicates the magnitude and direction of the apple's velocity when it passes through that point. (b) Make a second copy of the figure. This time, at each marked point, place an arrow indicating the magnitude and direction of any acceleration the apple experiences at that point.

FIGURE GA3.8

(Problem 8 is courtesy of E. F. Redish. For more problems of this type, visit http://www.physics.umd.edu/perg/)

The Laws of Motion

This surf boarder is lifted into the air under the action of forces exerted by the wind on the sail and by water waves on the surf board. *Peter Sterling/FPG/Getty Images*

*C*lassical mechanics describes the relationship between the motion of objects found in our everyday world and the forces acting on them. There are conditions under which classical mechanics does not apply or applies in only a limited way. Most often these conditions are encountered when dealing either with very tiny objects (of a size comparable to that of atoms or smaller, about 10^{-10} m) or with objects moving at speeds near that of light (3×10^8 m/s). Our study of relativity and quantum mechanics in later chapters will enable us to handle such situations. However, for large, slowly moving objects, we can make accurate calculations using the laws of classical mechanics discussed in this chapter.

We learn in this chapter that an object remains in motion with constant velocity if no external force acts on it. We also see that if an external force acts on an object, the object accelerates in response to this force. The acceleration can be determined if the net force acting on the object is known and if the object's mass is known. Finally we learn that there is no such thing as a single isolated force, but that forces arise in pairs from the interaction of two entities, each pushing or pulling on the other with forces of equal magnitude and opposite direction.

4.1   THE CONCEPT OF FORCE

When we think of a **force,** we usually imagine a push or a pull exerted on some object. For instance, you exert a force on a ball when you throw it or kick it, and you exert a force on a chair when you sit down on it. What happens to an object when it is acted on by a force depends on the magnitude and the direction of the force. Force is a vector quantity; thus, we denote it with a directed arrow, just as we do velocity and acceleration.

If you pull on a spring, as in Figure 4.1a, the spring stretches. If a child pulls hard enough on a wagon (Fig. 4.1b), the wagon moves. When a football is kicked (Fig. 4.1c), it is deformed and set in motion. These are all examples of **contact forces,** so named because they result from physical contact between two objects.

Another class of forces does not involve physical contact between two objects. Early scientists, including Newton, were uneasy with the concept of forces that act between two disconnected objects. To overcome this conceptual difficulty, Michael Faraday (1791–1867) introduced the concept of a *field*. The corresponding forces are called **field forces.** According to this approach, when an object of mass *m* is placed at some point *P* near a second object of mass *M*, we say that the first object interacts with the second by virtue of the gravitational field that exists at *P*. Thus, the force of gravitational attraction between two objects, illustrated in Figure 4.1d, is an example of a field force. This force keeps objects bound to Earth and gives rise to what we commonly call the *weight* of the object.

Another common example of a field force is the electric force that one electric charge exerts on another (Fig. 4.1e). A third example is the force exerted by a bar magnet on a piece of iron (Fig. 4.1f).

The known fundamental forces in nature are all field forces. These are, in order of decreasing strength, (1) strong nuclear forces between subatomic particles; (2) electromagnetic forces between electric charges; (3) weak nuclear forces, which arise in certain radioactive decay processes; and (4) gravitational attractions between objects. In classical physics, we are concerned only with gravitational and electromagnetic forces.

Whenever a force is exerted on an object, the object's shape can change. For example, when you kick a soccer ball or strike a tennis ball with a racquet, as in Figure 4.2, the objects deform to some extent. Even objects we usually think of as rigid and inflexible are deformed under the action of external forces. Often the deformations are permanent, as in the case of a collision between automobiles.

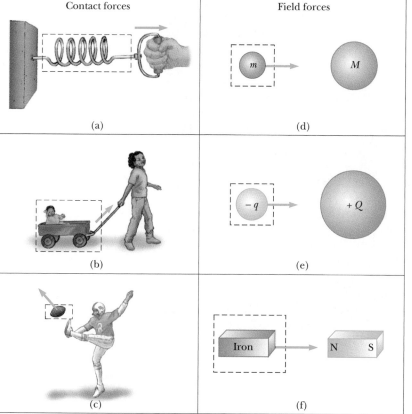

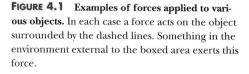

FIGURE 4.1 **Examples of forces applied to various objects.** In each case a force acts on the object surrounded by the dashed lines. Something in the environment external to the boxed area exerts this force.

(a)

(b)

FIGURE 4.2 (a) A soccer ball is set in motion by the contact force **F** exerted on it by the kicker's foot. The ball is deformed during the short time in contact with the foot. *(Loren Winters/Visuals Unlimited)* (b) A tennis ball being struck by a racquet. The ball experiences a large contact force as it is struck and a correspondingly large acceleration. *(Amoz Eckerson/Visuals Unlimited)*

4.2 NEWTON'S FIRST LAW

Consider the following simple experiment. A book is lying on a table. Obviously, the book remains at rest if left alone. Now imagine that you push the book with a horizontal force great enough to overcome the force of friction between book and table, so that the book is set in motion. Because the magnitude of your applied force exceeds the magnitude of the friction force, the book accelerates. If you stop applying the force, the book stops sliding after traveling a short distance because the force of friction retards its motion. Now imagine pushing the book across a smooth, waxed floor. The book again comes to rest once the force is no longer applied, but not as soon as before. Finally, imagine that the book is moving on a horizontal, frictionless surface. In this situation, the book continues to move in a straight line with constant velocity until it hits a wall or some other object.

Before about 1600, scientists felt that the natural state of matter was the state of rest. Galileo was the first to take a different approach. He devised thought experiments—such as an object moving on a frictionless surface, as just described—and concluded that it is not the nature of an object to *stop*, once set in motion; rather, it is an object's nature to *continue in its original state of motion*. This approach to motion was later formalized by Newton in a form that has come to be known as **Newton's first law of motion:**

> If the net force ΣF exerted on an object is zero, the object continues in its original state of motion. That is, if $\Sigma F = 0$, an object at rest remains at rest and an object moving with some velocity continues with that same velocity.

◀ Newton's first law

By *net force* we mean *the vector sum of all external forces exerted on the object.* An *external force* is any force that results from the interaction between the object and its envi-

Unless acted on by an external force, an object at rest will remain at rest and an object in motion will continue in motion with constant velocity. In this case, the wall of the building did not exert a large enough external force on the moving train to stop it. *(Roger Viollet, Mill Valley, CA, University Science Books, 1982)*

ronment or surroundings, such as the force exerted on an object when it is lifted.

As a special case, let us see what happens to an object if there are no external forces exerted on it. For example, consider a spaceship traveling in space, far from any planets or other matter. The spaceship requires a propulsion system to change its velocity. However, if the propulsion system is turned off when a velocity **v** is reached, the spaceship coasts in space with a constant velocity and the astronauts get a free ride.

MASS AND INERTIA

A bowling ball and a golf ball lie side by side on the ground. Newton's first law tells us that both remain at rest as long as no external force acts on them. Now imagine supplying a net force by striking each ball with a golf club. Both balls resist your attempt to change their state of motion. But you know from everyday experience that if the two are struck with equal force, the golf ball will travel much farther than the bowling ball. That is, the bowling ball is more successful in maintaining its original state of motion. The tendency of an object to continue in its original motion is called **inertia.**

While inertia is the tendency for an object to continue its motion in the absence of a force, **mass** is a measure of the resistance of an object to changes in its motion due to a force. The greater the mass of a body, the less it accelerates under the action of an applied force. The SI unit of mass is the kilogram. Mass is a scalar quantity that obeys the rules of ordinary arithmetic.

Inertia can be used to explain the operation of one type of seat belt mechanism. In the event of an accident, the purpose of the seat belt is to hold the passenger firmly in place relative to the car to prevent serious injury. Figure 4.3 illustrates how one type of shoulder harness operates. Under normal conditions, the ratchet turns freely to allow the harness to wind on or unwind from the pulley as the passenger moves. When an accident occurs, the car undergoes a large acceleration and rapidly comes to rest. The large block under the seat, because of its inertia, continues to slide forward along the tracks. The pin connection between the block and the rod causes the rod to pivot about its center and engage the ratchet wheel. At this point the ratchet wheel locks in place, and the harness no longer unwinds.

APPLICATION

SEAT BELTS

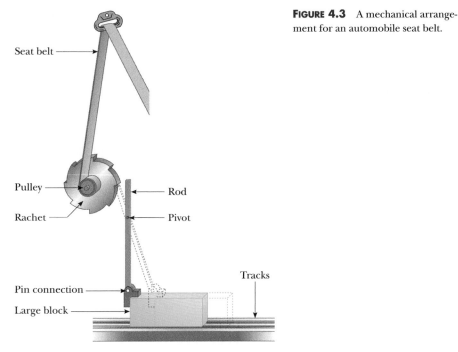

FIGURE 4.3 A mechanical arrangement for an automobile seat belt.

Seat belt

Pulley

Rachet

Rod

Pivot

Pin connection

Large block

Tracks

ISAAC NEWTON, ENGLISH PHYSICIST AND MATHEMATICIAN (1642–1727)

Newton was one of the most brilliant scientists in history. Before the age of 30, he formulated the basic concepts and laws of mechanics, discovered the law of universal gravitation, and invented the mathematical methods of calculus. As a consequence of his theories, Newton was able to explain the motions of the planets, the ebb and flow of the tides, and many special features of the motions of the Moon and Earth. He also interpreted many fundamental observations concerning the nature of light. His contributions to physical theories dominated scientific thought for two centuries and remain important today. *(Giraudon/Art Resource)*

4.3 NEWTON'S SECOND LAW

Newton's first law explains what happens to an object when no force acts on it: The object either remains at rest or continues moving in a straight line with constant speed. Newton's second law answers the question of what happens to an object that has a net force acting on it.

Imagine pushing a block of ice across a frictionless horizontal surface. When you exert some horizontal force on the block, it moves with an acceleration of, say, 2 m/s^2. If you apply a force twice as large, the acceleration doubles. Pushing three times as hard triples the acceleration, and so on. From such observations, we conclude that **the acceleration of an object is directly proportional to the net force acting on it.**

Common experience with pushing objects tells you that mass also affects acceleration. Imagine that you stack identical blocks of ice on top of each other while pushing the stack with constant force. If the force when you push one block produces an acceleration of 2 m/s^2, the acceleration will drop to half that value when two blocks are pushed, one-third that initial value when three blocks are pushed, and so on. We conclude that **the acceleration of an object is inversely proportional to its mass.** These observations are summarized in **Newton's second law:**

> The acceleration of an object is directly proportional to the net force acting on it and inversely proportional to its mass.
>
> $$\mathbf{a} \propto \frac{\sum \mathbf{F}}{m}$$

In vector equation form, and with the units of all quantities appropriately chosen, we can state Newton's second law as

$$\sum \mathbf{F} = m\mathbf{a} \qquad [4.1]$$

where **a** is the acceleration of the object, *m* is its mass, and $\sum \mathbf{F}$ represents the vec-

◄ Newton's second law

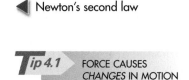

Tip 4.1 FORCE CAUSES *CHANGES* IN MOTION

Many times, students make the mistake of thinking that whenever there is motion, there must be an underlying force. But motion can occur in the absence of forces, as described in Newton's first law. Force is the cause of *changes* in motion.

Webnote 4.1

Find out more about Isaac Newton at:
http://www.newton.cam.ac.uk/ newtlife.html
http://userwww.sfsu.edu/~rsauzier/ Newton.html

tor sum of all forces acting on the object. You should note that, because this is a vector equation, it is equivalent to the following three component equations:

$$\sum F_x = ma_x \qquad \sum F_y = ma_y \qquad \sum F_z = ma_z \qquad \textbf{[4.2]}$$

UNITS OF FORCE AND MASS

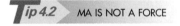 Definition of newton ▶

The SI unit of force is the **newton,** defined as the force that, when acting on an object that has a mass of 1 kg, produces an acceleration of 1 m/s². From this definition and Newton's second law, we see that the newton can be expressed in terms of the fundamental units of mass, length, and time:

$$1 \text{ N} \equiv 1 \text{ kg} \cdot \text{m/s}^2 \qquad \textbf{[4.3]}$$

In the U.S. customary system, the unit of force is the **pound.** The following conversion from pounds to newtons will be useful to you in many problems:

$$1 \text{ N} \equiv 0.225 \text{ lb} \qquad \textbf{[4.4]}$$

The units of mass, acceleration, and force in the SI and U.S. customary systems are summarized in Table 4.1.

 **Tip 4.2** MA IS NOT A FORCE

Equation 4.1 does *not* say that the product *ma* is a force. All forces exerted on an object are added vectorially to generate the net force on the left side of the equation. This net force is then equated to the product of the mass of the object and the acceleration that results from the net force. Do *not* include an "*ma* force" in your analysis. Also, be sure to use consistent units when applying Newton's second law.

Quick Quiz 4.1	True or false: (a) It is possible to have motion in the absence of a force. (b) If an object is not moving, no external force acts on it.
Quick Quiz 4.2	True or false: (a) If a single force acts on an object, the object accelerates. (b) If an object experiences an acceleration, a force acts on it. (c) If an object experiences no acceleration, no external force acts on it.
Quick Quiz 4.3	True or false: If the net force acting on an object is in the *x* direction, the object moves in the *x* direction.

THE GRAVITATIONAL FORCE

Tip 4.3 USING NEWTON'S SECOND LAW

The procedure for applying Newton's second law is to add *all* of the forces on the object vectorially and then find the resultant acceleration. Do *not* find individual accelerations for each force and then add the accelerations vectorially.

The **gravitational force** is the mutual force of attraction between any two objects in the Universe. It is interesting and rather curious that, although the gravitational force can be very strong between macroscopic objects, it is the weakest of the fundamental forces. For example, the gravitational force between the electron

TABLE 4.1	Units of Mass, Acceleration, and Force		
System	**Mass**	**Acceleration**	**Force**
SI	kg	m/s²	N = kg·m/s²
U.S. customary	slug	ft/s²	lb = slug·ft/s²
Note: 1 N = 0.225 lb.			

4.5 SOME APPLICATIONS OF NEWTON'S LAWS

This section applies Newton's laws to objects moving under the actions of constant external forces. We assume that objects behave as particles, and so we need not worry about rotational motion. We also neglect any friction effects. Finally, we neglect the masses of any ropes or strings involved; in these approximations, the magnitude of the force exerted along a rope (the *tension*) is the same at all points in the rope. This is illustrated by the rope in Figure 4.8, showing forces **T** and **T'** acting on it. If the rope has mass m, then Newton's second law applied to the rope gives $T - T' = ma$. However, if $m = 0$, as it will be in the upcoming examples, then $T = T'$.

When we apply Newton's law to an object, we are interested only in those forces that act *on the object*. For example, in Figure 4.6b, the only external forces acting on the TV are **n** and $\mathbf{F}_g$. The reactions to these forces, **n'** and $\mathbf{F}_g{}'$, act on the table and on Earth, respectively, and do not appear in Newton's second law applied to the TV.

Consider a crate being pulled to the right on a frictionless, horizontal surface, as in Figure 4.9a. Suppose you are asked to find the acceleration of the crate and the force the surface exerts on it. The horizontal force exerted on the crate acts through the rope. The force that the rope exerts on the crate is denoted by **T** (because it is a tension force). The magnitude of **T** is equal to the tension in the rope. What we mean by the words "tension in the rope" is just the force read by a spring scale when the rope in question has been cut and the scale inserted! A dashed circle is drawn around the crate in Figure 4.9a to remind you to isolate the crate from its surroundings.

Because we are interested only in the motion of the crate, we must be able to identify all forces acting on it. These are illustrated in Figure 4.9b. In addition to the force **T**, the force diagram for the crate includes the force of gravity $\mathbf{F}_g$ exerted by Earth and the normal force **n** exerted by the floor. Such a force diagram is referred to as a **free-body diagram.** We call it a "free-body diagram"

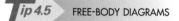

FIGURE 4.8 Newton's second law applied to the rope gives $T - T' = ma$. However, if $m = 0$, then $T = T'$. Thus, the tension in a massless rope is the same at all points in the rope.

*T*ip 4.5   FREE-BODY DIAGRAMS

The *most important* step in solving a problem using Newton's second law is to draw the free-body diagram. Be sure to draw only those forces that act on the object that you are isolating. An incorrect diagram will most likely lead to an incorrect solution.

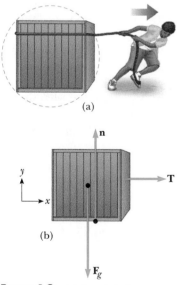

FIGURE 4.9 (a) A crate being pulled to the right on a frictionless surface. (b) The free-body diagram that represents the forces exerted on the crate.

PHYSICS *IN ACTION*

Forces and Motion

On the left is an athlete running at sunset. The external forces acting on the athlete, as described by the blue vectors, are (a) the force $\mathbf{F}_g$ exerted by the Earth (b) the force $\mathbf{F}$ exerted by the ground, and (c) the force $\mathbf{R}$ due to air resistance.

On the right is a multiflash exposure of a golfer striking a golf ball. The force exerted by the golf club on the ball is equal in magnitude and opposite in direction to the force exerted by the ball on the golf club.

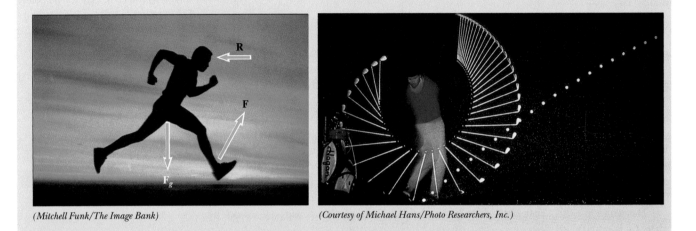

(Mitchell Funk/The Image Bank)   *(Courtesy of Michael Hans/Photo Researchers, Inc.)*

because the environment is replaced by a series of forces on an otherwise free body. The construction of a correct free-body diagram is an essential step in applying Newton's laws; *its importance cannot be overemphasized.*

The *reactions* to the forces we have listed—namely, the force exerted by the rope on the hand doing the pulling, the force exerted by the crate on Earth, and the force exerted by the crate on the floor—are not included in the free-body diagram because they act on other objects and not on the crate. Consequently, they do not determine the crate's motion.

Now let us apply Newton's second law to the crate. First we must choose an appropriate coordinate system. In this case it is convenient to use the one shown in Figure 4.9b, with the x axis horizontal and the y axis vertical. We can apply Newton's second law in the x direction, y direction, or both, depending on what we are asked to find in a problem. In addition, we may be able to use the equations of motion for constant acceleration that are found in Chapter 2. However, you should use these equations *only when the acceleration is constant.*

OBJECTS IN EQUILIBRIUM

Objects that are either at rest or moving with constant velocity are said to be in *equilibrium.* In equation form, because $\mathbf{a} = 0$, this condition of equilibrium can be expressed as

$$\sum \mathbf{F} = 0 \qquad\qquad [4.9]$$

This statement signifies that the *vector* sum of all the forces (the net force) acting on an object in equilibrium is zero.

Usually, equilibrium problems are solved more easily if we work with Equation 4.9 in terms of the components of the external forces acting on an object. By this

Webnote 4.3

For more on free-body diagrams including a tutorial and a self-test, see:
http://eta.physics.uoguelph.ca/ tutorials/fbd/Q.fbd.html

we mean that, in a two-dimensional problem, the sum of all the external forces in the x and y directions must separately equal zero; that is,

$$\sum F_x = 0 \quad \text{and} \quad \sum F_y = 0 \qquad \text{[4.10]}$$

We shall not consider three-dimensional problems in this book, but the extension of Equation 4.10 to a three-dimensional situation can be made by adding a third equation, $\sum F_z = 0$.

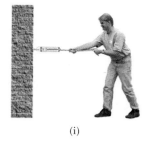

> ## Tip 4.6　A PARTICLE IN EQUILIBRIUM
>
> Zero net force on a particle does *not* mean that the particle is not moving. It means that the particle is not *accelerating*. If the particle has an initial velocity and experiences zero net force, it continues to move with the same velocity.

> **Quick Quiz 4.7**　Consider the two situations shown in Figure 4.10, in which there is no acceleration. In both cases, all individuals pull with a force of magnitude F. Is the reading on the scale in part (i) of the figure (a) greater than, (b) less than, or (c) equal to the reading in part (ii)?

PROBLEM-SOLVING *STRATEGY*　Objects in Equilibrium

The following procedure is recommended for problems involving objects in equilibrium:

1. Make a sketch of the situation described in the problem statement.
2. Draw a free-body diagram for the *isolated* object under consideration, and label all forces acting on the object.
3. Resolve all forces into x and y components, choosing a convenient coordinate system.
4. Use the equations $\sum F_x = 0$ and $\sum F_y = 0$. Keep track of the signs of the various force components.
5. Application of Step 4 leads to a set of equations with several unknowns. Solve the simultaneous equations for the unknowns in terms of the known quantities.

(i)

(ii)

FIGURE 4.10　(Quick Quiz 4.7) (i) A person pulls with a force of magnitude F on a spring scale attached to a wall. (ii) Two people pull with forces of magnitude F in opposite directions on a spring scale attached between two ropes.

Example 4.1　A Traffic Light at Rest

A traffic light weighing 100 N hangs from a vertical cable tied to two other cables that are fastened to a support, as in Figure 4.11a. The upper cables make angles of 37.0° and 53.0° with the horizontal. Find the tension in each of the three cables.

Reasoning　We must construct two free-body diagrams. The first is for the traffic light, shown in Figure 4.11b; the second is for the knot that holds the three cables together (Fig. 4.11c). The knot is a convenient point to choose because all forces in question act on this point.

Solution　From Figure 4.11b, we see that $\sum F_y = 0$ yields $T_3 - F_g = 0$, or $T_3 = F_g = 100$ N. Next, we choose the coordinate axes shown in Figure 4.10c and resolve all forces into their x and y components:

Force	x component	y component
$\mathbf{T}_1$	$-T_1 \cos 37.0°$	$T_1 \sin 37.0°$
$\mathbf{T}_2$	$T_2 \cos 53.0°$	$T_2 \sin 53.0°$
$\mathbf{T}_3$	0	-100 N

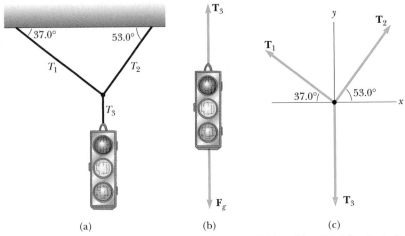

FIGURE 4.11 (Example 4.1) (a) A traffic light suspended by cables. (b) A free-body diagram for the traffic light. (c) A free-body diagram for the knot joining the cables.

The condition for equilibrium gives us the equations

$$\sum F_x = T_2 \cos 53.0° - T_1 \cos 37.0° = 0 \tag{1}$$

$$\sum F_y = T_1 \sin 37.0° + T_2 \sin 53.0° - 100\text{ N} = 0 \tag{2}$$

(In this case, equilibrium means a motionless light.) From (1) we see that the horizontal components of $\mathbf{T}_1$ and $\mathbf{T}_2$ must be equal in magnitude, and from (2) we see that the sum of the vertical components of $\mathbf{T}_1$ and $\mathbf{T}_2$ must balance the force of gravity acting on the light. We can solve (1) for T_2 in terms of T_1 to give

$$T_2 = T_1\left(\frac{\cos 37.0°}{\cos 53.0°}\right) = T_1\left(\frac{0.799}{0.602}\right) = 1.33T_1$$

This value for T_2 can be substituted into (2) to give

$$T_1 \sin 37.0° + (1.33T_1)(\sin 53.0°) - 100\text{ N} = 0$$

$$T_1 = 60.1\text{ N}$$

$$T_2 = 1.33T_1 = 1.33(60.0\text{ N}) = 79.9\text{ N}$$

EXERCISE When will $T_1 = T_2$?

ANSWER When the supporting cables make equal angles with the horizontal support.

EXERCISE Solve this problem using the rules of vector addition and the law of cosines and sines found in Appendix A.4.

Example 4.2 Sled on a Frictionless Hill

A child holds a sled at rest on a frictionless, snow-covered hill, as shown in Figure 4.12a. If the sled weighs 77.0 N, find the force $\mathbf{T}$ exerted by the rope on the sled and the force $\mathbf{n}$ exerted by the hill on the sled.

Reasoning Figure 4.12b shows the forces acting on the sled and a convenient coordinate system to use for this type of problem. Note that $\mathbf{n}$, the force exerted by the hill on the sled, is perpendicular (normal) to the hill. The hill can exert a component of force along the incline only if there is friction between the sled and the hill. Because the sled is at rest, we are able to apply the condition for equilibrium as $\Sigma F_x = 0$ and $\Sigma F_y = 0$.

Solution First, we replace the force of gravity $\mathbf{F}_g$ acting on the sled with its x and y components. The x component has magnitude $mg\sin\theta = (77.0\text{ N})(\sin 30.0°)$

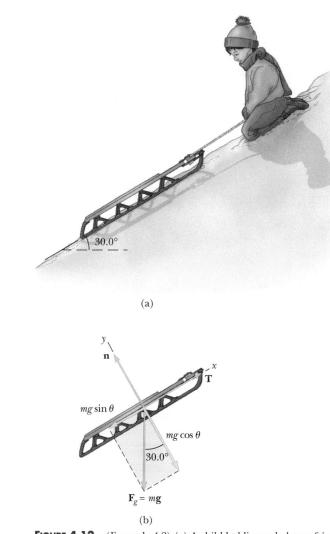

(a)

(b)

FIGURE 4.12　(Example 4.2) (a) A child holding a sled on a frictionless hill. (b) A free-body diagram for the sled.

and points along the negative *x* axis. The *y* component has magnitude
$mg \cos \theta = (77.0 \text{ N})(\cos 30.0°)$ and points in the negative *y* direction. Applying the condition for equilibrium to the sled, we find that

$$\sum F_x = T - (77.0 \text{ N})(\sin 30.0°) = 0$$

$$T = 38.5 \text{ N}$$

$$\sum F_y = n - (77.0 \text{ N})(\cos 30.0°) = 0$$

$$n = 66.7 \text{ N}$$

Note that *n* is less than the weight of the sled in this situation. This is so because the sled is on an incline and **n** is equal in magnitude and opposite in direction to the component of the force of gravity perpendicular to the incline.

EXERCISE　What happens to the normal force as the angle of incline increases?

ANSWER　It decreases.

EXERCISE　When is the magnitude of the normal force equal to the weight of the sled?

ANSWER　When the sled is on a horizontal surface and the force exerted by the rope on the sled is either zero or along the horizontal.

Quick Quiz 4.8

Consider the child being pulled on the toboggan in Figure 4.13. Is the magnitude of the normal force exerted by the ground on the toboggan (a) equal to the total weight of the child plus the toboggan, (b) greater than the total weight, (c) less than the total weight, or (d) possibly greater than or less than the total weight depending on the size of the weight relative to the tension in the rope?

FIGURE 4.13 (Quick Quiz 4.8)

ACCELERATING OBJECTS AND NEWTON'S SECOND LAW

In a situation in which a net force acts on an object, the object accelerates, and we use Newton's second law in order to analyze the motion. The examples and suggestions that follow should help you solve problems of this kind.

PROBLEM-SOLVING STRATEGY   **Newton's Second Law**

The following procedure is recommended for working problems that involve the application of Newton's second law:

1. Draw a diagram of the system.
2. Isolate the object of interest whose motion is being analyzed. Draw a free-body diagram for this object, showing all external forces acting on it. For systems containing more than one object, draw a *separate* free-body diagram for each object.
3. Establish convenient coordinate axes for each object, and find the components of the forces along these axes. It is convenient to choose a coordinate system where one of the axes is parallel to the motion of the objects. Apply Newton's second law in the *x* and *y* directions for each object.
4. Solve the component equations for the unknowns. Remember that, in order to obtain a complete solution, you must have as many independent equations as you have unknowns.
5. If necessary, use the equations of kinematics (motion with constant acceleration) from Chapter 2 to find all the unknowns.

n

$F = 20.0$ N

$F_g = w = 300$ N

FIGURE 4.14 (Example 4.3)

Example 4.3   Moving a Crate

The combined weight of the crate and dolly in Figure 4.14 is 300 N. If the person pulls on the rope with a constant force of 20.0 N, what is the acceleration of the system (crate plus dolly), and how far will it move in 2.00 s? Assume that the system starts from rest and that there are no friction forces opposing its motion.

Reasoning   We can find the acceleration of the system from Newton's second law. Because the force exerted on the system is constant, its acceleration is constant. Therefore, we can apply the equations of motion with constant acceleration to find the distance traveled in 2.00 s.

Solution   In order to apply Newton's second law to the system, we must first know the system's mass:

$$m = \frac{w}{g} = \frac{300 \text{ N}}{9.80 \text{ m/s}^2} = 30.6 \text{ kg}$$

Now we can find the acceleration of the system from the second law:

$$a_x = \frac{F_x}{m} = \frac{20.0 \text{ N}}{30.6 \text{ kg}} = \boxed{0.654 \text{ m/s}^2}$$

Because the acceleration is constant, we can find the distance the system moves in 2.00 s using the relation $x = v_0 t + \frac{1}{2}at^2$ with $v_0 = 0$:

$$x = \tfrac{1}{2}at^2 = \tfrac{1}{2}(0.654 \text{ m/s}^2)(2.00 \text{ s})^2 = \boxed{1.31 \text{ m}}$$

It is important to note that the constant applied force of 20.0 N is assumed to act on the system at all times during its motion. If the force were removed at some instant, the system would continue to move with constant velocity and hence zero acceleration.

Example 4.4   The Run-Away Car

A car of mass m is on an icy driveway inclined at an angle $\theta = 20.0°$, as in Figure 4.15a. Determine the acceleration of the car, assuming the incline is frictionless.

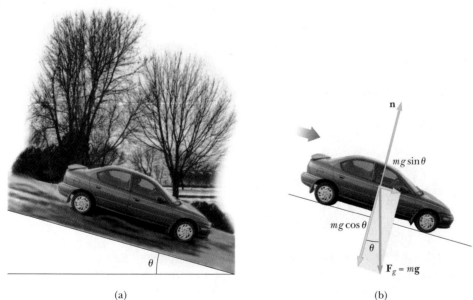

(a)   (b)

FIGURE 4.15   (Example 4.4)

Reasoning   The free-body diagram for the car is shown in Figure 4.15b. The only forces exerted on the car are the normal force **n** acting perpendicular to the driveway surface and the force of gravity $\mathbf{F}_g$ acting vertically downward. It is convenient to choose the coordinate axes with the x axis along the incline and the y axis perpendicular to it. Then we replace the force of gravity with a component of magnitude $mg \sin \theta$ along the positive x axis and a component of magnitude $mg \cos \theta$ along the negative y axis.

Solution   Applying Newton's second law in component form, with $a_y = 0$, gives

$$\sum F_x = mg \sin \theta = ma_x \tag{1}$$

$$\sum F_y = n - mg \cos \theta = 0 \tag{2}$$

From (1) we see that the acceleration of the car along the driveway is provided by the component of the force of gravity directed down the incline:

$$a_x = g \sin \theta \tag{3}$$

Note that the acceleration given by (3) is *constant* and *independent of the mass* of the car—it depends only on the angle of inclination and on *g*. In our example, $\theta = 20.0°$, and so we find that

$$a_x = \boxed{3.35 \text{ m/s}^2}$$

EXERCISE   A compact car and a large luxury sedan are at rest at the top of an ice-covered (frictionless) driveway. If they both slide down the driveway, which reaches the bottom first?

ANSWER   Because the acceleration is independent of the mass and thus is the same for both vehicles, they arrive at the bottom simultaneously.

EXERCISE   If the length of the driveway is 25.0 m and a car starts from rest at the top, how long does it take to travel to the bottom? What is the car's speed at the bottom?

ANSWER   3.86 s; 12.9 m/s

Example 4.5   Weighing a Fish in an Elevator

A person weighs a fish of mass *m* on a spring scale attached to the ceiling of an elevator, as shown in Figure 4.16. Show that if the elevator accelerates in either direction, the spring scale gives a reading different from the true weight of the fish.

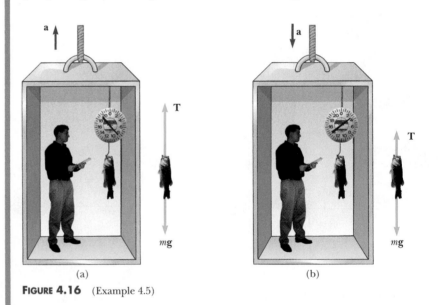

(a)   (b)

FIGURE 4.16   (Example 4.5)

Solution   The forces acting on the fish are the downward force of gravity $\mathbf{F}_g$ and the upward force $\mathbf{T}$ exerted by the spring scale. By Newton's third law, the tension T is also the reading of the scale. If the elevator is either at rest or moving at constant velocity, the fish is not accelerating, so $\Sigma F_y = T - F_g = 0$ or $T = F_g = mg$.

If the elevator moves with an acceleration $\mathbf{a}$, then Newton's second law applied to the fish gives the net force exerted on the fish:

$$\sum F_y = T - mg = ma_y \tag{1}$$

where we have chosen upward as the positive direction. Thus, we conclude from (1) that the scale reading T is *greater* than the weight *mg* if $\mathbf{a}$ is *upward*, so that a_y is positive as in

Figure 4.16a. The reading is *less* than *mg* if **a** is *downward,* so that a_y is negative as in Figure 4.16b.

For example, if the weight of the fish is 40.0 N and **a** is *upward,* so that $a_y = +2.00 \text{ m/s}^2$, then the scale reading from (1) is

$$T = ma_y + mg = mg\left(\frac{a_y}{g} + 1\right) \tag{2}$$

$$= (40.0 \text{ N})\left(\frac{+2.00 \text{ m/s}^2}{9.80 \text{ m/s}^2} + 1\right) = \boxed{48.2 \text{ N}}$$

If **a** is *downward,* so that $a_y = -2.00 \text{ m/s}^2$, then (2) gives us

$$T = (40.0 \text{ N})\left(\frac{-2.00 \text{ m/s}^2}{9.80 \text{ m/s}^2} + 1\right) = \boxed{31.8 \text{ N}}$$

Hence, if you buy a fish by weight in an elevator, make sure the fish is weighed while the elevator is either at rest or else accelerating downward!

Special Cases   If the elevator cable breaks, the elevator falls freely and $a_y = -g$. We see from (2) that the scale reading T is zero in this case; that is, the fish appears to be weightless. If the elevator accelerates downward with an acceleration greater than g, the fish (along with the person in the elevator) eventually hits the ceiling because the acceleration of fish and person is still that of a freely falling object.

4.6   FORCES OF FRICTION

When an object is in motion either on a surface or through a viscous medium, such as air or water, there is resistance to the motion because the object interacts with its surroundings. We call such resistance a **force of friction.** Forces of friction are very important in our everyday lives. They allow us to walk or run and are necessary for the motion of wheeled vehicles.

Imagine you are working in your garden and have filled a plastic trash can with yard clippings. You then try to drag the trash can across the surface of your concrete patio. If you apply an external horizontal force **F** to the trash can, acting to the right, as shown in Figure 4.17a, the trash can remains stationary if **F** is small. The force that counteracts **F** and keeps the trash can from moving acts to the left and is called the **force of static friction f_s.** As long as the trash can is not moving, $f_s = F$. Thus, if **F** is increased, f_s also increases. Likewise, if **F** decreases, f_s also decreases. Experiments show that the friction force arises from the nature of the two surfaces: Because of their roughness, contact is made only at a few points, as shown in the magnified view of the surfaces in Figure 4.17a.

If we increase the magnitude of **F**, as in Figure 4.17b, the trash can eventually slips. When the trash can is on the verge of slipping, f_s is a maximum as shown in Figure 4.17c. When F exceeds $f_{s, \max}$, the trash can moves and accelerates to the right. When the trash can is in motion, the friction force is less than $f_{s, \max}$ (Fig. 4.17c). We call the friction force for an object in motion the **force of kinetic friction f_k.** The net force $F - f_k$ in the x direction produces an acceleration to the right, according to Newton's second law. If $F = f_k$, the acceleration is zero, and the trash can moves to the right with constant speed. If the applied force is removed, the friction force acting to the left provides an acceleration of the trash can in the $-x$ direction and eventually brings it to rest, again consistent with Newton's second law.

Experimentally, we find that, to a good approximation, both $f_{s, \max}$ and f_k for an object on a surface are proportional to the normal force exerted by the surface on the object. Thus, the experimental observations can be summarized as follows:

FIGURE 4.17 (a) The force of friction $\mathbf{f}_s$ exerted by a concrete surface on a trash can is directed opposite the force $\mathbf{F}$ that you exert on the can. As long as the trash can is not moving, the magnitude of the force of static friction equals that of the applied force $\mathbf{F}$. (b) When the magnitude of $\mathbf{F}$ exceeds the magnitude of $\mathbf{f}_k$, the force of kinetic friction, the trash can accelerates to the right. (c) A graph of the magnitude of the friction force versus that of the applied force. Note that $f_{s,\,max} > f_k$.

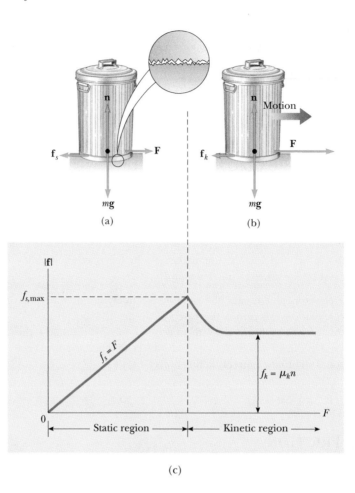

(a)

(b)

(c)

- The magnitude of the force of static friction between any two surfaces in contact can have the values

$$f_s \leq \mu_s n \qquad \text{[4.11]}$$

where the dimensionless constant μ_s is called the **coefficient of static friction** and n is the magnitude of the normal force exerted by one surface on the other. Equation 4.11 holds when an object is on the verge of slipping, that is, when $f_s = f_{s,\,max} \equiv \mu_s n$. This situation is called *impending motion*. The inequality holds when the component of the applied force parallel to the surfaces is less than this value.

- The magnitude of the force of kinetic friction acting between two surfaces is

$$f_k = \mu_k n \qquad \text{[4.12]}$$

where μ_k is the **coefficient of kinetic friction.**

- The values of μ_k and μ_s depend on the nature of the surfaces, but μ_k is generally less than μ_s. Table 4.2 lists some reported values.

- The direction of the friction force exerted by a surface on an object is opposite the actual motion (kinetic friction) or the impending motion (static friction) of the object relative to the surface.

- The coefficients of friction are nearly independent of the area of contact between the surfaces.

 USE THE EQUAL SIGN
IN LIMITED SITUATIONS

In Equation 4.11, the equal sign is used *only* when the surfaces are just about to break free and begin sliding. Do not fall into the common trap of using $f_s = \mu_s n$ in *any* static situation.

Although the coefficient of kinetic friction varies with speed, we shall neglect any such variations. The approximate nature of Equations 4.11 and 4.12 is easily demonstrated by trying to get an object to slide down an incline at constant speed. Especially at low speeds, the motion is likely to be characterized by alternate stick and slip episodes.

TABLE 4.2	Coefficients of Friction[a]	
	μ_s	μ_k
Steel on steel	0.74	0.57
Aluminum on steel	0.61	0.47
Copper on steel	0.53	0.36
Rubber on concrete	1.0	0.8
Wood on wood	0.25–0.5	0.2
Glass on glass	0.94	0.4
Waxed wood on wet snow	0.14	0.1
Waxed wood on dry snow	—	0.04
Metal on metal (lubricated)	0.15	0.06
Ice on ice	0.1	0.03
Teflon on Teflon	0.04	0.04
Synovial joints in humans	0.01	0.003

[a] All values are approximate.

Quick Quiz 4.9 You press your physics textbook flat against a vertical wall with your hand. What is the direction of the friction force exerted by the wall on the book? (a) downward (b) upward (c) out from the wall (d) into the wall.

Quick Quiz 4.10 A crate is sitting in the center of a flatbed truck. The truck accelerates toward the east, and the crate moves with it, not sliding on the bed of the truck. What is the direction of the friction force exerted by the bed of the truck on the crate? (a) To the west. (b) To the east. (c) There is no friction force because the crate is not sliding.

Quick Quiz 4.11 You are playing with your younger sister in the snow. She is sitting on a sled and asks you to move her across a flat, horizontal field. You have a choice of (a) pushing her from behind by applying a force downward on her shoulders at 30° below the horizontal (Fig. 4.18a), or (b) attaching a rope to the front of the sled and pulling with a force at 30° above the horizontal (Fig 4.18b). Which would be easier for you and why?

(a)   (b)

FIGURE 4.18 (Quick Quiz 4.11)

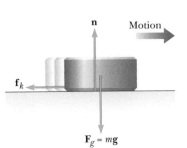

FIGURE 4.19   (Example 4.6) *After the puck is given an initial velocity to the right, the external forces acting on it are the force of gravity* $\mathbf{F}_g$, *the normal force* $\mathbf{n}$, *and the force of kinetic friction* $\mathbf{f}_k$.

Example 4.6   The Sliding Hockey Puck

The hockey puck in Figure 4.19 is given an initial speed of 20.0 m/s on a frozen pond. The puck remains on the ice and slides 120 m before coming to rest. Determine the coefficient of kinetic friction between puck and ice.

Reasoning   The puck slides to rest with a constant acceleration along the horizontal. Thus, we can use the kinematic equation $v^2 = v_0^2 + 2a\Delta x$ to find a. Newton's second law applied in the horizontal direction is $-f_k = -\mu_k n = ma$. To find μ_k, we first find the normal force $\mathbf{n}$ by applying $\Sigma F_y = 0$ in the vertical direction.

Solution   With the final speed $v = 0$; the initial speed $v_0 = 20.0$ m/s; and the displacement, $\Delta x = 120$ m:

$$v^2 = v_0^2 + 2a\Delta x$$

$$a = \frac{v^2 - v_0^2}{2\Delta x}$$

$$= \frac{0 - (20.0 \text{ m/s})^2}{2(120 \text{ m})} = -1.67 \text{ m/s}^2$$

The negative sign means that the acceleration is to the left in Figure 4.19, *opposite* the direction of the velocity.

The magnitude of the force of kinetic friction is found from $f_k = \mu_k n$, and n is found from $\Sigma F_y = 0$ as follows:

$$\sum F_y = n - F_g = 0$$

$$n = F_g = mg$$

Thus,

$$f_k = \mu_k n = \mu_k mg$$

Now we apply Newton's second law along the horizontal direction, taking the positive direction toward the right:

$$\sum F_x = -f_k = ma$$

$$-\mu_k mg = ma$$

$$\mu_k = -\frac{a}{g} = \frac{1.67 \text{ m/s}^2}{9.80 \text{ m/s}^2} = \boxed{0.170}$$

Example 4.7   Connected Objects

Two objects are connected by a light string that passes over a frictionless pulley, as in Figure 4.20a. The coefficient of kinetic friction between the cube and the surface is 0.300. Find the acceleration of the two objects and the tension in the string.

Reasoning   Connected objects are best handled by applying Newton's second law separately to each. The free-body diagrams for the cube and the ball are shown in Figure 4.20b. Assuming the string connecting the two objects does not stretch, the magnitude of the acceleration for both objects has the same value, a. We shall obtain two equations involving the unknowns T and a that can be solved simultaneously.

Solution   With the positive x direction to the right and the positive y direction upward, Newton's second law applied to the cube of mass m_1 gives

$$\sum F_x = T - f_k = m_1 a$$

$$\sum F_y = n - m_1 g = 0$$

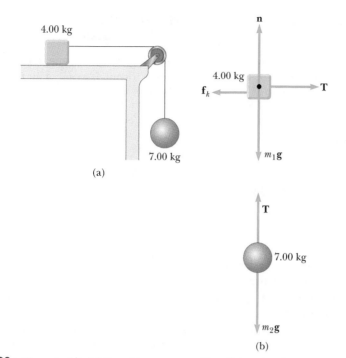

FIGURE 4.20   (Example 4.7) (a) Two objects connected by a light string that passes over a frictionless pulley. (b) Free-body diagrams for the objects.

Because $f_k = \mu_k n$ and $n = m_1 g$, we have $f_k = \mu_k m_1 g$ and the first equation above becomes

$$T - \mu_k m_1 g = m_1 a \qquad (1)$$

Now we apply Newton's second law to the ball, which is moving in the vertical direction and has mass m_2. Because the ball moves downward when the cube moves in the positive x direction (to the right), we define down as the positive direction for the ball:

$$\sum F_y = m_2 g - T = m_2 a \qquad (2)$$

Adding (1) and (2) eliminates T, leaving a single equation for a:

$$m_2 g - \mu_k m_1 g = (m_1 + m_2) a$$

$$a = \frac{m_2 g - \mu_k m_1 g}{m_1 + m_2}$$

Substituting the known values into this equation gives

$$a = \frac{(7.00\ \text{kg})(9.80\ \text{m/s}^2) - (0.300)(4.00\ \text{kg})(9.80\ \text{m/s}^2)}{(4.00\ \text{kg} + 7.00\ \text{kg})} = \boxed{5.17\ \text{m/s}^2}$$

When this value for the acceleration is substituted into (1), we get

$$T = \boxed{32.4\ \text{N}}$$

It is interesting that the result for a can be more directly obtained (and checked) by considering the two connected objects as a single system. In this approach, the total system mass is $m_1 + m_2$ because both objects are accelerated; the net external force exerted by the outside world on the two-object system is the difference between the gravitational force on m_2 and the friction force retarding m_1. That is, $\sum F_{ext} = m_2 g - \mu_k m_1 g = (m_1 + m_2) a$. Thus

$$a = \frac{m_2 g - \mu_k m_1 g}{m_1 + m_2}$$

Unfortunately the single system approach gives no information about **internal forces,**

Webnote 4.4

Pulleys such as the one in Example 4.7 are involved in several end-of-chapter problems. Practical pulley configurations are often more complicated than those shown in this book. If you are interested, you can study pulley support systems thoroughly at:

http://www.howstuffworks.com/pulley.htm

which are forces between adjacent parts of a system. Here tension is an internal force that does not contribute to the net external force on the system.

EXERCISE If the coefficient of kinetic friction were very large, the ball and cube could move with constant velocity. What coefficient of friction is required for this situation to exist?

ANSWER 1.75

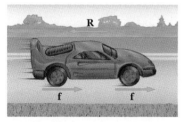

FIGURE 4.21 The horizontal forces acting on the car are the *forward* forces **f** exerted by the road on each tire and the force of air resistance **R**, which acts *opposite* the car's velocity. (The car's tires exert a rearward force on the road, not shown in the diagram.)

FRICTION AND THE MOTION OF A CAR

Forces of friction are important in the analysis of the motion of cars and other wheeled vehicles. There are several types of friction forces to consider, the main ones being the force of friction between tires and road surface and the retarding force produced by air resistance.

Assuming the car is a four-wheel-drive vehicle of mass m, as each wheel turns to propel the car forward, the tire exerts a rearward force on the road. The reaction to this rearward force is a forward force **f** exerted by the road on the tire (Fig. 4.21). If we assume that the same forward force **f** is exerted on each tire, the net forward force on the car is 4**f**, and the car's acceleration is therefore $\mathbf{a} = 4\mathbf{f}/m$.

When the car is in motion, we must also consider the force of air resistance **R**, which acts in the direction opposite the velocity of the car. The net force exerted on the car is therefore $4\mathbf{f} - \mathbf{R}$, and so the car's acceleration is $\mathbf{a} = (4\mathbf{f} - \mathbf{R})/m$. At

APPLICATION ⟩

SKYDIVING

APPLYING PHYSICS 4.2

Consider a sky diver falling through air before reaching her terminal speed, as in Figure 4.22. As her speed increases, what happens to her acceleration? What is her acceleration once she reaches terminal speed?

Explanation The forces exerted on the skydiver are the downward force of gravity $m\mathbf{g}$ and an upward force of air resistance **R**. Before she reaches terminal speed, the magnitude of **R** is less than her weight. As her downward speed increases, the force of air resistance increases. The vector sum of the force of gravity and the force of air resistance gives a total force that decreases with time, so her acceleration decreases. Once the two forces balance each other so that the net force is zero, the acceleration is consequently zero, and she has reached terminal speed.

FIGURE 4.22 (Applying Physics 4.2) *(Guy Sauvage, Photo Researchers, Inc.)*

normal driving speeds, the magnitude of **R** is proportional to the first power of the speed. That is, $R = bv$, where b is a constant. Thus, the force of air resistance increases with increasing speed. When R is equal to $4f$, the acceleration is zero, and the car moves at a constant speed.

A similar situation occurs when an object falls through air. When the upward force exerted by the air on the object balances the downward force of gravity exerted on the object, the net force on the object is zero and hence its acceleration is zero. Once this condition is reached, the object continues to move downward with some constant maximum speed called the **terminal speed.**

SUMMARY

Newton's first law states that, if the net force exerted on an object is zero, an object at rest remains at rest and an object in motion continues in motion with a constant velocity.

The tendency for an object to maintain its original state of motion is called **inertia. Mass** is the physical quantity that measures the resistance of an object to changes in its velocity.

Newton's second law states that the acceleration of an object is directly proportional to the net force acting on it and inversely proportional to its mass. In equation form, this is expressed as:

$$\sum \mathbf{F} = m\mathbf{a} \qquad \text{[4.1]}$$

The magnitude of the force of gravity exerted on an object is called the **weight** of an object, denoted by w. The weight of an object of mass m is equal to the product mg:

$$w = mg \qquad \text{[4.6]}$$

Newton's third law states that, if two objects interact, the force $\mathbf{F}_{12}$ exerted by object 1 on object 2 is equal in magnitude and opposite in direction to the force $\mathbf{F}_{21}$ exerted by object 2 on object 1. Thus, an isolated force can never occur in nature.

An **object in equilibrium** has no net external force acting on it, and the second law, in component form, implies that $\Sigma F_x = 0$ and $\Sigma F_y = 0$.

The magnitude of the maximum force of static friction, $f_{s,\,max}$, between an object and a surface is proportional to the magnitude of the normal force acting on the object. This maximum force occurs when the object is on the verge of slipping. In general,

$$f_s \le \mu_s n \qquad \text{[4.11]}$$

where μ_s is the **coefficient of static friction.** When an object slides over a surface, the direction of the force of kinetic friction on the object $\mathbf{f}_k$ is opposite the direction of the motion of the object relative to the surface, and the magnitude is proportional to that of the normal force. The magnitude of $\mathbf{f}_k$ is

$$f_k = \mu_k n \qquad \text{[4.12]}$$

where μ_k is the **coefficient of kinetic friction.** In general, $\mu_k < \mu_s$.

CONCEPTUAL QUESTIONS

1. A ball is held in a person's hand. (a) Identify all the external forces acting on the ball and the reaction to each. (b) If the ball is dropped, what force is exerted on it while it is falling? Identify the reaction force in this case. (Neglect air resistance.)

2. If a car is traveling westward with a constant speed of 20 m/s, what is the resultant force acting on it?

3. If a car moves with a constant acceleration, can you conclude there are no forces acting on it?

4. A rubber ball is dropped onto the floor. What force causes the ball to bounce?

5. If you push on a heavy box that is at rest, you must exert some force to start its motion. However, once the box is sliding, you can apply a smaller force to maintain that motion. Why?

6. If gold were sold by weight, would you rather buy it in Denver or in Death Valley? If it were sold by mass, in which of the two locations would you prefer to buy it? Why?

7. A passenger sitting in the rear of a bus claims that she was injured as the driver slammed on the brakes, causing a suitcase to come flying toward her from the front of the

bus. If you were the judge in this case, what disposition would you make? Why?

8. A space explorer is moving through space far from any planet or star. She notices a large rock, taken as a specimen from an alien planet, floating around the cabin of the ship. Should she push it gently or kick it toward the storage compartment? Why?

9. What force causes an automobile to move? A propeller-driven airplane? A rowboat?

10. Analyze the motion of a rock dropped in water in terms of its speed and acceleration as it falls. Assume that a resistive force is acting on the rock that increases as the velocity increases.

11. In the motion picture *It Happened One Night* (Columbia Pictures, 1934), Clark Gable is standing inside a stationary bus in front of Claudette Colbert, who is seated. The bus suddenly starts moving forward and Clark falls into Claudette's lap. Why did this happen?

12. A weightlifter stands on a bathroom scale. He pumps a barbell up and down. What happens to the reading on the bathroom scale as this is done? Suppose he is strong enough to actually *throw* the barbell upward. How does the reading on the scale vary now?

13. In a tug-of-war between two athletes, each pulls on the rope with a force of 200 N. What is the tension in the rope? If the rope does not move, what horizontal force does each athlete exert against the ground?

14. As a rocket is fired from a launching pad, its speed and ac-celeration increase with time as its engines continue to operate. Explain why this occurs even though the thrust of the engines remains constant.

15. Identify the action-reaction pairs in the following situations: (a) A man takes a step, (b) a snowball hits a girl in the back, (c) a baseball player catches a ball, (d) a gust of wind strikes a window.

16. The driver of a speeding empty truck slams on the brakes and skids to a stop through a distance *d.* (a) If the truck carried a load that doubled its mass, what would be the truck's "skidding distance"? (b) If the initial speed of the truck were halved, what would be the truck's "skidding distance"?

17. Suppose you are driving a car at a high speed. Why should you avoid "slamming on" your brakes when you want to stop in the shortest possible distance? (Newer cars have antilock brakes that avoid this problem.)

18. A truck loaded with sand accelerates along a highway. If the driving force on the truck remains constant, what happens to the truck's acceleration if its trailer leaks sand at a constant rate through a hole in its bottom?

19. A large crate is placed on the bed of a truck but not tied down. (a) As the truck accelerates forward, the crate remains at rest relative to the truck. What force causes the crate to accelerate forward? (b) If the driver slams on the brakes, what could happen to the crate?

20. Describe a few examples in which the force of friction exerted on an object is in the direction of motion of the object.

PROBLEMS

1, 2, 3 = straightforward, intermediate, challenging   ☐ = full solution available in Student Solutions Manual/Study Guide

web = solution posted at **http://info.brookscole.com/serway**   🔬 = biomedical application

Section 4.1   The Concept of Force

Section 4.2   Newton's First Law

Section 4.3   Newton's Second Law

Section 4.4   Newton's Third Law

1. A 6.0-kg object undergoes an acceleration of 2.0 m/s^2. (a) What is the magnitude of the resultant force acting on it? (b) If this same force is applied to a 4.0-kg object, what acceleration is produced?

2. A football punter accelerates a football from rest to a speed of 10 m/s during the time in which his toe is in contact with the ball (about 0.20 s). If the football has a mass of 0.50 kg, what average force does the punter exert on the ball?

3. The heaviest invertebrate is the giant squid, which is estimated to have a weight of about 2 tons spread out over its length of 70 feet. What is its weight in newtons?

4. The heaviest flying bird is the trumpeter swan, which weighs in at about 38 pounds at its heaviest. What is its weight in newtons?

5. A bag of sugar weighs 5.00 lb on Earth. What should it weigh in newtons on the Moon, where the free-fall acceleration is $\frac{1}{6}$ that on Earth? Repeat for Jupiter, where *g* is 2.64 times that on Earth. Find the mass of the bag of sugar in kilograms at each of the three locations.

6. A freight train has a mass of 1.5×10^7 kg. If the locomotive can exert a constant pull of 7.5×10^5 N, how long does it take to increase the speed of the train from rest to 80 km/h?

7. The air exerts a forward force of 10 N on the propeller of a 0.20-kg model airplane. If the plane accelerates forward at 2.0 m/s^2, what is the magnitude of the resistive force exerted by the air on the airplane?

8. A 5.0-g bullet leaves the muzzle of a rifle with a speed of 320 m/s. What total force (assumed constant) is exerted on the bullet while it is traveling down the 0.82-m-long barrel of the rifle?

9. A performer in a circus is fired from a cannon as a "human cannonball" and leaves the cannon with a speed of 18.0 m/s. The performer's mass is 80.0 kg. The cannon barrel is 9.20 m long. Find the average net force exerted on the performer while he is being accelerated inside the cannon.

10. To lift a patient, four nurses grip the sheet on which the patient is lying and lift upward. If each nurse exerts an upward force of 240 N and the patient has an upward acceleration of 0.504 m/s^2, what is the weight of the patient?

11. A boat moves through the water with two forces acting on it. One is a 2000-N forward push by the water on the propeller, and the other is an 1800-N resistive force due to the water around the bow. (a) What is the acceleration of

speed for 5.0 s, undergoes a uniform *negative* acceleration for 1.5 s, and comes to rest. What does the spring scale register (a) before the elevator starts to move? (b) during the first 0.80 s? (c) while the elevator is traveling at constant speed? (d) during the negative acceleration?

76. A sled weighing 60.0 N is pulled horizontally across snow so that the coefficient of kinetic friction between sled and snow is 0.100. A penguin weighing 70.0 N rides on the sled, as in Fig. P4.76. If the coefficient of static friction between penguin and sled is 0.700, find the maximum horizontal force that can be exerted on the sled before the penguin begins to slide off.

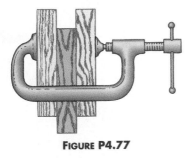

FIGURE P4.76

77. The board sandwiched between two other boards in Figure P4.77 weighs 95.5 N. If the coefficient of friction between the boards is 0.663, what must be the magnitude of the compression forces (assume horizontal) acting on both sides of the center board to keep it from slipping?

FIGURE P4.77

78. Determine the magnitude of the net force exerted by the cable on the leg in Figure P4.78.

79. An inventive child named Chris wants to reach an apple in a tree without climbing the tree. Sitting in a chair connected to a rope that passes over a frictionless pulley (Fig. P4.79), Chris pulls on the loose end of the rope with such a force that the spring scale reads 250 N. Chris's true weight is 320 N, and the chair weighs 160 N. (a) Show that the acceleration of the system is *upward* and find its magnitude. (b) Find the force Chris exerts on the chair.

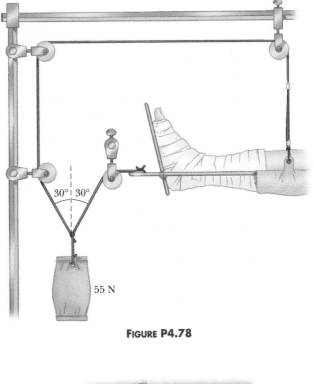

FIGURE P4.78

FIGURE P4.79

80. A fire helicopter carries a 620-kg bucket of water at the end of a 20.0-m long cable. When the helicopter is flying back from a fire at a constant speed of 40.0 m/s, the cable makes an angle of 40.0° with respect to the vertical. Determine the force exerted by air resistance on the bucket.

GROUP ACTIVITIES

G.1 There is a simple method for measuring the coefficients of static and kinetic friction between an object and some surface. For this investigation, you will need a few coins, your textbook or some other flat surface that can be inclined, a protractor, and some double-stick tape. Place a coin at one edge of the book as it lies on a table, and lift

that edge of the book until the coin just slips down the incline at constant speed after you give it a push to get it started as shown in Figure GA4.1. When this occurs, measure the incline angle using your protractor. Repeat the measurement five times, and find the average value of this critical angle θ_c. The coefficient of static friction between

FIGURE P4.64

65. Two boxes of fruit on a frictionless horizontal surface are connected by a light string as in Figure P4.65, where $m_1 = 10$ kg and $m_2 = 20$ kg. A force of 50 N is applied to the 20-kg box. (a) Determine the acceleration of each box and the tension in the string. (b) Repeat the problem for the case where a coefficient of kinetic friction between each box and the surface is 0.10.

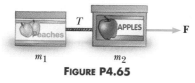

FIGURE P4.65

66. A high diver of mass 70.0 kg jumps off a board 10.0 m above the water. If his downward motion is stopped 2.00 s after he enters the water, what average upward force did the water exert on him?

67. Two people pull as hard as they can on ropes attached to a 200-kg boat. If they pull in the same direction, the boat has an acceleration of 1.52 m/s² to the right. If they pull in opposite directions, the boat has an acceleration of 0.518 m/s² to the left. What is the force exerted by each person on the boat? (Disregard any other forces on the boat.)

68. A 3.0-kg object hangs at one end of a rope that is attached to a support on a railroad car. When the car accelerates to the right, the rope makes an angle of 4.0° with the vertical, as shown in Figure P4.68. Find the acceleration of the car.

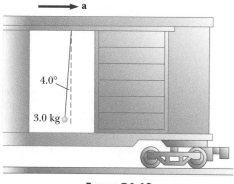

FIGURE P4.68

69. The three blocks of masses 10.0 kg, 5.00 kg, and 3.00 kg are connected by light strings that pass over frictionless pulleys as shown in Figure P4.69. The acceleration of the 5.00-kg block is 2.00 m/s² to the left, and the surfaces are rough. Find (a) the tension in each string and (b) the coefficient of kinetic friction between blocks and surfaces.

(Assume the same μ_k for both blocks in contact with surfaces.)

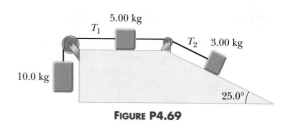

FIGURE P4.69

70. An inquisitive physics student, wishing to combine pleasure with scientific inquiry, rides on a roller coaster sitting on a bathroom scale. (Do not try this yourself on a roller coaster that forbids loose heavy packages.) The bottom of the seat in the roller coaster car is in a plane parallel to the track. The seat has a perpendicular back and a seat belt that fits around the student's chest in a plane parallel to the bottom of the seat. The student lifts his feet from the floor, so that the scale reads his weight, 200 lb, when the car is horizontal. At one point during the ride, the car zooms with negligible friction down a straight slope inclined at 30.0° below the horizontal. What does the scale read there?

71. A van accelerates down a hill (Fig. P4.71), going from rest to 30.0 m/s in 6.00 s. During the acceleration, a toy ($m = 0.100$ kg) hangs by a string from the van's ceiling. The acceleration is such that the string remains perpendicular to the ceiling. Determine (a) the angle θ and (b) the tension in the string.

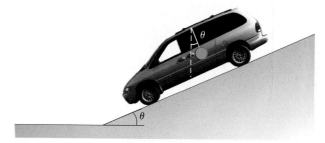

FIGURE P4.71

72. An 80-kg stuntman jumps from the top of a building 30 m above a catching net. Assuming that air resistance exerts a 100-N force on the stuntman as he falls, determine his velocity just before he hits the net.

73. The parachute on a race car of weight 8800 N opens at the end of a quarter-mile run when the car is traveling at 35 m/s. What total retarding force must be supplied by the parachute to stop the car in a distance of 1000 m?

74. On takeoff, the combined action of the air around the engines and wings of an airplane exerts an 8000-N force on the plane, directed upward at an angle of 65.0° above the horizontal. The plane rises with constant velocity in the vertical direction while continuing to accelerate in the horizontal direction. (a) What is the weight of the plane? (b) What is its horizontal acceleration?

75. A 72-kg man stands on a spring scale in an elevator. Starting from rest, the elevator ascends, attaining its maximum speed of 1.2 m/s in 0.80 s. It travels with this constant

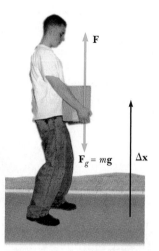

FIGURE 5.2 No work is done on a bucket when it is moved horizontally because the applied force **F** is perpendicular to the displacement.

FIGURE 5.3 Positive work is done by this person when the box is lifted because the applied force **F** is in the same direction as the displacement. When the box is lowered to the floor, the work done by the person is negative because his applied force is still directed upward but the displacement is downward.

Quick Quiz 5.1

Figure 5.4 shows four situations in which a force is exerted on an object. In all four cases, the force has the same magnitude, and the displacement of the object is to the right and of the same magnitude. Rank the situations in order of the work done by the force on the object, from most positive to most negative.

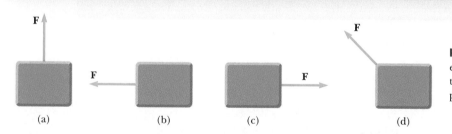

(a)   (b)   (c)   (d)

FIGURE 5.4 (Quick Quiz 5.1) A force **F** is exerted on an object that undergoes a displacement to the right. The magnitudes of the force and displacement are the same in all four cases.

APPLYING PHYSICS 5.1

A student lifts a heavy box of mass m a vertical height h and then walks horizontally a distance d while holding the box, as in Figure 5.5. Determine the work done by the student and by the force of gravity on the box.

Explanation Assuming that he lifts the box with a force of magnitude equal to the weight of the box mg, the work done by the student on the box during the vertical displacement is $W = F\Delta y = (mg)(h) = mgh$ because the force is in the direction of the displacement. For the horizontal displacement, we assume that the acceleration of the box is approximately zero. Hence, the horizontal force he exerts on the box is approximately zero. As a result, the work he does on the box as it moves horizontally is zero because the force he applies (which is holding up the box) is perpendicular to the displacement. Thus, the net work he does during the complete process is mgh.

The work done by the force of gravity on the box during the box's vertical displacement is $-mgh$, negative because the direction of this force is opposite that of the displacement. The work done by the force of gravity is zero during the horizontal displacement because this force is perpendicular to the displacement. Hence, the net work done by the force of gravity for the complete process is $-mgh$. The net work done by *all* forces on the box is zero because $+mgh + (-mgh) = 0$.

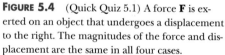

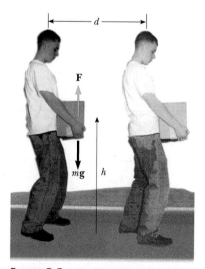

FIGURE 5.5 (Applying Physics 5.1) A student lifts a heavy box of mass m a distance h and then walks a horizontal distance d.

Example 5.1   Mr. Clean

A man cleaning his apartment pulls the canister of a vacuum cleaner with a force of magnitude $F = 55.0$ N. The force makes an angle of $30.0°$ with the horizontal, as shown in Figure 5.6. The canister is displaced 3.00 m to the right. Calculate the work done by the 55.0-N force on the vacuum cleaner.

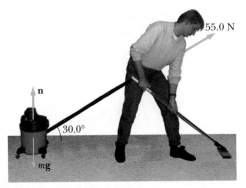

FIGURE 5.6   (Example 5.1) A vacuum cleaner being pulled at an angle of $30.0°$ above the horizontal.

Solution   Using the definition of work (Eq. 5.1), we have

$$W = (F\cos\theta)\,\Delta x = (55.0\text{ N})(\cos 30.0°)(3.00\text{ m}) = \boxed{143\text{ J}}$$

Note that the normal force **n**, the gravitational force $m\mathbf{g}$, and the upward component of the applied force do *no* work because they are perpendicular to the displacement.

EXERCISE   Find the work done by the man on the canister if he pulls it 3.00 m with a horizontal force of magnitude $F = 32.0$ N.

ANSWER   96.0 J

EXERCISE   A 65.0-kg woman climbs a flight of 20 stairs, each 23.0 cm high. How much work is done by the force of gravity on the woman in this process?

ANSWER -2.93 kJ

5.2   KINETIC ENERGY AND THE WORK-KINETIC ENERGY THEOREM

Now that we have seen how to evaluate the work done by a force on an object, let us explore the power of this approach. Solving problems using Newton's second law can be difficult if the forces involved are complicated. An alternative approach to such problems is to relate the speed of an object to the work done on the object by some net external force. If the work done by the net force on the object can be calculated for a given displacement, the change in the object's speed is easy to evaluate. Let's see how this is done.

Figure 5.7 shows an object of mass m moving to the right under the action of a constant net force $\mathbf{F}_{\text{net}}$, also directed to the right. Because the force is constant, we know from Newton's second law that the object moves with constant acceleration $\mathbf{a}$. If the object is displaced a distance of Δx, the work done by $\mathbf{F}_{\text{net}}$ on the object is

$$W_{\text{net}} = F_{\text{net}}\,\Delta x = (ma)\,\Delta x \qquad [5.3]$$

In Chapter 2 we found that the following relationship holds when an object undergoes constant acceleration:

$$v^2 = v_0^2 + 2a\,\Delta x \qquad \text{or} \qquad a\,\Delta x = \frac{v^2 - v_0^2}{2}$$

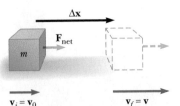

FIGURE 5.7   An object undergoing a displacement and a change in velocity under the action of a constant net force $\mathbf{F}_{\text{net}}$.

principle tells us that any increase (or decrease) in potential energy is accompanied by an equal decrease (or increase) in kinetic energy. **In any isolated system of objects that interact only through conservative forces, the total mechanical energy of the system remains constant.**

Because the total mechanical energy $E \equiv KE + PE$, we can state the principle of conservation of mechanical energy as $E_i = E_f$, and so we have

$$KE_i + PE_i = KE_f + PE_f \qquad [5.10]$$

If the force of gravity is the *only* force doing work within a system, then the total mechanical energy of the system remains constant, and the principle of conservation of mechanical energy takes the form

$$\tfrac{1}{2}mv_i^2 + mgy_i = \tfrac{1}{2}mv_f^2 + mgy_f \qquad [5.11]$$

Because there are other forms of energy besides kinetic and gravitational potential energy, we will modify Equation 5.11 to include these new forms as we learn about them.

 Conservation of mechanical energy

Webnote 5.2

A short but informative discussion of potential and kinetic energy in the context of a freely falling object may be found at:
http://theory.uwinnipeg.ca/mod_tech/ node32.html

Tip 5.4 CONSERVATION PRINCIPLES

Equation 5.10 represents conservation of the mechanical energy of an isolated system. In later chapters, we will generate conservation laws for such quantities as momentum, angular momentum, and electric charge. Conservation principles essentially say that a conserved quantity that either changes in form or moves out of one part of an isolated system must appear in exactly the same amount in either another form or in another part of the system.

Quick Quiz 5.2

Three identical balls are thrown from the top of a building, all with the same initial speed. The first ball is thrown horizontally, the second at some angle above the horizontal, and the third at some angle below the horizontal as in Figure 5.15. Neglecting air resistance, rank the speeds of the balls as they reach the ground.

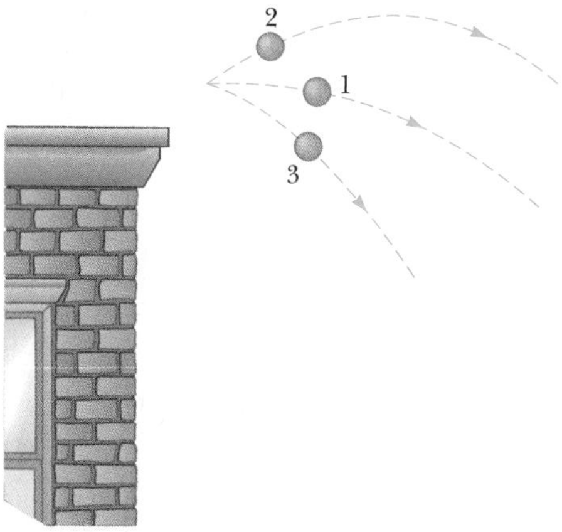

FIGURE 5.15 (Quick Quiz 5.2) Three identical balls are thrown with the same initial speed from the top of a building.

Quick Quiz 5.3

Bobby, of mass m, drops from a tree limb at the same time that Sally, also of mass m, begins her descent down a frictionless slide. The slide is in the shape of a quadrant of a circle. If they both start at the same height above the ground, which of the following is true about their kinetic energies as they reach the ground?
(a) The kinetic energy of Bobby is greater than that of Sally.
(b) The kinetic energy of Sally is greater than that of Bobby.
(c) They have the same kinetic energy.
(d) It is impossible to say from the information given.

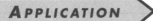

APPLICATION

ROLLER COASTER DESIGN

Webnote 5.3

You can practice designing a roller coaster at:
http://www.learner.org/exhibits/ parkphysics/coaster.html

APPLYING **PHYSICS** 5.4

You have graduated from college and now design roller coasters for a living. Your boss asks you to design a roller coaster that is supposed to release a car from rest at the top of a hill of height h and then roll freely down the hill and reach the peak of the next hill of height $1.1h$. What would you tell your boss in this situation?

Explanation   You should politely tell your boss to study some physics. It is impossible to design such a roller coaster because it would violate the principle of conservation of mechanical energy. At the starting point, the roller coaster has no kinetic energy and the Earth-coaster system has gravitational potential energy mgh relative to ground level. If the roller coaster were able to reach the top of the second hill, the system would have gravitational potential energy $1.1\ mgh$, which is greater than the initial potential energy. If this roller coaster were actually built, the car would move up the second hill to a height h (neglecting friction), stop short of the peak, and then start rolling backward, becoming trapped between the two peaks.

PROBLEM-SOLVING STRATEGY	**Applying Conservation of Mechanical Energy**

Take the following steps in applying the principle of conservation of mechanical energy:

1. Define your system.
2. Select a location for the zero of gravitational potential energy and *do not change* this location while solving the problem.
3. Determine whether or not nonconservative forces are present.
4. If mechanical energy is conserved (that is, if only conservative forces are present), you can write the total initial mechanical energy of the system $KE_i + PE_i$ at some point. Then, write an expression for the total final energy of the system $KE_f + PE_f$ at the final point of interest. Because mechanical energy is conserved, you can equate the two total energies and solve for the unknown.

Example 5.4   **The Daring Diver**

A diver of mass m drops from a board 10.0 m above the water surface, as in Figure 5.16.

A Use conservation of mechanical energy to find his speed 5.00 m above the water surface. Neglect air resistance.

Reasoning   The system consists of the diver and Earth. As the diver falls, only one force acts on him: the force of gravity. Therefore, only the force of gravity does any work on him, and the mechanical energy of the system is conserved. In order to find the diver's speed at the 5.00-m mark, we will choose the water surface as the zero level for potential energy. Also, note that the diver drops from the board; that is, he leaves with zero velocity and therefore zero kinetic energy.

Solution   Because mechanical energy is conserved and $v_i = 0$, Equation 5.11 gives

$$\tfrac{1}{2}mv_i^2 + mgy_i = \tfrac{1}{2}mv_f^2 + mgy_f$$

$$0 + mgy_i = \tfrac{1}{2}mv_f^2 + mgy_f$$

$$v_f = \sqrt{2g(y_i - y_f)}$$

$$= \sqrt{2(9.80\ \text{m/s}^2)(10.0\ \text{m} - 5.00\ \text{m})} = \boxed{9.90\ \text{m/s}}$$

Note that the result does not depend on the mass of the diver.

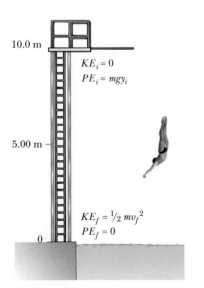

10.0 m

$KE_i = 0$
$PE_i = mgy_i$

5.00 m

$KE_f = \tfrac{1}{2}mv_f^2$
$PE_f = 0$

0

FIGURE 5.16   (Example 5.4) The zero of gravitational potential energy is taken to be when the diver is at the water surface.

B Find the speed of the diver just before he strikes the water.

Solution With the final position of the diver at the surface of the pool, $y_f = 0$ and Equation 5.11 gives

$$0 + mgy_i = \tfrac{1}{2}mv_f^2 + 0$$

$$v_f = \sqrt{2gy_i} = \sqrt{2(9.80 \ \text{m/s}^2)(10.0 \ \text{m})} = \boxed{14.0 \ \text{m/s}}$$

EXERCISE If the diver pushes off so that he leaves the board with an initial speed of 2.00 m/s, find his speed when he strikes the water. Use conservation of mechanical energy.

ANSWER 14.1 m/s

Example 5.5 Sliding Down a Frictionless Hill

A sled and its rider together weigh 800 N. They move down a frictionless hill through a vertical distance of 10.0 m, as shown in Figure 5.17a. Use conservation of mechanical energy to find the speed of the sled at the bottom of the hill, assuming the rider pushes off with an initial speed of 5.00 m/s. Neglect air resistance.

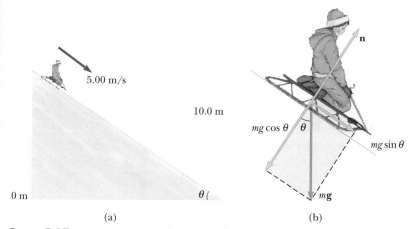

5.00 m/s

10.0 m

$mg \cos \theta$ θ

$mg \sin \theta$

0 m θ (

$m\mathbf{g}$

(a)   (b)

FIGURE 5.17 (Example 5.5) (a) A sled and rider start from the top of a frictionless hill with a speed of 5.00 m/s. (b) A free-body diagram for the sled plus rider.

Reasoning The forces acting on the sled and rider as they move down the hill are shown in Figure 5.17b. In the absence of a frictional force, the only forces acting are the normal force **n** and the gravitational force. At all points along the path, **n** is perpendicular to the direction of travel and hence does no work. Likewise, the component of weight perpendicular to the incline ($mg \cos \theta$) does no work. The only force that does any work is the component of the gravitational force, $mg \sin \theta$, along the slope of the hill. As a result, we are justified in using Equation 5.11.

Solution The initial energy of the sled-rider-Earth system includes kinetic energy because of the initial speed:

$$\tfrac{1}{2}mv_i^2 + mgy_i = \tfrac{1}{2}mv_f^2 + mgy_f$$

or, after canceling m throughout the equation,

$$\tfrac{1}{2}v_i^2 + gy_i = \tfrac{1}{2}v_f^2 + gy_f$$

If we set the origin of our coordinates at the bottom of the incline, the initial and final y

coordinates of the sled are $y_i = 10.0$ m and $y_f = 0$. Thus, we get

$$\tfrac{1}{2}v_i^2 + gy_i = \tfrac{1}{2}v_f^2 + 0$$

$$v_f^2 = v_i^2 + 2gy_i$$

$$= (5.00 \text{ m/s})^2 + 2(9.80 \text{ m/s}^2)(10.0 \text{ m})$$

$$v_f = \boxed{14.9 \text{ m/s}}$$

Notice that this solution does not depend on the shape of the hill. The only things we need to know are the rider's initial speed and the vertical distance traveled. If the hill had a varying curvature with bumps and troughs, the conservation of energy approach would yield the same result. If you try to solve the problem using Newton's laws, the varying slope of the hill would cause the acceleration to vary along the hill, making this approach very complicated. As you progress further with your study of physics, you will find many problems that are difficult or impossible to solve using Newton's laws, whereas conservation of energy provides a simple, yet powerful, alternative.

EXERCISE If the sled and rider start at the bottom of the incline and are given an initial speed of 5.00 m/s up the incline, how high will they rise vertically, and what will their speed be when they return to the bottom of the hill?

ANSWER 1.28 m; 5.00 m/s

Example 5.6 The Jumping Bug

A powerful grasshopper launches itself at an angle of 45° above the horizontal and rises vertically about 1.0 m. With what speed v does it leave the ground? Neglect air resistance.

Solution Because we are ignoring the nonconservative force of air resistance, we can use the principle of conservation of mechanical energy. We sketch the initial state of the grasshopper as it takes off and its final state when it is at a height of 1.0 m and its vertical speed is zero, as shown in Figure 5.18. The initial energy of the grasshopper-Earth system is all kinetic energy. We take the ground as the zero level for the gravitational potential energy. The final energy is partly kinetic energy and partly potential energy. Because mechanical energy is conserved,

$$KE_i + PE_i = KE_f + PE_f$$

$$\tfrac{1}{2}mv^2 + 0 = \tfrac{1}{2}mv_x^2 + mgh$$

$$v^2 = v_x^2 + 2gh \qquad (1)$$

We can eliminate v_x from (1) by noting that initially $v^2 = v_x^2 + v_y^2$ and that $v_x = v_y$ for 45°; these conditions yield $v^2 = 2v_x^2$ or $v_x^2 = v^2/2$. Substituting this result into (1) gives

$$v = 2\sqrt{gh}$$

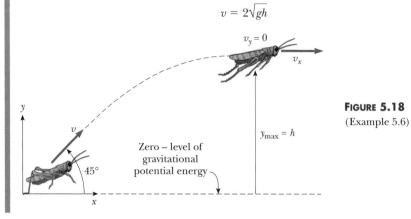

FIGURE 5.18
(Example 5.6)

light-emitting bacteria. The emitted light attracts other smaller creatures that become food for the fish.

Quick Quiz 5.4 A book of mass m is projected with a speed v across a horizontal surface. The book slides until it stops due to the friction force between the book and the surface. The surface is now tilted 30°, and the book is projected up the surface with the same initial speed v. When the book has come to rest, how does the decrease in mechanical energy of the book-Earth system compare to that when the book slid over the horizontal surface? (a) It is the same. (b) It is larger. (c) It is smaller. (d) The relationship cannot be determined.

FIGURE 5.23 This small plant, found in warm southern waters, exhibits bioluminescence, a process in which chemical energy is converted to light. The red areas are chlorophyll, which glows when excited by blue light. *(Jan Hinsch/Science Photo Library/Photo Researchers, Inc.)*

APPLYING PHYSICS 5.5

An automobile with kinetic energy plus potential energy carried within the gasoline strikes a tree and comes to rest. The total energy of the system (the automobile) is now less than before. How did the energy leave the system, and what is the form of the energy remaining in the system?

Explanation The energy left the system in several ways. The automobile did some *work* on the tree during the collision, causing the tree to deform. There was a large crash during the collision, representing transfer of energy by *sound*. After the automobile came to rest, there was no more kinetic energy. *Potential energy* is left in any gas remaining in the tank. There is also more *internal energy* in the automobile, as its temperature is likely to be higher after the collision than before.

PROBLEM-SOLVING STRATEGY | Conservation of Energy Revisited

If nonconservative forces such as friction act within a system, mechanical energy is not conserved. Hence, the problem-solving strategy given immediately before Example 5.4 must be modified as follows. First, write expressions for the total initial and total final mechanical energies. In this case, the difference between the two total energies is equal to the work done by the nonconservative force(s). These steps are summarized in Equation 5.15.

Example 5.9 Fun on the Water Slide

A child of mass $m = 20.0$ kg takes a ride on an irregularly curved water slide of height 2.00 m, as in Figure 5.24. He starts from rest at the top.

A Determine the speed of the child at the bottom, assuming no friction is present.

Reasoning The normal force does no work on the child because this force is always perpendicular to the displacement. Furthermore, because there is no friction, $W_{nc} = 0$ and we can apply the principle of conservation of mechanical energy for the child-Earth system.

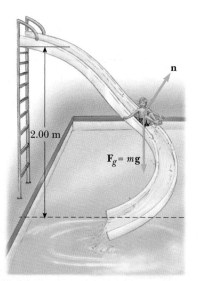

FIGURE 5.24   (Example 5.9) If the slide is frictionless, the speed of the child at the bottom depends only on the height of the slide.

Solution   If we measure the y coordinate for the child from the bottom of the slide, then $y_i = h = 2.00$ m, $y_f = 0$, and we have for the system,

$$KE_i + PE_i = KE_f + PE_f$$

$$0 + mgh = \tfrac{1}{2}mv_f^2 + 0$$

$$v_f = \sqrt{2gh}$$

$$= \sqrt{2(9.80 \text{ m/s}^2)(2.00 \text{ m})} = \boxed{6.26 \text{ m/s}}$$

Note that this speed is the same as if the child fell vertically through a distance $h = 2.00$ m!

B If a friction force acts on the child, and he arrives at the bottom of the slide with a speed $v_f = 3.00$ m/s, by how much does the mechanical energy of the system decrease due to this force?

Solution   We define the system as the child, Earth, and the slide. In this case, a non-conservative force acts within the system, and therefore mechanical energy is *not* conserved. We can find the change in mechanical energy ΔE due to friction, given that the final speed at the bottom is known:

$$\Delta E = KE_f + PE_f - KE_i - PE_i = \tfrac{1}{2}mv_f^2 + 0 - 0 - mgh$$

$$= \tfrac{1}{2}(20.0 \text{ kg})(3.00 \text{ m/s})^2 - (20.0 \text{ kg})(9.80 \text{ m/s}^2)(2.00 \text{ m})$$

$$= \boxed{-302 \text{ J}}$$

The change in mechanical energy is negative because friction reduces the mechanical energy of the system. The "lost" mechanical energy appears as an increase in internal energy in the system, $+302$ J.

Example 5.10   Let's Go Skiing

A skier starts from rest at the top of a frictionless incline of height 20.0 m as in Figure 5.25. At the bottom of the incline, she encounters a horizontal surface where the coefficient of kinetic friction between skis and snow is 0.210. How far does she travel on the horizontal surface before coming to rest?

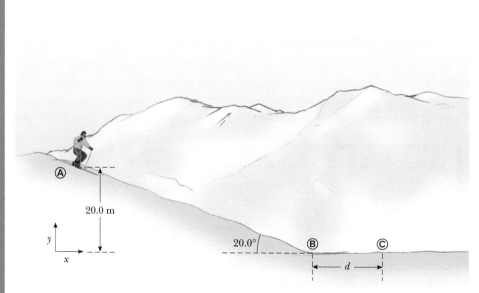

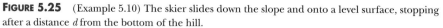

FIGURE 5.25   (Example 5.10) The skier slides down the slope and onto a level surface, stopping after a distance d from the bottom of the hill.

Solution   First, let's calculate her speed at the bottom of the incline, which we choose as our zero point of gravitational potential energy. Because the incline is frictionless, the mechanical energy of the skier-Earth system remains constant, and we find, as we did in the previous example,

$$v_B = \sqrt{2gh} = \sqrt{2(9.80 \text{ m/s}^2)(20.0 \text{ m})} = 19.8 \text{ m/s}$$

Now we apply Equation 5.15 as the skier moves along the rough horizontal surface from Ⓑ to Ⓒ. Note that the work done by the nonconservative force of friction is $W_{nc} = -f_k d$, so Equation 5.15 gives

$$-f_k d = \tfrac{1}{2}mv_C{}^2 - \tfrac{1}{2}mv_B{}^2$$

where d is the horizontal distance the skier travels before coming to rest. Because $v_B = 19.8$ m/s, $v_C = 0$, and the friction force is given by $f_k = \mu_k n = \mu_k mg$, we get

$$-\mu_k mgd = -\tfrac{1}{2}mv_B{}^2$$

$$d = \frac{v_B{}^2}{2\mu_k g} = \frac{(19.8 \text{ m/s})^2}{2(0.210)(9.80 \text{ m/s}^2)} = \boxed{95.2 \text{ m}}$$

EXERCISE   Find the horizontal distance the skier travels before coming to rest if the incline also has a coefficient of kinetic friction equal to 0.210.

ANSWER   40.3 m

5.7   POWER

From a practical viewpoint, it is interesting to know not only the amount of energy transferred to or from a system, but also the rate at which the transfer occurred. This is a particularly interesting issue for living creatures because the maximum work per second, or power output, of an animal varies greatly with output duration. To handle such problems in a systematic way, **power is defined as the time rate of energy transfer.**

If an external force is applied to an object and if the work done by this force is W in the time interval Δt, then the **average power** during this interval is defined as the ratio of the work done to the time interval:

$$\overline{\mathcal{P}} \equiv \frac{W}{\Delta t} \qquad\qquad [5.16] \qquad \blacktriangleleft \text{ Average power}$$

It is sometimes useful to rewrite Equation 5.16 by substituting $W = F\Delta x$ and noting that $\Delta x / \Delta t$ is the average speed of the object during the time Δt:

$$\overline{\mathcal{P}} = \frac{W}{\Delta t} = \frac{F\Delta x}{\Delta t} = F\overline{v} \qquad\qquad [5.17]$$

That is, the average power delivered to an object is equal to the product of the force acting on the object during some time interval and the object's average speed during this time interval. In Equation 5.17, F is the component of force in the direction of the average velocity.

A result of the same form as Equation 5.17 is obtained for instantaneous values. That is, the instantaneous power delivered to an object is equal to the product of the force on the object at that instant and the instantaneous speed, or $\mathcal{P} = Fv$.

The SI units of power are joules per second, which are also called **watts** (W) after James Watt:

$$1 \text{ W} = 1 \text{ J/s} = 1 \text{ kg} \cdot \text{m}^2/\text{s}^3 \qquad\qquad [5.18]$$

The unit of power in the U.S. customary system is the horsepower (hp), where

$$1 \text{ hp} \equiv 550 \, \frac{\text{ft} \cdot \text{lb}}{\text{s}} = 746 \text{ W} \qquad [5.19]$$

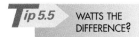

Tip 5.5   WATTS THE DIFFERENCE?

Do not confuse the nonitalic symbol W for the unit watt with the italic symbol *W* for work. Also, remember that the watt represents a rate of energy transfer—the watt is *the same as* a joule per second.

The horsepower was first defined by Watt, who needed a large power unit to rate the power output of his new invention, the steam engine.

The watt is commonly used in electrical applications, but it can be used in other scientific areas as well. For example, European sports-car engines are now rated in kilowatts.

We now define units of energy (or work) in terms of the unit of power, the watt. One kilowatt-hour (kWh) is the energy transferred in 1 h at the constant rate of 1 kW = 1 000 J/s. Therefore,

$$1 \text{ kWh} = (10^3 \text{ W})(3\,600 \text{ s}) = (10^3 \text{ J/s})(3\,600 \text{ s}) = 3.60 \times 10^6 \text{ J}$$

It is important to realize that a kilowatt-hour is a unit of energy, *not* power. When you pay your electric bill, you are buying energy, and that's why your bill lists a charge for electricity (no pun intended) of about 10 cents/kWh. The amount of electricity used by an appliance can be calculated by multiplying its power rating (usually expressed in watts and valid only for normal household electrical circuits) by the length of time the appliance is operated. For example, an electric bulb rated at 100 W (= 0.100 kW) "consumes" 3.6×10^5 J of energy in 1 h.

APPLYING **PHYSICS** 5.6

A lightbulb is described by some individuals as "having 60 watts." What is wrong with this phrase?

Reasoning   The number represents the power of the lightbulb, which is the rate at which energy passes through it, entering as electrical transmission from the power source. It is not a value that the lightbulb "possesses," however. It is the approximate power rating of the bulb when it is connected to a normal 120-volt household supply of electricity. If the bulb were connected to a 12-volt supply, energy would enter the bulb at a rate other than 60 watts. Thus, the 60 watts is not an intrinsic property of the bulb, such as density is for a given solid material.

Example 5.11   **Power Delivered by an Elevator Motor**

A 1 000-kg elevator carries a maximum load of 800 kg. A constant frictional force of 4 000 N retards its motion upward, as in Figure 5.26. What minimum power, in kilowatts

FIGURE 5.26   (Example 5.11) The motor exerts an upward force **T** on the elevator. A frictional force **f** and the force of gravity *M***g** act downward.

and in horsepower, must the motor deliver to lift the fully loaded elevator at a constant speed of 3.00 m/s?

Solution   The motor must supply the force **T** that pulls the elevator upward. From Newton's second law and from the fact that $\mathbf{a} = 0$ because **v** is constant, we get

$$T - f - Mg = 0$$

where M is the total mass (elevator plus load), equal to 1800 kg. Therefore,

$$
\begin{aligned}
T &= f + Mg \\
&= 4.00 \times 10^3 \, \text{N} + (1.80 \times 10^3 \, \text{kg})(9.80 \, \text{m/s}^2) \\
&= 2.16 \times 10^4 \, \text{N}
\end{aligned}
$$

From $\mathcal{P} = Fv$ and the fact that **T** is in the same direction as **v**, we have

$$
\begin{aligned}
\mathcal{P} &= Tv \\
&= (2.16 \times 10^4 \, \text{N})(3.00 \, \text{m/s}) = 6.48 \times 10^4 \, \text{W} \\
&= \boxed{64.8 \, \text{kW} = 86.9 \, \text{hp}}
\end{aligned}
$$

ENERGY AND POWER IN A VERTICAL JUMP

The stationary jump consists of two parts, extension and free flight.[2] In the extension phase the person jumps up from a crouch, straightening the legs and throwing up the arms; the free-flight phase occurs when the jumper leaves the ground. Because the body is an extended object and different parts move with different speeds, we describe the motion of the jumper in terms of the position and velocity of the **center of mass (CM),** which is the point in the body at which all the mass may be considered to be concentrated. Figure 5.27 shows the position and velocity of the CM at different stages of the jump.

Using the principle of the conservation of mechanical energy, we can find H, the maximum increase in height of the CM, in terms of the velocity v_{CM} of the CM at liftoff. Taking PE_i, the gravitational potential energy of the jumper-Earth system just as the jumper lifts off from the ground to be zero, and noting that the kinetic

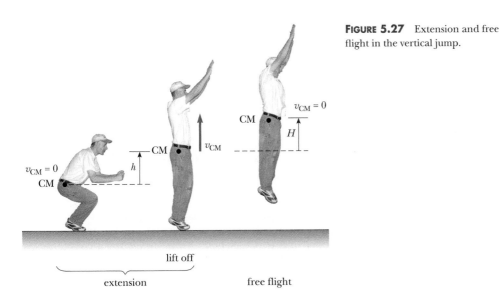

FIGURE 5.27   Extension and free flight in the vertical jump.

---

[2] For more information on this topic, see E. J. Offenbacher, *American Journal of Physics,* **38**, 829 (1969).

energy KE_f of the jumper at the peak is zero, we have

$$PE_i + KE_i = PE_f + KE_f$$

$$\tfrac{1}{2}mv_{CM}^2 = mgH \quad \text{or} \quad H = \frac{v_{CM}^2}{2g}$$

We can estimate v_{CM} by assuming that the acceleration of the CM is constant during the extension phase. If the depth of the crouch is h and the time for extension is Δt, we find $v_{CM} = 2\bar{v} = 2h/\Delta t$. Measurements on a group of male college students show typical values of $h = 0.40$ m and $\Delta t = 0.25$ s, the latter value being set by the fixed speed with which muscle can contract. Substituting, we find $v_{CM} = 2(0.40 \text{ m})/(0.25 \text{ s}) = 3.2$ m/s and

$$H = \frac{v_{CM}^2}{2g} = \frac{(3.2 \text{ m/s})^2}{2(9.80 \text{ m/s}^2)} = 0.52 \text{ m}$$

Measurements on this same group of students found that H was between 0.45 m and 0.61 m in all cases, confirming the basic validity of our simple calculation.

In order to relate the physical concepts of energy, power, and efficiency to humans, it is interesting to calculate these values for the vertical jump. The kinetic energy given to the body in a jump is $KE = \tfrac{1}{2}mv_{CM}^2$ and for a person of mass 68 kg, the kinetic energy is

$$KE = \tfrac{1}{2}(68 \text{ kg})(3.2 \text{ m/s})^2 = 3.5 \times 10^2 \text{ J}$$

Although this may seem like a large expenditure of energy, we can make a simple calculation to show that jumping and exercise in general, are not good ways to lose weight, in spite of their many health benefits. Since the muscles are at most 25% efficient at producing work from chemical energy (muscles always warm up a person as well as do work—that's why you sweat when you work out), they use up four times the 350 J (or 1 400 J) of chemical energy in one jump. This chemical energy ultimately comes from the food we eat, whose energy content is given in units of food calories, with 1 food calorie equal to 4 200 J. So the total energy supplied by the body in a vertical jump is only about $\tfrac{1}{3}$ of a food calorie! You are a lot better off not eating that piece of cheesecake than trying to work it off by jumping.

Finally, it is interesting to calculate the mechanical power that can be generated by the body in strenuous activity for brief periods. Here we find

$$\mathscr{P} = \frac{KE}{\Delta t} = \frac{3.5 \times 10^2 \text{ J}}{0.25 \text{ s}} = 1.4 \times 10^3 \text{ W}$$

or $(1\ 400 \text{ W})(1 \text{ hp}/746 \text{ W}) = 1.9$ hp. Thus, humans can produce about 2 hp of mechanical power for periods on the order of seconds. Table 5.2 shows the maximum power outputs from humans for various time intervals based on bicycling and rowing, activities in which it is possible to measure power output accurately.

TABLE 5.2	Maximum Power Output from Humans for Various Time Periods
Power	**Time**
2 hp or 1 500 W	6 s
1 hp or 750 W	60 s
0.35 hp or 260 W	35 min
0.2 hp or 150 W	5 h
0.1 hp or 75 W (Safe daily level)	8 h

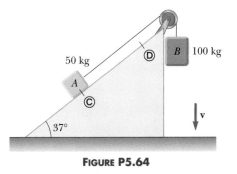

FIGURE P5.64

65. A 700-N Marine in basic training climbs a 10.0-m vertical rope at a constant speed in 8.00 s. What is his power output?

66. Energy is conventionally measured in Calories as well as in joules. One Calorie in nutrition is one kilocalorie, which we define in Chapter 11 as 1 kcal = 4 186 J. Metabolizing one gram of fat can release 9.00 kcal. A student decides to try to lose weight by exercising. She plans to run up and down the stairs in a football stadium as fast as she can and as many times as necessary. Is this in itself a practical way to lose weight? To evaluate the program, suppose she runs up a flight of 80 steps, each 0.150 m high, in 65.0 s. For simplicity ignore the energy she uses in coming down (which is small). Assume that a typical efficiency for human muscles is 20.0%. This means that when your body converts 100 J from metabolizing fat, 20 J goes into doing mechanical work (here, climbing stairs). The remainder goes into internal energy. Assume the student's mass is 50.0 kg. (a) How many times must she run the flight of stairs to lose one pound of fat? (b) What is her average power output, in watts and in horsepower, as she is running up the stairs?

67. For saving energy, bicycling and walking are far more efficient means of transportation than is travel by automobile. For example, when riding at 10.0 mi/h, a cyclist uses food energy at a rate of about 400 kcal/h above what he would use if merely sitting still. (In exercise physiology, power is often measured in kcal/h rather than in watts. Here 1 kcal = 1 nutritionist's Calorie = 4 186 J.) Walking at 3.00 mi/h requires about 220 kcal/h. It is interesting to compare these values with the energy consumption required for travel by car. Gasoline yields about 1.30×10^8 J/gal. Find the fuel economy in equivalent miles per gallon for a person (a) walking, and (b) bicycling.

68. A catcher "gives" with the ball when he catches a 0.15-kg baseball moving at 25 m/s. (a) If he moves his glove a distance of 2.0 cm, what is the average force acting on his hand? (b) Repeat for the case in which his glove and hand move 10 cm.

69. A ski jumper starts from rest 50.0 m above the ground on a frictionless track, and flies off the track at an angle of 45.0° above the horizontal and at a height of 10.0 m above the level ground. Neglect air resistance. (a) What is his speed when he leaves the track? (b) What is the maximum altitude he attains after leaving the track? (c) Where does he land relative to the end of the track?

70. A 5.0-kg block is pushed 3.0 m up a vertical wall with constant speed by a constant force of magnitude F applied at an angle of $\theta = 30°$ with the horizontal, as shown in Figure P5.70. If the coefficient of kinetic friction between block and wall is 0.30, determine the work done by (a) **F**, (b) the force of gravity, and (c) the normal force between block and wall. (d) By how much does the gravitational potential energy increase during this motion?

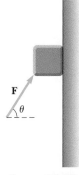

FIGURE P5.70

71. The ball launcher in a pinball machine has a spring that has a force constant of 1.20 N/cm (Fig. P5.71). The surface on which the ball moves is inclined 10.0° with respect to the horizontal. If the spring is initially compressed 5.00 cm, find the launching speed of a 0.100-kg ball when the plunger is released. Friction and the mass of the plunger are negligible.

FIGURE P5.71

72. The masses of the javelin, discus, and shot are 0.80 kg, 2.0 kg, and 7.2 kg, respectively, and record throws in the corresponding track events are about 89 m, 69 m, and 21 m, respectively. Neglecting air resistance, (a) calculate the minimum initial kinetic energies that would produce these throws, and (b) estimate the average force exerted on each object during the throw, assuming the force acts over a distance of 2.0 m. (c) Do your results suggest that air resistance is an important factor?

73. Jane, whose mass is 50.0 kg, needs to swing across a river filled with man-eating crocodiles in order to rescue Tarzan. However, she must swing into a *constant* horizontal wind force **F** on a vine that is initially at an angle of θ with the vertical (see Fig. P5.73). $D = 50.0$ m, $F = 110$ N, $L = 40.0$ m, and $\theta = 50.0°$. (a) With what minimum speed must Jane begin her swing in order to just make it to the other side? (*Hint:* First determine the potential energy that can be associated with the wind force. Because the wind force is constant, use an analogy with the constant gravitational force.) (b) Once the rescue is complete, Tarzan and Jane must swing back across the river. With what minimum speed must they begin their swing? Assume that Tarzan has a mass of 80.0 kg.

74. As it plows a parking lot, a snowplow pushes an ever-growing pile of snow in front of it. Suppose a car moving through air is similarly modeled as a cylinder pushing a growing plug of air in front of it. The originally stationary air is set into motion at the constant speed v of the

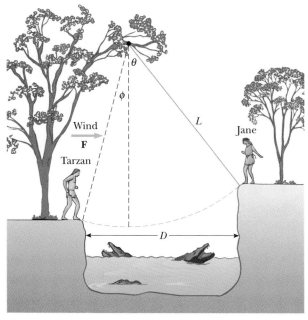

FIGURE P5.73

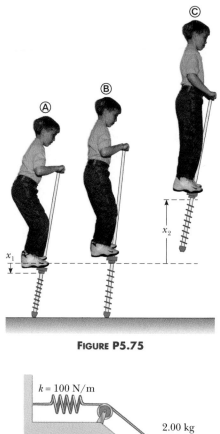

FIGURE P5.75

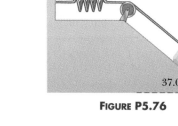

FIGURE P5.76

cylinder, as in Figure P5.74. In a time Δt, a new disk of air of area A and mass Δm must be moved a distance $v\Delta t$ and hence must be given kinetic energy $\frac{1}{2}(\Delta m)v^2$. Using this model, show that the power loss due to air resistance is $\rho A v^3/2$ and the resistive force is $\rho A v^2/2$, where ρ is the density of air.

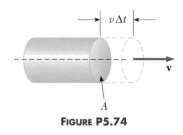

FIGURE P5.74

75. A child's pogo stick (Fig. P5.75) stores energy in a spring ($k = 2.50 \times 10^4$ N/m). At position Ⓐ ($x_1 = -0.100$ m), the spring compression is a maximum and the child is momentarily at rest. At position Ⓑ ($x = 0$), the spring is relaxed and the child is moving upward. At position Ⓒ, the child is again momentarily at rest at the top of the jump. Assuming that the combined mass of child and pogo stick is 25.0 kg, (a) calculate the total energy of the system if both potential energies are zero at $x = 0$, (b) determine x_2, (c) calculate the speed of the child at $x = 0$, (d) determine the value of x for which the kinetic energy of the system is a maximum, and (e) obtain the child's maximum upward speed.

76. A 2.00-kg block situated on a rough incline is connected to a spring of negligible mass having a spring constant of 100 N/m (Fig. P5.76). The block is released from rest when the spring is unstretched, and the pulley is frictionless. The block moves 20.0 cm down the incline before coming to rest. Find the coefficient of kinetic friction between the block and incline.

77. In the dangerous "sport" of bungee jumping, a daring student jumps from a balloon with a specially designed elastic cord attached to his waist, as shown in Figure P5.77. The

unstretched length of the cord is 25.0 m, the student weighs 700 N, and the balloon is 36.0 m above the surface of a river below. Calculate the required force constant of the cord if the student is to stop safely 4.00 m above the river.

FIGURE P5.77 Bungee jumping. *(Gamma)*

78. An object of mass m is suspended from the top of a cart by a string of length L as in Figure P5.78a. The cart and object are initially moving to the right at constant speed v_0. The cart comes to rest after colliding and sticking to a bumper, as in Figure P5.78b, and the suspended object swings through an angle θ. (a) Show that the initial speed is $v_0 = \sqrt{2gL(1 - \cos\theta)}$. (b) If $L = 1.20$ m and $\theta = 35.0°$, find the initial speed of the cart. (*Hint:* The force exerted by the string on the object does no work on the object.)

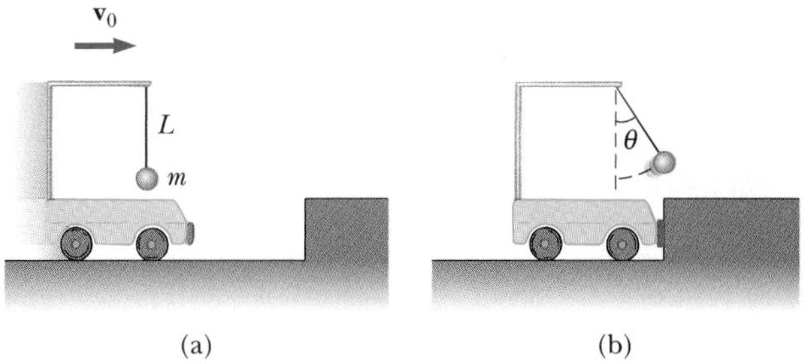

(a)   (b)

FIGURE P5.78

GROUP ACTIVITIES

G.1 Suspend a rubber band from a support and borrow some weights from your instructor to measure the rubber band's spring constant for small extensions. Calculate how much elastic potential energy is stored in the rubber band for a given extension. Use conservation of energy to predict how high a paper wad will go into the air when released from a given extension of the band. Try it to test your prediction.

G.2 Tightly wrap a rubber band around a tennis ball. Now fasten one end of a string through the rubber band and the other end to the top of a door frame to construct a pendulum. Pull the pendulum to the side at a variety of angles to observe that the energy of the pendulum-Earth system is always conserved as the pendulum swings (almost) to the same height of its arc as the height from which it was released. The word "almost" in the last sentence occurs because some energy is lost to friction at the point of support and to air resistance. You can observe this slight loss of energy by pulling the ball to the side and letting it go from a point about half an inch from your chin. Let it go (but don't push it) and test your belief in conservation of energy by seeing if you can avoid flinching when it swings back toward your chin.

G.3 While you have your pendulum of the preceding activity set up, predict what will happen in the following situation and then test your guess. When a pendulum is released from a given height, it swings to the same height at the other end of its arc as you noted above. However, suppose you place a meter stick across the door opening such that the pendulum string strikes the stick about halfway up the string when it moves through the door opening. How high will the ball swing in this case? Will it return to the same height as that at which it started, swing to a lower height, or swing to a greater height? Explain your answer.

G.4 Measure your pulse rate while at rest. Now slowly walk up a flight of stairs and measure your pulse rate at the top of the stairs. Repeat this activity starting with about the same rest pulse rate at the bottom of the stairs. This time run up the stairs. Based on your pulse rate readings, what can you conclude about the amount of work and power expended in each case? Repeat this experiment with a series of 10 pushups.

G.5 (a) A block with a mass m is pulled along a horizontal surface for a distance x by a constant force $\mathbf{F}$ at an angle θ with respect to the horizontal. The coefficient of kinetic friction between block and table is μ_k. Is the force exerted by friction equal to $\mu_k mg$? If not, what is the force exerted by friction? (b) How much work is done by the friction force and by $\mathbf{F}$? Don't forget the signs. (c) Identify all the forces that do no work on the block. (d) Let $m = 2.00$ kg, $x = 4.00$ m, $\theta = 37.0°$, $F = 15.0$ N, and $\mu_k = 0.400$ to find the answers to the questions posed above. Do the answers agree with your initial judgments?

G.6 (a) A child slides down a waterslide at an amusement park from an initial height h. The slide can be considered frictionless because of the water flowing down it. Can the equation for conservation of mechanical energy be used for the child? Defend your answer. (b) Is the mass of the child a factor in determining the speed at the bottom of the slide? Defend your answer. (c) The child drops straight down rather than following the curved ramp of the slide. In which case will he be traveling faster at ground level? Defend your answer. (d) If friction is present, how would the conservation of energy equation be modified? Defend your choice for the equation to be used. (e) Find the maximum speed of the child if the initial height of the slide is 25.0 m.

G.7 A child's toy consists of a base that is clamped firmly to a table, a spring and launching ramp attached to the base, and a ball. By compressing the spring, the child can launch the ball up the ramp to fire it off the table. The spring has a spring constant k, the ball has a mass m, and the ramp rises a height h. The spring is compressed a distance Δr in order to launch the ball. When the ball leaves the launching ramp, its velocity makes an angle θ with respect to the horizontal. (a) Assuming that friction and air resistance can be ignored for the purposes of this problem, describe the changes in the forms of energy in the system from the time the spring is compressed until the ball hits the ground. (b) Calculate the vector velocity of the ball when it leaves the launching ramp. Be sure to specify your coordinate system. (c) The spring constant is 32 N/m, the spring's compression is 5.0 cm, the ball's mass is 20 g, the height of the ramp is 10 cm, and the top of the table is 1.0 m above the floor. With what speed will the ball hit the floor?

(Problem 7 is courtesy of E. F. Redish. For more problems of this type, visit http://www.physics.umd.edu/ripe/perg/)

Momentum and Collisions

Chapter Outline

A bowling ball rolling down an alley has momentum, a quantity that is proportional to the ball's velocity. As the ball collides with the pins, part of the ball's momentum before the collision is transferred to the pins after the collision.

(Mark Cooper/Corbis Stock Market)

*C*onsider what happens when you strike a golf ball with a club. As you well know, the ball undergoes a large change in velocity and travels a long distance because of the force exerted by the club on that ball. But why doesn't the club also undergo a large change in velocity? It has the same force acting on it. Why does a good follow-through on the swing help the ball to go farther? Does the deformation of the ball upon contact have any effect on its subsequent motion?

As a first step in answering such questions, we shall introduce the term *momentum,* a term that has a slightly different meaning in physics than it does in everyday life. For example, a candidate moving rapidly in the opinion polls is said to be gaining momentum, but from the standpoint of physics, is he? As another example, a very massive football player is said to have a great deal of momentum as he runs down the field. Can a much less massive player have the same momentum? What must he do to achieve this momentum? Can a Ping-Pong ball have the same momentum as the expertly released bowling ball in the chapter opening photo?

The concept of momentum will lead us to a second conservation law: conservation of momentum. This law is especially useful for treating problems that involve collisions between objects because it gives us important information about the objects' motion without having to know the complicated variation of collision forces with time.

6.1 MOMENTUM AND IMPULSE

The linear momentum of an object of mass *m* moving with a velocity v is defined as the product of the mass and the velocity:

$$\mathbf{p} \equiv m\mathbf{v} \qquad [6.1]$$

◀ Linear momentum

As its definition shows, momentum is a vector quantity, with its direction matching that of the velocity. Momentum has dimensions ML/T, and its SI units are kilogram-meters per second (kg· m/s).

Often we find it advantageous to work with the components of momentum. For two-dimensional motion, these are

$$p_x = mv_x \qquad p_y = mv_y$$

where p_x represents the momentum of an object in the x direction and p_y its momentum in the y direction.

Quick Quiz 6.1	Two objects have equal kinetic energy. How do the magnitudes of their momenta compare? (a) $p_1 < p_2$, (b) $p_1 = p_2$, (c) $p_1 > p_2$, (d) not enough information to determine the answer.

The definition of momentum in Equation 6.1 coincides with our everyday use of the word. When we think of a massive object moving with a high velocity, we often say that the object has a large momentum, in accordance with Equation 6.1. Likewise, a small object moving slowly is said to have a small momentum. On the other hand, a small object moving with a high velocity can have a large momentum.

In order to change the momentum of an object, a force must be applied to it. In modern terms, Newton's second law of motion can be stated as follows:

The time rate of change of momentum of an object is equal to the net force acting on the object.

Newton's second law ▶

$$\mathbf{F}_{net} = \frac{\text{change in momentum}}{\text{time interval}} = \frac{\Delta \mathbf{p}}{\Delta t} \qquad [6.2]$$

where Δt is the time interval during which the momentum changes by $\Delta \mathbf{p}$ and $\mathbf{F}_{net}$ is the net force acting on the object. To see that this is equivalent to $\mathbf{F}_{net} = m\mathbf{a}$ for an object of constant mass, consider a single constant force $\mathbf{F}$ acting on a particle and producing a constant acceleration. In this case, we can write Equation 6.2 as

$$\mathbf{F} = \frac{\Delta \mathbf{p}}{\Delta t} = \frac{m\mathbf{v}_f - m\mathbf{v}_i}{\Delta t} = \frac{m(\mathbf{v}_f - \mathbf{v}_i)}{\Delta t} \qquad [6.3]$$

Now recall that the average acceleration of an object as given by Equation 3.8 is $\mathbf{a} = \Delta \mathbf{v}/\Delta t$. Thus, we see that Equation 6.3 reduces to the familiar equation $\mathbf{F} = m\mathbf{a}$.

Note from Equation 6.2 that if the net force is zero, the momentum of the object does not change. In other words, the linear momentum and velocity of a particle are conserved when $\mathbf{F}_{net} = 0$.

When a single constant force $\mathbf{F}$ acts on a particle, Equation 6.3 can be written as $\mathbf{F}\,\Delta t = \Delta \mathbf{p}$ or

Impulse-momentum theorem ▶

$$\mathbf{F}\,\Delta t = \Delta \mathbf{p} = m\mathbf{v}_f - m\mathbf{v}_i \qquad [6.4]$$

The product $\mathbf{F}\,\Delta t$ is defined to be the **impulse** of a constant force for the time interval Δt:

$$\text{Impulse} \equiv \mathbf{F}\,\Delta t \qquad [6.5]$$

Note that impulse is a vector quantity whose direction is the same as the constant force on the object. Equation 6.4 is called the **impulse-momentum theorem** and shows that **the impulse of the force acting on an object equals the change in momentum of that object.** This is true even if the force is not constant, although we need to modify the definition of impulse in this case, as discussed below.

The concept of impulse is most useful in describing collisions that last for a short time, for example, a bat hitting a ball, but a word of caution is in order here. If you tried to solve a problem such as this using Newton's second law, you would encounter some difficulty in measuring the value to be used for $\mathbf{F}$, because the force exerted on the ball is large, of short duration, and not constant. Instead, it might be represented by a curve like that in Figure 6.1a. The force starts out small as the bat comes in contact with the ball, rises to a maximum value when they are firmly in contact, and then drops off as the ball leaves the bat. In such instances, it is necessary to define an **average force** $\overline{\mathbf{F}}$ shown as a dashed line in Figure 6.1b. This average force can be thought of as the constant force that would give the same impulse to the object in the time interval Δt as the actual time-varying force

FIGURE 6.1 (a) A force acting on an object may vary in time. The impulse is the area under the force-time curve. (b) The average force (horizontal dashed line) gives the same impulse to the object in the time interval Δt as the real time-varying force described in (a).

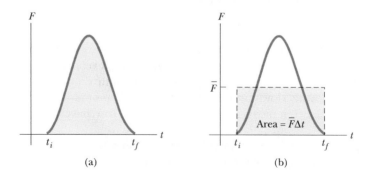

gives in this interval. In this situation, we can write the impulse-momentum theorem as

$$\overline{\mathbf{F}} \, \Delta t = \Delta \mathbf{p} \qquad \qquad [6.6]$$

The impulse imparted by a force during the time interval Δt is equal to the area under the force-time graph from the beginning to the end of the time interval or, equivalently, to $\overline{F} \, \Delta t$ as shown in Figure 6.1b. The brief collision between a bullet and an apple is illustrated in Figure 6.2.

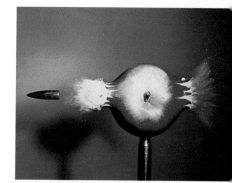

FIGURE 6.2 An apple being pierced by a 30-caliber bullet traveling at a supersonic speed of 900 m/s. This collision was photographed with a microflash stroboscope using an exposure time of 0.33 μs. Shortly after the photograph was taken, the apple disintegrated completely. Note that the points of entry and exit of the bullet are visually explosive. (© *Harold and Esther Edgerton Foundation, 2002, courtesy of Palm Press, Inc.*)

APPLYING PHYSICS 6.1

In boxing matches of the 19th century, bare fists were used. In modern boxing, fighters wear padded gloves. How does this protect the brain of the boxer from injury? Boxers often "roll with the punch." How does this protect their health?

Explanation The brain is immersed in a cushioning fluid inside the skull. If the head is struck suddenly by a bare fist, there is a rapid acceleration of the skull. The brain matches this acceleration only because of the large impulsive force exerted by the skull on the brain. This large and sudden force (large $\overline{F}$ and small Δt) can cause severe brain injury. Padded gloves extend the time Δt over which the force is applied to the head. Thus, for a given impulse $\overline{F} \, \Delta t$, the gloves provide a longer time interval than the bare fist, decreasing the average force. Because the average force is decreased, the acceleration of the skull is decreased, reducing (but not eliminating) the chance for brain injury. The same argument can be made for "rolling with the punch." If the head is held steady while being struck, the time interval over which the force is applied is relatively short and the average force is large. If the head is allowed to move in the same direction of the punch, the time interval is lengthened, and the average force reduced.

APPLICATION

BOXING AND BRAIN INJURY

Example 6.1 Teeing Off

A 50-g golf ball is struck with a club as in Figure 6.3. The force on the ball varies from zero when contact is made up to some maximum value (when the ball is deformed) and then back to zero when the ball leaves the club. Thus, the force-time graph is somewhat like that in Figure 6.1. Assume that the ball leaves the club face with a velocity of $+44$ m/s.

A Estimate the size and direction of the impulse due to the collision.

Solution Before the club hits the ball, the ball is at rest on the tee, and as a result its initial momentum is zero. Taking the positive direction in Figure 6.3 to be that of the motion of the ball (to the right), the momentum of the ball immediately after the collision is

$$\mathbf{p}_f = m\mathbf{v}_f = (50 \times 10^{-3} \, \text{kg})(44 \, \text{m/s}) = +2.2 \, \text{kg} \cdot \text{m/s}$$

The positive sign means the final momentum is directed to the right in Figure 6.3. Thus, the impulse imparted to the ball, which equals its change in momentum, is

$$\Delta \mathbf{p} = m\mathbf{v}_f - m\mathbf{v}_i = \boxed{+2.2 \, \text{kg} \cdot \text{m/s}}$$

where the positive sign means that the change in momentum and the impulse are both directed to the right.

B Estimate the duration of the collision and the average force on the ball.

Solution From Figure 6.3, it appears that a reasonable estimate of the distance the ball travels while in contact with the club is the radius of the ball, about 2.0 cm. The time

FIGURE 6.3 (Example 6.1) A golf ball being struck by a club. (© *Harold and Esther Edgerton Foundation, 2002, courtesy of Palm Press, Inc.*)

it takes the club to move this distance (the contact time) is then

$$\Delta t = \frac{\Delta x}{\overline{v}} = \frac{2.0 \times 10^{-2} \text{ m}}{22 \text{ m/s}} = \boxed{9.1 \times 10^{-4} \text{ s}}$$

Finally, the average force is estimated to be

$$\overline{\mathbf{F}} = \frac{\Delta \mathbf{p}}{\Delta t} = \frac{2.2 \text{ kg} \cdot \text{m/s}}{9.1 \times 10^{-4} \text{ s}} = \boxed{+2.4 \times 10^3 \text{ N}}$$

and the direction of $\overline{\mathbf{F}}$ is in the same direction as $\Delta \mathbf{p}$, the positive x direction.

Example 6.2   How Good Are the Bumpers?

In a particular crash test, a car of mass 1.50×10^3 kg collides with a wall as in Figure 6.4a. The initial and final velocities of the car are $\mathbf{v}_i = -15.0$ m/s and $\mathbf{v}_f = +2.60$ m/s, respectively. If the collision lasts for 0.150 s, find the impulse on the car due to the collision and the size and direction of the average force exerted on the car.

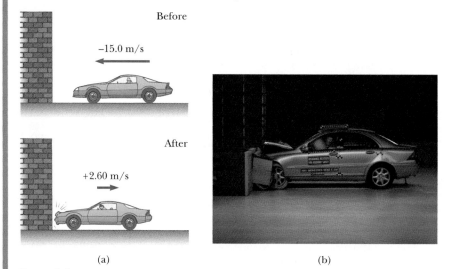

Before

−15.0 m/s

After

+2.60 m/s

(a)   (b)

FIGURE 6.4   (Example 6.2) (a) This car's momentum changes as a result of its collision with the wall. (b) In a crash test (an inelastic collision), much of the car's initial kinetic energy is transformed into the energy it took to damage the vehicle. *(b, courtesy of the Insurance Institute for Highway Safety)*

Solution   The initial and final momenta of the car are

$$\mathbf{p}_i = m\mathbf{v}_i = (1.50 \times 10^3 \text{ kg})(-15.0 \text{ m/s}) = -2.25 \times 10^4 \text{ kg} \cdot \text{m/s}$$

$$\mathbf{p}_f = m\mathbf{v}_f = (1.50 \times 10^3 \text{ kg})(+2.60 \text{ m/s}) = +0.390 \times 10^4 \text{ kg} \cdot \text{m/s}$$

where the minus sign means the initial momentum is to the left in Figure 6.4a and the plus sign means the final momentum is to the right. Hence, the impulse, which equals the change in momentum, is

$$\overline{\mathbf{F}} \, \Delta t = \Delta \mathbf{p} = \mathbf{p}_f - \mathbf{p}_i$$
$$= +0.390 \times 10^4 \text{ kg} \cdot \text{m/s} - (-2.25 \times 10^4 \text{ kg} \cdot \text{m/s})$$
$$= \boxed{+2.64 \times 10^4 \text{ kg} \cdot \text{m/s}}$$

The average force exerted on the car is

$$\overline{\mathbf{F}} = \frac{\Delta \mathbf{p}}{\Delta t} = \frac{2.64 \times 10^4 \text{ kg} \cdot \text{m/s}}{0.150 \text{ s}} = \boxed{+1.76 \times 10^5 \text{ N}}$$

where the plus sign indicates the average force is in the positive x direction.

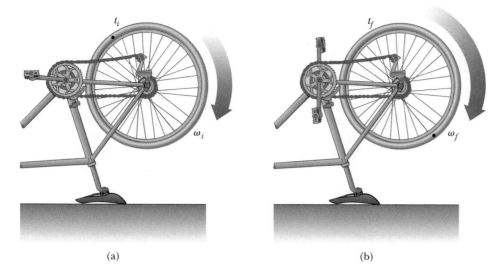

FIGURE 7.3 An accelerating bicycle wheel rotates with (a) angular speed ω_i at time t_i and (b) angular speed ω_f at time t_f.

Δt it takes the object to undergo the change:

$$\overline{\alpha} \equiv \frac{\omega_f - \omega_i}{t_f - t_i} = \frac{\Delta\omega}{\Delta t} \qquad [7.4]$$

The instantaneous angular acceleration is defined as the limit of the ratio $\Delta\omega/\Delta t$ as Δt approaches zero. Angular acceleration has the units radians per second per second (rad/s^2). **When a rigid object rotates about a fixed axis, as does the bicycle wheel, every portion of the object has the same angular speed and the same angular acceleration.** This, in fact, is precisely what makes these variables so useful for describing rotational motion. Note that the linear speed and acceleration have a different value at every point along a radius.

◀ Instantaneous angular acceleration

The following argument should convince you that ω and α are the same for every point on the wheel. If a point on the rim of the wheel had a greater angular speed than a point nearer the center, the shape of the wheel would be changing. The wheel remains circular (symmetrically distributed about the axle) only if all points have the same angular speed and the same angular acceleration.

7.2 ROTATIONAL MOTION UNDER CONSTANT ANGULAR ACCELERATION

Let us pause for a moment to consider some similarities between the equations we have found thus far for rotational motion and those we found for linear motion in earlier chapters. For example, compare the defining equation for average angular speed,

$$\overline{\omega} \equiv \frac{\theta_f - \theta_i}{t_f - t_i} = \frac{\Delta\theta}{\Delta t}$$

with the defining equation for average linear speed,

$$\overline{v} \equiv \frac{x_f - x_i}{t_f - t_i} = \frac{\Delta x}{\Delta t}$$

The equations are similar in the sense that θ replaces x and ω replaces v. Take careful note of such similarities as you study rotational motion, because virtually every linear quantity we have encountered thus far has a corresponding "twin" in rotational motion. Once you are adept at recognizing such analogies, you will find it unnecessary to memorize many of the equations in this chapter. In addition, the techniques for solving rotational motion problems are quite similar to those you have already learned for linear motion. For example, problems concerned with objects that rotate with constant angular acceleration can be solved in much the same manner as those dealing with linear motion under constant acceleration. If you understand problems involving objects that move with constant linear acceleration, these rotational motion problems should be little more than a review for you.

An additional analogy between linear motion and rotational motion is revealed when we compare the defining equation for average angular acceleration,

$$\bar{\alpha} \equiv \frac{\omega_f - \omega_i}{t_f - t_i} = \frac{\Delta\omega}{\Delta t}$$

with the defining equation for average linear acceleration,

$$\bar{a} \equiv \frac{v_f - v_i}{t_f - t_i} = \frac{\Delta v}{\Delta t}$$

In light of the analogies between variables in linear motion and those in rotational motion, it should not surprise you that the equations of rotational motion involve the variables θ, ω, and α. In Section 2.6, we developed a set of kinematic equations for linear motion under constant acceleration. The same procedure can be used to derive a similar set of equations for rotational motion under constant angular acceleration. The resulting equations of rotational kinematics, along with the corresponding equations for linear motion under constant acceleration, are as follows:

Tip 7.2   JUST LIKE TRANSLATION?

Equations 7.5–7.7 suggest that rotational kinematics is just like translational kinematics. That is almost true, but remember the following two differences. (1) In rotational kinematics, you must specify a rotation axis. (2) In rotational motion, the object keeps returning to its original orientation—thus, you may be asked for the number of revolutions made by a rigid body, which is something that has no meaning in translational motion.

Rotational Motion About a Fixed Axis with α Constant (Variables: θ and ω)	Linear Motion with a Constant (Variables: x and v)
$\omega = \omega_i + \alpha t$ [7.5]	$v = v_i + at$
$\Delta\theta = \omega_i t + \frac{1}{2}\alpha t^2$ [7.6]	$\Delta x = v_i t + \frac{1}{2}at^2$
$\omega^2 = \omega_i^2 + 2\alpha\,\Delta\theta$ [7.7]	$v^2 = v_i^2 + 2a\,\Delta x$

Again, note the one-to-one correspondence between the rotational equations involving the angular variables θ, ω, and α and the equations of linear motion involving the variables x, v, and a.

Example 7.2   A Rotating Wheel

A wheel rotates with a constant angular acceleration of 3.50 rad/s^2.

A If the angular speed of the wheel is 2.00 rad/s at $t_i = 0$, through what angle does the wheel rotate between $t = 0$ and $t = 2.00$ s?

Solution   Because the angular acceleration in the problem is given as constant, we can use the rotational kinematic equations. We use Equation 7.6, setting $\theta_i = 0$,

$$\Delta\theta = \omega_i t + \frac{1}{2}\alpha t^2$$
$$= (2.00 \text{ rad/s})(2.00 \text{ s}) + \frac{1}{2}(3.50 \text{ rad/s}^2)(2.00 \text{ s})^2$$
$$= \boxed{11.0 \text{ rad}}$$

This is equivalent to $11.0 \text{ rad}/(2\pi \text{ rad/rev}) = 1.75$ rev.

B What is the angular speed of the wheel at $t = 2.00$ s?

Solution   We use Equation 7.5:

$$\omega = \omega_i + \alpha t = 2.00 \text{ rad/s} + (3.50 \text{ rad/s}^2)(2.00 \text{ s}) = \boxed{9.00 \text{ rad/s}}$$

We could also obtain this result using Equation 7.7 and the results of part A. Try it!

EXERCISE   Find the angle through which the wheel rotates between $t = 2.00$ s and $t = 3.00$ s.

ANSWER   10.8 rad

7.3 RELATIONS BETWEEN ANGULAR AND LINEAR QUANTITIES

In this section we derive some useful relations between the angular speed and acceleration of a rotating object and the linear speed and acceleration of an arbitrary point in the object.

Consider the arbitrarily shaped object in Figure 7.4, rotating about the z axis through the point O. Assume that the object rotates through the angle $\Delta\theta$, and hence point P moves through the arc length Δs in the interval Δt. We know from the defining equation for radian measure that

$$\Delta\theta = \frac{\Delta s}{r}$$

Let us now divide both sides of this equation by Δt, the interval during which the rotation occurred:

$$\frac{\Delta\theta}{\Delta t} = \frac{1}{r}\frac{\Delta s}{\Delta t}$$

If Δt is very small, then the angle $\Delta\theta$ through which the object rotates is small and the ratio $\Delta\theta/\Delta t$ is the instantaneous angular speed ω. Also, Δs is very small when Δt is very small, and the ratio $\Delta s/\Delta t$ equals the instantaneous linear speed v for small values of Δt. Hence, the preceding equation is equivalent to

$$\omega = \frac{v}{r}$$

Figure 7.4 allows us to interpret this equation. The distance Δs is traversed along an arc of the circular path followed by the point P as it moves during the time Δt. Thus, v must be the linear speed of a point lying along this arc, where the direction of the point's vector velocity **v** is *tangent to the circular path*. As a result, we often refer to this linear speed as the *tangential speed* of a particle moving in a circular path, and write

$$\boxed{v_t = r\omega} \qquad\qquad [7.8] \qquad \blacktriangleleft \text{ Tangential speed}$$

That is, **the tangential speed of a point on a rotating object equals the distance of that point from the axis of rotation multiplied by the angular speed.** Equation 7.8 shows that the linear speed of a point on the rotating object increases with movement outward from the center of rotation toward the rim, as we would intuitively expect. **Although every point on the rotating object has the same angular speed, not every point has the same linear (tangential) speed.**

Exercise caution when you use Equation 7.8. It has been derived using the defining equation for radian measure and hence is valid only when ω is measured

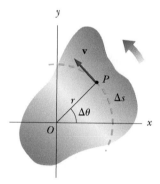

FIGURE 7.4   Rotation of an object about an axis through O that is perpendicular to the plane of the figure (the z axis). Note that a point P on the object rotates in a circle of radius r centered at O.

in radians per unit time. Other measures of angular speed, such as degrees per second and revolutions per second, will not give correct answers in Equation 7.8.

To find a second equation relating linear and angular quantities, imagine that an object rotating about a fixed axis (Fig. 7.4) changes its angular speed by $\Delta\omega$ in the interval Δt. At the end of this interval, the speed of a point on the object, such as P, has changed by the amount Δv_t. From Equation 7.8 we have

$$\Delta v_t = r\,\Delta\omega$$

Dividing by Δt gives

$$\frac{\Delta v_t}{\Delta t} = r\frac{\Delta\omega}{\Delta t}$$

If the time interval Δt is very small, then the ratio $\Delta v_t/\Delta t$ is the tangential acceleration of that point and $\Delta\omega/\Delta t$ is the angular acceleration. Therefore, we see that

Tangential acceleration

$$a_t = r\alpha \qquad\qquad [7.9]$$

That is, **the tangential acceleration of a point on a rotating object equals the distance of that point from the axis of rotation multiplied by the angular acceleration.** Again, radian measure must be used for the angular acceleration in this equation.

There is one more equation that relates linear quantities to angular quantities, but we shall defer its derivation to the next section.

Quick Quiz 7.3   When a wheel of radius R rotates about a fixed axis as in Figure 7.3, which of the following statements are false? (a) All points on the wheel have the same angular speed. (b) All points on the wheel have the same tangential speed. (c) All points on the wheel have the same angular acceleration. (d) All points on the wheel have the same tangential acceleration.

APPLYING PHYSICS 7.1

The launch area for the European Space Agency is not in Europe—it is in South America. Why?

Explanation   Placing a satellite in Earth orbit requires providing a large tangential speed to the satellite. This is the task of the rocket propulsion system. Anything that reduces the requirements on the propulsion system is a welcome contribution. The surface of Earth is already traveling toward the east at a high speed, due to Earth's rotation. Thus, if rockets are launched toward the east, the rotation of Earth provides some initial tangential speed, reducing somewhat the requirements on the propulsion system. If rockets were launched from Europe, which is at a relatively large latitude, the contribution of Earth's rotation is relatively small, because the distance between Europe and the rotation axis of Earth is relatively small. The ideal place for launching is at the Equator, which is as far as one can be from the rotation axis of Earth and still be on Earth's surface. This results in the largest possible tangential speed due to Earth's rotation. The European Space Agency exploits this advantage by launching from French Guiana, which is only a few degrees north of the Equator.

A second advantage of this launch location is that launching toward the east takes the spacecraft over water. In the event of an accident or a failure, the wreckage will fall into the ocean rather than falling into populated areas as it would if launched to the east from Europe. This is the primary reason why the United States launches spacecraft from Florida rather than California, despite the more favorable weather conditions in California.

APPLYING PHYSICS 7.2

When a tall smokestack falls over, it often breaks somewhere along its length before it hits the ground as in Figure 7.5. Why does this happen?

Explanation As the smokestack rotates around its base, each higher portion of the smokestack falls with an increasing tangential acceleration. The tangential acceleration of a given point on the smokestack is proportional to the distance of that portion from the base. As the acceleration increases, eventually higher portions of the smokestack will experience an acceleration greater than that resulting from gravity alone. This can only happen if these portions are being pulled downward by a force in addition to the gravitational force. The force that causes this to occur is the cohesive force from lower portions of the smokestack. Eventually, the cohesive force that provides this acceleration is larger than the smokestack can withstand, and it breaks.

FIGURE 7.5 (Applying Physics 7.2) A falling smokestack.

Example 7.3 Compact Discs

A compact disc in a computer rotates from rest up to an angular speed of 31.4 rad/s in a time of 0.892 s.

A What is the angular acceleration of the disc, assuming the angular acceleration is uniform?

Solution If we use $\omega = \omega_i + \alpha t$ and the fact that $\omega_i = 0$ at $t = 0$, we get

$$\alpha = \frac{\omega}{t} = \frac{31.4 \text{ rad/s}}{0.892 \text{ s}} = \boxed{35.2 \text{ rad/s}^2}$$

B How many rotations does the disc make while coming up to speed?

Solution Equation 7.6 enables us to find the angular displacement during the 0.892-s time interval. Taking $\omega_i = 0$ and $\theta_i = 0$ gives

$$\Delta\theta = \omega_i t + \tfrac{1}{2}\alpha t^2 = \tfrac{1}{2}(35.2 \text{ rad/s}^2)(0.892 \text{ s})^2 = \boxed{14.0 \text{ rad}}$$

Because 2π rad = 1 rev, this angular displacement corresponds to 2.23 rev.

C If the radius of the disc is 4.45 cm, find the final linear speed of a microbe riding on the rim of the disc.

Solution The relation $v_t = r\omega$ and the given values lead to

$$v_t = r\omega = (0.044\ 5 \text{ m})(31.4 \text{ rad/s}) = \boxed{1.40 \text{ m/s}}$$

D What is the magnitude of the tangential acceleration of the microbe at this time?

Solution We use $a_t = r\alpha$, which gives

$$a_t = r\alpha = (0.044\ 5 \text{ m})(35.2 \text{ rad/s}^2) = \boxed{1.57 \text{ m/s}^2}$$

EXERCISE What is the angular speed and angular displacement of the disc 0.300 s after it begins to rotate?

ANSWER 10.6 rad/s; 1.58 rad

Long before compact discs became the medium of choice for recorded music, phonographs were popular. There are similarities between the rotational motion of phonograph records and compact discs, and there are some interesting differences. A phonograph record, for instance, rotates at a constant angular speed. Popular angular speeds were $33\tfrac{1}{3}$ rev/min for long-playing albums (hence the nickname "LP"), 45 rev/min for "singles," and 78 rev/min used in very early

APPLICATION

PHONOGRAPH RECORDS AND COMPACT DISCS

recordings. At the outer edge of the record, the pickup needle (stylus) was moving over the vinyl material at a faster tangential speed than when the needle was close to the center of the record. As a result, the sound information was compressed into a smaller length of track near the center of the record than near the outer edge.

Compact discs, on the other hand, are designed such that the disc moves past the laser pickup at a constant tangential speed. Because the pickup moves radially as it follows the tracks of information, this requires that the angular speed of the compact disc must vary according to the radial position of the laser. Because the tangential speed is fixed, the information density (per length of track) anywhere on the disc is the same. Example 7.4 demonstrates numerical calculations for both compact discs and phonograph records.

Example 7.4 Length of a CD Track

A compact disc player is designed such that as the read head moves out from the center of the disc, the angular speed of the disc changes so that the linear speed at the position of the head will always be at a constant value of about 1.3 m/s.

A Find the angular speed of the disc when the read head is at a distance of 2.0 cm from the center of the disc and at 5.6 cm.

Solution   From $v_t = r\omega$, we have at $r = 2.0$ cm,

$$\omega = \frac{v_t}{r} = \frac{1.3 \text{ m/s}}{2.0 \times 10^{-2} \text{ m}} = \boxed{65 \text{ rad/s}}$$

and at 5.6 cm,

$$\omega = \frac{v_t}{r} = \frac{1.3 \text{ m/s}}{5.6 \times 10^{-2} \text{ m}} = \boxed{23 \text{ rad/s}}$$

B On the other hand, an old-fashioned record player rotated at a constant angular speed so that the linear speed of the record groove moving under the detector (the stylus) changed. Find the linear speed of a 45.0-rpm record at points 2.0 and 5.6 cm from the center.

Solution   Because 45.0 rev/min = 4.7 rad/s, the linear speed at $r = 2.0$ cm is

$$v_t = r\omega = (2.0 \times 10^{-2} \text{ m})(4.7 \text{ rad/s}) = \boxed{0.094 \text{ m/s}}$$

At $r = 5.6$ cm, we have

$$v_t = r\omega = (5.6 \times 10^{-2} \text{ m})(4.7 \text{ rad/s}) = \boxed{0.26 \text{ m/s}}$$

C In both the CD and phonograph record, information is recorded in a continuous spiral track as one moves over the surface of the disc from the edge. Use the information above to calculate the total length of the track for a CD designed to play for 1 hour.

Solution   The CD moves under the read head with a constant linear speed of 1.3 m/s. Thus, the total length of the track that moves under it in 3 600 seconds is

$$\Delta x = v_t t = (1.3 \text{ m/s})(3\,600 \text{ s}) = \boxed{4\,680 \text{ m}}$$

7.4 CENTRIPETAL ACCELERATION

Figure 7.6a shows a car moving in a circular path with *constant linear speed v.* Students are often surprised to find that **even though the car moves at a constant speed, it still has an acceleration.** To see why this occurs, consider the defining equation for average acceleration:

$$\bar{\mathbf{a}} = \frac{\mathbf{v}_f - \mathbf{v}_i}{t_f - t_i} \qquad [7.10]$$

Note that the average acceleration depends on the *change in the velocity vector*. Because velocity is a vector, there are two ways in which an acceleration can be produced: by a change in the *magnitude* of the velocity and by a change in the *direction* of the velocity. It is the latter situation that occurs for the car moving in a circular path with constant speed (see Fig. 7.6b). We shall show that the acceleration vector in this case is perpendicular to the path and always points toward the center of the circle. An acceleration of this nature is called a **centripetal** (center-seeking) **acceleration.** Its magnitude is given by

$$a_c = \frac{v^2}{r} \qquad [7.11]$$

To derive Equation 7.11, consider Figure 7.7a. Here an object is seen first at point Ⓐ with velocity $\mathbf{v}_i$ at time t_i and then at point Ⓑ with velocity $\mathbf{v}_f$ at a later time t_f. Let us assume that here $\mathbf{v}_i$ and $\mathbf{v}_f$ differ only in direction; their magnitudes are the same (that is, $v_i = v_f = v$). To calculate the acceleration, we begin with Equation 7.10, which indicates that we must vectorially subtract $\mathbf{v}_i$ from $\mathbf{v}_f$:

$$\bar{\mathbf{a}} = \frac{\mathbf{v}_f - \mathbf{v}_i}{t_f - t_i} = \frac{\Delta\mathbf{v}}{\Delta t} \qquad [7.12]$$

where $\Delta\mathbf{v} = \mathbf{v}_f - \mathbf{v}_i$ is the change in velocity. This can be accomplished graphically, as shown by the vector triangle in Figure 7.7b. Note that when Δt is very small, Δs and $\Delta\theta$ will also be very small. In this case, $\mathbf{v}_f$ will be almost parallel to $\mathbf{v}_i$, and the vector $\Delta\mathbf{v}$ will be approximately perpendicular to them, pointing toward the center of the circle. In the limiting case where Δt becomes vanishingly small, $\Delta\mathbf{v}$ will point exactly toward the center of the circle. Furthermore, in this limiting case the acceleration will also be directed toward the center of the circle because it is in the direction of $\Delta\mathbf{v}$.

Now consider the triangle in Figure 7.7a, which has sides Δs and r. This triangle and the one formed by the vectors in Figure 7.7b are similar. This enables us to write a relationship between the lengths of the sides:

$$\frac{\Delta v}{v} = \frac{\Delta s}{r}$$

or

$$\Delta v = \frac{v}{r}\Delta s$$

This can be substituted into Equation 7.12 for Δv to give

$$a = \frac{v}{r}\frac{\Delta s}{\Delta t} \qquad [7.13]$$

In this situation, Δs is a small distance measured along the arc of the circle (a tangential distance), so $v = \Delta s/\Delta t$, where v is the tangential speed. Therefore, Equation 7.13 reduces to Equation 7.11:

$$a_c = \frac{v^2}{r}$$

Because the tangential speed is related to the angular speed through the relation $v_t = r\omega$ (Eq. 7.8), an alternative form of Equation 7.11 is

$$a_c = \frac{r^2\omega^2}{r} = r\omega^2 \qquad [7.14]$$

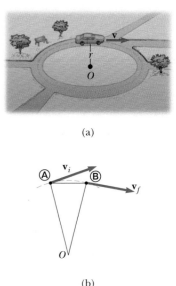

(a)

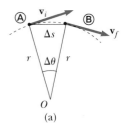

(b)

FIGURE 7.6 (a) Circular motion of a car moving with constant speed. (b) As the car moves along the circular path from Ⓐ to Ⓑ, the direction of its velocity vector changes, so the car undergoes a centripetal acceleration.

(a)

(b)

FIGURE 7.7 (a) As the particle moves from Ⓐ to Ⓑ, the direction of its velocity vector changes from $\mathbf{v}_i$ to $\mathbf{v}_f$. (b) The construction for determining the direction of the change in velocity $\Delta\mathbf{v}$, which is toward the center of the circle.

Thus, we conclude that

Centripetal acceleration ▶

In circular motion the centripetal acceleration is directed inward toward the center of the circle and has a magnitude given by either v^2/r or $r\omega^2$.

You should show that the dimensions of a_c are L/T^2, as required.

Let us clear up any misconceptions that might exist concerning centripetal and tangential acceleration by considering a car moving around a circular race track. In order to move in a circular path, the car always has a centripetal component of acceleration because its direction of travel, and hence the direction of its velocity, is continuously changing. If the car's speed is increasing or decreasing, it also has a tangential component of acceleration. To summarize, the tangential component of acceleration is due to changing speed; the centripetal component of acceleration is due to changing direction.

When both components of acceleration exist simultaneously, the component vector associated with the tangential acceleration is tangent to the circular path, and the component vector associated with the centripetal acceleration points toward the center of the circular path. Because these component vectors are perpendicular to each other, we can find the magnitude of the **total acceleration** using the Pythagorean theorem:

Total acceleration ▶

$$a = \sqrt{a_t^2 + a_c^2}$$ [7.15]

Quick Quiz 7.4

A race track is constructed such that two arcs of radius 80 m at Ⓐ and 40 m at Ⓑ are joined by two stretches of straight track as in Figure 7.8. In a particular trial run, a driver traveled at a constant speed of 50 m/s for one complete lap.

1. The ratio of the tangential acceleration at Ⓐ to that at Ⓑ is
   (a) $\frac{1}{2}$ (b) $\frac{1}{4}$ (c) 2   (d) 4
   (e) The tangential acceleration is zero at both points.
2. The ratio of the centripetal acceleration at Ⓐ to that at Ⓑ is
   (a) $\frac{1}{2}$ (b) $\frac{1}{4}$ (c) 2   (d) 4
   (e) The centripetal acceleration is zero at both points.
3. The angular speed is greatest at
   (a) Ⓐ   (b) Ⓑ   (c) It is equal at both Ⓐ and Ⓑ.

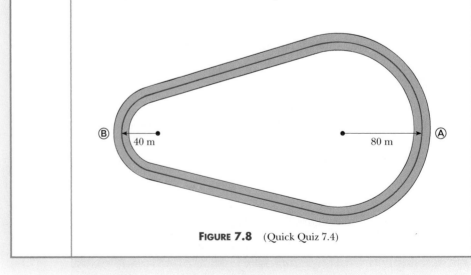

FIGURE 7.8   (Quick Quiz 7.4)

Quick Quiz 7.5 An object moves in a circular path with constant speed v. Which of the following statements is true concerning the object? (a) Its velocity is constant but its acceleration is changing. (b) Its acceleration is constant but its velocity is changing. (c) Both its velocity and acceleration are changing. (d) Its velocity and acceleration remain constant.

Example 7.5   Let's Go for a Spin

A test car moves at a constant speed of 10 m/s around a circular road of radius 50 m. Find the car's (a) centripetal acceleration and (b) angular speed.

Solution

A From Equation 7.11, the magnitude of the centripetal acceleration of the car is found to be

$$a_c = \frac{v^2}{r} = \frac{(10 \text{ m/s})^2}{50 \text{ m}} = 2.0 \text{ m/s}^2$$

By definition, the direction of $\mathbf{a}_c$ is toward the center of curvature of the road.

B The angular speed of the car can be found using the expression $v_t = r\omega$ which gives

$$\omega = \frac{v_t}{r} = \frac{10 \text{ m/s}}{50 \text{ m}} = 0.20 \text{ rad/s}$$

Note that we can also find the centripetal acceleration by using this value of ω, in an alternative method to part A. From Equation 7.14 we have

$$a_c = r\omega^2 = (50 \text{ m})(0.20 \text{ rad/s})^2 = 2.0 \text{ m/s}^2$$

As expected, the two methods for finding the centripetal acceleration give the same answer.

EXERCISE   Find the car's tangential acceleration and total acceleration.

ANSWER   The tangential acceleration is zero because the speed of the car remains constant. Because $a_t = 0$, the total acceleration in this example equals the centripetal acceleration, 2.0 m/s^2, found in part A.

7.5   THE VECTOR NATURE OF ANGULAR QUANTITIES

Before continuing our discussion of centripetal acceleration, a brief digression is in order. In Sections 7.1–7.4 we treated angular displacement, speed, and acceleration as though they were scalar quantities with linear analogs. You should not be surprised to learn that, like their linear analogs, these rotational concepts generalize to vector quantities that have both magnitude and direction. The first hint of this came just before Example 7.1, when we assigned positive and negative values to angular speed ω. That assignment was analogous to our practice in Chapter 2 of choosing a positive or negative direction for one-dimensional linear motion.

Let us describe how to find the direction of the **angular velocity** vector $\boldsymbol{\omega}$. When a rigid object such as a disk rotates about a fixed axis, the angular velocity $\boldsymbol{\omega}$ points along this axis. A convenient right-hand rule, illustrated in Figure 7.9a, can

(a)

(b)

FIGURE 7.10 **An end view of a disk rotating about an axis through its center perpendicular to the page.** (a) When the disk rotates clockwise, **ω** is into the page. (b) When the disk rotates counterclockwise, **ω** is out of the page.

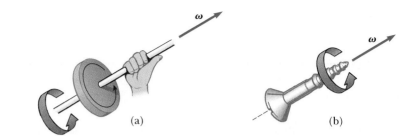

FIGURE 7.9 (a) The right-hand rule for determining the direction of the angular velocity vector **ω**. (b) The direction of **ω** is in the direction of advance of a right-handed screw.

be used to determine the direction of **ω**. Grasp the axis of rotation with your right hand so that your fingers wrap in the direction of rotation. Your extended thumb points in the direction of **ω**. Figure 7.9b illustrates that **ω** is also in the direction of advance of a rotating right-handed screw.

We can apply this rule to a rotating disk when viewed along the axis of rotation as in Figure 7.10. When the disk rotates clockwise as in Figure 7.10a, applying the right-hand rule shows that the direction of **ω** is into the page. When the disk rotates counterclockwise as in Figure 7.10b, the direction of **ω** is out of the page.

Finally, the direction of the angular acceleration vector **α** is in the direction of **ω** if the angular speed (the magnitude of **ω**) is increasing in time and antiparallel to **ω** if the angular speed is decreasing in time.

Although we will not work problems involving vector angular velocity and acceleration, the vector nature of rotational quantities plays an important role in a number of phenomena we shall consider later in this book.

7.6 FORCES CAUSING CENTRIPETAL ACCELERATION

Consider a ball of mass m tied to a string of length r and being whirled in a horizontal circular path, as in Figure 7.11. Let us assume that the ball moves with constant speed. Because the velocity vector **v** changes its direction continuously during the motion, the ball experiences a centripetal acceleration directed toward the center of motion, as described in Section 7.4, with magnitude

$$a_c = \frac{v_t^2}{r}$$

The inertia of the ball tends to maintain motion in a straight-line path; however, the string exerts a force on the ball that makes it follow a circular path instead. This force (equal to the force of tension) is directed along the length of the string toward the center of the circle in Figure 7.11. Thus, the equation for Newton's second law along the radial direction is

$$F = ma_c = m\frac{v_t^2}{r} \qquad [7.16]$$

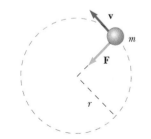

FIGURE 7.11 A ball attached to a string of length r, rotating in a circular path at constant speed.

Here F stands for any force—tension in a string, gravity, the force of friction—acting in a given situation to keep an object following a circular path.

Because it acts at right angles to the motion, a force that produces a centripetal acceleration causes a change in the direction of the velocity. Beyond this, such a force is no different from any of the other forces we have studied. For example, friction between tires and road provides the force that enables a race car to travel in a circular path on a flat road, and the gravitational force exerted by Earth on the Moon provides the force necessary to keep the Moon in its orbit. (We shall discuss the gravitational force in more detail in Section 7.8.)

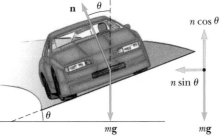

FIGURE 7.12   Front view of a car rounding a banked roadway.

Regardless of the example used, if the force that produces a centripetal acceleration vanishes, the object does not continue to move in its circular path; instead, it moves along a straight-line path tangent to the circle. As the race car in the preceding paragraph travels faster and faster on the curved, flat road, the centripetal acceleration increases, as does the friction force needed to cause this acceleration. At some speed, the friction force needed to keep the car on a circular path will be greater than the maximum possible static friction force allowed by the static coefficient of friction. At this point, the car will skid off the road.

In order to prevent this from happening, roadways are often banked instead of being built with flat curves. Banking the roadway toward the inside of the circular path as in Figure 7.12 creates a component of the normal force exerted by the road on the car that is directed toward the center of the path. This component of the normal force adds to the centrally directed friction force. Thus, the car can negotiate a banked curve at a higher speed than it can an unbanked curve. The banking of roadways is especially evident at high-speed race tracks, in bicycle tracks in velodromes, and in bobsled tracks.

> **Tip 7.3   CENTRIPETAL FORCE**
>
> The force causing centripetal acceleration is called a *centripetal force* in some textbooks. Giving the force causing circular motion a name—centripetal force—leads many students to consider this as a new *kind* of force. Furthermore, a common mistake in drawing force diagrams is to include all of the usual forces and then add another vector for the centripetal force. But it is not a separate force— *it is simply one of our familiar forces that causes circular motion.* For Earth's motion around the Sun, the "centripetal force" is the force of *gravity*. For a rock whirled on the end of a string, the "centripetal force" is the *tension* in the string. For an amusement park patron pressed against the inner wall of a rapidly rotating circular room, the "centripetal force" is the *normal force* exerted by the wall.

◄ APPLICATION

BANKED ROADWAYS

PROBLEM-SOLVING STRATEGY   Forces That Cause Centripetal Acceleration

Use the following steps when dealing with centripetal accelerations and the forces that produce them:

1. Draw a free-body diagram of the object(s) under consideration, showing and labeling all forces that act on it (them).
2. Choose a coordinate system that has one axis perpendicular to the circular path followed by the object and one axis tangent to the circular path.
3. Find the net force toward the center of the circular path. This is the force that causes the centripetal acceleration.
4. From here on the steps are virtually identical to those encountered when solving Newton's second-law problems with $\Sigma F_x = ma_x$ and $\Sigma F_y = ma_y$. However, instead of the x and y directions, we apply Newton's second law in the radial and tangential directions by writing $\Sigma F_{\text{along radius}} = ma_c$ and $\Sigma F_t = ma_t$. Also, note that the magnitude of the centripetal acceleration can always be written $a_c = v_t^2/r$.

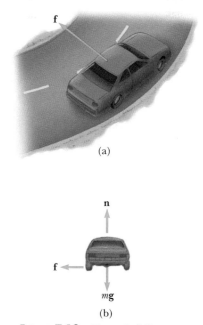

(a)

(b)

FIGURE 7.13   (Example 7.6) (a) Top view of a car on a curved path. (b) A free-body diagram for the car, showing an end view.

Example 7.6   Buckle Up for Safety

A car travels at a constant speed of 30.0 mi/h (13.4 m/s) on a level circular turn of radius 50.0 m, as shown in the bird's-eye view in Figure 7.13a. What is the minimum coefficient of static friction between the tires and roadway so that the car makes the circular turn without sliding?

Reasoning   The force in the radial direction acting on the car is the force of static friction directed toward the center of the circular path as in Figure 7.13a. Thus, our objective in this example is to find an expression for the maximum frictional force $f = \mu_s n$ and use it in Equation 7.16.

Solution   Equation 7.16 applied to this situation becomes

$$f = m\frac{v_t^2}{r} \tag{1}$$

All the forces acting on the car are shown in Figure 7.13b. Because the car is in equilibrium in the vertical direction, the normal force upward is balanced by the force of gravity downward, so that

$$n = mg$$

From this expression we can find the maximum force of static friction, as follows:

$$f = \mu_s n = \mu_s mg \tag{2}$$

By setting the right-hand sides of (1) and (2) equal to each other, we find that

$$\mu_s mg = m\frac{v_t^2}{r}$$

or

$$\mu_s = \frac{v_t^2}{rg} \tag{3}$$

Therefore, the minimum coefficient of static friction required for the car to make the turn without sliding outward is

$$\mu_s = \frac{v_t^2}{rg} = \frac{(13.4 \text{ m/s})^2}{(50.0 \text{ m})(9.80 \text{ m/s}^2)} = \boxed{0.366}$$

The value of μ_s for rubber on dry concrete is very close to 1; thus, the car can negotiate the curve with ease. If the road were wet or icy, however, the value for μ_s could be 0.2 or lower. Under such conditions, the radial force provided by static friction would not be great enough to enable the car to follow the circular path, and it would slide off the roadway.

Example 7.7   Fun with a Yo-Yo

A child swings a yo-yo of weight mg in a horizontal circle so that the cord makes an angle of θ with the vertical, as in Figure 7.14a. Find the centripetal acceleration of the yo-yo.

Reasoning   Two forces act on the yo-yo: the force of gravity and the force **T** associated with the tension in the cord, as in Figure 7.14b. The condition for equilibrium from Chapter 4 (Equation 4.9) applied in the y direction, $\Sigma F_y = 0$, enables us to find T. The component of **T** toward the center of the circular path is the force that produces the centripetal acceleration, and we can therefore find a_c from $F = ma_c$.

Solution   Applying ΣF_y gives

$$T\cos\theta - mg = 0$$

$$T = \frac{mg}{\cos\theta}$$

The *horizontal component* of **T** causes the centripetal acceleration.

$$F = T\sin\theta = \frac{mg\sin\theta}{\cos\theta} = mg\tan\theta$$

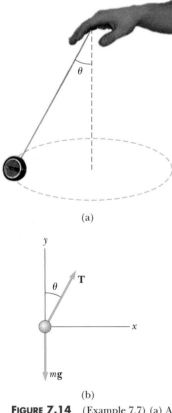

(a)

(b)

Figure 7.14   (Example 7.7) (a) A yo-yo swinging in a horizontal circle so that the cord makes a constant angle θ with the vertical. (b) A free-body diagram for the yo-yo.

The **average angular acceleration** $\bar{\alpha}$ of a rotating object is defined as the ratio of the change in angular speed $\Delta\omega$ to the time interval Δt.

$$\bar{\alpha} \equiv \frac{\omega_f - \omega_i}{t_f - t_i} = \frac{\Delta\omega}{\Delta t} \qquad \text{[7.4]}$$

where $\bar{\alpha}$ is in radians per second per second (rad/s^2).

If an object undergoes rotational motion about a fixed axis under constant angular acceleration α, we can describe its motion with the following set of equations:

$$\omega = \omega_i + \alpha t \qquad \text{[7.5]}$$

$$\Delta\theta = \omega_i t + \tfrac{1}{2}\alpha t^2 \qquad \text{[7.6]}$$

$$\omega^2 = \omega_i^2 + 2\alpha\,\Delta\theta \qquad \text{[7.7]}$$

When an object rotates about a fixed axis, the angular speed and angular acceleration are related to the tangential speed and tangential acceleration through the relationships

$$v_t = r\omega \qquad \text{[7.8]}$$

$$a_t = r\alpha \qquad \text{[7.9]}$$

Any object moving in a circular path has an acceleration directed toward the center of the circular path, called a **centripetal acceleration.** Its magnitude is given by

$$a_c = \frac{v^2}{r} = r\omega^2 \qquad \text{[7.11, 7.14]}$$

Any object moving in a circular path must have a net force exerted on it that is directed toward the center of the circular path. From Newton's second law, this force and the centripetal acceleration are related by

$$F = m\frac{v_t^2}{r} \qquad \text{[7.16]}$$

Some examples of forces that cause centripetal acceleration are the force of gravity (as in the motion of a satellite) and the force of tension in a string.

Newton's law of universal gravitation states that every particle in the Universe attracts every other particle with a force that is directly proportional to the product of their masses and inversely proportional to the square of the distance r between them:

$$F = G\frac{m_1 m_2}{r^2} \qquad \text{[7.17]}$$

where $G = 6.673 \times 10^{-11} \text{ N·m}^2/\text{kg}^2$ is the **constant of universal gravitation.**

Kepler's laws of planetary motion state that

1. All planets move in elliptical orbits with the Sun at one of the focal points.
2. A line drawn from the Sun to any planet sweeps out equal areas in equal time intervals.
3. The square of the orbital period of a planet is proportional to the cube of the average distance from the planet to the Sun:

$$T^2 = \left(\frac{4\pi^2}{GM_S}\right) r^3 \qquad \text{[7.22]}$$

CONCEPTUAL QUESTIONS

1. In a race like the Indianapolis 500, a driver circles the track counterclockwise and feels his head pulled toward one shoulder. To relieve his neck muscles from having to hold his head erect, the driver fastens a strap to one wall of the car and the other to his helmet. The length of the strap is adjusted to keep his head vertical. (a) Which shoulder does his head tend to lean toward? What force(s) produce the centripetal acceleration (b) when there is no strap, and (c) when there is a strap?

2. Consider a compact disc rotating at constant angular speed. Does a scratch on the rim have a radial component of velocity? A tangential component? A radial component

of acceleration? A tangential component? Repeat these questions for the brief interval of time during which the CD moves from rest to its final angular speed.

3. An object executes circular motion with a constant speed whenever a net force of constant magnitude acts perpendicular to the velocity. What happens to the speed if the force is not perpendicular to the velocity?

4. Explain why Earth is not spherical in shape, but bulges at the Equator.

5. If a car's wheels are replaced with wheels of greater diameter, will the reading of the speedometer change? Explain.

6. At night, you are farther from the Sun than during the day. What's more, the force exerted by the Sun on you is downward into Earth at night, and upward into the sky during the day. If you had a sensitive enough bathroom scale, would you appear to weigh more at night than during the day?

7. Correct the following statement: "The race car rounds the turn at a constant velocity of 90 miles per hour."

8. Why does an astronaut in a spacecraft orbiting Earth experience a feeling of weightlessness?

9. Explain why it is easier to determine the mass of a planet when it has a moon than to find its mass if it has no moon.

10. Because of Earth's rotation about its axis, you will weigh slightly less at the Equator than at the poles. Why?

11. It has been suggested that rotating cylinders about 10 miles long and 5 miles in diameter be placed in space for colonies. The purpose of their rotations is to simulate gravity for the inhabitants. Explain the concept behind this proposal.

12. Describe the path of a moving object in the event that the object's acceleration is constant in magnitude at all times and (a) perpendicular to its velocity; (b) parallel to its velocity.

13. A pail of water can be whirled in a vertical circular path such that no water is spilled. Why does the water remain in the pail, even when the pail is upside down above your head?

14. The orbital planes of all the planets must pass through the center of the Sun. Why?

15. Is it possible for a car to move in a circular path in such a way that it has a tangential acceleration but no centripetal acceleration?

16. Use Kepler's second law to convince yourself that Earth moves faster in its orbit during the northern-hemisphere winter, when it is closest to the Sun, than during the summer, when it is farthest from the Sun.

17. Why does a pilot sometimes black out when pulling out of a steep dive?

PROBLEMS

1, 2, 3 = straightforward, intermediate, challenging   ☐ = full solution available in Student Solutions Manual/Study Guide

web = solution posted at **http://info.brookscole.com/serway**   🐾 = biomedical application

Section 7.1 Angular Speed and Angular Acceleration

1. The tires on a new compact car have a diameter of 2.0 ft and are warranted for 60 000 miles. (a) Determine the angle (in radians) through which one of these tires will rotate during the warranty period. (b) How many revolutions of the tire are equivalent to your answer in (a) above?

2. A wheel has a radius of 4.1 m. How far (path length) does a point on the circumference travel if the wheel is rotated through angles of $30°$, 30 rad, and 30 rev, respectively?

3. Find the angular speed of Earth about the Sun in radians per second and degrees per day.

4. A potter's wheel moves from rest to an angular speed of 0.20 rev/s in 30 s. Find its angular acceleration in radians per second per second.

Section 7.2 Rotational Motion Under Constant Angular Acceleration

Section 7.3 Relations Between Angular and Linear Quantities

5. A dentist's drill starts from rest. After 3.20 s of constant angular acceleration, it turns at a rate of 2.51×10^4 rev/min. (a) Find the drill's angular acceleration. (b) Determine the angle (in radians) through which the drill rotates during this period.

6. An electric motor rotating a workshop grinding wheel at a rate of 100 rev/min is switched off. Assume constant negative angular acceleration of magnitude 2.00 rad/s². (a) How long does it take for the grinding wheel to stop? (b) Through how many radians has the wheel turned during the interval found in (a)?

7. A car is traveling with a velocity of 17.0 m/s on a straight horizontal highway. The wheels of the car have a radius of 48.0 cm. If the car then speeds up with an acceleration of 2.00 m/s² for 5.00 s, find the number of revolutions of the wheels during this period.

8. The car in the preceding problem runs out of gas and the rotation of the wheels slows with an angular acceleration of magnitude 1.35 rad/s². If the car has a speed of 24.0 m/s when this happens, through how many revolutions do the wheels turn as the car comes to rest?

9. The diameters of the main rotor and tail rotor of a single-engine helicopter are 7.60 m and 1.02 m, respectively. The respective rotational speeds are 450 rev/min and 4 138 rev/min. Calculate the speeds of the tips of both rotors. Compare these speeds with the speed of sound, 343 m/s.

10. The tub of a washer goes into its spin-dry cycle, starting from rest and reaching an angular speed of 5.0 rev/s in 8.0 s. At this point the person doing the laundry opens the lid, and a safety switch turns off the washer. The tub slows to rest in 12.0 s. Through how many revolutions does the tub turn during this 20-s interval? Assume constant angular acceleration while it is starting and stopping.

11. A standard cassette tape is placed in a standard cassette player. Each side lasts for 30 minutes. The two tape wheels of the cassette fit onto two spindles in the player. Suppose

that a motor drives one spindle at constant angular velocity ~ 1 rad/s and the other spindle is free to rotate at any angular speed. Find the order of magnitude of the tape's thickness. Specify any other quantities you estimate and the values you take for them.

12. A coin with a diameter of 2.40 cm is dropped on edge onto a horizontal surface. The coin starts out with an initial angular speed of 18.0 rad/s and rolls in a straight line without slipping. If the rotation slows with an angular acceleration of magnitude 1.90 rad/s^2, how far does the coin roll before coming to rest?

13. A pulsar is a celestial object that emits light in short bursts. A pulsar in the Crab Nebula flashes at a rate of 30 times/s. Suppose the light pulses are caused by the rotation of a spherical object that emits light from a pair of diametrically opposed "flashlights" on its equator. What is the maximum radius of the pulsar if no part of its surface can move faster than the speed of light (3.00×10^8 m/s)?

Section 7.4 Centripetal Acceleration

14. As noted in Conceptual Question 11, it has been suggested that rotating cylinders about 10 mi long and 5.0 mi in diameter be placed in space and used as colonies. What angular speed must such a cylinder have so that the centripetal acceleration at its surface equals the free-fall acceleration?

15. Find the centripetal accelerations of (a) a point on the Equator and (b) the North Pole, due to the rotation of Earth about its axis.

16. In civil aviation, a "standard turn" for level flight of a propeller-driven airplane is one in which the airplane makes a complete circular turn in 2.0 min. If the plane's speed is 180 m/s, what is the radius of the circle? What is the centripetal acceleration of the plane?

17. (a) What is the tangential acceleration of a bug on the rim of a 10-in.-diameter disk if the disk moves from rest to an angular speed of 78 rev/min in 3.0 s? (b) When the disk is at its final speed, what is the tangential velocity of the bug? (c) One second after the bug starts from rest, what are its tangential acceleration, centripetal acceleration, and total acceleration?

18. A race car starts from rest on a circular track of radius 400 m. The car's speed increases at the constant rate of 0.500 m/s^2. At the point where the magnitudes of the centripetal and tangential accelerations are equal, determine (a) the speed of the race car, (b) the distance traveled, and (c) the elapsed time.

Section 7.6 Forces Causing Centripetal Acceleration

Section 7.7 Describing Forces in Accelerated Reference Frames

19. A 55.0-kg ice skater is moving at 4.00 m/s when she grabs the loose end of a rope, the opposite end of which is tied to a pole. She then moves in a circle of radius 0.800 m around the pole. (a) Determine the force exerted by the horizontal rope on her arms. (b) Compare this force with her weight.

20. A sample of blood is placed in a centrifuge of radius 15.0 cm. The mass of a red blood cell is 3.0×10^{-16} kg, and the magnitude of the force acting on it as it settles out

of the plasma is 4.0×10^{-11} N. At how many revolutions per second should the centrifuge be operated?

21. A 2 000-kg car rounds a circular turn of radius 20 m. If the road is flat and the coefficient of friction between tires and road is 0.70, how fast can the car go without skidding?

22. The cornering performance of an automobile is evaluated on a skidpad, where the maximum speed that a car can maintain around a circular path on a dry, flat surface is measured. Then the centripetal acceleration, also called the lateral acceleration, is calculated as a multiple of the free-fall acceleration g. The main factors affecting the performance are the tire characteristics and the suspension system of the car. A Dodge Viper GTS can negotiate a skidpad of radius 61.0 m at 86.5 km/h. Calculate its maximum lateral acceleration.

23. A 50.0-kg child stands at the rim of a merry-go-round of radius 2.00 m, rotating with an angular speed of 3.00 rad/s. (a) What is the child's centripetal acceleration? (b) What is the minimum force between his feet and the floor of the carousel that is required to keep him in the circular path? (c) What minimum coefficient of static friction is required? Is the answer you found reasonable? In other words, is he likely to stay on the merry-go-round?

24. An engineer wishes to design a curved exit ramp for a toll road in such a way that a car will not have to rely on friction to round the curve without skidding. She does so by banking the road in such a way that the force causing the centripetal acceleration will be supplied by the component of the normal force toward the center of the circular path. (a) Show that for a given speed v and a radius r, the curve must be banked at the angle θ such that $\tan \theta = v^2/rg$. (b) Find the angle at which the curve should be banked if a typical car rounds it at a 50.0-m radius and a speed of 13.4 m/s.

25. An air puck of mass 0.25 kg is tied to a string and allowed to revolve in a circle of radius 1.0 m on a frictionless horizontal table. The other end of the string passes through a hole in the center of the table, and a mass of 1.0 kg is tied to it (Fig. P7.25). The suspended mass remains in equilibrium while the puck on the tabletop revolves. (a) What is the tension in the string? (b) What is the horizontal force acting on the puck? (c) What is the speed of the puck?

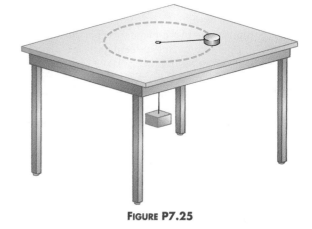

FIGURE P7.25

26. Tarzan ($m = 85$ kg) tries to cross a river by swinging from a 10-m-long vine. His speed at the bottom of the swing (as he just clears the water) is 8.0 m/s. Tarzan doesn't know

that the vine has a breaking strength of 1 000 N. Does he make it safely across the river? Justify your answer.

27. A 40.0-kg child takes a ride on a Ferris wheel that rotates four times each minute and has a diameter of 18.0 m. (a) What is the centripetal acceleration of the child? (b) What force (magnitude and direction) does the seat exert on the child at the lowest point of the ride? (c) What force does the seat exert on the child at the highest point of the ride? (d) What force does the seat exert on the child when she is halfway between the top and bottom?

28. A roller-coaster vehicle has a mass of 500 kg when fully loaded with passengers (Fig. P7.28). (a) If the vehicle has a speed of 20.0 m/s at Ⓐ, what is the magnitude of the force that the track exerts on the vehicle at this point? (b) What is the maximum speed the vehicle can have at Ⓑ in order for gravity to hold it on the track?

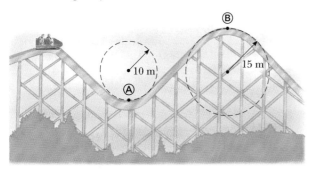

FIGURE P7.28

Section 7.8   Newton's Law of Universal Gravitation

Section 7.9   Gravitational Potential Energy Revisited

Section 7.10   Kepler's Laws

29. The average distance separating Earth and the Moon is 384 000 km. Use the data in Table 7.3 to find the net gravitational force exerted by Earth and the Moon on a spaceship with mass 3.00×10^4 kg located halfway between them.

30. During a solar eclipse, the Moon, Earth, and Sun all lie on the same line, with the Moon between Earth and the Sun. (a) What force is exerted by the Sun on the Moon? (b) What force is exerted by Earth on the Moon? (c) What force is exerted by the Sun on Earth? (See Table 7.3 and Problem 29 above.)

31. A coordinate system (in meters) is constructed on the surface of a pool table, and three objects are placed on the coordinate system as follows: a 2.0-kg object at the origin, a 3.0-kg object at (0, 2.0), and a 4.0-kg object at (4.0, 0). Find the resultant gravitational force exerted by the other two objects on the object at the origin.

32. Use the data of Table 7.3 to find the point between Earth and the Sun at which an object can be placed so that the net gravitational force exerted by Earth and Sun on this object is zero.

33. Objects with masses of 200 kg and 500 kg are separated by 0.400 m. (a) Find the net gravitational force exerted by these objects on a 50.0-kg object placed midway between them. (b) At what position (other than infinitely remote ones) can the 50.0-kg object be placed so as to experience a net force of zero?

34. Two objects attract each other with a gravitational force of

magnitude 1.00×10^{-8} N when separated by 20.0 cm. If the total mass of the two objects is 5.00 kg, what is the mass of each?

35. A satellite moves in a circular orbit around Earth at a speed of 5 000 m/s. Determine (a) the satellite's altitude above Earth's surface and (b) the period of the satellite's orbit.

36. A 600-kg satellite is in a circular orbit about Earth at a height above Earth equal to Earth's mean radius. Find (a) the satellite's orbital speed, (b) the period of its revolution, and (c) the gravitational force acting on it.

37. Io, a satellite of Jupiter, has an orbital period of 1.77 days and an orbital radius of 4.22×10^5 km. From these data, determine the mass of Jupiter.

38. A satellite has a mass of 100 kg and is located at 2.00×10^6 m above Earth's surface. (a) What is the potential energy associated with the satellite at this location? (b) What is the magnitude of the gravitational force on the satellite?

39. A satellite of mass 200 kg is launched from a site on the Equator into an orbit at 200 km above Earth's surface. (a) If the orbit is circular, what is the orbital period of this satellite? (b) What is the satellite's speed in orbit? (c) What is the minimum energy necessary to place this satellite in orbit, assuming no air friction?

40. Neutron stars are extremely dense objects that are formed from the remnants of supernova explosions. Many rotate very rapidly. Suppose that the mass of a certain spherical neutron star is twice the mass of the Sun and its radius is 10.0 km. Determine the greatest possible angular speed it can have so that the matter at the surface of the star on its equator is just held in orbit by the gravitational force.

ADDITIONAL PROBLEMS

41. (a) Find the angular speed of Earth's rotation on its axis. As Earth turns toward the east, we see the sky turning toward the west at this same rate.

(b) *The rainy Pleiads wester*
   And seek beyond the sea
   The head that I shall dream of
   That shall not dream of me.
      —A. E. Housman (© Robert E. Symons)

Cambridge, England, is at longitude 0° and Saskatoon, Saskatchewan, is at longitude 107° west. How much time elapses after the Pleiades set in Cambridge until these stars fall below the western horizon in Saskatoon?

42. The Mars probe *Pathfinder* is designed to drop the instrument package from a height of 20 m above the surface, after the speed of the probe has been brought to zero by a combination parachute-rocket system at that height. To cushion the landing, giant airbags surround the package. The mass of Mars is 0.107 4 times that of Earth and the radius of Mars is 0.528 2 that of Earth. Find (a) the acceleration due to gravity at the surface of Mars and (b) how long it takes for the instrument package to fall the last 20 meters.

43. An athlete swings a 5.00-kg ball horizontally on the end of a rope. The ball moves in a circle of radius 0.800 m at an angular speed of 0.500 rev/s. What are (a) the tangential speed of the ball and (b) its centripetal acceleration?

(c) If the maximum tension the rope can withstand before breaking is 100 N, what is the maximum tangential speed the ball can have?

44. A digital audio compact disc carries data, with each bit occupying 0.6 μm, along a continuous spiral track from the inner circumference of the disc to the outside edge. A CD player turns the disc to carry the track counterclockwise above a lens at a constant speed of 1.30 m/s. Find the required angular speed (a) at the beginning of the recording, where the spiral has a radius of 2.30 cm and (b) at the end of the recording, where the spiral has a radius of 5.80 cm. (c) A full-length recording lasts for 74 min 33 s. Find the average angular acceleration of the disc. (d) Assuming that the acceleration is constant, find the total angular displacement of the disc as it plays. (e) Find the total length of the track.

45. The Solar Maximum Mission Satellite was placed in a circular orbit about 150 mi above Earth. Determine (a) the orbital speed of the satellite and (b) the time required for one complete revolution.

46. A car rounds a banked curve where the radius of curvature of the road is R, the banking angle is θ, and the coefficient of static friction is μ. (a) Determine the range of speeds the car can have without slipping up or down the road. (b) What is the range of speeds possible if $R = 100$ m, $\theta = 10°$, and $\mu = 0.10$ (slippery conditions)?

47. A car moves at speed v across a bridge made in the shape **web** of a circular arc of radius r. (a) Find an expression for the normal force acting on the car when it is at the top of the arc. (b) At what minimum speed will the normal force become zero (causing occupants of the car to seem weightless) if $r = 30.0$ m?

48. A 0.400-kg pendulum bob passes through the lowest part of its path at a speed of 3.00 m/s. (a) What is the tension in the pendulum cable at this point if the pendulum is 80.0 cm long? (b) When the pendulum reaches its highest point, what angle does the cable make with the vertical? (c) What is the tension in the pendulum cable when the pendulum reaches its highest point?

49. Because of Earth's rotation about its axis, a point on the Equator experiences a centripetal acceleration of 0.034 0 m/s^2 while a point at the poles experiences no centripetal acceleration. (a) Show that at the Equator the gravitational force on an object (the true weight) must exceed the object's apparent weight. (b) What are the apparent weights at the Equator and at the poles of a 75.0-kg person? (Assume Earth is a uniform sphere, and take $g = 9.800$ m/s^2.)

50. A stunt man whose mass is 70 kg swings from the end of a 4.0-m-long rope along the arc of a vertical circle. Assuming he starts from rest when the rope is horizontal, find the tensions in the rope that are required to make him follow his circular path, (a) at the beginning of his motion, (b) at a height of 1.5 m above the the bottom of the circular arc, and (c) at the bottom of his arc.

51. In a popular amusement park ride, a rotating cylinder of radius 3.00 m is set in rotation at an angular speed of 5.00 rad/s, as in Figure P7.51. The floor then drops away, leaving the riders suspended against the wall in a vertical position. What minimum coefficient of friction between a rider's clothing and the wall is needed to keep the rider from slipping? (*Hint*: Recall that the magnitude of the

maximum force of static friction is equal to μn, where n is the normal force—in this case, the force causing the centripetal acceleration.

FIGURE P7.51

52. A 0.50-kg ball that is tied to the end of a 1.5-m light cord is revolved in a horizontal plane with the cord making a 30° angle with the vertical (see Fig. 7.14). (a) Determine the ball's speed. (b) If instead the ball is revolved so that its speed is 4.0 m/s, what angle does the cord make with the vertical? (c) If the cord can withstand a maximum tension of 9.8 N, what is the highest speed at which the ball can move?

53. A skier starts at rest at the top of a large hemispherical hill (Fig. P7.53). Neglecting friction, show that the skier will leave the hill and become airborne at a distance of $h = R/3$ below the top of the hill. (*Hint*: At this point, the normal force goes to zero.)

FIGURE P7.53

54. After consuming all of its nuclear fuel, a massive star can collapse to form a black hole, which is an immensely dense object for which the escape speed is greater than the speed of light. Newton's law of universal gravitation still describes the force that a black hole exerts on objects

outside it. A spacecraft in the shape of a long cylinder has a length of 100 m and its mass with occupants is 1 000 kg. It has strayed too close to a 1.0-m-radius black hole having a mass 100 times that of the Sun (Fig. P7.54). (a) If the nose of the spacecraft points toward the center of the black hole, and if distance between the nose of the spacecraft and the black hole's center is 10 km, determine the total force on the spacecraft. (b) What is the difference in the force per kilogram of mass felt by the occupants in the nose of the ship and those in the rear of the ship farthest from the black hole?

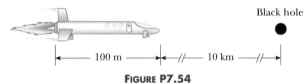

FIGURE P7.54

55. In Robert Heinlein's *The Moon Is a Harsh Mistress,* the colonial inhabitants of the Moon threaten to launch rocks down onto Earth if they are not given independence (or at least representation). Assuming that a gun could launch a rock of mass *m* at twice the lunar escape speed, calculate the speed of the rock as it enters Earth's atmosphere.

56. Show that the escape speed from the surface of a planet of **web** uniform density is directly proportional to the radius of the planet.

57. A massless spring of constant $k = 78.4$ N/m is fixed on the left side of a level track. A block of mass $m = 0.50$ kg is pressed against the spring and compresses it a distance *d*, as in Figure P7.57. The block (initially at rest) is then released and travels toward a circular loop-the-loop of radius $R = 1.5$ m. The entire track and the loop-the-loop are frictionless except for the section of track between points *A* and *B*. Given that the coefficient of kinetic friction between the block and the track along *AB* is $\mu_k = 0.30$, and that the length of *AB* is 2.5 m, determine the minimum compression *d* of the spring that enables the block to just make it through the loop-the-loop at point *C*. (*Hint:* The force exerted by the track on the block will be zero if the block barely makes it through the loop-the-loop.)

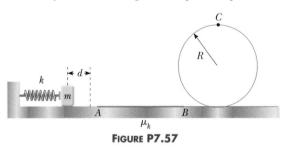

FIGURE P7.57

58. A small block of mass $m = 0.50$ kg is fired with an initial speed of $v_0 = 4.0$ m/s along a horizontal section of frictionless track, as shown in the top portion of Figure P7.58. The block then moves along the frictionless semicircular, *vertical* tracks of radius $R = 1.5$ m. (a) Determine the force exerted by the track on the block at points Ⓐ and Ⓑ. (b) The bottom of the track consists of a section ($L = 0.40$ m) with friction. Determine the coefficient of kinetic friction between the block and that portion of the bottom track if the block just makes it to point Ⓒ on the first trip. (*Hint:* If the block just makes it to point Ⓒ, the force of contact exerted by the track on the block at that point should be zero.)

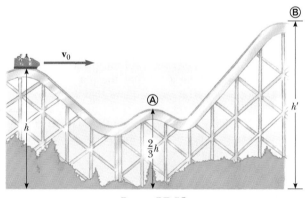

FIGURE P7.58

59. A frictionless roller coaster is given an initial velocity of v_0 at height *h*, as in Figure P7.59. The radius of curvature of the track at point Ⓐ is *R*. (a) Find the maximum value of v_0 so that the roller coaster stays on the track at Ⓐ solely because of gravity. (b) Using the value of v_0 calculated in (a), determine the value of h' that is necessary if the roller coaster just makes it to point Ⓑ.

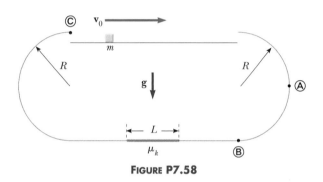

FIGURE P7.59

GROUP ACTIVITIES

G.1 Experiment with a bicycle wheel, like the one in Figure 7.3. Wrap tape around the middle of one spoke so that you can easily see it and draw a chalk mark on a point on the rim at the end of the spoke. Make the tire rotate by always gently and steadily turning the crank by hand. Compare the linear speed of the tape with that of the chalk mark. Why do these two points move at different speeds?

What happens to the speed of the tape as you slide it closer to the center of the wheel?

With the tape at a given distance from the center, measure the time it takes for the tape to make five rotations. From this, find the angular speed and the linear speed. Repeat at least five times and average your results. Move the tape to a different point and repeat the observations.

G.2 Tie a tennis ball to string of about 1.0 m length. At a safe distance from other students, whirl the ball in a horizontal circle. Note the increasing tension in the string as you whirl it faster. Now, whirl the ball in a vertical circle and observe the difference in tension at the top of the path and at the bottom. Why does this occur? While whirling the ball, release it and observe that it flies off in a direction tangent to its circular path.

G.3 On a clear night spend a few hours outside observing the stars. Choose a pattern of stars toward the north, and others toward the east, south, and west. As time passes, notice how these patterns move. Obtain a star chart of the constellations from your instructor and also a list of the planets visible for your date of observation. Locate as many of these as possible.

G.4 A roller coaster travels in a circular path in a vertical plane. (a) Identify the forces on a passenger at the top of the circular loop that cause centripetal acceleration. Show the direction of all forces in a sketch. (b) Identify the forces on the passenger at the bottom of the loop that produce centripetal acceleration. Show these in a sketch. (c) Based on your answers to (a) and (b) at what point, top or bottom, should the seat exert the greatest force on the passenger? (d) Assume the speed of the roller coaster is 12.0 m/s at the top of the loop, which has radius 8.00 m. Find the force exerted by the seat on a 70.0-kg passenger. Then, assume the speed remains the same at the bottom of the path and find the force exerted by the seat on the passenger at this point. Are your answers consistent with your choice of answers for (a) and (b)?

G.5 (a) If Earth's mass were doubled at the same time its radius was doubled, what would happen to free-fall acceleration? Would it increase, decrease, or stay the same? (b) Use the mass of Earth and its radius as given in Table 7.3 to find precise values for g at these values of M and R. Are your answers consistent with your predictions?

G.6 Figure GA7.6 shows the elliptical orbit of a spacecraft around Earth. Take the origin of your coordinate system to be at the center of Earth.
  (a) On a copy of the figure (enlarged if necessary), draw vectors representing
   (i) the position of the spacecraft when it is at Ⓐ and Ⓑ;
   (ii) the velocity of the spacecraft when it is at Ⓐ and Ⓑ;

   (iii) the acceleration of the spacecraft when it is at Ⓐ and Ⓑ.
Make sure that each type of vector can be distinguished. (You might use the color scheme found on the inside front cover of this book.) Provide a legend that shows how each type is represented.
  (b) Have you drawn the velocity vector at Ⓐ longer, shorter, or the same length as the one at Ⓑ? Explain why. Have you drawn the acceleration vector at Ⓐ longer, shorter, or the same length as the one at Ⓑ? Explain why you have done so.

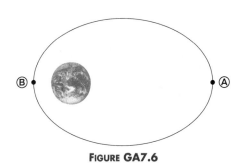

FIGURE GA7.6

G.7 Two students are driving a Porsche of mass M at a constant speed v around the traffic circle in front of the physics building. Assume that the road is level and has a radius R. The coefficient of friction between the car's wheels and the road is μ.
  (a) In terms of the quantities given, what is the magnitude and direction of the car's acceleration?
  (b) At the instant when it passes directly in front of the physics building, is there a net force on the car? If there is, what is responsible for it and how big must it be? Express your answer in terms of the quantities given and explain your reasoning.
  (c) The mass of the car and passengers is 1 700 kg, the circle has a radius $R = 25$ m, and the coefficient of friction between rubber and a dry road is $\mu = 0.60$. What is the fastest speed the car can go without slipping?
(Problems 6 and 7 are courtesy of E. F. Redish. For more problems of this type, visit http://www.physics.umd.edu/perg/)

8

Rotational Equilibrium and Rotational Dynamics

Chapter Outline

This adventurous child shifts his body weight around the center of the steering wheel to change the direction of the ship. As the center of his body shifts to the left of the wheel's rotation axis, the wheel rotates counterclockwise, causing the ship to move toward the left as indicated by the tilt from the horizontal. If he were to shift his body to the right of the wheel's axis, the ship would move to the right. *(Christoph Wilhelm/FPG/Getty Images)*

his chapter completes our study of mechanical equilibrium begun in Chapter 4. An understanding of equilibrium problems is important in a variety of fields. For example, students of architecture or industrial technology benefit from understanding the forces that act on buildings or large machines, and biology students should understand the forces at work in muscles and on bones and joints. In order to fully understand objects in equilibrium, we must consider torque. This concept will also play a key role in completing our discussion of rotational motion.

In this chapter we build on the definitions of angular speed and angular acceleration encountered in Chapter 7 by examining the relationship between these concepts and the forces that produce rotational motion. Specifically, we shall find the rotational analog of Newton's second law and define a term that needs to be added to our equation for conservation of mechanical energy: rotational kinetic energy. One of the central purposes of this chapter is to develop the concept of angular momentum, a quantity that plays a key role in rotational motion. Finally, just as we found that linear momentum is conserved for an isolated system, we shall also find that the angular momentum of any isolated system is always conserved.

8.1 TORQUE

Consider Figure 8.1, an overhead view of a door hinged at point O. From this viewpoint, the door is free to rotate about a line perpendicular to the page and passing through O. When the force **F** is applied at the outer edge, as shown, the door easily rotates counterclockwise; that is, the rotational effect of the force is quite large. On the other hand, the same force applied at a point nearer the hinges has a smaller rotational effect on the door. (Observe for yourself that a half-open door rotates with difficulty if you push it near the inner edge, close to the hinges, but turns easily if you push near the doorknob.)

The tendency of a force to rotate an object about some axis is measured by a quantity called the torque τ. The magnitude of the torque due to a force **F** is given by

$$\tau = Fd \qquad [8.1]$$

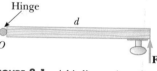

FIGURE 8.1 A bird's-eye view of a door hinged at O, with a force applied perpendicularly to the door.

In this equation, τ (the Greek letter tau) is the torque, and the distance d is the **lever arm** (or moment arm) of the force **F**. **The lever arm is the perpendicular distance from the axis of rotation to a line drawn along the direction of the force.** Note that **the value of τ depends on the axis of rotation.** Torque is a vector perpendicular to the plane determined by the lever arm and the force. For the two-dimensional problems used in this book, the torque is either into or out of the plane of the paper. We shall use the clockwise or counterclockwise turning tendency of an applied force to determine whether the torque is positive or negative.

As an example, consider the wrench pivoted about the axis O in Figure 8.2a.

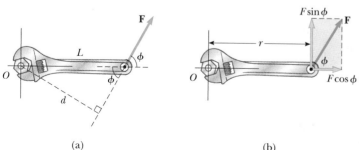

(a)   (b)

FIGURE 8.2 (a) A force **F** acting at an angle ϕ with the horizontal produces a torque of magnitude Fd about the pivot O. (b) The component $F\sin\phi$ tends to rotate the system about O. The component $F\cos\phi$ produces no torque about O.

In this case, the applied force **F** acts at an angle ϕ with the horizontal. If you examine the definition of lever arm just given, you will see that the lever arm is the distance d shown in the figure and not L, the length of the wrench. That is, d is the perpendicular distance from the axis of rotation to the line along which the applied force acts. Here, d is related to L by the expression $d = L \sin \phi$. Thus, the net torque produced by the force **F** is given by $\tau = FL \sin \phi$. You can also find the torque in situations such as this by resolving the force into components, as shown in Figure 8.2b. The torque about the axis of rotation produced by the component $F \cos \phi$ is zero because the lever arm of this component is zero. That is, the distance from the pivot to the line along which the force acts is zero, because the line along which the force acts passes through the axis of rotation. The component $F \sin \phi$ has a lever arm of L, and the torque produced by this component is, as before, $\tau = FL \sin \phi$.

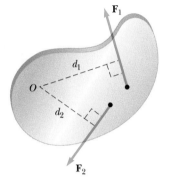

FIGURE 8.3   The force **F**$_1$ tends to rotate the object counterclockwise about *O* and **F**$_2$ tends to rotate the object clockwise.

If two or more forces act on an object, as in Figure 8.3, then each has a tendency to produce a rotation about the pivot *O*. In this example, **F**$_2$ tends to rotate the object clockwise and **F**$_1$ tends to rotate the object counterclockwise. We shall use the convention that **the sign of the torque is positive if its turning tendency is counterclockwise and negative if its turning tendency is clockwise.** In Figure 8.3, the torque associated with **F**$_1$, which has a moment arm of d_1, is positive and equal to $F_1 d_1$; the torque associated with **F**$_2$ is negative and equal to $-F_2 d_2$. The *net torque* acting on the object about *O* is found by summing the torques:

$$\sum \tau = \tau_1 + \tau_2 = F_1 d_1 - F_2 d_2$$

Notice that the units of torque are units of force times length, such as the newton-meter (N·m) or the foot-pound (ft·lb).

Example 8.1   The Spinning Crate

Figure 8.4 is a top view of a packing crate being pushed by two forces of equal magnitude acting in opposite directions as shown. Find the net torque exerted on the crate if its width is 1.0 m. Assume an axis of rotation through the center of the crate.

$F_1 = 500$ N   **FIGURE 8.4**   (Example 8.1) A top view of a packing crate being pushed by forces of equal magnitude and opposite direction.

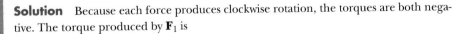

$F_2 = 500$ N

Solution   Because each force produces clockwise rotation, the torques are both negative. The torque produced by **F**$_1$ is

$$\tau_1 = F_1 d_1 = -(500 \text{ N})(0.50 \text{ m}) = -250 \text{ N} \cdot \text{m}$$

and the torque produced by **F**$_2$ is

$$\tau_2 = F_2 d_2 = -(500 \text{ N})(0.50 \text{ m}) = -250 \text{ N} \cdot \text{m}$$

Thus, the net torque is exerted on the crate $-500 \text{ N} \cdot \text{m}$.

Example 8.2 The Swinging Door

Find the torque produced by the 300-N force applied at an angle of 60° to the door of Figure 8.5a.

Reasoning   In Figure 8.5b, the 300-N force has been replaced by its horizontal and vertical components:

$$F_x = (300 \text{ N})\cos 60° = 150 \text{ N}$$

$$F_y = (300 \text{ N})\sin 60° = 260 \text{ N}$$

A convenient and obvious location for the axis of rotation is the hinge of the door.

Solution   In this case, the 150-N force produces zero torque about the axis of rotation, because the line along which the force acts passes through the axis of rotation and hence the lever arm is zero. The 260-N force has a lever arm of 2.0 m and thus produces a torque of

$$\tau = (260 \text{ N})(2.0 \text{ m}) = \boxed{+520 \text{ N} \cdot \text{m}}$$

where the plus sign indicates the door rotates in the counterclockwise direction.

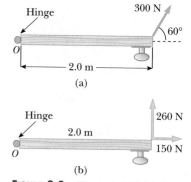

FIGURE 8.5   (Example 8.2) (a) A top view of a door being pulled by a 300-N force. (b) The components of the 300-N force.

8.2 TORQUE AND THE TWO CONDITIONS FOR EQUILIBRIUM

In Chapter 4 we examined some situations in which an object has no net external force acting on it; that is, $\Sigma F_x = 0$ and $\Sigma F_y = 0$. We found there that if such an object is initially at rest with respect to an observer, it remains at rest. If the object is initially moving with respect to the observer, the object's velocity remains constant (does not change). Such an object was said to be in equilibrium. However, the condition that $\Sigma \mathbf{F} = 0$ (Equation 4.9), although necessary, is not sufficient to ensure that an object is in complete mechanical equilibrium. This can be understood by considering Figure 8.4. In this situation, the two applied forces acting on the crate are equal in magnitude and opposite in direction. Nevertheless, the crate will still accelerate—it will begin to rotate clockwise because the forces do not act through a common point.

This case illustrates that, to fully understand the effect of a force or group of forces on an object, we must know not only the magnitude and direction of the force(s) but also their point(s) of application. That is, the net torque acting on an object must be considered.

We shall call the requirement that $\Sigma \mathbf{F} = 0$ the first condition for equilibrium. **The second condition for equilibrium asserts that if an object is in rotational equilibrium, the net torque acting on it about any axis must be zero.** That is,

$$\sum \tau = 0$$

In summary, an object in mechanical equilibrium must satisfy both conditions:

1. **The net external force must be zero.** $\sum \mathbf{F} = 0$

2. **The net external torque must be zero.** $\sum \tau = 0$

[8.2]   ◀ The two conditions for mechanical equilibrium

The first condition is a statement of translational equilibrium; the second is a statement of rotational equilibrium. We shall discuss torque and its relation to rotational motion in more detail later in the chapter, but for now we simply point out

This large balanced rock at the Garden of the Gods in Colorado Springs, Colorado is in mechanical equilibrium. (*David Serway*)

Tip 8.1   SPECIFY YOUR AXIS

In solving rotation problems, you must specify an axis of rotation. The choice is arbitrary, but once you make it, you must maintain that choice consistently throughout the problem. For example, there is no unique value of the torque. Its value depends on your choice of rotation axis.

Webnote 8.1

To see the classic case of equilibrium for a seesaw, go to:
http://theory.uwinnipeg.ca/mod_tech/node48.html

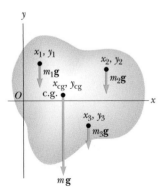

FIGURE 8.7   The center of gravity of an object is the point where all the weight of the object can be considered to be concentrated.

that just as an object in translational equilibrium has **a** = 0, an object in rotational equilibrium has $\boldsymbol{\alpha}$ = 0, meaning no rotational acceleration.

POSITION OF THE AXIS OF ROTATION

In the cases we have described so far, the axes of rotation for calculating torques were selected without explanation. Often the nature of a problem suggests a convenient location for the axis, but just as often no single location stands out as being preferable. You might ask, "What axis should I choose in calculating the net torque?" The answer is as follows: **If the object is in equilibrium, it does not matter where you put the axis of rotation for calculating the net torque; the location of the axis is completely arbitrary.**

APPLYING PHYSICS 8.1

When an automobile driver steps on the accelerator, the nose of the car moves upward. When the driver brakes, the nose moves downward. Why do these effects occur?

Explanation   When the driver steps on the accelerator, there is an increased force exerted by the roadway on the tires. This force is parallel to the roadway and directed toward the front of the automobile, as suggested in Figure 8.6a. This force provides a torque that causes the car to rotate in the clockwise direction. The result of this rotation is a "nosing up" of the car. When the driver steps on the brake, there is also an increased force exerted by the roadway on the tires, but now directed toward the rear of the automobile, as suggested in Figure 8.6b. This force results in a torque that causes a counterclockwise rotation and the subsequent "nosing down" of the automobile.

(a)   (b)

FIGURE 8.6   (Applying Physics 8.1)

8.3   THE CENTER OF GRAVITY

One of the forces that must be considered in dealing with a rigid object is that of gravity acting on the object. To compute the torque due to the force of gravity, all of the weight can be thought of as concentrated at a single point.

Consider an object of arbitrary shape lying in the *xy* plane, as in Figure 8.7. The object is divided into a large number of very small particles of weight m_1g, m_2g, m_3g, . . . having coordinates (x_1, y_1), (x_2, y_2), (x_3, y_3), ... If the object is free to rotate about the origin, each particle contributes a torque about the origin that is equal to its weight multiplied by its lever arm. For example, the torque due to the weight m_1g is m_1gx_1, and so forth.

We wish to locate the point of application of the single force of magnitude $w = F_g$ (the total weight of the object) whose effect on the rotation of the object is the same as that of the individual particles. This point is called the **center of gravity** of the object. Equating the torque exerted by w at the center of gravity to the sum of the torques acting on the individual particles gives

$$(m_1g + m_2g + m_3g + \cdots)x_{\text{cg}} = m_1gx_1 + m_2gx_2 + m_3gx_3 + \cdots$$

If we assume that g is uniform over the object (which will be the case in all the situations we examine), then the g terms in the preceding equation cancel and we get

$$x_{cg} = \frac{m_1 x_1 + m_2 x_2 + m_3 x_3 + \cdots}{m_1 + m_2 + m_3 + \cdots} = \frac{\Sigma m_i x_i}{\Sigma m_i}$$ **[8.3]**

Similarly, the y coordinate of the center of gravity of the system can be found from

$$y_{cg} = \frac{\Sigma m_i y_i}{\Sigma m_i}$$ **[8.4]**

The center of gravity of a homogeneous, symmetric body must lie on the axis of symmetry. For example, the center of gravity of a homogeneous rod must lie midway between the ends of the rod. The center of gravity of a homogeneous sphere or a homogeneous cube must lie at the geometric center of the object. We can determine the center of gravity of an irregularly shaped object, such as a wrench, experimentally by suspending the wrench from two different arbitrary points (Fig. 8.8). The wrench is first hung from point A, and a vertical line AB (which can be established with a plumb bob) is drawn when the wrench is in equilibrium. The wrench is then hung from point C, and a second vertical line CD is drawn. The center of gravity coincides with the intersection of these two lines. In fact, if the wrench is hung freely from any point, the center of gravity always lies straight below the point of support. Furthermore, the vertical line through that point must pass through the center of gravity.

In several examples in Section 8.4, we shall be concerned with homogeneous, symmetric objects whose centers of gravity coincide with their geometric centers. A rigid object in a uniform gravitational field can be balanced by a single force equal in magnitude to the weight of the object, as long as the force is directed upward through the object's center of gravity.

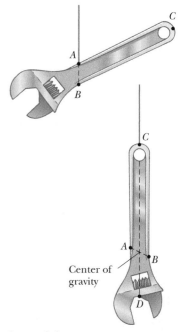

FIGURE 8.8 An experimental technique for determining the center of gravity of a wrench. The wrench is hung freely from two different pivots, A and C. The intersection of the two lines, AB and CD, locates the center of gravity.

Example 8.3   Where Is the Center of Gravity?

Three particles are located in a coordinate system as shown in Figure 8.9. Find the center of gravity.

Reasoning and Solution   The y coordinate of the center of gravity is zero because all the particles are on the x axis. To find the x coordinate of the center of gravity, we use Equation 8.3:

$$x_{cg} = \frac{\Sigma m_i x_i}{\Sigma m_i}$$

For the numerator, we find

$$\Sigma m_i x_i = m_1 x_1 + m_2 x_2 + m_3 x_3$$
$$= (5.00 \text{ kg})(-0.500 \text{ m}) + (2.00 \text{ kg})(0 \text{ m}) + (4.00 \text{ kg})(1.00 \text{ m})$$
$$= 1.50 \text{ kg} \cdot \text{m}$$

The denominator is $\Sigma m_i = 11.0$ kg; therefore,

$$x_{cg} = \frac{1.50 \text{ kg} \cdot \text{m}}{11.0 \text{ kg}} = \boxed{0.136 \text{ m}}$$

EXERCISE   If a fourth particle of mass 2.00 kg is placed at $x = 0$, $y = 0.250$ m, find the x and y coordinates of the center of gravity for this system of four particles.

ANSWER $x_{cg} = 0.115$ m; $y_{cg} = 0.038$ 0 m

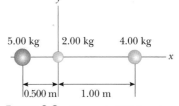

FIGURE 8.9   (Example 8.3) Locating the center of gravity for a system of three particles.

Example 8.4 Locating Your Lab Partner's Center of Gravity

In this example we show how to find the location of a person's center of gravity. Suppose your lab partner has a height L of 173 cm (5 ft 8 in) and a weight w of 715 N (160 lb). You can easily determine the position of his center of gravity by having him stretch out on a uniform board supported at one end by a scale as shown in Figure 8.10. If the board's weight w_b is 49 N and the scale reading F is 350 N, find the distance of your lab partner's center of gravity from the left end of the board.

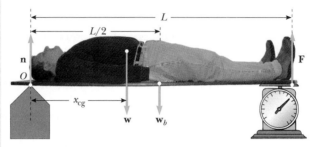

FIGURE 8.10 (Example 8.4) Determining your lab partner's center of gravity.

Reasoning The center of gravity must lie somewhere on the long symmetry axis of the body. To find the position x_{cg} of the center of gravity, we set the sum of the torques about point O equal to zero.

Solution Referring to Figure 8.10,

$$\sum \tau_O = 0$$

$$- wx_{cg} - w_b(L/2) + FL = 0$$

$$x_{cg} = \frac{FL - w_b(L/2)}{w} = \frac{(350 \text{ N})(173 \text{ cm}) - (49 \text{ N})(86.5 \text{ cm})}{715 \text{ N}} = \boxed{79 \text{ cm}}$$

Thus x_{cg}, measured from the head, is about 46% of the person's height.

8.4 EXAMPLES OF OBJECTS IN EQUILIBRIUM

In Chapter 4 we discussed some techniques for solving problems concerned with objects in equilibrium. Recall that when the objects are treated as geometric points, it is sufficient to simply apply the condition that the net force on the object must be zero. In this chapter we have shown that for objects of finite dimensions, a second condition for equilibrium must be satisfied—namely that the net torque on the object must also be zero. The following general procedure is recommended for solving problems that involve objects in equilibrium.

Tip 8.2 ZERO TORQUE

Zero net torque does not mean an absence of rotational motion. An object that rotates at a constant angular speed can be under the influence of zero net torque. This is analogous to the translational situation—zero net force on an object does not necessarily mean that it is not in motion.

PROBLEM-SOLVING *STRATEGY*	Objects in Equililbrium

1. Draw a simple, neat diagram of the system.
2. Isolate the object that is being analyzed. Draw a free-body diagram showing all external forces that are acting on this object. For systems that contain more than one object, draw a *separate* diagram for each object. Do not include forces that the object exerts on its surroundings.
3. Establish convenient coordinate axes for each object and find the components of the forces along these axes. Now apply the first condition of equilibrium (the net force on the object in the x and y directions must be zero) for each object under consideration.

4. Choose a convenient rotation axis for calculating the net torque on the object. Now apply the second condition of equilibrium (the net torque on the object about any axis must be zero). Remember that the choice of the axis for the torque equation is arbitrary; therefore, choose an axis that will simplify your calculation as much as possible. Note that a force that acts along a line passing through the axis of rotation gives zero contribution to the torque.

5. The first and second conditions for equilibrium will give a set of simultaneous equations with several unknowns. All that is left to complete your solution is to solve for the unknowns in terms of the known quantities.

Example 8.5 Walking a Horizontal Beam

A uniform horizontal 300-N beam, 5.00 m long, is attached to a wall by a pin connection that allows the beam to rotate. Its far end is supported by a cable that makes an angle of 53.0° with the horizontal (Fig. 8.11a). If a 600-N person stands 1.50 m from the wall, find the tension in the cable and the force exerted by the wall on the beam.

Reasoning   First we must identify all the external forces acting on the beam and sketch them on a free-body diagram. This is shown in Figure 8.11b. The forces on the beam consist of the downward force of gravity, which has a magnitude of 300 N; the downward force exerted by the man, which is equal in magnitude to his weight, 600 N; the force **T** exerted by the cable; and the force **R** exerted by the wall. We then resolve the forces **T** and **R** into their horizontal and vertical components, as shown in Figure 8.11c. Note that the *x* component of the force exerted by the cable ($T \cos 53.0°$) is to the left, whereas the *y* component ($T \sin 53.0°$) is upward. The horizontal and vertical components of **R** are denoted by R_x and R_y, respectively. The first condition for equilibrium can now be applied in the *x* and *y* directions to give us two equations in terms of our unknowns R_x, R_y, and T. The necessary third equation can be found from the second condition of equilibrium.

Solution   From the first condition for equilibrium, we find

$$\sum F_x = R_x - T \cos 53.0° = 0 \tag{1}$$

$$\sum F_y = R_y + T \sin 53.0° - 600 \text{ N} - 300 \text{ N} = 0 \tag{2}$$

The unknowns are R_x, R_y, and T. Because there are three unknowns and only two equations, we cannot find a solution from just the first condition of equilibrium.

Now let us use the second condition of equilibrium. The axis that passes through the pivot at the wall is a convenient one to choose for the torque equation because the components R_x, R_y, and $T \cos 53.0°$ all have lever arms of zero and hence have zero torque about this axis. Recalling our sign convention for the torque about an axis and noting that the lever arms of the 600-N, 300-N, and $T \sin 53.0°$ components are 1.50 m, 2.50 m, and 5.00 m, respectively, we get

$$\sum \tau_O = (T \sin 53.0°)(5.00 \text{ m}) - (300 \text{ N})(2.50 \text{ m}) - (600 \text{ N})(1.50 \text{ m}) = 0 \tag{3}$$

$$T = \boxed{413 \text{ N}}$$

Thus, the torque equation using this axis gives us one of the unknowns immediately! This value for T is then substituted into (1) and (2) to give

$$R_x = \boxed{249 \text{ N}} \qquad R_y = \boxed{570 \text{ N}}$$

If we selected some other axis for the torque equation, the solution would be the same. For example, if the axis were to pass through the center of gravity of the beam, the torque equation would involve both T and R_y; together with (1) and (2), however, it could still be solved for the unknowns. Try it!

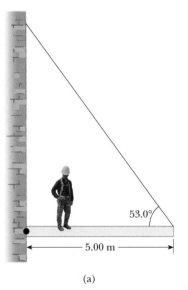

(a)

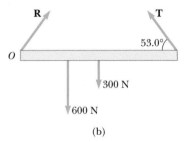

(b)

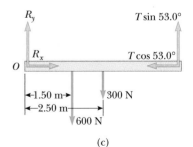

(c)

FIGURE 8.11   (Example 8.5) (a) A uniform beam attached to a wall and supported by a cable. (b) A free-body diagram for the beam. (c) The component form of the free-body diagram.

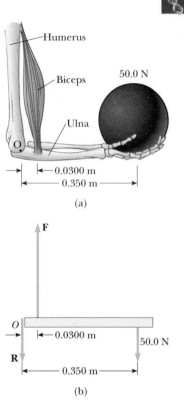

FIGURE 8.12 (Example 8.6) (a) A weight held with the forearm horizontal. (b) The mechanical model for the system.

Example 8.6   A Weighted Forearm

A 50.0-N (11-lb) weight is held in a person's hand with the forearm horizontal, as in Figure 8.12a. The biceps muscle is attached 0.030 0 m from the joint, and the weight is 0.350 m from the joint. Find the upward force exerted by the biceps on the forearm (the ulna) and the downward force exerted by the humerus on the forearm, acting at the joint. Neglect the weight of the forearm.

Solution   The forces acting on the forearm are equivalent to those acting on a bar of length 0.350 m, as shown in Figure 8.12b, where **F** is the upward force exerted by the biceps and **R** is the downward force at the joint. From the first condition for equilibrium, we have

$$\sum F_y = F - R - 50.0 \text{ N} = 0 \tag{1}$$

From the second condition for equilibrium, we know that the sum of the torques about any axis must be zero. With the joint O as the axis, we have

$$F(0.030\ 0 \text{ m}) - (50.0 \text{ N})(0.350 \text{ m}) = 0$$

$$F = \boxed{583 \text{ N} \ (131 \text{ lb})}$$

This value for F can be substituted into (1) to give $R = \boxed{533 \text{ N} \ (120 \text{ lb})}$. Clearly, the forces at joints and in muscles can be extremely large compared to the weight of the lifted object.

Example 8.7   Don't Climb the Ladder

A uniform 10-m-long, 50-N ladder rests against a smooth vertical wall as in Figure 8.13a. If the ladder is just on the verge of slipping when it makes a 50° angle with the ground, find the coefficient of static friction between the ladder and ground.

Reasoning   Figure 8.13b is the free-body diagram for the ladder, showing all external forces acting on it. At the base of the ladder, Earth exerts an upward normal force **n**, and a force of static friction **f** acts to the right. The wall exerts the force **P** to the left. Note that **P** is horizontal because the wall is smooth. (If the wall were rough, an upward frictional force would be exerted on the ladder.) The first condition for equilibrium can now be applied in the x and y directions to give us two equations in terms of our unknowns f, P, and n. The necessary third equation can be found from the second condition of equilibrium.

Solution   From the first condition for equilibrium applied to the ladder, we have

$$\sum F_x = f - P = 0 \tag{1}$$

$$\sum F_y = n - 50 \text{ N} = 0 \tag{2}$$

From (2) we see that $n = 50$ N. Furthermore, when the ladder is on the verge of slipping, the magnitude of the force of static friction must be maximum and given by the relation $f_{s,max} = \mu_s n = \mu_s(50 \text{ N})$. Thus, (1) reduces to

$$\mu_s(50 \text{ N}) = P \tag{3}$$

Let us now apply the second condition of equilibrium and take the torques about the axis through O at the bottom of the ladder, as in Figure 8.13c. The force **P** and the force of gravity acting on the ladder are the only forces that contribute to the torque about this axis, and their lever arms are shown in Figure 8.13c. Note that, because the length of the ladder is 10 m, the lever arm for **P** is $d_1 = (10 \text{ m})\sin 50°$. Likewise, the lever arm for

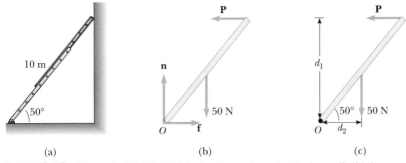

FIGURE 8.13 (Example 8.7) (a) A ladder leaning against a frictionless wall. (b) A free-body diagram for the ladder. (c) Lever arms for the force of gravity and **P**.

the 50-N force of gravity is $d_2 = (5.0 \text{ m}) \cos 50°$, where the force of gravity acts through the center because the ladder is uniform. Thus, we find that

$$\sum \tau_O = P(10 \text{ m}) \sin 50° - (50 \text{ N})(5.0 \text{ m}) \cos 50° = 0$$

$$P = 21 \text{ N}$$

Now that P is known, we can substitute its value into (3) to find μ_s:

$$\mu_s = \frac{P}{50 \text{ N}} = \frac{21 \text{ N}}{50 \text{ N}} = \boxed{0.42}$$

EXERCISE Determine the direction of the total force acting on the ladder at its base. Does this force act along the ladder?

ANSWER $\tan^{-1}(n/P) = 67°$; no. The force at the base of the ladder is generally *not* in the direction parallel to the ladder.

8.5 RELATIONSHIP BETWEEN TORQUE AND ANGULAR ACCELERATION

Earlier in this chapter we considered the situation in which both the net force and the net torque acting on an object are zero. Objects in this situation are said to be in equilibrium. We now examine the behavior of an object when the net torque acting on it is not zero. As you shall see, when a rigid object is subject to a net torque, it undergoes an angular acceleration. Furthermore, the angular acceleration is directly proportional to the net torque. The end result of our investigation will be an expression that is analogous to $\sum \mathbf{F} = m\mathbf{a}$ in translational motion.

Let us begin by considering the system shown in Figure 8.14, which consists of an object of mass m connected to a very light rod of length r. The rod is pivoted at the point O, and its movement is confined to rotation on a frictionless *horizontal* table. Now let us assume that a force $\mathbf{F}_t$ perpendicular to the rod and hence tangent to the circular orbit is acting on m. Because there is no force to oppose this tangential force, the object undergoes a tangential acceleration according to Newton's second law:

$$F_t = ma_t$$

Multiplying the left and right sides of this equation by r gives

$$F_t r = mra_t$$

In Chapter 7 (Equation 7.9) we found that the tangential acceleration and angular acceleration for a particle rotating in a circular path are related by the expression

$$a_t = r\alpha$$

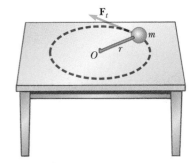

FIGURE 8.14 An object of mass m attached to a light rod of length r moves in a circular path on a frictionless horizontal surface while the tangential force $\mathbf{F}_t$ acts on it.

so now we find that

$$F_t r = mr^2 \alpha \qquad [8.5]$$

The left side of Equation 8.5, which should be familiar to you, is the torque acting on the object about its axis of rotation. That is, the torque is equal in magnitude to the force on the object multiplied by the perpendicular distance from the pivot to the line along which the force acts, or $\tau = F_t r$. Hence, we can write Equation 8.5 as

$$\tau = mr^2 \alpha \qquad [8.6]$$

Equation 8.6 shows that the torque on the system is proportional to angular acceleration, where the constant of proportionality mr^2 is called the **moment of inertia** of the object of mass m. (Because the rod is very light, its moment of inertia can be neglected.)

Quick Quiz 8.1	Using a screwdriver, you try to loosen a stubborn screw from a piece of wood and you fail. In order to succeed, you should find a screwdriver that (a) is longer, (b) is shorter, (c) has a narrower handle, or (d) has a fatter handle.

TORQUE ON A ROTATING OBJECT

Now consider a solid disk rotating about its axis as in Figure 8.15a. The disk consists of many particles at various distances from the axis of rotation, as in Figure 8.15b. The torque on each one of these particles is given by Equation 8.6. The *total* torque on the disk is given by the sum of the individual torques on all the particles:

$$\sum \tau = \left(\sum mr^2 \right) \alpha \qquad [8.7]$$

Note that, because the disk is rigid, all particles have the *same* angular acceleration, so α is not involved in the sum. If the masses and distances of the particles are labeled with subscripts as in Figure 8.15b, then

$$\sum mr^2 = m_1 r_1^2 + m_2 r_2^2 + m_3 r_3^2 + \cdots$$

This quantity is the moment of inertia of the whole body and is given the symbol I:

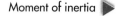
Moment of inertia

$$I \equiv \sum mr^2 \qquad [8.8]$$

The moment of inertia has the SI units kg·m². Using this result in Equation 8.7, we see that the total torque on a rigid body rotating about a fixed axis is given by

Relationship between net torque and angular acceleration

$$\sum \tau = I\alpha \qquad [8.9]$$

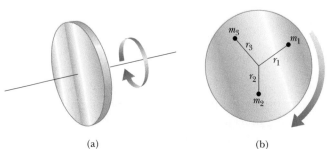

FIGURE 8.15 (a) A solid disk rotating about its axis. (b) The disk consists of many particles, all with the same angular acceleration.

(a)   (b)

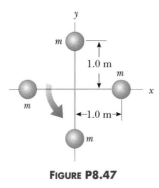

FIGURE P8.47

10.0 rev/min about a frictionless vertical axle. Facing the axle, a 25.0-kg child hops onto the merry-go-round, and manages to sit down on the edge. What is the new angular speed of the merry-go-round?

49. A solid, horizontal cylinder of mass 10.0 kg and radius 1.00 m rotates with an angular speed of 7.00 rad/s about a fixed vertical axis through its center. A 0.250-kg piece of putty is dropped vertically onto the cylinder at a point 0.900 m from the center of rotation, and sticks to the cylinder. Determine the final angular speed of the system.

50. A student sits on a rotating stool holding two 3.0-kg objects. When his arms are extended horizontally, the objects are 1.0 m from the axis of rotation, and he rotates with an angular speed of 0.75 rad/s. The moment of inertia of the student plus stool is 3.0 kg·m² and is assumed to be constant. The student then pulls the objects horizontally to 0.30 m from the rotation axis. (a) Find the new angular speed of the student. (b) Find the kinetic energy of the student before and after the objects are pulled in.

51. The puck in Figure P8.51 has a mass of 0.120 kg. Its original distance from the center of rotation is 40.0 cm, and the puck is moving with a speed of 80.0 cm/s. The string is pulled downward 15.0 cm through the hole in the frictionless table. Determine the work done on the puck. (*Hint:* Consider the change of kinetic energy of the puck.)

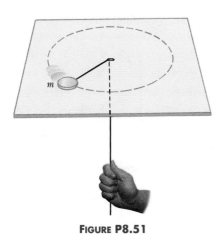

FIGURE P8.51

52. A merry-go-round rotates at the rate of 0.20 rev/s with an 80-kg man standing at a point 2.0 m from the axis of rotation. (a) What is the new angular speed when the man walks to a point 1.0 m from the center? Assume that the merry-go-round is a solid 25-kg cylinder of radius 2.0 m.

(b) Calculate the change in kinetic energy due to this movement. How do you account for this change in kinetic energy?

53. A 60.0-kg woman stands at the rim of a horizontal turntable having a moment of inertia of 500 kg·m² and a radius of 2.00 m. The turntable is initially at rest and is free to rotate about a frictionless, vertical axle through its center. The woman then starts walking around the rim clockwise (as viewed from above the system) at a constant speed of 1.50 m/s relative to Earth. (a) In what direction and with what angular speed does the turntable rotate? (b) How much work does the woman do to set herself and the turntable into motion?

54. A space station shaped like a giant wheel has a radius 100 m and a moment of inertia of 5.00×10^8 kg·m². A crew of 150 are living on the rim, and the station is rotating so that the crew experience an apparent acceleration of 1g (Fig. P8.54). When 100 people move to the center of the station for a union meeting, the angular speed changes. What apparent acceleration is experienced by the managers remaining at the rim? Assume an average mass of 65.0 kg for all the inhabitants.

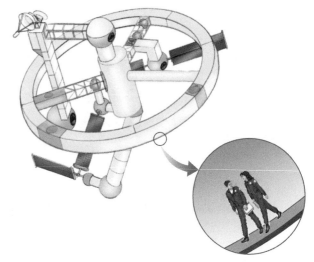

FIGURE P8.54

ADDITIONAL PROBLEMS

55. A cylinder with moment of inertia I_1 rotates with angular velocity ω_0 about a frictionless vertical axle. A second cylinder, with moment of inertia I_2, initially not rotating, drops onto the first cylinder (Fig. P8.55). Since the surfaces are rough, the two eventually reach the same angular speed ω. (a) Calculate ω. (b) Show that kinetic energy is lost in this situation, and calculate the ratio of the final to the initial kinetic energy.

56. A 0.100-kg meter stick is supported at its 40.0-cm mark by a string attached to the ceiling. A 0.700-kg object hangs vertically from the 5.00-cm mark. An object of mass m is attached somewhere on the meter stick to keep it horizontal and in rotational and translational equilibrium. If the tension in the string attached to the ceiling is 19.6 N, determine (a) the value of m and (b) its point of attachment on the stick.

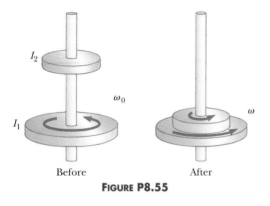

FIGURE P8.55

57. Show that the kinetic energy of an object rotating about a fixed axis with angular momentum $L = I\omega$ can be written as $KE = L^2/2I$.

58. Figure P8.58 shows a claw hammer as it is being used to pull a nail out of a horizontal board. If a force of magnitude 150 N is exerted horizontally as shown, find (a) the force exerted by the hammer claws on the nail and (b) the force exerted by the surface on the point of contact with the hammer head. Assume that the force the hammer exerts on the nail is parallel to the nail.

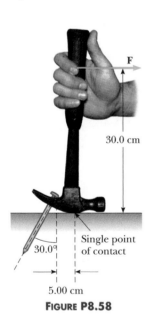

FIGURE P8.58

59. An electric motor turns a flywheel through a drive belt that joins a pulley on the motor and a pulley that is rigidly attached to the flywheel, as shown in Figure P8.59. The flywheel is a solid disk with a mass of 80.0 kg and a diameter of 1.25 m. It turns on a frictionless axle. Its pulley has much smaller mass and a radius of 0.230 m. If the tension in the upper (taut) segment of the belt is 135 N and the flywheel has a clockwise angular acceleration of 1.67 rad/s², find the tension in the lower (slack) segment of the belt.

60. A 12.0-kg object is attached to a cord that is wrapped around a wheel of radius $r = 10.0$ cm (Fig. P8.60). The acceleration of the object down the frictionless incline is measured to be 2.00 m/s². Assuming the axle of the wheel to be frictionless, determine (a) the tension in the rope,

(b) the moment of inertia of the wheel, and (c) the angular speed of the wheel 2.00 s after it begins rotating, starting from rest.

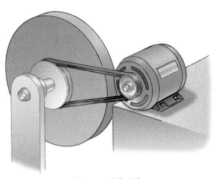

FIGURE P8.59

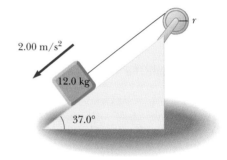

FIGURE P8.60

61. A uniform ladder of length L and weight w is leaning against a vertical wall. The coefficient of static friction between the ladder and the floor is the same as that between the ladder and the wall. If this coefficient of static friction is $\mu_s = 0.500$, determine the smallest angle the ladder can make with the floor without slipping.

62. A uniform 10.0-N picture frame is supported as shown in Figure P8.62. Find the tension in the cords and the magnitude of the horizontal force at P that are required to hold the frame in the position shown.

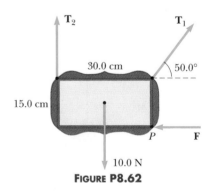

FIGURE P8.62

63. A solid 2.0-kg ball of radius 0.50 m starts at a height of 3.0 m above the surface of Earth and *rolls* down a 20° slope. A solid disk and a ring start at the same time and the same height. The ring and disk each have the same mass and radius as the ball. Which of the three wins the race to the bottom if all roll without slipping?

64. A common physics demonstration (Figure P8.64) consists of a ball resting at the end of a board of length ℓ that is elevated at an angle θ with the horizontal. A light cup is attached to the board at r_c so that it will catch the ball when the support stick is suddenly removed. (a) Show that the ball will lag behind the falling board when $\theta < 35.3°$, and (b) the ball will fall into the cup when the board is supported at this limiting angle and the cup is placed at

$$r_c = \frac{2\ell}{3 \cos \theta}$$

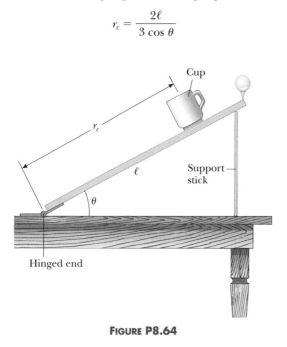

FIGURE P8.64

65. In Figure P8.65 the sliding block has a mass of 0.850 kg, the counterweight has a mass of 0.420 kg, and the pulley is a uniform solid cylinder with a mass of 0.350 kg and an outer radius of 0.0300 m. The coefficient of kinetic friction between the block and the horizontal surface is 0.250. The pulley turns without friction on its axle. The light cord does not stretch and does not slip on the pulley. The block has a velocity of 0.820 m/s toward the pulley when it passes through a photogate. (a) Use energy methods to predict its speed after it has moved to a second photogate, 0.700 m away. (b) Find the angular speed of the pulley at the same moment.

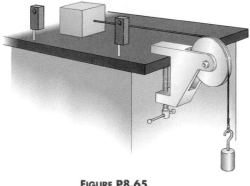

FIGURE P8.65

66. (a) Without the wheels, a bicycle frame has a mass of 8.44 kg. Each of the wheels can be roughly modeled as a

uniform solid disk with a mass of 0.820 kg and a radius of 0.343 m. Find the kinetic energy of the whole bicycle when it is moving forward at 3.35 m/s. (b) Before the invention of a wheel turning on an axle, ancient people moved heavy loads by placing rollers under them. (Modern people use rollers too. Any hardware store will sell you a roller bearing for a lazy susan.) A stone block of mass 844 kg moves forward at 0.335 m/s, supported by two uniform cylindrical tree trunks, each of mass 82.0 kg and radius 0.343 m. There is no slipping between the block and the rollers or between the rollers and the ground. Find the total kinetic energy of the moving objects.

67. In exercise physiology studies it is sometimes important to determine the location of a person's center of gravity. This can be done with the arrangement shown in Figure P8.67. A light plank rests on two scales, which read $F_{g1} = 380$ N and $F_{g2} = 320$ N. The scales are separated by a distance of 2.00 m. How far from the woman's feet is her center of gravity?

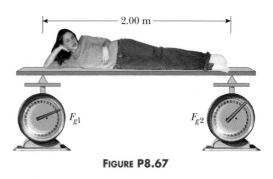

FIGURE P8.67

68. Two astronauts (Fig. P8.68), each having a mass of 75.0 kg, are connected by a 10.0-m rope of negligible mass. They are isolated in space, moving in circles around the point halfway between them at speeds of 5.00 m/s. Treating the astronauts as particles, calculate (a) the magnitude of the angular momentum and (b) the rotational energy of the system. By pulling on the rope, the astronauts shorten the distance between them to 5.00 m. (c) What is the new angular momentum of the system? (d) What are their new speeds? (e) What is the new rotational energy of the system? (f) How much work is done by the astronauts in shortening the rope?

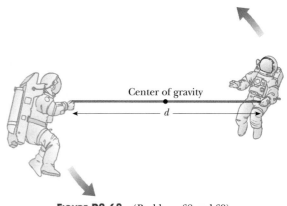

FIGURE P8.68   (Problems 68 and 69)

69. Two astronauts (Fig. P8.68), each having a mass M, are connected by a rope of length d having negligible mass. They are isolated in space, moving in circles around the point halfway between them at speeds v. Calculate (a) the magnitude of the angular momentum of the system by treating the astronauts as particles and (b) the rotational energy of the system. By pulling on the rope, the astronauts shorten the distance between them to $d/2$. (c) What is the new angular momentum of the system? (d) What are their new speeds? (e) What is the new rotational energy of the system? (f) How much work is done by the astronauts in shortening the rope?

70. Two window washers, Bob and Joe, are on a 3.00-m-long, 345-N scaffold supported by two cables attached to its ends. Bob weighs 750 N and stands 1.00 m from the left end, as shown in Figure P8.70. Two meters from the left end is the 500-N washing equipment. Joe is 0.500 m from the right end and weighs 1000 N. Given that the scaffold is in rotational and translational equilibrium, what are the forces on each cable?

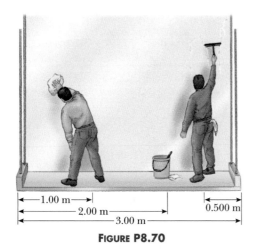

|←—1.00 m—→| |←0.500 m|
|←——2.00 m——→|
|←———3.00 m———→|

FIGURE P8.70

71. We have all complained that there aren't enough hours in a day. In an attempt to change that, suppose that all the people in the world lined up at the Equator, and all started running east at 2.5 m/s relative to the surface of Earth. By how much would the length of a day increase? (Assume that there are 5.5×10^9 people in the world with an average mass of 70 kg each, and that Earth is a solid homogeneous sphere. In addition, you may use the result $1/(1 - x) \approx 1 + x$ for small x.)

72. In a circus performance, a large 5.0-kg hoop of radius 3.0 m rolls without slipping. If the hoop is given an angular speed of 3.0 rad/s while rolling on the horizontal and allowed to roll up a ramp inclined at 20° with the horizontal, how far (measured along the incline) does the hoop roll?

73. A uniform, solid cylinder of mass M and radius R rotates on a frictionless horizontal axle (Fig. P8.73). Two objects with equal masses hang from light cords wrapped around the cylinder. If the system is released from rest, find (a) the tension in each cord and (b) the acceleration of each object after the objects have descended a distance h.

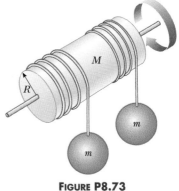

FIGURE P8.73

74. Figure P8.74 shows a vertical force applied tangentially to a uniform cylinder of weight w. The coefficient of static friction between the cylinder and all surfaces is 0.500. Find, in terms of w, the maximum force **F** that can be applied without causing the cylinder to rotate. (*Hint:* When the cylinder is on the verge of slipping, both friction forces are at their maximum values. Why?)

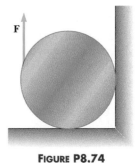

FIGURE P8.74

GROUP ACTIVITIES

G.1 Compare the motion of an empty soup can and a filled soup can down the same incline such as a tilted table. If they are released from rest at the same height on the incline, which one reaches the bottom first? Repeat your observations with different kinds of soup (tomato, chicken noodle, etc.) and a can of beans. Compare their motions, and try to explain your observations. Finally, compare the motion of a filled soup can with a tennis ball, and explain your results to a friend.

G.2 Before attempting this exercise, review Example 8.13 dealing with the spinning stool. The techniques used here are similar to those used there. (a) First, make an estimate of the moment of inertia of your body. One way to do this would be to model your body as a solid cylinder and find I from $I = \frac{1}{2}MR^2$. You would have to determine your mass and to estimate your average "radius" for this approach. Can you think of an alternative way to estimate I? (b) Now, use the approach of Example 8.13 to measure

I. Sit on a rotating stool holding two weights (say two books), and determine the angular speed of rotation with the books extended and after they are pulled in. The angular speed can be found by counting the time for a given number of rotations. Use conservation of angular momentum to determine *I*. Do this five times to determine an average value for *I*. How well do your results for (a) and (b) compare, and if they differ greatly, what might cause the discrepancy?

G.3 This experiment demonstrates a simple way to find the center of gravity of an irregularly shaped object. Cut out an irregular shape from a piece of cardboard, and punch three to five holes around the edge of the shape. Put a pushpin through one of the holes, and tack the shape to a corkboard so that the shape can rotate freely. Now tie a weight to one end of a string and hang the other end of the string from the pushpin. When the string stops moving, trace a line on the cardboard that follows the string. Repeat this for each hole in the cardboard. You will find that there is a point where all the lines intersect. This is the center of gravity.

G.4 A painter climbs a ladder leaning against a smooth wall. At a certain height, the ladder is on the verge of slipping. (a) Explain why the force exerted by the vertical wall on the ladder is horizontal. (b) If the ladder of length *L* leans at an angle θ with the horizontal, what is the lever arm for this horizontal force with the axis of rotation taken at the base of the ladder? (c) If the ladder is uniform, what is the lever arm for the weight of the ladder? (d) Let the mass of the painter be 80 kg, *L* = 4.0 m, the ladder's mass be 30 kg, $\theta = 53°$, and the coefficient of friction between ground and ladder be 0.45. Find the maximum distance the painter can climb up the ladder.

G.5 A horizontal cylinder is fixed in place and allowed to rotate about an axis through its center. The cylinder has a radius of 1.00 m and moment of inertia 350 kg·m². It is given an angular acceleration by wrapping several turns of rope around it and pulling on the rope with a constant force of 120 N. Then, the rope is again wrapped around the cylinder, but this time the rope is passed over a pulley and attached to a hanging weight of 120 N. The weight is released and the cylinder again receives an angular acceleration. (a) Without doing any calculations, in which case is the angular acceleration the greatest, or will it be the same in both cases? Defend your answer. (b) Do the calculations to find the acceleration in each case.

G.6 The polar ice caps contain about 2.3×10^{19} kg of ice. This mass contributes essentially nothing to the moment of inertia of Earth because it is located at the poles, close to the axis of rotation. Estimate the change in the length of the day that would be expected if the polar ice caps were to melt and the water were distributed uniformly over the surface of Earth. (Note that the moment of inertia of a thin spherical shell of radius *r* and mass *m* is $\frac{2}{3}mr^2$.)

G.7 (a) Estimate the angular momentum your body has as a result of Earth's daily rotation about its north-south axis. (b) Estimate the angular momentum an automobile tire has about its axis of rotation while the car is driving along a highway at 55 miles per hour.

G.8 In the movie *Jurassic Park*, there is a scene in which some members of the visiting group are trapped in the kitchen with dinosaurs outside. The paleontologist is pressing against the center of the door, trying to keep out the dinosaurs who are on the other side. The botanist throws herself against the door at the edge near the hinge. A pivotal point in the film is that she cannot reach a gun on the floor because she is trying to hold the door closed. If the paleontologist is pressing at the center of the door, and the botanist at the edge about 8 cm from the hinge, estimate how far the paleontologist would have to relocate in order to have a greater effect on keeping the door closed than both of them pushing together have in their original positions.

(Problems 6, 7, and 8 are courtesy of Edward F. Redish. For more problems of this type, see http://www.physics.umd.edu/perg/)

9 Solids and Fluids

Chapter Outline

This bather is floating in the Dead Sea, a highly saline body of water. The bather floats in the Dead Sea with ease, because the upward buoyant force exerted by the salt water on the bather is much greater than it would be if the bather were floating in fresh water. The Dead Sea is actually a landlocked lake located between Israel and Jordan. At 395 m below sea level, it is the lowest body of water on Earth. Its salinity near the surface is about 30% and increases to about 33% at depths of 150 ft (46 m). The salinity of the Dead Sea is so high that except for bacteria, it cannot sustain animal or vegetable life. *(© Carl Purcell/Photo Researchers, Inc.)*

*I*n this chapter we consider some properties of solids and fluids (both liquids and gases). We spend some time looking at properties that are peculiar to solids, but much of our emphasis is on the properties of fluids. We take this approach because an understanding of the behavior of fluids is fundamentally important to students in the life sciences. We open the fluids part of the chapter with a study of fluids at rest and finish with a discussion of the properties of fluids in motion. Additional topics discussed in this chapter include surface tension, viscosity, and transport phenomena (diffusion and osmosis).

Crystals of natural quartz (SiO_2), one of the most common minerals on Earth. Quartz crystals are used to make special lenses and prisms and in certain electronic applications. *(Charles Winters)*

9.1 STATES OF MATTER

Matter is normally classified as being in one of three states—solid, liquid, or gas. Often this classification system is extended to include a fourth state, referred to as a plasma. When matter is heated to high temperatures, many of the electrons surrounding each atom are freed from the nucleus. The resulting substance is a collection of free, electrically charged particles—negatively charged electrons and positively charged ions. Such a highly ionized substance containing equal amounts of positive and negative charges is a **plasma.** Plasmas exist inside stars, for example. If we were to take a grand tour of our Universe, we would find that there is far more matter in the plasma state than in the more familiar solid, liquid, and gaseous states because there are far more stars around than any other form of celestial matter. In this chapter, however, we ignore plasmas and concentrate on the more familiar solid, liquid, and gaseous forms that make up the environment on our planet.

Everyday experience tells us that a solid has definite volume and shape. A brick, for example, maintains its familiar shape and size day in and day out. We also know that a liquid has a definite volume but no definite shape. For instance, when you fill the tank on a lawn mower, the gasoline changes its shape from that of the original container to that of the tank on the mower. If there is a gallon of gasoline before you pour, however, there still is a gallon after. Finally, a gas has neither definite volume nor definite shape.

All matter consists of some distribution of atoms or molecules. The atoms in a solid are held, by forces that are mainly electrical, at specific positions with respect to one another and vibrate about these equilibrium positions. At low temperatures, however, the vibrating motion is slight and the atoms can be considered essentially fixed. As energy is added to the material, the amplitude of the vibrations increases. A vibrating atom can be viewed as being bound in its equilibrium position by springs attached to neighboring atoms. A collection of such atoms and imaginary springs is shown in Figure 9.1. If a solid is compressed by external forces, we can picture the forces as compressing these tiny internal springs. When the external forces are removed, the solid tends to return to its original shape and size. Consequently, a solid is said to have *elasticity*.

Solids can be classified as either crystalline or amorphous. A **crystalline solid** is one in which the atoms have an ordered structure. For example, in the sodium chloride crystal (common table salt), sodium and chlorine atoms occupy alternate corners of a cube, as in Figure 9.2a. In an **amorphous solid,** such as glass, the atoms are arranged randomly, as in Figure 9.2b.

For any given substance, the liquid state exists at a higher temperature than the solid state. The intermolecular forces in a liquid are not strong enough to keep the molecules in fixed positions, and they wander through the liquid in a random fashion (Fig. 9.2c). Solids and liquids have the following property in common: When an attempt is made to compress them, strong repulsive atomic forces act internally to resist compression.

In the gaseous state, molecules are in constant random motion and exert only weak forces on each other. The average distance between the molecules of a gas is

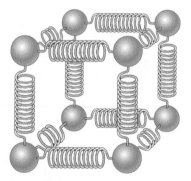

FIGURE 9.1 A model of a portion of a solid. The atoms (spheres) are imagined as being attached to each other by springs, which represent the elastic nature of the interatomic forces. A solid consists of trillions of segments like this with springs connecting all of them.

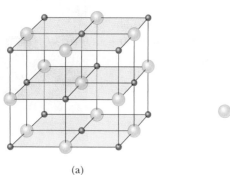

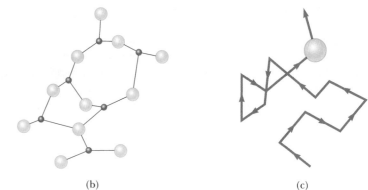

(a)   (b)   (c)

FIGURE 9.2 (a) The NaCl structure, with the Na$^+$ (red) and Cl$^-$ (blue) ions at alternate corners of a cube. (b) In an amorphous solid, the atoms are arranged randomly. (c) Erratic motion of a molecule in a liquid.

Webnote 9.1
For a good visualization of solids, liquids, and gases from a macro-scopic and microscopic viewpoint, take a look at
http://intro.chem.okstate.edu/1225/ Lecture/Chapter13/State.html
http://intro.chem.okstate.edu/1225/ Lecture/Chapter13/Microstates.html

quite large compared with the size of the molecules. Occasionally the molecules collide with each other, but most of the time they move as nearly free, noninteracting particles. We shall say more about gases in subsequent chapters.

9.2 THE DEFORMATION OF SOLIDS

As mentioned in Section 9.1, we usually think of a solid as an object having definite shape and volume. In our study of mechanics, to keep things simple, we assumed that objects remain undeformed when external forces act on them. In reality, all objects are deformable. That is, it is possible to change the shape or size of an object (or both) through the application of external forces. When the forces are removed, the object tends to return to its original shape and size, which means that the deformation exhibits an elastic behavior.

The elastic properties of solids are discussed in terms of stress and strain. **Stress** is related to the force causing a deformation; **strain** is a measure of the degree of deformation. For sufficiently small stresses, stress is proportional to strain, and the constant of proportionality depends on the material being deformed and on the nature of the deformation. We call this proportionality constant the **elastic modulus**:

$$\text{Elastic modulus} = \frac{\text{stress}}{\text{strain}} \tag{9.1}$$

The elastic modulus can be thought of as the stiffness of a material: A material having a large elastic modulus is very stiff and is difficult to deform.

YOUNG'S MODULUS: ELASTICITY IN LENGTH

Consider a long bar of cross-sectional area A and length L_0, clamped at one end (Fig. 9.3). When an external force **F** is applied along the bar, perpendicularly to the cross section, internal forces in the bar resist the distortion ("stretching") that **F** tends to produce, but the bar nevertheless attains an equilibrium in which (1) its length is greater than L_0 and (2) the external force is balanced by internal forces. In such a situation the bar is said to be stressed. We define the **tensile stress** as the ratio of the magnitude of the external force F to the cross-sectional area A. The word "tensile" has the same root as the word "tension" and is used because the bar is under tension. The SI units of stress are newtons per square meter (N/m^2), which are also given a special name, the **pascal** (Pa):

FIGURE 9.3 A long bar clamped at one end is stretched by the amount ΔL under the action of a force **F**.

The pascal ▶

$$1 \text{ Pa} \equiv 1 \text{ N/m}^2 \tag{9.2}$$

TABLE 9.1	Typical Values for Elastic Moduli		
Substance	Young's Modulus (Pa)	Shear Modulus (Pa)	Bulk Modulus (Pa)
Aluminum	7.0×10^{10}	2.5×10^{10}	7.0×10^{10}
Bone	1.8×10^{10}	8.0×10^{10}	—
Brass	9.1×10^{10}	3.5×10^{10}	6.1×10^{10}
Copper	11×10^{10}	4.2×10^{10}	14×10^{10}
Steel	20×10^{10}	8.4×10^{10}	16×10^{10}
Tungsten	35×10^{10}	14×10^{10}	20×10^{10}
Glass	$6.5-7.8 \times 10^{10}$	$2.6-3.2 \times 10^{10}$	$5.0-5.5 \times 10^{10}$
Quartz	5.6×10^{10}	2.6×10^{10}	2.7×10^{10}
Rib Cartilage	1.2×10^{7}	—	—
Rubber	0.1×10^{7}	—	—
Tendon	2×10^{7}	—	—
Water	—	—	0.21×10^{10}
Mercury	—	—	2.8×10^{10}

The **tensile strain** in this case is defined as the ratio of the change in length ΔL to the original length L_0 and is therefore a dimensionless quantity. Thus, we can use Equation 9.1 to define **Young's modulus** Y:

$$Y \equiv \frac{\text{tensile stress}}{\text{tensile strain}} = \frac{F/A}{\Delta L/L_0} = \frac{FL_0}{A \, \Delta L} \qquad [9.3]$$

A material having a large Young's modulus is difficult to stretch. This quantity is typically used to characterize a rod or wire stressed under *either tension or compression*. Note that because the strain is a dimensionless quantity, Y is in pascals. Typical values are given in Table 9.1. Experiments show that (1) the change in length for a fixed external force is proportional to the original length and (2) the force necessary to produce a given strain is proportional to the cross-sectional area.

It is possible to exceed the *elastic limit* of a substance by applying a sufficiently great stress (Fig. 9.4). At the *elastic limit,* the stress-strain curve departs from a straight line. A material subjected to a stress beyond this level ordinarily does not return to its original length when the external force is removed. As the stress is increased further, the material ultimately breaks.

SHEAR MODULUS: ELASTICITY OF SHAPE

Another type of deformation occurs when an object is subjected to a force **F** parallel to one of its faces while the opposite face is held fixed (Fig. 9.5a). If the object is originally a rectangular block, such a parallel force results in a shape whose cross section is a parallelogram. In this situation, the stress is called a shear stress. A book pushed sideways as in Figure 9.5b is under a shear stress. There is no change in volume with this deformation. We define the **shear stress** as F/A, the ratio of the magnitude of the parallel force to the area A of the face being sheared. The **shear strain** is the ratio $\Delta x/h$, where Δx is the horizontal distance the sheared face moves and h is the height of the object. In terms of these quantities, the **shear modulus** S is defined as

$$S \equiv \frac{\text{shear stress}}{\text{shear strain}} = \frac{F/A}{\Delta x/h} \qquad [9.4]$$

A material having a large shear modulus is difficult to bend.

Shear moduli for some representative materials are listed in Table 9.1. Note that the units of shear modulus are force per unit area (Pa).

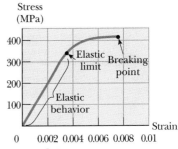

FIGURE 9.4 Stress-versus-strain curve for an elastic solid.

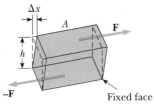

FIGURE 9.5 (a) A shear deformation in which a rectangular block is distorted by a force applied tangent to one of its faces. (b) A book under shear stress.

BULK MODULUS: VOLUME ELASTICITY

Bulk modulus characterizes the response of a substance to uniform squeezing. Suppose that the external forces acting on an object are all perpendicular to the surface on which the force acts and distributed uniformly over the surface of the object (Fig. 9.6). This occurs when an object is immersed in a fluid. An object subject to this type of deformation undergoes a change in volume but no change in shape. The **volume stress** ΔP is defined as the ratio of the magnitude of the change in the applied force ΔF to the surface area A. (When dealing with fluids, we shall refer to the quantity F/A as the **pressure.**) The volume strain is equal to the change in volume ΔV divided by the original volume V. Thus, from Equation 9.1 we can characterize a volume compression in terms of the **bulk modulus** B, defined as

Bulk modulus ▶

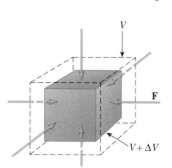

$$B \equiv \frac{\text{volume stress}}{\text{volume strain}} = -\frac{\Delta F/A}{\Delta V/V} = -\frac{\Delta P}{\Delta V/V} \qquad [9.5]$$

A material having a large bulk modulus does not compress easily. Note that a negative sign is included in this defining equation so that B is always positive. An increase in pressure (positive ΔP) causes a decrease in volume (negative ΔV), and vice versa.

Table 9.1 lists bulk modulus values for some materials. If you look up such values in a different source, you will often find that the reciprocal of the bulk modulus, called the **compressibility** of the material, is listed. Note from Table 9.1 that both solids and liquids have bulk moduli. However, there is no shear modulus and no Young's modulus for liquids because a liquid will not sustain a shearing stress or a tensile stress. (It flows instead.)

FIGURE 9.6 When a solid is under uniform pressure, it undergoes a change in volume but no change in shape. This cube is compressed on all sides by forces normal to its six faces.

Example 9.1 Built to Last

A vertical steel beam in a building supports a load of 6.0×10^4 N. If the length of the beam is 4.0 m and its cross-sectional area is 8.0×10^{-3} m², find the distance it is compressed along its length.

Solution Because the beam is under compression, we can use Equation 9.3. Taking Young's modulus for steel, $Y = 20 \times 10^{10}$ Pa, from Table 9.1, we have

$$Y = \frac{FL_0}{A\,\Delta L}$$

or

$$\Delta L = \frac{FL_0}{YA} = \frac{(6.0 \times 10^4 \text{ N})(4.0 \text{ m})}{(20 \times 10^{10} \text{ Pa})(8.0 \times 10^{-3} \text{ m}^2)} = \boxed{1.5 \times 10^{-4} \text{ m}}$$

Example 9.2 Squeezing a Lead Sphere

A solid lead sphere of volume 0.50 m³ is dropped in the ocean to a depth of about 2 000 m (about 1 mile) where the pressure increases by 2.0×10^7 Pa. Lead has a bulk modulus of 7.7×10^9 Pa. What is the change in volume of the sphere?

Solution From the definition of bulk modulus (Eq. 9.5), we have

$$B = -\frac{\Delta P}{\Delta V/V}$$

$$\Delta V = -\frac{V\,\Delta P}{B}$$

When the sphere is at the surface, where its volume is 0.50 m^3, the pressure exerted on it is atmospheric pressure. The increase in pressure ΔP when the sphere is submerged is 2.0×10^7 Pa. Therefore, the change in its volume when it is submerged is

$$\Delta V = -\frac{(0.50 \text{ m}^3)(2.0 \times 10^7 \text{ Pa})}{7.7 \times 10^9 \text{ Pa}} = \boxed{-1.3 \times 10^{-3} \text{ m}^3}$$

The negative sign indicates a *decrease* in volume.

ARCHES AND THE ULTIMATE STRENGTH OF MATERIALS

APPLICATION

ARCH STRUCTURES IN
BUILDINGS

The **ultimate strength** of a material is the maximum force per unit area the material can withstand before it breaks or fractures. Such values are of great importance, particularly in the construction of buildings, bridges, and roads. Table 9.2 gives the ultimate strength of a variety of materials under both tension and compression. Note that both bone and a variety of building materials (concrete, brick and marble) are stronger in compression than in tension. The greater ability of brick and stone to resist compression is the basis of the semicircular arch, developed and used extensively by the Romans in everything from memorial arches, to expansive temples, to aqueduct supports.

Before the development of the arch, the principal method of spanning a space was the simple post-and-beam construction (Fig. 9.7a), in which a horizontal beam is supported by two columns. This type of construction was used to build the great Greek temples. The columns of these temples are closely spaced because of the limited length of available stones and also because of the low ultimate tensile strength of a sagging stone beam.

TABLE 9.2	**Ultimate Strength of Materials**	
Material	**Tensile Strength (N/m²)**	**Compressive Strength (N/m²)**
Iron	1.7×10^8	5.5×10^8
Steel	5.0×10^8	5.0×10^8
Aluminum	2.0×10^8	2.0×10^8
Bone	1.2×10^8	1.5×10^8
Marble	—	8.0×10^7
Brick	1×10^6	3.5×10^7
Concrete	2×10^6	2×10^7

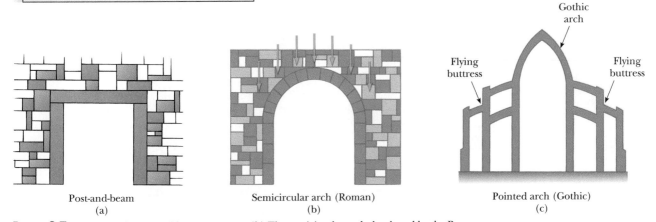

Post-and-beam
(a)

Semicircular arch (Roman)
(b)

Pointed arch (Gothic)
(c)

Gothic arch

Flying buttress

Flying buttress

FIGURE 9.7 (a) A simple post-and-beam structure. (b) The semicircular arch developed by the Romans. (c) Gothic arch with flying buttresses to provide lateral support.

The semicircular arch (Fig. 9.7b) developed by the Romans was a great technological achievement in architectural design. It effectively allowed the heavy load of a wide roof span to be channeled into horizontal and vertical forces on narrow supporting columns. The stability of this arch depends on the compression between its wedge-shaped stones. The stones are forced to squeeze against each other by the uniform loading as shown in Figure 9.7b. This results in horizontal outward forces at the base of the arch where it starts curving away from the vertical, which must be balanced by the stone walls shown on the sides of the arch. It is common to use very heavy walls (buttresses) on either side of the arch to provide horizontal stability. If the foundation of the arch should move, the compressive forces between the wedge-shaped stones may decrease to the extent that the arch collapses. The stone surfaces used in the arches constructed by the Romans were cut to make a very tight joint; it is interesting to note that mortar was usually not used in these joints. The resistance to slipping between stones was provided by the compression force and the friction between the stone faces.

Another important architectural innovation was the pointed Gothic arch shown in Figure 9.7c. This type of structure was first used in Europe beginning in the 12th century, followed by the construction of several magnificent Gothic cathedrals in France in the 13th century. One of the most striking features of these cathedrals is their extreme height. For example, the cathedral at Chartres rises to 118 ft and the one at Reims has a height of 137 ft. Such magnificent buildings evolved over a very short period of time, without the benefit of any mathematical theory of structures. However, Gothic arches required flying buttresses to prevent the spreading of the arch supported by the tall, narrow columns.

9.3   DENSITY AND PRESSURE

The density of a substance of uniform composition is defined as its mass per unit volume.

In symbolic form, a substance of mass M and volume V has a density ρ (Greek rho), given by

Density ▶

$$\rho \equiv \frac{M}{V} \qquad [9.6]$$

The units of density are kilograms per cubic meter in SI units and grams per cubic centimeter in the cgs system. Table 9.3 lists the densities of some substances.

TABLE 9.3	Density of Some Common Substances		
Substance	$\rho(\text{kg/m}^3)^a$	**Substance**	$\rho(\text{kg/m}^3)^a$
Ice	0.917×10^3	Water	1.00×10^3
Aluminum	2.70×10^3	Glycerin	1.26×10^3
Iron	7.86×10^3	Ethyl alcohol	0.806×10^3
Copper	8.92×10^3	Benzene	0.879×10^3
Silver	10.5×10^3	Mercury	13.6×10^3
Lead	11.3×10^3	Air	1.29
Gold	19.3×10^3	Oxygen	1.43
Platinum	21.4×10^3	Hydrogen	8.99×10^{-2}
Uranium	18.7×10^3	Helium	1.79×10^{-1}

[a] All values are at standard atmospheric temperature and pressure (STP), defined as 0°C (273 K) and 1 atm (1.013×10^5 Pa). To convert to grams per cubic centimeter, multiply by 10^{-3}.

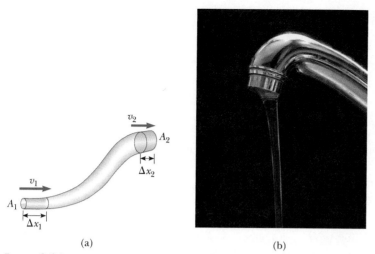

(a)   (b)

FIGURE 9.26 (a) A fluid moving with streamline flow through a pipe of varying cross-sectional area. The volume of fluid flowing through A_1 in a time interval Δt must equal the volume flowing through A_2 in the same time interval. Therefore, $A_1 v_1 = A_2 v_2$. (b) Water flowing slowly out of a faucet. The width of the stream narrows as the water falls and speeds up in accord with the continuity equation. *(George Semple)*

time interval Δt, the fluid entering the bottom end of the pipe moves a distance of $\Delta x_1 = v_1 \Delta t$, where v_1 is the speed of the fluid at this location. If A_1 is the cross-sectional area in this region, then the mass contained in the bottom blue region is $\Delta M_1 = \rho_1 A_1 \Delta x_1 = \rho_1 A_1 v_1 \Delta t$, where ρ_1 is the density of the fluid at A_1. Similarly, the fluid that moves out of the upper end of the pipe in the same time interval Δt has a mass of $\Delta M_2 = \rho_2 A_2 v_2 \Delta t$. However, **because mass is conserved** and because the flow is steady, the mass that flows into the bottom of the pipe through A_1 in the time Δt must equal the mass that flows out through A_2 in the same interval. Therefore, $\Delta M_1 = \Delta M_2$, or

$$\rho_1 A_1 v_1 = \rho_2 A_2 v_2 \qquad \text{[9.13]}$$

For the case of an incompressible fluid, $\rho_1 = \rho_2$ and Equation 9.13 reduces to

$$A_1 v_1 = A_2 v_2 \qquad \text{[9.14]}$$

◀ Equation of continuity

This expression is called the **equation of continuity.** From this result, we see that **the product of the cross-sectional area of the pipe and the fluid speed at that cross section is a constant.** Therefore, the speed is high where the tube is constricted and low where the tube has a larger diameter. The product Av, which has dimensions of volume per unit time, is called the **flow rate. The condition $Av =$ constant is equivalent to the fact that the amount of fluid that enters one end of the tube in a given time interval equals the amount of fluid leaving the tube in the same interval, assuming that the fluid is incompressible and that there are no leaks.** Figure 9.26b is a one example of an application of the equation of continuity. As the stream of water flows continuously from a faucet, the width of the stream narrows as it falls and speeds up.

There are many instances in your everyday experiences when you have seen the equation of continuity in action. For example, you place your thumb over the open end of a garden hose to make the water spray out farther and with greater speed. As you reduce the cross-sectional area of the nozzle, the water drops leave the nozzle at a higher speed, causing them to move a longer distance. Similar reasoning explains why rising smoke from a smoldering piece of wood behaves as it does. The smoke first rises in a streamline pattern, getting thinner as it rises, and eventually breaks up into a swirling, turbulent pattern. The smoke rises because it is less dense than air, and the buoyant force of the air accelerates it upward. As the

Tip 9.4 CONTINUITY EQUATIONS

Equation 9.14 states that the rate of flow of fluid into the system equals the rate of flow out of the system, because fluid cannot be stored in the system once it is full of fluid. Example 9.4, which analyzed a car lift, represents an analogous result for energy conservation—the energy transferred into the lift is equal to that transferred out.

speed of the smoke stream increases, the cross-sectional area of the stream decreases, as seen from the equation of continuity. However, the stream soon reaches a speed so great that streamline flow is not possible. We shall study the relationship between speed of fluid flow and turbulence in a later discussion of the Reynolds number.

Example 9.7 Watering a Garden

A water hose 2.50 cm in diameter is used by a gardener to fill a 30.0-liter bucket (1 liter = 1 000 cm^3, exactly). The gardener notes that it takes 1.00 min to fill the bucket. A nozzle with an opening of cross-sectional area 0.500 cm^2 is then attached to the hose. The nozzle is held so that water is projected horizontally from a point 1.00 m above the ground. Over what horizontal distance can the water be projected?

Solution We identify point 1 within the hose and point 2 at the exit of the nozzle. We first use the continuity equation for fluids to find the speed $v_2 = v_{x0}$ with which the water exits the nozzle:

$$A_1 v_1 = A_2 v_2 = A_2 v_{x0}$$

$$v_{x0} = \frac{A_1 v_1}{A_2}$$

The numerator of this expression, $A_1 v_1$, is the volume flow rate in the hose, which can be evaluated from the bucket-filling information. Thus,

$$v_{x0} = \frac{A_1 v_1}{A_2} = \left(\frac{30.0 \ \text{L/min}}{0.500 \ \text{cm}^2} \right)\left(\frac{10^3 \ \text{cm}^3}{1 \ \text{L}} \right)\left(\frac{1 \ \text{min}}{60 \ \text{s}} \right)\left(\frac{1 \ \text{m}}{100 \ \text{cm}} \right) = 10.0 \ \text{m/s}$$

Because the water is in free fall once it exits the nozzle, it exhibits projectile motion after leaving the hose. A particle of the water falls through a vertical distance of 1.00 m starting from rest. Setting $y_0 = 0$ and $v_{y0} = 0$ at $t = 0$, the time required to fall this distance can be calculated from Equation 3.12:

$$\Delta y = v_{y0}t - \tfrac{1}{2}gt^2$$

$$t = \sqrt{\frac{2y}{g}}$$

$$= \sqrt{\frac{2(1.00 \ \text{m})}{9.80 \ \text{m/s}^2}} = 0.452 \ \text{s}$$

In the horizontal direction, we apply Equation 3.10 to a particle of water to find the horizontal distance, setting $x_0 = 0$ at $t = 0$:

$$x = v_{x0}t = (10.0 \ \text{m/s})(0.452 \ \text{s}) = \boxed{4.52 \ \text{m}}$$

EXERCISE The Garfield Thomas water tunnel at Pennsylvania State University has a circular cross-section that constricts from a diameter of 3.6 m to the test section, which is 1.2 m in diameter. If the speed of flow is 3.0 m/s in the larger-diameter pipe, determine the speed of flow in the test section.

ANSWER 27 m/s

DANIEL BERNOULLI, SWISS PHYSICIST AND MATHEMATICIAN (1700–1782)

In his most famous work, *Hydrodynamica,* Bernoulli showed that, as the velocity of fluid flow increases, its pressure decreases. In this same publication, Bernoulli also attempted the first explanation of the behavior of gases with changing pressure and temperature; this was the beginning of kinetic theory of gases. *(Corbis-Bettmann)*

BERNOULLI'S EQUATION

As a fluid moves through a pipe of varying cross section and elevation, the pressure changes along the pipe. In 1738 the Swiss physicist Daniel Bernoulli (1700–1782) derived an expression that relates pressure to fluid speed and elevation. Bernoulli's equation is not a freestanding law of physics. As we shall see in this section, **Bernoulli's equation is a consequence of energy conservation as applied to an ideal fluid.**

In deriving Bernoulli's equation, we again assume that the fluid is incompressible and nonviscous and flows in a nonturbulent, steady-state manner. Consider the flow through a nonuniform pipe in the time Δt, as illustrated in Figure 9.27. The force on the lower end of the fluid is $P_1 A_1$, where P_1 is the pressure at the lower end. The work done on the lower end of the fluid by the fluid behind it is

$$W_1 = F_1\,\Delta x_1 = P_1 A_1\,\Delta x_1 = P_1 V$$

where V is the volume of the lower blue region in Figure 9.27. In a similar manner, the work done on the fluid on the upper portion in the time Δt is

$$W_2 = -P_2 A_2\,\Delta x_2 = -P_2 V$$

(Remember that the volume of fluid that passes through A_1 in the time Δt equals the volume that passes through A_2 in the same interval.) The work W_2 is negative because the force on the fluid at the top is opposite its displacement. The net work done by these forces in the time Δt equals the net work done by the fluid:

$$W_{\text{fluid}} = P_1 V - P_2 V$$

Part of this work goes into changing the fluid's kinetic energy, and part goes into changing the gravitational potential energy of the fluid-earth system. If m is the mass of the fluid passing through the pipe in the time interval Δt, then the change in kinetic energy of the volume of fluid is

$$\Delta KE = \tfrac{1}{2}mv_2^{\,2} - \tfrac{1}{2}mv_1^{\,2}$$

The change in the gravitational potential energy is

$$\Delta PE = mgy_2 - mgy_1$$

Because the net work done by the fluid on the segment of fluid shown in Figure 9.27 changes the kinetic energy and the potential energy of the nonisolated system, we have

$$W_{\text{fluid}} = \Delta KE + \Delta PE$$

The three terms in this equation are those we have just evaluated. Substituting expressions for each of the terms gives

$$P_1 V - P_2 V = \tfrac{1}{2}mv_2^{\,2} - \tfrac{1}{2}mv_1^{\,2} + mgy_2 - mgy_1$$

If we divide each term by V and recall that $\rho = m/V$, this expression reduces to

$$P_1 - P_2 = \tfrac{1}{2}\rho v_2^{\,2} - \tfrac{1}{2}\rho v_1^{\,2} + \rho gy_2 - \rho gy_1$$

Let us move those terms that refer to point ① to one side of the equation and those that refer to point ② to the other side:

$$P_1 + \tfrac{1}{2}\rho v_1^{\,2} + \rho gy_1 = P_2 + \tfrac{1}{2}\rho v_2^{\,2} + \rho gy_2 \qquad \text{[9.15]}$$

◀ Bernoulli's equation

This is **Bernoulli's equation.** It is often expressed as

$$P + \tfrac{1}{2}\rho v^2 + \rho gy = \text{constant} \qquad \text{[9.16]}$$

> **Bernoulli's equation states that the sum of the pressure (P), the kinetic energy per unit volume ($\tfrac{1}{2}\rho v^2$), and the potential energy per unit volume (ρgy) has the same value at all points along a streamline.**

An important consequence of Bernoulli's equation can be demonstrated by considering Figure 9.28, which shows water flowing through a horizontal constricted pipe from a region of large cross-sectional area into a region of smaller cross-sectional area. This device, called a *Venturi tube,* can be used to measure the

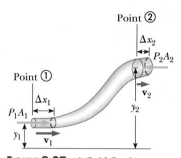

FIGURE 9.27 A fluid flowing through a constricted pipe with streamline flow. The fluid in the section with a length of Δx_1 moves to the section with a length of Δx_2. The volumes of fluid in the two sections are equal.

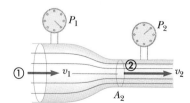

FIGURE 9.28 The pressure P_1 is greater than the pressure P_2, because $v_1 < v_2$. This device can be used to measure the speed of fluid flow.

BERNOULLI'S PRINCIPLE FOR GASES

For gases, we cannot use the assumption that the fluid is incompressible. Thus, Equation 9.16 is not true for gases. The qualitative behavior is the same, however—as the speed of the gas increases, its pressure decreases.

speed of fluid flow. Let us compare the pressure at point ① to the pressure at point ②. Because the pipe is horizontal, $y_1 = y_2$ and Equation 9.15 applied to points ① and ② gives

$$P_1 + \tfrac{1}{2}\rho v_1{}^2 = P_2 + \tfrac{1}{2}\rho v_2{}^2 \qquad \textbf{[9.17]}$$

Because the water is not backing up in the pipe, its speed v_2 in the constriction must be greater than its speed v_1 in the region of greater diameter. From Equation 9.17, $v_2 > v_1$ means that P_2 must be less than P_1. This result is often expressed by the statement that **swiftly moving fluids exert less pressure than do slowly moving fluids.** As we shall see in the next section, this important result enables us to understand a wide range of everyday phenomena.

Quick Quiz 9.7

You observe two helium balloons floating next to each other at the ends of strings secured to a table. The facing surfaces of the balloons are separated by 1–2 cm. You blow through the opening between the balloons. What happens to the balloons? (a) They move toward each other. (b) They move away from each other. (c) They are unaffected.

Example 9.8 Shoot-Out at the Old Water Tank

A nearsighted sheriff fires at a cattle rustler with his trusty six-shooter. Fortunately for the cattle rustler, the bullet misses him and penetrates the town water tank and causes a leak (Fig. 9.29). If the top of the tank is open to the atmosphere, determine the speed at which the water leaves the hole when the water level is 0.500 m above the hole.

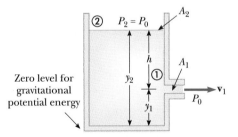

FIGURE 9.29 (Example 9.8) The water speed v_1 from the hole in the side of the container is given by $v_1 = \sqrt{2gh}$.

Reasoning If we assume that the cross-sectional area of the tank is large relative to that of the hole ($A_2 \gg A_1$), then the water level drops very slowly and we can assume $v_2 \approx 0$. Let us apply Bernoulli's equation to points ① and ②. If we note that $P_1 = P_0$ at the hole, we get

$$P_0 + \tfrac{1}{2}\rho v_1{}^2 + \rho g y_1 = P_0 + \rho g y_2$$

$$v_1 = \sqrt{2g(y_2 - y_1)} = \sqrt{2gh}$$

Solution This says that the speed of the water emerging from the hole is equal to the speed acquired by a body falling freely through the vertical distance h. This is known as **Torricelli's law.** If the height h is 0.500 m, for example, the speed of the stream is

$$v = \sqrt{2(9.80 \text{ m/s}^2)(0.500 \text{ m})} = \boxed{3.13 \text{ m/s}}$$

EXERCISE If the head of the cattle rustler is 3.00 m below the level of the hole in the tank, where must he stand to get doused with water?

ANSWER 2.45 m from the base of the tank

9.8 OTHER APPLICATIONS OF FLUID DYNAMICS

In this section we describe some common phenomena that can be explained, at least in part, by Bernoulli's equation.

In general, an object moving through a fluid experiences a net upward force as the result of any effect that causes the fluid to change its direction as it flows past the object. For example, a golf ball struck with a club is given a rapid back-spin, as shown in Figure 9.30. The dimples on the ball help "entrain" the air to follow the curvature of the ball's surface. Figure 9.30 shows a thin layer of air wrapping part way around the ball and being deflected downward as a result. Because the ball pushes the air down, by Newton's third law the air must push up on the ball and cause it to rise. Without the dimples, the air is not as well entrained and the golf ball does not travel as far. For the same reason, a tennis ball's fuzz helps the spinning ball "grab" the air rushing by and helps deflect it.

Many devices operate in the manner illustrated in Figure 9.31. A stream of air passing over an open tube reduces the pressure above the tube. This causes the liquid to rise into the airstream. The liquid is then dispersed into a fine spray of droplets. You might recognize that this so-called atomizer is used in perfume bottles and paint sprayers. The same principle is used in the carburetor of a gasoline engine. In that case, the low-pressure region in the carburetor is produced by air drawn in by the piston through the air filter. The gasoline vaporizes, mixes with the air, and enters the cylinder of the engine for combustion.

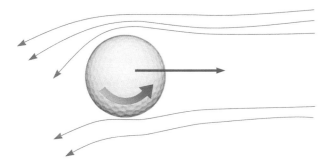

FIGURE 9.30 A spinning golf ball experiences a lifting force that allows it to travel much farther than it would if it were not spinning.

In a person with advanced arteriosclerosis, the Bernoulli effect produces a symptom called *vascular flutter*. In this situation, the artery is constricted as a result of accumulated plaque on its inner walls, as in Figure 9.32. To maintain a constant flow rate, the blood must travel faster than normal through the constriction. If the blood speed is sufficiently high in the constricted region, its pressure is low, and the artery may collapse under external pressure, causing a momentary interruption in blood flow. At this moment there is no Bernoulli effect, and the vessel reopens under arterial pressure. As the blood rushes through the constricted artery, the internal pressure drops and again the artery closes. Such variations in blood flow can be heard with a stethoscope. If the plaque becomes dislodged and ends up in a smaller vessel that delivers blood to the heart, the person can suffer a heart attack.

An aneurysm is a weakened spot on an artery where the artery walls have ballooned outward. Blood flows more slowly through this region, as can be seen from the equation of continuity, resulting in an increase in pressure in the vicinity of the aneurysm relative to the pressure in other parts of the artery. This condition is dangerous because the excess pressure can cause the artery to rupture.

The lift on an aircraft wing can also be explained, in part, by the Bernoulli effect. Airplane wings are designed so that the air speed above the wing is greater than that below. As a result, the air pressure above the wing is less than the pressure below, and there is a net upward force on the wing, called the "lift." Another

▶ **APPLICATION**

LIFT EXPERIENCED BY A SPINNING BALL

▶ **APPLICATION**

"ATOMIZERS" IN PERFUME BOTTLES AND PAINT SPRAYERS

FIGURE 9.31 A stream of air passing over a tube dipped in a liquid causes the liquid to rise in the tube as shown. This effect is used in perfume bottles and paint sprayers.

▶ **APPLICATION**

VASCULAR FLUTTER AND ANEURYSMS

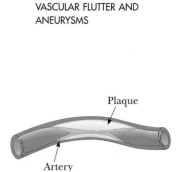

FIGURE 9.32 Blood must travel faster than normal through a constricted region of an artery.

▶ **APPLICATION**

LIFT ON AIRCRAFT WINGS

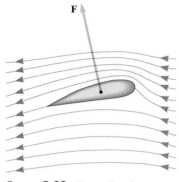

FIGURE 9.33 Streamline flow around an airplane wing. The pressure above is less than the pressure below, and there is a dynamic lift force upward.

factor influencing the lift on a wing is shown in Figure 9.33. The wing has a slight upward tilt that causes air molecules striking the bottom to be deflected downward. The air molecules bouncing off the wing at the bottom produce an upward force on the wing and a significant lift on the aircraft. Finally, turbulence also has an effect. If the wing is tilted too much, the flow of air across the upper surface becomes turbulent, and the pressure difference across the wing is not as great as that predicted by the Bernoulli effect. In an extreme case, this turbulence may cause the aircraft to stall.

APPLYING **PHYSICS** 9.4

The Bernoulli effect can be used to partially explain how a sailboat can accomplish the seemingly impossible task of sailing into the wind. How can this be done?

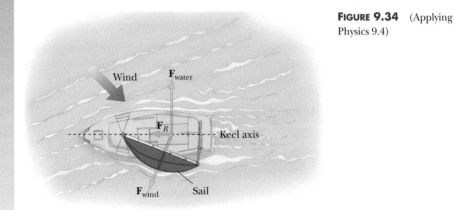

FIGURE 9.34 (Applying Physics 9.4)

Explanation As shown in Figure 9.34, the wind blowing in the direction of the arrow causes the sail to billow out, taking on a shape similar to that of an airplane wing. For the same reason as for an airplane wing, there is a force acting on the sail in the direction shown. The component of force perpendicular to the boat tends to make the boat move sideways in the water, but the keel prevents this sideways motion. The component of the force in the forward direction drives the boat almost against the wind. The word *almost* is used because a sailboat can move forward only when the wind direction is about $10-15°$ with respect to the forward direction. This means that in order to sail directly against the wind, a boat must follow a zigzag path, a procedure called *tacking*, so that the wind is always at some angle with respect to the direction of travel.

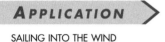

APPLICATION

SAILING INTO THE WIND

APPLICATION

HOME PLUMBING

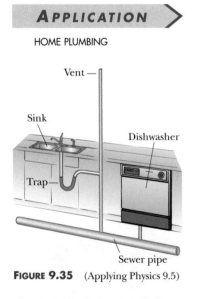

FIGURE 9.35 (Applying Physics 9.5)

APPLYING **PHYSICS** 9.5

Consider the portion of a home plumbing system shown in Figure 9.35. The water trap in the pipe below the sink captures a plug of water that prevents sewer gas from finding its way from the sewer pipe, up the sink drain, and into the home. Suppose the dishwasher is draining, so that water is moving to the left in the sewer pipe. What is the purpose of the vent, which is open to the air above the roof of the house? In which direction is air moving at the opening of the vent, upward or downward?

Explanation Let us imagine that the vent is not present, so that the drain pipe for the sink is simply connected through the trap to the sewer pipe. As water from the dishwasher moves to the left in the sewer pipe, the pressure in the sewer pipe is reduced below atmospheric pressure, according to Bernoulli's principle. The pressure at the drain in the sink is still at atmospheric pressure. Thus, this pressure differential can push the plug of water in the water trap of the sink down the drain pipe into the sewer pipe, removing it as a barrier to sewer gas. With the addition of the vent to the roof, the reduced pressure of the dishwasher water results in air entering the vent pipe at the roof. This keeps the pressure in the vent pipe and the right-hand side of the sink drain pipe at a pressure close to atmospheric, so that the plug of water in the water trap remains in place.

9.9 SURFACE TENSION, CAPILLARY ACTION, AND VISCOUS FLUID FLOW

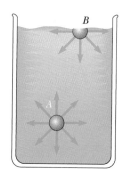

FIGURE 9.36 The net force on a molecule at *A* is zero because such a molecule is completely surrounded by other molecules. The net force on a surface molecule at *B* is downward because it is not completely surrounded by other molecules.

If you look closely at a dewdrop sparkling in the morning sunlight, you will find that the drop is spherical. The drop takes this shape because of a property of liquid surfaces called **surface tension.** In order to understand the origin of surface tension, consider a molecule at point *A* in a container of water, as in Figure 9.36. Although nearby molecules exert forces on this molecule, the net force on it is zero because it is completely surrounded by other molecules and hence is attracted equally in all directions. The molecule at *B*, however, is not attracted equally in all directions. Because there are no molecules above it to exert upward forces, the molecule at *B* is pulled toward the interior of the liquid. The contraction at the surface of the liquid ceases when the inward pull exerted on the surface molecules is balanced by the outward repulsive forces that arise from collisions with molecules in the interior of the liquid. **The net effect of this pull on all the surface molecules is to make the surface of the liquid contract and consequently to make the surface area of the liquid as small as possible.** Drops of water take on a spherical shape because a sphere has the smallest surface area for a given volume.

If you place a sewing needle very carefully on the surface of a bowl of water, you will find that the needle floats even though the density of steel is about eight times that of water. This can also be explained by surface tension. A close examination of the needle shows that it actually rests in a depression in the liquid surface, as shown in Figure 9.37. The water surface acts like an elastic membrane under tension. The weight of the needle produces a depression, thus increasing the surface area of the film. Molecular forces now act at all points along the depression in an attempt to restore the surface to its original horizontal position. The vertical components of these forces act to balance the force of gravity acting on the needle.

The **surface tension** γ in a film of liquid is defined as the ratio of the magnitude of the surface tension force **F** to the length *L* along which the force acts:

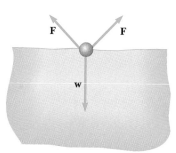

FIGURE 9.37 End view of a needle resting on the surface of water. The components of surface tension balance the force of gravity.

$$\gamma \equiv \frac{F}{L} \qquad\qquad [9.18]$$

The SI units of surface tension are newtons per meter, and values for a few representative materials are given in Table 9.4.

The concept of surface tension can be thought of as the energy content of the fluid at its surface per unit surface area. To see that this is reasonable, we can

TABLE 9.4	Surface Tensions for Various Liquids	
Liquid	**$T(°C)$**	**Surface Tension (N/m)**
Ethyl alcohol	20	0.022
Mercury	20	0.465
Soapy water	20	0.025
Water	20	0.073
Water	100	0.059

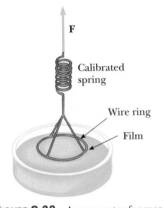

FIGURE 9.38 An apparatus for measuring the surface tension of liquids. The force on the wire ring is measured just before it breaks free of the liquid.

Webnote 9.4

Surface tension plays an important part in the formation of bubbles. Read more at *http://www.exploratorium.edu/ronh/bubbles/bubbles.html*

manipulate the units of surface tension γ as

$$\frac{N}{m} = \frac{N \cdot m}{m^2} = \frac{J}{m^2}$$

In general, **any equilibrium configuration of an object is one in which the energy is a minimum.** Consequently, a fluid will take on a shape such that its surface area is as small as possible. For a given volume, a spherical shape is the one that has the smallest surface area; therefore, a drop of water takes on a spherical shape.

An apparatus used to measure the surface tension of liquids is shown in Figure 9.38. A circular wire with a circumference L is lifted from a body of liquid. The surface film clings to the inside and outside edges of the wire, holding back the wire and causing the spring to stretch. If the spring is calibrated, we can measure the force required to overcome the surface tension of the liquid. In this case, the surface tension is given by

$$\gamma = \frac{F}{2L}$$

We must use $2L$ for the length because the surface film exerts forces on the inside and outside of the ring.

The surface tension of liquids decreases with increasing temperature. This occurs because the faster moving molecules of a hot liquid are not bound together as strongly as are those in a cooler liquid. Furthermore, certain ingredients called surfactants decrease surface tension when added to liquids. For example, soap or detergent decreases the surface tension of water. This reduction in surface tension makes it easier for soapy water to penetrate the cracks and crevices of your clothes to clean them better than plain water. An effect similar to this occurs in our lungs. The surface tissue of the air sacs in the lungs contains a fluid that has a surface tension of about 0.050 N/m. A liquid with a surface tension this high would make it very difficult for the lungs to expand as we inhale. However, as the area of the lungs increases with inhalation, the body secretes into the tissue a substance that gradually reduces the surface tension of the liquid. At full expansion, the surface tension of the lung fluid can drop to as low as 0.005 N/m.

Example 9.9   Walking on Water

In this example, we illustrate how an insect is supported on the surface of water by surface tension. Let us assume that the insect's "foot" is spherical. When the insect steps onto the water with all six legs, a depression is formed in the water around each foot, as shown in Figure 9.39a. The surface tension of the water produces upward forces on the water that tend to restore the water surface to its normally flat shape. If the insect has a mass of 2.0×10^{-5} kg and if the radius of each foot is 1.5×10^{-4} m, find the angle θ.

Solution   From the definition of surface tension, we can find the magnitude of the net force **F** directed tangential to the depressed part of the water surface:

$$F = \gamma L$$

The length L along which this force acts is equal to the distance around the insect's foot, $2\pi r$. (It is assumed that the insect depresses the water surface such that the radius of the depression is equal to the radius of the foot.) Thus, $F = \gamma(2\pi r)$, and the net vertical force is

$$F_v = \gamma(2\pi r)(\cos \theta)$$

Because the insect has six legs, this upward force must equal one sixth the weight of the insect, assuming its weight is equally distributed on all 6 feet. Thus,

$$\gamma(2\pi r)(\cos\theta) = \tfrac{1}{6}w = \tfrac{1}{6}mg$$

$$\cos\theta = \frac{mg}{12\pi r\gamma} = \frac{(2.0\times10^{-5}\text{ kg})(9.80\text{ m/s}^2)}{12\pi(1.5\times10^{-4}\text{ m})(0.073\text{ N/m})} \qquad \textbf{(1)}$$

$$\theta = \boxed{62°}$$

Note that if the weight of the insect were great enough to make the right side of (1) greater than unity, a solution for θ would be impossible because the cosine can never be greater than unity. Under these conditions, the insect would sink.

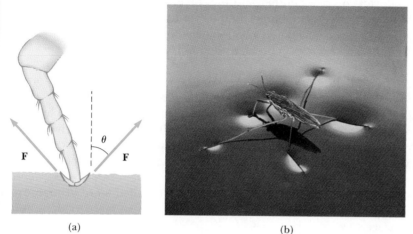

(a)   (b)

FIGURE 9.39 (Example 9.9) (a) One foot of an insect resting on the surface of water. (b) This water strider resting on the surface of a lake remains on the surface, rather than sinking, because an upward surface tension force acts on each leg, which balances the force of gravity acting on the insect. *(Herman Eisenbeiss/Photo Researchers, Inc.)*

THE SURFACE OF LIQUIDS

If you have ever closely examined the surface of water in a glass container, you may have noticed that the surface of the liquid near the walls of the glass curves upward as you move from the center to the edge, as shown in Figure 9.40a. However, if mercury is placed in a glass container, the mercury surface curves downward, as in Figure 9.40b. These surface effects can be explained by considering the forces between molecules. In particular, we must consider the forces that the molecules

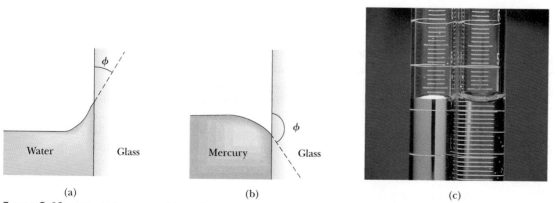

(a)   (b)   (c)

FIGURE 9.40 A liquid in contact with a solid surface. (a) For water, the adhesive force is greater than the cohesive force. (b) For mercury, the adhesive force is less than the cohesive force. (c) The surface of mercury *(left)* curves downward in a glass container, whereas the surface of water *(right)* curves upward as you move from the center to the edge. *(Charles D. Winters)*

FIGURE 9.41 (a) The contact angle between water and paraffin is about 107°. In this case, the cohesive force is greater than the adhesive force. (b) When a chemical called a wetting agent is added to the water, it wets the paraffin surface, and $\phi < 90°$. In this case, the adhesive force is greater than the cohesive force.

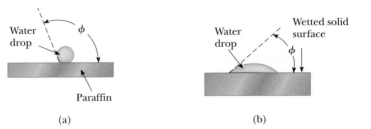

(a)

(b)

of the liquid exert on one another and the forces that the molecules of the glass surface exert on those of the liquid. In general terms, forces between like molecules, such as the forces between water molecules, are called *cohesive forces* and forces between unlike molecules, such as those exerted by glass on water, are called *adhesive forces*.

Water tends to cling to the walls of the glass because the adhesive forces between the liquid molecules and the glass molecules are *greater* than the cohesive forces between the liquid molecules. In effect, the liquid molecules cling to the surface of the glass rather than fall back into the bulk of the liquid. When this condition prevails, the liquid is said to "wet" the glass surface. The surface of the mercury curves downward near the walls of the container because the cohesive forces between the mercury atoms are greater than the adhesive forces between mercury and glass. That is, a mercury atom near the surface is pulled more strongly toward other mercury atoms than toward the glass surface; hence mercury does not wet the glass surface.

The angle ϕ between the solid surface and a line drawn tangent to the liquid at the surface is called *the angle of contact* (Fig. 9.40). Note that ϕ is less than 90° for any substance in which adhesive forces are stronger than cohesive forces and greater than 90° if cohesive forces predominate. For example, if a drop of water is placed on paraffin, the contact angle is approximately 107° (Fig. 9.41a). If certain chemicals, called wetting agents or detergents, are added to the water, the contact angle becomes less than 90°, as shown in Figure 9.41b. The addition of such substances to water is of value when one wants to ensure that water makes intimate contact with a surface and penetrates it. For this reason, detergents are added to water to wash clothes or dishes. On the other hand, it is often necessary to keep water from making intimate contact with a surface, as in waterproofing clothing, where a situation somewhat the reverse of that shown in Figure 9.41 is called for. The clothing is sprayed with a waterproofing agent, which changes ϕ from less than 90° to greater than 90°. Thus, the water beads up on the surface and does not easily penetrate the clothing.

APPLICATION

DETERGENTS AND
WATERPROOFING AGENTS

CAPILLARY ACTION

Capillary tubes are tubes in which the diameter of the opening is very small, on the order of a hundredth of a centimeter. In fact, the word *capillary* means "hairlike." If such a tube is inserted into a fluid for which adhesive forces dominate over cohesive forces, the liquid will rise into the tube, as shown in Figure 9.42. The rising of the liquid in the tube can be explained in terms of the shape of the surface of the liquid and in terms of the surface tension effects in the liquid. At the point of contact between liquid and solid, the upward force of surface tension is directed as shown in Figure 9.42. From Equation 9.18, the magnitude of this force is

$$F = \gamma L = \gamma(2\pi r)$$

We use $L = 2\pi r$ here because the liquid is in contact with the surface of the tube at all points around its circumference. The vertical component of this force due to surface tension is

$$F_v = \gamma(2\pi r)(\cos \phi) \qquad [9.19]$$

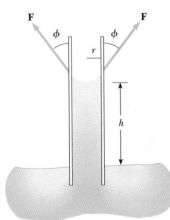

FIGURE 9.42 A liquid rises in a narrow tube because of capillary action, a result of surface tension and adhesive forces.

For the liquid in the capillary tube to be in equilibrium, this upward force must be equal to the weight of the cylinder of water of height h inside the capillary tube. The weight of this water is

$$w = Mg = \rho Vg = \rho g \pi r^2 h \qquad \text{[9.20]}$$

Equating F_v in Equation 9.19 to w in Equation 9.20, we have

$$\gamma(2\pi r)(\cos\phi) = \rho g \pi r^2 h$$

Thus, the height to which water is drawn into the tube is

$$h = \frac{2\gamma}{\rho g r}\cos\phi \qquad \text{[9.21]}$$

If a capillary tube is inserted into a liquid in which cohesive forces dominate over adhesive forces, the level of the liquid in the capillary tube will be below the surface of the surrounding fluid, as shown in Figure 9.43. An analysis similar to that done above would show that the distance h the surface is depressed is given by Equation 9.21.

Capillary tubes are often used to draw small samples of blood from a needle prick in the skin. Capillary action must also be considered in the construction of concrete-block buildings because water seepage through capillary pores in the blocks or the mortar may cause damage to the inside of the building. To prevent this, the blocks are usually coated with a waterproofing agent either outside or inside the building. Water seepage through a wall is an undesirable effect of capillary action, but paper towels use capillary action in a useful manner to absorb spilled fluids.

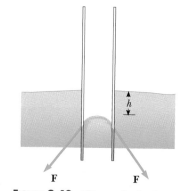

FIGURE 9.43 When cohesive forces between molecules of the liquid exceed adhesive forces, the liquid level in the capillary tube is below the surface of the surrounding fluid.

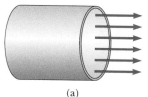

(a)

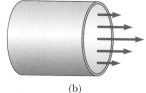

(b)

FIGURE 9.44 (a) The particles in an ideal (nonviscous) fluid all move through the pipe with the same velocity. (b) In a viscous fluid, the velocity of the fluid particles is zero at the surface of the pipe and increases to its maximum value at the center of the pipe.

Example 9.10 How High Does the Water Rise?

Find the height to which water would rise in a capillary tube with a radius equal to 5.0×10^{-5} m. Assume that the angle of contact between the water and the material of the tube is small enough to be considered zero.

Solution From Table 9.4, the surface tension of water is 0.073 N/m. For a contact angle of 0°, we have $\cos\phi = \cos 0° = 1$, so that Equation 9.21 gives

$$h = \frac{2\gamma}{\rho g r} = \frac{2(0.073 \text{ N/m})}{(1.00 \times 10^3 \text{ kg/m}^3)(9.80 \text{ m/s}^2)(5.0 \times 10^{-5} \text{ m})} = \boxed{0.29 \text{ m}}$$

VISCOUS FLUID FLOW

It is considerably easier to pour water out of a container than to pour honey. This is because honey has a higher viscosity than water. In a general sense, *viscosity refers to the internal friction of a fluid.* It is very difficult for layers of a viscous fluid to slide past one another. Likewise, it is difficult for one solid surface to slide past another if there is a highly viscous fluid, such as soft tar, between them.

When an ideal (nonviscous) fluid flows through a pipe, the fluid layers slide past one another with no resistance. If the pipe has a uniform cross section, each layer has the same velocity as in Figure 9.44a. In contrast, the layers of a viscous fluid have different velocities as in Figure 9.44b. The fluid has the greatest velocity at the center of the pipe; the layer next to the wall does not move because of cohesive forces between molecules and the wall surface.

To better understand the concept of viscosity, consider a liquid layer placed between two solid surfaces, as in Figure 9.45. The lower surface is fixed in position, and the top surface moves to the right with a velocity **v** under the action of an external force **F**. Because of this motion, a portion of the liquid is distorted

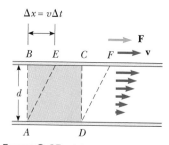

FIGURE 9.45 A layer of liquid between two solid surfaces in which the lower surface is fixed and the upper surface moves to the right with a velocity **v**.

from its original shape, *ABCD,* at one instant to the shape *AEFD* a moment later. The force required to move the upper plate and distort the liquid depends on several factors. The force is proportional to both the area *A* in contact with the fluid and the speed *v* of the fluid. Furthermore, the force is inversely proportional to the distance *d* between the two plates. We can express these proportionalities as $F \propto Av/d$. Thus, the force required to move the upper plate at a fixed speed *v* is

$$F = \eta \frac{Av}{d}$$ [9.22]

▶ Coefficient of viscosity

where η (lowercase Greek letter *eta*) is the **coefficient of viscosity** of the fluid.

The SI units of viscosity are $N \cdot s/m^2$. The units of viscosity in many reference sources are often expressed in $dyne \cdot s/cm^2$, called 1 **poise** in honor of the French scientist J. L. Poiseuille (1799–1869). The relationship between the SI unit of viscosity and the poise is

$$1 \text{ poise} = 10^{-1} \text{ N} \cdot s/m^2$$ [9.23]

Small viscosities are often expressed in centipoise (cp), where 1 cp = 10^{-2} poise. The coefficients of viscosity for some common substances are listed in Table 9.5.

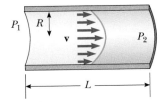

FIGURE 9.46 Velocity profile of a fluid flowing through a uniform pipe of circular cross section. The rate of flow is given by Poiseuille's law. Note that the fluid velocity is greatest at the middle of the pipe.

POISEUILLE'S LAW

Figure 9.46 shows a section of a tube of length *L* and radius *R* containing a fluid under a pressure P_1 at the left end and a pressure P_2 at the right. Because of this pressure difference, the fluid flows through the tube. The rate of flow (volume per unit time) depends on the pressure difference $(P_1 - P_2)$, the dimensions of the tube, and the viscosity of the fluid. The result, known as **Poiseuille's law,** is

▶ Poiseuille's law

$$\text{Rate of flow} = \frac{\Delta V}{\Delta t} = \frac{\pi R^4 (P_1 - P_2)}{8 \eta L}$$ [9.24]

where η is the coefficient of viscosity of the fluid. We shall not attempt to derive this equation here because the methods of integral calculus are required. However, you should note that the equation does agree with common sense. That is, it is reasonable that the rate of flow should increase if the pressure difference across the tube or the tube radius increases. Likewise, the flow rate should decrease if the viscosity of the fluid or the length of the tube increases. Thus, the presence of *R* and the pressure difference in the numerator of Equation 9.24 and of *L* and η in the denominator makes sense.

From Poiseuille's law, we see that in order to maintain a constant flow rate, the pressure difference across the tube has to increase if the viscosity of the fluid increases. This is important when we consider the flow of blood through the circulatory system. The viscosity of blood increases as the number of red blood cells rises. Thus, blood with a high concentration of red blood cells requires greater pumping pressure from the heart to keep it circulating than does blood of lower red blood cell concentration.

Webnote 9.5

A calculator version of Poiseuille's law can be found at the following URL. Make sure you understand how to use Equation 9.24 on your own at exam time.
http://hyperphysics.phy-astr.gsu.edu/hbase/ppois.html
If you want to know how Poiseuille came up with his some-what complicated relationship, work through the site below. Keep going at least until you get to the animation.
http://grad.math.arizona.edu/~walton/biomath/history.htm

TABLE 9.5	The Viscosities of Various Fluids	
Fluid	$T(°C)$	**Viscosity $\eta(N \cdot s/m^2)$**
Water	20	1.0×10^{-3}
Water	100	0.3×10^{-3}
Whole blood	37	2.7×10^{-3}
Glycerin	20	$1\,500 \times 10^{-3}$
10-wt motor oil	30	250×10^{-3}

Note that the flow rate varies as the radius of the tube raised to the fourth power. Consequently, if a constriction occurs in a vein or artery, the heart will have to work considerably harder in order to produce a higher pressure drop and hence to maintain the required flow rate.

Example 9.11   A Blood Transfusion

A patient receives a blood transfusion through a needle of radius 0.20 mm and length 2.0 cm. The density of blood is 1 050 kg/m^3. The bottle supplying the blood is 0.50 m above the patient's arm. What is the rate of flow through the needle?

Solution   The pressure differential between the level of the blood and the patient's arm is

$$P_1 - P_2 = \rho g h = (1\,050 \text{ kg/m}^3)(9.80 \text{ m/s}^2)(0.50 \text{ m}) = 5.15 \times 10^3 \text{ Pa}$$

The viscosity of whole blood is given in Table 9.5. Thus, the rate of flow, from Poiseuille's law, is

$$\frac{\Delta V}{\Delta t} = \frac{\pi R^4 (P_1 - P_2)}{8\eta L}$$

$$= \frac{\pi (2.0 \times 10^{-4} \text{ m})^4 (5.15 \times 10^3 \text{ Pa})}{8(2.7 \times 10^{-3} \text{ N} \cdot \text{s/m}^2)(2.0 \times 10^{-2} \text{ m})} = \boxed{6.0 \times 10^{-8} \text{ m}^3/\text{s}}$$

EXERCISE   How long will it take to inject 1 pint (500 cm^3) of blood into the patient?

ANSWER   140 min

REYNOLDS NUMBER

As we mentioned earlier, at sufficiently high velocities, fluid flow changes from simple streamline flow to turbulent flow, that is, flow characterized by a highly irregular motion of the fluid. It is found experimentally that the onset of turbulence in a tube is determined by a dimensionless factor called the **Reynolds number** (*RN*), given by

$$RN = \frac{\rho v d}{\eta} \qquad\qquad \text{[9.25]}$$

◀ Reynolds number

where ρ is the density of the fluid, v is the average speed of the fluid along the direction of flow, d is the diameter of the tube, and η is the viscosity of the fluid. If *RN* is below about 2 000, the flow of fluid through a tube is streamline; turbulence occurs if *RN* is above 3 000. In the region between 2 000 and 3 000, the flow is unstable, meaning that the fluid can move in streamline flow but any small disturbance will cause its motion to change to turbulent flow.

Example 9.12   Turbulent Flow of Blood

Determine the speed at which blood flowing through an artery of diameter 0.20 cm would become turbulent. Assume that the density of blood is 1.05 × 10^3 kg/m^3 and that its viscosity is 2.7 × 10^{-3} N · s/m^2.

Solution   At the onset of turbulence, the Reynolds number is 3 000. Thus, the speed of the blood would have to be

$$v = \frac{\eta (RN)}{\rho d} = \frac{(2.7 \times 10^{-3} \text{ N} \cdot \text{s/m}^2)(3\,000)}{(1.05 \times 10^3 \text{ kg/m}^3)(0.20 \times 10^{-2} \text{ m})} = \boxed{3.9 \text{ m/s}}$$

9.10   TRANSPORT PHENOMENA

When a fluid flows through a tube, the basic mechanism that produces the flow is a difference in pressure across the ends of the tube. This pressure difference is responsible for the transport of a mass of fluid from one location to another. The fluid may also move from place to place because of a second mechanism, one that depends on a concentration difference between two points in the fluid, as opposed to a pressure difference. When the concentration (the number of molecules per unit volume) is higher at one location than at another, molecules will flow from the point where the concentration is high to the point where it is lower. The two fundamental processes involved in fluid transport resulting from concentration differences are called *diffusion* and *osmosis*. The following sections examine the nature and importance of these processes.

DIFFUSION

In a diffusion process, molecules move from a region where their concentration is high to a region where their concentration is lower. To understand why diffusion occurs, consider Figure 9.47, which represents a container in which a high concentration of molecules has been introduced into the left side. The dashed line in Figure 9.47 represents an imaginary barrier separating the region of high concentration from the region of lower concentration. Because the molecules are moving with high speeds in random directions, many of them cross the imaginary barrier moving from left to right. Very few molecules pass through this area moving from right to left simply because very few of them are on the right side of the container at any instant. Thus, there will always be a *net* movement from the region with many molecules to the region with fewer molecules. For this reason, the concentration on the left side of the container decreases in time and that on the right side increases. Once a concentration equilibrium has been reached, there will be no *net* movement across the cross-sectional area. That is, when the concentration is the same on both sides, the number of molecules diffusing from right to left in a given time interval equals the number moving from left to right in the same time interval.

The basic equation for diffusion is **Fick's law,** which in equation form is

Fick's law ▶

$$\text{Diffusion rate} = \frac{\text{mass}}{\text{time}} = \frac{\Delta M}{\Delta t} = DA\left(\frac{C_2 - C_1}{L}\right) \qquad \text{[9.26]}$$

where D is a constant of proportionality. The left side of this equation is called the diffusion rate and is a measure of the mass being transported per unit time. This equation says that the rate of diffusion is proportional to the cross-sectional area A and to the change in concentration per unit distance, $(C_2 - C_1)/L$, which is called the concentration gradient. The concentrations C_1 and C_2 are measured in kilograms per cubic meter. The proportionality constant D is called the **diffusion coefficient** and has units of square meters per second. Table 9.6 lists diffusion coefficients for a few substances.

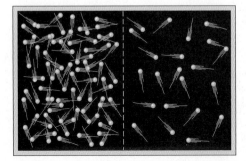

FIGURE 9.47   When the concentration of gas molecules on the left side of the container exceeds the concentration on the right side, there will be a net motion (diffusion) of molecules from left to right.

TABLE 9.6	Diffusion Coefficients for Various Substances at 20°C
Substance	D **(m²/s)**
Oxygen through air	6.4×10^{-5}
Oxygen through tissue	1×10^{-11}
Oxygen through water	1×10^{-9}
Sucrose through water	5×10^{-10}
Hemoglobin through water	76×10^{-11}

THE SIZE OF CELLS AND OSMOSIS

Diffusion through cell membranes is vital for carrying oxygen to the cells of the body and for removing carbon dioxide and other waste products from them. Cells require oxygen for those metabolic processes in which substances are either synthesized or broken down. In such metabolic processes, the cell uses up oxygen and produces carbon dioxide as a by-product. A fresh supply of oxygen diffuses from the blood, where its concentration is high, into the cell, where its concentration is low. Likewise, carbon dioxide diffuses from the cell into the blood, where it is in lower concentration. Water, ions, and other nutrients also pass into and out of cells by diffusion.

A common characteristic of cells in all plants and animals is their extremely small size. The adult human body contains literally trillions of cells. In order to understand why cells are so small, we must consider the relationship between the surface area of an object and its volume.

Let us consider a cube 2 cm on a side. The area of one of its faces is 2 cm $\times$ 2 cm = 4 cm², and because a cube has six sides, the total surface area is 24 cm². Its volume is 2 cm $\times$ 2 cm $\times$ 2 cm = 8 cm³. Hence, the ratio of surface area to volume is 24/8 = 3. Now consider a larger cube, one measuring 3 cm on a side. Repeating the calculations gives us a surface area of 54 cm² and a volume of 27 cm³. In this case, the ratio of surface area to volume is 54/27 = 2. Thus, we see that as the size of an object decreases, the ratio of its surface area to its volume increases. This, of course, says that a small cell has a larger surface-area-to-volume ratio than a large cell. But how does this pertain to the operation of a cell?

A cell can function properly only if it can (1) rapidly receive vital substances such as oxygen and (2) rapidly eliminate waste products. If such substances are to move readily into and out of cells, the cells should have a large surface area. However, if the volume of the cell is too large, it could take a considerable period of time for the nutrients to diffuse into the interior of the cell where they are needed. Under optimum conditions, the surface area of the cell should be large enough so that the exposed membrane area can exchange materials effectively, while at the same time the volume should be small enough so that materials can reach or leave particular locations rapidly. To reach these optimum conditions, a small cell with its high surface-area-to-volume ratio is necessary.

As we have seen, the movement of material through cell membranes is necessary for the efficient functioning of cells. The diffusion of material through a membrane is partially determined by the size of the pores (holes) in the membrane wall. That is, small molecules, such as water, pass through the pores easily and larger molecules, such as sugar, pass through only with difficulty or not at all. A membrane that allows passage of some molecules but not others is called a **selectively permeable** membrane.

Osmosis is defined as the movement of water from a region where its concentration is high, across a selectively permeable membrane, into a region where its concentration is lower. As in the case of diffusion, osmosis continues until the concentrations on the two sides of the membrane are equal. Osmosis is often described simply as the diffusion of water across a membrane.

To understand the effect of osmosis on living cells, let us consider a particular cell in the body that contains a sugar concentration of 1%. (That is, 1 g of sugar is dissolved in enough water to make 100 mL of solution; "mL" is the abbreviation for milliliters, 10^{-3} L = 1 cm^3.) Now assume that this cell is immersed in a 5% sugar solution (5 g of sugar dissolved in enough water to make 100 mL). In such a situation, water diffuses from inside the cell, where its concentration is higher, across the cell wall membrane, to the outside solution, where the concentration of water is lower. This loss of water from the cell causes it to shrink and perhaps become damaged through dehydration. If the concentrations are reversed, water diffuses into the cell, causing it to swell and perhaps burst. It should be obvious from this description that normal osmotic relationships must be maintained in the body. If solutions are introduced into the body intravenously, care must be taken to ensure that these solutions do not disturb the osmotic balance of the body because such a disturbance can lead to cell damage. For example, if 9% saline solution surrounds a red blood cell, the cell will shrink. On the other hand, if the saline solution is about 1%, the cell will eventually burst.

In the body, blood is cleansed of impurities by osmosis as it flows through the kidneys (see Fig. 9.48a). Arterial blood first passes through a bundle of capillaries known as a glomerulus where most of the waste products along with some essential salts and minerals are removed. From the glomerulus, a narrow tube emerges that is in intimate contact with other capillaries throughout its length. As blood passes through the tubules, most of the essential elements are returned to it; waste products are not allowed to reenter and are eventually removed as urine.

If the kidneys fail, an artificial kidney or a dialysis machine can filter the blood. Figure 9.48b shows how this is done. Blood from an artery in the arm is mixed with heparin, a blood thinner, and allowed to pass through a tube covered with a semipermeable membrane. This tubing is immersed in a bath of a dialysate fluid with the same chemical composition as purified blood. Waste products from the blood enter the dialysate by diffusion through the membrane walls. The filtered blood is then returned to a vein.

APPLICATION

KIDNEY FUNCTION AND DIALYSIS

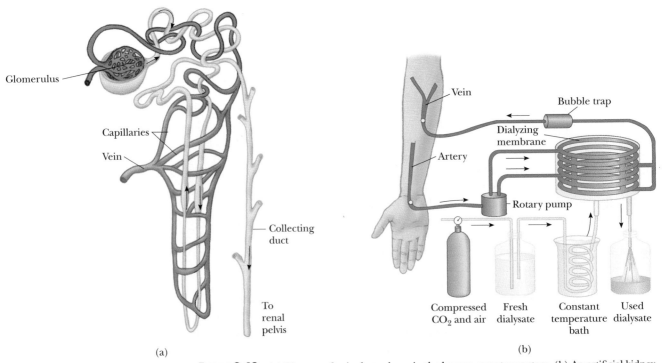

(a)   (b)

FIGURE 9.48   (a) Diagram of a single nephron in the human excretory system. (b) An artificial kidney.

umn. The thermometer can be calibrated by placing it in thermal contact with some environments that remain at constant temperature. One such environment is a mixture of water and ice in thermal equilibrium at atmospheric pressure. Another commonly used system is a mixture of water and steam in thermal equilibrium at atmospheric pressure. Once we mark the ends of the liquid column for our chosen environment on our thermometer, we need to define a scale of numbers associated with various temperatures. An example of such a scale is the **Celsius temperature scale.** On the Celsius scale, the temperature of the ice-water mixture is defined to be zero degrees Celsius, written 0°C; this temperature is called the **ice point** or **freezing point** of water. Likewise, on the Celsius scale, the temperature of the water-steam mixture is defined as 100°C, called the **steam point** or **boiling point** of water. Once the ends of the liquid column in the thermometer are marked at these two points, the distance between marks is divided into 100 equal segments, each corresponding to a change in temperature of one degree Celsius.

Thermometers calibrated in this way present problems when extremely accurate readings are needed. For instance, an alcohol thermometer calibrated at the ice and steam points of water might agree with a mercury thermometer only at the calibration points. Because mercury and alcohol have different thermal expansion properties, when one indicates a temperature of 50°C, say, the other may indicate a slightly different value. The discrepancies between different types of thermometers are especially large when the temperatures to be measured are far from the calibration points.

THE CONSTANT-VOLUME GAS THERMOMETER AND THE KELVIN SCALE

We can construct practical thermometers such as the mercury thermometer, but these types of thermometers do not define temperature in a fundamental way. One thermometer, however, is more fundamental, and it offers a way to define temperature and relate it to internal energy directly—the **gas thermometer.** In a gas thermometer, the temperature readings are nearly independent of the substance used in the thermometer. One type of gas thermometer is the constant-volume unit shown in Figure 10.3. The behavior observed in this device is the pressure variation with temperature of a fixed volume of gas. When the constant-volume gas thermometer was developed, it was calibrated using the ice and steam points of water, as follows. (A different calibration procedure, to be discussed shortly, is now used.) The gas flask was inserted into an ice-water bath, and mercury reservoir B was raised or lowered until the volume of the confined gas was at some value, indicated by the zero point on the scale. The height h, the difference between the levels in the reservoir and column A, indicated the pressure in the flask at 0°C. The flask was inserted into water at the steam point, and reservoir B was readjusted until the height in column A was again brought to zero on the scale, ensuring that the gas volume was the same as it had been in the ice bath (hence the designation "constant volume"). A measure of the new value for h gave a value for the pressure at 100°C. These pressure and temperature values were then plotted on a graph, as in Figure 10.4. The line connecting the two points served as a calibration curve for measuring unknown temperatures. If we want to measure the temperature of a substance, we place the gas flask in thermal contact with the substance and adjust the column of mercury until the level in column A returns to zero. The height of the mercury column tells us the pressure of the gas, and we then find the temperature of the substance from the calibration curve.

Now suppose that temperatures are measured with various gas thermometers containing different gases. Experiments show that the thermometer readings are nearly independent of the type of gas used, as long as the gas pressure is low and the temperature is well above the point at which the gas liquifies.

We can also perform the temperature measurements with the gas in the flask at different starting pressures at 0°C. As long as the pressure is low, we generate

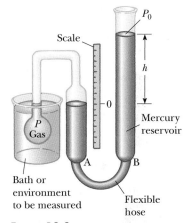

FIGURE 10.3 A constant-volume gas thermometer measures the pressure of the gas contained in the flask immersed in the bath. The volume of gas in the flask is kept constant by raising or lowering reservoir B to keep the mercury level constant.

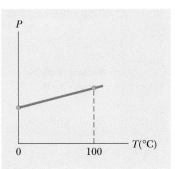

FIGURE 10.4 A typical graph of pressure versus temperature taken with a constant-volume gas thermometer. The dots represent known reference temperatures (the ice point and the steam point).

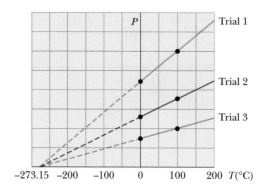

FIGURE 10.5 **Pressure versus temperature for dilute gases.** Note that, for all three experimental trials, the pressure extrapolates to zero at the unique temperature of −273.15°C.

straight-line calibration curves for each starting pressure, as shown for three experimental trials (solid lines) in Figure 10.5.

If the curves in Figure 10.5 are extended back toward negative temperatures, we find a startling result. In every case, regardless of the type of gas or the value of the low starting pressure, **the pressure extrapolates to zero when the temperature is −273.15°C.** This suggests that this particular temperature is universal in its importance, because it does not depend on the substance used in the thermometer. In addition, because the lowest possible pressure is $P = 0$, which would be a perfect vacuum, this temperature must represent a lower bound for physical processes. Thus, we define this temperature as **absolute zero.**

This significant temperature is used as the basis for the **Kelvin temperature scale,** which sets −273.15°C as its zero point (0 K). The size of a "degree" on the Kelvin scale (called a kelvin) is chosen to be identical to the size of a degree on the Celsius scale. Thus, the relationship that enables us to convert between these temperatures is

$$T_C = T - 273.15 \qquad [10.1]$$

where T_C is the Celsius temperature and T is the Kelvin temperature (sometimes called the **absolute temperature**).

Early gas thermometers made use of ice and steam points according to the procedure just described. However, these points are experimentally difficult to duplicate because they are pressure-sensitive. Consequently, a procedure based on two new points was adopted in 1954 by the International Committee on Weights and Measures. The first point is absolute zero. The second point is **the triple point of water, which is the single temperature and pressure at which water, water vapor, and ice can coexist in equilibrium.** This point is a convenient and reproducible reference temperature for the Kelvin scale. It occurs at a temperature of 0.01°C and a pressure of 4.58 mm of mercury. The temperature at the triple point of water on the Kelvin scale occurs at 273.16 K. Thus, **the SI unit of temperature, the kelvin, is defined as 1/273.16 of the temperature of the triple point of water.**

Figure 10.6 shows the Kelvin temperatures for various physical processes and structures. As the figure reveals, absolute zero has never been achieved, although laboratory experiments have come very close to absolute zero.

What would happen to a substance if its temperature could reach 0 K? As Figure 10.5 indicates, the substance would exert zero pressure on the walls of its container (assuming that the gas does not liquefy or solidify all the way to absolute zero). In Section 10.6 we will show that the pressure of a gas is proportional to the kinetic energy of the molecules of that gas. Thus, according to classical physics, the kinetic energy of the gas would go to zero, and there would be no motion at all of the individual components of the gas; hence, the molecules would settle out on the bottom of the container. Quantum theory, to be discussed in Chapter 27, modifies this statement to indicate that there would be some residual energy, called the *zero-point energy,* at this low temperature.

Webnote 10.1

The SI unit of temperature is described at *http://physics.nist.gov/cuu/Units/kelvin.html*

The kelvin ▶

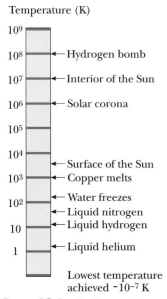

FIGURE 10.6 Absolute temperatures at which various selected physical processes take place. Note that the scale is logarithmic.

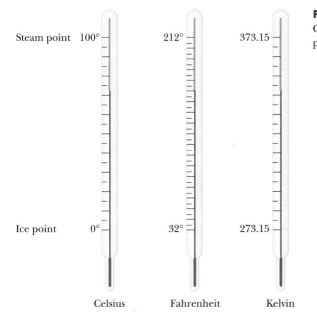

FIGURE 10.7 A comparison of the Celsius, Fahrenheit, and Kelvin temperature scales.

THE CELSIUS, KELVIN, AND FAHRENHEIT TEMPERATURE SCALES

Equation 10.1 shows that the Celsius temperature T_C is shifted from the absolute (Kelvin) temperature T by 273.15. Because the size of a Celsius degree is the same as that of a kelvin, a temperature difference of 5°C is equal to a temperature difference of 5 K. The two scales differ only in the choice of zero point. Thus, the ice point (273.15 K) corresponds to 0.00°C, and the steam point (373.15 K) is equivalent to 100.00°C.

The most common temperature scale in use in the United States is the Fahrenheit scale. It sets the temperature of the ice point at 32°F and the temperature of the steam point at 212°F. The relationship between the Celsius and Fahrenheit temperature scales is

$$T_F = \tfrac{9}{5}T_C + 32$$ **[10.2]**

For example, you should show that a temperature of 50.0°F corresponds to a Celsius temperature of 10.0°C and an absolute temperature of 283 K.

Equation 10.2 can easily be used to find a relationship between changes in temperature on the Celsius and Fahrenheit scales. In an end-of-chapter problem you will be asked to show that if the Celsius temperature changes by ΔT_C, the Fahrenheit temperature changes by the amount ΔT_F, given by

$$\Delta T_F = \tfrac{9}{5}\Delta T_C$$ **[10.3]**

Figure 10.7 compares the three temperature scales that we have discussed.

APPLYING **PHYSICS** *10.1*

A group of future astronauts lands on an inhabited planet. They strike up a conversation with the life forms about temperature scales. It turns out that the inhabitants of this planet have a temperature scale based on the freezing and boiling points of water. Would these two temperatures on this planet be the same as those on Earth? Would the planet's inhabitants use degrees that are the same size as ours? Suppose that the inhabitants have also devised a scale similar to the Kelvin scale. Would their absolute zero be the same as ours?

Explanation The values of 0°C and 100°C for the freezing and boiling points of water are defined at atmospheric pressure. On another planet, it is unlikely

Webnote 10.2

For a convenient calculator that converts temperature readings between all three scales, go to the URL below. (You should be able to convert temperatures on your own during exams!)
http://hyperphysics.phy-str.gsu.edu/ hbase/thermo/temper.html#c3

that atmospheric pressure will be exactly the same as that on Earth. Thus, water will freeze and boil at different temperatures on the other planet. The inhabitants may call these temperatures 0° and 100°, but their degrees will not be the same size as our Celsius degrees (unless their atmospheric pressure was the same as ours). However, this planet's version of the Kelvin scale will use the same absolute zero as ours, because it is based on a natural, universal definition, rather than being associated with a particular substance or a given atmospheric pressure.

Example 10.1   Heating a Pan of Water

A pan of water is heated from 25°C to 80°C. What is the change in its temperature on the Kelvin scale and on the Fahrenheit scale?

Solution   From Equation 10.1 we see that the change in temperature on the Celsius scale equals the change on the Kelvin scale. Therefore,

$$\Delta T = \Delta T_C = 80 - 25 = 55°C = \boxed{55 \text{ K}}$$

From Equation 10.3 we find that the change in temperature on the Fahrenheit scale is $\frac{9}{5}$ as great as the change on the Celsius scale. That is,

$$\Delta T_F = \frac{9}{5}\Delta T_C = \frac{9}{5}(80 - 25) = \boxed{99°F}$$

EXERCISE   On a hot summer day, the temperature is reported as 30.0°C. What is this temperature in Fahrenheit degrees and in kelvins?

ANSWER   86.0°F; 303 K

10.3   THERMAL EXPANSION OF SOLIDS AND LIQUIDS

Our discussion of the liquid thermometer made use of one of the best-known changes that occur in most substances: as temperature increases, volume increases. This phenomenon, known as **thermal expansion,** plays an important role in numerous applications. For example, thermal expansion joints must be included in buildings, concrete highways, and bridges to compensate for changes in dimensions with temperature variations (Figure 10.8).

The overall thermal expansion of an object is a consequence of the change in the average separation between its constituent atoms or molecules. To understand this, consider how the atoms in a solid substance behave. These atoms are located at fixed equilibrium positions; if an atom is pulled away from its position, a restoring force pulls it back. We can imagine that the atoms are particles connected by springs to their neighboring atoms. (See Figure 9.1.) If an atom is pulled away from its equilibrium position, the distortion of the springs provides a restoring force.

At ordinary temperatures, the atoms vibrate about their equilibrium positions with an amplitude (maximum distance from the center of vibration) of about 10^{-11} m, and the average spacing between the atoms is about 10^{-10} m. As the temperature of the solid increases, the atoms vibrate with greater amplitudes and the average separation between them increases. Consequently, the solid as a whole expands.

If the thermal expansion of an object is sufficiently small compared with the object's initial dimensions, then the change in any dimension is, to a good approximation, proportional to the first power of the temperature change. Suppose an object has an initial length L_0 along some direction at some temperature. The length increases by ΔL for the change in temperature ΔT. Thus, for small changes

FIGURE 10.8   Thermal expansion joints are used to separate sections of roadways on bridges. Without these joints, the surfaces would buckle due to thermal expansion on very hot days or crack due to contraction on very cold days. (© *Frank Siteman, Stock/Boston*)

TABLE 10.1	**Average Coefficients of Expansion for Some Materials Near Room Temperature**		
Material	**Average Coefficient of Linear Expansion $[(°C)^{-1}]$**	**Material**	**Average Coefficient of Volume Expansion $[(°C)^{-1}]$**
Aluminum	24×10^{-6}	Ethyl alcohol	1.12×10^{-4}
Brass and bronze	19×10^{-6}	Benzene	1.24×10^{-4}
Copper	17×10^{-6}	Acetone	1.5×10^{-4}
Glass (ordinary)	9×10^{-6}	Glycerin	4.85×10^{-4}
Glass (Pyrex®)	3.2×10^{-6}	Mercury	1.82×10^{-4}
Lead	29×10^{-6}	Turpentine	9.0×10^{-4}
Steel	11×10^{-6}	Gasoline	9.6×10^{-4}
Invar (Ni-Fe alloy)	0.9×10^{-6}	Air	3.67×10^{-3}
Concrete	12×10^{-6}	Helium	3.665×10^{-3}

in temperature,

$$\Delta L = \alpha L_0 \Delta T \qquad \text{[10.4]}$$

or

$$L - L_0 = \alpha L_0 (T - T_0)$$

where L is the final length, T is the final temperature, and the proportionality constant α is called the **coefficient of linear expansion** for a given material and has units of $(°C)^{-1}$.

Table 10.1 lists the coefficients of linear expansion for various materials. Note that for these materials α is positive, indicating an increase in length with increasing temperature.

It is helpful to picture a thermal expansion as a magnification or a photographic enlargement. For example, as the temperature of a metal washer increases (Fig. 10.9), all dimensions, including the radius of the hole, increase according to Equation 10.4.

One practical application of thermal expansion is the common technique of using hot water to loosen a metal lid that is stuck on a glass jar. This succeeds because the circumference of the lid expands more than that of the glass jar.

Because the linear dimensions of an object change with temperature, it follows that surface area and volume also change with temperature. Consider a square having an initial length L_0 on a side and therefore an initial area $A_0 = L_0^2$. As the temperature is increased, the length of each side increases to

$$L = L_0 + \alpha L_0 \Delta T$$

We are now able to calculate the change in the area of the square object as follows. The new area $A = L^2$ is

$$L^2 = (L_0 + \alpha L_0 \Delta T)(L_0 + \alpha L_0 \Delta T) = L_0^2 + 2\alpha L_0^2 \Delta T + \alpha^2 L_0^2 (\Delta T)^2$$

The last term in this expression contains the quantity $\alpha \Delta T$ raised to the second power. Because $\alpha \Delta T$ is much less than unity, squaring it makes it even smaller. Therefore, we can neglect this term to get a simpler expression:

$$A = L^2 = L_0^2 + 2\alpha L_0^2 \Delta T$$
$$= A_0 + 2\alpha A_0 \Delta T$$

or

$$\Delta A = A - A_0 = \gamma A_0 \Delta T \qquad \text{[10.5]}$$

Tip 10.1 COEFFICIENTS OF EXPANSION ARE NOT CONSTANTS

Note that the coefficients of expansion can vary somewhat with temperature, so the coefficients are actually averages.

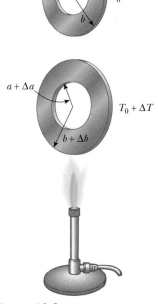

FIGURE 10.9 Thermal expansion of a homogeneous metal washer. As the washer is heated, all dimensions increase. (Note that the expansion is exaggerated in this figure.)

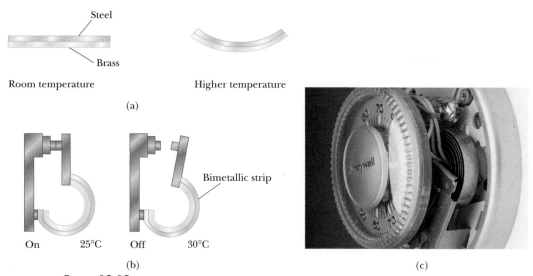

Steel

Brass

Room temperature   Higher temperature

(a)

Bimetallic strip

On   25°C   Off   30°C

(b)   (c)

FIGURE 10.10 (a) A bimetallic strip bends as the temperature changes because the two metals have different expansion coefficients. (b) A bimetallic strip used in a thermostat to break or make electrical contact. (c) The interior of a thermostat, showing the coiled bimetallic strip. Why do you suppose the strip is coiled? *(c, Charles D. Winters)*

where $\gamma = 2\alpha$. The quantity γ (Greek letter gamma) is called the **coefficient of area expansion.**

By a similar procedure we can show that the *increase in volume* of an object accompanying a change in temperature is

$$\Delta V = \beta V_0 \, \Delta T \qquad [10.6]$$

where β, the **coefficient of volume expansion,** is equal to 3α. (Note that $\gamma = 2\alpha$ and $\beta = 3\alpha$ only if the coefficient of linear expansion of the object is the same in all directions.)

As Table 10.1 indicates, each substance has its own characteristic coefficients of expansion. For example, when the temperatures of a brass rod and a steel rod of equal length are raised by the same amount from some common initial value, the brass rod expands more than the steel rod because brass has a larger coefficient of expansion than steel. A simple device that uses this principle is a *bimettalic strip*. Certain thermostats in home furnace systems, for example, make use of a bimetallic strip. The strip is made by securely bonding two different metals together. As the temperature of the strip increases, the two metals expand by different amounts and the strip bends, as in Figure 10.10.

Thermal expansion affects the choice of glassware used in kitchens and laboratories. If hot liquid is poured into a cold container made of ordinary glass, the container may well break due to thermal stress. The inside surface of the glass becomes hot and expands while the outside surface remains at room temperature, and ordinary glass may not withstand the differential expansion without breaking. Pyrex glass has a coefficient of linear expansion of about one third that of ordinary glass, so that the thermal stresses are smaller. Kitchen measuring cups and laboratory beakers are often made of Pyrex so that they can be used with hot liquids.

The thermal expansion of water also has profound influence on the rising ocean levels. At current rates of global warming, scientists predict that about one half of the expected rise in sea level will be caused by thermal expansion; the remainder will be due to the melting of polar ice.

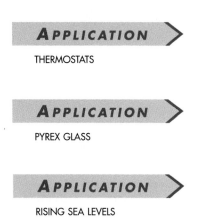

A P P L I C A T I O N

THERMOSTATS

A P P L I C A T I O N

PYREX GLASS

A P P L I C A T I O N

RISING SEA LEVELS

Quick Quiz 10.5 A helium-filled balloon is released into the atmosphere. As the balloon rises, it (a) expands, (b) contracts, or (c) remains unchanged in size.

APPLYING **PHYSICS** 10.4

On some canned foods, you can test the seal by observing the lid of the container when it is opened. When sealed, the lid is indented, but when opened, the lid springs back out. How does this work?

Explanation   The food is sealed inside the container while it is still hot. As the food and the air above it cools, the pressure of the trapped air is lowered. Thus, the atmospheric pressure pressing against the outside of the lid causes it to indent inward. When you open the container, breaking the seal allows outside air to rush in, and the lid snaps back to its original shape.

Example 10.4   How Much Gas Is in a Container?

An ideal gas occupies a volume of 100 cm^3 at 20°C and a pressure of 100 Pa. Determine the number of moles of gas in the container.

Solution   The quantities given are volume, pressure, and temperature:

$$V = 100 \text{ cm}^3 = 1.00 \times 10^{-4} \text{ m}^3; \qquad P = 100 \text{ Pa}; \qquad T = 20°C = 293 \text{ K}$$

From the ideal gas law $PV = nRT$, we solve for n and find

$$n = \frac{PV}{RT} = \frac{(100 \text{ Pa})(1.00 \times 10^{-4} \text{ m}^3)}{(8.31 \text{ J/mol} \cdot \text{K})(293 \text{ K})} = \boxed{4.11 \times 10^{-6} \text{ mol}}$$

EXERCISE   Show that the universal gas constant $R = 0.082\ 1 \text{ L} \cdot \text{atm/mol} \cdot \text{K}$ is also equal to the value $R = 8.31 \text{ J/mol K}$.

Example 10.5   Squeezing a Tank of Gas

Pure helium gas is admitted into a leakproof cylinder containing a movable piston. The initial volume, pressure, and temperature of the gas are 15 L, 2.0 atm, and 300 K. If the volume is decreased to 12 L and the pressure increased to 3.5 atm, find the final temperature of the gas. (Assume that helium behaves as an ideal gas.)

Solution   From the ideal gas law, $PV = nRT$, we divide both sides by T to find

$$\frac{PV}{T} = nR \qquad\qquad (1)$$

Because no gas escapes from the cylinder, the number of moles remains constant, so the right side of (1) is a constant. Applying (1) at the initial and final points gives

$$\frac{P_i V_i}{T_i} = \frac{P_f V_f}{T_f} \qquad\qquad (2)$$

where i and f refer to the initial and final values. Solving for T_f, we get

$$T_f = \left(\frac{P_f V_f}{P_i V_i}\right)(T_i) = \frac{(3.5 \text{ atm})(12 \text{ L})}{(2.0 \text{ atm})(15 \text{ L})}(300 \text{ K}) = \boxed{420 \text{ K}}$$

Example 10.6   Heating a Bottle of Air

A sealed glass bottle at 27°C contains air at atmospheric pressure and has a volume of 30 cm³. The bottle is then tossed into an open fire. When the temperature of the air in the bottle reaches 200°C, what is the pressure inside the bottle? Assume that any volume changes of the bottle are small enough to be negligible.

Solution   As in the previous example, the number of moles of gas remains constant, so we can start with Equation (2) of Example 10.5:

$$\frac{P_i V_i}{T_i} = \frac{P_f V_f}{T_f}$$

Because the initial and final volumes of the gas are assumed equal, this expression reduces to

$$\frac{P_i}{T_i} = \frac{P_f}{T_f}$$

Before evaluating the final pressure, we must convert the given temperatures to kelvins: $T_i = 27°C = 300$ K and $T_f = 200°C = 473$ K. Thus,

$$P_f = \left(\frac{T_f}{T_i}\right)(P_i) = \left(\frac{473\text{ K}}{300\text{ K}}\right)(1.0\text{ atm}) = \boxed{1.6\text{ atm}}$$

Obviously, the higher the temperature, the higher the pressure exerted by the trapped air. Of course, if the pressure rises high enough, the bottle will shatter.

Exercise   In this example we neglected the change in volume of the bottle. If the coefficient of volume expansion for glass is 27×10^{-6} (°C)$^{-1}$, find the magnitude of this volume change.

Answer   0.14 cm³

10.5   AVOGADRO'S NUMBER AND THE IDEAL GAS LAW

In the early 1800s an important field of experimental investigation was created to determine the relative masses of molecules. The equipment used in one such investigation is shown in Figure 10.13. The lower section of the glass vessel contains two electrodes that are connected to a battery so that an electric current can be passed through lightly salted water. (The salt improves the electrical conductivity of the water.) Bubbles of gas are produced at each electrode and become trapped in the column above it. An analytical examination of the bubbles reveals that the column above the positive electrode contains oxygen gas and the column above the negative electrode contains hydrogen gas. Obviously, the current decomposes the water into its constituent parts.

Experimentally it is found that the volume of hydrogen gas collected is always exactly twice the volume of oxygen gas collected, as we now know in light of the chemical composition of water, H_2O. In addition, it is found that if 9 g of water is decomposed, 8 g of oxygen and 1 g of hydrogen are collected. From this information it is possible to determine the relative masses of oxygen and hydrogen molecules.

Following such an experimental investigation, Amedeo Avogadro in 1811 stated the following hypothesis:

Equal volumes of gas at the same temperature and pressure contain the same numbers of molecules.

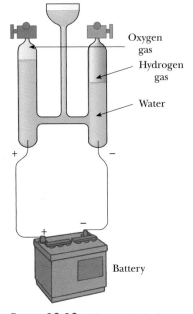

FIGURE 10.13   When water is decomposed by an electric current, the volume of hydrogen gas collected is twice that of oxygen.

Oxygen gas

Hydrogen gas

Water

+   –

–

+

Battery

7. Objects deep below the surface of the ocean are subjected to extremely high pressures, as we saw in Chapter 9. Some bacteria in these environments have adapted to pressures as much as a thousand times atmospheric. How might these bacteria be affected if they were rapidly moved to the surface of the ocean?

8. Why is a power line more likely to break in winter than in summer even if it is loaded with the same weight?

9. Although the average speed of gas molecules in thermal equilibrium at some temperature is greater than zero, the average velocity is zero. Explain.

10. After food is cooked in a pressure cooker, why is it very important to cool the container with cold water before attempting to remove the lid?

11. Some picnickers stop at a convenience store to buy food, including bags of potato chips. They then drive up into the mountains to their picnic site. When they unload the food, they notice that the chip bags are puffed up like balloons. Why did this happen?

12. Markings to indicate length are placed on a steel tape in a room that is at a temperature of 22°C. Measurements are then made with the same tape on a day when the temperature is 27°C. Are the measurements too long, too short, or accurate?

13. Why do vapor bubbles in a pot of boiling water get larger as they approach the surface?

14. Small planets tend to have little or no atmosphere. Why is this?

PROBLEMS

1, 2, 3 = straightforward, intermediate, challenging   ☐ = full solution available in Student Solutions Manual/Study Guide

web = solution posted at **http://info.brookscole.com/serway**   ▨ = biomedical application

Section 10.1 Temperature and the Zeroth Law of Thermodynamics

Section 10.2 Thermometers and Temperature Scales

1. For each of the following temperatures, find the equivalent temperature on the indicated temperature scale: (a) $-273.15°C$ on the Fahrenheit scale, (b) $98.6°F$ on the Celsius scale, and (c) 100 K on the Fahrenheit scale.

2. The highest recorded temperature on Earth was 136°F, at Azizia, Libya, in 1922. The lowest recorded temperature was $-127°F$, at Vostok Station, Antarctica, in 1960. Express these temperature extremes in degrees Celsius.

3. Convert the following temperatures to their values on the Fahrenheit and Kelvin scales. (a) the boiling point of liquid hydrogen $-252.87°C$; (b) the temperature of a room at 20°C.

4. Death Valley holds the record for the highest recorded temperature in the United States. On July 10, 1913, at a place called Furnace Creek Ranch, the temperature rose to 134°F. The lowest occurred at Prospect Creek Camp in Alaska on January 23, 1971, when the temperature plummeted to $-79.8°$ F. Convert these temperatures to the Celsius scale.

5. Show that the temperature $-40°$ is unique in that it has the same numerical value on the Celsius and Fahrenheit scales.

6. A constant-volume gas thermometer is calibrated in dry ice ($-80.0°C$) and in boiling ethyl alcohol (78.0°C). The two pressures are 0.900 atm and 1.635 atm. (a) What value of absolute zero does the calibration yield? (b) What pressures would be found at the freezing and boiling points of water? (Note that we have a linear relationship between P and T as $P = A + BT$, where A and B are constants.)

7. Show that if the temperature on the Celsius scale changes by ΔT_C, the Fahrenheit temperature changes by the amount $\Delta T_F = \frac{9}{5}\Delta T_C$.

8. The temperature difference between the inside and the outside of an automobile engine is 450°C. Express this temperature difference on the (a) Fahrenheit scale (b) Kelvin scale.

9. The melting point of gold is 1 064°C, and the boiling point is 2 660°C. (a) Express these temperatures in kelvins. (b) Compute the difference between these temperatures in Celsius degrees and kelvins.

Section 10.3 Thermal Expansion of Solids and Liquids

10. A copper telephone wire has essentially no sag between poles 35.0 m apart on a winter day when the temperature is $-20.0°C$. How much longer is the wire on a summer day when $T_C = 35.0°C$? Assume that the thermal coefficient of copper is constant throughout this range at its room temperature value.

11. The New River Gorge bridge in West Virginia is a 518-m-long steel arch. How much will its length change between temperature extremes of $-20°C$ and 35°C?

12. A grandfather clock is controlled by a swinging brass pendulum that is 1.3 m long at a temperature of 20°C. (a) What is the length of the pendulum rod when the temperature drops to 0.0°C? (b) If a pendulum's period is given by $T = 2\pi\sqrt{L/g}$, where L is its length, does the change in length of the rod cause the clock to run fast or slow?

13. A pair of eyeglass frames are made of epoxy plastic [coefficient of linear expansion $= 130 \times 10^{-6}$ (°C)$^{-1}$]. At room temperature (assume 20.0°C) the frames have circular lens holes 2.20 cm in radius. To what temperature must the frames be heated in order to insert lenses 2.21 cm in radius?

14. A cube of solid aluminum has a volume of 1.00 m^3 at 20°C. What temperature change is required to produce a 100-cm^3 increase in the volume of the cube?

15. A brass ring of diameter 10.00 cm at 20.0°C is heated and web slipped over an aluminum rod of diameter 10.01 cm at 20.0°C. Assuming the average coefficients of linear expansion are constant, (a) to what temperature must this combination be cooled to separate them? Is this attainable? (b) What if the aluminum rod is 10.02 cm in diameter?

16. Show that the coefficient of volume expansion β is related to the coefficient of linear expansion α through the expression $\beta = 3\alpha$.

17. The inside diameter of a steel lid and the outside diameter of a glass peanut butter jar are both precisely 11.500 cm at room temperature, 21.0°C. If the lid is stuck and you run 80.0°C hot water over it until the lid and the jar both come to 80.0°C, what will the new diameters be?

18. A construction worker uses a steel tape to measure the length of an aluminum support column. If the measured length is 18.700 m when the temperature is 21.2°C, what is the measured length when the temperature rises to 29.4°C? (*Note:* Don't neglect the expansion of the steel tape.)

19. The band in Figure P10.19 is stainless steel (coefficient of linear expansion = 17.3×10^{-6} (°C)$^{-1}$; Young's modulus = 18×10^{10} N/m^2). It is essentially circular with an initial mean radius of 5.0 mm, a height of 4.0 mm, and a thickness of 0.50 mm. If the band just fits snugly over the tooth when heated to 80°C, what is the tension in the band when it cools to 37°C?

FIGURE P10.19

20. The Trans-Alaskan pipeline is 1 300 km long, reaching from Prudhoe Bay to the port of Valdez, and experiences temperatures from -73°C to $+35$°C. How much does the steel pipeline expand due to the difference in temperature? How can this expansion be compensated for?

21. An automobile fuel tank is filled to the brim with 45 L (12 gal) of gasoline at 10°C. Immediately afterward, the vehicle is parked in the Sun, where the temperature is 35°C. How much gasoline overflows from the tank as a result of the expansion? (Neglect the expansion of the tank.)

22. An underground gasoline tank at 54°F can hold 1 000 gallons of gasoline. If the driver of a tanker truck fills the underground tank on a day when the temperature is 90°F, how many gallons, according to his measure on the truck, can he pour in? Assume that the temperature of the gasoline cools to 54°F upon entering the tank.

23. A disk has moment of inertia I_0. Its temperature increases by ΔT. Show that its moment of inertia increases by $\Delta I = 2\alpha I_0 \Delta T$.

24. A concrete walk is poured on a day when the temperature is 20.0°C in such a way that the ends are unable to move. (a) What is the stress in the cement when its temperature is 50.0°C on a hot sunny day? (b) Does the concrete fracture? Take Young's modulus for concrete to be 7.00×10^9 N/m^2 and the compressive strength to be 2.00×10^7 N/m^2.

25. Figure P10.25 shows a circular steel casting with a gap. If the casting is heated, (a) does the width of the gap increase or decrease? (b) The gap width is 1.600 cm when the temperature is 30.0°C. Determine the gap width when the temperature is 190°C.

FIGURE P10.25

26. A hollow aluminum cylinder 20.0 cm deep has an internal capacity of 2.000 L at 20.0°C. It is completely filled with turpentine and then warmed to 80.0°C. (a) How much turpentine overflows? (b) If it is then cooled back to 20.0°C, how far below the surface of the cylinder's rim is the turpentine surface?

Section 10.4 Macroscopic Description of an Ideal Gas

27. The ideal gas constant in the SI is 8.315 J/mol·K. In chemistry classes, the mixed units L·atm/mol·K are used for this constant. Find the ideal gas constant in the mixed units to four significant figures.

28. Gas is contained in an 8.0-L vessel at a temperature of 20°C and a pressure of 9.0 atm. (a) Determine the number of moles of gas in the vessel. (b) How many molecules are in the vessel?

29. (a) An ideal gas occupies a volume of 1.0 cm^3 at 20°C and atmospheric pressure. Determine the number of molecules of gas in the container. (b) If the pressure of the 1.0-cm^3 volume is reduced to 1.0×10^{-11} Pa (an extremely good vacuum) while the temperature remains constant, how many moles of gas remain in the container?

30. A tank having a volume of 0.100 m^3 contains helium gas at 150 atm. How many balloons can the tank blow up if each filled balloon is a sphere 0.300 m in diameter at an absolute pressure of 1.20 atm?

31. Gas is confined in a tank at a pressure of 10.0 atm and a temperature of 15.0°C. If half of the gas is withdrawn and the temperature is raised to 65.0°C, what is the new pressure in the tank?

32. A rigid tank contains 0.40 mol of oxygen (O$_2$). Determine the mass (in kg) of oxygen that must be withdrawn from the tank to lower the pressure of the gas from 40 atm to 25 atm. Assume that the volume of the tank and the temperature of the oxygen are constant during this operation.

33. A weather balloon is designed to expand to a maximum radius of 20 m when in flight at its working altitude where the air pressure is 0.030 atm and the temperature is 200 K. If the balloon is filled at atmospheric pressure and 300 K, what is its radius at lift-off?

34. A cylindrical diving bell, 3.00 m in diameter and 4.00 m tall with an open bottom, is submerged to a depth of 220 m in the ocean. The surface temperature is 25.0°C, and the temperature 220 m down is 5.00°C. The density of sea water is 1 025 kg/m^3. How high does the sea water rise in the bell when it is submerged?

35. An air bubble has a volume of 1.50 cm³ when it is released by a submarine 100 m below the surface of a lake. What is the volume of the bubble when it reaches the surface? Assume that the temperature and the number of air molecules in the bubble remains constant during ascent.

Section 10.5 Avogadro's Number and the Ideal Gas Law

Section 10.6 The Kinetic Theory of Gases

36. A sealed cubical container 20.0 cm on a side contains three times Avogadro's number of molecules at a temperature of 20.0°C. Find the force exerted by the gas on one of the walls of the container.

37. What is the average kinetic energy of a molecule of oxygen at a temperature of 300 K?

38. (a) What is the total random kinetic energy of all the molecules in 1 mol of hydrogen at a temperature of 300 K? (b) With what speed would a mole of hydrogen have to move so that the kinetic energy of the mass as a whole would be equal to the total random kinetic energy of its molecules?

39. Use Avogadro's number to find the mass of a helium atom.

40. The temperature near the top of the atmosphere on Venus is 240 K. (a) Find the rms speed of hydrogen (H_2) at this point in the atmosphere. (b) Repeat for carbon dioxide (CO_2). (c) It has been found that if the rms speed exceeds one sixth of the planet's escape velocity, the gas eventually leaks out of the atmosphere and into outer space. If the escape velocity on Venus is 10.3 km/s, does hydrogen escape? Does carbon dioxide?

41. A cylinder contains a mixture of helium and argon gas in equilibrium at a temperature of 150°C. (a) What is the average kinetic energy of each type of molecule? (b) What is the rms speed of each type of molecule?

42. If 2.0 mol of a gas is confined to a 5.0-L vessel at a pressure of 8.0 atm, what is the average kinetic energy of a gas molecule?

43. Superman leaps in front of Lois Lane to save her from a **web** volley of bullets. In a 1-min interval, an automatic weapon fires 150 bullets, each of mass 8.0 g, at 400 m/s. The bullets strike his mighty chest, which has an area of 0.75 m². Find the average force exerted on Superman's chest if the bullets bounce back after an elastic, head-on collision.

44. In a period of 1.0 s, 5.0×10^{23} nitrogen molecules strike a wall of area 8.0 cm². If the molecules move at 300 m/s and strike the wall head-on in a perfectly elastic collision, find the pressure exerted on the wall. (The mass of one N_2 molecule is 4.68×10^{-26} kg.)

ADDITIONAL PROBLEMS

45. Inside the wall of a house, an L-shaped section of hot water pipe consists of a straight horizontal piece 28.0 cm long, an elbow, and a straight vertical piece 134 cm long (Figure P10.45). A stud and a second-story floorboard hold stationary the ends of this section of copper pipe. Find the magnitude and direction of the displacement of the pipe elbow when the water flow is turned on, raising the temperature of the pipe from 18.0°C to 46.5°C.

FIGURE P10.45

46. The active element of a certain laser is an ordinary glass rod 20 cm long and 1.0 cm in diameter. If the temperature of the rod increases by 75°C, find its increases in (a) length, (b) diameter, and (c) volume.

47. A popular brand of cola contains 6.50 g of carbon dioxide dissolved in 1.00 L of soft drink. If the evaporating carbon dioxide is trapped in a cylinder at 1.00 atm and 20.0°C, what volume does the gas occupy?

48. A 1.5-m-long glass tube, closed at one end, is weighted and lowered to the bottom of a freshwater lake. When the tube is recovered, an indicator mark shows that water rose to within 0.40 m of the closed end. Determine the depth of the lake. Assume constant temperature.

49. Long-term space missions require reclamation of the oxygen in the carbon dioxide exhaled by the crew. In one method of reclamation, 1.00 mol of carbon dioxide produces 1.00 mol of oxygen and 1.00 mol of methane as a byproduct. The methane is stored in a tank under pressure and is available to control the attitude of the spacecraft by controlled venting. A single astronaut exhales 1.09 kg of carbon dioxide each day. If the methane generated in the respiration recycling of three astronauts during one week of flight is stored in an originally empty 150-L tank at −45.0°C, what is the final pressure in the tank?

50. A vertical cylinder of cross-sectional area 0.050 m² is fitted with a tight-fitting, frictionless piston of mass 5.0 kg (Fig. P10.50). If there is 3.0 mol of an ideal gas in the cylinder at 500 K, determine the height h at which the piston will be in equilibrium under its own weight.

51. A liquid with coefficient of volume expansion β just fills a spherical flask of volume V_0 at temperature T (Fig. P10.51). The flask is made of a material that has a coefficient of linear expansion α. The liquid is free to expand into a capillary of cross-sectional area A at the top. (a) If the temperature increases by ΔT, show that the liquid rises in the capillary by the amount $\Delta h = (V_0/A)(\beta - 3\alpha)\Delta T$. (b) For a typical system, such as a mercury thermometer, why is it a good approximation to neglect the expansion of the flask?

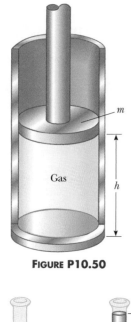

FIGURE P10.50

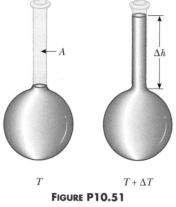

FIGURE P10.51

52. A hollow aluminum cylinder is to be fitted over a steel piston. At 20°C the inside diameter of the cylinder is 99% of the outside diameter of the piston. To what common temperature should the two pieces be heated in order that the cylinder just fit over the piston?

53. A steel measuring tape was designed to read correctly at 20°C. A parent uses the tape to measure the height of a 1.1-m-tall child. If the measurement is made on a day when the temperature is 25°C, is the tape reading longer or shorter than the actual height, and by how much?

54. Before beginning a long trip on a hot day, a driver inflates an automobile tire to a gauge pressure of 1.80 atm at 300 K. At the end of the trip the gauge pressure has increased to 2.20 atm. (a) Assuming the volume has remained constant, what is the temperature of the air inside the tire? (b) What percentage of the original mass of air in the tire should be released so the pressure returns to the original value? Assume the temperature remains at the value found in (a), and the volume of the tire remains constant as air is released.

55. Two concrete spans of a 250-m-long bridge are placed end to end so that no room is allowed for expansion (Fig. P10.55a). If the temperature increases by 20.0°C, what is

the height y to which the spans rise when they buckle (Fig. P10.55b)?

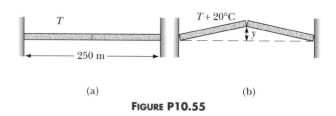

(a)   (b)

FIGURE P10.55

56. A copper rod and steel rod are heated. At 0°C the copper rod has a length L_C and the steel one has a length L_S. When the rods are being heated or cooled, a difference of 5.00 cm is maintained between their lengths. Determine the values of L_C and L_S.

57. If 9.00 g of water is placed in a 2.00-L pressure cooker and heated to 500°C, what is the pressure inside the container?

58. An expandable cylinder has its top connected to a spring of constant 2.00×10^3 N/m (see Fig. P10.58). The cylinder is filled with 5.00 L of gas with the spring relaxed at a pressure of 1.00 atm and a temperature of 20.0°C. (a) If the lid has a cross-sectional area of 0.010 0 m² and negligible mass, how high will the lid rise when the temperature is raised to 250°C? (b) What is the pressure of the gas at 250°C?

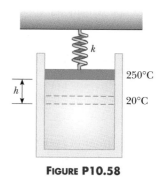

FIGURE P10.58

59. A swimmer has 0.820 L of dry air in his lungs when he dives into a lake. Assuming the pressure of the dry air is 95% of the external pressure at all times, what is the volume of the dry air at a depth of 10.0 m? Assume that atmospheric pressure at the surface is 1.013×10^5 Pa.

60. Two small containers of equal volume, 100 cm³, each contain helium gas at 0°C and 1.00 atm pressure. The two containers are joined by a small open tube of negligible volume, allowing gas to flow from one container to the other. What common pressure will exist in the two containers if the temperature of one container is raised to 100°C while the other container is kept at 0°C?

61. A bimetallic bar is made of two thin strips of dissimilar metals bonded together. As they are heated, the one with the larger average coefficient of expansion expands more than the other, forcing the bar into an arc, with the outer

perature difference of 30°C is established across the block. Repeat the calculation assuming the material is (b) a block of stagnant air with these dimensions; (c) a block of wood with these dimensions.

34. A window has a glass surface of 1.6×10^3 cm^2 and a thickness of 3.0 mm. (a) Find the rate of energy transfer by conduction through this pane when the temperature of the inside surface of the glass is 70°F and the outside temperature is 90°F. (b) Repeat for the same inside temperature and an outside temperature of 0°F.

35. A steam pipe is covered with 1.50-cm-thick insulating material of thermal conductivity 0.200 cal/cm·°C·s. How much energy is lost every second when the steam is at 200°C and the surrounding air is at 20.0°C? The pipe has a circumference of 800 cm and a length of 50.0 m. Neglect losses through the ends of the pipe.

36. A box with a total surface area of 1.20 m^2 and a wall thickness of 4.00 cm is made of an insulating material. A 10.0-W electric heater inside the box maintains the inside temperature at 15.0°C above the outside temperature. Find the thermal conductivity k of the insulating material.

37. Determine the R value for a wall constructed as follows: The outside of the house consists of lapped wood shingles placed over 0.50-in.-thick sheathing, over 3.0 in. of cellulose fiber, over 0.50 in. of dry wall.

38. A Thermopane window consists of two glass panes, each 0.50 cm thick, with a 1.0-cm-thick sealed layer of air between. If the inside surface temperature is 23°C and the outside surface temperature is 0.0°C, determine the rate of energy transfer through 1.0 m^2 of the window. Compare this with the rate of energy transfer through 1.0 m^2 of a single 1.0-cm-thick pane of glass.

39. A copper rod and an aluminum rod of equal diameter are joined end to end in good thermal contact. The temperature of the free end of the copper rod is held constant at 100°C, and that of the far end of the aluminum rod is held at 0°C. If the copper rod is 0.15 m long, what must be the length of the aluminum rod so that the temperature at the junction is 50°C?

40. A Styrofoam box has a surface area of 0.80 m^2 and a wall thickness of 2.0 cm. The temperature of the inner surface is 5.0°C, and that outside is 25°C. If it takes 8.0 h for 5.0 kg of ice to melt in the container, determine the thermal conductivity of the Styrofoam.

Section 11.7 Energy Transfer by Radiation

41. A sphere that is a perfect blackbody radiator has a radius of 0.060 m and is at 200°C in a room where the temperature is 22°C. Calculate the net rate at which the sphere radiates energy.

42. The surface temperature of the Sun is about 5 800 K. Taking the Sun's radius to be 6.96×10^8 m, calculate the total energy radiated by the Sun each second. (Assume $e = 0.965$.)

43. A large hot pizza floats in outer space. It is 70 cm in diameter and 2.0 cm thick, with a temperature of 100°C. Assume its emissivity is 0.8. What is the order of magnitude of its rate of energy loss?

44. Calculate the temperature at which a tungsten filament that has an emissivity of 0.25 and a surface area of 2.5×10^{-5} m^2 will radiate energy at the rate of 25 W in a room where the temperature is 22°C.

45. Measurements on two stars indicate that Star X has a surface temperature of 5 727°C and Star Y has a surface temperature of 11 727°C. If both stars have the same radius, what is the ratio of the luminosity (total power output) of Star Y to the luminosity of Star X? Both stars can be considered to have an emissivity of 1.0.

46. At high noon, the Sun delivers 1.00 kW to each square meter of a blacktop road. If the hot asphalt loses energy only by radiation, what is its equilibrium temperature?

ADDITIONAL PROBLEMS

47. The bottom of a copper kettle has a 10-cm radius and is 2.0 mm thick. The temperature of the outside surface is 102°C, and the water inside the kettle is boiling at 1 atm of pressure. Find the rate at which energy is being transferred through the bottom of the kettle.

48. A 1.00-m^2 solar collector collects radiation from the Sun and focuses it on 250 g of water that is initially at 23.0°C. The average intensity of radiation arriving from the Sun at Earth's surface at this location is 550 W/m^2, and we assume that this is collected with 100% efficiency. Find the time required for the collector to raise the temperature of the water to 100°C.

49. Solar energy can be the primary source of winter space heating for a typical house (with floor area 130 m^2 = 1 400 ft^2) in the north central United States. If the house has very good insulation, you may model it as losing energy by heat steadily at the rate of 1 000 W during the winter, when the average exterior temperature is -5°C. The passive solar-energy collector can consist simply of large windows facing south. Sunlight shining in during the daytime is absorbed by the floor, interior walls, and objects in the house, raising their temperature to 30°C. As the Sun goes down, insulating draperies or shutters are closed over the windows. During the period between 4 P.M. and 8 A.M., the temperature of the house will drop, and a sufficiently large "thermal mass" is required to keep it from dropping too far. The thermal mass can be a large quantity of stone (with specific heat 800 J/kg·°C) in the floor and the interior walls exposed to sunlight. What mass of stone is required if the temperature is not to drop below 18°C overnight?

50. A water heater is operated by solar power. If the solar collector has an area of 6.00 m^2, and the intensity delivered by sunlight is 550 W/m^2, how long does it take to increase the temperature of 1.00 m^3 of water from 20.0°C to 60.0°C?

51. A 40-g ice cube floats in 200 g of water in a 100-g copper cup; all are at a temperature of 0°C. A piece of lead at 98°C is dropped into the cup, and the final equilibrium temperature is 12°C. What is the mass of the lead?

52. Assume that a hailstone at 0°C falls through air at a uniform temperature of 0°C and lands on a sidewalk also at this temperature. From what initial height must the hailstone fall in order to entirely melt on impact?

53. A 200-g block of copper at a temperature of 90°C is dropped into 400 g of water at 27°C. The water is contained in a 300-g glass container. What is the final temperature of the mixture?

54. A class of 10 students taking an exam has a power output per student of about 200 W. Assume that the initial

temperature of the room is 20°C and that its dimensions are 6.0 m by 15.0 m by 3.0 m. What is the temperature of the room at the end of 1.0 h if all the energy remains in the air in the room and none is added by an outside source? The specific heat of air is 837 J/kg·°C, and its density is about 1.3×10^{-3} g/cm^3.

55. The human body must maintain its core temperature inside a rather narrow range around 37°C. Metabolic processes, notably muscular exertion, convert chemical energy into internal energy deep in the interior. From the interior, energy must flow out to the skin or lungs to be lost by heat to the environment. During moderate exercise, an 80-kg man can metabolize food energy at the rate 300 kcal/h, do 60 kcal/h of mechanical work, and put out the remaining 240 kcal/h of energy by heat. Most of the energy is carried from the body interior out to the skin by forced convection (as a plumber would say): Blood is warmed in the interior and then cooled at the skin, which is a few degrees cooler than the body core. Without blood flow, living tissue is a good thermal insulator, with thermal conductivity about 0.210 W/m·°C. Show that blood flow is essential to cool the body by calculating the rate of energy conduction in kcal/h through the tissue layer under the skin. Assume that its area is 1.40 m^2, its thickness is 2.50 cm, and it is maintained at 37.0°C on one side and at 34.0°C on the other side.

56. An aluminum rod and an iron rod are joined end to end in good thermal contact. The two rods have equal lengths and radii. The free end of the aluminum rod is maintained at a temperature of 100°C, and the free end of the iron rod is maintained at 0°C. (a) Determine the temperature of the interface between the two rods. (b) If each rod is 15 cm long and each has a cross-sectional area of 5.0 cm^2, what quantity of energy is conducted across the combination in 30 min?

57. Water is being boiled in an open kettle that has a 0.500-cm-thick circular aluminum bottom with a radius of 12.0 cm. If the water boils away at rate of 0.500 kg/min, what is the temperature of the lower surface of the bottom of the kettle? Assume that the top surface of the bottom of the kettle is at 100°C.

58. A 3.00-g copper penny at 25.0°C drops 50.0 m to the ground. (a) If 60.0% of the initial potential energy associated with the penny goes into increasing the internal energy, determine the final temperature. (b) Does the result depend on the mass of the penny? Explain.

59. A bar of gold is in thermal contact with a bar of silver of the same length and area (Fig. P11.59). One end of the compound bar is maintained at 80.0°C, and the opposite end is at 30.0°C. When the energy flow reaches steady state, find the temperature at the junction.

60. An iron plate is held against an iron wheel so that a sliding frictional force of 50 N acts between the two pieces of metal. The relative speed at which the two surfaces slide over each other is 40 m/s. (a) Calculate the rate at which

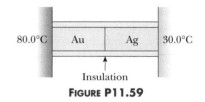

FIGURE P11.59

mechanical energy is converted to internal energy. (b) The plate and the wheel have masses of 5.0 kg each, and each receives 50% of the internal energy. If the system is run as described for 10 s and each object is then allowed to reach a uniform internal temperature, what is the resultant temperature increase?

61. An automobile has a mass of 1 500 kg, and its aluminum brakes have an overall mass of 60 kg. (a) Assuming that all of the internal energy transformed by friction when the car stops is deposited in the brakes, and neglecting energy transfer, how many times could the car be braked to rest starting from 25 m/s (56 mi/h) before the brakes would begin to melt? (Assume an initial temperature of 20°C.) (b) Identify some effects that are neglected in part (a) but are likely to be important in a more realistic assessment of the temperature increase of the brakes.

62. A 1.0-m-long aluminum rod of cross-sectional area 2.0 cm^2 is inserted vertically into a thermally insulated vessel containing liquid helium at 4.2 K. The rod is initially at 300 K. If half of the rod is inserted into the helium, how many liters of helium boil off in the very short time during which the inserted half cools to 4.2 K? The density of liquid helium at 4.2 K is 122 kg/m^3.

63. A *flow calorimeter* is an apparatus used to measure the specific heat of a liquid. The technique is to measure the temperature difference between the input and output points of a flowing stream of the liquid while adding energy at a known rate. (a) Start with the equations $Q = mc \, \Delta T$ and $m = \rho V$ and show that the rate at which energy is added to the liquid is given by the expression

$$\frac{\Delta Q}{\Delta t} = \rho c \, \Delta T \left(\frac{\Delta V}{\Delta t} \right)$$

(b) In a particular experiment, a liquid of density 0.72 g/cm^3 flows through the calorimeter at the rate of 3.5 cm^3/s. At steady state, a temperature difference of 5.8°C is established between the input and output points when energy is supplied at the rate of 40 J/s. What is the specific heat of the liquid?

64. Three liquids are at temperatures of 10°C, 20°C, and 30°C, respectively. Equal masses of the first two liquids are mixed, and the equilibrium temperature is 17°C. Equal masses of the second and third are then mixed, and the equilibrium temperature is 28°C. Find the equilibrium temperature when equal masses of the first and third are mixed.

GROUP ACTIVITIES

G.1 A plot of the temperature of a substance versus time as its temperature decreases is called a cooling curve and has the same shape and basic explanation as the curve shown in Figure 11.2. You can plot such a curve using observations on some water in a container in the freezer compartment of a refrigerator. Place a thermometer in the liq-

12.5 REVERSIBLE AND IRREVERSIBLE PROCESSES

In Section 12.6 we shall discuss a theoretical heat engine that is the most efficient engine possible. To understand its nature, we must first examine the meaning of reversible and irreversible processes. A **reversible** process is one in which every state along some path is an equilibrium state, and one for which the system can be returned to its initial conditions along the same path. A process that does not satisfy these requirements is **irreversible.**

Most natural processes are known to be irreversible—the reversible process is an idealization. Although real processes are always irreversible, some are *almost* reversible. If a real process occurs very slowly so that the system is virtually always in equilibrium, the process can be considered reversible. For example, imagine compressing a gas very slowly by dropping some grains of sand onto a frictionless piston, as in Figure 12.11. The temperature can be kept constant by placing the gas in thermal contact with an energy reservoir. The pressure, volume, and temperature of the gas are well defined during this isothermal compression. Each added grain of sand represents a change to a new equilibrium state. The process can be reversed by slowly removing grains of sand from the piston.

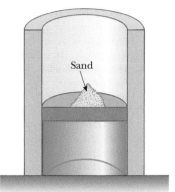

Energy reservoir

FIGURE 12.11 A gas in thermal contact with an energy reservoir is compressed slowly by grains of sand dropped onto the piston. The compression is isothermal and reversible.

12.6 THE CARNOT ENGINE

In 1824, a French engineer named Sadi Carnot (1796–1832), in an effort to understand the efficiency of real engines, described a theoretical engine, now called a *Carnot engine,* that is of great importance from both practical and theoretical viewpoints. He showed that a heat engine operating in an ideal, reversible cycle—called a **Carnot cycle**—between two energy reservoirs is the most efficient engine possible. Such an engine establishes an upper limit on the efficiencies of all real engines. **Carnot's theorem** can be stated as follows:

> No real engine operating between two energy reservoirs can be more efficient than a Carnot engine operating between the same two reservoirs.

To describe the Carnot cycle, we assume that the substance whose temperature varies between T_c and T_h is an ideal gas contained in a cylinder with a movable piston at one end. The cylinder walls and the piston are thermally nonconducting. Figure 12.12 shows four stages of the Carnot cycle, and Figure 12.13 is the *PV* diagram for the cycle. The cycle consists of two adiabatic and two isothermal processes, all reversible.

1. The process $A \rightarrow B$ is an isothermal expansion at temperature T_h, in which the gas is placed in thermal contact with a hot reservoir (that is, placed in a large oven) at temperature T_h (Fig. 12.12a). During the process, the gas absorbs energy Q_h from the reservoir and does work W_{AB} in raising the piston.
2. In the process $B \rightarrow C$, the base of the cylinder is replaced by a thermally nonconducting wall and the gas expands adiabatically; that is, no energy enters or leaves the system by heat (Fig. 12.12b). During the process, the temperature falls from T_h to T_c and the gas does work W_{BC} in raising the piston.
3. In the process $C \rightarrow D$, the gas is placed in thermal contact with a cold reservoir at temperature T_c (Fig. 12.12c) and is compressed isothermally at temperature T_c. During this time, the gas expels energy Q_c to the reservoir, and the work done on the gas is W_{CD}.
4. In the final process $D \rightarrow A$, the base of the cylinder is again replaced by a thermally nonconducting wall (Fig. 12.12d) and the gas is compressed adiabatically. The temperature of the gas increases to T_h, and the work done on the gas is W_{DA}.

SADI CARNOT, FRENCH ENGINEER (1796–1832)

Carnot is considered to be the founder of the science of thermodynamics. Some of his notes found after his death indicate that he was the first to recognize the relationship between work and heat. (*J.-L. Charmet/Science Photo Library/Photo Researchers, Inc.*)

FIGURE 12.12 The Carnot cycle.
(a) In process $A \rightarrow B$, the gas expands isothermally while in contact with a reservoir at T_h. (b) In process $B \rightarrow C$, the gas expands adiabatically ($Q = 0$). (c) In process $C \rightarrow D$, the gas is compressed isothermally while in contact with a reservoir at $T_c < T_h$. (d) In process $D \rightarrow A$, the gas is compressed adiabatically. The upward arrows on the piston indicate removal of sand during the expansions, and the downward arrows indicate addition of sand during the compressions.

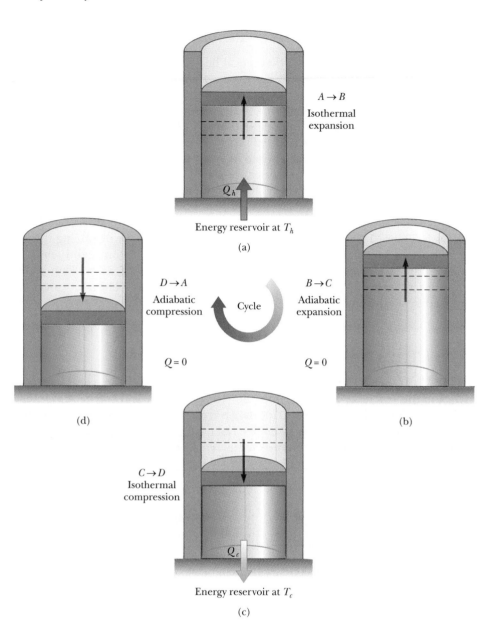

$A \rightarrow B$
Isothermal expansion

Q_h

Energy reservoir at T_h

(a)

$D \rightarrow A$
Adiabatic compression

Cycle

$B \rightarrow C$
Adiabatic expansion

$Q = 0$

$Q = 0$

(d)

(b)

$C \rightarrow D$
Isothermal compression

Q_c

Energy reservoir at T_c

(c)

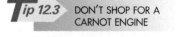

Tip 12.3 DON'T SHOP FOR A CARNOT ENGINE

The Carnot engine is an idealization—do not expect a Carnot engine to be developed for commercial use. If a Carnot engine were developed in an effort to maximize the efficiency, it would have zero power output, because, in order for all of the processes to be reversible, the engine would have to run infinitely slowly.

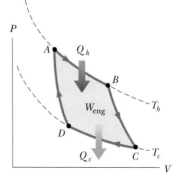

FIGURE 12.13 The *PV* diagram for the Carnot cycle. The net work done W equals the net energy received by heat in one cycle, $Q_h - Q_c$.

Carnot showed that the thermal efficiency of a Carnot engine is

$$e_c = 1 - \frac{T_c}{T_h} \qquad [12.8]$$

where T must be in kelvins. From this result, we see that all Carnot engines operating reversibly between the same two temperatures have the same efficiency.

Equation 12.8 can be applied to any working substance operating in a Carnot cycle between two energy reservoirs. According to this result, the efficiency is zero if $T_c = T_h$. The efficiency increases as T_c is lowered and as T_h is increased. The efficiency can be unity (100%), however, only if $T_c = 0$ K. It is impossible to reach absolute zero, so such reservoirs are not available. Thus, the maximum efficiency is always less than unity. In most practical cases, the cold reservoir is near room temperature, about 300 K. Therefore, we usually strive to increase the efficiency by raising the temperature of the hot reservoir. **All real engines are less efficient than the Carnot engine because they operate irreversibly because of friction and because they complete a cycle in a brief time period.**

Section 12.4 Heat Engines and the Second Law of Thermodynamics

Section 12.5 Reversible and Irreversible Processes

Section 12.6 The Carnot Engine

23. A heat engine operates between two reservoirs at temperatures of 20°C and 300°C. What is the maximum efficiency possible for this engine?

24. The efficiency of a Carnot engine is 30.0%. The engine absorbs 800 J of energy per cycle by heat from a hot reservoir at 500 K. Determine (a) the energy expelled per cycle and (b) the temperature of the cold reservoir.

25. The energy absorbed by an engine is three times greater than the work it performs. (a) What is its thermal efficiency? (b) What fraction of the energy absorbed is expelled to the cold reservoir?

26. A particular engine has a power output of 5.00 kW and an efficiency of 25.0%. If the engine expels 8 000 J of energy as heat in each cycle, find (a) the energy absorbed in each cycle and (b) the time for each cycle.

27. One of the most efficient engines ever built is a coal-fired steam turbine in the Ohio valley, driving an electric generator as it operates between 1 870°C and 430°C. (a) What is its maximum theoretical efficiency? (b) Its actual efficiency is 42.0%. How much mechanical power does the engine deliver if it absorbs 1.40×10^5 J of energy each second from the hot reservoir?

28. A heat engine performs 200 J of work in each cycle and has an efficiency of 30%. For each cycle of operation, (a) how much energy is absorbed by heat and (b) how much energy is expelled by heat?

29. An engine absorbs 1 700 J from a hot reservoir and expels 1 200 J to a cold reservoir in each cycle. (a) What is the engine's efficiency? (b) How much work is done in each cycle? (c) What is the mechanical power output of the engine if each cycle lasts for 0.300 s?

30. A power plant has been proposed that would make use of the temperature gradient in the ocean. The system is to operate between 20.0°C (surface-water temperature) and 5.00°C (water temperature at a depth of about 1 km). (a) What is the maximum efficiency of such a system? (b) If the useful power output of the plant is 75.0 MW, how much energy is absorbed per hour? (c) In view of your answer to (a), do you think such a system is worthwhile (considering that there is no charge for fuel)?

31. In one cycle, a heat engine absorbs 500 J from the high-temperature reservoir and expels 300 J to a low-temperature reservoir. If the efficiency of this engine is 60% of the efficiency of a Carnot engine, what is the ratio of the low temperature to the high temperature in the Carnot engine?

32. A heat engine operates in a Carnot cycle between 80.0°C and 350°C. It absorbs 21 000 J of energy per cycle from the hot reservoir. The duration of each cycle is 1.00 s. (a) What is the mechanical power output of this engine? (b) How much energy does it expel in each cycle by heat?

33. A nuclear power plant has an electrical power output of 1 000 MW and operates with an efficiency of 33%. If excess energy is carried away from the plant by a river with a flow rate of 1.0×10^6 kg/s, what is the rise in temperature of the flowing water?

34. A 20.0%-efficient real engine is used to speed up a train from rest to 5.00 m/s. It is known that an ideal (Carnot) engine using the same cold and hot reservoirs would accelerate the same train from rest to a speed of 6.50 m/s using the same amount of fuel. If the engines use air at 300 K as a cold reservoir, find the temperature of the steam serving as the hot reservoir.

Section 12.7 Entropy

Section 12.8 Entropy and Disorder

35. A freezer is used to freeze 1.0 L of water completely into ice. The water and the freezer remain at a constant temperature of $T = 0°C$. Determine (a) the change in the entropy of the water and (b) the change in the entropy of the freezer.

36. What is the change in entropy of 1.00 kg of liquid water at 100°C as it changes to steam at 100°C?

37. A 70-kg log falls from a height of 25 m into a lake. If the log, the lake, and the air are all at 300 K, find the change in entropy of the Universe for this process.

38. Two 2 000-kg cars, both traveling at 20 m/s, undergo a head-on collision and stick together. Find the entropy change of the Universe resulting from the collision if the temperature is 23°C.

39. The surface of the Sun is approximately at 5 700 K, and the temperature of Earth's surface is approximately 290 K. What entropy change occurs when 1 000 J of energy is transferred by heat from the Sun to Earth?

40. Repeat the procedure used to construct Table 12.4 (a) for the case in which you draw three marbles rather than four from your bag and (b) for the case in which you draw five rather than four.

41. Prepare a table like Table 12.4 for the following occurrence. You toss four coins into the air simultaneously. Record all the possible results of the toss in terms of the numbers of heads and tails that can result. (For example, HHTH and HTHH are two possible ways in which three heads and one tail can be achieved.) (a) On the basis of your table, what is the most probable result of a toss? (b) In terms of entropy, what is the most ordered state and (c) what is the most disordered?

42. Consider a standard deck of 52 playing cards that has been thoroughly shuffled. (a) What is the probability of drawing the ace of spades in one draw? (b) What is the probability of drawing any ace? (c) What is the probability of drawing any spade?

ADDITIONAL PROBLEMS

43. A student claims that she has constructed a heat engine that operates between the temperatures of 200 K and 100 K with 60% efficiency. The professor does not give her credit for the project. Why not?

44. A Carnot heat engine uses a steam boiler at 100°C as the high-temperature reservoir. The low-temperature reservoir is the outside environment at 20.0°C. Energy is exhausted to the low-temperature reservoir at the rate of 15.4 W. (a) Determine the useful power output of the heat engine. (b) How much steam will it cause to condense in the high-temperature reservoir in one hour?

45. A Carnot heat engine extracts energy Q_h from a hot reservoir at constant temperature T_h and rejects energy Q_c to a cold reservoir at constant temperature T_c. Find the entropy changes of the (a) hot reservoir, (b) the cold reservoir, (c) the engine, and (d) the complete system.

46. One end of a copper rod is in thermal contact with a hot reservoir at $T = 500$ K, and the other end is in thermal contact with a cooler reservoir at $T = 300$ K. If 8 000 J of energy is transferred from one end to the other, with no change in the temperature distribution, find the entropy change of each reservoir and the total entropy change of the Universe.

47. Find the Celsius temperature change of a river due to the exhausted energy from a nuclear power plant. Assume that the input power to the boiler in the plant is 25×10^8 W, the efficiency of the use of this power is 30%, and the river flow rate is 9.0×10^6 kg/min.

48. Every second at Niagara Falls, some 5 000 m³ of water falls a distance of 50 m. What is the increase in entropy per second due to the falling water? (Assume a 20°C environment.)

49. One object is at a temperature of T_h and another is at a lower temperature T_c. Use the second law of thermodynamics to show that energy transfer by heat can only occur from the hotter to the colder object. Assume that the heat capacity of each object is large enough that its temperature is not measurably changed.

50. When a gas follows path 123 on the *PV* diagram in Figure P12.50, 418 J of energy flows into the system by heat, and − 167 J of work is done *on the gas*. (a) What is the change in the internal energy of the system? (b) How much energy Q flows into the system if the gas follows path 143? The work done on the gas along this path is − 63.0 J. What net work would be done on or by the system if the system followed (c) path 12341? (d) path 14321? (e) What is the change in internal energy of the system in the processes described in parts (c) and (d)?

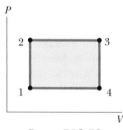

FIGURE P12.50

51. A substance undergoes the cyclic process shown in Figure P12.51. Work output occurs along path *AB*, whereas work input is required along path *BC*, and no work is involved in the constant-volume process *CA*. Energy transfers by heat occur during each process involved in the cycle. (a) What is the work output during process *AB*? (b) How much work input is required during process *BC*? (c) What is the net energy input Q during this cycle?

52. A power plant, having a Carnot efficiency, produces 1 000 MW of electrical power from turbines that take in steam at 500 K and reject water at 300 K into a flowing river. The water downstream is 6.00 K warmer due to the output of the power plant. Determine the flow rate of the river.

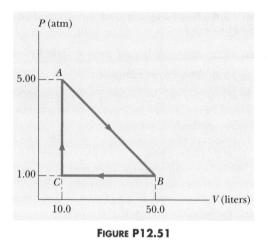

FIGURE P12.51

53. A 100-kg steel support rod in a building has a length of 2.0 m at a temperature of 20°C. The rod supports a hanging load of 6 000 kg. Find (a) the work done *on* the rod as the temperature increases to 40°C, (b) the energy Q added to the rod (assume the specific heat of steel is the same as that for iron), and (c) the change in internal energy of the rod.

54. An ideal gas initially at pressure P_0, volume V_0, and temperature T_0 is taken through the cycle described in Figure P12.54. (a) Find the net work done *by* the gas per cycle in terms of P_0 and V_0. (b) What is the net energy Q added to the system per cycle? (c) Obtain a numerical value for the net work done per cycle for 1.00 mol of gas initially at 0°C. (*Hint:* Recall that work done by the system equals the area under a *PV* curve.)

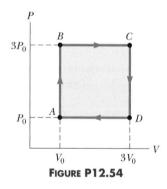

FIGURE P12.54

55. One mole of neon gas is heated from 300 K to 420 K at constant pressure. Calculate (a) the energy Q transferred to the gas, (b) the change in the internal energy of the gas, and (c) the work done on the gas. Note that neon has a molar specific heat of $c = 20.79$ J/mol·K for a constant-pressure process.

56. A 1.0-kg block of aluminum is heated at atmospheric pressure so that its temperature increases from 22°C to 40°C. Find (a) the work done *on* the aluminum, (b) the energy Q added to the aluminum, and (c) the change in internal energy of the aluminum.

57. Suppose a heat engine is connected to two energy reservoirs, one a pool of molten aluminum (660°C) and the other a block of solid mercury (− 38.9°C). The engine runs by freezing 1.00 g of aluminum and melting 15.0 g of mercury during each cycle. The latent heat of fusion of aluminum is 3.97×10^5 J/kg, and that of mercury is 1.18×10^4 J/kg. (a) What is the efficiency of this engine? (b) How does the efficiency compare with that of a Carnot engine?

58. One mole of an ideal gas is taken through the cycle shown in Figure P12.58. At point A, the pressure, volume, and temperature are P_0, V_0, and T_0. In terms of R and T_0, find (a) the total energy entering the system by heat per cycle, (b) the total energy leaving the system by heat per cycle, (c) the efficiency of an engine operating in this cycle, and (d) the efficiency of an engine operating in a Carnot cycle between the temperature extremes for this process. (*Hint:* Recall that work done on the gas is the negative of the area under a *PV* curve.)

59. An electrical power plant has an overall efficiency of 15%. The plant is to deliver 150 MW of electrical power to a city, and its turbines use coal as fuel. The burning coal produces steam at $190°C$, which drives the turbines. This steam is then condensed into water at $25°C$ by passing through coils in contact with river water. (a) How many

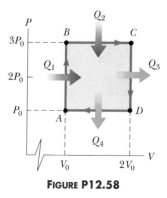

FIGURE P12.58

metric tons of coal does the plant consume each day (1 metric ton = 1×10^3 kg)? (b) What is the total cost of the fuel per year if the delivered price is \$8 per metric ton? (c) If the river water is delivered at $20°C$, at what minimum rate must it flow over the cooling coils in order that its temperature not exceed $25°C$? (*Note:* The heat of combustion of coal is 7.8×10^3 cal/g.)

GROUP ACTIVITIES

G.1 Bend a paper clip back and forth several times and touch it to your lip. Why does it warm up? Repeat this with a rubber band after it is stretched several times. You should notice that it also warms up after stretching. Now take the stretched rubber band from your lip and wait for it to come to equilibrium with the surrounding air. Touch the stretched band to your lip, let it relax to its unstretched length, and note that it becomes cooler. Explain how the second law of thermodynamics helps you understand these results.

G.2 You should be able to locate one of the following in either the chemistry or physics department of your university: a helium tank, a nitrogen tank, or a CO_2 tank. Ask to use one of these for a short period for the following observation. Learn to open and close the valve or valves correctly. Touch the tanks and nozzle to verify that they are at room temperature. Open a tank for 2–3 seconds and then close it. Touch the nozzle again, and you will find that it is very cold. Why? As a hint, the expansion of the gas is nearly adiabatic.

G.3 You can study order and disorder and verify the answer to Quick Quiz 12.5 by rolling a pair of dice 100 times and recording the number of spots appearing on the dice for each throw. Which total comes up most frequently?

G.4 Another way to study order and disorder is to toss pennies. Predict in advance which is the more likely occurrence when tossing three pennies: ordered with all the same orientation (all heads or all tails), or disordered (not all the same). Show that the ordered result includes two out of eight possibilities, whereas the disordered re-

sult includes six out of eight. Toss the three pennies 100 times to see how close you come to this prediction. Repeat this process for four pennies. (The all-the-same result should be expected two times in sixteen throws).

G.5 For each of the situations described below, the system considered is undergoing some changes. Among the possibilities, you should consider these changes: (*Q*) The object is absorbing or giving off energy by heat. (*T*) The object's temperature is changing. (*U*) The object's internal energy is changing. (*W*) The object is doing mechanical work or having work done on it. For each of the situations described below, identify which of the four changes are taking place and write as many of the letters *Q*, *T*, *U*, or *W* (or none) as are appropriate.

(a) A cylinder with a piston on top contains a compressed gas and is sitting on an energy reservoir (a large iron block). After everything has come to thermal equilibrium, the piston is moved upward somewhat (very slowly). The system to be considered is the gas in the cylinder.

(b) A cylinder with a piston on top contains a compressed gas and is wrapped in styrofoam, a very good thermal insulator. The piston is pressed downward (again, very slowly), compressing the gas. The system to be considered is the gas in the cylinder.

(c) An ice cube sitting in the open air is melting.

(Problem 5 is courtesy of Edward F. Redish. For more problems of this type, see http://www.physics.umd.edu/perg/)

PART Three

Vibrations and Waves

As we look around us, we find many examples of objects that vibrate or oscillate: a pendulum, the strings of a guitar, an object suspended on a spring, the piston of an engine, the head of a drum, the reed of a saxophone. Most elastic objects vibrate when an impulse is applied to them; that is, once they are distorted, their shape tends to be restored to some equilibrium configuration. Even at the atomic level, the atoms in a solid vibrate about some position as if they were connected to their neighbors by imaginary springs.

Wave motion is closely related to the phenomenon of vibration. Sound waves, earthquake waves, waves on stretched strings, and water waves are all produced by vibrations. As a sound wave travels through a medium (such as air), the elements of the medium vibrate back and forth; as a water wave travels across a pond, the water elements vibrate up and down. When any wave travels through a medium, the elements of the medium move in repetitive cycles. Therefore, the motion of the elements bears a strong resemblance to the periodic motion of a vibrating pendulum or an object attached to a spring.

Many other natural phenomena occur whose explanations require an understanding of vibrations and waves. Although many large structures, such as skyscrapers and bridges, appear to be rigid, they actually vibrate—a fact that must be taken into account by the architects and engineers who design and build them. To understand how radio and television work, we must comprehend the origin and nature of electromagnetic waves and how they propagate through space. Finally, much of what scientists have learned about atomic structure has come from information carried by waves; therefore, to understand the concepts and theories of atomic physics, we must first study waves and vibrations.

As these water droplets fall into a pond, a circular pattern of waves spread outward from the entry point of the droplets. (© Peter K. Ziminski/Visuals Unlimited)

13

Vibrations and Waves

Chapter Outline

When sand is abundant and evenly distributed, transverse dunes are formed perpendicular to the direction of prevailing winds. These transverse dunes were formed in Great Sand Dunes National Monument, Colorado. *(Arthur C. Smith III/Grant Heilman Photography, Inc.)*

This chapter constitutes a brief return to the subject of mechanics as we examine various forms of periodic motion. We concentrate especially on motion that occurs when the force on an object is proportional to the displacement of the object from its equilibrium position. When such a force acts only toward the equilibrium position, the result is a back-and-forth periodic motion—an oscillation, or vibration, between two extreme positions. You may be familiar with several types of periodic motion, such as the oscillations of an object attached to a spring, the motion of a pendulum, and the vibrations of a stringed musical instrument.

Because vibrations can cause disturbances that move through a medium, we also study wave motion in this chapter. Many kinds of waves occur in nature, including sound waves, water waves, seismic waves, and electromagnetic waves. We conclude with a brief discussion of some terms and concepts that are common to all types of waves, and in later chapters we shall focus our attention on specific categories of waves.

13.1 HOOKE'S LAW

One of the simplest types of vibrational motion is that of an object attached to a spring, which we discussed briefly in Chapter 5. Let us assume that the object moves on a frictionless horizontal surface. If the spring is stretched or compressed a small distance x from its unstretched or equilibrium position and then released, it exerts a force on the object as shown in Figure 13.1. From experiment, this spring force is found to obey the equation

$$F_s = - kx \qquad \text{[13.1]}$$

◀ Hooke's law

where x is the displacement of the object from its equilibrium ($x = 0$) position and k is a positive constant called the **spring constant.** This force law for springs was discovered by Robert Hooke in 1678 and is known as **Hooke's law.** The value of k is a measure of the stiffness of the spring. Stiff springs have large k values, and soft springs have small k values.

The negative sign in Equation 13.1 signifies that the force exerted by the spring is always directed *opposite* the displacement of the object. When the object is to the right of the equilibrium position, as in Figure 13.1a, x is positive and F_s is negative. This means that the force is in the negative direction, to the left. When the object is to the left of the equilibrium position, as in Figure 13.1c, x is negative and F_s is positive, indicating that the direction of the force is to the right. Of course, when $x = 0$, as in Figure 13.1b, the spring is unstretched and $F_s = 0$. Because the spring force always acts toward the equilibrium position, it is sometimes called a restoring force. **The direction of the restoring force is such that the object is being either pulled or pushed toward the equilibrium position.**

Let us examine the motion of the object if it is initially pulled a distance A to the right and released from rest. The force exerted by the spring on the object pulls the object back toward the equilibrium position. As the object moves toward $x = 0$, the magnitude of the force decreases (because x decreases) and reaches zero at $x = 0$. However, the object gains speed as it moves toward the equilibrium position. In fact, the speed reaches its maximum value when $x = 0$. The momentum achieved by the object causes it to overshoot the equilibrium position and to compress the spring. As the object moves to the left of the equilibrium position (negative x values), the force acting on it begins to increase to the right and the speed of the object begins to decrease. The object finally comes to a stop at $x = - A$. The process is then repeated, and the object continues to oscillate back and forth over the same path. This type of motion is called **simple harmonic motion. Simple harmonic motion occurs when the net force along the direction**

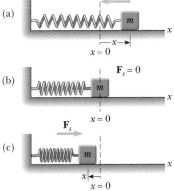

FIGURE 13.1 The force exerted by a spring on an object varies with the displacement of the object from the equilibrium position, $x = 0$.
(a) When x is positive (stretched spring), the spring force is to the left.
(b) When x is zero (unstretched spring), the spring force is zero.
(c) When x is negative (compressed spring), the spring force is to the right.

of motion is a Hooke's law type of force—that is, when the net force is proportional to the displacement and in the opposite direction.

Not all periodic motion over the same path can be classified as simple harmonic motion. For example, consider a ball being tossed back and forth between a parent and a child. The ball moves repetitively, but the motion is not simple harmonic. Unless the force acting on an object along the direction of motion has the form of Equation 13.1, the object does not exhibit simple harmonic motion.

The motion of an object suspended from a vertical spring is also simple harmonic. In this case, the force of gravity acting on the attached object stretches the spring until equilibrium is reached (the object is suspended and at rest.) This position of equilibrium establishes the position of the object for which $x = 0$. When the object is moved away from equilibrium by a distance x and then released, a net force acts (in the form of Hooke's law) toward the equilibrium position. Because the net force is proportional to x, the motion is simple harmonic.

In order to discuss the motion of an object oscillating in simple harmonic motion in some detail, it is necessary to define the following terms.

- The **amplitude** A is the magnitude of the maximum position of the object relative to its equilibrium position. In the absence of friction, an object continues in simple harmonic motion and reaches a maximum position equal to $\pm A$ on each side of the equilibrium position during each cycle.
- The **period** T is the time it takes the object to execute one complete cycle of motion—that is, from $x = A$ to $x = -A$ and back to $x = A$.
- The **frequency** f is the number of complete cycles or vibrations per unit of time.

Example 13.1 Measuring the Spring Constant

A common technique used to evaluate the spring constant is illustrated in Figure 13.2. The spring is hung vertically (Fig. 13.2a), and an object of mass m is attached to the lower end of the spring (Fig. 13.2b). The spring is stretched by a distance d from its initial position under the action of the "load" mg. Because the spring force is upward, it must balance the weight mg downward *when the system is at rest* (see Fig. 13.2c). In this case, we can apply Hooke's law to give

$$F_s = -kd = -mg$$

$$k = \frac{mg}{d}$$

For example, if a spring is displaced 2.0 cm by a mass of 0.55 kg, the spring constant is

$$k = \frac{mg}{d} = \frac{(0.55 \text{ kg})(9.80 \text{ m/s}^2)}{2.0 \times 10^{-2} \text{ m}} = \boxed{2.7 \times 10^2 \text{ N/m}}$$

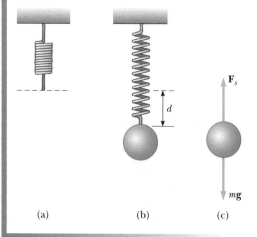

FIGURE 13.2 (Example 13.1) Determining the spring constant. The elongation d of the spring is due to the suspended weight, mg. Because the upward spring force balances the weight when the system is in equilibrium, it follows that $k = mg/d$.

(a)   (b)   (c)

Example 13.2 Simple Harmonic Motion on a Frictionless Surface

A 0.35-kg object attached to a spring of spring constant 130 N/m is free to move on a frictionless horizontal surface, as in Figure 13.1. If the object is released from rest at $x = 0.10$ m, find the force on it and its acceleration at (a) $x = 0.10$ m, (b) $x = 0.050$ m, (c) $x = 0$ m, and (d) $x = -0.050$ m.

Solution

A The point at which the object is released at rest defines the amplitude of the motion. In this case, $A = 0.10$ m. The object moves continuously between the limits 0.10 m and -0.10 m. When x is a maximum ($x = A$), the force on the object is a maximum and is calculated as follows:

$$F_s = -kx$$
$$F_{max} = -kA = -(130 \text{ N/m})(0.10 \text{ m}) = \boxed{-13 \text{ N}}$$

The negative sign indicates that the force acts to the left, in the negative x direction. We can use Newton's second law to calculate the acceleration at this position:

$$F_{max} = ma_{max}$$
$$a_{max} = \frac{F_{max}}{m} = -\frac{13 \text{ N}}{0.35 \text{ kg}} = \boxed{-37 \text{ m/s}^2}$$

Again, the negative sign indicates that the acceleration is to the left.

B We can use the same approach to find the force and acceleration at other positions. At $x = 0.05$ m, we have

$$F_s = -kx = -(130 \text{ N/m})(0.050 \text{ m}) = \boxed{-6.5 \text{ N}}$$
$$a = \frac{F}{m} = -\frac{6.5 \text{ N}}{0.35 \text{ kg}} = \boxed{-19 \text{ m/s}^2}$$

Note that the acceleration of an object moving with simple harmonic motion *is not constant* because F_s is not constant.

C At $x = 0$, the spring force is zero (because $F_s = -kx = 0$) and the acceleration is zero. In other words, when the spring is unstretched, it exerts no force on the object attached to it.

D At $x = -0.050$ m, we have

$$F_s = -kx = -(130 \text{ N/m})(-0.050 \text{ m}) = \boxed{6.5 \text{ N}}$$
$$a = \frac{F_s}{m} = \frac{6.5 \text{ N}}{0.35 \text{ kg}} = \boxed{19 \text{ m/s}^2}$$

This result shows that the force and acceleration are positive when the object is on the negative side of the equilibrium position. This indicates that the force exerted by the spring on the object is acting to the right as the spring is being compressed. At the same time, the object is slowing down as it moves from $x = 0$ to $x = -0.050$ m.

EXERCISE Find the force and acceleration when $x = -0.10$ m.

ANSWER 13 N; 37 m/s^2

As indicated in Example 13.2, the acceleration of an object moving with simple harmonic motion can be found by using Hooke's law to specify the force in the equation for Newton's second law, $F = ma$. This gives

$$-kx = ma$$

$$a = -\frac{k}{m}x \qquad \text{[13.2]} \qquad \blacktriangleleft \text{ Acceleration in simple harmonic motion}$$

 **A NONCONSTANT ACCELERATION**

The acceleration of a particle in simple harmonic motion is *not* constant. Equation 13.2 shows that *a* is proportional to the position *x*. Thus, *a* varies with *x* and we *cannot* apply the constant acceleration kinematic equations of Chapter 2 in this situation.

Because the maximum value of *x* is defined to be the amplitude, *A*, we see that the acceleration ranges over the values $-kA/m$ to $+kA/m$. Equation 13.2 enables us to find the acceleration of the object as a function of its position. In subsequent sections, we will find equations for velocity as a function of position and position as a function of time.

Earlier we stated that an object moves with simple harmonic motion when the net force acting on it is proportional to its displacement from equilibrium and is directed toward the equilibrium position. Equation 13.2 provides an alternative definition of simple harmonic motion. An object moves with simple harmonic motion if its acceleration is proportional to its displacement and is in the opposite direction to it.

Quick Quiz 13.1 A block on the end of a spring is pulled to position $x = A$ and released. Through what total distance does it travel in one full cycle of its motion? (a) $A/2$ (b) A (c) $2A$ (d) $4A$

Quick Quiz 13.2 Which of the following combination of quantities cannot be in the same direction for a simple harmonic oscillator? (a) position and velocity, (b) velocity and acceleration, (c) position and acceleration.

13.2 ELASTIC POTENTIAL ENERGY

For the most part, we have worked with three types of mechanical energy: gravitational potential energy, translational kinetic energy, and rotational kinetic energy. In Chapter 5 we briefly discussed a fourth type of mechanical energy, elastic potential energy. We now consider this form of energy in more detail.

A system of interacting objects has potential energy associated with the configuration of the system. As we learned in Chapter 5, a system of an object and Earth has potential energy of *mgh* if the object is at a height *h* above Earth's surface and the zero of potential energy is chosen as that configuration in which the object is located on the surface. This potential energy can be transformed to kinetic energy of the object as the object falls after being released. Likewise, a compressed spring has potential energy. The compressed spring, when allowed to expand, can apply a force on an object such that the potential energy in the spring is transformed to kinetic energy of the object. As an example, Figure 13.3 shows a ball being projected from a spring-loaded toy gun, where the spring is compressed a distance *x*. As the gun is fired, the compressed spring does work on the ball and imparts kinetic energy to it. **The energy stored in a stretched or compressed spring or other elastic material is called elastic potential energy PE_s given by**

FIGURE 13.3 A ball projected from a spring-loaded gun. The elastic potential energy stored in the spring is transformed into the kinetic energy of the ball.

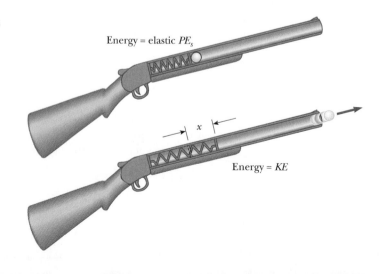

Energy = elastic PE_s

Energy = KE

A **longitudinal wave** is one in which the elements of the medium move parallel to the direction of the wave velocity. An example is a sound wave.

The relationship of the speed, wavelength, and frequency of a wave is

$$v = f\lambda \qquad\qquad [13.16]$$

The speed of a wave traveling on a stretched string of mass per unit length μ and under tension F is

$$v = \sqrt{\frac{F}{\mu}} \qquad\qquad [13.17]$$

The **superposition principle** states that if two or more traveling waves are moving through a medium, the resultant wave is found by adding the individual waves together point by point. When waves meet crest to crest and trough to trough, they undergo **constructive interference.** When crest meets trough, the waves undergo **destructive interference.**

When a wave pulse reflects from a rigid boundary, the pulse is inverted. When the boundary is free, the reflected pulse is not inverted.

CONCEPTUAL QUESTIONS

1. If one end of a heavy rope is attached to one end of a light rope, the speed of a wave will change as the wave goes from the heavy rope to the light one. Will the speed increase or decrease? What happens to the frequency? To the wavelength?

2. If a spring is cut in half, what happens to its spring constant?

3. An object-spring system undergoes simple harmonic motion with an amplitude A. Does the total energy change if the mass is doubled but the amplitude is not changed? Are the kinetic and potential energies at a given point in its motion affected by the change in mass? Explain.

4. The speed of sound in air as given in this chapter (343 m/s) is an enormous speed compared to the speed of common objects. Yet the speed of sound is of the same order of magnitude as the rms speed of air molecules at one atmosphere and 20°C as given by the kinetic theory of gases. Is this just a remarkable coincidence? Explain.

5. An object is hung on a spring and the frequency of the oscillation of the system f is measured. The object, a second identical object, and the spring are carried in the Space Shuttle to space. The two objects are attached to the ends of the spring, and the system is taken out into space on a space walk. The spring is extended, and the system is released to oscillate while floating in space. The coils of the spring do not bump into one another. What is the frequency of oscillation for this system, in terms of f?

6. If an object-spring system is hung vertically and set into oscillation, why does the motion eventually stop?

7. Is a bouncing ball an example of simple harmonic motion? Is the daily movement of a student from home to school and back simple harmonic motion?

8. If a pendulum clock keeps perfect time at the base of a mountain, will it also keep perfect time when moved to the top of the mountain? Explain.

9. A pendulum bob is made from a sphere filled with water. What would happen to the frequency of vibration of this pendulum if the sphere had a hole in it that allowed the water to leak out slowly?

10. If a grandfather clock were running slow, how could we adjust the length of the pendulum to correct the time?

11. A grandfather clock depends on the period of a pendulum to keep correct time. Suppose a grandfather clock is calibrated correctly and then the temperature of the room in which it resides increases. Does the grandfather clock run slow, fast, or correctly? (*Hint:* A material expands when its temperature increases.)

12. If you stretch a rubber hose and pluck it, you can observe a pulse traveling up and down the hose. What happens to the speed of the pulse if you stretch the hose more tightly? What happens to the speed if you fill the hose with water?

13. In a long line of people waiting to buy tickets at a movie theater, when the first person leaves, a pulse of motion occurs as people step forward to fill in the gap. The gap moves through the line of people. What determines the speed of this pulse? Is it transverse or longitudinal? How

FIGURE Q13.13 (*Gregg Adams/Stone/Getty Images*)

about the "wave" at a baseball game, where people in the stands stand up and shout as the wave arrives at their location (Fig. Q13.13), and this pulse moves around the stadium—what determines the speed of this pulse? Is it transverse or longitudinal?

14. How would you create a longitudinal wave in a stretched spring? Would it be possible to create a transverse wave in a spring?

15. In mechanics, massless strings are often assumed. Why is this not a good assumption when discussing waves on strings?

16. What happens to the wavelength of a wave on a string when the frequency is doubled? Assume that the tension in the string remains the same.

17. Explain why the kinetic and potential energies of an object-spring system can never be negative.

18. What happens to the speed of a wave on a string when the frequency is doubled? Assume that the tension in the string remains the same.

19. By what factor would you have to multiply the tension in a stretched spring in order to double the wave speed?

20. The motion of Earth going around the Sun is periodic with a period of 1 year. Is this motion simple harmonic? Explain.

PROBLEMS

1, 2, 3 = straightforward, intermediate, challenging   ☐ = full solution available in Student Solutions Manual/Study Guide

web = solution posted at **http://info.brookscole.com/serway**   🔬 = biomedical application

Section 13.1   Hooke's Law

1. A 0.40-kg object is attached to a spring with a spring constant 160 N/m so that the object is allowed to move on a horizontal frictionless surface. The object is released from rest when the spring is compressed 0.15 m. Find (a) the force on the object and (b) its acceleration at this instant.

2. A load of 50 N attached to a spring hanging vertically stretches the spring 5.0 cm. The spring is now placed horizontally on a table and stretched 11 cm. (a) What force is required to stretch the spring by this amount? (b) Plot a graph of force (on the *y* axis) versus spring displacement from the equilibrium position along the *x* axis.

3. A ball dropped from a height of 4.00 m makes a perfectly elastic collision with the ground. Assuming no mechanical energy is lost due to air resistance, (a) show that the motion is periodic and (b) determine the period of the motion. (c) Is the motion simple harmonic? Explain.

4. The four springs of the device shown in Figure P13.4 each have a spring constant of 65.0 N/m. The hands holding the unstretched device are 43.5 cm apart. If the person cannot move his hands more than 80.0 cm apart, what is the maximum lateral force each hand can exert?

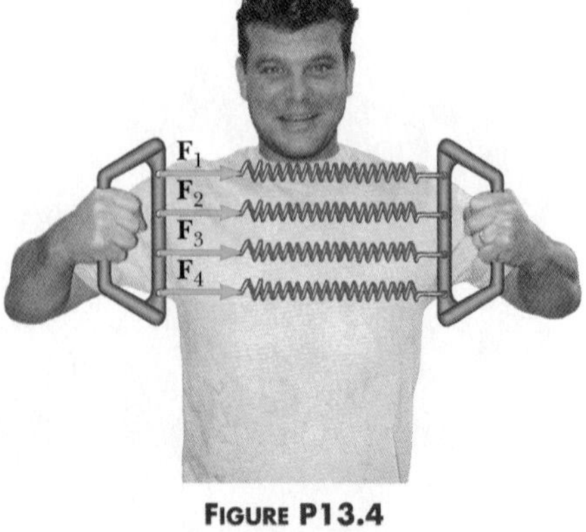

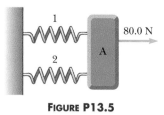

FIGURE P13.4

5. The springs 1 and 2 in Figure P13.5 have spring constants of 40.0 N/cm and 25.0 N/cm, respectively. The object *A* remains at rest, and both springs are stretched equally. Determine the stretch.

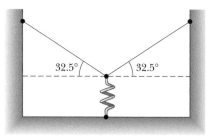

FIGURE P13.5

6. (a) The spring in Figure P13.6 has a force constant of 5.60×10^4 N/m. If the tension in either side of the rope is 210 N, how far is the spring stretched? (b) Suppose the rope is replaced by two springs which make the same V shape as the rope. If the stretch of each of these springs is twice that of the lower spring, what must be the spring constant of these springs?

FIGURE P13.6

Section 13.2   Elastic Potential Energy

7. A slingshot consists of a light leather cup, containing a stone, that is pulled back against two parallel rubber bands. It takes a force of 15 N to stretch either one of these bands 1.0 cm. (a) What is the potential energy stored in the two bands together when a 50-g stone is placed in the cup and pulled back 0.20 m from the equilibrium position? (b) With what speed does the stone leave the slingshot?

8. An archer pulls her bow string back 0.400 m by exerting a force that increases uniformly from zero to 230 N. (a) What is the equivalent spring constant of the bow? (b) How much work is done in pulling the bow?

14.2 CHARACTERISTICS OF SOUND WAVES

As already noted, the general motion of elements of air near a vibrating object is back and forth between regions of compression and rarefaction. Back-and-forth motion of elements of the medium in the direction of the disturbance is characteristic of a longitudinal wave. **The motion of the elements of the medium in a longitudinal sound wave is back and forth along the direction in which the wave travels.** In contrast, **in a transverse wave the vibrations of the elements of the medium are at right angles to the direction of travel of the wave.**

CATEGORIES OF SOUND WAVES

Sound waves fall into three categories that cover different ranges of frequencies. **Audible waves** are longitudinal waves that lie within the range of sensitivity of the human ear, approximately 20 to 20 000 Hz. **Infrasonic waves** are longitudinal waves with frequencies below the audible range. Earthquake waves are an example. **Ultrasonic waves** are longitudinal waves with frequencies above the audible range for humans. For example, certain types of whistles produce ultrasonic waves. Some animals, such as dogs, can hear the waves emitted by these whistles, even though humans cannot.

APPLICATIONS OF ULTRASOUND

Ultrasonic waves are sound waves with frequencies greater than 20 kHz. Because of their high frequency and corresponding short wavelengths, ultrasonic waves can be used to produce images of small objects and are currently in wide use in medical applications, both as a diagnostic tool and in certain treatments. Internal organs are examined via the images produced by the reflection and absorption of ultrasonic waves. Although ultrasonic waves are far safer than x-rays, their images do not always have as much detail. Certain organs, however, such as the liver and the spleen, are invisible to x-rays but can be imaged with ultrasonic waves.

Medical workers can measure the speed of the blood flow in the body using a device called an ultrasonic flow meter, which makes use of the Doppler effect. The Doppler effect is discussed in Section 14.6. By comparing the frequency of the waves scattered by flowing blood with the incident frequency, we can obtain the flow speed.

Figure 14.3 illustrates the technique used to produce ultrasonic waves for clinical use. Electrical contacts are made to the opposite faces of a crystal, such as quartz or strontium titanate. If an alternating voltage of high frequency is applied to these contacts, the crystal vibrates at the same frequency as the applied voltage, emitting a beam of ultrasonic waves. At one time, this was how almost all headphones produced sound. This method of transforming electrical energy into mechanical energy, called the **piezoelectric effect,** is also reversible. If some external source causes the crystal to vibrate, an alternating voltage is produced across the crystal. Hence, a single crystal both generates and receives ultrasonic waves.

The primary physical principle that makes ultrasound imaging possible is the fact that a sound wave is partially reflected whenever it is incident on a boundary between two materials having different densities. If a sound wave is traveling in a material of density ρ_i and strikes a material of density ρ_t, the percentage of the incident sound wave intensity reflected, *PR*, is given by

$$PR = \left(\frac{\rho_i - \rho_t}{\rho_i + \rho_t} \right)^2 \times 100$$

This equation assumes that the incident sound wave travels perpendicularly to the boundary and that the speed of sound is approximately the same in the two

▶ **APPLICATION**

MEDICAL USES OF ULTRASOUND

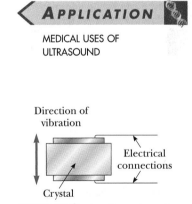

FIGURE 14.3 An alternating voltage applied to the faces of a piezoelectric crystal causes the crystal to vibrate.

Webnote 14.1

The March of Dimes maintains a Web site related to the obstetrical use of ultrasound:
http://www.modimes.org/ HealthLibrary2/factsheets/Ultrasound. htm

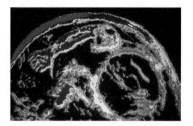

An ultrasound image of a human fetus in the womb, showing the head, right arm, and hand. *(Howard Sochurek/Woodfin Camp & Associates, Inc.)*

APPLICATION

ULTRASONIC RANGE UNIT FOR CAMERAS

Webnote 14.2

For some interesting information on autofocus cameras, go to
http://www.howstuffworks.com/ autofocus.htm

materials. The latter assumption holds very well for the human body because the speed of sound does not vary much in the organs of the body.

Physicians commonly use ultrasonic waves to observe fetuses. This technique presents far less risk than x-rays, which deposit more energy in cells and can produce birth defects. First the abdomen of the mother is coated with a liquid, such as mineral oil. If this were not done, most of the incident ultrasonic waves from the piezoelectric source would be reflected at the boundary between the air and the mother's skin. Mineral oil has a density similar to that of skin, and a very small fraction of the incident ultrasonic wave is reflected when $\rho_i \approx \rho_t$. The ultrasound energy is emitted in pulses rather than as a continuous wave, so the same crystal can be used as a detector as well as a transmitter. An image of the fetus is obtained using an array of transducers placed on the abdomen. The reflected sound waves picked up by the transducers are converted to an electric signal, which is used to form an image on a fluorescent screen. Difficulties such as the likelihood of spontaneous abortion or of breech birth are easily detected with this technique. Also, such fetal abnormalities as spina bifida and water on the brain are readily observable.

A relatively new medical application of ultrasonics is the cavitron ultrasonic surgical aspirator (CUSA). This device has made it possible to surgically remove brain tumors that were previously inoperable. The probe of this device emits ultrasonic waves (about 23 kHz) at its tip. When the tip of the probe touches a tumor, the part of the tumor near the probe is shattered, and the residue can be sucked up (aspirated) through the hollow probe. Using this technique, neurosurgeons are able to remove brain tumors without causing serious damage to healthy surrounding tissue.

Another interesting application of ultrasound is the ultrasonic ranging unit used in some cameras to provide an almost instantaneous measurement of the distance between the camera and object to be photographed. The principal component of this device is a crystal that acts as both a loudspeaker and a microphone. A pulse of ultrasonic waves is transmitted from the transducer to the object to be photographed. The object reflects part of the signal, producing an echo that is detected by the device. The time interval between the outgoing pulse and the detected echo is then electronically converted to a distance value, because the speed of sound is a known quantity.

14.3 THE SPEED OF SOUND

The speed of a sound wave in a liquid depends on the liquid's compressibility and inertia. If the liquid has a bulk modulus B and an equilibrium density ρ, the speed of sound is

Speed of sound in a liquid

$$v = \sqrt{\frac{B}{\rho}}$$ [14.1]

Recall from Chapter 9 that bulk modulus is defined as the ratio of the change in pressure ΔP to the resulting fractional change in volume $\Delta V/V$:

$$B \equiv -\frac{\Delta P}{\Delta V/V}$$ [14.2]

Note that B is always positive because an increase in pressure (positive ΔP) results in a decrease in volume. Hence the ratio $\Delta P/\Delta V$ is always negative.

It is interesting to compare Equation 14.1 with Equation 13.17 for the speed of transverse waves on a string $v = \sqrt{F/\mu}$ discussed in Chapter 13. In both cases, the wave speed depends on an elastic property of the medium (B or F) and on an inertial property of the medium (ρ or μ). In fact, the speed of all mechanical waves follows an expression of the general form

$$v = \sqrt{\frac{\text{elastic property}}{\text{inertial property}}}$$

Another example of this general form is the **speed of a longitudinal wave in a solid rod,** which is

$$v = \sqrt{\frac{Y}{\rho}}$$ [14.3]

where Y is the Young's modulus of the solid, defined as the longitudinal stress divided by the longitudinal strain (Equation 9.3), and ρ is the density of the solid.

Table 14.1 lists the speeds of sound in various media. As you can see, the speed of sound is much higher in solids than in gases. This makes sense because the molecules in a solid interact more strongly with each other than those in a gas. In general, sound travels more slowly in liquids than in solids because liquids are more compressible and hence have smaller bulk moduli. To see how sound travels faster in a solid than in air, suppose that a student places her ear against a long metal object, such as a long steel rail, and listens as her distant friend strikes the rail with a hammer. The student will hear two sounds. The first is produced by the sound traveling rapidly through the rail, and the second is the sound taking the slower path through the air.

The speed of sound also depends on the temperature of the medium. For sound traveling through air, the relationship between the speed of sound and temperature is

$$v = (331 \text{ m/s}) \sqrt{\frac{T}{273 \text{ K}}}$$ [14.4]

where 331 m/s is the speed of sound in air at 0°C and T is the absolute (Kelvin) temperature. Using this equation, we find that at 293 K, the speed of sound in air is approximately 343 m/s.

Quick Quiz 14.1	The speed of sound in air is affected by changes in (a) wavelength, (b) frequency, (c) temperature, (d) amplitude, or (e) none of these.

TABLE 14.1	**Speeds of Sound in Various Media**
Medium	v **(m/s)**
Gases	
Air (0°C)	331
Air (100°C)	386
Hydrogen (0°C)	1 290
Oxygen (0°C)	317
Helium (0°C)	972
Liquids at 25°C	
Water	1 490
Methyl alcohol	1 140
Sea water	1 530
Solids	
Aluminum	5 100
Copper	3 560
Iron	5 130
Lead	1 320
Vulcanized rubber	54

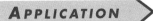

APPLICATION

THE SOUNDS HEARD DURING
A STORM

Why does thunder produce an extended "rolling" sound? And, how does lightning produce thunder in the first place?

Explanation Let us assume that we are at ground level and neglect ground reflections. When lightning strikes, a channel of ionized air carries a large electric current from a cloud to the ground. This results in a rapid temperature increase of this channel of air as the current moves through it. The temperature increase causes a sudden expansion of the air. This expansion is so sudden and so intense that a tremendous disturbance—thunder—is produced in the air. The thunder rolls due to the fact that the lightning channel is a long extended source—the entire length of the channel produces the sound at essentially the same instant of time. Sound produced at the bottom of the channel reaches you first, if you are on the ground, because that is the point closest to you, and then sounds from progressively higher portions of the channel reach you. If the lightning channel were a perfectly straight line, the resulting sound might be a steady roar, but the zigzagged shape of the path results in the rolling variation in loudness, with sound from some portions of the channel arriving at your location simultaneously. The result is a particularly loud sound, interspersed with other instants when the net sound reaching you is low in intensity.

Example 14.1 Sound Waves in a Solid Bar

If a solid bar is struck at one end with a hammer, a longitudinal pulse propagates down the bar. Find the speed of sound in a bar of aluminum, which has a Young's modulus of 7.0×10^{10} Pa and a density of 2.7×10^3 kg/m^3.

Solution From Equation 14.3 we find that

$$v_{\text{Al}} = \sqrt{\frac{Y}{\rho}} = \sqrt{\frac{7.0 \times 10^{10}\ \text{Pa}}{2.7 \times 10^3\ \text{kg/m}^3}} \approx \boxed{5\ 100\ \text{m/s or about 11 000 mi/h!}}$$

This is a typical value for the speed of sound in solids (see Table 14.1).

14.4 Energy and Intensity of Sound Waves

As the tines of a tuning fork move back and forth through the air, they exert a force on a layer of air and cause it to move. In other words, the tines do work on the layer of air. The fact that the fork pours energy into the air as sound is one of the reasons that the vibration of the fork slowly dies out. (Other factors, such as the energy lost to friction as the tines bend, also are responsible for the diminution of movement.)

> We define the **intensity** I of a wave to be the rate at which energy flows through a unit area, A, oriented perpendicular to the direction of travel of the wave.

In equation form this is

$$I \equiv \frac{1}{A}\frac{\Delta E}{\Delta t} \tag{14.5}$$

Equation 14.5 can be written in an alternative form if you recall that the rate of transfer of energy is defined as power. Thus,

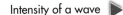 Intensity of a wave ▶

$$I \equiv \frac{\text{power}}{\text{area}} = \frac{\mathcal{P}}{A} \tag{14.6}$$

where $\mathcal{P}$ is the sound power passing through A, measured in watts, and the intensity has units of watts per square meter.

The faintest sounds the human ear can detect at a frequency of 1 000 Hz have an intensity of about 1×10^{-12} W/m^2. This intensity is called the **threshold of hearing.** The loudest sounds the ear can tolerate have an intensity of about 1 W/m^2 (the **threshold of pain**). At the threshold of hearing, the increase in pressure in the ear is approximately 3×10^{-5} Pa over normal atmospheric pressure. Because atmospheric pressure is about 1×10^5 Pa, this means the ear can detect pressure fluctuations as small as about 3 parts in 10^{10}! Also, at the threshold of hearing, the maximum displacement of an air molecule is about 1×10^{-11} m. This is a remarkably small number! If we compare this result with the diameter of a molecule (about 10^{-10} m), we see that the ear is an extremely sensitive detector of sound waves.

In a similar manner, we find that the loudest sounds the human ear can tolerate at 1 kHz correspond to a pressure variation of about 29 Pa away from normal atmospheric pressure, with a maximum displacement of air molecules of 1×10^{-5} m.

INTENSITY LEVEL IN DECIBELS

As was just mentioned, the human ear can detect a wide range of intensities, with the loudest tolerable sounds having intensities about 1.0×10^{12} times greater than those of the faintest detectable sounds. However, the most intense sound is not perceived as being 1.0×10^{12} times louder than the faintest sound. This is because the sensation of loudness is approximately logarithmic in the human ear. (If the word *logarithmic* strikes fear into your heart, look at Section A.3, heading G in Appendix A.) The relative intensity of a sound is called the **intensity level** or **decibel level** β, and is defined as

$$\beta \equiv 10 \log \left(\frac{I}{I_0} \right) \qquad [14.7]$$

◀ Intensity level

The constant I_0 is the reference intensity, taken to be the sound intensity at the threshold of hearing ($I_0 = 1.0 \times 10^{-12}$ W/m^2), I is any intensity, and β is the corresponding intensity level measured in decibels (dB). (The word *decibel*, which is one tenth of a *bel*, comes from the name of the inventor of the telephone, Alexander Graham Bell, 1847–1922.) On this scale, the threshold of pain ($I = 1.0$ W/m^2) corresponds to an intensity level of $\beta = 10 \log(1/1 \times 10^{-12}) = 10 \log(10^{12}) = 120$ dB. Nearby jet airplanes can create intensity levels of 150 dB, and subways and riveting machines have levels of 90–100 dB. The electronically amplified sound heard at rock concerts can be at levels of up to 120 dB, the threshold of pain. Exposure to such high intensity levels can seriously damage the ear. Earplugs are recommended whenever prolonged intensity levels exceed 90 dB. Recent evidence suggests that noise pollution, which is common in most large cities and in some industrial environments, may be a contributing factor to high blood pressure, anxiety, and nervousness. Table 14.2 gives some idea of the intensity levels of various sounds.

Webnote 14.3

To see a decibel calculator, try the following Web site. (You must be able to do calculations on your own for exams!)
http://hyperphysics.phy-astr.gsu.edu/hbase/sound/db.html#c3

Quick Quiz 14.2	A violin plays a melody line and is then joined by nine other violins, all playing at the same intensity as the first violin, in a repeat of the same melody. (a) When all of the violins are playing together, by how many decibels does the sound level increase? (b) If ten more violins join in, by how many decibels does the sound increase over that for the single violin?

TABLE 14.2	Intensity Levels in Decibels for Different Sources
Source of Sound	β **(dB)**
Nearby jet airplane	150
Jackhammer, machine gun	130
Siren, rock concert	120
Subway, power mower	100
Busy traffic	80
Vacuum cleaner	70
Normal conversation	50
Mosquito buzzing	40
Whisper	30
Rustling leaves	10
Threshold of hearing	0

Example 14.2   Intensity Levels of Sound

Calculate the intensity level of a sound wave having an intensity of
(a) 1.0×10^{-12} W/m²; (b) 1.0×10^{-11} W/m²; (c) 1.0×10^{-10} W/m².

Solution

A For an intensity of 1.0×10^{-12} W/m², the intensity level, in decibels, is

$$\beta = 10 \log\left(\frac{1.0 \times 10^{-12} \text{ W/m}^2}{1.0 \times 10^{-12} \text{ W/m}^2}\right) = 10 \log(1) = \boxed{0 \text{ dB}}$$

This answer should have been obvious without calculation, because an intensity of 1.0×10^{-12} W/m² corresponds to the threshold of hearing.

B In this case, the intensity is exactly ten times that in part (a). The intensity level is

$$\beta = 10 \log\left(\frac{1.0 \times 10^{-11} \text{ W/m}^2}{1.0 \times 10^{-12} \text{ W/m}^2}\right) = 10 \log(10) = \boxed{10 \text{ dB}}$$

C Here the intensity is 100 times greater than at the threshold of hearing, and the intensity level is

$$\beta = 10 \log\left(\frac{1.0 \times 10^{-10} \text{ W/m}^2}{1.0 \times 10^{-12} \text{ W/m}^2}\right) = 10 \log(100) = \boxed{20 \text{ dB}}$$

Note the pattern in these answers. A sound with an intensity level of 10 dB is ten times more intense than the 0-dB sound, and a sound with an intensity level of 20 dB is 100 times more intense than a 0-dB sound. This pattern is continued throughout the decibel scale. In short, on the decibel scale **an increase of 10 dB means that the intensity of the sound is multiplied by a factor of 10.** For example, a 50-dB sound is 10 times as intense as a 40-dB sound and a 60-dB sound is 100 times as intense as a 40-dB sound.

EXERCISE Determine the intensity level of a sound wave with an intensity of 5.0×10^{-7} W/m².

ANSWER 57 dB

Example 14.3   A Noisy Grinding Machine

A noisy grinding machine in a factory produces a sound intensity of 1.0×10^{-5} W/m². Find the decibel level of this machine, and calculate the new intensity level when a second, identical machine is added to the factory.

Solution   The intensity level of the single grinder is

$$\beta = 10 \log\left(\frac{1.0 \times 10^{-5} \text{ W/m}^2}{1.0 \times 10^{-12} \text{ W/m}^2}\right) = 10 \log(10^7) = \boxed{70 \text{ dB}}$$

Adding the second grinder doubles the energy input into sound and hence doubles the intensity. The new intensity level is

$$\beta = 10 \log\left(\frac{2.0 \times 10^{-5} \text{ W/m}^2}{1.0 \times 10^{-12} \text{ W/m}^2}\right) = \boxed{73 \text{ dB}}$$

Federal regulations now demand that no office or factory worker be exposed to noise levels that average more than 90 dB over an 8-h day. The results in this example read like one of the old jokes that start "There is some good news and some bad news." First the good news. Imagine that you are a manager analyzing the noise conditions in your factory. One machine in the factory produces a noise level of 70 dB. When you add a second machine, the noise level increases by only 3 dB. Because of the logarithmic nature of intensity levels, doubling the intensity does not double the intensity level; in fact, it alters it by a surprisingly small amount. This means that additional equipment can be added to the factory without appreciably altering the intensity level of the environment.

Now the bad news. The results also work in reverse. As you remove noisy machinery, the intensity level is not lowered appreciably. For example, consider a factory with 60 machines producing a noise level of 93 dB, which is 3 dB above the maximum allowed. In order to reduce the noise level by 3 dB, half the machines would have to be removed! That is, you would have to remove 30 machines to reduce the noise level to 90 dB. To reduce the level another 3 dB, you would have to remove half of the remaining machines, and so on.

APPLICATION

OSHA NOISE LEVEL
REGULATIONS

14.5   SPHERICAL AND PLANE WAVES

If a small spherical object oscillates so that its radius changes periodically with time, a spherical sound wave is produced (Fig. 14.4). The wave moves outward from the source at a constant speed.

Because all points on the vibrating sphere behave in the same way, we conclude that the energy in a spherical wave propagates equally in all directions. That is, no one direction is preferred over any other. If $\mathcal{P}_{av}$ is the average power emitted

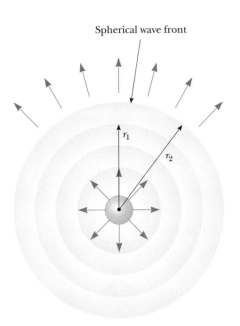

Spherical wave front

FIGURE 14.4   A spherical wave propagating radially outward from an oscillating sphere. The intensity of the spherical wave varies as $1/r^2$.

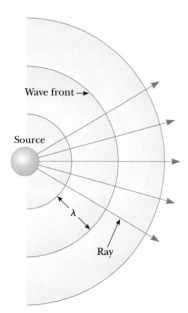

FIGURE 14.5 Spherical waves emitted by a point source. The circular arcs represent the spherical wavefronts concentric with the source. The rays are radial lines pointing outward from the source, perpendicular to the wavefronts.

Webnote 14.4

For a diagram clarifying the inverse-square behavior of sound intensity, see
http://hyperphysics.phy-astr.gsu.edu/ hbase/acoustic/invsqs.html#c1

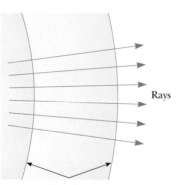

Wave fronts

FIGURE 14.6 Far away from a point source, the wavefronts are nearly parallel planes and the rays are nearly parallel lines perpendicular to the planes. Hence, a small segment of a spherical wavefront is approximately a plane wave.

by the source, then at any distance r from the source, this power must be distributed over a spherical surface of area $4\pi r^2$, assuming no absorption in the medium. (Recall that $4\pi r^2$ is the surface area of a sphere.) Hence, the **intensity** of the sound at a distance of r from the source is

$$I = \frac{\text{average power}}{\text{area}} = \frac{\mathscr{P}_{\text{av}}}{A} = \frac{\mathscr{P}_{\text{av}}}{4\pi r^2} \qquad [14.8]$$

This shows that the intensity of a wave decreases with increasing distance from its source, as you might expect. The fact that I varies as $1/r^2$ is a result of the assumption that the small source (sometimes called a **point source**) emits a spherical wave. (In fact, light waves also obey this so-called inverse-square relationship.) Because the average power is the same through any spherical surface centered at the source, we see that the intensities at distances r_1 and r_2 (Fig. 14.4) from the center of the source are

$$I_1 = \frac{\mathscr{P}_{\text{av}}}{4\pi r_1{}^2} \qquad I_2 = \frac{\mathscr{P}_{\text{av}}}{4\pi r_2{}^2}$$

Therefore, the ratio of intensities at these two spherical surfaces is

$$\frac{I_1}{I_2} = \frac{r_2{}^2}{r_1{}^2}$$

It is useful to represent spherical waves graphically with a series of circular arcs (lines of maximum intensity) concentric with the source representing part of a spherical surface, as in Figure 14.5. We call such an arc a **wave front.** The distance between adjacent wave fronts equals the wavelength, λ. The radial lines pointing outward from the source and cutting the arcs perpendicularly are called **rays.**

Now consider a small portion of a wave front that is at a *great* distance (great relative to λ) from the source, as in Figure 14.6. In this case, the rays are nearly parallel to each other and the wave fronts are very close to being planes. Therefore, at distances from the source that are great relative to the wavelength, we can approximate the wave front with parallel planes. We call such waves plane waves. Any small portion of a spherical wave that is far from the source can be considered a **plane wave.** Figure 14.7 illustrates a plane wave propagating along the x axis. If x is taken to be the direction of the wave motion (or ray) in this figure, then the wave fronts are parallel to the plane containing the y and z axes.

Example 14.4 Intensity Variations of a Point Source

A small source emits sound waves with a power output of 80 W.

A Find the intensity 3.0 m from the source.

Solution A small source emits energy in the form of spherical waves (see Fig. 14.4). Let $\mathscr{P}_{\text{av}}$ be the average power output of the source. At a distance of r from the source, the power is distributed over the surface area of a sphere, $4\pi r^2$. Therefore, the intensity at distance r from the source is given by Equation 14.8. Because $\mathscr{P}_{\text{av}} = 80$ W and $r = 3.0$ m, we find that

$$I = \frac{\mathscr{P}_{\text{av}}}{4\pi r^2} = \frac{80 \text{ W}}{4\pi (3.0 \text{ m})^2} = 0.71 \text{ W/m}^2$$

which is close to the threshold of pain.

B At what distance would the intensity be one fourth as much as it is at $r = 3.0$ m?

Solution Taking $I = (0.71 \text{ W/m}^2)/4$, and solving for r in Equation 14.8 gives

$$r = \left(\frac{\mathscr{P}_{\text{av}}}{4\pi I}\right)^{1/2} = \left(\frac{80 \text{ W}}{4\pi (0.71 \text{ W/m}^2)/4}\right)^{1/2} = 6.0 \text{ m}$$

C Find the distance at which the sound level is 40 dB.

Solution We can find the intensity at the 40-dB intensity level by using Equation 14.7 with $I_0 = 1.0 \times 10^{-12}$ W/m^2:

$$40 = 10 \log(I/I_0)$$

$$4 = \log(I/I_0)$$

$$10^4 = I/I_0$$

$$I = 10^4 I_0 = 1.0 \times 10^{-8} \text{ W/m}^2$$

When this value for I is used in Equation 14.8, solving for r gives

$$r = \left(\frac{\mathcal{P}_{\text{av}}}{4\pi I}\right)^{1/2} = \left(\frac{80 \text{ W}}{4\pi \times 10^{-8} \text{ W/m}^2}\right)^{1/2} = \boxed{2.5 \times 10^4 \text{ m}}$$

which is approximately 15 mi!

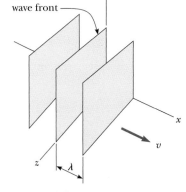

FIGURE 14.7 A representation of a plane wave moving in the positive x direction with a speed of v. The wave fronts are planes parallel to the yz plane.

14.6 THE DOPPLER EFFECT

If a car or truck is moving while its horn is blowing, the frequency of the sound you hear is higher as the vehicle approaches you and lower as it moves away from you. This is one example of the Doppler effect, named for the Austrian physicist Christian Doppler (1803–1853), who discovered it.

> In general, a **Doppler effect** is experienced whenever there is relative motion between a source of waves and an observer. When the source and observer are moving toward each other, the observer hears a frequency higher than the frequency of the source in the absence of relative motion. When the source and observer are moving away from each other, the observer hears a frequency lower than the source frequency.

Although the Doppler effect is most commonly experienced with sound waves, it is a phenomenon common to all waves. For example, the relative motion of source and observer also shifts the frequencies of light waves.

First let us consider the case in which the observer is moving and the sound source is stationary. For simplicity, we assume that the air is also stationary and that all velocity measurements are made relative to this stationary medium. Figure 14.8 describes the situation in which the observer is moving with a speed of v_O toward the source (considered a point source), which is at rest ($v_S = 0$).

We shall take the frequency of the source to be f, the wavelength to be λ, and the speed of sound in air to be v. Clearly, if both observer and source were stationary, the observer would detect f wave fronts per second. (That is, when $v_O = 0$ and $v_S = 0$, the observed frequency equals the source frequency.) When the observer is moving toward the source, he or she moves a distance of $v_O t$ in t seconds. During this interval, **the observer detects an additional number of wave fronts.** The number of extra wave fronts detected is equal to the distance traveled, $v_O t$, divided by the wavelength, λ. Thus,

$$\text{Additional wave fronts detected} = \frac{v_O t}{\lambda}$$

The number of additional wave fronts detected *per second* is v_O/λ. Hence, the frequency f' heard by the observer is *increased* to

$$f' = f + \frac{v_O}{\lambda}$$

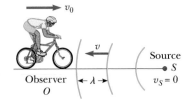

FIGURE 14.8 An observer moving with a speed of v_O *toward* a stationary point source (S) hears a frequency, f', that is *greater* than the source frequency, f.

Substituting $\lambda = v/f$, into this expression for f' we find

Observed frequency—
observer in motion and
source at rest

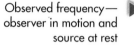

$$f' = f\left(\frac{v + v_O}{v}\right) \qquad [14.9]$$

An observer traveling *away* from the source, as in Figure 14.9, **detects fewer wave fronts per second.** In this situation, we simply take v_O to be *negative* in Equation 14.9. Thus, it follows that the observed frequency in this case is *lower* than *f*.

In general, when an observer is moving with velocity v_O relative to a stationary source, we take v_O to be positive when the observer is moving toward the source, and negative when the observer is moving away from the source.

Now consider the situation in which the source is in motion and the observer is at rest. If the source is moving directly toward observer A in Figure 14.10a, the wave fronts heard by A are closer together because the source is moving in the direction of the outgoing wave. As a result, the wavelength λ' measured by observer A is shorter than the wavelength λ of the source at rest. During each vibration, which lasts for an interval of T (the period), the source moves a distance of $v_S T = v_S/f$ and **the wavelength is shortened by this amount.** Therefore, the observed wavelength λ' is given by

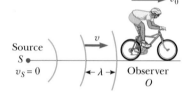

FIGURE 14.9 An observer moving with a speed of v_O *away* from a stationary source hears a frequency, f', that is *lower* than the source frequency.

$$\lambda' = \lambda - \frac{v_S}{f}$$

Because $\lambda = v/f$, the frequency heard by observer A is

$$f' = \frac{v}{\lambda'} = \frac{v}{\lambda - \dfrac{v_S}{f}} = \frac{v}{\dfrac{v}{f} - \dfrac{v_S}{f}}$$

or

Observed frequency—
source in motion and
observer at rest

$$f' = f\left(\frac{v}{v - v_S}\right) \qquad [14.10]$$

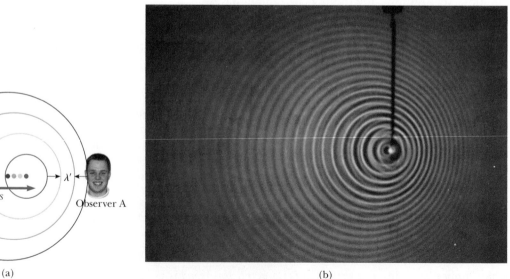

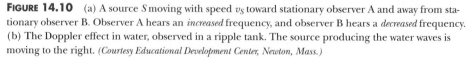

FIGURE 14.10   (a) A source *S* moving with speed v_S toward stationary observer A and away from stationary observer B. Observer A hears an *increased* frequency, and observer B hears a *decreased* frequency. (b) The Doppler effect in water, observed in a ripple tank. The source producing the water waves is moving to the right. *(Courtesy Educational Development Center, Newton, Mass.)*

That is, **the observed frequency increases when the source is moving toward the observer.**

In Figure 14.10a, the source is moving away from observer B, who is at rest to the left of the source. In this situation, we take v_O to be negative in Equation 14.10. Thus, observer B hears a frequency that is lower than f.

Finally, if both the source and the observer are in motion, we find the following general relationship for the observed frequency:

$$f' = f\left(\frac{v + v_O}{v - v_S}\right)$$ [14.11]

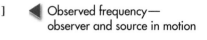 Observed frequency—
observer and source in motion

In this expression, the signs for the values substituted for v_O and v_S depend on the direction of the velocity. A positive value is used for motion of the observer or the source *toward* the other, and a negative sign for motion of one *away* from the other.

When working with any Doppler-effect problem, remember the following rules concerning signs. The word *toward* is associated with an *increase* in the observed frequency. The words *away from* are associated with a *decrease* in the observed frequency.

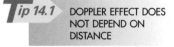 **Tip 14.1** DOPPLER EFFECT DOES NOT DEPEND ON DISTANCE

The Doppler effect does not depend on the distance between the source and the observer. Although the intensity of a sound varies as the distance changes, the observed frequency does not—the frequency depends only on the speed. As you listen to an approaching source, you will hear a gradual increase in intensity but a constant frequency. As the source passes, you will hear a sudden decrease in frequency to a new constant value and a gradual decrease in intensity.

Quick Quiz 14.3 You are a passenger on a hot-air balloon that is rising with constant velocity, and you are carrying a buzzer that emits a sound of frequency f. If you accidentally drop the buzzer from the balloon, what can you conclude about the sound you hear as it falls towards the ground? (a) The frequency and intensity increase, (b) the frequency decreases and the intensity increases, (c) the frequency decreases and the intensity decreases, or (d) the frequency remains the same but the intensity decreases.

APPLYING **PHYSICS** 14.2

Suppose you place your stereo speakers far apart and run past them from right to left or left to right. If you run rapidly enough and have excellent pitch discrimination, you may notice that the music that is playing seems to be out of tune when you are between the speakers. Why?

Explanation When you are between the speakers, you are running away from one of them and toward the other. Thus, there is a Doppler shift downward for the sound from the speaker behind you and a Doppler shift upward for the sound from the speaker ahead of you. As a result, the sound from the two speakers will not be in tune. A calculation shows that a world-class sprint runner could run fast enough to generate about a semitone difference in the sound from the two speakers.

Example 14.5 Listen, but Don't Stand on the Track

A train moving at a speed of 40.0 m/s sounds its whistle, which has a frequency of 500 Hz. Determine the frequency heard by a stationary observer as the train approaches the observer.

Solution We can use Equation 14.10 to get the observed frequency as the train approaches the observer. Taking $v = 345$ m/s as the speed of sound in air, we have

$$f' = f\left(\frac{v}{v - v_S}\right) = (500 \text{ Hz})\left(\frac{345 \text{ m/s}}{345 \text{ m/s} - 40.0 \text{ m/s}}\right) = \boxed{566 \text{ Hz}}$$

EXERCISE   Determine the frequency heard by the stationary observer as the train recedes from the observer.

ANSWER   448 Hz

Example 14.6   The Noisy Siren

An ambulance travels down a highway at a speed of 75.0 mi/h, its siren emitting sound at a frequency of 400 Hz. What frequency is heard by a passenger in a car traveling at 55.0 mi/h in the opposite direction as the car (a) approaches? (b) moves away from the ambulance?

Solution   Let us take the velocity of sound in air to be $v = 345$ m/s and use the conversion 1.00 mi/h = 0.447 m/s. Therefore, $v_S = 75.0$ mi/h = 33.5 m/s and $v_O = 55.0$ mi/h = 24.6 m/s. We can use Equation 14.11 in both cases.

A As the ambulance and car approach each other, the observed frequency is

$$f' = f\left(\frac{v + v_O}{v - v_S}\right) = (400 \text{ Hz})\left(\frac{345 \text{ m/s} + 24.6 \text{ m/s}}{345 \text{ m/s} - 33.5 \text{ m/s}}\right) = \boxed{475 \text{ Hz}}$$

B As the two vehicles recede from each other, we must take $v_O = -24.6$ m/s and $v_S = -33.5$ m/s in Equation 14.11. Thus, the passenger in the car hears a frequency of

$$f' = f\left(\frac{v + v_O}{v - v_S}\right) = (400 \text{ Hz})\left[\frac{345 \text{ m/s} + (-24.6 \text{ m/s})}{345 \text{ m/s} - (-3.5 \text{ m/s})}\right] = \boxed{339 \text{ Hz}}$$

SHOCK WAVES

Now let us consider what happens when the source speed v_S *exceeds* the wave velocity v. Figure 14.11a describes this situation graphically. The circles represent spherical wave fronts emitted by the source at various times during its motion. At $t = 0$, the source is at point S_0, and at some later time t, the source is at point S_n. In the interval t, the wave front centered at S_0 reaches a radius of vt. In this same interval, the source travels to S_n, a distance of $v_S t$. At the instant the source is at S_n, the waves just beginning to be generated at this point have wave fronts of zero radius. The line drawn from S_n to the wave front centered on S_0 is tangent to all other wave fronts generated at intermediate times. All such tangent lines lie on the surface of a cone. The angle θ between one of these tangent lines and the direction of travel is given by

$$\sin\theta = \frac{v}{v_S}$$

The ratio v_S/v is referred to as the **Mach number.** The conical wave front produced when $v_S > v$ (supersonic speeds) is known as a **shock wave.** Figure 14.11b is a photograph of a bullet traveling at supersonic speed through the hot air rising above a candle. Note the shock waves in the vicinity of the bullet. Another interesting example of shock waves is the V-shaped wave front produced by a canoe (the bow wave) when the canoe's speed exceeds the speed of the water waves (Fig. 14.12).

Jet aircraft and space shuttles traveling at supersonic speeds produce shock waves, which are responsible for the loud explosion, or sonic boom, heard on the

Webnote 14.5

Take a look at an amazing and varied gallery of shock wave images. Explore
http://www.eng.vt.edu/fluids/msc/gallery/gall.htm
Start by clicking on "Shock Waves."

Webnote 14.6

To interact with an excellent animation illustrating both the Doppler effect and shock waves, visit
http://www.phy.ntnu.edu.tw/~hwang/Doppler/Doppler.html

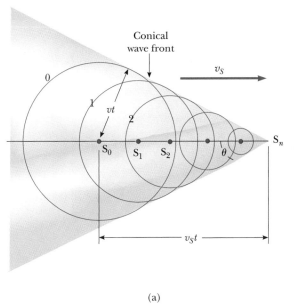

(a)

(b)

FIGURE 14.11 (a) A representation of a shock wave, produced when a source moves from S_0 to S_n with a speed, v_S, that is *greater* than the wave speed, v, in that medium. The envelope of the wave fronts forms a cone whose half-angle is $\sin \theta = v/v_S$. (b) A stroboscopic photograph of a bullet moving at supersonic speed through the hot air above a candle. *(Harold Edgerton, Courtesy of Palm Press, Inc.)*

ground. A shock wave carries a great deal of energy concentrated on the surface of the cone, with correspondingly great pressure variations. Shock waves are unpleasant to hear and can damage buildings when aircraft fly supersonically at low altitudes. In fact, an airplane flying at supersonic speeds produces a double boom because two shock waves are formed, one from the nose of the plane and one from the tail (Fig. 14.13).

> **Quick Quiz 14.4** As an airplane flying with constant velocity moves from a cold air mass into a warm air mass, does the Mach number (a) increase, (b) decrease, or (c) remain the same?

APPLICATION

SONIC BOOMS

FIGURE 14.12 The V-shaped bow wave of a canoe is formed because the canoe travels at a speed greater than the speed of water waves. A bow wave is analogous to a shock wave formed by an airplane traveling faster than sound. *(© 1994, Comstock)*

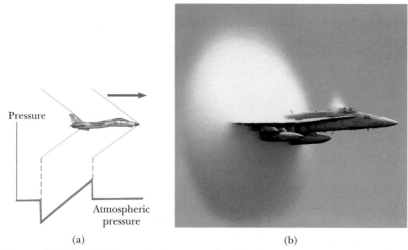

FIGURE 14.13   (a) The two shock waves produced by the nose and tail of a jet airplane traveling at supersonic speed. (b) A shock wave due to a jet traveling at the speed of sound is made visible as a fog of water vapor. The large pressure variation in the shock wave causes the water in the air to condense into water droplets. *(U.S. Navy. Photo by Ensign John Gay)*

14.7   INTERFERENCE OF SOUND WAVES

Sound waves can be made to interfere with each other. This can be demonstrated with the device shown in Figure 14.14. Sound from a loudspeaker at S is sent into a tube at *P*, where there is a T-shaped junction. The sound splits and follows two separate pathways indicated by the red arrows. Half of the sound travels upward, and half downward. Finally, the two sounds merge at an opening where a listener places her ear. If the two paths r_1 and r_2 have the same length, the waves that enter the junction will separate into two halves, travel the two paths, and then combine again at the ear. This reuniting of the two waves produces constructive interference, and thus the listener hears a loud sound. If the upper path is adjusted to be one full wavelength longer than the lower path, constructive interference of the two waves occurs again, and a loud sound is detected at the receiver. In general, **if the path difference $r_2 - r_1$ is zero or some integer multiple of wavelengths, constructive interference occurs:**

Condition for constructive interference ▶

$$r_2 - r_1 = n\lambda \qquad (n = 0, 1, 2, \ldots) \qquad \text{[14.12]}$$

Suppose, however, that path length r_2 is adjusted so that the upper path is half a wavelength *longer* than the lower path r_1. In this case, an entering sound wave splits and travels the two paths as before, but now the wave along the upper path must travel a distance equivalent to half a wavelength farther than the wave traveling along the lower path. As a result, the crest of one wave meets the trough of the other when they merge at the receiver. Because this is the condition for destructive

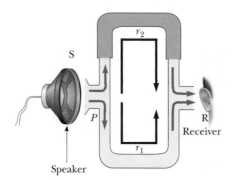

FIGURE 14.14   An acoustical system for demonstrating interference of sound waves. Sound from the speaker enters the tube and splits into two parts at *P*. The two waves combine at the opposite side and are detected at *R*. The upper path length is varied by the sliding section.

interference, no sound is detected at the receiver. In general, **if the path difference $r_2 - r_1$ is $\frac{1}{2}$, $1\frac{1}{2}$, $2\frac{1}{2}$, . . . wavelengths, destructive interference occurs:**

$$r_2 - r_1 = (n + \tfrac{1}{2})\lambda \qquad (n = 0, 1, 2, \ . \ . \ .) \qquad \text{[14.13]}$$

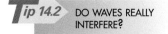

◀ Condition for destructive interference

Nature provides many other examples of interference phenomena. In Chapter 24 we shall describe several interesting interference effects involving light waves.

When connecting the wires between your stereo system and your loudspeakers, you may notice that the wires are color-coded and that the speakers have positive and negative signs on the connections. The reason for this is that the speakers need to be connected with the same "polarity." If they are not, then the same electrical signal fed to both speakers will result in one speaker cone moving outward at the same time that the other speaker cone is moving inward. In this case, the sound leaving the two speakers will be 180° out of phase with each other. If you are sitting midway between the speakers, the sounds from both speakers travel the same distance and preserve the phase difference they had when they left. In an ideal situation, for a 180° phase difference you would get complete destructive interference and no sound! In reality, the cancellation is not complete and is much more significant for bass notes (which have a long wavelength) than for the shorter wavelength treble notes. Nevertheless, to avoid significant reduction in the intensity of bass notes, the color-coded wires and the signs on the speaker connections should be noted carefully.

APPLICATION

CONNECTING YOUR STEREO SPEAKERS

Tip 14.2 DO WAVES REALLY INTERFERE?

In popular usage, the term *interfere* implies that an agent affects the outcome of some event. For example, in American football, *pass interference* means that a defending player has made contact with the receiver so that he is unable to catch the ball. This is very different than its use in physics, where waves pass through each other and interfere, but they do not affect each other in any way.

Example 14.7 Two Speakers Driven by the Same Source

Two speakers placed 3.00 m apart are driven by the same oscillator (Fig. 14.15). A listener is originally at point O, which is located 8.00 m from the center of the line connecting the two speakers. The listener then walks to point P, which is a perpendicular distance 0.350 m from O, before reaching the *first minimum* in sound intensity. What is the frequency of the oscillator?

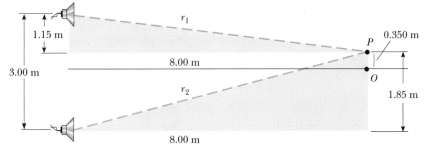

FIGURE 14.15 (Example 14.7) Two loudspeakers driven by the same source can produce interference.

Solution In order to find the frequency of the oscillator, we need to know the wavelength of the sound coming from the speakers. This information, combined with our knowledge of the speed of sound, allows us find the frequency. We can determine the wavelength from the information given in the problem about the interference of the sound waves from the speakers. The first minimum occurs when the two waves reaching the listener at P are 180° out of phase—in other words, when their path difference Δr equals $\lambda/2$. In order to calculate the path difference, we must first find the path lengths r_1 and r_2. Figure 14.15 shows the physical arrangement of the speakers, along with two shaded right triangles, drawn using the lengths described in the problem. Making use of these triangles in Figure 14.15, we find the path lengths to be

$$r_1 = \sqrt{(8.00 \text{ m})^2 + (1.15 \text{ m})^2} = 8.08 \text{ m}$$

$$r_2 = \sqrt{(8.00 \text{ m})^2 + (1.85 \text{ m})^2} = 8.21 \text{ m}$$

Hence, the path difference is $r_2 - r_1 = 0.13$ m. Because we require that this path difference be equal to $\lambda/2$ for the first minimum, we find that $\lambda = 0.26$ m.

To obtain the oscillator frequency, we use $v = \lambda f$, where v is the speed of sound in air, 343 m/s:

$$f = \frac{v}{\lambda} = \frac{343 \text{ m/s}}{0.26 \text{ m}} = \boxed{1.3 \text{ kHz}}$$

EXERCISE If the oscillator frequency is adjusted such that the first location at which a listener hears no sound is at a distance of 0.75 m from O, what is the new frequency?

ANSWER 0.63 kHz

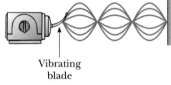

Vibrating blade

FIGURE 14.16 Standing waves can be set up in a stretched string by connecting one end of the string to a vibrating blade. When the blade vibrates at one of the natural frequencies of the string, large-amplitude standing waves are created.

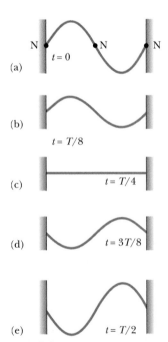

FIGURE 14.17 A standing-wave pattern in a stretched string, shown by snapshots of the string during one half of a cycle.

14.8 STANDING WAVES

Standing waves can be set up in a stretched string by connecting one end of the string to a stationary clamp and connecting the other end to a vibrating object, such as the end of a tuning fork, or by shaking your hand up and down at a steady rate (Fig. 14.16). In this situation, traveling waves reflect from the ends, creating waves traveling in both directions on the string. The incident and reflected waves combine according to the **superposition principle.** (See section 13.12). If the string is vibrated at exactly the right frequency, the wave appears to stand still— hence its name, **standing wave.** A **node** occurs where the two traveling waves always have the same magnitude of displacement but have opposite sign, so that the net displacement is zero at that point. There is no motion in the string at the nodes but midway between two adjacent nodes, at an **antinode,** the string vibrates with the largest amplitude.

Figure 14.17 shows snapshots of the oscillation of a standing wave during half of a cycle. Notice that **all points on the string oscillate together vertically with the same frequency but that different points have different amplitudes of motion.** The nodes are stationary and the points at which the string is attached to the wall are also nodes, labeled N in Figure 14.17a. From the figure, observe that the distance between adjacent nodes is one half of the wavelength of the wave:

$$d_{\text{NN}} = \tfrac{1}{2}\lambda$$

Consider a string of length L that is fixed at both ends, as in Figure 14.18a. For a string, we can set up standing-wave patterns at many frequencies—the more loops, the higher the frequency. Three of these are pictured in Figures 14.18b, 14.18c, and 14.18d. Each has a characteristic frequency, which we will now calculate.

First, note that **the ends of the string must be nodes because these points are fixed.** If the string is displaced at its midpoint and released, the vibration shown in Figure 14.18b can be produced, in which case the center of the string is an antinode. For this standing-wave pattern, the length of the string equals $\lambda/2$ (the distance between nodes). Thus,

$$L = \frac{\lambda_1}{2} \quad \text{or} \quad \lambda_1 = 2L$$

and the frequency of this vibration is

$$f_1 = \frac{v}{\lambda_1} = \frac{v}{2L} \qquad [14.14]$$

In Chapter 13 (Equation 13.17), the speed of a wave on a string was given as $v = \sqrt{F/\mu}$, where F is the tension in the string and μ is its mass per unit length.

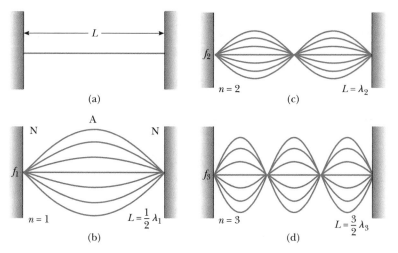

FIGURE 14.18 (a) Standing waves in a stretched string of length L, fixed at both ends. The characteristic frequencies of vibration form a harmonic series: (b) the fundamental frequency, or first harmonic; (c) the second harmonic; and (d) the third harmonic.

Thus, we can express Equation 14.14 as

$$f_1 = \frac{1}{2L}\sqrt{\frac{F}{\mu}} \qquad \text{[14.15]}$$

This lowest frequency of vibration is called the **fundamental frequency** of the vibrating string.

The next natural frequency of vibration occurs when the length of the string equals one wavelength λ_2—that is, when $L = \lambda_2$ (Fig. 14.18c). Hence,

$$f_2 = \frac{v}{L} = \frac{2v}{2L} = 2f_1 \qquad \text{[14.16]}$$

Note that this frequency is equal to *twice* the fundamental frequency. You should convince yourself that the next highest frequency of vibration, shown in Figure 14.18d, is

$$f_3 = \frac{3v}{2L} = 3f_1 \qquad \text{[14.17]}$$

In general, the characteristic frequencies are given by

$$f_n = nf_1 = \frac{n}{2L}\sqrt{\frac{F}{\mu}} \qquad \text{[14.18]}$$

Tip 14.3   JUMPING ROPE

If you played jump rope as a child, you set up a pattern on the rope something like that in Figure 14.18b, although the elements of the rope also exhibited a circular motion, rather than a simple up-and-down motion. If you had moved your hand faster, you could have set up a pattern something like Figure 14.18c, and you could have had two friends jumping rope—one in each of the two loops!

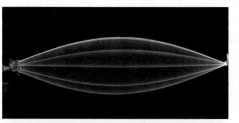

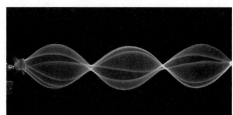

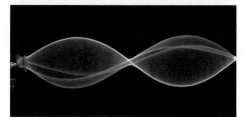

Multiflash photographs of standing-wave patterns in a cord driven by a vibrator at the left end. The single-loop pattern at the top left represents the fundamental ($n = 1$), the two-loop pattern at the right represents the second harmonic ($n = 2$), and the three-loop pattern at the lower left represents the third harmonic ($n = 3$).
(© *Richard Megna, Fundamental Photographs, NYC*)

where $n = 1, 2, 3, \ldots$. In other words, the frequencies are integral multiples of the fundamental frequency. The frequencies f_1, $2f_1$, $3f_1$, and so on form a **harmonic series.** The fundamental f_1 corresponds to the **first harmonic;** the frequency $f_2 = 2f_1$, to the **second harmonic;** and so on.

When a stretched string is distorted to a shape that corresponds to any one of its harmonics, after being released it vibrates only at the frequency of that harmonic. If the string is struck or bowed, however, the resulting vibration includes different amounts of various harmonics, including the fundamental. Waves not in the harmonic series are quickly damped out on a string fixed at both ends. In effect, when disturbed, the string "selects" the standing-wave frequencies. As we shall see later, the presence of several harmonics on a string gives stringed instruments their characteristic sound, which enables us to distinguish one from another even when they produce identical fundamental frequencies.

The frequency of a string on a musical instrument can be changed either by varying the tension or by changing the length. For example, the tension in guitar and violin strings is varied by turning pegs on the neck of the instrument. As the tension is increased, the frequency of the harmonic series increases according to Equation 14.18. Once the instrument is tuned, the musician varies the frequency by pressing the strings against the neck at a variety of positions, thereby changing the effective lengths of the vibrating portions of the strings. As the length is reduced, the frequency increases, as Equation 14.18 indicates.

Finally, Equation 14.18 shows that a string of fixed length can be made to vibrate at a lower fundamental frequency by increasing its mass per unit length. This is achieved in the bass strings of guitars and pianos by wrapping them with metal windings.

APPLICATION

TUNING A MUSICAL
INSTRUMENT

A concert-style harp. *(Lyon & Healy Harps, Chicago)*

Example 14.8   Harmonics of a Stretched String

Find the first four harmonics of a 1.0-m-long string if the string has a mass per unit length of 2.0×10^{-3} kg/m and is under a tension of 80 N.

Solution   The speed of the wave on the string is

$$v = \sqrt{\frac{F}{\mu}} = \sqrt{\frac{80 \text{ N}}{2.0 \times 10^{-3} \text{ kg/m}}} = 200 \text{ m/s}$$

The fundamental frequency can be found using Equation 14.14:

$$f_1 = \frac{v}{2L} = \frac{200 \text{ m/s}}{2(1.0 \text{ m})} = \boxed{100 \text{ Hz}}$$

The frequencies of the next three modes are $f_2 = 2f_1$, $f_3 = 3f_1$, and $f_4 = 4f_1$. Thus, $f_2 = \boxed{200 \text{ Hz}}$, $f_3 = \boxed{300 \text{ Hz}}$, and $f_4 = \boxed{400 \text{ Hz}}$.

EXERCISE   Find the tension in the string if the fundamental frequency is increased to 120 Hz.

ANSWER   120 N

14.9   FORCED VIBRATIONS AND RESONANCE

In Chapter 13 we learned that the energy of a damped oscillator decreases in time because of friction. It is possible to compensate for this energy loss by applying an external force that does positive work on the system.

For example, suppose an object-spring system having some natural frequency of vibration f_0 is pushed back and forth with a periodic force whose frequency is f. The system vibrates at the frequency f of the driving force. This type of motion is referred to as a **forced vibration.** Its amplitude reaches a maximum when the frequency of the driving force equals the natural frequency of the system f_0 called the **resonant frequency** of the system. Under this condition, the system is said to be in **resonance.**

In Section 14.8 we learned that a stretched string can vibrate in one or more of its natural modes. Here again, if a periodic force is applied to the string, the amplitude of vibration increases as the frequency of the applied force approaches one of the natural frequencies of vibration.

Resonance vibrations occur in a wide variety of circumstances. Figure 14.19 illustrates one experiment that demonstrates a resonance condition. Several pendulums of different lengths are suspended from a flexible beam. If one of them, such as *A*, is set in motion, the others begin to oscillate because of vibrations in the flexible beam. Pendulum *C*, the same length as *A*, oscillates with the greatest amplitude because its natural frequency matches that of pendulum *A* (the driving force).

Another simple example of resonance is a child being pushed on a swing, which is essentially a pendulum with a natural frequency that depends on the length. The swing is kept in motion by a series of appropriately timed pushes. For its amplitude to be increased, the swing must be pushed each time it returns to the person's hands. This corresponds to a frequency equal to the natural frequency of the swing. If the energy put into the system per cycle of motion exactly equals the energy lost due to friction, the amplitude remains constant.

Opera singers have been known to set crystal goblets in audible vibration with their powerful voices, as shown in Figure 14.20. This is yet another example of resonance: The sound waves emitted by the singer can set up large-amplitude vibrations in the glass. If a highly amplified sound wave has the right frequency, the amplitude of forced vibrations in the glass increases to the point where the glass becomes heavily strained and shatters.

A striking example of structural resonance occurred in 1940, when the Tacoma Narrows Bridge in the state of Washington was set in oscillation by the wind (Fig. 14.21). The amplitude of the oscillations increased rapidly and reached a high value until the bridge ultimately collapsed (probably because of metal fatigue).

A more recent example of destruction by structural resonance occurred during the Loma Prieta earthquake near Oakland, California, in 1989. In one section—almost a mile long—of the double-decker Nimitz Freeway, the upper deck collapsed onto the lower deck, killing several people. The collapse occurred because this particular section was built on mudfill, whereas other parts were built on bedrock. As seismic waves pass through mudfill or other loose soil, their wave speed decreases and the amplitude of the wave increases. Thus, the section of the freeway that collapsed oscillated at the same frequency as other sections but at a much larger amplitude.

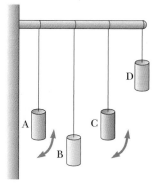

FIGURE 14.19 **Resonance.** If pendulum A is set in oscillation, only pendulum C, whose length matches that of A, will eventually oscillate with a large amplitude, or resonate. The arrows indicate motion perpendicular to the page.

APPLICATION

SHATTERING GOBLETS WITH THE VOICE

APPLICATION

STRUCTURAL RESONANCE IN BRIDGES AND BUILDINGS

FIGURE 14.20 *(Left)* Standing-wave pattern in a vibrating wine glass. The wine glass will shatter if the amplitude of vibration becomes too large. *(Courtesy of Professor Thomas D. Rossing, Northern Illinois University)* *(Right)* A wine glass shattered by the amplified sound of a human voice. *(Ben Rose/The IMAGE Bank)*

FIGURE 14.21 The collapse of the Tacoma Narrows suspension bridge in 1940 was a vivid demonstration of mechanical resonance. High winds set up standing waves in the bridge, causing it to oscillate at one of its natural frequencies. Once established, this resonance condition led to the bridge's collapse. *(United Press International Photo)*

14.10 STANDING WAVES IN AIR COLUMNS

Standing longitudinal waves can be set up in a tube of air, such as an organ pipe, as the result of interference between sound waves traveling in opposite directions. The relationship between the incident wave and the reflected wave depends on whether the reflecting end of the tube is open or closed. A portion of the sound wave is reflected back into the tube even at an open end. **If one end is closed, a node must exist at this end because the movement of air is restricted. If the end is open, the elements of air have complete freedom of motion, and an antinode exists.**

Figure 14.22a shows the first three modes of vibration of a pipe open at both ends. When air is directed against an edge at the left, longitudinal standing waves are formed and the pipe vibrates at its natural frequencies. Note that for the fundamental frequency the wavelength is twice the length of the pipe and hence

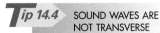

SOUND WAVES ARE NOT TRANSVERSE

The standing longitudinal waves in Figure 14.22 are drawn as transverse waves. This is because it is difficult to draw longitudinal displacements—they are in the same direction as the propagation. Thus, in Figure 14.22, the vertical axis represents either pressure or horizontal displacement of the elements of the medium.

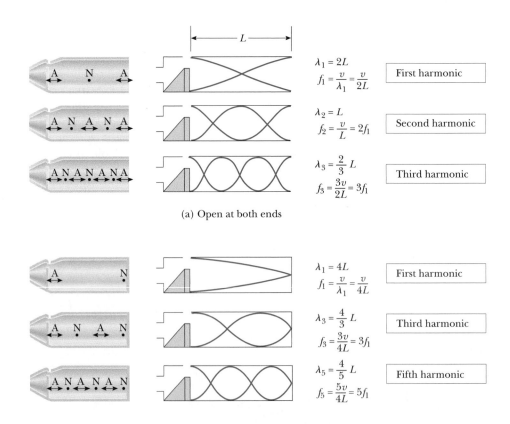

(a) Open at both ends

(b) Closed at one end, open at the other

FIGURE 14.22 (a) Standing longitudinal waves in an organ pipe open at both ends. The natural frequencies f_1, $2f_1$, $3f_1$, . . . form a harmonic series. (b) Standing longitudinal waves in an organ pipe closed at one end. Only *odd* harmonics are present, and the natural frequencies are f_1, $3f_1$, $5f_1$, and so on.

$f_1 = v/2L$. Similarly, we find that the frequencies of the second and third harmonics are $2f_1$, $3f_1$, ... Thus, **in a pipe open at both ends, the natural frequencies of vibration form a series in which all harmonics are present. These harmonics are equal to integral multiples of the fundamental frequency.** We can express this harmonic series as

$$f_n = n\,\frac{v}{2L} \qquad n = 1, 2, 3, \ldots \qquad \text{[14.19]}$$

 Pipe open at both ends; all harmonics present

where v is the speed of sound in air.

If a pipe is closed at one end and open at the other, the closed end is a node (Fig. 14.22b). In this case, the wavelength of the fundamental mode is four times the length of the tube. Hence, $f_1 = v/4L$ and the frequencies of the third and fifth harmonics are $3f_1$, $5f_1$, ... That is, **in a pipe closed at one end and open at the other, only odd harmonics are present.** These are given by

$$f_n = n\,\frac{v}{4L} \qquad n = 1, 3, 5, \ldots \qquad \text{[14.20]}$$

 Pipe closed at one end; odd harmonics present

APPLYING **PHYSICS** 14.3

Passing ocean waves sometimes cause the water in a harbor to undergo very large oscillations, called a *seiche* (pronounced *saysh*). Why does this happen?

Explanation Water in a harbor is enclosed and possesses a natural frequency based on the size of the harbor. This is similar to the natural frequency of the enclosed air in a bottle, which can be excited by blowing across the edge of the opening. Ocean waves pass by the opening of the harbor at a certain frequency. If this frequency matches that of the enclosed harbor, then a large standing wave can be set up in the water by resonance. This can be simulated by carrying a fish tank with water. If your walking frequency matches the natural frequency of the water as it sloshes back and forth, a large standing wave in the fish tank can be established.

APPLICATION

OSCILLATIONS IN A HARBOR

APPLYING **PHYSICS** 14.4

If an orchestra doesn't warm up before a performance, the strings go flat and the wind instruments go sharp during the performance. Why?

Explanation Without warming up, all the instruments will be at room temperature at the beginning of the concert. As the wind instruments are played, they fill with warm air from the player's exhalation. The increase in temperature of the air in the instrument causes an increase in the speed of sound, which raises the resonance frequencies of the air columns. As a result, the instruments go sharp. The strings on the stringed instruments also increase in temperature due to the friction of rubbing with the bow. This results in thermal expansion, which causes a decrease in tension in the strings. With a decrease in tension, the wave speed on the strings drops, and the fundamental frequencies decrease. Thus, the stringed instruments go flat.

APPLICATION

WHY ARE INSTRUMENTS WARMED UP?

APPLYING **PHYSICS** 14.5

A bugle has no valves, keys, slides, or finger holes—how can it play a song?

Explanation Songs for the bugle are limited to harmonics of the fundamental frequency, because there is no control over frequencies without valves, keys, slides, or finger holes. The player obtains different notes by changing the tension

APPLICATION

HOW DO BUGLES WORK?

in the lips as the bugle is played, in order to excite different harmonics. The normal playing range of a bugle is among the third, fourth, fifth, and sixth harmonics of the fundamental. For example, "Reveille" is played with just the three notes G, C, and F. And "Taps" is played with these three notes and the G one octave above the lower G.

Example 14.9 Harmonics of a Pipe

A pipe is 2.46 m long.

A Determine the frequencies of the first three harmonics if the pipe is open at both ends. Take 345 m/s as the speed of sound in air.

Solution   The fundamental frequency of a pipe open at both ends can be found from Equation 14.19, with $n = 1$.

$$f_1 = \frac{v}{2L} = \frac{345 \text{ m/s}}{2(2.46 \text{ m})} = \boxed{70.0 \text{ Hz}}$$

Because all harmonics are present in a pipe open at both ends, the second and third harmonics have frequencies of $f_2 = 2f_1 = \boxed{140 \text{ Hz}}$ and $f_3 = 3f_1 = \boxed{210 \text{ Hz}}$.

B What are the three lowest possible frequencies if the pipe is closed at one end and open at the other?

Solution   The fundamental frequency of a pipe closed at one end can be found from Equation 14.20, with $n = 1$.

$$f_1 = \frac{v}{4L} = \frac{345 \text{ m/s}}{4(2.46 \text{ m})} = \boxed{35.0 \text{ Hz}}$$

In this case, only odd harmonics are present, and so the third and fifth harmonics have frequencies of $f_3 = 3f_1 = \boxed{105 \text{ Hz}}$ and $f_5 = 5f_1 = \boxed{175 \text{ Hz}}$.

EXERCISE   If the pipe is open at one end, how many harmonics are possible in the normal hearing range, 20 to 20 000 Hz?

ANSWER   286

Example 14.10 Resonance in a Tube of Variable Length

Figure 14.23 shows a simple apparatus for demonstrating resonance in a tube. A long tube open at both ends is partially submerged in a beaker of water, and a vibrating tuning fork of unknown frequency is placed near the top of the tube. The length of the air column L is adjusted by moving the tube vertically. The sound waves generated by the

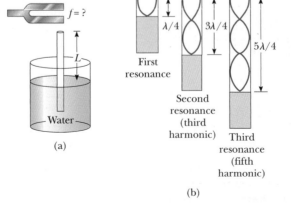

FIGURE 14.23   (Example 14.10) (a) Apparatus for demonstrating the resonance of sound waves in a tube closed at one end. The length, L, of the air column is varied by moving the tube vertically while it is partially submerged in water. (b) The first three resonances of the system.

$f = ?$

$\lambda/4$

$3\lambda/4$

$5\lambda/4$

First resonance

Second resonance (third harmonic)

Third resonance (fifth harmonic)

Water

(a)

(b)

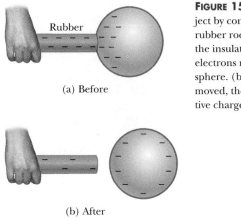

(a) Before

(b) After

FIGURE 15.3 Charging a metal object by conduction. (a) The charged rubber rod is placed in contact with the insulated metal sphere. Some electrons move from the rod onto the sphere. (b) When the rod is removed, the sphere is left with a negative charge.

Conductors are materials in which electric charges move freely, and insulators are materials in which electric charges do not move freely.

Glass and rubber are insulators. When such materials are charged by rubbing, only the rubbed area becomes charged, and there is no tendency for the charge to move into other regions of the material. In contrast, materials such as copper, aluminum, and silver are good conductors. When such materials are charged in some small region, the charge readily distributes itself over the entire surface of the material. If you hold a copper rod in your hand and rub the rod with wool or fur, it will not attract a piece of paper. This might suggest that a metal cannot be charged. However, if you hold the copper rod with an insulator and then rub it with wool or fur, the rod remains charged and attracts the paper. In the first case, the electric charges produced by rubbing readily move from the copper through your body and finally to ground. In the second case, the insulating handle prevents the flow of charge to ground.

Semiconductors are a third class of materials, and their electrical properties are somewhere between those of insulators and those of conductors. Silicon and germanium are well-known semiconductors that are widely used in the fabrication of a variety of electronic devices.

CHARGING BY CONDUCTION

Consider a negatively charged rubber rod brought into contact with a neutral conducting sphere that is insulated so that there is no conducting path for charges to leave the sphere. Some electrons on the rubber rod are now able to move onto the sphere, as in Figure 15.3. When the rubber rod is removed, the sphere is left with a negative charge. This process is referred to as charging by **conduction.** The object being charged in such a process (the sphere) is always left with a charge having the same sign as the object doing the charging (the rubber rod).

CHARGING BY INDUCTION

When an object is connected to a conducting wire or copper pipe buried in Earth, it is said to be **grounded.** Earth can be considered an infinite reservoir for electrons; this means that it can accept or supply an unlimited number of electrons. With this in mind, we can understand the charging of a conductor by a process known as **induction.**

Consider a negatively charged rubber rod brought near a neutral (uncharged) conducting sphere that is insulated so that there is no conducting path to ground (Fig. 15.4a). The repulsive force between the electrons in the rod and those in the sphere causes a redistribution of charge on the sphere so that some

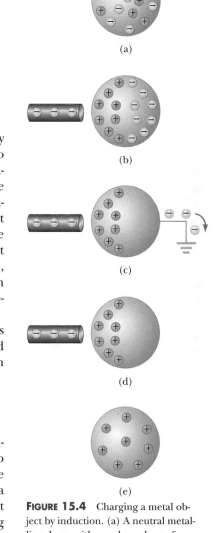

(a)

(b)

(c)

(d)

(e)

FIGURE 15.4 Charging a metal object by induction. (a) A neutral metallic sphere, with equal numbers of positive and negative charge. (b) The charge on the neutral sphere is redistributed when a charged rubber rod is placed near the sphere. (c) When the sphere is grounded, some of its electrons leave through the ground wire. (d) When the ground connection is removed, the sphere has excess positive charge that is nonuniformly distributed. (e) When the rubber rod is removed, the excess positive charge becomes uniformly distributed over the surface of the sphere.

FIGURE 15.5 (a) The charged object on the left induces charges on the surface of an insulator. (b) A charged comb attracts bits of paper because charges are displaced in the paper. *(© 1968 Fundamental Photographs)*

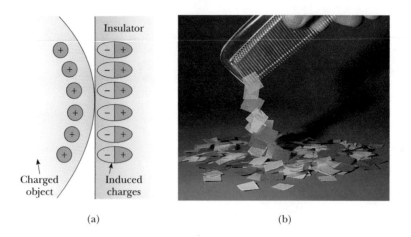

(a)   (b)

electrons move to the side of the sphere farthest away from the rod (Fig. 15.4b). The region of the sphere nearest the negatively charged rod has an excess of positive charge because of the migration of electrons away from this location. If a grounded conducting wire is then connected to the sphere, as in Figure 15.4c, some of the electrons leave the sphere and travel to ground. If the wire to ground is then removed (Fig. 15.4d), the conducting sphere is left with an excess of induced positive charge. Finally, when the rubber rod is removed from the vicinity of the sphere (Fig. 15.4e), the induced positive charge remains on the ungrounded sphere. Even though the positively charged atomic nuclei remain fixed, this excess positive charge becomes uniformly distributed over the surface of the ungrounded sphere because of the repulsive forces among the like charges and the high mobility of electrons in a metal.

In the process of inducing a charge on the sphere, the charged rubber rod loses none of its negative charge because it never came in contact with the sphere. Furthermore, the sphere is left with a charge opposite to that of the rubber rod. Note that **charging an object by induction requires no contact with the object inducing the charge.**

A process very similar to charging by induction in conductors also takes place in insulators. In most neutral atoms or molecules, the center of positive charge coincides with the center of negative charge. However, in the presence of a charged object, these centers may separate slightly, resulting in more positive charge on one side of the molecule than on the other. This effect is known as **polarization.** The realignment of charge within individual molecules produces an induced charge on the surface of the insulator, as shown in Figure 15.5a. Knowing about induction in insulators, we can explain why a comb that has been rubbed through hair attracts bits of electrically neutral paper (Fig. 15.5b) and why a balloon that has been rubbed against our clothing is able to stick to an electrically neutral wall.

APPLYING **PHYSICS** 15.1

A positively charged ball hanging from a nonconducting string is brought near a nonconducting object. Based on the behavior of the ball-string combination, the ball is seen to be attracted to the object. From this experiment, it is not possible to determine whether the object is negatively charged or neutral. Why not? What additional experiment would help you decide between these two possibilities?

Explanation The attraction between the ball and the object could be an attraction of unlike charges, or it could be an attraction between a charged object and a neutral object as a result of polarization of the molecules of the neutral object. Two additional experiments help us determine whether the object is charged. First, a known neutral ball is brought near the object—if there is an attraction, the object is negatively charged. Another possibility is to bring a known negatively charged ball near the object—if there is a repulsion, then the object is negatively charged. If there is no attraction, then the object is neutral.

Quick Quiz 15.1

If a suspended object A is attracted to object B, which is charged, we can conclude that (a) object A is uncharged, (b) object A is charged, (c) object B is positively charged, or (d) object A may be either charged or uncharged.

15.3 COULOMB'S LAW

In 1785 Charles Coulomb (1736–1806) established the fundamental law of electric force between two stationary charged particles. Experiments show that

An electric force has the following properties:

1. It is inversely proportional to the square of the separation r between the two particles and is along the line joining them.
2. It is proportional to the product of the magnitudes of the charges $|q_1|$ and $|q_2|$ on the two particles.
3. It is attractive if the charges are of opposite sign and repulsive if the charges have the same sign.

From these observations, we can express the magnitude of the electric force between two charges separated by a distance r as

$$F = k_e \frac{|q_1||q_2|}{r^2} \tag{15.1}$$

where k_e is a constant called the *Coulomb constant*. Note that Equation 15.1, known as **Coulomb's law,** applies only to point charges and to spherical distributions of charges.

The value of the Coulomb constant in Equation 15.1 depends on the choice of units. The SI unit of charge is the **coulomb** (C). From experiment, we know that the **Coulomb constant** in SI units has the value

$$k_e = 8.987\ 5 \times 10^9\ \text{N} \cdot \text{m}^2/\text{C}^2 \tag{15.2}$$

To simplify our calculations, we shall use the approximate value

$$k_e \approx 8.99 \times 10^9\ \text{N} \cdot \text{m}^2/\text{C}^2 \tag{15.3}$$

The charge on the proton has a magnitude of $e = 1.6 \times 10^{-19}$ C. Therefore, it would take $1/e = 6.3 \times 10^{18}$ protons to create a total charge of $+1$ C. Likewise, 6.3×10^{18} electrons would have a total charge of -1 C. Compare this with the number of free electrons in 1 cm³ of copper, which is on the order of 10^{23}. Even so, 1 C is a large amount of charge. In typical electrostatic experiments, where a rubber or glass rod is charged by friction, a net charge on the order of 10^{-6} C ($= 1\ \mu$C) is obtained. In other words, only a very small fraction of the total available charge is transferred between the rod and rubbing material. Table 15.1 lists the charges and masses of the electron, proton, and neutron.

When dealing with Coulomb's force law, remember that force is a vector quantity and must be treated accordingly. Figure 15.6a shows the electric force of repulsion between two positively charged particles. Electric forces obey Newton's third law, and hence the forces $\mathbf{F}_{12}$ and $\mathbf{F}_{21}$ are equal in magnitude but opposite in direction. (The notation $\mathbf{F}_{12}$ denotes the force exerted by particle 1 on particle 2. Likewise, $\mathbf{F}_{21}$ is the force exerted by particle 2 on particle 1.) It bears repeating

CHARLES COULOMB (1736–1806)

Coulomb's major contribution to science was in the field of electrostatics and magnetism. During his lifetime, he also investigated the strengths of materials and determined the forces that affect objects on beams, thereby contributing to the field of structural mechanics. (*Photo courtesy of AIP Niels Bohr Library, E. Scott Barr Collection*)

◀ Coulomb's law

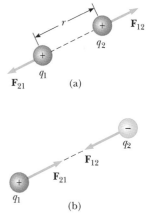

FIGURE 15.6 Two point charges separated by a distance r exert a force on each other given by Coulomb's law. The force on q_1 is equal in magnitude and opposite in direction to the force on q_2. (a) When the charges are of the same sign, the force is repulsive. (b) When the charges are of opposite sign, the force is attractive.

TABLE 15.1	Charge and Mass of the Electron, Proton, and Neutron	
Particle	**Charge (C)**	**Mass (kg)**
Electron	-1.60×10^{-19}	9.11×10^{-31}
Proton	$+1.60 \times 10^{-19}$	1.67×10^{-27}
Neutron	0	1.67×10^{-27}

that F_{12} and F_{21} are always equal regardless of whether q_1 and q_2 have the same magnitude.

The Coulomb force is the second example we have seen of a *field force*—a force exerted by one object on another even though there is no physical contact between them. Recall that another example of a field force is gravitational attraction. The mathematical form of the Coulomb force is the same as that of the gravitational force. That is, they are both inversely proportional to the square of the distance of separation. However, there are some important differences between electric and gravitational forces. Electric forces can be either attractive or repulsive, but gravitational forces are always attractive.

Quick Quiz 15.2	Object A has a charge of $+2$ μC, and object B has a charge of $+6$ μC. Which statement is true: (a) $\mathbf{F}_{AB} = -3\mathbf{F}_{BA}$, (b) $\mathbf{F}_{AB} = -\mathbf{F}_{BA}$, or (c) $3\mathbf{F}_{AB} = -\mathbf{F}_{BA}$

Example 15.1   The Electric Force and the Gravitational Force

The electron and proton of a hydrogen atom are separated (on the average) by a distance of about 5.3×10^{-11} m. Find the magnitudes of the electric force and the gravitational force that each particle exerts on the other.

Solution   From Coulomb's law, we find that the attractive electric force has the magnitude

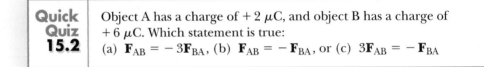

$$F_e = k_e \frac{|e|^2}{r^2} = \left(8.99 \times 10^9 \, \frac{\text{N} \cdot \text{m}^2}{\text{C}^2}\right) \frac{(1.6 \times 10^{-19} \, \text{C})^2}{(5.3 \times 10^{-11} \, \text{m})^2} = \boxed{8.2 \times 10^{-8} \, \text{N}}$$

From Newton's law of universal gravitation and Table 15.1, we find that the gravitational force has the magnitude

$$F_g = G \frac{m_e m_p}{r^2} = \left(6.67 \times 10^{-11} \, \frac{\text{N} \cdot \text{m}^2}{\text{kg}^2}\right) \frac{(9.11 \times 10^{-31} \, \text{kg})(1.67 \times 10^{-27} \, \text{kg})}{(5.3 \times 10^{-11} \, \text{m})^2}$$

$$= \boxed{3.6 \times 10^{-47} \, \text{N}}$$

Because $F_e/F_g \approx 2 \times 10^{39}$, the gravitational force between the charged atomic particles is negligible compared with the electric force.

The Superposition Principle

Frequently, more than two charges are present and it is necessary to find the resultant electric force on one of them. This can be accomplished by noting that the electric force between any pair of charges is given by Equation 15.1. Therefore, the resultant force on any one charge equals the vector sum of the forces exerted by the other individual charges that are present. This is another example of the **superposition principle.** For example, if you have three charges and you want to find the force exerted by charges 2 and 3 on charge 1, you first find the force exerted by charge 2 on charge 1 and the force exerted by charge 3 on charge 1. You then add these two forces together *vectorially* to get the resultant force on charge 1. The following example illustrates this procedure.

Webnote 15.2

You can check your answer to a Coulomb's law calculation at *http://hyperphysics.phy-astr.gsu.edu/hbase/electric.elefor.html#c1*

Example 15.2   Using the Superposition Principle

Consider three point charges at the corners of a triangle, as in Figure 15.7, where $q_1 = 6.00 \times 10^{-9}$ C, $q_3 = 5.00 \times 10^{-9}$ C, $q_2 = -2.00 \times 10^{-9}$ C, and the distances of separation are as shown in the figure. Find the resultant force on q_3.

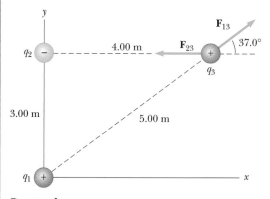

FIGURE 15.7 (Example 15.2) The force exerted by q_1 on q_3 is $\mathbf{F}_{13}$. The force exerted by q_2 on q_3 is $\mathbf{F}_{23}$. The *resultant force* exerted on q_3 is the *vector* sum $\mathbf{F}_{13} + \mathbf{F}_{23}$.

Reasoning   It is first necessary to find the direction of the forces exerted by q_1 and q_2 on q_3. The force $\mathbf{F}_{23}$ exerted by q_2 on q_3 is attractive because q_2 and q_3 have opposite signs. The force $\mathbf{F}_{13}$ exerted by q_1 on q_3 is repulsive because both q_1 and q_3 are positive. To find the resultant force on q_3, it is necessary to find the magnitudes of $\mathbf{F}_{23}$ and $\mathbf{F}_{13}$ by use of Coulomb's law and then add the two forces vectorially.

Solution   The magnitude of the force exerted by q_2 on q_3 is

$$F_{23} = k_e \frac{|q_2||q_3|}{r^2} = (8.99 \times 10^9 \text{ N} \cdot \text{m}^2/\text{C}^2) \frac{(2.00 \times 10^{-9} \text{ C})(5.00 \times 10^{-9} \text{ C})}{(4.00 \text{ m})^2}$$

$$= 5.62 \times 10^{-9} \text{ N}$$

The magnitude of the force exerted by q_1 on q_3 is

$$F_{13} = k_e \frac{|q_1||q_3|}{r^3} = (8.99 \times 10^9 \text{ N} \cdot \text{m}^2/\text{C}^2) \frac{(6.00 \times 10^{-9} \text{ C})(5.00 \times 10^{-9} \text{ C})}{(5.00 \text{ m})^2}$$

$$= 1.08 \times 10^{-8} \text{ N}$$

The force $\mathbf{F}_{13}$ makes an angle of 37.0° with the x axis and is directed along the line connecting q_1 and q_3. Therefore, the x component of this force has the magnitude $F_{13} \cos 37.0° = 8.63 \times 10^{-9}$ N, and the y component of $\mathbf{F}_{13}$ has the magnitude $F_{13} \sin 37.0° = 6.50 \times 10^{-9}$ N. The force $\mathbf{F}_{23}$ is in the negative x direction. Hence, the x and y components of the resultant force on q_3 are

$$F_x = 8.63 \times 10^{-9} \text{ N} - 5.62 \times 10^{-9} \text{ N} = 3.01 \times 10^{-9} \text{ N}$$

$$F_y = 6.50 \times 10^{-9} \text{ N}$$

The magnitude of the resultant force on the charge q_3 is therefore

$$\sqrt{(3.01 \times 10^{-9} \text{ N})^2 + (6.50 \times 10^{-9} \text{ N})^2} = \boxed{7.16 \times 10^{-9} \text{ N}}$$

and the force vector makes an angle with the x axis of

$$\theta = \tan^{-1}\left(\frac{F_y}{F_x}\right) = \tan^{-1}\left(\frac{6.50 \times 10^{-9} \text{ N}}{3.01 \times 10^{-9} \text{ N}}\right) = \boxed{65.2°}$$

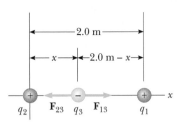

FIGURE 15.8 (Example 15.3)
Three point charges are placed along
the x axis. The charge q_3 is negative,
whereas q_1 and q_2 are positive. If the
resultant force on q_3 is zero, then the
force $\mathbf{F}_{13}$ exerted by q_1 on q_3 must be
equal in magnitude and opposite the
force $\mathbf{F}_{23}$ exerted by q_2 on q_3.

Example 15.3   Where Is the Resultant Force Zero?

Three charges lie along the x axis as in Figure 15.8. The positive charge $q_1 = 15\ \mu C$ is at
$x = 2.0$ m, and the positive charge $q_2 = 6.0\ \mu C$ is at the origin. Where must a *negative*
charge q_3 be placed on the x axis so that the resultant force on it is zero?

Reasoning   Because q_3 is negative and q_1 and q_2 are positive, the forces $\mathbf{F}_{13}$ and $\mathbf{F}_{23}$
are both attractive, as indicated in Figure 15.8. The only location where the force ex-
erted by q_2 on q_3 is opposite the force exerted by q_1 on q_3 lies on the x axis between q_1
and q_2. Because we require that the resultant force on q_3 be zero, then F_{13} must
equal F_{23}.

Solution   If we let x be the coordinate of q_3, then the forces $\mathbf{F}_{13}$ and $\mathbf{F}_{23}$ have the
magnitudes

$$F_{13} = k_e \frac{(15 \times 10^{-6}\ \text{C})|q_3|}{(2.0 - x)^2} \quad \text{and} \quad F_{23} = k_e \frac{(6.0 \times 10^{-6}\ \text{C})|q_3|}{x^2}$$

If the resultant force on q_3 is zero, then $F_{13} = F_{23}$, or

$$k_e \frac{(15 \times 10^{-6}\ \text{C})|q_3|}{(2.0 - x)^2} = k_e \frac{(6.0 \times 10^{-6}\ \text{C})|q_3|}{x^2}$$

Because k_e, 10^{-6}, and q_3 are common to both sides, they can be cancelled from the equation, and we have (after some reduction)

$$(2.0 - x)^2(6.0) = x^2(15)$$

This can be expanded to a quadratic equation, which can then be solved for x. An easier
approach is first to take the positive square root of both sides:

$$(2.0 - x)\sqrt{6.0} = x\sqrt{15}$$

$$2.0 - x = x(1.58)$$

$$x = \boxed{0.78\ \text{m}}$$

15.4   THE ELECTRIC FIELD

Two different field forces have been introduced into our discussions so far: the
gravitational force and the electrostatic force. As pointed out earlier, these forces
are capable of acting through space, producing an effect even when there is no
physical contact between the objects involved. Field forces can be discussed in a variety of ways, but an approach developed by Michael Faraday (1791–1867) is of
such practical value that we shall devote much attention to it in the next few chapters. In this approach, an **electric field** is said to exist in the region of space
around a charged object. When another charged object enters this electric field,
the field is what exerts a force on the second charged object. As an example, consider Figure 15.9, which shows an object with a small positive charge q_0 placed
near a second object with a larger positive charge Q.

We define the electric field at the location of the small "test" charge to be the
electric force acting on it divided by the charge q_0 of the test charge:

$$\mathbf{E} \equiv \frac{\mathbf{F}}{q_0} \qquad\qquad [15.4]$$

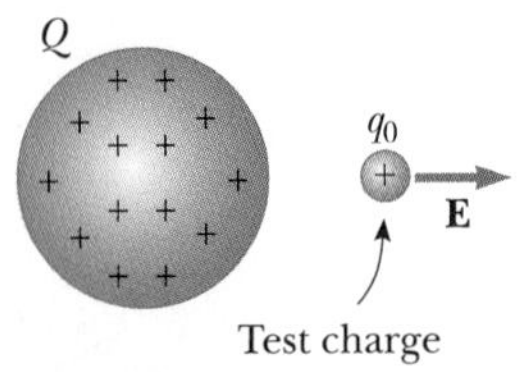

FIGURE 15.9   A small object with a
positive charge q_0 placed near an object with a larger positive charge Q experiences an electric field $\mathbf{E}$ directed
as shown. The magnitude of the elec-
tric field at the location of q_0 is defined as the electric force on q_0 divided by the charge q_0.

Note that this is the electric field at the location of q_0 produced by the charge Q,
not the field produced by q_0. The electric field is a vector quantity having the SI
units newtons per coulomb (N/C). **The direction of E at a point is defined to
be the direction of the electric force that would be exerted on a small posi-**

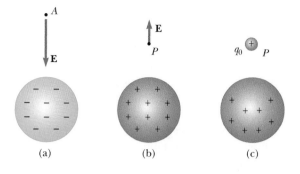

(a) (b) (c)

FIGURE 15.10 (a) The electric field at *A* due to the negatively charged sphere is downward, toward the negative charge. (b) The electric field at *P* due to the positively charged conducting sphere is upward, away from the positive charge. (c) A test charge q_0 placed at *P* will cause a rearrangement of charge on the sphere unless q_0 is very small compared with the charge on the sphere.

tive test charge placed at that point. Thus, in Figure 15.9, the direction of the electric field is horizontal and to the right. The electric field at point *A* in Figure 15.10a is vertical and downward because at this point a positive charge would experience a force of attraction toward the negatively charged sphere. Note that the test charge q_0 is required to be a small charge, in order to cause no significant rearrangement of the charge on the large sphere as shown in Figure 15.10c.

Once the electric field is known at some point, the force on *any* particle with charge *q* placed at that point can be calculated from a rearrangement of Equation 15.4:

$$\mathbf{F} = q\mathbf{E} \qquad [15.5]$$

Consider a point charge *q* located a distance *r* from a test charge q_0. According to Coulomb's law, the *magnitude* of the force on the test charge is

$$F = k_e \frac{|q||q_0|}{r^2}$$

Because the magnitude of the electric field at the position of the test charge is defined as $E = F/q_0$, we see that the *magnitude* of the electric field due to the charge *q* at the position of q_0 is

$$E = k_e \frac{|q|}{r^2} \qquad [15.6]$$

◀ Electric field due to a charge *q*

If *q* is *positive,* as in Figure 15.11a, the field at *P* due to this charge is *radially outward* from *q*. If *q* is *negative,* as in Figure 15.11b, the field at *P* is directed *toward q*. Equation 15.6 points out an important property of electric fields that makes them useful quantities for describing electrical phenomena. As the equation indicates, an electric field at a given point depends only on the charge *q* on the object setting up the field and the distance *r* from that object to a specific point in space. As a result, we can say that an electric field exists at point *P* in Figure 15.11 whether or not there is a test charge at *P*.

The principle of superposition holds when the electric field due to a group of point charges is calculated. We first use Equation 15.6 to calculate the electric field produced by each charge individually at a point, and then add these electric fields together as vectors.

PROBLEM-SOLVING STRATEGY **Electric Forces and Fields**

1. **Units.** When performing calculations that use the Coulomb constant k_e, charges must be in coulombs and distances in meters. If they are given in other units, you must convert them.
2. **Applying Coulomb's law to point charges.** It is important to use the superposition principle properly when dealing with a collection of

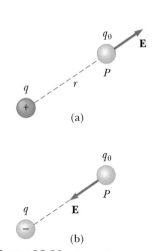

FIGURE 15.11 A test charge q_0 at *P* is a distance *r* from a point charge *q*. (a) If *q* is positive, the electric field at *P* points radially *outward* from *q*. (b) If *q* is negative, the electric field at *P* points radially *inward* toward *q*.

interacting point charges. If several charges are present, the resultant force on any one of them is found by determining the individual force exerted on it by every other charge and then determining the vector sum of all these forces. The force that any charged object exerts on another is given by Coulomb's law, and the direction of the force is found by noting that the forces are repulsive between like charges and attractive between unlike charges.

3. **Calculating the electric field of point charges.** Because the electric field is a vector quantity, the superposition principle can be applied. To find the resultant electric field at a given point due to several charges, first calculate the electric field at the point due to each individual charge. The vector sum of the fields due to all the individual charges is the resultant field at the point.

APPLYING PHYSICS 15.2

An electron moving horizontally passes between two horizontal plates, the upper charged negatively, the lower positively. A uniform, upward-directed electric field exists in this region, and this field exerts an electric force downward on the electron. Describe the movement of the electron in this region.

Explanation The magnitude of the electric force on the electron of charge e due to a uniform electric field $\mathbf{E}$ is $F = eE$. Thus, the force is constant. Compare this to the force on a projectile of mass m moving in the gravitational field of Earth. The magnitude of the gravitational force is mg. In both cases, the particle is subject to a constant force in the vertical direction and has an initial velocity in the horizontal direction. Thus, the path will be the same in each case — the electron will move as a projectile with an acceleration in the vertical direction and constant velocity in the horizontal direction. Once the electron leaves the region between the plates, the electric field disappears, and the electron continues moving in a straight line according to Newton's first law.

Example 15.4 Electric Force on a Proton

Find the electric force on a proton placed in an upward electric field of magnitude 2.0×10^4 N/C.

Solution Because the charge on a proton is $+q = +1.6 \times 10^{-19}$ C, the electric force acting on the proton is

$$F = qE = (1.6 \times 10^{-19} \text{ C})(2.0 \times 10^4 \text{ N/C}) = \boxed{3.2 \times 10^{-15} \text{ N}}$$

where the force is upward, in the positive y direction. The magnitude of the gravitational force acting downward on the proton has the value $m_p g = (1.67 \times 10^{-27} \text{ kg})(9.80 \text{ m/s}^2) = 1.64 \times 10^{-26}$ N. Hence, the magnitude of the gravitational force is negligible compared with that of the electric force.

Example 15.5 Electric Field Due to Two Point Charges

Charge $q_1 = 7.00 \ \mu$C is at the origin, and charge $q_2 = -5.00 \ \mu$C is on the x axis, 0.300 m from the origin (Fig. 15.12). Find the electric field at point P, which has coordinates (0, 0.400) m.

Reasoning   It is first necessary to find the direction of the field at P set up by each charge. The field $\mathbf{E}_1$ at P due to q_1 is vertically upward as in Figure 15.12 because $\mathbf{E}_1$ is in the direction that a positive test charge would move if it were placed at P. Likewise, the field $\mathbf{E}_2$ at P due to q_2 is directed toward q_2 as in Figure 15.12. The magnitudes of the fields can be found from $E = k_e |q|/r^2$ and then added together vectorially.

Solution   The magnitudes of $\mathbf{E}_1$ and $\mathbf{E}_2$ are

$$E_1 = k_e \frac{|q_1|}{r_1{}^2} = (8.99 \times 10^9 \text{ N} \cdot \text{m}^2/\text{C}^2) \frac{(7.00 \times 10^{-6} \text{ C})}{(0.400 \text{ m})^2} = 3.93 \times 10^5 \text{ N/C}$$

$$E_2 = k_e \frac{|q_2|}{r_2{}^2} = (8.99 \times 10^9 \text{ N} \cdot \text{m}^2/\text{C}^2) \frac{(5.00 \times 10^{-6} \text{ C})}{(0.500 \text{ m})^2} = 1.80 \times 10^5 \text{ N/C}$$

The vector $\mathbf{E}_1$ has an x component of zero. The vector $\mathbf{E}_2$ has an x component given by $E_2 \cos\theta = \frac{3}{5}E_2 = 1.08 \times 10^5$ N/C and a negative y component given by $-E_2 \sin\theta = -\frac{4}{5}E_2 = -1.44 \times 10^5$ N/C. Hence, the resultant component in the x direction is

$$E_x = 1.08 \times 10^5 \text{ N/C}$$

and the resultant component in the y direction is

$$E_y = E_{y1} + E_{y2} = 3.93 \times 10^5 \text{ N/C} - 1.44 \times 10^5 \text{ N/C} = 2.49 \times 10^5 \text{ N/C}$$

From the Pythagorean theorem ($E = \sqrt{E_x{}^2 + E_y{}^2}$), we find that $\mathbf{E}$ has a magnitude 2.72×10^5 N/C and makes an angle ϕ with the positive x axis given by

$$\phi = \tan^{-1}\left(\frac{E_y}{E_x}\right) = \tan^{-1}\left(\frac{2.49 \times 10^5 \text{ N/C}}{1.08 \times 10^5 \text{ N/C}}\right) = 66.5°$$

EXERCISE   Find the force on a positive test charge of 2.00×10^{-8} C placed at P.

ANSWER 5.44×10^{-3} N in the same direction as $\mathbf{E}$.

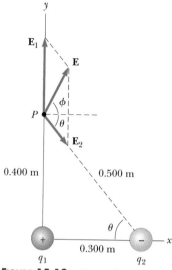

FIGURE 15.12   (Example 15.5) The resultant electric field $\mathbf{E}$ at P equals the vector sum $\mathbf{E}_1 + \mathbf{E}_2$, where $\mathbf{E}_1$ is the field due to the positive charge q_1 and $\mathbf{E}_2$ is the field due to the negative charge q_2.

Quick Quiz 15.3   A test charge of $+3 \ \mu$C is at a point P where the electric field due to other charges is directed to the right and has a magnitude of 4×10^6 N/C. If the test charge is replaced with a charge of $-3 \ \mu$C, the electric field at P (a) has the same magnitude but changes direction, (b) increases in magnitude and changes direction, (c) remains the same, or (d) decreases in magnitude and changes direction.

Quick Quiz 15.4   A Styrofoam ball covered with a conducting paint has a mass of 5.0×10^{-3} kg and a charge of $4.0 \ \mu$C. What electric field directed upward will produce an electric force on the ball that will balance the weight of the ball? (a) 8.2×10^2 N/C   (b) 1.2×10^4 N/C   (c) 2.0×10^{-2} N/C   (d) 5.1×10^6 N/C

Quick Quiz 15.5   A circular ring of radius b has a total charge q uniformly distributed around it. The magnitude of the electric field at the center of the ring is (a) 0, (b) $k_e q/b^2$, (c) $k_e q^2/b^2$, (d) $k_e q^2/b$, or (e) none of these.

Quick Quiz 15.6 A "free" electron and "free" proton are placed in an identical electric field. Which of the following statements are true? (a) Each particle experiences the same electric force and the same acceleration. (b) The electric force on the proton is greater in magnitude than the force on the electron but in the opposite direction. (c) The electric force on the proton is equal in magnitude to the force on the electron, but in the opposite direction. (d) The magnitude of the acceleration of the electron is greater than that of the proton. (e) Both particles experience the same acceleration.

15.5 ELECTRIC FIELD LINES

A convenient aid for visualizing electric field patterns is to draw lines pointing in the direction of the electric field vector at any point. These lines introduced by Michael Faraday—called **electric field lines**—are related to the electric field in any region of space in the following manner:

1. The electric field vector **E** is tangent to the electric field lines at each point.
2. The number of lines per unit area through a surface perpendicular to the lines is proportional to the strength of the electric field in a given region.

Thus, **E** is large when the field lines are close together and small when they are far apart.

Figure 15.13a shows some representative electric field lines for a single positive point charge. Note that this two-dimensional drawing contains only the field lines that lie in the plane containing the point charge. The lines are actually directed radially outward from the charge in *all* directions, somewhat like the quills of an angry porcupine. Because a positive test charge placed in this field would be repelled by the charge *q*, the lines are directed radially away from the positive charge. In a similar way, the electric field lines for a single negative point charge are directed toward the charge (Fig. 15.13b). In either case, the lines are radial and extend all the way to infinity. Note that the lines are closer together as they get near the charge, indicating that the strength of the field is increasing. Equation 15.6 verifies that this should indeed be the case.

The rules for drawing electric field lines for any charge distribution are as follows:

1. The lines for a group of point charges must begin on positive charges and end on negative charges. In the case of an excess of charge, some lines will begin or end infinitely far away.

FIGURE 15.13 The electric field lines for a point charge. (a) For a positive point charge, the lines radiate outward. (b) For a negative point charge, the lines converge inward. Note that the figures show only those field lines that lie in the plane containing the charge. (c) The dark lines are small pieces of thread suspended in oil, which align with the electric field produced by a small charged conductor at the center. *(Photo courtesy of Harold M. Waage, Princeton University)*

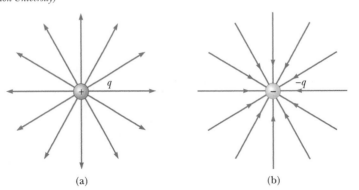

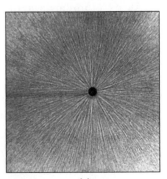

(a)   (b)   (c)

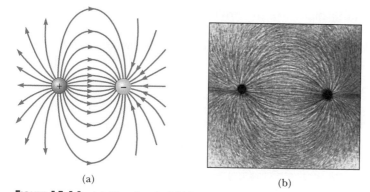

(a)   (b)

FIGURE 15.14   (a) The electric field lines for two equal and opposite point charges (an electric dipole). Note that the number of lines leaving the positive charge equals the number terminating at the negative charge. (b) The dark lines are small pieces of thread suspended in oil, which align with the electric field produced by two charged conductors. *(Photo courtesy of Harold M. Waage, Princeton University)*

2. The number of lines drawn leaving a positive charge or ending on a negative charge is proportional to the magnitude of the charge.

3. No two field lines can cross each other.

Figure 15.14 shows the beautifully symmetric electric field lines for two point charges of equal magnitude but opposite sign. This charge configuration is called an **electric dipole.** In this case, the number of lines that begin at the positive charge must equal the number that terminate at the negative charge. At points very near either charge, the lines are nearly radial. The high density of lines between the charges indicates a strong electric field in this region.

Figure 15.15 shows the electric field lines in the vicinity of two equal positive point charges. Again, close to either charge the lines are nearly radial. The same number of lines emerges from each charge because the charges are equal in magnitude. At great distances from the charges, the field is approximately equal to that of a single point charge of magnitude $2q$. The bulging out of the electric field lines between the charges indicates the repulsive nature of the electric force between like charges. Also, the low density of field lines between the charges indicates a weak field in this region, unlike the dipole.

Finally, Figure 15.16 is a sketch of the electric field lines associated with the positive charge $+2q$ and the negative charge $-q$. In this case, the number of lines leaving charge $+2q$ is twice the number terminating on charge $-q$. Hence, only

Tip 15.1   ELECTRIC FIELD LINES ARE NOT PATHS OF PARTICLES

Electric field lines are *not* material objects. They are used only as a pictorial representation of the electric field at various locations. Except in special cases, they *do not* represent the path of a charged particle released in an electric field.

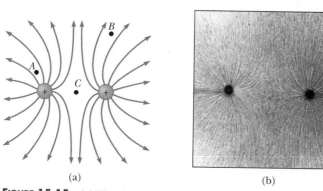

(a)   (b)

FIGURE 15.15   (a) The electric field lines for two positive point charges. The points *A*, *B*, and *C* will be discussed in Quick Quiz 15.7. (b) The dark lines are small pieces of thread suspended in oil, which align with the electric field produced by two charged conductors. *(Photo courtesy of Harold M. Waage, Princeton University)*

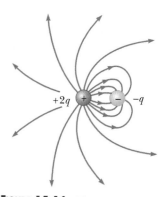

FIGURE 15.16   The electric field lines for a point charge $+2q$ and a second point charge $-q$. Note that two lines leave the charge $+2q$ for every line that terminates on $-q$.

half of the lines that leave the positive charge end at the negative charge. The remaining half terminate on negative charges that we assume to be located at infinity. At great distances from the charges (great compared with the charge separation), the electric field lines are equivalent to those of a single charge $+q$.

Quick Quiz 15.7   Rank the magnitudes of the electric field at points A, B, and C in Figure 15.15, largest magnitude first.

APPLICATION

MEASURING ATMOSPHERIC ELECTRIC FIELDS

APPLYING **PHYSICS** 15.3

The electric field near Earth's surface in fair weather is about 100 N/C downward. Under a thundercloud, the electric field can be very large, on the order of 20 000 N/C. How are these electric fields measured?

Explanation   A device for measuring these fields is called the *field mill*. Figure 15.17 shows the fundamental components of a field mill—two metal plates parallel to the ground. Each plate is connected to ground with a wire, with an ammeter (a low-resistance device for measuring flow of charge, to be discussed in Section 19.6) in one path. Consider first just the lower plate. Because it is connected to ground, and the ground happens to carry a negative charge, the plate is negatively charged. Thus, electric field lines, directed downward, end on the plate, as in Figure 15.17a. Now, imagine that the upper plate is suddenly moved over the lower plate, as in Figure 15.17b. This plate is also connected to ground, and is also negatively charged, so the field lines now end on the upper plate. The negative charges in the lower plate are repelled by those on the upper plate and must pass through the ammeter, registering a flow of charge. The amount of charge that was on the lower plate is related to the strength of the electric field. Thus, the flow of charge through the ammeter can be calibrated to measure the electric field. The plates are normally designed like the blades of a fan, with the upper plate rotating so that the lower plate is alternately covered and uncovered. As a result, charges flow back and forth continually through the ammeter, and the reading can be related to the electric field strength.

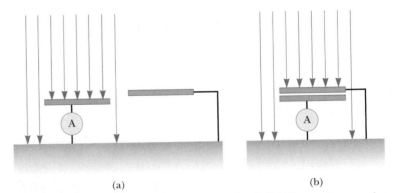

(a)   (b)

FIGURE 15.17   (Applying Physics 15.3) In (a), electric field lines end on negative charges on the lower plate. In (b), the second plate is moved above the lower plate. Electric field lines now end on the upper plate, and the negative charges in the lower plate are repelled through the ammeter.

15.6   CONDUCTORS IN ELECTROSTATIC EQUILIBRIUM

A good electric conductor, such as copper, contains charges (electrons) that are not bound to any atom and are free to move about within the material. When no net motion of charge occurs within a conductor, the conductor is said to be in

electrostatic equilibrium. An isolated conductor (one that is insulated from ground) has the following properties:

1. **The electric field is zero everywhere inside the conducting material.**
2. **Any excess charge on an isolated conductor resides entirely on its surface.**
3. **The electric field just outside a charged conductor is perpendicular to the conductor's surface.**
4. **On an irregularly shaped conductor, the charge accumulates at locations where the radius of curvature of the surface is smallest—that is, at sharp points.**

◀ Properties of an isolated conductor

The first property can be understood by examining what would happen if it were *not* true. If there were an electric field inside a conductor, the free charge there would move and a flow of charge, or current, would be created. However, if there were a net movement of charge, the conductor would no longer be in electrostatic equilibrium.

Property 2 is a direct result of the $1/r^2$ repulsion between like charges described by Coulomb's law. If by some means an excess of charge is placed inside a conductor, the repulsive forces between the like charges push them as far apart as possible, causing them to quickly migrate to the surface. (We do not prove it here, but it is of interest to note that the excess charge resides on the surface due to the fact that Coulomb's law is an inverse-square law. With any other power law, an excess of charge would exist on the surface, but there would be a distribution of charge, of either the same or opposite sign, inside the conductor.)

Property 3 can be understood by again considering what would happen if it were not true. If the electric field in Figure 15.18a were not perpendicular to the surface, the electric field would have a component along the surface, which would cause the free charges of the conductor to move (to the left in the figure). If the charges moved, however, a current would be created and the conductor would no longer be in electrostatic equilibrium. Hence, **E** must be perpendicular to the surface.

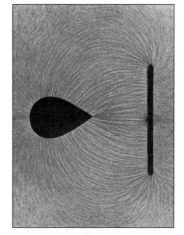

Electric field pattern of a charged conducting plate near an oppositely charged pointed conductor. Small pieces of thread suspended in oil align with the electric field lines. Note that the electric field is most intense near the pointed part of the conductor where the radius of curvature is the smallest. Also, the lines are perpendicular to the conductors. *(Courtesy of Harold M. Waage, Princeton University)*

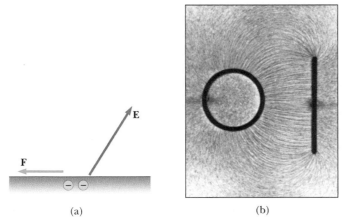

(a) (b)

FIGURE 15.18 (a) Negative charges at the surface of a conductor. If the electric field were at an angle to the surface as shown, an electric force would be exerted on the charges along the surface and they would move to the left. Because the conductor is assumed to be in electrostatic equilibrium, **E** cannot have a component along the surface and hence must be perpendicular to it. (b) The electric field pattern of a charged conducting plate near an oppositely charged conducting cylinder. Small pieces of thread suspended in oil align with the electric field lines. Note that (1) the electric field lines are perpendicular to the conductors and (2) there are no lines inside the cylinder (**E** = 0). *(Courtesy of Harold M. Waage, Princeton University)*

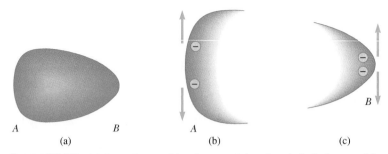

FIGURE 15.19 (a) A conductor with a flatter end A and a relatively sharp end B. Excess charge placed on a conductor resides entirely at its surface and is distributed so that (b) there is less charge per unit area on the flatter end and (c) there is a large charge per unit area on the sharper end.

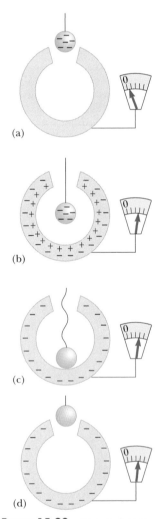

FIGURE 15.20 An experiment showing that any charge transferred to a conductor resides on its surface in electrostatic equilibrium. The hollow conductor is insulated from ground, and the small metal ball is supported by an insulating thread.

To see why property 4 must be true, consider Figure 15.19a, which shows a conductor that is fairly flat at one end and relatively pointed at the other. Any excess charge placed on the object moves to its surface. Figure 15.19b shows the forces between two such charges at the flatter end of the object. These forces are predominantly directed parallel to the surface. Thus, the charges move apart until repulsive forces from other nearby charges create an equilibrium situation. At the sharp end, however, the forces of repulsion between two charges are directed predominantly away from the surface, as in Figure 15.19c. As a result, there is less tendency for the charges to move apart along the surface here, and the amount of charge per unit area is greater than at the flat end. The cumulative effect of many such outward forces from nearby charges at the sharp end produces a large resultant force directed away from the surface that can be great enough to cause charges to leap from the surface into the surrounding air.

Many experiments have shown that the net charge on a conductor resides on its surface. The experiment described here was first performed by Michael Faraday. A metal ball having a negative charge was lowered at the end of a silk thread (an insulator) into an uncharged hollow conductor insulated from ground, as in Figure 15.20a. (This experiment is referred to as **Faraday's ice-pail experiment** because he used a metal ice pail as the hollow conductor.) As the ball was lowered into the pail, the needle on an electrometer attached to the outer surface of the pail was observed to deflect. (An electrometer is a device used to measure charge.) The needle deflected because the charged ball induced a positive charge on the inner wall of the pail, which left an equal negative charge on the outer wall (Fig. 15.20b).

Faraday next touched the inner surface of the pail with the ball and noted that the needle deflection did not change, either when the ball touched the inner surface of the pail (Fig. 15.20c) or when it was removed (Fig. 15.20d). Furthermore, he found that the ball was now completely uncharged. Apparently, when the ball touched the inside of the pail, the excess negative charge on the ball was neutralized by the induced positive charge on the inner surface of the pail. Thus, Faraday discovered the useful result that *all* the excess charge on an object can be transferred to an already charged metal shell if the object is touched to the *inside* of the shell. As we shall shortly see, this is the principle of operation of the Van de Graff generator.

Faraday concluded that because the electrometer deflection did not change when the charged ball touched the inside of the pail, the positive charge induced on the inside surface of the pail was just enough to neutralize the negative charge on the ball. As a result of his investigations, he concluded that a charged object suspended inside a metal container causes a rearrangement of charge on the container in such a manner that the sign of the charge on the inside surface of the container is *opposite* the sign of the charge on the suspended object. This produces a charge on the outside surface of the container of the same sign as that on the suspended object.

The potential difference across the 12.0-μF capacitor (and all other capacitors in this case) is equal to the voltage of the battery, and so

$$Q = C\Delta V = (12.0 \times 10^{-6}\,\text{F})(18.0\,\text{V}) = 216 \times 10^{-6}\,\text{C} = \boxed{216\,\mu\text{C}}$$

SERIES COMBINATION

Now consider two capacitors connected in *series*, as illustrated in Figure 16.18a. **For a series combination of capacitors, the magnitude of the charge must be the same on all the plates.** To see why this must be true, let us consider the charge transfer process in some detail. We start with uncharged capacitors. When a battery is connected to the circuit, electrons are transferred from the left plate of C_1 to the right plate of C_2 through the battery. As this negative charge accumulates on the right plate of C_2, an equivalent amount of negative charge is removed from the left plate of C_2, leaving it with an excess positive charge. The negative charge leaving the left plate of C_2 accumulates on the right plate of C_1, where again an equivalent amount of negative charge is removed from the left plate. The result of this is that **all of the right plates gain charges of $-Q$ and all the left plates have charges of $+Q$.** (This is a consequence of the conservation of charge.)

We can find an equivalent capacitor that performs the same function as the series combination. After it is fully charged, **the equivalent capacitor must end up with a charge of $-Q$ on its right plate and a charge of $+Q$ on its left plate.** By applying the definition of capacitance to the circuit in Figure 16.18b, we have

$$\Delta V = \frac{Q}{C_{\text{eq}}}$$

◀ Q is the same for all capacitors connected in series

where ΔV is the potential difference between the terminals of the battery and C_{eq} is the equivalent capacitance. From Figure 16.18a we see that

$$\Delta V = \Delta V_1 + \Delta V_2 \qquad\qquad \textbf{[16.14]}$$

where ΔV_1 and ΔV_2 are the potential differences across capacitors C_1 and C_2. (This is a consequence of the conservation of energy.) The potential difference across any number of capacitors (or other circuit elements) in series equals the sum of the potential differences across the individual capacitors. Because $Q = C\Delta V$ can be applied to each capacitor, the potential differences across them

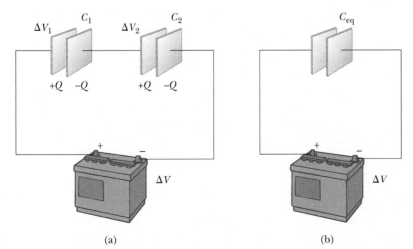

(a)   (b)

FIGURE 16.18 A series combination of two capacitors. The charges on the capacitors are the same, and the equivalent capacitance can be calculated from the reciprocal relationship $1/C_{\text{eq}} = (1/C_1) + (1/C_2)$.

are given by

$$\Delta V_1 = \frac{Q}{C_1} \qquad \Delta V_2 = \frac{Q}{C_2}$$

Substituting these expressions into Equation 16.14, and noting that $\Delta V = Q/C_{eq}$, we have

$$\frac{Q}{C_{eq}} = \frac{Q}{C_1} + \frac{Q}{C_2}$$

Canceling Q, we arrive at the relationship

$$\frac{1}{C_{eq}} = \frac{1}{C_1} + \frac{1}{C_2} \qquad \left(\begin{array}{c} \text{series} \\ \text{combination} \end{array} \right) \qquad \textbf{[16.15]}$$

If this analysis is applied to three or more capacitors connected in series, the equivalent capacitance is found to be

$$\frac{1}{C_{eq}} = \frac{1}{C_1} + \frac{1}{C_2} + \frac{1}{C_3} + \cdots \qquad \left(\begin{array}{c} \text{series} \\ \text{combination} \end{array} \right) \qquad \textbf{[16.16]}$$

As we shall demonstrate in Example 16.6, this implies that **the equivalent capacitance of a series combination is always less than any individual capacitance in the combination.**

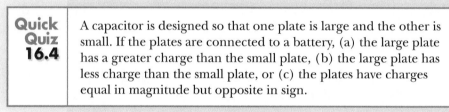

Quick Quiz 16.4

A capacitor is designed so that one plate is large and the other is small. If the plates are connected to a battery, (a) the large plate has a greater charge than the small plate, (b) the large plate has less charge than the small plate, or (c) the plates have charges equal in magnitude but opposite in sign.

PROBLEM-SOLVING STRATEGY   **Capacitors**

1. Be careful with your choice of units. To calculate the capacitance of a device in farads, make sure that distances are in meters and use the SI value of ϵ_0.

2. When two or more unequal capacitors are connected in *series*, they carry the same charge, but the potential differences across them are not the same. Their capacitances add as reciprocals, and the equivalent capacitance of the combination is always *less* than the smallest individual capacitor.

3. When two or more capacitors are connected in *parallel*, the potential differences across them are the same. The charge on each capacitor is proportional to its capacitance; hence, the capacitances add directly to give the equivalent capacitance of the parallel combination.

4. A complicated circuit consisting of capacitors can often be reduced to a simple circuit containing only one capacitor. To do this, examine your initial circuit and replace any capacitors in series or any in parallel with equivalent capacitors, using the rules in 2 and 3. After making these changes, sketch your new circuit. Examine it and replace any series or parallel combinations again. Continue this process until a single equivalent capacitor is found.

5. To find the charge on, or the potential difference across, one of the capacitors in the complicated circuit, start with the final circuit found in 4, and gradually work your way back through the circuits using $C = Q/\Delta V$ and the information in 2 and 3 above.

Example 16.6 Four Capacitors Connected in Series

Four capacitors are connected in series with a battery, as in Figure 16.19.

A Find the capacitance of the equivalent capacitor.

Solution The equivalent capacitance is found from Equation 16.16:

$$\frac{1}{C_{eq}} = \frac{1}{3.0 \ \mu F} + \frac{1}{6.0 \ \mu F} + \frac{1}{12 \ \mu F} + \frac{1}{24 \ \mu F}$$

$$C_{eq} = \boxed{1.6 \ \mu F}$$

Note that the equivalent capacitance is less than the capacitance of any of the individual capacitors in the combination.

B Find the charge on the 12-μF capacitor.

Solution We find the charge on the equivalent capacitor:

$$Q = C_{eq} \Delta V = (1.6 \times 10^{-6} \ \text{F})(18 \ \text{V}) = \boxed{29 \ \mu C}$$

This is also the charge on each of the capacitors it replaced. Thus, the charge on the 12-μF capacitor in the original circuit is 29 μC.

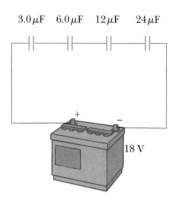

3.0μF 6.0μF 12μF 24μF

18 V

FIGURE 16.19 (Example 16.6)
Four capacitors connected in series.

Example 16.7 Equivalent Capacitance

Find the equivalent capacitance between a and b for the combination of capacitors shown in Figure 16.20a. All capacitances are in microfarads.

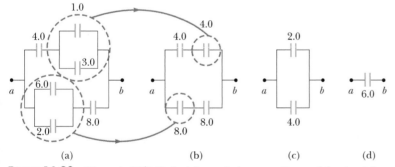

(a)   (b)   (c)   (d)

FIGURE 16.20 (Example 16.7) To find the equivalent capacitance of the circuit in (a), the circuit is reduced in steps—as indicated in (b), (c), and (d)—using the series and parallel rules described in the text.

Solution Using Equations 16.13 and 16.16, we reduce the combination step by step as indicated in the figure. The 1.0-μF and 3.0-μF capacitors are in *parallel* and combine according to $C_{eq} = C_1 + C_2$. Their equivalent capacitance is 4.0 μF. Likewise, the 2.0-μF and 6.0-μF capacitors are also in *parallel* and have an equivalent capacitance of 8.0 μF. The upper branch in Figure 16.20b now consists of two 4.0-μF capacitors in *series*, which combine according to

$$\frac{1}{C_{eq}} = \frac{1}{C_1} + \frac{1}{C_2} = \frac{1}{4.0 \ \mu F} + \frac{1}{4.0 \ \mu F} = \frac{1}{2.0 \ \mu F}$$

$$C_{eq} = 2.0 \ \mu F$$

Likewise, the lower branch in Figure 16.20b consists of two 8.0-μF capacitors in *series* and has an equivalent capacitance of 4.0 μF. Finally, the 2.0-μF and 4.0-μF capacitors in Figure 16.20c are in *parallel* and have an equivalent capacitance of 6.0 μF. Hence, the equivalent capacitance of the circuit is $\boxed{6.0 \ \mu F}$.

16.9   ENERGY STORED IN A CHARGED CAPACITOR

Almost everyone who works with electronic equipment has at some time verified that a capacitor can store energy. If the plates of a charged capacitor are connected by a conductor such as a wire, charge transfers from one plate to the other until the two are uncharged. The discharge can often be observed as a visible spark. If you accidentally touched the opposite plates of a charged capacitor, your fingers would act as a pathway by which the capacitor could discharge, inflicting an electric shock. The degree of shock would depend on the capacitance and voltage applied to the capacitor. Where high voltages and large quantities of charge are present, as in the power supply of a television set, such a shock can be fatal.

If a capacitor is initially uncharged (both plates neutral), so that the plates are at the same potential, almost no work is required to transfer a small amount of charge ΔQ from one plate to the other. However, once this charge has been transferred, a small potential difference $\Delta V = \Delta Q / C$ appears between the plates. Therefore, work must be done to transfer additional charge against this potential difference. We must look at the way the voltage changes at each moment. As more charge is transferred from one plate to the other, the potential difference increases in proportion to the charge. If the potential difference at any instant during the charging process is ΔV, the work required to move more charge ΔQ through this potential difference is $\Delta V \Delta Q$; that is,

$$\Delta W = \Delta V \Delta Q$$

We know that $\Delta V = Q / C$ for a capacitor that has a total charge of Q. Therefore, a plot of voltage versus total charge gives a straight line with a slope of $1/C$, as shown in Figure 16.21. Because the work ΔW is the area of the shaded rectangle, the total work done in charging the capacitor to a final voltage ΔV is the area under the voltage-charge curve, which in this case equals the area under the straight line. Because the area under this line is the area of a triangle (which is one half of the product of the base and height), the total work done is

$$W = \tfrac{1}{2} Q \Delta V \qquad [16.17]$$

Note that this is also the energy stored in the capacitor, because the work required to charge the capacitor equals the energy stored in the capacitor after it is charged. From the definition of capacitance, we find $Q = C \Delta V$; hence, we can express the energy stored as

$$\text{Energy stored} = \tfrac{1}{2} Q \Delta V = \tfrac{1}{2} C (\Delta V)^2 = \frac{Q^2}{2C} \qquad [16.18]$$

For example, the amount of energy stored in a 5.0-μF capacitor when it is connected across a 120-V battery is

$$\text{Energy stored} = \tfrac{1}{2} C (\Delta V)^2 = \tfrac{1}{2} (5.0 \times 10^{-6}\,\text{F})(120\,\text{V})^2 = 3.6 \times 10^{-2}\,\text{J}$$

In practice, there is a limit to the maximum energy (or charge) that can be stored in a capacitor, because electrical breakdown ultimately occurs between the plates of the capacitor at a sufficiently large value of ΔV. For this reason, capacitors are usually labeled with a maximum operating voltage.

Large capacitors can store enough electrical energy to cause severe burns or even death if they are discharged so that the flow of charge can pass through the heart. Under the proper conditions, however, they can be used to sustain life by stopping cardiac fibrillation in heart attack victims. When fibrillation occurs, the heart produces a rapid, irregular pattern of beats. A fast discharge of electrical energy through the heart can return the organ to its normal beat pattern. Emergency medical teams use portable defibrillators that contain batteries capable of charging a capacitor to a high voltage. (The circuitry actually permits the capacitor to be charged to a much higher voltage than the battery.) In this case and others (camera flash units and lasers used for fusion experiments), capacitors serve as

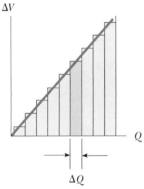

FIGURE 16.21   A plot of voltage versus charge for a capacitor is a straight line with the slope $1/C$. The work required to move a charge of ΔQ through a potential difference of ΔV across the capacitor plates is $\Delta W = \Delta V \Delta Q$, which equals the area of the blue rectangle. The *total work* required to charge the capacitor to a final charge of Q is the area under the straight line, which equals $Q \Delta V / 2$.

APPLICATION

DEFIBRILLATORS

energy reservoirs that can be slowly charged and then discharged quickly to provide large amounts of energy in a short pulse. The stored electrical energy is released through the heart by conducting electrodes, called paddles, that are placed on both sides of the victim's chest. The paramedics must wait between applications of the electrical energy due to the time necessary for the capacitors to become fully charged. The high voltage on the capacitor can be obtained from a low-voltage battery in a portable machine with the phenomenon of *electromagnetic induction,* to be studied in Chapter 20.

Webnote 16.2

Researchers report evidence suggesting that the heart's erratic shivering during cardiac fibrillation is a form of chaos. Read more in
http://www.sciencenews.org/sn_arc97/ 1_25_97/fob1.htm

Example 16.8   Typical Voltage, Energy, and Discharge Time for a Defibrillator

A fully charged defibrillator contains 1.2 kJ of energy stored in a 110-μF capacitor. In a discharge through a patient, 600 J of electrical energy are delivered in 2.5 ms. (This is impossible for a battery, which cannot supply such a large amount of energy in such a short time.)

A Find the voltage needed to store 1.2 kJ in the unit.

Solution   Because we know the energy stored and the capacitance, we can use Equation 16.18 to find the required voltage:

$$\text{Energy stored} = \tfrac{1}{2}C\,\Delta V^2$$

$$\Delta V = \sqrt{\frac{2 \times (\text{energy stored})}{C}} = \sqrt{\frac{2(1.2 \times 10^3 \, \text{J})}{1.1 \times 10^{-4} \, \text{F}}} = \boxed{4.7 \times 10^3 \, \text{V}}$$

B What power is delivered to the patient in watts?

Solution   Because 600 J of energy are delivered in 2.5 ms, we can find the power delivered to the patient as follows:

$$\mathcal{P} = \frac{\text{Energy delivered}}{\Delta t} = \frac{600 \, \text{J}}{2.5 \times 10^{-3} \, \text{s}} = \boxed{2.4 \times 10^5 \, \text{W}}$$

APPLYING PHYSICS 16.2

You have three capacitors and two batteries. How should you connect the batteries to all three capacitors so that the capacitors will store the maximum possible energy?

Explanation   The energy stored in the capacitor is proportional to the capacitance and the square of the potential difference. Thus, we would like to maximize each of these quantities. We can do this by connecting the three capacitors in parallel (so that the capacitances add) across the two batteries in series (so that the potential differences add).

Quick Quiz 16.5   You charge a parallel-plate capacitor, remove it from the battery, and prevent the wires connected to the plates from touching each other. When you pull the plates farther apart, do the following quantities increase, decrease, or stay the same? (a) C, (b) Q, (c) E between the plates, (d) ΔV, (e) energy stored in the capacitor.

16.10   CAPACITORS WITH DIELECTRICS

A **dielectric** is an insulating material, such as rubber, plastic, or waxed paper. When a dielectric is inserted between the plates of a capacitor, the capacitance increases. If the dielectric completely fills the space between the plates, the capacitance is multiplied by the factor κ, called the **dielectric constant.**

The following experiment illustrates the effect of a dielectric in a capacitor. Consider a parallel-plate capacitor of charge Q_0 and capacitance C_0 in the absence of a dielectric. The potential difference across the capacitor plates can be measured, and it is given by $\Delta V_0 = Q_0/C_0$ (Fig. 16.22a). Because the capacitor is not connected to an external circuit, there is no pathway for charge to leave or be added to the plates. If a dielectric is now inserted between the plates as in Figure 16.22b, it is found that the voltage across the plates is *reduced* by the factor κ to the value ΔV, where

$$\Delta V = \frac{\Delta V_0}{\kappa}$$

Because $\kappa > 1$, ΔV is less than ΔV_0. Because the charge Q_0 on the capacitor does not change, we conclude that the capacitance C in the presence of the dielectric must change to the value

$$C = \frac{Q_0}{\Delta V} = \frac{Q_0}{\Delta V_0/\kappa} = \frac{\kappa Q_0}{\Delta V_0}$$

or

$$C = \kappa C_0 \qquad\qquad \text{[16.19]}$$

According to this result, the capacitance is *multiplied* by the factor κ when the dielectric completely fills the region between the plates. For a parallel-plate capacitor, where the capacitance in the absence of a dielectric is $C_0 = \epsilon_0 A/d$, we can express the capacitance in the presence of a dielectric as

$$C = \kappa \epsilon_0 \frac{A}{d} \qquad\qquad \text{[16.20]}$$

From this result it appears that the capacitance could be made very large by decreasing d, the plate separation. In practice, the lowest value of d is limited by the electric discharge that can occur through the dielectric material separating the plates. For any given plate separation, there is a maximum electric field that can be produced in the dielectric before it breaks down and begins to conduct. This maximum electric field is called the **dielectric strength,** and for air its value is

FIGURE 16.22   (a) With air between the plates, the voltage across the capacitor is ΔV_0, the capacitance is C_0, and the charge is Q_0. (b) With a dielectric between the plates, the charge remains at Q_0, but the voltage and capacitance both change.

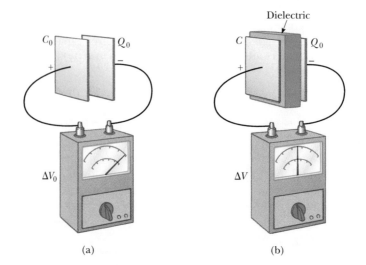

(a)   (b)

TABLE 16.1	Dielectric Constants and Dielectric Strengths of Various Materials at Room Temperature	
Material	**Dielectric Constant, κ**	**Dielectric Strength (V/m)**
Vacuum	1.000 00	—
Air	1.000 59	3×10^6
Bakelite	4.9	24×10^6
Fused quartz	3.78	8×10^6
Pyrex glass	5.6	14×10^6
Polystyrene	2.56	24×10^6
Teflon	2.1	60×10^6
Neoprene rubber	6.7	12×10^6
Nylon	3.4	14×10^6
Paper	3.7	16×10^6
Strontium titanate	233	8×10^6
Water	80	—
Silicone oil	2.5	15×10^6

Dielectric breakdown in air. Sparks are produced when a large alternating voltage is applied across the wires using a high-voltage induction coil power supply. *(© Loren Winters/Visuals Unlimited)*

about 3×10^6 V/m. Most insulating materials have dielectric strengths greater than that of air, as indicated by the values in Table 16.1.

Commercial capacitors are often made using metal foil interlaced with thin sheets of paraffin-impregnated paper or mylar, which serves as the dielectric material. These alternating layers of metal foil and dielectric are then rolled into a small cylinder (Fig. 16.23a). A high-voltage capacitor commonly consists of a number of interwoven metal plates immersed in silicone oil (Fig. 16.23b). Small capacitors are often constructed from ceramic materials. Variable capacitors (typically 10–500 pF) usually consist of two interwoven sets of metal plates, one fixed and the other movable, with air as the dielectric.

An electrolytic capacitor (Fig. 16.23c) is often used to store large amounts of charge at relatively low voltages. It consists of a metal foil in contact with an electrolyte—a solution that conducts charge by virtue of the motion of the ions contained in it. When a voltage is applied between the foil and the electrolyte, a thin layer of metal oxide (an insulator) is formed on the foil, and this layer serves as the dielectric. Enormous capacitances can be attained because the dielectric layer is very thin.

When electrolytic capacitors are used in circuits, the polarity (the plus and minus signs on the device) must be observed. If the polarity of the applied voltage is opposite that intended, the oxide layer will be removed and the capacitor will conduct rather than store charge. Furthermore, reversing the polarity can result in such a large current that the capacitor may either burn or produce steam and explode.

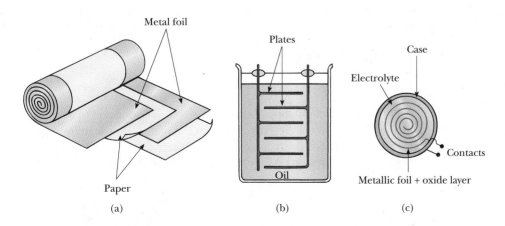

FIGURE 16.23 Three commercial capacitor designs. (a) A tubular capacitor whose plates are separated by paper and then rolled into a cylinder, (b) a high-voltage capacitor consisting of many parallel plates separated by oil, and (c) an electrolytic capacitor.

Metal foil

Paper

Plates

Oil

Case

Electrolyte

Contacts

Metallic foil + oxide layer

(a)　　　　(b)　　　　(c)

A collection of capacitors used in a variety of applications. *(Paul Silverman/Fundamental Photographs)*

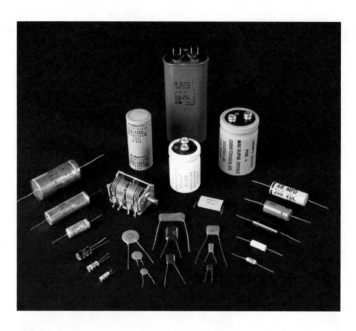

APPLYING **PHYSICS** **16.3**

APPLICATION

STUD FINDERS

If you have ever tried to hang a picture on a wall securely, you know that it can be difficult to locate a wooden stud in which to anchor your nail or screw. The principles discussed in this section can be used to detect a stud electronically. The primary element of an electronic stud finder is a capacitor with its plates arranged side by side instead of facing one another, as in Figure 16.24. How does this device work?

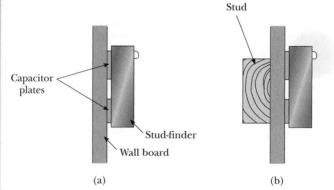

(a)   (b)

FIGURE 16.24   (Applying Physics 16.3) A stud finder. (a) The materials between the plates of the capacitor are the drywall and the air behind it. (b) The materials become drywall and wood when the detector moves across a stud in the wall. The change in the dielectric constant causes a signal light to illuminate.

Explanation   As the detector is moved along a wall, its capacitance changes when it passes across a stud because the dielectric constant of the material "between" the plates changes. The change in capacitance can be used to cause a light to come on, signaling the presence of the stud.

Webnote 16.3

At home you might have a touch-sensitive lamp (turns on when you touch it). This device elegantly makes use of capacitance. Visit *http://www.howstuffworks.com/question42.htm*

Quick Quiz 16.6

A fully charged parallel-plate capacitor remains connected to a battery while you slide a dielectric between the plates. Do the following quantities increase, decrease, or stay the same? (a) C, (b) Q, (c) E between the plates, (d) ΔV, (e) energy stored in the capacitor.

Example 16.9 A Paper-Filled Capacitor

A parallel-plate capacitor has plates 2.0 cm by 3.0 cm. The plates are separated by a 1.0-mm thickness of paper.

A Find the capacitance of this device.

Solution   Because $\kappa = 3.7$ for paper (Table 16.1), we get

$$C = \kappa\epsilon_0 \frac{A}{d} = 3.7\left(8.85 \times 10^{-12}\,\frac{C^2}{N\cdot m^2}\right)\left(\frac{6.0 \times 10^{-4}\,m^2}{1.0 \times 10^{-3}\,m}\right)$$

$$= 20 \times 10^{-12}\,F = \boxed{20\ pF}$$

B Find the maximum charge that can be placed on the capacitor.

Solution   From Table 16.1 we see that the dielectric strength of paper is equal to 16×10^6 V/m. Because the paper thickness is 1.0 mm, the maximum voltage that can be applied before electrical breakdown occurs can be calculated using Equation 16.4:

$$\Delta V_{max} = E_{max}d = (16 \times 10^6\,V/m)(1.0 \times 10^{-3}\,m) = 16 \times 10^3\,V$$

Hence, the maximum charge that can be placed on the capacitor is

$$Q_{max} = C\Delta V_{max} = (20 \times 10^{-12}\,F)(16 \times 10^3\,V) = \boxed{0.32\ \mu C}$$

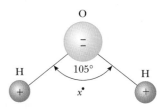

FIGURE 16.25   The water molecule, H_2O, has a permanent polarization resulting from its bent geometry. The point labeled x is the center of positive charge.

AN ATOMIC DESCRIPTION OF DIELECTRICS

The explanation of why a dielectric increases the capacitance of a capacitor is based on an atomic description of the material, which in turn involves a property of some molecules called **polarization.** A molecule is said to be polarized when there is a separation between the "centers of gravity" of its negative charge and its positive charge. In some molecules, such as water, this condition is always present. To see why, consider the geometry of a water molecule (Fig. 16.25).

The molecule is arranged so that the negative oxygen atom is bonded to the positively charged hydrogen atoms with a 105° angle between the two bonds. The center of negative charge is at the oxygen atom, and the center of positive charge lies at a point midway along the line joining the hydrogen atoms (point x in the diagram). Materials composed of molecules that are permanently polarized in this fashion have large dielectric constants, and indeed Table 16.1 shows that the dielectric constant of water is quite large ($\kappa = 80$).

A symmetric molecule (Fig. 16.26a) can have no permanent polarization, but a polarization can be induced by an external electric field. A field directed to the left, as in Figure 16.26b, would cause the center of positive charge to shift to the left from its initial position, and the center of negative charge to shift to the right. This *induced polarization* is the effect that predominates in most materials used as dielectrics in capacitors.

To understand why the polarization of a dielectric can affect capacitance, consider Figure 16.27, which shows a slab of dielectric placed between the plates of a parallel-plate capacitor. The dielectric becomes polarized as shown because it is in the electric field that exists between the metal plates. Notice that a net positive charge appears on the dielectric surface adjacent to the negatively charged metal plate. The presence of this positive charge on the dielectric effectively reduces some of the negative charge on the metal, allowing more negative charge to be stored on the capacitor plates for a given applied voltage. From the definition of capacitance, $C = Q\Delta V$, we see that, because the plates can store more charge for a given voltage, the capacitance must increase.

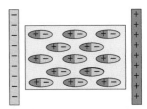

FIGURE 16.26   (a) A symmetric molecule has no permanent polarization. (b) An external electric field induces a polarization in the molecule.

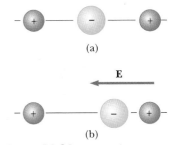

FIGURE 16.27   When a dielectric is placed between the plates of a charged parallel-plate capacitor, the dielectric becomes polarized. This creates a net positive induced charge on the left side of the dielectric and a net negative induced charge on the right side. As a result, the capacitance of the device is multiplied by the factor κ.

APPLYING PHYSICS 16.4

Consider a parallel-plate capacitor with a dielectric material between the plates. Is the capacitance higher on a cold day or a hot day?

Explanation   The polarization of the molecules in the dielectric increases the capacitance when the dielectric is added. As the temperature increases, there is more vibrational motion of the polarized molecules. This disturbs the orderly arrangement of the polarized molecules, and the net polarization decreases. Thus, the capacitance must decrease as the temperature increases.

SUMMARY

The **difference in electric potential** ΔV between two points A and B is

$$\Delta V \equiv V_B - V_A = \frac{\Delta PE}{q} \qquad \text{[16.2]}$$

where ΔPE is the *change* in electrical potential energy experienced by a charge q as it moves between A and B. The units of potential difference are joules per coulomb, or **volts;** $1\,\text{J/C} = 1\,\text{V}$.

The **electric potential difference** between two points A and B in a uniform electric field $\mathbf{E}$ is

$$V_B - V_A = -Ed \qquad \text{[16.4]}$$

where d is the distance between A and B and E is the strength of the electric field in that region.

The **electric potential** due to a point charge q at distance r from the point charge is

$$V = k_e \frac{q}{r} \qquad \text{[16.5]}$$

The **electrical potential energy** of a pair of point charges separated by distance r is

$$PE = k_e \frac{q_1 q_2}{r} \qquad \text{[16.6]}$$

Every point on the surface of a charged conductor in electrostatic equilibrium is at the same potential. Furthermore, the potential is constant everywhere inside the conductor and equals its value on the surface.

The **electron volt** is defined as the energy that an electron (or proton) gains when accelerated through a potential difference of 1 V. The conversion between electron volts and joules is

$$1\,\text{eV} = 1.60 \times 10^{-19}\,\text{C}\cdot\text{V} = 1.60 \times 10^{-19}\,\text{J} \qquad \text{[16.8]}$$

A **capacitor** consists of two metal plates with charges that are equal in magnitude but opposite in sign. The capacitance (C) of any capacitor is the ratio of the magnitude of the charge Q on either plate to the potential difference ΔV between them:

$$C \equiv \frac{Q}{\Delta V} \qquad \text{[16.9]}$$

Capacitance has the units coulombs per volt, or farads; $1\,\text{C/V} = 1\,\text{F}$.

The capacitance of two parallel metal plates of area A separated by distance d is

$$C = \epsilon_0 \frac{A}{d} \qquad \text{[16.10]}$$

where ϵ_0 is a constant called the **permittivity of free space,** with the value $\epsilon_0 = 8.85 \times 10^{-12}\,\text{C}^2/\text{N}\cdot\text{m}^2$.

The **equivalent capacitance of a parallel combination** of capacitors is

$$C_{eq} = C_1 + C_2 + C_3 + \cdots \qquad \text{[16.13]}$$

If two or more capacitors are connected in series, the **equivalent capacitance of the series combination** is

$$\frac{1}{C_{eq}} = \frac{1}{C_1} + \frac{1}{C_2} + \frac{1}{C_3} + \cdots \qquad \text{[16.16]}$$

Three equivalent expressions for calculating the **energy stored** in a charged capacitor are

$$\text{Energy stored} = \tfrac{1}{2}Q\,\Delta V = \tfrac{1}{2}C(\Delta V)^2 = \frac{Q^2}{2C} \qquad \text{[16.18]}$$

When a nonconducting material, called a **dielectric,** is placed between the plates of a capacitor, the capacitance is multiplied by the factor κ, which is called the **dielectric constant** and is a property of the dielectric material. The capacitance of a parallel-plate capacitor filled with a dielectric is

$$C = \kappa \epsilon_0 \frac{A}{d} \qquad \text{[16.20]}$$

CONCEPTUAL QUESTIONS

1. Criticize this statement: "There is no potential in a region where the electric field is uniform."

2. "The equipotentials of a uniformly charged conducting sphere are spheres." Is this statement true? Discuss.

3. A parallel-plate capacitor is charged by a battery, and the battery is then disconnected from the capacitor. Because the charges on the capacitor plates are opposite in sign, they attract each other. Hence, it takes positive work to increase the plate separation. Show that the external work done when the plate separation is increased leads to an increase in the energy stored in the capacitor.

4. Distinguish between electric potential and electrical potential energy.

5. Suppose you are sitting in a car and a 20-kV power line drops across the car. Should you stay in the car or get out? The power line potential is 20 kV compared to the potential of the ground.

6. Why is it important to avoid sharp edges or points on conductors used in high-voltage equipment?

7. Explain why, under static conditions, all points in a conductor must be at the same electric potential.

8. If you are given three different capacitors C_1, C_2, and C_3, how many different combinations of capacitance can you produce, using all capacitors in your circuits?

9. Why is it dangerous to touch the terminals of a high-voltage capacitor even after the voltage source that charged the battery is disconnected from the capacitor? What can be done to make the capacitor safe to handle after the voltage source has been removed?

10. The plates of a capacitor are connected to a battery. What happens to the charge on the plates if the connecting wires are removed from the battery? What happens to the charge if the wires are removed from the battery and connected to each other?

11. Can electric field lines ever cross? Why or why not? Can equipotentials ever cross? Why or why not?

12. What happens to the energy stored in a capacitor (a) if the potential difference across the plates is doubled? (b) if the plate separation is doubled while the capacitor is connected to a battery?

13. (a) What happens to the *charge* stored in a capacitor if the potential difference across the plates is doubled? (b) What happens to the *energy* stored in a capacitor if the plate separation is doubled while the charge on the capacitor remains constant (as it would after being disconnected from a battery)?

14. Is it always possible to reduce a combination of capacitors to one equivalent capacitor with the rules developed in this chapter? Explain.

15. If you were asked to design a capacitor where small size and large capacitance were required, what factors would be important in your design?

16. Explain why a dielectric increases the maximum operating voltage of a capacitor although the physical size of the capacitor does not change.

17. What happens to the energy stored in a capacitor (a) if a dielectric is inserted into the capacitor while it is connected to a battery? (b) if a dielectric is inserted into the capacitor while the charge on the capacitor remains constant (as it would after being disconnected from a battery)?

PROBLEMS

1, 2, 3 = straightforward, intermediate, challenging ☐ = full solution available in Student Solutions Manual/Study Guide

web = solution posted at **http://info.brookscole.com/serway** 🖼 = biomedical application

Section 16.1 Potential Difference and Electric Potential

1. A proton moves 2.00 cm parallel to a uniform electric field with $E = 200$ N/C. (a) How much work is done by the field on the proton? (b) What change occurs in the potential energy of the proton? (c) Through what potential difference did the proton move?

2. A uniform electric field of magnitude 250 V/m is directed in the positive *x* direction. A positive 12-μC charge moves from the origin to the point $(x, y) = (20$ cm, 50 cm). (a) What is the change in the potential energy of this charge? (b) Through what potential difference does the charge move?

3. A potential difference of 90 mV exists between the inner and outer surfaces of a cell membrane. The inner surface is negative relative to the outer surface. How much work is required to eject a positive sodium ion (Na$^+$) from the interior of the cell?

4. An ion accelerated through a potential difference of 115 V experiences an increase in kinetic energy of 7.37×10^{-17} J. Calculate the charge on the ion.

5. The difference in potential between the accelerating plates of a TV set is about 25 kV. If the distance between these plates is 1.5 cm, find the magnitude of the uniform electric field in this region.

6. To recharge a 12-V battery, a battery charger must move 3.6×10^5 C of charge from the negative terminal to the positive terminal. How much work is done by the battery charger? Express your answer in joules.

7. A pair of oppositely charged parallel plates are separated
web by 5.33 mm. A potential difference of 600 V exists between the plates. (a) What is the magnitude of the electric field between the plates? (b) What is the magnitude of the force on an electron between the plates? (c) How much work must be done on the electron to move it to the negative plate if it is initially positioned 2.90 mm from the positive plate?

8. Suppose an electron is released from rest in a uniform electric field whose strength is 5.90×10^3 V/m. (a) Through what potential difference will it have passed after moving 1.00 cm? (b) How fast will the electron be moving after it has traveled 1.00 cm?

9. A 4.00-kg block carrying a charge $Q = 50.0$ μC is connected to a spring for which $k = 100$ N/m. The block lies on a frictionless horizontal track, and the system is immersed in a uniform electric field of magnitude $E = 5.00 \times 10^5$ V/m directed as in Figure P16.9. (a) If the block is released at rest when the spring is unstretched (at

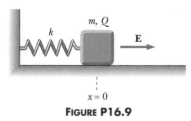

$x = 0$

FIGURE P16.9

$x = 0$), by what maximum amount does the spring expand? (b) What is the equilibrium position of the block?

10. On planet Tehar, the free-fall acceleration is the same as that on Earth, but there is also a strong downward electric field that is uniform close to the planet's surface. A 2.00-kg ball having a charge of 5.00 μC is thrown upward at a speed of 20.1 m/s, and it hits the ground after an interval of 4.10 s. What is the potential difference between the starting point and the top point of the trajectory?

Section 16.2 Electric Potential and Potential Energy Due to Point Charges

Section 16.3 Potentials and Charged Conductors

Section 16.4 Equipotential Surfaces

11. (a) Find the potential 1.00 cm from a proton. (b) What is the potential difference between two points that are 1.00 cm and 2.00 cm from a proton?

12. Find the potential at point *P* for the rectangular grouping of charges shown in Figure P16.12.

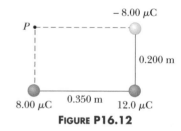

FIGURE P16.12

13. (a) Find the electric potential, taking zero at infinity, at the upper-right corner (the corner without a charge) of the rectangle in Figure P16.13. (b) Repeat if the 2.00-μC charge is replaced with a charge of -2.00 μC.

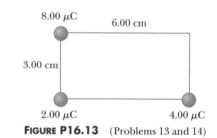

FIGURE P16.13 (Problems 13 and 14)

14. Three charges are situated at corners of a rectangle as in Figure P16.13. How much energy would be expended in moving the 8.00-μC charge to infinity?

15. Two point charges $Q_1 = +5.00$ nC and $Q_2 = -3.00$ nC are separated by 35.0 cm. (a) What is the electric potential at a point midway between the charges? (b) What is the potential energy of the pair of charges? What is the significance of the algebraic sign of your answer?

16. Calculate the speed of (a) an electron that has a kinetic energy of 1.00 eV and (b) a proton that has a kinetic energy of 1.00 eV.

17. The three charges in Figure P16.17 are at the vertices of an isosceles triangle. Calculate the electric potential at the midpoint of the base, taking $q = 7.00$ nC.

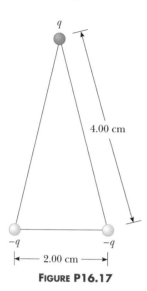

FIGURE P16.17

18. An electron starts from rest 3.00 cm from the center of a uniformly charged sphere of radius 2.00 cm. If the sphere carries a total charge of 1.00×10^{-9} C, how fast will the electron be moving when it reaches the surface of the sphere?

19. In Rutherford's famous scattering experiments (which led **web** to the planetary model of the atom), alpha particles (having charges of $+2e$ and masses of 6.64×10^{-27} kg) were fired toward a gold nucleus with charge $+79e$. An alpha particle, initially very far from the gold nucleus, is fired at 2.00×10^{7} m/s directly toward the gold nucleus as in Figure P16.19. How close does the alpha particle get to the gold nucleus before turning around? Assume the gold nucleus remains stationary.

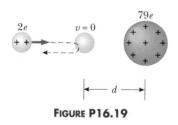

FIGURE P16.19

20. Starting with the definition of work, prove that at every point on an equipotential surface the surface must be perpendicular to the local electric field.

21. A small spherical object carries a charge of 8.00 nC. At what distance from the center of the object is the potential equal to 100 V? 50.0 V? 25.0 V? Is the spacing of the equipotentials proportional to the change in potential?

Section 16.6   Capacitance

Section 16.7   The Parallel-Plate Capacitor

22. (a) How much charge is on each plate of a 4.00-μF capacitor when it is connected to a 12.0-V battery? (b) If this same capacitor is connected to a 1.50-V battery, what charge is stored?

23. Consider Earth and a cloud layer 800 m above Earth to be the plates of a parallel-plate capacitor. (a) If the cloud layer has an area of 1.0 km^2 = 1.0×10^{6} m^2, what is the capacitance? (b) If an electric field strength greater than 3.0×10^{6} N/C causes the air to break down and conduct charge (lightning), what is the maximum charge the cloud can hold?

24. The potential difference between a pair of oppositely charged parallel plates is 400 V. (a) If the spacing between the plates is doubled without altering the charge on the plates, what is the new potential difference between the plates? (b) If the plate spacing is doubled while the potential difference between the plates is kept constant, what is the ratio of the final charge on one of the plates to the original charge?

25. An air-filled capacitor consists of two parallel plates, each with an area of 7.60 cm^2, separated by a distance of 1.80 mm. If a 20.0-V potential difference is applied to these plates, calculate (a) the electric field between the plates, (b) the capacitance, and (c) the charge on each plate.

26. A 1-megabit computer memory chip contains many 60.0×10^{-15} F capacitors. Each capacitor has a plate area of 21.0×10^{-12} m^2. Determine the plate separation of such a capacitor (assume a parallel-plate configuration). The diameter of an atom is on the order of 10^{-10} m = 1 Å. Express the plate separation in angstroms.

27. The plates of a parallel-plate capacitor are separated by 0.100 mm. If the material between the plates is air, what plate area is required to provide a capacitance of 2.00 pF?

28. A small object with a mass of 350 mg carries a charge of 30.0 nC and is suspended by a thread between the vertical plates of a parallel-plate capacitor. The plates are separated by 4.00 cm. If the thread makes an angle of 15.0° with the vertical, what is the potential difference between the plates?

Section 16.8   Combinations of Capacitors

29. A series circuit consists of a 0.050-μF capacitor, a 0.100-μF capacitor, and a 400-V battery. Find the charge (a) on each of the capacitors and (b) on each of the capacitors if they are reconnected in parallel across the battery.

30. Three capacitors $C_1 = 5.00$ μF, $C_2 = 4.00$ μF, and $C_3 = 9.00$ μF are connected together. Find the effective capacitance of the group (a) if they are all in parallel, and (b) if they are all in series.

31. (a) Find the equivalent capacitance of the group of capacitors in Figure P16.31. (b) Find the charge on and the potential difference across each.

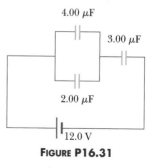

FIGURE P16.31

32. Two capacitors when connected in parallel give an equivalent capacitance of 9.00 pF and give an equivalent capacitance of 2.00 pF when connected in series. What is the capacitance of each capacitor?

33. Four capacitors are connected as shown in Figure P16.33. (a) Find the equivalent capacitance between points *a* and *b*. (b) Calculate the charge on each capacitor if a 15.0-V battery is connected across points *a* and *b*.

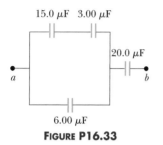

FIGURE P16.33

34. Consider the combination of capacitors in Figure P16.34. (a) What is the equivalent capacitance of the group? (b) Determine the charge on each capacitor.

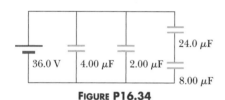

FIGURE P16.34

35. Find the charge on each of the capacitors in Figure P16.35.

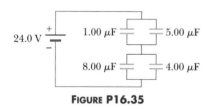

FIGURE P16.35

36. To repair a power supply for a stereo amplifier, an electronics technician needs a 100-μF capacitor capable of withstanding a potential difference of 90 V between the plates. The only available supply is a box of five 100-μF capacitors, each having a maximum voltage capability of 50 V. Can the technician substitute a combination of these capacitors that has the proper electrical characteristics, and if so, what will be the maximum voltage across any of the capacitors used? (*Hint:* The technician may not have to use all the capacitors in the box.)

37. A 25.0-μF capacitor and a 40.0-μF capacitor are charged web by being connected across separate 50.0-V batteries. (a) Determine the resulting charge on each capacitor. (b) The capacitors are then disconnected from their batteries and connected to each other, with each negative plate connected to the other positive plate. What is the final charge of each capacitor, and what is the final potential difference across the 40.0-μF capacitor?

38. A 10.0-μF capacitor is fully charged across a 12.0-V battery. The capacitor is then disconnected from the battery and connected across an initially uncharged capacitor, *C*. The resulting voltage across each capacitor is 3.00 V. What is the capacitance *C*?

39. A 1.00-μF capacitor is first charged by being connected across a 10.0-V battery. It is then disconnected from the battery and connected across an uncharged 2.00-μF capacitor. Determine the resulting charge on each capacitor.

40. Find the equivalent capacitance between points *a* and *b* for the group of capacitors connected as shown in Figure P16.40 if $C_1 = 5.00$ μF, $C_2 = 10.0$ μF, and $C_3 = 2.00$ μF.

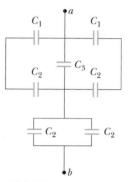

FIGURE P16.40 (Problems 40 and 41)

41. For the network described in the previous problem if the potential between points *a* and *b* is 60.0 V, what charge is stored on C_3?

42. Find the equivalent capacitance between points *a* and *b* in the combination of capacitors shown in Figure P16.42.

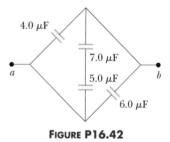

FIGURE P16.42

Section 16.9 Energy Stored in a Charged Capacitor

43. A parallel-plate capacitor has 2.00-cm² plates that are separated by 5.00 mm with air between them. If a 12.0-V battery is connected to this capacitor, how much energy does it store?

44. Two capacitors $C_1 = 25.0$ μF and $C_2 = 5.00$ μF are connected in parallel and charged with a 100-V power supply. (a) Calculate the total energy stored in the two capacitors. (b) What potential difference would be required across the same two capacitors connected in *series* in order that the combination store the same energy as in (a)?

45. Consider the parallel-plate capacitor formed by Earth and a cloud layer as described in Problem 16.23. Assume this capacitor will discharge (that is, lightning occurs) when the electric field strength between the plates reaches 3.0×10^6 N/C. What is the energy released if the capacitor discharges completely during a lightning strike?

46. A certain storm cloud has a potential difference of 1.00×10^8 V relative to a tree. If, during a lightning storm, 50.0 C of charge is transferred through this potential difference and 1.00% of the energy is absorbed by the tree, how much water (sap in the tree) initially at 30.0°C can be boiled away? Water has a specific heat of 4 186 J/kg · °C, a boiling point of 100°C, and a heat of vaporization of 2.26×10^6 J/kg.

Section 16.10   Capacitors with Dielectrics

47. A capacitor with air between its plates is charged to 100 V and then disconnected from the battery. When a piece of glass is placed between the plates, the voltage across the capacitor drops to 25 V. What is the dielectric constant of this glass? (Assume the glass completely fills the space between the plates.)

48. Two parallel plates, each of area 2.00 cm², are separated by 2.00 mm with purified nonconducting water between them. A voltage of 6.00 V is applied between the plates. Calculate (a) the magnitude of the electric field between the plates, (b) the charge stored on each plate, and (c) the charge stored on each plate if the water is removed and replaced with air.

49. Determine (a) the capacitance and (b) the maximum voltage that can be applied to a Teflon-filled parallel-plate capacitor having a plate area of 175 cm² and insulation thickness of 0.040 0 mm.

50. A commercial capacitor is constructed as in Figure 16.23a. This particular capacitor is made from two strips of aluminum separated by a strip of paraffin-coated paper. Each strip of foil and paper is 7.00 cm wide. The foil is 0.004 00 mm thick, and the paper is 0.025 0 mm thick and has a dielectric constant of 3.70. What length should the strips be if a capacitance of 9.50×10^{-8} F is desired before the capacitor is rolled up? (Use the parallel-plate formula. Adding a second strip of paper and rolling up the capacitor doubles its capacitance by allowing both surfaces of each strip of foil to store charge.)

51. A model of a red blood cell portrays the cell as a spherical capacitor—a positively charged liquid sphere of surface area A, separated by a membrane of thickness t from the surrounding negatively charged fluid. Tiny electrodes introduced into the interior of the cell show a potential difference of 100 mV across the membrane. The membrane's thickness is estimated to be 100 nm and its dielectric constant to be 5.00. (a) If an average red blood cell has a mass of 1.00×10^{-12} kg, estimate the volume of the cell and thus find its surface area. The density of blood is 1 100 kg/m³. (b) Estimate the capacitance of the cell. (c) Calculate the charge on the surface of the membrane. How many electronic charges does this represent?

ADDITIONAL PROBLEMS

52. Three parallel-plate capacitors are constructed, each having the same plate spacing d, and with C_1 having plate area A_1, C_2 having area A_2, and C_3 having area A_3. Show that the total capacitance C of these three capacitors connected in parallel is the same as a capacitor having plate spacing d and plate area $A = A_1 + A_2 + A_3$.

53. Three parallel-plate capacitors are constructed, each having the same plate area A, and with C_1 having plate

web

spacing d_1, C_2 having plate spacing d_2, and C_3 having plate spacing d_3. Show that the total capacitance C of these three capacitors connected in series is the same as a capacitor of plate area A and with plate spacing $d = d_1 + d_2 + d_3$.

54. Charges of equal magnitude 1.00×10^{-15} C and opposite sign are distributed over the inner and outer surfaces of the cell wall in Figure P16.54. Find the force on the potassium ion (K$^+$) if the ion is (a) 2.70 μm from the center of the cell, (b) 2.92 μm from the center, and (c) 4.00 μm from the center.

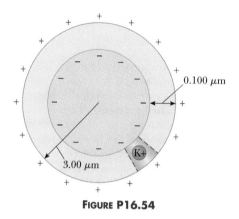

FIGURE P16.54

55. A virus rests on the bottom plate of oppositely charged parallel plates in the vacuum chamber of an electron microscope. The electric field strength between the plates is 2.00×10^5 N/C, and the bottom plate is negative. If the virus has a mass of 1.00×10^{-15} kg and suddenly acquires a charge of -1.60×10^{-19} C, what are its velocity and position 75.0 ms later? Do not disregard gravity.

56. A plastic pellet has a mass of 7.50×10^{-10} kg and carries a charge of 2.00×10^{-9} C. The pellet is fired horizontally between a pair of oppositely charged parallel plates, as shown in Figure P16.56. If the electric field strength between the plates is 9.20×10^4 N/C and the pellet enters the plates with an initial speed of 150 m/s, determine the deflection angle δ. (*Hint:* This exercise is similar to those involving the motion of a projectile under gravity.)

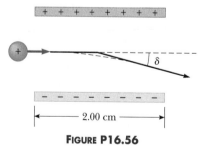

FIGURE P16.56

57. Find the equivalent capacitance of the group of capacitors in Figure P16.57.

58. A spherical capacitor consists of a spherical conducting shell of radius b and charge $-Q$ concentric with a smaller conducting sphere of radius a and charge Q. (a) Find the capacitance of this device. (b) Show that as the radius b of

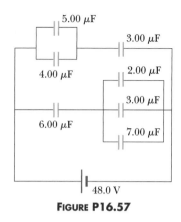

FIGURE P16.57

the outer sphere approaches infinity, the capacitance approaches the value of $a/k_e = 4\pi\epsilon_0 a$.

59. The immediate cause of many deaths is ventricular fibrillation, uncoordinated quivering of the heart as opposed to proper beating. An electric shock to the chest can cause momentary paralysis of the heart muscle, after which the heart will sometimes start organized beating again. A *defibrillator* is a device that applies a strong electric shock to the chest over a time of a few milliseconds. The device contains a capacitor of several microfarads, charged to several thousand volts. Electrodes called paddles, about 8 cm across and coated with conducting paste, are held against the chest on both sides of the heart. Their handles are insulated to prevent injury to the operator, who calls, "Clear!" and pushes a button on one paddle to discharge the capacitor through the patient's chest. Assume that an energy of 300 W · s is to be delivered from a 30.0 μF capacitor. To what potential difference must it be charged?

60. When a certain air-filled parallel-plate capacitor is connected across a battery, it acquires a charge (on each plate) of 150 μC. While the battery connection is maintained, a dielectric slab is inserted into and fills the region between the plates. This results in the accumulation of an additional charge of 200 μC on each plate. What is the dielectric constant of the dielectric slab?

61. Capacitors $C_1 = 6.0$ μF and $C_2 = 2.0$ μF are charged as a parallel combination across a 250-V battery. The capacitors are disconnected from the battery and from each other. They are then connected positive plate to negative plate and negative plate to positive plate. Calculate the resulting charge on each capacitor.

62. Capacitors $C_1 = 4.0$ μF and $C_2 = 2.0$ μF are charged as a series combination across a 100-V battery. The two capacitors are disconnected from the battery and from each other. They are then connected positive plate to positive plate and negative plate to negative plate. Calculate the resulting charge on each capacitor.

63. The charge distribution shown in Figure P16.63 is referred to as a linear quadrupole. (a) Show that the electric poten-

tial at a point on the *x* axis where $x > d$ is

$$V = \frac{2k_e Q d^2}{x^3 - x d^2}$$

(b) Show that the expression obtained in (a) when $x \gg d$ reduces to

$$V = \frac{2k_e Q d^2}{x^3}$$

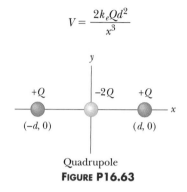

Quadrupole
FIGURE P16.63

64. The energy stored in a 52.0-μF capacitor is used to melt a 6.00-mg sample of lead. To what voltage must the capacitor be initially charged, assuming that the initial temperature of the lead is 20.0°C? (Lead has a specific heat of 128 J/kg · °C, a melting point of 327.3°C, and a latent heat of fusion of 24.5 kJ/kg.)

65. Consider a parallel-plate capacitor with charge Q and area A, filled with dielectric material having dielectric constant κ. It can be shown that the magnitude of the attractive force exerted by each plate on the other is given by $F = Q^2/(2\kappa\epsilon_0 A)$. When a potential difference of 100 V exists between the plates of an air-filled 20-μF parallel-plate capacitor, what force does each plate exert on the other if they are separated by 2.0 mm?

66. An electron is fired at a speed of $v_0 = 5.6 \times 10^6$ m/s and at an angle of $\theta_0 = -45°$ between two parallel conducting plates that are $D = 2.0$ mm apart, as in Figure P16.66. If the potential difference between the plates is $\Delta V = 100$ V, determine (a) how close d the electron will get to the bottom plate and (b) where the electron will strike the top plate.

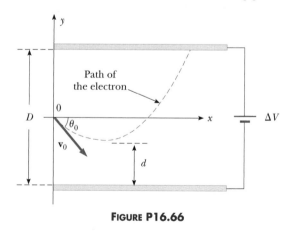

FIGURE P16.66

GROUP ACTIVITIES

G.1 It takes an electric field of about 30 kV/cm to cause a spark in dry air. Shuffle across a rug and reach toward a doorknob. By estimating the length of the spark, determine the electric potential difference between your finger and the doorknob that existed just before you touched the knob. Try this experiment again on a very humid day and you will find that the spark is much shorter or is imperceptible. Why?

G.2 Suppose you are given a battery, a capacitor, two switches, a lightbulb and several pieces of connecting wire. On a sheet of paper, design a circuit that will do the following. (1) When switch 1 is closed and switch 2 is open, the capacitor charges but no current moves through the lightbulb. (2) Then when switch 1 is opened and switch 2 closed, the lightbulb is connected to the capacitor but not to the battery. Describe the motion of charge in the circuit when switch 1 is closed and switch 2 is open. Is energy being stored in the capacitor? What measurements would you have to make to determine how much energy is stored if any? What happens to the lightbulb when switch 1 is opened after the capacitor has charged and switch 2 then closed? Will the bulb light and stay lit? What happens to the charge on the capacitor when switch 2 is closed in this way?

G.3 (a) Under what conditions can the equation $V_B - V_A = -Ed$ be used? (b) Can this equation be used to find the difference in potential between two points in an electric field set up by a point charge? Defend your answer. (c) Can this equation be used to find the difference in potential between two points in the electric field between the plates of a charged, parallel-plate capacitor? Defend your answer. (d) What is the strength of the electric field between two parallel metal plates attached to a battery of 12 V when the plates are separated by 5 mm?

G.4 Two charges of 1.0 μC and -2.0 μC are 0.5 m apart at two vertices of an equilateral triangle. (a) What is the direction of the potential set up by the 1.0-μC charge at the third vertex of the triangle? Be careful, and defend your answer. (b) What is the direction of the potential set up by the charge of -2.0 μC at this third vertex? Again be very careful before you defend your answer too strongly. (c) Find the electric potential at this point.

G.5 (a) A group of capacitors connected in parallel have the same (i) charge on them, (ii) potential difference across them, or (iii) neither of the above.

(b) A group of capacitors connected in series have the same (i) charge on them, (ii) potential difference across them, or (iii) neither of the above.

(c) The equivalent capacitance for a group of capacitors connected in parallel is (i) greater than any of the capacitors in the group, (ii) less than any of the capacitors in the group, or (iii) neither of the above.

(d) The equivalent capacitance for a group of capacitors connected in series is (i) greater than any of the capacitors in the group, (ii) less than any of the capacitors in the group, or (iii) neither of the above.

(e) Defend your answers by examining a circuit consisting of a 3.0-μF and a 5.0-μF capacitor used with a 12-V battery. Be sure that the answers you find are consistent with your answers to parts (a) through (d).

G.6 Figure GA16.6 shows a contour map of a hilly island. The outer part of the figure is at sea level (marked 0). Each contour line from the region marked zero shows a level 10 m higher than the previous. The maximum height is 70 m and is shown by the number 70. Answer the following questions by giving the pair of grid markers (a letter and a number) closest to the point being requested.

(a) Where is there a steep cliff?

(b) Where is there a pass between two hills?

(c) Where is the easiest slope to climb?

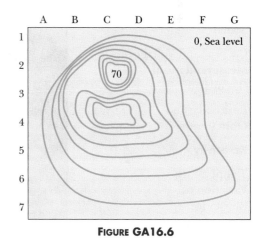

FIGURE GA16.6

Group Activity 6 is courtesy of Edward F. Redish. For more problems of this type, see
http://www.physics.umd.edu/perg/

17 Current and Resistance

This lightbulb emits light when its tungsten filaments carry sufficient current. Tungsten is an ideal filament material because it is strong and has a high melting point (about 3 410°C). The interior of the bulb contains an inert gas, such as a mixture of argon and nitrogen. Air must be avoided because the oxygen it contains would corrode the tungsten filaments. *(Tek Image/Science Photo Library/Photo Researchers, Inc.)*

chapter, let us digress briefly to discuss what a circuit is and how you measure current in and voltage across some circuit element.

Look at the circuit shown in Figure 17.5a. This is the actual circuit that would have to be set up to measure the current in Example 17.1. Figure 17.5b shows a stylized figure called a **circuit diagram** (as first introduced in Chapter 16) that represents the actual circuit of Figure 17.5a. This circuit consists of only a battery and a lightbulb. The word *circuit* means that current circulates around a closed loop of some sort. The battery pumps charge through the bulb and around the loop. No charge would flow if you did not have a complete conducting path from the positive terminal of the battery into one side of the bulb, out the other, and through the copper conducting wires back to the negative terminal of the battery. The most important quantities that characterize how the bulb will work in different situations are the current I in the bulb and the potential difference ΔV *across* the bulb. To measure the current in the bulb, we place an ammeter, the device for measuring current, in line with the bulb so that there is no path for the current to sneak around the meter; all of the charge passing through the bulb must also pass through the ammeter. The voltmeter measures the potential difference or voltage between the two ends of the bulb's filament. If we use two meters simultaneously as in Figure 17.5a, we can remove the voltmeter and see if its presence effects the current reading. Figure 17.5c shows a digital **multimeter,** a convenient device with a digital readout that can be used to measure voltage, current, or resistance. An advantage of using a digital multimeter as a voltmeter is that it usually will not affect the current, because a digital meter has enormous resistance to the flow of charge in the voltmeter mode.

At this point, you can measure the current as a function of voltage (the I-ΔV curve) of various devices in the lab. All you need is a variable voltage supply (an adjustable battery) capable of supplying potential differences from about -5 V to $+5$ V, a bulb, a resistor, some wires and alligator clips, and a couple of multimeters. Always start your measurements using the highest multimeter scales, that is, 10 A and 1 000 V, and increase the sensitivity one scale at a time to obtain the highest accuracy without overloading the meters. (To increase the sensitivity means that we lower the maximum current or voltage that the scale reads.) Also note that the meters must be connected with the proper polarity with respect to the voltage supply, as shown in Figure 17.5b. Finally, follow your instructor's directions carefully to avoid damaged meters and a soaring lab fee.

(a) (b) (c)

FIGURE 17.5   (a) A sketch of an actual circuit used to measure the current in a flashlight bulb and the potential difference across it. (b) A schematic diagram of the circuit shown in part (a). (c) A digital multimeter that can be used to measure both currents and potential differences. Here, the meter is measuring the potential difference across the terminals of a battery. *(c, Michael Dalton, Fundamental Photographs)*

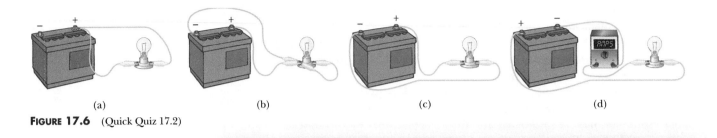

(a) (b) (c) (d)

FIGURE 17.6 (Quick Quiz 17.2)

> **Quick Quiz 17.2**
>
> Look at the four "circuits" shown in Figure 17.6 and select those that will light the bulb.

17.4 RESISTANCE AND OHM'S LAW

When a voltage (potential difference) ΔV is applied across the ends of a metallic conductor as in Figure 17.7, the current in the conductor is found to be proportional to the applied voltage; that is, $I \propto \Delta V$. If the proportionality holds, we can write $\Delta V = IR$, where the proportionality constant R is called the resistance of the conductor. In fact, we define the **resistance** as the ratio of the voltage across the conductor to the current it carries:

Resistance ▶

$$R \equiv \frac{\Delta V}{I} \qquad\qquad [17.4]$$

Resistance has the SI units volts per ampere, called **ohms** (Ω). Thus, if a potential difference of 1 V across a conductor produces a current of 1 A, the resistance of the conductor is 1 Ω. For example, if an electrical appliance connected to a 120-V source carries a current of 6 A, its resistance is 20 Ω.

It is useful to compare the concepts of electric current, voltage, and resistance with the flow of water in a river. As water flows downhill in a river of constant width and depth, the flow rate (water current) depends on the steepness of descent of the river and the effects of rocks, the river bank, and other obstructions. Based on this analogy, it seems reasonable that increasing the voltage applied to a circuit should increase the current in the circuit just as increasing the steepness of descent increases the water current. Also, increasing the obstructions in the river's path will reduce the water current just as increasing the resistance in a circuit will lower the electric current. Resistance in a circuit arises due to collisions between the electrons carrying the current with fixed atoms inside the conductor. These collisions inhibit the movement of charges in much the same way as would a force of friction. For many materials, including most metals, experiments show that **the resistance remains constant over a wide range of applied voltages or currents.** This statement is known as **Ohm's law** after Georg Simon Ohm (1789–1854), who was the first to conduct a systematic study of electrical resistance.

It is common practice to express Ohm's law as

Ohm's law ▶

$$\Delta V = IR \qquad\qquad [17.5]$$

where R is understood to be independent of ΔV, the potential drop across the resistor, and I, the current in the resistor. We shall continue to use this traditional form of Ohm's law when discussing electrical circuits. A **resistor** is a conductor that provides a specified resistance in an electric circuit. The symbol for a resistor in circuit diagrams is a zigzag line, .

Ohm's law is an empirical relationship that is valid only for certain materials. Materials that obey Ohm's law, and hence have a constant resistance over a wide

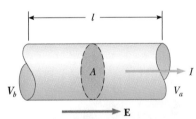

FIGURE 17.7 A uniform conductor of length l and cross-sectional area A. The current I in the conductor is proportional to the applied voltage $\Delta V = V_b - V_a$. The electric field **E** set up in the conductor is also proportional to the current.

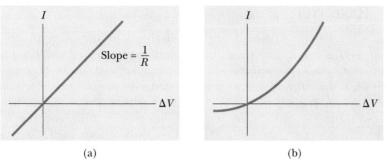

FIGURE 17.8 (a) The current-voltage curve for an ohmic material. The curve is linear, and the slope gives the resistance of the conductor. (b) A nonlinear current-voltage curve for a semiconducting diode. This device does not obey Ohm's law.

range of voltages, are said to be **ohmic.** Materials whose resistance changes with voltage or current are **nonohmic.** Ohmic materials have a linear current-voltage relationship over a large range of applied voltages (Fig. 17.8a). Nonohmic materials have a nonlinear current-voltage relationship (Fig. 17.8b). One common semiconducting device that is nonohmic is the **diode,** a circuit element that acts like a one-way valve for current (see Problem 62). Its resistance is small for currents in one direction (positive ΔV) and large for currents in the reverse direction (negative ΔV). Most modern electronic devices, such as transistors, have nonlinear current-voltage relationships; their operation depends on the particular ways in which they violate Ohm's law.

GEORG SIMON OHM (1787–1854)

Ohm, a high school teacher in Cologne and later a professor at Munich, formulated the concept of resistance and discovered the proportionalities expressed in Equation 17.6. But Henry Cavendish had earlier determined the proportionality of current to voltage that we call Ohm's law. (*© Bettmann/CORBIS*)

Quick Quiz 17.3 In Figure 17.8b, does the resistance of the diode (a) increase or (b) decrease as the positive voltage ΔV increases?

Example 17.3   The Resistance of a Steam Iron

All electric devices are required to have identifying plates that specify their electrical characteristics. The plate on a certain steam iron states that the iron carries a current of 6.4 A when connected to a 120-V source. What is the resistance of the steam iron?

Solution   From Ohm's law, we find the resistance to be

$$R = \frac{\Delta V}{I} = \frac{120\ \text{V}}{6.4\ \text{A}} = \boxed{19\ \Omega}$$

EXERCISE   The resistance of a hot plate is 48 Ω. How much current does the plate carry when connected to a 120-V source?

ANSWER   2.5 A

An assortment of resistors used for a variety of applications in electronic circuits. (*Courtesy of Henry Leap and Jim Lehman*)

17.5 RESISTIVITY

In an earlier section we pointed out that electrons do not move in straight-line paths through a conductor. Instead, they undergo repeated collisions with the metal atoms. Consider a conductor with a voltage applied across its ends. An electron gains speed as the electric force associated with the internal electric field accelerates it, giving it a velocity in the direction opposite that of the electric field. A collision with an atom randomizes the electron's velocity, thus reducing its velocity in the direction opposite the field. The process then repeats itself. Together these collisions affect the electron somewhat as a force of internal friction would. This is the origin of a material's resistance. The resistance of an ohmic conductor is proportional to its length l and inversely proportional to its cross-sectional area A. That is,

$$R = \rho \frac{l}{A} \qquad [17.6]$$

where the constant of proportionality ρ is called the **resistivity** of the material.[1] Every material has a characteristic resistivity that depends on its electronic structure and on temperature. Good electric conductors have very low resistivities, and good insulators have very high resistivities. Table 17.1 lists the resistivities of a variety of materials at 20°C. Because resistance values are in ohms, resistivity values must be in ohm-meters.

Equation 17.6 states that the resistance of a cylindrical conductor is proportional to its length and inversely proportional to its cross-sectional area. This may be understood by analogy to the flow of liquid through a pipe. As the length of the pipe is increased, the resistance to liquid flow increases because of a gain in friction between the fluid and the walls of the pipe. As its cross-sectional area is in-

TABLE 17.1	Resistivities and Temperature Coefficients of Resistivity for Various Materials (at 20°C)	
Material	**Resistivity ($\Omega \cdot$ m)**	**Temperature Coefficient of Resistivity [(°C)$^{-1}$]**
Silver	1.59×10^{-8}	3.8×10^{-3}
Copper	1.7×10^{-8}	3.9×10^{-3}
Gold	2.44×10^{-8}	3.4×10^{-3}
Aluminum	2.82×10^{-8}	3.9×10^{-3}
Tungsten	5.6×10^{-8}	4.5×10^{-3}
Iron	10.0×10^{-8}	5.0×10^{-3}
Platinum	11×10^{-8}	3.92×10^{-3}
Lead	22×10^{-8}	3.9×10^{-3}
Nichrome*	150×10^{-8}	0.4×10^{-3}
Carbon	3.5×10^{5}	-0.5×10^{-3}
Germanium	0.46	-48×10^{-3}
Silicon	640	-75×10^{-3}
Glass	$10^{10} - 10^{14}$	
Hard rubber	$\approx 10^{13}$	
Sulfur	10^{15}	
Quartz (fused)	75×10^{16}	

* A nickel-chromium alloy commonly used in heating elements.

[1] The symbol ρ used for resistivity should not be confused with the same symbol used earlier in the book for density. Very often, a single symbol is used to represent different quantities.

creased, the pipe can transport more fluid in a given time interval, so its resistance drops.

APPLYING PHYSICS 17.3

It is a common observation that as a lightbulb ages, it gives off less light than when new. Why?

Explanation   There are two reasons for this, one electrical and one optical, but both are related to the same phenomenon occurring within the bulb. The filament of a lightbulb is made of a tungsten wire that, in an old lightbulb, has been kept at a high temperature for many hours. These high temperatures cause tungsten to be evaporated from the filament, decreasing its radius. From $R = \rho\, l/A$, we see that a decreased cross-sectional area leads to an increase in resistance of the filament. This increasing resistance with age means that the filament will carry less current for the same applied voltage. With less current in the filament, there is less light output, and the filament glows more dimly.

At the high operating temperature of the filament, tungsten atoms leave the surface of the filament, much as water molecules evaporate from a puddle of water. These atoms are carried away by convection currents in the gas in the bulb and are deposited on the inner surface of the glass. In time, the glass becomes less transparent because of this tungsten coating, which decreases the amount of light that passes through the glass.

◀ **APPLICATION**

DIMMING OF AGING
LIGHTBULBS

Example 17.4   The Resistance of Nichrome Wire

A Calculate the resistance per unit length of a 22-gauge nichrome wire of radius 0.321 mm.

Solution   The cross-sectional area of this wire is

$$A = \pi r^2 = \pi(0.321 \times 10^{-3}\,\text{m})^2 = 3.24 \times 10^{-7}\,\text{m}^2$$

The resistivity of nichrome is $1.5 \times 10^{-6}\ \Omega \cdot \text{m}$ (Table 17.1). Thus, we can use Equation 17.6 to find the resistance per unit length:

$$\frac{R}{l} = \frac{\rho}{A} = \frac{1.5 \times 10^{-6}\ \Omega \cdot \text{m}}{3.24 \times 10^{-7}\ \text{m}^2} = \boxed{4.6\ \Omega/\text{m}}$$

B If a potential difference of 10.0 V is maintained across a 1.0-m length of the nichrome wire, what is the current in the wire?

Solution   Because a 1.0-m length of this wire has a resistance of 4.6 Ω, Ohm's law gives

$$I = \frac{\Delta V}{R} = \frac{10.0\ \text{V}}{4.6\ \Omega} = \boxed{2.2\ \text{A}}$$

Note from Table 17.1 that the resistivity of nichrome is about 100 times that of copper, a typical good conductor. Therefore, a copper wire of the same radius would have a resistance per unit length of only 0.052 Ω/m, and a 1.0-m length of copper wire of the same radius would carry the same current (2.2 A) with an applied voltage of only 0.11 V.

Because of its resistance to oxidation, nichrome is often used for heating elements in toasters, irons, and electric heaters.

EXERCISE   What is the resistance of a 6.0-m length of 22-gauge nichrome wire? How much current does it carry when connected to a 120-V source?

ANSWER   28 Ω; 4.3 A

**Quick
Quiz
17.4**

Aliens with strange powers visit Earth and double every linear dimension of every object on the surface of the Earth. Does the electrical cord from the wall socket to your floor lamp now have
(a) more resistance than before, (b) less resistance, or (c) the same resistance? Does the lightbulb filament glow (d) more brightly than before, (e) less brightly, or (f) the same? (Assume the resistivities of materials remain the same before and after the doubling.)

**Quick
Quiz
17.5**

A voltage ΔV is applied across the ends of a nichrome heater wire having a cross-sectional area A and length L. The same voltage is applied across the ends of a second heater wire having a cross-sectional area A and length $2L$. Which wire gets hotter? (a) The shorter wire, (b) the longer wire, or (c) not enough information to say.

17.6 TEMPERATURE VARIATION OF RESISTANCE

The resistivity, and hence the resistance, of a conductor depends on a number of factors. One of the most important is the temperature of the metal. For most metals, resistivity increases with increasing temperature. This correlation can be understood as follows. As the temperature of the material increases, its constituent atoms vibrate with increasingly greater amplitudes. Just as it is more difficult to weave one's way through a crowded room when the people are in motion than when they are standing still, so do the electrons find it more difficult to pass atoms vibrating with large amplitudes. The increased electron scattering with increasing temperature results in increased resistivity.

An old-fashioned carbon-filament incandescent lamp. The resistance of such a lamp is typically 10 Ω, but it changes with temperature. *(Courtesy of Central Scientific Company)*

For most metals, resistivity increases approximately linearly with temperature over a limited temperature range, according to the expression

$$\rho = \rho_0[1 + \alpha(T - T_0)] \qquad \text{[17.7]}$$

where ρ is the resistivity at some temperature T (in Celsius degrees), ρ_0 is the resistivity at some reference temperature T_0 (usually taken to be 20°C), and α is a parameter called the **temperature coefficient of resistivity.** The temperature coefficients for various materials are provided in Table 17.1. The interesting negative values of α for semiconductors arise because these materials possess weakly bound charge carriers that become free to move and contribute to the current as the temperature is raised.

Because the resistance of a conductor with uniform cross section is proportional to the resistivity, according to Equation 17.6 ($R = \rho l/A$), the temperature variation of resistance can be written

$$R = R_0[1 + \alpha(T - T_0)] \qquad \text{[17.8]}$$

Precise temperature measurements are often made using this property, as shown by the following example.

Example 17.5   A Platinum Resistance Thermometer

A resistance thermometer, which measures temperature by measuring the change in resistance of a conductor, is made of platinum and has a resistance of 50.0 Ω at 20.0°C. When the device is immersed in a vessel containing melting indium, its resistance increases to 76.8 Ω. From this information, find the melting point of indium.

Solution If we solve Equation 17.8 for $T - T_0$ and get α for platinum from Table 17.1, we obtain

$$T - T_0 = \frac{R - R_0}{\alpha R_0} = \frac{76.8 \ \Omega - 50.0 \ \Omega}{[3.92 \times 10^{-3} \ (°\text{C})^{-1}][50.0 \ \Omega]} = 137°\text{C}$$

Because $T_0 = 20.0°\text{C}$, we find that the melting point of indium is

$$T = \boxed{157°\text{C}}$$

17.7 SUPERCONDUCTORS

There is a class of metals and compounds whose resistances fall virtually to *zero* below a certain temperature T_c called the *critical temperature*. These materials are known as **superconductors.** The resistance-temperature graph for a superconductor follows that of a normal metal at temperatures above T_c (Fig. 17.9). When the temperature is at or below T_c, the resistance suddenly drops to zero. This phenomenon was discovered in 1911 by the Dutch physicist H. Kamerlingh Onnes as he and a graduate student worked with mercury, which is a superconductor below 4.1 K. Recent measurements have shown that the resistivities of superconductors below T_c are less than $4 \times 10^{-25} \ \Omega \cdot \text{m}$—around 10^{17} times smaller than the resistivity of copper and in practice considered to be zero.

Today, thousands of superconductors are known, including such common metals as aluminum, tin, lead, zinc, and indium. Table 17.2 lists the critical temperatures of several superconductors. The value of T_c is sensitive to chemical composition, pressure, and crystalline structure. Interestingly, copper, silver, and gold, which are excellent conductors, do not exhibit superconductivity.

One of the truly remarkable features of superconductors is the fact that, once a current is set up in them, it persists *without any applied voltage* (because $R = 0$). In fact, steady currents in superconducting loops have been observed to persist for many years with no apparent decay!

An important development in physics that created much excitement in the scientific community was the discovery of high-temperature copper-oxide-based superconductors. The excitement began with a 1986 publication by J. Georg Bednorz and K. Alex Müller, scientists at the IBM Zurich Research Laboratory in Switzerland, in which they reported evidence for superconductivity at a temperature near 30 K in an oxide of barium, lanthanum, and copper. Bednorz and Müller were awarded the Nobel prize for physics in 1987 for their discovery. This discovery was remarkable in view of the fact that the critical temperature was significantly higher than that of any previously known superconductor. Shortly thereafter, a new family of compounds was investigated, and research activity in the field of superconductivity proceeded vigorously. In early 1987, groups at the University of Alabama at Huntsville and the University of Houston announced the discovery of superconductivity at about 92 K in an oxide of yttrium, barium, and copper ($\text{YBa}_2\text{Cu}_3\text{O}_7$), shown as the gray disk in Figure 17.10. Late in 1987, teams of scientists from Japan and the United States reported superconductivity at 105 K in an oxide of bismuth, strontium, calcium, and copper. More recently, scientists have reported superconductivity at temperatures as high as 150 K in an oxide containing mercury. At this point one cannot rule out the possibility of room-temperature superconductivity, and the search for novel superconducting materials continues. It is an important search both for scientific reasons and because practical applications become more probable and widespread as the critical temperature is raised and the expense of cooling materials decreases.

An important and useful application is superconducting magnets in which the magnetic field intensities are about ten times greater than those of the best normal electromagnets. Such magnets are being considered as a means of storing

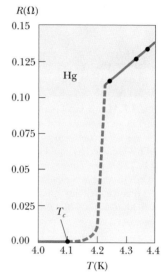

FIGURE 17.9 Resistance versus temperature for a sample of mercury. The graph follows that of a normal metal above the critical temperature T_c. The resistance drops to zero at the critical temperature, which is 4.1 K for mercury, and remains at zero for lower temperatures.

TABLE 17.2	
Critical Temperatures for Various Superconductors	
Material	**T_c (K)**
Zn	0.88
Al	1.19
Sn	3.72
Hg	4.15
Pb	7.18
Nb	9.46
Nb_3Sn	18.05
Nb_3Ge	23.2
$\text{YBa}_2\text{Cu}_3\text{O}_7$	90
Bi-Sr-Ca-Cu-O	105
Tl-Ba-Ca-Cu-O	125
$\text{HgBa}_2\text{Ca}_2\text{Cu}_3\text{O}_8$	134

Webnote 17.1

The Department of Energy has a good Web site on superconductivity at *http://www.eren.doe.gov/ superconductivity/*

FIGURE 17.10 A small permanent magnet floats freely above a ceramic disk of the superconductor YBa$_2$Cu$_3$O$_7$ cooled by liquid nitrogen at 77 K. The superconductor has zero electric resistance at temperatures below 92 K and expels any applied magnetic field. *(Courtesy of IBM Research Laboratory)*

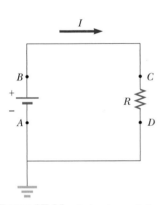

FIGURE 17.11 A circuit consisting of a battery and resistance *R*. Positive charge flows clockwise from the positive to the negative terminal of the battery. Point A is grounded.

Tip 17.3   MISCONCEPTION ABOUT CURRENT

Note that current is *not* "used up" in the resistor. The current is the same *everywhere* in the circuit. Charges flow in the same rotational sense at *all* points in the circuit.

Power delivered to a resistor ▶

energy. The idea of using superconducting power lines for transmitting power efficiently is also receiving serious consideration. Modern superconducting electronic devices consisting of two thin-film superconductors separated by a thin insulator have been constructed. They include magnetometers (magnetic-field measuring devices) and various microwave devices.

17.8 ELECTRICAL ENERGY AND POWER

If a battery is used to establish an electric current in a conductor, chemical energy stored in the battery is continuously transformed into kinetic energy of the charge carriers. This kinetic energy is quickly lost as a result of collisions between the charge carriers and fixed atoms in the conductor, causing an increase in the temperature of the conductor. Thus, the chemical energy stored in the battery is continuously transformed into internal energy.

In order to understand the process of energy transfer in a simple circuit, consider a battery whose terminals are connected to a resistor (Fig. 17.11). (Remember that the positive terminal of the battery is always at the higher potential.) Now imagine following a quantity of positive charge ΔQ around the circuit from point A through the battery and resistor and back to A. Point A is a reference point that is grounded (the ground symbol is ⏚), and its potential is taken to be zero. As the charge moves from A to B through the battery, whose potential difference is ΔV, the electrical potential energy of the system increases by the amount $\Delta Q \, \Delta V$, and the chemical potential energy in the battery decreases by the same amount. (Recall from Chapter 16 that $\Delta PE = q \, \Delta V$.) However, as the charge moves from C to D through the resistor, it loses this electrical potential energy during collisions with atoms in the resistor. In this process, the energy is transformed to internal energy corresponding to increased vibrational motion of the atoms in the resistor. Because we have ignored the resistance of the interconnecting wires, no energy transformation occurs for paths BC and DA. When the charge returns to point A, the net result is that some of the chemical energy in the battery has been delivered to the resistor and causes its temperature to rise. Feeling the resistor, we would often notice that it was warmer than its surroundings.

The charge ΔQ loses energy $\Delta Q \, \Delta V$ as it passes through the resistor. If Δt is the time it takes the charge to pass through the resistor, then the rate at which it loses electrical potential energy is

$$\frac{\Delta Q}{\Delta t} \Delta V = I \, \Delta V$$

where I is the current in the resistor and ΔV is the potential difference across it. Of course, the charge regains this energy when it passes through the battery, at the expense of chemical energy in the battery. The rate at which the system loses potential energy as the charge passes through the resistor is equal to the rate at which the system gains internal energy in the resistor. Thus, the power $\mathcal{P}$ representing the rate at which energy is delivered to the resistor is

$$\mathcal{P} = I \, \Delta V \tag{17.9}$$

We have developed this result by considering a battery delivering energy to a resistor. However, Equation 17.9 can be used to determine the power transferred from a voltage source to *any* device carrying a current I and having a potential difference ΔV between its terminals.

Using Equation 17.9 and the fact that $\Delta V = IR$ for a resistor, we can express the power delivered to the resistor in the alternative forms

$$\mathcal{P} = I^2 R = \frac{(\Delta V)^2}{R} \tag{17.10}$$

When *I* is in amperes, ΔV in volts, and *R* in ohms, the SI unit of power is the watt (introduced in Chapter 5). The power delivered to a conductor of resistance *R* is often referred to as an I^2R *loss*. Note that Equation 17.10 applies only to resistors and not to nonohmic devices like lightbulbs and diodes.

Regardless of the ways in which you use electrical energy in your home, you ultimately must pay for it or risk having your power turned off. The unit of energy used by electric companies to calculate consumption, the **kilowatt-hour,** is defined in terms of the unit of power and the amount of time it is supplied. One kilowatt-hour (kWh) is the energy converted or consumed in 1 h at the constant rate of 1 kW. It has the numerical value

$$1 \text{ kWh} = (10^3 \text{ W})(3\,600 \text{ s}) = 3.60 \times 10^6 \text{ J} \qquad [17.11]$$

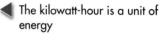 ◀ The kilowatt-hour is a unit of energy

On an electric bill, the amount of electricity used in a given period is usually stated in multiples of kilowatt-hours.

APPLYING PHYSICS 17.4

When is more power delivered to a lightbulb—just after it is turned on and the glow of the filament is increasing or after it has been on for a few seconds and the glow is steady?

Explanation Once the switch is closed, the line voltage is applied across the lightbulb. As the voltage is applied across the cold filament when first turned on, the resistance of the filament is low, the current is high, and a relatively large amount of power is delivered to the bulb. As the filament warms, its resistance rises and the current decreases. As a result, the power delivered to the bulb decreases. The large current spike at the beginning of operation is the reason that lightbulbs often fail just after they are turned on.

Quick Quiz 17.6

For the two resistors shown in Figure 17.12, rank the currents at points *a* through *f*, from largest to smallest.

Quick Quiz 17.7

Two resistors, A and B, are connected across the same potential difference. The resistance of A is twice that of B. (a) Which resistor dissipates more power? (b) Which carries the greater current?

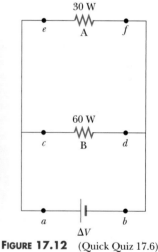

FIGURE 17.12 (Quick Quiz 17.6)

Example 17.6 The Power Converted by an Electric Heater

An electric heater is operated by applying a potential difference of 50.0 V to a nichrome wire of total resistance 8.00 Ω. Find the current carried by the wire and the power rating of the heater.

Solution Because $\Delta V = IR$, we have

$$I = \frac{\Delta V}{R} = \frac{50.0 \text{ V}}{8.00 \ \Omega} = \boxed{6.25 \text{ A}}$$

We can find the power rating using $\mathcal{P} = I^2R$:

$$\mathcal{P} = I^2R = (6.25 \text{ A})^2(8.00 \ \Omega) = \boxed{313 \text{ W}}$$

EXERCISE If we doubled the applied voltage to the heater, what would happen to the current and power?

ANSWER The current would double, and the power would quadruple.

PHYSICS *IN ACTION*

The photo shows electrical workers at dusk, trying to restore power to the eastern Ontario town of St. Isadore that was without power for several days in January 1998 because of a severe ice storm. When power transmission lines fall because of lightning, ice storms, or earthquakes, it is very dangerous to touch them due to the high electric potential, possibly hundreds of thousands of volts relative to the ground. When transporting electrical energy through power lines, utility companies seek to minimize the power transformed into internal energy in the lines and maximize the energy delivered to the consumer. Because $\mathcal{P} = I\,\Delta V$, the same amount of power can be transported either at high currents and low potential differences or at low currents and high potential differences. Utility companies choose to transport electrical energy at low currents and high potential differences primarily for economic reasons. Copper wire is very expensive, and so it is cheaper to use high-resistance wire (that is, wire having a small cross-sectional area; see Eq. 17.6). In the expression for power delivered to a resistor, $\mathcal{P} = I^2 R$, the resistance of the transmission wire is fixed at a relatively high value for economic considerations. The $I^2 R$ loss can be reduced by keeping the current I as low as possible.

(AP/Wide World Photos/Fred Chartrand)

Example 17.7 **Lighting Up Your Life**

A circuit provides a current of 20.0 A at an operating voltage of 120 V. How many 75-W bulbs can operate with this voltage source?

Solution The total power that can be delivered by the voltage source is given by Equation 17.9:

$$\mathcal{P}_{\text{total}} = I\,\Delta V = (20.0 \text{ A})(120 \text{ V}) = 2.40 \times 10^3 \text{ W}$$

Because each bulb requires a power of 75 W, the number of bulbs that can operate from this voltage source is

$$\text{Number of bulbs} = \frac{\mathcal{P}_{\text{total}}}{\mathcal{P}_{\text{bulb}}} = \frac{2.40 \times 10^3 \text{ W}}{75 \text{ W}} = \boxed{32}$$

Example 17.8   The Cost of Operating a Lightbulb

How much does it cost to burn a 100-W lightbulb for 24 h if electric energy costs 12 cents or $0.12 per kilowatt-hour?

Solution   A 100-W lightbulb is equivalent to a 0.10-kW bulb. Because energy consumed equals power × time, the amount of energy you must pay for, expressed in kilowatt-hours, is

$$\text{Energy} = (0.10 \text{ kW})(24 \text{ h}) = 2.4 \text{ kWh}$$

If energy is purchased at 12 cents or $0.12 per kilowatt-hour, the 24-h cost is

$$\text{Cost} = (2.4 \text{ kWh})(\$0.12/\text{kWh}) = \boxed{\$0.29}$$

This is a small amount of money, but when larger and more complex electric devices are used, the cost goes up rapidly.

EXERCISE   If electric energy costs $0.12/kWh, what does it cost to operate an electric oven, which operates at 20.0 A and 220 V, for 5.0 h?

ANSWER $2.70

17.9   ELECTRICAL ACTIVITY IN THE HEART

ELECTROCARDIOGRAMS

Every action involving the body's muscles is initiated by electrical activity. The voltages produced by muscular action in the heart are particularly important to physicians. Voltage pulses cause the heart to beat, and the waves of electrical excitation that sweep across the heart associated with the heartbeat are conducted through the body via the body fluids. These voltage pulses are large enough to be detected by suitable monitoring equipment attached to the skin. A sensitive voltmeter making good electrical contact with the skin by means of contacts attached with conducting paste can be used to measure heart pulses that are typically of the order of 1 mV at the surface of the body. The voltage pulses can be recorded on an instrument called an **electrocardiograph,** and the pattern recorded by this instrument is called an **electrocardiogram** (EKG). In order to understand the information contained in an EKG pattern, it is necessary first to describe the underlying principles concerning electrical activity in the heart.

The right atrium of the heart contains a specialized set of muscle fibers called the SA (sinoatrial) node, which initiate the heartbeat (Fig. 17.13). Electric impulses that originate in these fibers gradually spread from cell to cell throughout the right and left atrial muscles, causing them to contract. The pulse that passes through the muscle cells is often called a *depolarization wave* because of its effect on individual cells. If an individual muscle cell were examined in its resting state, a double-layered electric charge distribution would be found on its surface, as shown in Figure 17.14a. The impulse generated by the SA node momentarily and locally allows positive charge on the outside of the cell to flow in and neutralize the negative charge on the inside layer. This changes the cell's charge distribution to that shown in Figure 17.14b. Once the depolarization wave has passed through a cell, an individual heart muscle cell recovers the resting state charge distribution (positive out, negative in) shown in Figure 17.14a in about 250 ms. When the impulse reaches the AV (atrioventricular) node (Fig. 17.13), the muscles of the atria begin to relax, and the pulse is directed by the AV node to the ventricular muscles. The muscles of the ventricles contract as the depolarization wave spreads through the ventricles along a group of fibers called the *Purkinje fibers*. The ventricles then

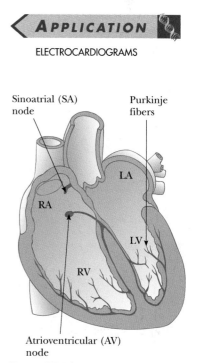

APPLICATION

ELECTROCARDIOGRAMS

FIGURE 17.13   The electrical conduction system of the human heart. (RA: right atrium; LA: left atrium; RV: right ventricle; LV: left ventricle.)

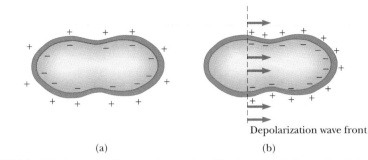

FIGURE 17.14   (a) Charge distribution of a muscle cell in the atrium before a depolarization wave has passed through the cell. (b) Charge distribution as the wave passes.

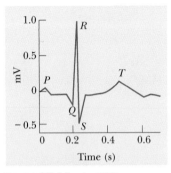

FIGURE 17.15   An EKG response for a normal heart.

Webnote 17.2

Electrocardiograms have unique signatures associated with malfunctions. To see and learn about these signatures, visit the home page of the following EKG library: *http://www.ecglibrary.com/ ecghome.html*

relax after the pulse has passed through. At this point, the SA node is again triggered and the cycle is repeated.

A sketch of the electrical activity registered on an EKG for one beat of a normal heart is shown in Figure 17.15. The pulse indicated by *P* occurs just before the atria begin to contract. The *QRS* pulse occurs in the ventricles just before they contract, and the *T* pulse occurs when the cells in the ventricles begin to recover. EKGs for an abnormal heart are shown in Figure 17.16. The *QRS* portion of the pattern shown in Figure 17.16a is wider than normal. This indicates that the patient may have an enlarged heart. (Why?) Figure 17.16b indicates that there is no constant relationship between the *P* pulse and the *QRS* pulse. This suggests a blockage in the electrical conduction path between the SA and AV nodes, which results in the atria and ventricles beating independently, leading to inefficient heart pumping. Finally, Figure 17.16c shows a situation in which there is no *P* pulse and an irregular spacing between the *QRS* pulses. This is symptomatic of irregular atrial contraction, which is called *fibrillation*. In this situation, the atrial and ventricular contractions are irregular.

As noted previously, the sinoatrial node directs the heart to beat at the appropriate rate, usually about 72 beats per minute. However, disease or the aging process can damage the heart and slow its beating, and a medical assist may be

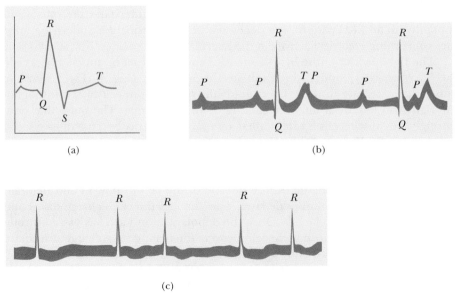

FIGURE 17.16   Abnormal EKGs.

necessary in the form of a *cardiac pacemaker* attached to the heart. This matchbox-sized electrical device implanted under the skin has a lead that is connected to the wall of the right ventricle. Pulses from this lead stimulate the heart to maintain its proper rhythm. In general, a pacemaker is designed to produce pulses at a rate of about 60 per minute, slightly slower than the normal beats per minute but sufficient to maintain life. The circuitry basically consists of a capacitor charging up from a lithium battery to a certain voltage and then discharging. The design of the circuit is such that if the heart is beating normally the capacitor is never allowed to charge completely and send pulses to the heart.

APPLICATION

CARDIAC PACEMAKERS

AN EMERGENCY ROOM IN YOUR CHEST

The operation in June 2001 on Vice President Dick Cheney focused attention on the progress in treating heart problems with tiny implanted electrical devices. Aptly termed "an emergency room in your chest" by Cheney's attending physician, current devices called <u>I</u>mplanted <u>C</u>ardioverter <u>D</u>efibrillators **(ICDs)** can monitor, record, and logically process heart signals and then supply different corrective signals to hearts beating too slowly, too rapidly, or irregularly, even monitoring and sending independent signals to atria and ventricles! Figure 17.17a shows a sketch of an ICD with conducting leads that are implanted in the heart. Figure 17.17b shows an actual titanium-encapsulated dual chamber ICD.

APPLICATION

IMPLANTED CARDIOVERTER DEFIBRILLATORS

The latest ICDs are quite sophisticated devices capable of

1. monitoring both atrial and ventricular chambers to differentiate between atrial and potentially fatal ventricular arrhythmias, which require prompt regulation;
2. storing about $\frac{1}{2}$ hour of heart signals, which can easily be read out by a physician;
3. being easily reprogrammed with an external magnetic wand;
4. performing complicated signal analysis and comparison;

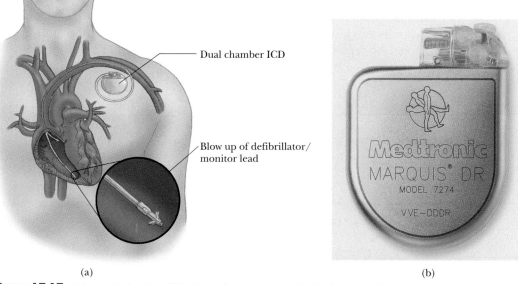

(a)   (b)

FIGURE 17.17 (a) A dual chamber ICD with leads in the heart. One lead monitors/stimulates the right atrium and the other monitors/stimulates the right ventricle. (b) Medtronic Dual Chamber ICD. *(Courtesy of Medtronic, Inc.)*

TABLE 17.3	Properties of Implanted Cardioverter Defibrillators[a]
Physical Specifications	
Mass (g)	85
Size (cm)	7.3 × 6.2 × 1.3 (about 5 stacked silver dollars)
Antitachycardia Pacing	ICD delivers a burst of critically timed, low-energy pulses
Number of bursts	1–15
Burst cycle length (ms)	200–552
Number of pulses per burst	2–20
Pulse amplitude (V)	7.5 or 10
Pulsewidth (ms)	1.0 or 1.9
High-Voltage Defibrillation	
Pulse energy (J)	37 stored/33 delivered
Pulse amplitude (V)	801
Bradycardia Pacing	A dual chamber ICD can steadily deliver repetitive pulses to both the atrium and ventricle
Base frequency (beats/min)	40–100
Pulse amplitude (V)	0.25–7.5
Pulsewidth (ms)	0.05, 0.1–1.5, 1.9

[a] For more information, see www.photonicd.com/specs.html

5. supplying 0.25- to 10-V repetitive pacing signals to speed up or slow down a malfunctioning heart, or a high-voltage pulse of about 800 V to halt the potentially fatal condition of ventricular fibrillation, in which the heart rapidly quivers rather than beats (people who have experienced such a high-voltage jolt say that it feels like a kick or a bomb going off in the chest);

6. automatically adjusting the number of pacing pulses per minute to match the patient's activity.

ICDs are powered by lithium batteries and have implanted lifetimes of 4–6 years. Some basic properties of these adjustable ICDs are given in Table 17.3. In this table, *tachycardia* means rapid heartbeat and *bradycardia* means slow heartbeat. A key factor in developing tiny electrical implants that serve as defibrillators is the development of capacitors with relatively large capacitance (125 μF) and small physical size.

SUMMARY

The **electric current** I in a conductor is defined as

$$I \equiv \frac{\Delta Q}{\Delta t}$$ [17.1]

where ΔQ is the charge that passes through a cross section of the conductor in time Δt. The SI unit of current is the **ampere** (A); 1 A = 1 C/s. By convention, the direction of current is in the direction of flow of positive charge.

The current in a conductor is related to the motion of the charge carriers by

$$I = nqv_d A$$ [17.3]

where n is the number of mobile charge carriers per unit volume, q is the charge on each carrier, v_d is the drift speed of the charges, and A is the cross-sectional area of the conductor.

The **resistance** R of a conductor is defined as the ratio of the potential difference across the conductor to the current:

$$R \equiv \frac{\Delta V}{I} \qquad [17.4]$$

The SI units of resistance are volts per ampere, or **ohms** (Ω); $1\ \Omega = 1\ \text{V/A}$.

Ohm's law describes many conductors for which the applied voltage is directly proportional to the current it causes. The proportionality constant is the resistance:

$$\Delta V = IR \qquad [17.5]$$

If a conductor has length l and cross-sectional area A, its **resistance** is

$$R = \rho\,\frac{l}{A} \qquad [17.6]$$

where ρ is an intrinsic property of the conductor called the **resistivity.** The SI unit of resistivity is the **ohm-meter** ($\Omega \cdot \text{m}$).

The resistivity of a conductor varies with temperature over a limited temperature range, according to the expression

$$\rho = \rho_0[1 + \alpha(T - T_0)] \qquad [17.7]$$

where α is the **temperature coefficient of resistivity** and ρ_0 is the resistivity at some reference temperature T_0 (usually taken to be 20°C).

The resistance of a conductor varies with temperature according to the expression

$$R = R_0[1 + \alpha(T - T_0)] \qquad [17.8]$$

If a potential difference ΔV is maintained across an electrical device, the **power,** or rate at which energy is supplied to the device, is

$$\mathcal{P} = I\,\Delta V \qquad [17.9]$$

Because the potential difference across a resistor is $\Delta V = IR$, the **power delivered to a resistor** can be expressed as

$$\mathcal{P} = I^2 R = \frac{(\Delta V)^2}{R} \qquad [17.10]$$

A **kilowatt-hour** is the amount of energy converted or consumed in 1 h by a device supplied with power at the rate of 1 kW. This is equivalent to

$$1\ \text{kWh} = 3.60 \times 10^6\ \text{J} \qquad [17.11]$$

CONCEPTUAL QUESTIONS

1. A coulomb is a very large unit of charge, and yet currents of several amperes are quite common. How is this possible?

2. Edison's original lightbulb contained a carbon filament. How should carbon be described electrically?

3. Why don't the free electrons in a metal fall to the bottom of the metal due to gravity? And charges in a conductor are supposed to reside on the surface—why don't the free electrons all go to the surface?

4. In an analogy between traffic flow and electrical current, what would correspond to the charge Q? What would correspond to the current I?

5. Newspaper articles often have statements such as, "10 000 volts of electricity surged *through* the victim's body." What is wrong with this statement?

6. "A microvoltmeter measured the resistance of the sciatic nerve." Criticize this statement. How might it be corrected?

7. When the voltage across a certain conductor is doubled, the current is observed to triple. What can you conclude about the conductor?

8. There is an old admonition given to experimenters to "keep one hand in the pocket" when working around high voltages. Why might this be a good idea?

9. What factors affect the resistance of a conductor?

10. Some homes have light dimmers that are operated by rotation of a knob. What is being changed in the electric circuit when the knob is rotated?

11. Two wires A and B with circular cross section are made of the same metal and have equal lengths, but the resistance of wire A is three times greater than that of wire B. What is the ratio of their cross-sectional areas? How do the radii compare?

12. What single experimental requirement makes superconducting devices expensive to operate? In principle, can this limitation be overcome?

13. What could happen to the drift velocity of the electrons in a wire and to the current in the wire if the electrons could move freely without resistance through this metallic conductor?

14. Use the atomic theory of matter to explain why the resistance of a material should increase as its temperature increases.

PROBLEMS

1, **2**, **3** = straightforward, intermediate, challenging   ☐ = full solution available in Student Solutions Manual/Study Guide

web = solution posted at **http://info.brookscole.com/serway**   🔬 = biomedical application

Section 17.1   Electric Current

Section 17.2   A Microscopic View: Current and Drift Speed

1. If a current of 80.0 mA exists in a metal wire, how many electrons flow past a given cross section of the wire in 10.0 min? Sketch the directions of the current and the electrons' motion.

2. A small sphere that carries a charge q is whirled in a circle at the end of an insulating string. The angular frequency of rotation is ω. What average current does this rotating charge represent?

3. A total charge of 6.0 mC passes through a cross-sectional area of a wire in 2.0 s. What is the current in the wire?

4. In a particular television picture tube, the measured beam current is 60.0 μA. How many electrons strike the screen every second?

5. In the Bohr model of the hydrogen atom, an electron in the lowest energy state moves at a speed of 2.19×10^6 m/s in a circular path having a radius of 5.29×10^{-11} m. What is the effective current associated with this orbiting electron?

6. If 3.25×10^{-3} kg of gold is deposited on the negative electrode of an electrolytic cell in a period of 2.78 h, what is the current through the cell in this period? Assume that the gold ions carry one elementary unit of positive charge.

7. A 200-km-long high-voltage transmission line 2.0 cm in diameter carries a steady current of 1 000 A. If the conductor is copper with a free charge density of 8.5×10^{28} electrons per cubic meter, how long (in years) does it take one electron to travel the full length of the cable?

8. An aluminum wire with a cross-sectional area of **web** 4.0×10^{-6} m^2 carries a current of 5.0 A. Find the drift speed of the electrons in the wire. The density of aluminum is 2.7 g/cm^3. (Assume that one electron is supplied by each atom.)

9. If the current carried by a conductor is doubled, what happens to the (a) charge carrier density and (b) electron drift velocity?

Section 17.4   Resistance and Ohm's Law

Section 17.5   Resistivity

10. A lightbulb has a resistance of 240 Ω when operating at a voltage of 120 V. What is the current through the lightbulb?

11. A person notices a mild shock if the current along a path 🔬 through the thumb and index finger exceeds 80 μA. Compare the maximum allowable voltage without shock across the thumb and index finger with a dry-skin resistance of 4.0×10^5 Ω and a wet-skin resistance of 2 000 Ω.

12. Suppose that you wish to fabricate a uniform wire out of 1.00 g of copper. If the wire is to have a resistance of $R = 0.500$ Ω, and if all of the copper is to be used, what will be (a) the length and (b) the diameter of this wire?

13. Calculate the diameter of a 2.0-cm length of tungsten filament in a small lightbulb if its resistance is 0.050 Ω.

14. Eighteen-gauge wire has a diameter of 1.024 mm. Calculate the resistance of 15 m of 18-gauge copper wire at 20°C.

15. A potential difference of 12 V is found to produce a current of 0.40 A in a 3.2-m length of wire with a uniform radius of 0.40 cm. What is (a) the resistance of the wire and (b) the resistivity of the wire?

16. A length L_0 of copper wire has a resistance R_0. The wire is cut into three pieces of equal length. The pieces are then connected as parallel lengths between points A and B. What resistance will this new "wire" of length $L_0/3$ have between points A and B?

17. A wire 50.0 m long and 2.00 mm in diameter is connected to a source with a potential difference of 9.11 V, and the current is found to be 36.0 A. Assume a temperature of 20°C and, using Table 17.1, identify the metal of the wire.

18. A rectangular block of copper has sides of length 10 cm, 20 cm, and 40 cm. If the block is connected to a 6.0-V source across opposite faces of the rectangular block, what are (a) the maximum current and (b) minimum current that can be carried?

19. The breathing monitor shown in Figure P17.19 girds the 🔬 patient with a mercury-filled rubber tube and measures the variation of the tube resistance. The tube has an unstretched length of 1.25 m and an inside diameter of 2.51 mm. The monitor is connected to a 100-mV power supply, and the total resistance of the circuit is that due to

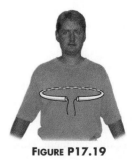

FIGURE P17.19

the mercury *plus* 1.00 Ω (an internal resistance of the power supply). Determine the change of current through the monitor as a patient draws in a breath and stretches the hose by 10.0 cm. Take $\rho_{Hg} = 9.40 \times 10^{-7}$ Ω·m.

Section 17.6 Temperature Variation of Resistance

20. A certain lightbulb has a tungsten filament with a resistance of 19 Ω when cold and 140 Ω when hot. Assume that Equation 17.8 can be used over the large temperature range involved here, and find the temperature of the filament when it is hot. Assume an initial temperature of 20°C.

21. While taking photographs in Death Valley on a day when the temperature is 58.0°C, Bill Hiker finds that a certain voltage applied to a copper wire produces a current of 1.000 A. Bill then travels to Antarctica and applies the same voltage to the same wire. What current does he register there if the temperature is −88.0°C? Assume that no change occurs in the wire's shape and size.

22. If a silver wire has a resistance of 10.0 Ω at 20.0°C, what resistance does it have at 40.0°C? Neglect any change in length or cross-sectional area resulting from the change in temperature.

23. At 20°C the carbon resistor in an electric circuit, connected to a 5.0-V battery, has a resistance of 200 Ω. What is the current in the circuit when the temperature of the carbon rises to 80°C?

24. At 40.0°C, the resistance of a segment of gold wire is 100.0 Ω. When the wire is placed in a liquid bath, the resistance decreases to 97.0 Ω. What is the temperature of the bath? (*Hint:* First determine the resistance of the gold wire at room temperature, 20°C.)

25. The copper wire used in a house has a cross-sectional area of 3.00 mm². If 10.0 m of this wire is used to wire a circuit in the house at 20.0°C, find the resistance of the wire at temperatures of (a) 30.0°C and (b) 10.0°C.

26. An aluminum rod has a resistance of 1.234 Ω at 20.0°C. Calculate the resistance of the rod at 120°C by accounting for the changes in both the resistivity and the dimensions of the rod.

27. (a) A 34.5-m length of copper wire at 20.0°C has a radius
web of 0.25 mm. If a potential difference of 9.0 V is applied across the length of the wire, determine the current in the wire. (b) If the wire is heated to 30.0°C while the 9.0-V potential difference is maintained, what is the resulting current in the wire?

28. A toaster rated at 1 050 W operates on a 120-V household circuit and has a 4.00-m length of nichrome wire as its heating element. The operating temperature of this element is 320°C. What is the cross-sectional area of the wire?

29. In one form of plethysmograph (a device for measuring volume), a rubber capillary tube with an inside diameter of 1.00 mm is filled with mercury at 20°C. The resistance of the mercury is measured with the aid of electrodes sealed into the ends of the tube. If 100.00 cm of the tube is wound in a spiral around a patient's upper arm, the blood flow during a heartbeat causes the arm to expand, stretching the tube to a length of 100.04 cm. From this observation (assuming cylindrical symmetry) you can find the change in volume of the arm, which gives an indication of blood flow. (a) Calculate the resistance of the mercury. (b) Calculate the fractional change in resistance dur-

ing the heartbeat. (*Hint:* The fraction by which the cross-sectional area of the mercury thread decreases is the fraction by which the length increases, since the volume of mercury is constant.) Take $\rho_{Hg} = 9.4 \times 10^{-7}$ Ω·m.

30. A platinum resistance thermometer has resistances of 200.0 Ω when placed in a 0°C ice bath and 253.8 Ω when immersed in a crucible containing melting potassium. What is the melting point of potassium? (*Hint:* First determine the resistance of the platinum resistance thermometer at room temperature, 20°C.)

Section 17.8 Electrical Energy and Power

31. A toaster is rated at 600 W when connected to a 120-V source. What current does the toaster carry, and what is its resistance?

32. The output power of the Sun is 4.0×10^{26} W. Calculate at eight cents per kilowatt-hour the cost of running the Sun for one second.

33. How many 100-W lightbulbs can you use in a 120-V circuit without tripping a 15-A circuit breaker? (The bulbs are connected in parallel, which means that the potential difference across each lightbulb is 120 V.)

34. A high-voltage transmission line with a resistance of 0.31 Ω/km carries a current of 1 000 A. The line is at a potential of 700 kV at the power station and carries the current to a city located 160 km from the power station. (a) What is the power loss due to resistance in the line? (b) What fraction of the transmitted power does this loss represent?

35. The heating element of a coffee maker operates at 120 V and carries a current of 2.00 A. Assuming that the water absorbs all of the energy converted by the resistor, calculate how long it takes to heat 0.500 kg of water from room temperature (23.0°C) to the boiling point.

36. The power supplied to a typical black-and-white television set is 90 W when the set is connected to 120 V. (a) How much electric energy does this set consume in one hour? (b) A color television set draws about 2.5 A when connected to 120 V. How much time is required for it to consume the same energy as the black-and-white model consumes in one hour?

37. What is the required resistance of an immersion heater that will increase the temperature of 1.50 kg of water from 10.0°C to 50.0°C in 10.0 min while operating at 120 V?

38. A certain toaster has a heating element made of nichrome resistance wire. When the toaster is first connected to a 120-V source of potential difference (and the wire is at a temperature of 20.0°C) the initial current is 1.80 A. However, the current begins to decrease as the resistive element warms up. When the toaster has reached its final operating temperature, the current has dropped to 1.53 A. (a) Find the power the toaster converts when it is at its operating temperature. (b) What is the final temperature of the heating element?

39. A copper cable is designed to carry a current of 300 A with a power loss of 2.00 W/m. What is the required radius of this cable?

40. A small motor draws a current of 1.75 A from a 120-V line. The output power of the motor is 0.20 hp. (a) At a rate of $0.060/kWh, what is the cost of operating the motor for 4.0 h? (b) What is the efficiency of the motor?

41. We estimate that there are 270 million plug-in electric clocks in the United States, approximately one clock for each person. The clocks convert energy at the average rate of 2.50 W. To supply this energy, how many metric tons of coal are burned per hour in coal-fired electric-generating plants that are, on average, 25.0% efficient? The heat of combustion for coal is 33.0 MJ/kg.

42. The cost of electricity varies widely throughout the United States; $0.120/kWh is one typical value. At this unit price, calculate the cost of (a) leaving a 40.0-W porch light on for two weeks while you are on vacation, (b) making a piece of dark toast in 3.00 min with a 970-W toaster, and (c) drying a load of clothes in 40.0 min in a 5 200-W dryer.

43. How much does it cost to watch a complete 21-hour-long World Series on a 180-W television set? Assume that electricity costs $0.070/kWh.

44. A house is heated by a 24-kW electric furnace using resistance heating. The rate for electrical energy is $0.080/kWh. If the heating bill for January is $200, how long must the furnace have been running on an average January day?

45. An 11-W energy-efficient fluorescent lamp is designed to produce the same illumination as a conventional 40-W lamp. How much does the energy-efficient lamp save during 100 hours of use? Assume a cost of $0.080/kWh for electrical energy.

46. An electric resistance heater is to deliver 1 500 kcal/h to a room using 110-V electricity. If fuses come in 10-A, 20-A, and 30-A sizes, what is the smallest fuse that can safely be used in the heater circuit?

47. The heating coil of a hot water heater has a resistance of
web 20 Ω and operates at 210 V. If electrical energy costs $0.080/kWh, what does it cost to raise the 200 kg of water in the tank from 15°C to 80°C? (See Chapter 11.)

ADDITIONAL PROBLEMS

48. One lightbulb is marked "25 W 120 V" and another "100 W 120 V"; this means that each converts its respective power when plugged into a constant 120-V potential difference. (a) Find the resistance of each bulb. (b) How long does it take for 1.00 C to pass through the dim bulb? How is this charge different upon its exit versus its entry into the bulb? (c) How long does it take for 1.00 J to pass through the dim bulb? How is this energy different upon its exit versus its entry into the bulb? (d) Find the cost of running the dim bulb continuously for 30.0 days if the electric company sells its product at $0.070 0 per kWh. What physical quantity *does* the electric company sell? What is its price for one SI unit of this quantity?

49. A particular wire has a resistivity of 3.0×10^{-8} Ω·m and a cross-sectional area of 4.0×10^{-6} m². A length of this wire is to be used as a resistor that will develop 48 W of power when connected across a 20-V battery. What length of wire is required?

50. A steam iron draws 6.0 A from a 120-V line. (a) How many joules of internal energy are produced in 20 min? (b) How much does it cost, at $0.080/kWh, to run the steam iron for 20 min?

51. An experiment is conducted to measure the electrical resistivity of nichrome in the form of wires with different lengths and cross-sectional areas. For one set of measurements, a student uses 30-gauge wire, which has a cross-sectional area of 7.30×10^{-8} m². The student measures the potential difference across the wire and the current in the wire with a voltmeter and an ammeter, respectively. For each of the measurements given in the table taken on wires of three different lengths, calculate the resistance of the wires and the corresponding values of the resistivity. What is the average value of the resistivity, and how does this value compare with the value given in Table 17.1?

L (m)	ΔV (V)	I (A)	R (Ω)	ρ (Ω·m)
0.540	5.22	0.500		
1.028	5.82	0.276		
1.543	5.94	0.187		

52. Birds resting on high-voltage power lines are a common sight. The copper wire on which a bird stands is 2.2 cm in diameter and carries a current of 50 A. If the bird's feet are 4.0 cm apart, calculate the potential difference across its body.

53. A small sphere that carries a charge of 8.00 nC is whirled in a circle at the end of an insulating string. The angular speed is 100π rad/s. What average current does this rotating charge represent?

54. The current in a conductor varies in time as shown in Figure P17.54. (a) How many coulombs of charge pass through a cross section of the conductor in the interval $t = 0$ to $t = 5.0$ s? (b) What constant current would transport the same total charge during the 5.0-s interval as does the actual current?

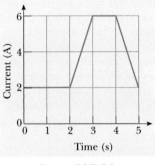

FIGURE P17.54

55. An electric car is designed to run off a bank of 12.0-V batteries with a total energy storage of 2.00×10^7 J. (a) If the electric motor draws 8.00 kW, what is the current delivered to the motor? (b) If the electric motor draws 8.00 kW as the car moves at a steady speed of 20.0 m/s, how far will the car travel before it is "out of juice"?

56. (a) A 115-g mass of aluminum is formed into a right circular cylinder, shaped so that its diameter equals its height. Calculate the resistance between the top and bottom faces of the cylinder at 20°C. (b) Calculate the resistance between opposite faces if the same mass of aluminum is formed into a cube.

57. A length of metal wire has a radius of 5.00×10^{-3} m and a resistance of 0.100 Ω. When the potential difference across the wire is 15.0 V, the electron drift speed is found to be 3.17×10^{-4} m/s. Based on these data, calculate the density of free electrons in the wire.

58. A carbon wire and a nichrome wire are connected one after the other. If the combination has total resistance of 10.0 kΩ at 20°C, what is the resistance of each wire at 20°C so that the resistance of the combination does not change with temperature?

59. (a) Determine the resistance of a lightbulb marked 100 W @ 120 V. (b) Assuming that the filament is tungsten and has a cross-sectional area of 0.010 mm^2, determine the length of the wire inside the bulb when the bulb is operating. (c) Why do you think the wire inside the bulb is tightly coiled? (d) If the temperature of the tungsten wire is 2 600°C when the bulb is operating, what is the length of the wire after the bulb is turned off and has cooled to 20°C? (See Chapter 10, and use 4.5×10^{-6}/°C as the coefficient of linear expansion for tungsten.)

60. In a certain stereo system, each speaker has a resistance of 4.00 Ω. The system is rated at 60.0 W in each channel. Each speaker circuit includes a fuse rated at a maximum current of 4.00 A. Is this system adequately protected against overload?

61. A resistor is constructed by forming a material of resistivity 3.5×10^5 Ω·m into the shape of a hollow cylinder of length 4.0 cm and inner and outer radii of 0.50 cm and 1.2 cm, respectively. In use, a potential difference is applied between the ends of the cylinder, producing a current parallel to the length of the cylinder. Find the resistance of the cylinder.

62. The graph in Figure P17.62a shows the current I in a diode as a function of potential difference ΔV across the diode. Figure P17.62b shows the circuit used to make the measurements. The symbol —▶|— represents the diode. (a) Using Equation 17.4, make a table of the resistance of the diode for different values of ΔV in the range from -1.5 V to $+1.0$ V. (b) Based on your results, what amazing electrical property does a diode possess?

63. An x-ray tube used for cancer therapy operates at 4.0 MV, with a beam current of 25 mA striking the metal target. Nearly all the power in this beam is transferred to a stream of water flowing through holes drilled in the target. What rate of flow, in kilograms per second, is needed if the temperature rise (ΔT) of the water is not to exceed 50°C?

64. A 50.0-g sample of a conducting material is all that is available. The resistivity of the material is measured to be 11×10^{-8} Ω·m, and the density is 7.86 g/cm^3. The material is to be shaped into a solid cylindrical wire that has a total resistance of 1.5 Ω. (a) What length is required? (b) What must be the diameter of the wire?

65. (a) A sheet of copper ($\rho = 1.7 \times 10^{-8}$ Ω·m) is 2.0 mm thick and has surface dimensions of 8.0 cm × 24 cm. If the long edges are joined to form a tube 24 cm in length, what is the resistance between the ends? (b) What mass of copper is required to manufacture a 1 500-m-long spool of copper cable with a total resistance of 4.5 Ω?

web

66. When a straight wire is heated, its resistance changes according to the equation

$$R = R_0[1 + \alpha(T - T_0)]$$

where α is the temperature coefficient of resistivity. (a) Show that a more precise result, which includes the fact that the length and area of a wire change when it is heated, is

$$R = \frac{R_0[1 + \alpha(T - T_0)][1 + \alpha'(T - T_0)]}{[1 + 2\alpha'(T - T_0)]}$$

where α' is the coefficient of linear expansion (see Chapter 10). (b) Compare these two results for a 2.00-m-long copper wire of radius 0.100 mm, starting at 20.0°C and heated to 100.0°C.

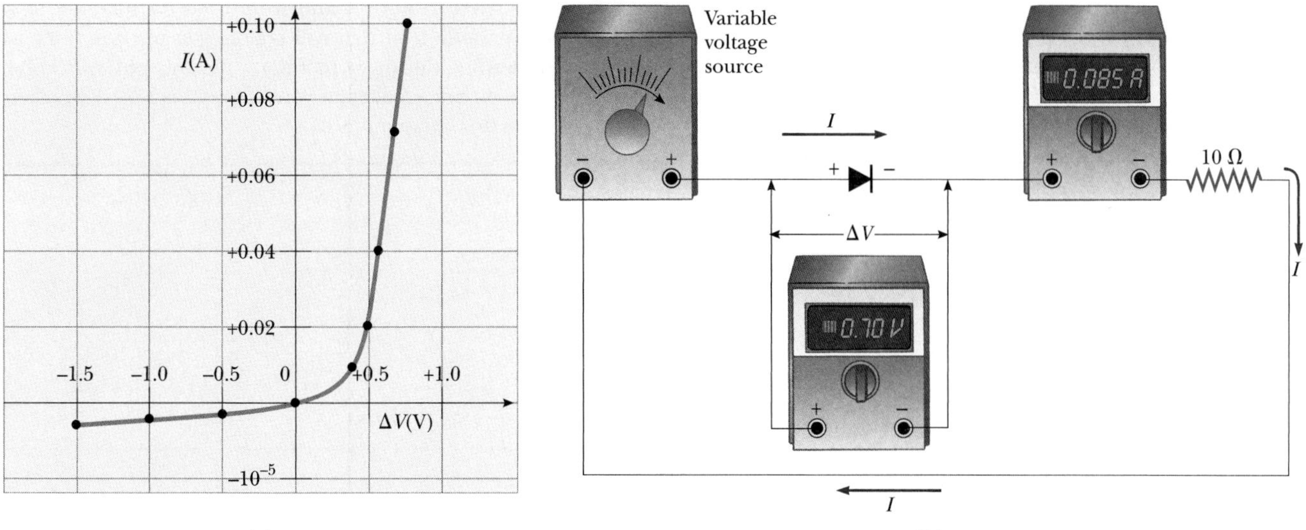

(a) (b)

FIGURE P17.62

GROUP ACTIVITIES

G.1 Connect one terminal of a D-cell battery to the base of a flashlight bulb using insulated wire, tape a second wire to the other battery terminal, and tape a third wire to the center conductor of the bulb as in Figure GA17.1. Make sure to remove about 1 cm of insulation from the ends of all wires before making connections. Now bridge the gap between the open wires with different objects, such as a plastic pen, an aluminum can, a penny, a rubber band, and a spoon. Which objects make the bulb light up? Explain your observations.

Touch objects with these wires

FIGURE GA17.1

G.2 When the lightbulbs in your home are used, they are always connected across the same potential difference. Which do you believe has a filament with the highest resistance when cool, a 60-W bulb or a 100-W bulb? To check your prediction, ask your instructor to lend you a device called an ohmmeter and to instruct you in its use. A resistor must always be disconnected from a circuit when its resistance is measured with an ohmmeter.

G.3 Examine the label on several household appliances such as a toaster, television, lamp, stereo, air conditioner, and clock. From the label, determine the power rating of the device in watts. Check the billing statement from your electric utility company to find the cost of electrical energy per kilowatt-hour. (Prices usually range from about a nickel to twenty cents.) Calculate the cost of running each appliance for 1 h. Estimate how many hours per day each appliance is used. Then calculate the monthly cost of using each appliance based on your daily estimate.

G.4 A piece of material is connected to a battery. All of the following factors affect the current in the wire. Explain why these influence the current and discuss whether the current is directly proportional or inversely proportional to each factor, or has another pattern of dependence. (a) The cross-sectional area of the material, the length of the material, the temperature of the material, the number of free charge carriers in the material. (b) "When I turn on a wall switch, the current in a room seems to come on instantly. Thus, I know that the drift velocity of the electrons in a wire is extremely fast." Argue for or against this statement. (c) Assume a certain material has 9.00×10^{28} free electrons per cubic meter, a cross-sectional area of 3.00×10^{-6} m², and a current of 0.300 A. Find the drift speed of the electrons in the material.

G.5 (a) Explain why temperature affects the resistance of a material. (b) Explain why length affects the resistance of a material. (c) Explain why cross-sectional area affects the resistance of a material. (d) A tungsten-filament lightbulb has a resistance of 240 Ω when connected to a 120-V outlet, and it is found that the temperature of the filament is about 2 000°C under these conditions. What is the current in the bulb when it is connected to a 2.00-V battery?

G.6 Two identical, parallel copper wires used as cable TV lines are placed underground between two points 1.00 mile apart. These wires are usually not connected, but a construction accident shorts the wires together at some point. Problem: Where is that point? To try to isolate the spot so that repair can be initiated, a technician goes to end A of the lines and finds that a 12.0-V battery connected across the wires at that end produces a current of 1.00 A. Doing the same at the other end, end B, produces a current of 0.20 A. (a) Is the break closer to end A or to end B? (b) How far from the 0.20-A end should digging begin?

G.7 You are cooking breakfast for yourself and a friend using a 1 200-W waffle iron and a 500-W coffee pot. Usually, you operate these appliances from a 110-V outlet for 0.50 h each day. (a) At eight cents per kWh, how much do you spend to cook breakfast each month? (b) You find yourself addicted to waffles and would like to upgrade to a 2 400-W waffle iron that will enable you to cook twice as many waffles during a half-hour period, but you know that the circuit breaker in your kitchen is a 20-A breaker. Can you do the upgrade?

Direct Current Circuits

18

Chapter Outline

18.1 Sources of emf

18.2 Resistors in Series

18.3 Resistors in Parallel

18.4 Kirchhoff's Rules and Complex DC Circuits

18.5 *RC* Circuits

18.6 Household Circuits

18.7 Electrical Safety

18.8 Conduction of Electrical Signals by Neurons

This versatile circuit enables the experimenter to examine the properties of circuit elements such as capacitors and resistors and their effects on circuit behavior. *(Courtesy of Central Scientific Company)*

This chapter analyzes some simple circuits whose elements include batteries, resistors, and capacitors in varied combinations. Such analysis is simplified by the use of two rules known as Kirchhoff's rules, which follow from the principle of conservation of energy and the law of conservation of charge. Most of the circuits are assumed to be in *steady state,* which means that the currents are constant in magnitude and direction. We close the chapter with a discussion of circuits containing resistors and capacitors, in which current varies with time.

18.1   SOURCES OF emf

The source that maintains the current in a closed circuit is called a source of emf.[1] Any devices (such as batteries and generators) that increase the potential energy of charges circulating in circuits are sources of emf. One can think of such a source as a "charge pump" that forces electrons to move in a direction opposite the electrostatic field inside the source. The emf $\mathcal{E}$ of a source is the work done per unit charge, and hence the SI unit of emf is the volt.

Consider the circuit in Figure 18.1a, consisting of a battery connected to a resistor. We assume that the connecting wires have no resistance. If we neglect the internal resistance of the battery, the potential difference across the battery (the terminal voltage) equals the emf of the battery. However, because a real battery always has some internal resistance r, the terminal voltage is not equal to the emf. The circuit of Figure 18.1a can be described schematically by the diagram in Figure 18.1b. The battery, represented by the dashed rectangle, consists of a source of emf $\mathcal{E}$ in series with an internal resistance r. Now imagine a positive charge moving from a to b in Figure 18.1b. As the charge passes from the negative to the positive terminal of the battery, the potential of the charge increases by $\mathcal{E}$. However, as the charge moves through the resistance r its potential decreases by the amount Ir, where I is the current in the circuit. Thus, the terminal voltage of the battery, $\Delta V = V_b - V_a$, is

$$\Delta V = \mathcal{E} - Ir \qquad [18.1]$$

Note from this expression that **$\mathcal{E}$ is equal to the terminal voltage when the current is zero,** called the **open-circuit voltage.** By inspecting Figure 18.1b, we see that the terminal voltage ΔV must also equal the potential difference across the external resistance R, often called the **load resistance;** that is, $\Delta V = IR$. Combining this with Equation 18.1, we see that

$$\mathcal{E} = IR + Ir \qquad [18.2]$$

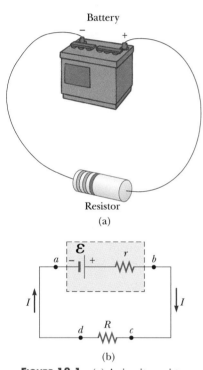

Battery

Resistor

(a)

(b)

FIGURE 18.1   (a) A circuit consisting of a resistor connected to the terminals of a battery. (b) A circuit diagram of a source of emf $\mathcal{E}$ having internal resistance r connected to an external resistor R.

An assortment of batteries. *(George Semple)*

[1] The term was originally an abbreviation for electromotive force, but emf is not really a force, so the long form is discouraged.

Solving for the current gives

$$I = \frac{\mathcal{E}}{R + r}$$

This shows that the current in this simple circuit depends on both the resistance external to the battery and the internal resistance. If R is much greater than r, we can neglect r in our analysis and we do, for many circuits. If we multiply Equation 18.2 by the current I, we get

$$I\mathcal{E} = I^2R + I^2r$$

This equation tells us that the total power output $I\mathcal{E}$ of the source of emf is converted to the rate I^2R at which energy is delivered to the load resistance *plus* the rate I^2r at which energy is delivered to the internal resistance. Again, if $r \ll R$, most of the power delivered by the battery is transferred to the load resistance.

Unless otherwise stated, we will assume in our examples and end-of-chapter problems that the internal resistance of a battery in a circuit is negligible.

18.2 RESISTORS IN SERIES

When two or more resistors are connected end-to-end as in Figure 18.2, they are said to be in *series*. The resistors could be simple devices such as lightbulbs or heating elements. When the two resistors R_1 and R_2 are connected to a battery as in Figure 18.2, **the current is the same in the two resistors because any charge that flows through R_1 must also flow through R_2.** This is analogous to water flowing through a pipe with two constrictions, corresponding to R_1 and R_2. Whatever volume of water flows in one end in a given time interval must exit the opposite end in the same time interval.

Because the potential difference between a and b in Figure 18.2b equals IR_1 and the potential difference between b and c equals IR_2, the potential difference between a and c is

$$\Delta V = IR_1 + IR_2 = I(R_1 + R_2)$$

Regardless of how many resistors we have in series, the sum of the potential differences across the resistors is equal to the total potential difference across the combination. As we will show later, this is a consequence of the conservation of energy. Figure 18.2c shows an equivalent resistor R_{eq} that can replace the two resistors of the original circuit. The equivalent resistance has the same effect on the circuit as

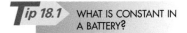

Tip 18.1 WHAT IS CONSTANT IN A BATTERY?

Equation 18.2 shows that the current in a circuit depends on the resistance of the battery. Note that the battery is not a source of constant current. Furthermore, the terminal voltage of a battery given by Equation 18.1 is also not constant. A battery is a source of constant emf.

◀ For a series connection of resistors, the current is the same in all the resistors.

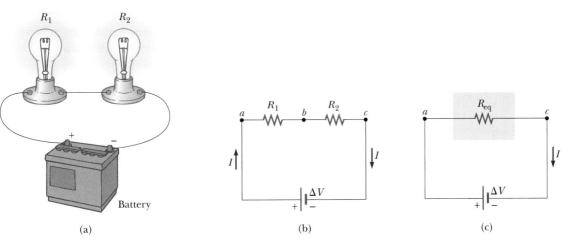

(a)   (b)   (c)

FIGURE 18.2 A series connection of two resistors R_1 and R_2. The currents in the resistors are the same, and the equivalent resistance of the combination is given by $R_{eq} = R_1 + R_2$.

the combination of resistors because it results in the same current in the circuit. Applying Ohm's law to this resistor, we have

$$\Delta V = IR_{eq}$$

Equating the preceding two expressions, we have

$$IR_{eq} = I(R_1 + R_2)$$

or

$$R_{eq} = R_1 + R_2 \qquad [18.3]$$

An extension of this analysis shows that the equivalent resistance of three or more resistors connected in series is

$$R_{eq} = R_1 + R_2 + R_3 + \cdots \qquad [18.4]$$

Equivalent resistance of a series combination of resistors.

Therefore, **the equivalent resistance of a series combination of resistors is the algebraic sum of the individual resistances and is always greater than any individual resistance.**

Note that if the filament of one lightbulb in Figure 18.2 were to fail, the circuit would no longer be complete (an open-circuit condition would exist) and the second bulb would also go out.

APPLYING **PHYSICS** *18.1*

A new design for Christmas tree lights allows them to be connected in series. One might expect that a failed bulb in such a string would result in an open circuit and all of the bulbs would go out. How can the bulbs be designed to prevent this from happening?

Explanation   If the string of lights contained normal bulbs, a failed bulb would be hard to locate. Each bulb would have to be replaced with a good bulb, one by one, until the failed bulb is found. If there happened to be two or more failed bulbs in the string of lights, finding them would be a formidable task.

Christmas lights use specially designed bulbs that have an insulated loop of wire (jumper) across the conducting supports to the bulb filaments (Fig. 18.3). If the filament breaks and the bulb fails, the resistance of this bulb increases dramatically. As a result, most of the applied voltage appears across the loop of wire. This voltage causes the insulation around the loop of wire to melt, causing the metal wire to make electrical contact with the supports. This produces a conducting path through the bulb, enabling the other bulbs to remain lit.

APPLICATION

CHRISTMAS LIGHTS IN SERIES

Webnote 18.1

Holiday lights are a tradition at the end of the year. Learn how these colorful and festive decorations are constructed at
http://www.howstuffworks.com/christmas-lights.htm

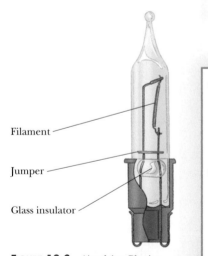

Filament

Jumper

Glass insulator

FIGURE 18.3   (Applying Physics 18.1) Diagram of a modern miniature holiday lightbulb, with a jumper connection to provide a current if the filament breaks.

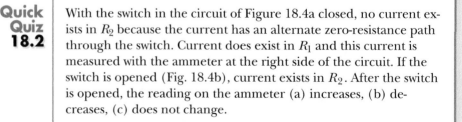

Quick Quiz 18.1   When a piece of wire is used to connect points b and c in Figure 18.2b, the brightness of bulb R_1 (a) increases, (b) decreases, or (c) stays the same. The brightness of bulb R_2 (a) increases, (b) decreases, or (c) stays the same.

Quick Quiz 18.2   With the switch in the circuit of Figure 18.4a closed, no current exists in R_2 because the current has an alternate zero-resistance path through the switch. Current does exist in R_1 and this current is measured with the ammeter at the right side of the circuit. If the switch is opened (Fig. 18.4b), current exists in R_2. After the switch is opened, the reading on the ammeter (a) increases, (b) decreases, (c) does not change.

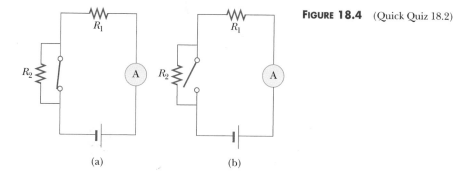

FIGURE 18.4 (Quick Quiz 18.2)

(a)   (b)

Example 18.1   Four Resistors in Series

Four resistors are arranged as shown in Figure 18.5a. Find (a) the equivalent resistance and (b) the current in the circuit if the emf of the battery is 6.0 V.

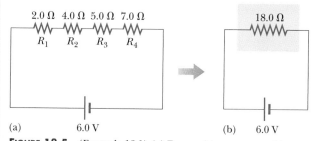

(a)   6.0 V   (b)   6.0 V

FIGURE 18.5 (Example 18.1) (a) Four resistors connected in series. (b) The equivalent resistance of the circuit in (a).

Solution

A The equivalent resistance is found from Equation 18.4:

$$R_{eq} = R_1 + R_2 + R_3 + R_4 = 2.0\ \Omega + 4.0\ \Omega + 5.0\ \Omega + 7.0\ \Omega = \boxed{18.0\ \Omega}$$

B If we apply Ohm's law to the equivalent resistor in Figure 18.5b, we find the current in the circuit to be

$$I = \frac{\Delta V}{R_{eq}} = \frac{6.0\ \text{V}}{18.0\ \Omega} = \boxed{\tfrac{1}{3}\ \text{A}}$$

EXERCISE Because the current in the equivalent resistor is $\tfrac{1}{3}$ A, this must also be the current in each resistor of the original circuit. Find the voltage drop across each resistor.

ANSWER $\Delta V_{2\Omega} = \tfrac{2}{3}$ V; $\Delta V_{4\Omega} = \tfrac{4}{3}$ V; $\Delta V_{5\Omega} = \tfrac{5}{3}$ V; $\Delta V_{7\Omega} = \tfrac{7}{3}$ V

18.3 RESISTORS IN PARALLEL

Now consider two resistors connected in parallel, as in Figure 18.6. In this case, **the potential differences across the resistors are the same because each is connected directly across the battery terminals.** The currents are generally not the same. When the charge reaches point a (called a junction) in Figure 18.6b, the current splits into two parts, I_1 going through R_1 and I_2 going through R_2. If R_1 is greater than R_2, then I_1 is less than I_2. That is, the charge tends to follow the path of least resistance. **Because charge is conserved, the current I that enters point a must equal the total current leaving that point, $I_1 + I_2$.** That is,

$$I = I_1 + I_2$$

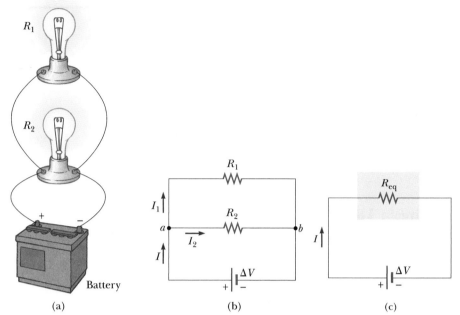

FIGURE 18.6   (a) A parallel connection of two resistors. (b) A circuit diagram for the parallel combination. (c) The voltages across the resistors are the same, and the equivalent resistance of the combination is given by the reciprocal relationship $1/R_{eq} = 1/R_1 + 1/R_2$.

The potential drop must be the same for the two resistors and must also equal the potential drop across the battery. Ohm's law applied to each resistor gives

$$I_1 = \frac{\Delta V}{R_1} \qquad I_2 = \frac{\Delta V}{R_2}$$

Ohm's law applied to the equivalent resistor in Figure 18.6c gives

$$I = \frac{\Delta V}{R_{eq}}$$

When these expressions for the current are substituted into the equation $I = I_1 + I_2$, and ΔV is cancelled, we obtain

$$\frac{1}{R_{eq}} = \frac{1}{R_1} + \frac{1}{R_2} \qquad \text{(parallel combination)} \qquad \textbf{[18.5]}$$

An extension of this analysis to three or more resistors in parallel produces the following general expression for the equivalent resistance:

Equivalent resistance of a ▶
parallel combination of
resistors.

$$\frac{1}{R_{eq}} = \frac{1}{R_1} + \frac{1}{R_2} + \frac{1}{R_3} + \cdots \qquad \textbf{[18.6]}$$

From this expression, it can be seen that **the inverse of the equivalent resistance of two or more resistors connected in parallel is the algebraic sum of the inverses of the individual resistances. The equivalent resistance is always less than the smallest resistance in the group.**

Household circuits are always wired so that the electrical devices are connected in parallel, as in Figure 18.6a. In this manner, each device operates independently of the others, so that if one is switched off, the others remain on. For example, if one of the lightbulbs in Figure 18.6 were removed from its socket, the other would continue to operate. Equally important, each device operates on the same voltage. If the devices were connected in series, the voltage across any one device would depend on how many devices were in the combination and on their individual resistances.

In many household circuits, circuit breakers are used in series with other circuit elements for safety purposes. A circuit breaker is designed to switch off and open the circuit at some maximum current (typically 15 A or 20 A) whose value depends on the nature of the circuit. If a circuit breaker were not used, excessive currents caused by operating several devices simultaneously could result in excessive wire temperatures and perhaps cause a fire. In older home construction, fuses were used in place of circuit breakers. When the current in a circuit exceeds some value, the conductor in a fuse melts and opens the circuit. The disadvantage of fuses is that they are destroyed in the process of opening the circuit, whereas circuit breakers can be reset.

◀ **APPLICATION**

CIRCUIT BREAKERS

PROBLEM-SOLVING *STRATEGY*   Resistors

1. When two or more unequal resistors are connected in *series*, they carry the same current, but the potential differences across them are not the same. The resistors add directly to give the equivalent resistance of the series combination.

2. When two or more unequal resistors are connected in *parallel*, the potential differences across them are the same. Because the current is inversely proportional to the resistance, the currents through them are not the same. The equivalent resistance of a parallel combination of resistors is found through reciprocal addition, and the equivalent resistance is always *less* than the smallest individual resistor in the combination.

3. A complicated circuit consisting of several resistors and batteries can often be reduced to a simple circuit with only one resistor. To do so, examine the initial circuit and replace any resistors in series or any in parallel using the procedures outlined in Steps 1 and 2. Sketch the new circuit after these changes have been made. Examine the new circuit and replace any series or parallel combinations. Continue this process until a single equivalent resistance is found.

4. If the current in or the potential difference across a resistor in the complicated circuit is to be identified, start with the final circuit found in Step 3 and gradually work back through the circuits, using $\Delta V = IR$ and the procedures of Steps 1 and 2.

APPLYING **PHYSICS** *18.2*

Compare the brightness of the four identical bulbs shown in Figure 18.7. What happens if bulb A fails, so that it cannot conduct current? What if C fails? What if D fails?

Explanation   Bulbs A and B are connected in series across the emf of the battery, whereas bulb C is connected by itself across the battery. Thus, the emf is split between bulbs A and B. As a result, bulb C will be brighter than bulbs A and B, which should be equally as bright as each other. Bulb D has a wire connected across it. Thus, there is no potential difference across D, and it does not glow at all. If bulb A fails, B goes out, but C stays lit. If C fails, there is no effect on the other bulbs. If D fails, the event is undetectable, because D was not glowing initially.

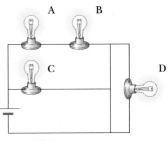

FIGURE 18.7   (Applying Physics 18.2)

APPLICATION

THREE-WAY LIGHTBULBS

FIGURE 18.8   (Applying Physics 18.3)

APPLYING **PHYSICS** 18.3

Figure 18.8 illustrates how a three-way lightbulb is constructed to provide three levels of light intensity. The socket of the lamp is equipped with a three-way switch for selecting different light intensities. The bulb contains two filaments. Why are the filaments connected in parallel? Explain how the two filaments are used to provide three different light intensities.

Explanation   If the filaments were connected in series and one of them were to fail, there would be no current in the bulb and the bulb would give no illumination, regardless of the switch position. However, when the filaments are connected in parallel and one of them (say the 75-W filament) fails, the bulb will still operate in one of the switch positions because there is current in the other (100-W) filament. The three light intensities are made possible by selecting one of three values of filament resistance using a single value of 120 V for the applied voltage. The 75-W filament offers one value of resistance, the 100-W filament offers a second value, and the third resistance is obtained by combining the two filaments in parallel. When switch S_1 is closed and switch S_2 is opened, only the 75-W filament carries current. When switch S_1 is open and switch S_2 is closed, only the 100-W filament carries current. When both switches are closed, both filaments carry current and a total illumination corresponding to 175 W is obtained.

Quick Quiz 18.3   With the switch in the circuit of Figure 18.9a open, there is no current in R_2. There is current in R_1 and this current is measured with the ammeter at the right side of the circuit. If the switch is closed (Fig. 18.9b), there is current in R_2. When the switch is closed, the reading on the ammeter (a) increases, (b) decreases, or (c) remains the same.

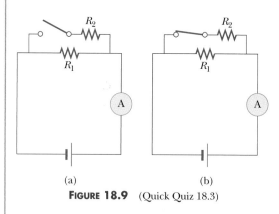

(a)   (b)

FIGURE 18.9   (Quick Quiz 18.3)

Quick Quiz 18.4   You have a large supply of lightbulbs and a battery. You start with one lightbulb connected to the battery and notice its brightness. You then add one lightbulb at a time, each new bulb being added in parallel to the previous bulbs. As the lightbulbs are added, what happens (a) to the brightness of the bulbs, (b) to the current in the bulbs, (c) to the power delivered by the battery, (d) to the lifetime of the battery, (e) to the terminal voltage of the battery? *Hint:* Do not ignore the internal resistance of the battery.

Example 18.2   **Three Resistors in Parallel**

Three resistors are connected in parallel as in Figure 18.10. A potential difference of 18 V is maintained between points a and b.

A Find the current in each resistor.

Solution   Because the resistors are in parallel, the potential difference across each is 18 V. Let us apply $\Delta V = IR$ to find the current in each resistor:

$$I_1 = \frac{\Delta V}{R_1} = \frac{18 \text{ V}}{3.0 \text{ }\Omega} = \boxed{6.0 \text{ A}}$$

$$I_2 = \frac{\Delta V}{R_2} = \frac{18 \text{ V}}{6.0 \text{ }\Omega} = \boxed{3.0 \text{ A}}$$

$$I_3 = \frac{\Delta V}{R_3} = \frac{18 \text{ V}}{9.0 \text{ }\Omega} = \boxed{2.0 \text{ A}}$$

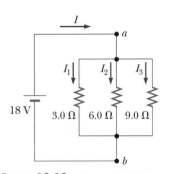

FIGURE 18.10   (Example 18.2) Three resistors connected in parallel. The voltage across each resistor is 18 V.

There is a saying that holds that current, like politicians, follows the path of least resistance. This is only partially correct, as our example demonstrates. The smallest (3.0-Ω) resistor carries the largest current, while the other larger resistors of 6.0 Ω and 9.0 Ω carry smaller currents.

B Calculate the power delivered to each resistor and the total power delivered to the three resistors.

Solution   Applying $\mathcal{P} = I^2R$ to each resistor gives

$$3 \text{ }\Omega: \quad \mathcal{P}_1 = I_1{}^2R_1 = (6.0 \text{ A})^2(3.0 \text{ }\Omega) = \boxed{110 \text{ W}}$$

$$6 \text{ }\Omega: \quad \mathcal{P}_2 = I_2{}^2R_2 = (3.0 \text{ A})^2(6.0 \text{ }\Omega) = \boxed{54 \text{ W}}$$

$$9 \text{ }\Omega: \quad \mathcal{P}_3 = I_3{}^2R_3 = (2.0 \text{ A})^2(9.0 \text{ }\Omega) = \boxed{36 \text{ W}}$$

(Note that you can also use $\mathcal{P} = (\Delta V)^2/R$ to find the power dissipated by each resistor.) Summing the three quantities gives a total power of 200 W.

EXERCISE   Calculate the equivalent resistance of the three resistors, and from this result find the total power dissipated.

ANSWER $\frac{18}{11}$ Ω;   200 W

Example 18.3   **Equivalent Resistance**

Four resistors are connected as shown in Figure 18.11a.

A Find the equivalent resistance between points a and c.

Solution   The circuit can be reduced in steps, as shown in Figures 18.11b and 18.11c. The 8.0-Ω and 4.0-Ω resistors are in series, and so the equivalent resistance between a and b is 12 Ω (Eq. 18.4). The 6.0-Ω and 3.0-Ω resistors are in parallel, and so from Equation 18.6 we find that the equivalent resistance from b to c is 2.0 Ω. Hence, the equivalent resistance from a to c is $\boxed{14 \text{ }\Omega}$.

B What is the current in each resistor if the terminals of a 42-V battery are connected between a and c?

Solution   The current I is the same in the 8.0-Ω and 4.0-Ω resistors because they are in series. Using Ohm's law and the results of (a), we get

$$I = \frac{\Delta V_{ac}}{R_{eq}} = \frac{42 \text{ V}}{14 \text{ }\Omega} = \boxed{3.0 \text{ A}}$$

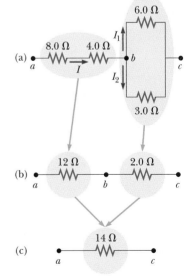

FIGURE 18.11   (Example 18.3) The four resistors shown in (a) can be reduced in steps to an equivalent 14-Ω resistor.

When this current enters the junction at *b*, it splits; part of it passes through the 6.0-Ω resistor (I_1), and part passes through the 3.0-Ω resistor (I_2). Because the potential difference ΔV_{bc} across these resistors is the *same* (they are in parallel), $(6\ \Omega)I_1 = (3\ \Omega)I_2$, or $I_2 = 2I_1$. Using this result and the fact that $I_1 + I_2 = 3.0$ A, we find that $I_1 =$ 1.0 A and $I_2 =$ 2.0 A. We could have guessed this from the start by noting that the current in the 3.0-Ω resistor has to be twice the current in the 6.0-Ω resistor in view of their relative resistances and the fact that the same voltage is applied to both.

As a final check, note that $\Delta V_{bc} = (6\ \Omega)I_1 = (3\ \Omega)I_2 = 6.0$ V and $\Delta V_{ab} = (12\ \Omega)I_1 = 36$ V; therefore, $\Delta V_{ac} = \Delta V_{ab} + \Delta V_{bc} = 42$ V, as expected.

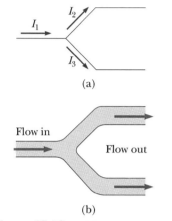

(a)

(b)

FIGURE 18.12 (a) A schematic diagram illustrating Kirchhoff's junction rule. Conservation of charge requires that whatever current enters a junction must leave that junction. Therefore, in this case, $I_1 = I_2 + I_3$. (b) A mechanical analog of the junction rule: The net flow in must equal the net flow out.

18.4 KIRCHHOFF'S RULES AND COMPLEX DC CIRCUITS

As demonstrated in the preceding section, we can analyze simple circuits using Ohm's law and the rules for series and parallel combinations of resistors. However, there are many ways in which resistors can be connected so that the circuits formed cannot be reduced to a single equivalent resistor. The procedure for analyzing more complex circuits is greatly simplified by the use of two simple rules called **Kirchhoff's rules:**

1. **The sum of the currents entering any junction must equal the sum of the currents leaving that junction. (This rule is often referred to as the junction rule.)**
2. **The sum of the potential differences across all the elements around any closed-circuit loop must be zero. (This rule is usually called the loop rule.)**

The junction rule is a statement of *conservation of charge*. Because charge must be conserved, whatever current enters a point in a circuit must equal the total current leaving that point. If we apply this rule to the junction in Figure 18.12a, we get

$$I_1 = I_2 + I_3$$

Figure 18.12b represents a mechanical analog to this situation, in which water flows through a branched pipe with no leaks. The flow rate into the pipe equals the total flow rate out of the two branches.

The loop rule is equivalent to the principle of *conservation of energy*. Any charge that moves around any closed loop in a circuit (starting and ending at the same point) must gain as much energy as it loses. It gains energy as it is pumped through a source of emf. Its energy may decrease in the form of a potential drop, $-IR$, across a resistor or as a result of flowing backward through a source of emf— that is, from the positive to the negative terminal inside the battery. In the latter case, electrical energy is converted to chemical energy as the battery is charged.

When applying Kirchhoff's rules, you must make two decisions at the beginning of a problem.

1. Assign symbols and directions to the currents in all branches of the circuit. Do not be alarmed that you might guess the direction of a current incorrectly; the resulting answer will be negative, but *its magnitude will be correct.*
2. When applying the loop rule, you must choose a direction for traversing the loop and be consistent in going either clockwise or counterclockwise. As you traverse the loop, record voltage drops and rises according to the rules stated next. They are summarized in Figure 18.13, where it is assumed that the traversal is from point *a* toward point *b*:

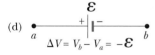

(a) $\Delta V = V_b - V_a = -IR$

(b) $\Delta V = V_b - V_a = +IR$

(c) $\Delta V = V_b - V_a = +\mathcal{E}$

(d) $\Delta V = V_b - V_a = -\mathcal{E}$

FIGURE 18.13 Rules for determining the potential changes across a resistor and a battery, assuming the battery has no internal resistance.

(a) If a resistor is traversed in the direction of the current, the change in electric potential across the resistor is $-IR$ (Fig. 18.13a).

(b) If a resistor is traversed in the direction opposite the current, the change in electric potential across the resistor is $+IR$ (Fig. 18.13b).

(c) If a source of emf is traversed in the direction of the emf (from $-$ to $+$ on the terminals), the change in electric potential is $+\mathcal{E}$ (Fig. 18.13c).

(d) If a source of emf is traversed in the direction opposite the emf (from $+$ to $-$ on the terminals), the change in electric potential is $-\mathcal{E}$ (Fig. 18.13d).

There are limits to the numbers of times the junction rule and the loop rule can be used. You can use the junction rule as often as needed so long as, each time you write an equation, you include in it a current that has not been used in a previous junction-rule equation. (If this procedure is not followed, a new equation will not be produced.) In general, the number of times the junction rule can be used is one fewer than the number of junction points in the circuit. The loop rule can be used as often as needed so long as a new circuit element (resistor or battery) or a new current appears in each new equation. In general, **to solve a particular circuit problem, you need as many independent equations as you have unknowns.**

GUSTAV KIRCHHOFF, GERMAN PHYSICIST (1824–1887)

Kirchhoff, a professor at Heidelberg, together with Robert Bunsen invented the spectroscopy that we study in Chapter 28. He also formulated another rule that states "a cool substance will absorb light of the same wavelengths that it emits when hot." *(AIP ESVA, W. F. Meggers Collection)*

PROBLEM-SOLVING *STRATEGY*	**Kirchhoff's Rules**

1. First, draw the circuit diagram and assign labels and symbols to all the known and unknown quantities. Assign *directions* to the currents in each part of the circuit. Although the assignment of current directions is arbitrary, you must stick with them throughout as you apply Kirchhoff's rules.

2. Apply the junction rule to any junction in the circuit. The junction rule may be applied as many times as a new current (one not used in a previous application) appears in the resulting equation.

3. Now apply Kirchhoff's loop rule to as many loops in the circuit as are needed to solve for the unknowns. In order to apply this rule, you must correctly identify the change in electric potential as you cross each element in traversing the closed loop. Watch out for signs! (Use the conventions in Fig. 18.13.)

4. Solve the equations simultaneously for the unknown quantities. Be careful in your algebraic steps, and check your numerical answers for consistency.

Example 18.4   Applying Kirchhoff's Rules

Find the currents in the circuit shown in Figure 18.14.

Reasoning   There are three unknown currents in this circuit, and so we must obtain three independent equations. We can find the equations with one application of the junction rule and two applications of the loop rule.

Solution   The first step is to assign a current to each branch of the circuit; these are our unknowns and are labeled I_1, I_2, and I_3 in Figure 18.14. It is also necessary to guess directions for the currents. Your experience with circuits such as this should tell you that the directions of all three have been chosen correctly. (However, recall that if a current direction is chosen incorrectly, the numerical answer will turn out negative, but the magnitude will be correct. This point will be demonstrated in the next example.)

We now apply Kirchhoff's rules. First we can apply the junction rule using either c or d, the only two junctions in the circuit. Let us choose junction c. The net current into

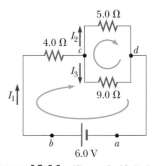

FIGURE 18.14   (Example 18.4) A multiloop circuit.

this junction is I_1, and the net current leaving it is $I_2 + I_3$. Thus, the junction rule applied to c gives

$$I_1 = I_2 + I_3$$

Recall that you may apply the junction rule over and over until you reach a situation in which no new currents appear in an equation. In this example we have reached that point with one application. If we apply the junction rule at d, we find that $I_1 = I_2 + I_3$, exactly the same equation.

We have three unknowns in our problem, I_1, I_2, and I_3; thus, we need two more independent equations before we can find a solution. We obtain these equations by applying the loop rule to the two loops indicated in the figure. Note that there are actually three loops in the circuit, but these two are sufficient to complete the problem. (Where is the loop that we do not use?)

When applying the loop rule, we must first choose the loops to be traversed, and then the directions in which to traverse them. We have selected the two loops indicated in the figure and have decided to traverse both of them clockwise. Other choices could be made, but the final result would be the same.

Starting at point a and moving clockwise around the large loop, we encounter the following voltage changes (see Fig. 18.13 for the basic sign conventions):

- From a to b, we encounter a voltage change of 6.0 V.
- From b to c through the 4.0-Ω resistor, we encounter a voltage change of $- (4.0\ \Omega)I_1$.
- From c to d through the 9.0-Ω resistor, we encounter a voltage change of $- (9.0\ \Omega)I_3$.

No voltage change occurs from d back to a. Now that we have made a complete traversal of the loop, we can equate the sum of the voltage changes to zero:

$$6.0\ \text{V} - (4.0\ \Omega)I_1 - (9.0\ \Omega)I_3 = 0$$

Moving clockwise around the small loop from point c, we encounter the following:

- From c to d through the 5.0-Ω resistor, a voltage change of $- (5.0\ \Omega)I_2$
- From d to c through the 9.0-Ω resistor, a voltage change of $+ (9.0\ \Omega)I_3$

We find that

$$- (5.0\ \Omega)I_2 + (9.0\ \Omega)I_3 = 0$$

Thus, we have the following three equations to be solved for the three unknowns:

$$I_1 - I_2 - I_3 = 0$$

$$6.0 - 4.0I_1 - 9.0I_3 = 0$$

$$-5.0I_2 + 9.0I_3 = 0$$

where we have dropped units of volts and ohms and realize that the currents will have units of amperes.

If you need help in solving three equations with three unknowns, see Example 18.5. You should be able to obtain the following answers:

$$I_1 = \boxed{0.83\ \text{A}} \qquad I_2 = \boxed{0.53\ \text{A}} \qquad I_3 = \boxed{0.30\ \text{A}}$$

Exercise Solve this same problem by using the methods learned earlier for series and parallel combinations of resistors. First find the equivalent resistance of the circuit, which you can then use to obtain I_1.

Example 18.5 Another Application of Kirchhoff's Rules

Find I_1, I_2, and I_3 in Figure 18.15.

Reasoning To find the three unknown currents, we apply the junction rule once and the loop rule twice.

Tip 18.2 CURRENT DOES NOT TAKE THE PATH OF LEAST RESISTANCE

The statement "current takes the path of least resistance" is incorrect in reference to a parallel combination of current paths. In this context, the current takes all paths. Those paths with lower resistance have larger currents, whereas those with higher resistance have lower currents.

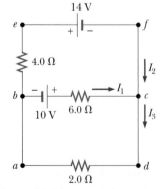

FIGURE 18.15 (Example 18.5) A circuit containing three loops.

Solution We choose the directions of the currents as shown in the figure. Applying Kirchhoff's first rule to junction c gives

$$I_1 + I_2 = I_3 \tag{1}$$

The circuit has three loops: *abcda*, *befcb*, and *aefda*. We need only two loop equations to determine the unknown currents. The third loop equation would give no new information. Applying Kirchhoff's second rule to loops *abcda* and *befcb* and traversing these loops clockwise, we obtain the following expressions:

Loop *abcda*: $\qquad 10 \text{ V} - (6.0 \ \Omega)I_1 - (2.0 \ \Omega)I_3 = 0 \tag{2}$

Loop *befcb*: $\qquad -14 \text{ V} + (6.0 \ \Omega)I_1 - 10 \text{ V} - (4.0 \ \Omega)I_2 = 0 \tag{3}$

Note that in loop *befcb*, a positive sign is obtained when the 6.0-Ω resistor is traversed, because the direction of the path is opposite the direction of the current I_1. A third loop equation for *aefda* gives $-14 \text{ V} - (2.0 \ \Omega)I_3 - (4.0 \ \Omega)I_2 = 0$, which is just the sum of (2) and (3). Expressions (1), (2), and (3) represent three linear, independent equations with three unknowns.

We can solve the problem as follows: substitution of (1) into (2) gives, with units ignored for the moment,

$$10 - 6.0I_1 - 2.0(I_1 + I_2) = 0$$

$$10 = 8.0I_1 + 2.0I_2 \tag{4}$$

Dividing each term in (3) by 2 and rearranging the equation gives

$$-12 = -3.0I_1 + 2.0I_2 \tag{5}$$

Subtracting (5) from (4) eliminates I_2, giving

$$22 = 11I_1$$

$$I_1 = 2.0 \text{ A}$$

Using this value of I_1 in (5) yields a value for I_2:

$$2.0I_2 = 3.0I_1 - 12 = 3.0(2.0) - 12 = -6.0$$

$$I_2 = -3.0 \text{ A}$$

Finally, $I_3 = I_1 + I_2 = -1.0$ A. Hence, the currents have the values

$$I_1 = \boxed{2.0 \text{ A}} \qquad I_2 = \boxed{-3.0 \text{ A}} \qquad I_3 = \boxed{-1.0 \text{ A}}$$

The fact that I_2 and I_3 are both negative merely indicates that we chose the wrong directions for these currents. The numerical values are correct.

EXERCISE Find the potential difference between junctions b and c.

ANSWER $V_b - V_c = 2.0$ V

18.5 *RC* CIRCUITS

So far, we have been concerned with circuits with constant currents. We now consider direct-current circuits containing capacitors, in which the currents vary with time. Consider the series circuit in Figure 18.16a. Let us assume that the capacitor is initially uncharged with the switch opened. After the switch is closed, the battery begins to charge the plates of the capacitor and a current passes through the resistor. The charging process continues until the capacitor is charged to its maximum equilibrium value $Q = C\mathcal{E}$, where $\mathcal{E}$ is the maximum voltage across the capacitor. Once the capacitor is fully charged, the current in the circuit is zero. If we assume that the capacitor is uncharged before the switch is closed, and the switch is closed

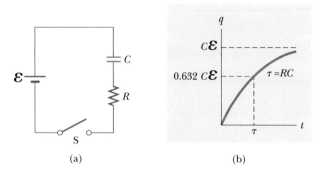

FIGURE 18.16 (a) A capacitor in series with a resistor, a battery, and a switch. (b) A plot of the charge on the capacitor versus time after the switch for the circuit is closed. After one time constant τ, the charge is 63% of the maximum value $C\mathcal{E}$. The charge approaches its maximum value as t approaches infinity.

at $t = 0$, we find that the charge on the capacitor varies with time according to the expression

$$q = Q(1 - e^{-t/RC})$$ [18.7]

where $e = 2.718$ . . . is Euler's constant, the base of the natural logarithms. Figure 18.16b is a graph of this expression. Note that the charge is zero at $t = 0$ and approaches its maximum value Q as t approaches infinity. The voltage ΔV across the capacitor at any time is obtained by dividing the charge by the capacitance. That is, $\Delta V = q/C$.

As you can see from Equation 18.7, it takes an infinite amount of time for the capacitor to become fully charged. The term RC that appears in Equation 18.7, is called the **time constant,** τ (Greek letter tau), so

$$\tau = RC$$ [18.8]

The time constant represents the time required for the charge to increase from zero to 63.2% of its maximum equilibrium value. That is, in one time constant, the charge on the capacitor increases from zero to $0.632Q$. This can be seen by substituting $t = \tau = RC$ in Equation 18.7 and solving for q. (Note that $1/e = 0.632$.) It is important to note that a capacitor charges very slowly in a circuit with a long time constant, whereas it charges very rapidly in a circuit with a short time constant.

Now consider the circuit in Figure 18.17a, consisting of a capacitor with an initial charge Q, a resistor, and a switch. Before the switch is closed, the potential difference across the charged capacitor is Q/C. Once the switch is closed, the charge begins to flow through the resistor from one capacitor plate to the other until the capacitor is fully discharged. If the switch is closed at $t = 0$, it can be shown that the charge q on the capacitor varies with time according to the expression

$$q = Qe^{-t/RC}$$ [18.9]

That is, the charge decreases exponentially with time as shown in Figure 18.17b. In the interval $t = \tau = RC$, the charge decreases from its initial value Q to $0.368Q$. In other words, in one time constant, the capacitor loses 63.2% of its initial charge. Because $\Delta V = q/C$, we see that the voltage across the capacitor also decreases ex-

Webnote 18.2

You can see the effects of charging and discharging the capacitor in an *RC* circuit by studying the following java applet:
http://www.phy.ntnu.edu.tw/java/rc/rc.html

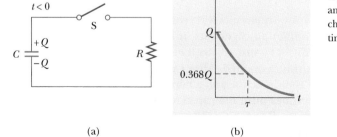

FIGURE 18.17 (a) A charged capacitor connected to a resistor and a switch. (b) A graph of the charge on the capacitor versus time after the switch is closed.

ponentially with time according to the expression $\Delta V = \mathcal{E}e^{-t/RC}$, where $\mathcal{E}$ (which equals Q/C) is the initial voltage across the fully charged capacitor.

APPLYING **PHYSICS** 18.4

Many automobiles are equipped with windshield wipers that can be used intermittently during a light rainfall. How does the operation of this feature depend on the charging and discharging of a capacitor?

Explanation   The wipers are part of an *RC* circuit whose time constant can be varied by selecting different values of *R* through a multiposition switch. The brief time that the wipers remain on and the time they are off are determined by the value of the time constant of the circuit.

APPLICATION

TIMED WINDSHIELD WIPERS

APPLYING **PHYSICS** 18.5

In biological applications concerned with population growth, an equation is used that is similar to the exponential equations encountered in the analysis of *RC* circuits. It is

$$N_f = N_i \, 2^n$$

where N_f is the number of bacteria present after n doubling times, N_i is the number present initially, and n is the number of growth cycles or doubling times. Doubling times vary according to the organism. The doubling time for the bacteria responsible for leprosy is about 30 days, and that for the salmonella bacteria responsible for food poisoning is about 20 minutes. For those of you with a biology inclination, consider that only 10 salmonella bacteria find their way onto a turkey leg after your Thanksgiving meal. Four hours later you come back for a midnight snack. How many bacteria are present now?

Explanation   The number of doubling times is 240 min/20 min = 12. Thus, we have

$$N_f = N_i \, 2^n = (10 \text{ bacteria})(2^{12}) = 40\,960 \text{ bacteria}$$

So, your system will have to deal with an invading host of about 41 000 bacteria, which are going to continue to double in a very promising environment.

APPLICATION

BACTERIAL GROWTH

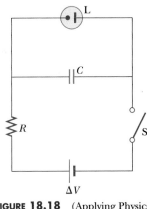

FIGURE 18.18   (Applying Physics 18.6)

APPLYING **PHYSICS** 18.6

Many roadway construction sites have flashing yellow lights to warn motorists of possible dangers. What causes the lights to flash?

Explanation   A typical circuit for such a flasher is shown in Figure 18.18. The lamp L is a gas-filled lamp that acts as an open circuit until a large potential difference causes a discharge, which gives off a bright light. During this discharge, charge flows through the gas between the electrodes of the lamp. When the switch is closed, the battery charges up the capacitor. At the beginning, the current is high and the charge on the capacitor is low, so most of the potential difference appears across the resistance *R*. As the capacitor charges, more potential difference appears across it, reflecting the lower current and, thus, lower potential difference across the resistor. Eventually, the potential difference across the capacitor reaches a value at which the lamp will conduct, causing a flash. This discharges the capacitor through the lamp and the process of charging begins again. The period between flashes can be adjusted by changing the time constant of the *RC* circuit.

APPLICATION

ROADWAY FLASHERS

Example 18.6   Charging a Capacitor in an *RC* Circuit

An uncharged capacitor and a resistor are connected in series to a battery, as in Figure 18.16a. If $\mathcal{E} = 12$ V, $C = 5.0\ \mu$F, and $R = 8.0 \times 10^5\ \Omega$, find the time constant of the circuit, the maximum charge on the capacitor, and the charge on the capacitor after one time constant — that is, when $t = RC$.

Solution   The time constant of the circuit is

$$\tau = RC = (8.0 \times 10^5\ \Omega)(5.0 \times 10^{-6}\ \text{F}) = \boxed{4.0\ \text{s}}$$

The maximum charge on the capacitor is

$$Q = C\mathcal{E} = (5.0 \times 10^{-6}\ \text{F})(12\ \text{V}) = \boxed{60\ \mu\text{C}}$$

After one time constant, the charge on the capacitor is 63.2% of its maximum value:

$$q = 0.632Q = 0.632(60 \times 10^{-6}\ \text{C}) = \boxed{38\ \mu\text{C}}$$

EXERCISE   Find the charge on the capacitor and the voltage across the capacitor after time t has elapsed.

ANSWER $q = (60\ \mu\text{C})(1 - e^{-t/4})$; $\Delta V = (12\ \text{V})(1 - e^{-t/4})$

Example 18.7   Discharging a Capacitor in an *RC* Circuit

Consider a capacitor C being discharged through a resistor R as in Figure 18.17a. After how many time constants does the charge on the capacitor drop to one fourth of its initial value?

Solution   The charge on the capacitor varies with time according to Equation 18.9:

$$q = Qe^{-t/RC}$$

where Q is the initial charge on the capacitor. To find the time it takes the charge q to drop to one fourth of its initial value, we substitute $q = Q/4$ into this expression and solve for t:

$$\tfrac{1}{4}Q = Qe^{-t/RC}$$

$$\tfrac{1}{4} = e^{-t/RC}$$

Taking logarithms of both sides, we find that

$$\ln\!\left(\tfrac{1}{4}\right) = -\ln 4 = -\frac{t}{RC}$$

$$t = RC \ln 4 = \boxed{1.39RC}$$

EXERCISE   If $R = 8.0 \times 10^5\ \Omega$, $C = 5.0\ \mu$F, and the initial voltage across the capacitor is 6.0 V, what is the voltage across the capacitor after time t has elapsed?

ANSWER $\Delta V = (6.0\ \text{V})e^{-t/4}$

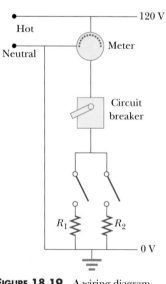

FIGURE 18.19 A wiring diagram for a household circuit. The resistances R_1 and R_2 represent appliances or other electrical devices that operate at an applied voltage of 120 V.

18.6   HOUSEHOLD CIRCUITS

Household circuits are a very practical application of some of the ideas presented in this chapter. In a typical installation, the utility company distributes electric power to individual houses with a pair of wires, or power lines. Electrical devices in a house are then connected in parallel to these lines, as shown in Figure 18.19.

The potential drop between the two wires is about 120 V. (These are actually alternating currents and voltages, but for the present discussion we shall assume that they are direct, constant currents and voltages.) One of the wires is connected to ground, and the other wire, sometimes called the "hot" wire, is at a potential of 120 V. A meter and a circuit breaker or fuse are connected in series with the wire entering the house, as indicated in Figure 18.19.

In modern homes, circuit breakers are used in place of fuses. When the current in a circuit exceeds some value (typically 20 A), the circuit breaker acts as a switch and opens the circuit. Figure 18.20 shows one design for a circuit breaker. Current passes through a bimetallic strip, the top of which bends to the left when excessive current heats it. If the strip bends far enough to the left, it settles into a groove in the spring-loaded metal bar. When this occurs, the bar drops enough to open the circuit at the contact point. The bar also flips a switch that indicates that the circuit breaker has interrupted the current. (After the overload is removed, the switch can be flipped back on.) Circuit breakers based on this design have the disadvantage that some time is required for the heating of the strip, and thus the circuit is not opened rapidly enough when it is overloaded. As a consequence, many circuit breakers are now designed to use electromagnets, which we shall discuss in Chapter 19.

The wire and circuit breaker are carefully selected to meet the current demands of a circuit. If the circuit is to carry currents as large as 30 A, a heavy-duty wire and appropriate circuit breaker must be used. Household circuits that are normally used to power lamps and small appliances often require only 20 A. Each circuit has its own circuit breaker to accommodate its maximum safe current.

As an example, consider a circuit that powers a toaster, a microwave oven, and a heater (represented by R_1, R_2, . . . in Fig. 18.19). We can calculate the current carried by each appliance using the equation $\mathcal{P} = I\,\Delta V$. The toaster, rated at 1 000 W, draws a current of $1\ 000/120 = 8.33$ A. The microwave oven, rated at 800 W, draws a current of 6.67 A, and the heater, rated at 1 300 W, draws a current of 10.8 A. If the three appliances are operated simultaneously, they draw a total current of 25.8 A. Therefore, the breaker should be able to handle at least this much current, or else it will be tripped. As an alternative, one could operate the toaster and microwave oven on one 20-A circuit and the heater on a separate 20-A circuit.

Many heavy-duty appliances, such as electric ranges and clothes dryers, require 240 V to operate. The power company supplies this voltage by providing, in addition to a live wire that is 120 V above ground potential, a wire, also considered live, that is 120 V below ground potential (Fig. 18.21). Therefore, the potential drop across the two live wires is 240 V. An appliance operating from a 240-V line requires half the current of one operating from a 120-V line; therefore, smaller wires can be used in the higher-voltage circuit without becoming overheated.

APPLICATION

FUSES AND CIRCUIT BREAKERS

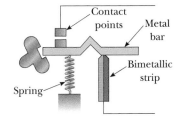

FIGURE 18.20 A circuit breaker that uses a bimetallic strip for its operation.

Webnote 18.3

At home you probably have at least one light that is controlled by two switches. Flipping either switch changes the light from off to on, or from on to off. Discover how this works at
http://www.howstuffworks.com/three-way.htm

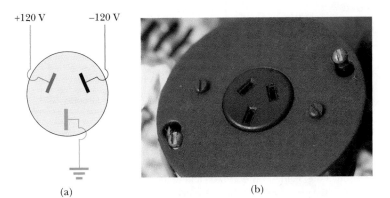

(a)   (b)

FIGURE 18.21 Power connections for a 240-V appliance. *(b, George Semple)*

18.7 ELECTRICAL SAFETY

A person can be electrocuted by touching a live wire (which commonly is live because of a frayed cord and exposed conductors) while the person is in contact with ground. The ground contact might be made by touching a water pipe (which is normally at ground potential) or by standing on the ground with wet feet, because impure water is a good conductor. Obviously, such situations should be avoided at all costs.

Electric shock can result in fatal burns, or it can cause the muscles of vital organs, such as the heart, to malfunction. The degree of damage to the body depends on the magnitude of the current, the length of time it acts, and the part of the body through which it passes. Currents of 5 mA or less can cause a sensation of shock but ordinarily do little or no damage. If the current is larger than about 10 mA, the hand muscles contract and the person may be unable to let go of the live wire. If a current of about 100 mA passes through the body for just a few seconds, it can be fatal. Such large currents paralyze the respiratory muscles. In some cases, currents of about 1 A through the body produce serious burns.

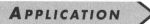

APPLICATION

THIRD WIRE ON CONSUMER APPLIANCES

As an additional safety feature for consumers, electrical equipment manufacturers now use electrical cords that have a third wire, called a ground. To understand how this works, consider the drill being used in Figure 18.22. A two-wire device has one wire, called the "hot" wire, connected to the high-potential (120-V)

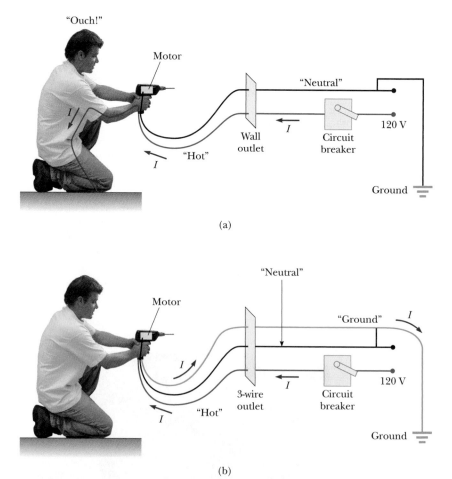

(a)

(b)

FIGURE 18.22   The "hot" wire, at 120 V, always includes a circuit breaker for safety. (a) When the drill is operated with two wires, the normal current path is from the "hot" wire through the motor connections and back to ground through the "neutral" wire. However, here the high-voltage wire has come in contact with the drill case so that the person holding the drill receives an electrical shock. (b) Shock can be prevented by a third wire running from the drill case to the ground.

side of the input power line, and the second wire is connected to ground (0 V). If the high-voltage wire comes in contact with the case of the drill (Fig. 18.22a), a "short circuit" occurs. In this undesirable circumstance, the pathway for the current is from the high-voltage wire through the person holding the drill and to Earth—a pathway that can kill. Protection is provided by a third wire, connected to the case of the drill (Fig. 18.22b). In this situation, if a short occurs, the path of least resistance for the current is from the high-voltage wire through the case and back to ground through the third wire. The resulting high current produced will trip a circuit breaker before the person is injured.

Special power outlets called ground-fault interrupters (GFIs) are now being used in kitchens, bathrooms, basements, and other hazardous areas of new homes. They are designed to protect people from electrical shock by sensing small currents—approximately 5 mA and greater—leaking to ground. When current above this level is detected, the device shuts off (interrupts) the current in less than a millisecond. Ground-fault interrupters will be discussed further in Chapter 19.

Webnote 18.4

Read more about the electrical safety features of three-prong cords at
http://www.howstuffworks.com/question110.htm

18.8 CONDUCTION OF ELECTRICAL SIGNALS BY NEURONS[2]

The most remarkable use of electrical phenomena in living organisms is found in the nervous system of animals. Specialized cells in the body called **neurons** form a complex network that receives, processes, and transmits information from one part of the body to another. The center of this network is located in the brain, which has the ability to store and analyze information. Based on this information, the nervous system controls parts of the body.

The nervous system is very complex and consists of about 10^{10} interconnected neurons. Neurons are the basic units of the nervous system. Some aspects of the nervous system are well known. Over the past 45 years, the method of signal propagation through the nervous system has been established. The messages are voltage pulses called *action potentials* transmitted by neurons. When a neuron receives a strong enough stimulus, it produces repetitive voltage pulses that are actively propagated along its structure. The strength of the stimulus is conveyed by the number of pulses produced. When the pulses reach the end of the neuron, they activate either muscle cells or other neurons. It is interesting that there is a "firing threshold" for neurons. That is, action potentials will propagate along a neuron only if the stimulus is sufficiently strong.

Neurons can be divided into three classes: sensory neurons, motor neurons, and interneurons. The sensory neurons receive stimuli from sensory organs that monitor the external and internal environment of the body. Depending on their specialized functions, the sensory neurons convey messages about factors such as light, temperature, pressure, muscle tension, and odor to higher centers in the nervous system. The motor neurons carry messages that control the muscle cells. These messages are based on the information provided by the sensory neurons and by the brain. The interneurons transmit information from one neuron to another.

Each neuron consists of a cell body to which are attached input ends called **dendrites** and a long tail called the **axon,** which transmits the signal away from the cell (Fig. 18.23). The far end of the axon branches into nerve endings that transmit the signal across small gaps to other neurons or to muscle cells. A simple sensory-motor neuron circuit is shown in Figure 18.24. A stimulus from a muscle produces nerve impulses that travel to the spine. Here the signal is transmitted to a motor neuron, which in turn sends impulses to control the muscle. Figure 18.25 shows an electron microscope image of neurons in the brain.

---

[2] This section is based upon an essay by Paul Davidovits of Boston College.

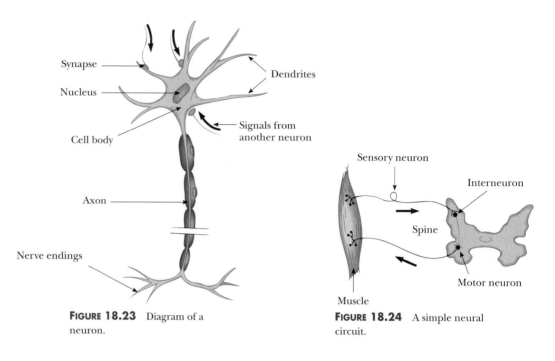

FIGURE 18.23 Diagram of a neuron.

FIGURE 18.24 A simple neural circuit.

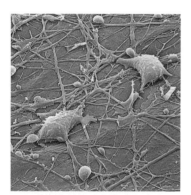

FIGURE 18.25 Stellate neuron from human cortex. (*Juergen Berger, Max-Planck Institute/Science Photo Library/Photo Researchers, Inc.*)

The axon, which is an extension of the neuron cell, conducts the electric impulses away from the cell body. Some axons are extremely long. In humans, for example, the axons connecting the spine with the fingers and toes are more than 1 m long. The neuron can transmit messages because of the special active electrical characteristics of the axon. (The axon acts as an *active* source of energy like a battery rather than like a *passive* stretch of resistive wire.) Much of the information about the electrical and chemical properties of the axon is obtained by inserting small needle-like probes into the axon. Figure 18.26 shows an experimental setup. Note that the outside of the axon is grounded so that all measured voltages are with respect to a zero potential on the outside of the axon. With such probes it is possible to inject current into the axon, measure the resulting action potential as a function of time at a fixed point, and sample the cell's chemical composition. Such experiments are usually difficult to run because the diameter of most axons is very small. Even the largest axons in the human nervous system have a diameter of only about 20×10^{-4} cm. The giant squid, however, has an axon with a diameter of about 0.5 mm, which is large enough for the convenient insertion of probes. Much of the information about signal transmission in the nervous system has come from experiments with the squid axon.

In the aqueous environment of the body, salts and other molecules dissociate into positive and negative ions. As a result, body fluids are relatively good conductors of electricity. The inside of the axon is filled with an ionic fluid that is sepa-

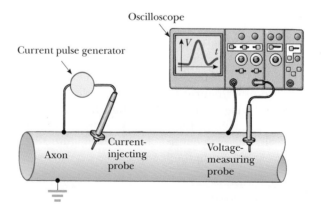

FIGURE 18.26 An axon stimulated electrically. The left probe injects a short current pulse and the right probe measures the resulting action potential as a function of time.

rated from the surrounding body fluid by a thin membrane that is only about 5 to 10 nm thick. The resistivities of the internal and external fluids are about the same, but their chemical compositions are substantially different. The external fluid is similar to sea water. Its ionic solutes are mostly positive sodium ions and negative chloride ions. Inside the axon, the positive ions are mostly potassium ions and the negative ions are mostly large organic ions.

Because there is a large concentration of sodium ions outside the axon and a large concentration of potassium ions inside, one might wonder why the concentrations are not equalized by diffusion. In other words, why don't the sodium ions leak into the axon and the potassium ions leak out of it? The answer lies in the properties of the axon membrane, which, as part of a living cell with an energy supply, can change its permeability on the time scale of milliseconds!

When the axon is not conducting an electric pulse, the axon membrane is highly permeable to potassium ions, slightly permeable to sodium ions, and impermeable to large organic ions. Thus, although sodium ions cannot easily enter the axon, potassium ions can leave it. As the positive potassium ions leave the axon, however, they leave behind large negative organic ions, which cannot follow them through the membrane. As a result, a negative potential builds up inside the axon with respect to the outside. The final negative potential reached, which has been measured at about -70 mV, holds back the outflow of potassium ions so that at equilibrium, the concentration of ions is as we have stated.

The mechanism for the production of an electric signal by the neuron is conceptually simple but was experimentally difficult to sort out. When a neuron changes its resting potential because of an appropriate stimulus, the properties of its membrane change locally. As a result, there is a sudden flow of sodium ions into the cell that lasts for about 2 ms. This produces the $+30$-mV peak in the action potential shown in Figure 18.27a. Immediately after, there is an increase in potassium ion flow out of the cell, which restores the resting action potential of -70 mV in an additional 3 ms. Both the Na^+ and K^+ ion flows have been

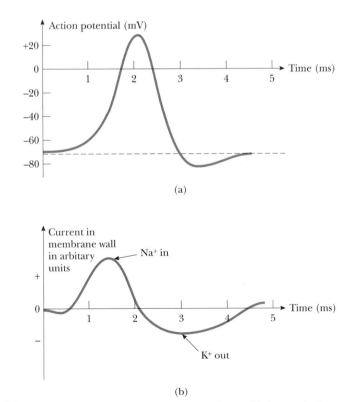

(a)

(b)

FIGURE 18.27   (a) Typical action potential as a function of time. (b) Current in the axon membrane wall as a function of time.

measured by using radioactive Na and K tracers. The nerve signal has been measured to travel along the axon at speeds of 50 m/s to about 150 m/s. This flow of charged particles (or signal transmission) in a nerve axon is *unlike* signal transmission in a metal wire. In an axon, charges move in a direction perpendicular to the direction of travel of the nerve signal, and the nerve signal moves much slower than a voltage pulse traveling along a metallic wire.

Although the axon is a complex structure, and much of how ion channels selectively pass Na^+ and K^+ is not understood,[3] standard electric circuit concepts of current and capacitance can be used to analyze axons. It is left as a problem to show that the axon, having equal and opposite charges separated by a thin dielectric membrane, acts like a capacitor (see Problem 18.41).

SUMMARY

A **source of emf** is any device that transforms nonelectrical energy into electrical energy.
The **equivalent resistance** of a set of resistors connected in **series** is

$$R_{eq} = R_1 + R_2 + R_3 + \cdots \qquad [18.4]$$

The **equivalent resistance** of a set of resistors connected in **parallel** is

$$\frac{1}{R_{eq}} = \frac{1}{R_1} + \frac{1}{R_2} + \frac{1}{R_3} + \cdots \qquad [18.6]$$

Complex circuits are conveniently analyzed by using **Kirchhoff's rules:**

1. The sum of the currents entering any junction must equal the sum of the currents leaving that junction.
2. The sum of the potential differences across all the elements around any closed circuit loop must be zero.

The first rule is a statement of **conservation of charge.** The second is a statement of **conservation of energy.**

As a capacitor is charged by a battery through a resistor, the charge increases from zero to some maximum value. The **time constant** $\tau = RC$ represents the time it takes the charge on the capacitor to increase from zero to 63% of its maximum value.

CONCEPTUAL QUESTIONS

1. Is the direction of current through a battery always from the negative terminal to the positive one? Explain.
2. Given three lightbulbs and a battery, sketch as many different circuits as you can.
3. If the energy transferred to a dead battery during charging is *W*, is the total energy transferred out of the battery to an external circuit during use in which it completely discharges also *W*?
4. How would you connect resistors so that the equivalent resistance is larger than the individual resistances? Give an example involving two or three resistors.
5. If you have your headlights on while you start your car, why do they dim while the car is starting?
6. How would you connect resistors so that the equivalent resistance is smaller than the individual resistances? Give an example involving two or three resistors.

7. Electrical devices are often rated with a voltage and a current—for example, 120 V, 5 A. Batteries, however, are only rated with a voltage—for example, 1.5 V. Why?
8. A "short circuit" is a circuit containing a path of very low resistance in parallel with some other part of the circuit. Discuss the effect of a short circuit on the portion of the circuit it parallels. Use a lamp with a frayed cord as an example.
9. Connecting batteries in series increases the emf applied to a circuit. What advantage might there be to connecting them in parallel?
10. If electrical power is transmitted over long distances, the resistance of the wires becomes significant. Why? Which mode of transmission would result in less energy loss—high current and low voltage or low current and high voltage? Discuss.

[3] For recent developments on this topic, see the November 1, 2001 issue of *Nature.*

11. In Figure Q18.11, describe what happens to the lightbulb after the switch is closed. Assume the capacitor has a large capacitance and is initially uncharged, and assume that the light illuminates when connected directly across the battery terminals.

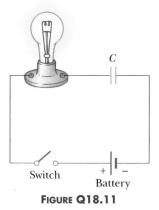

Switch

Battery

FIGURE Q18.11

12. Two sets of Christmas tree lights are available. For set A, when one bulb is removed, the remaining bulbs remain illuminated. For set B, when one bulb is removed, the remaining bulbs do not operate. Explain the difference in wiring for the two sets.

13. Why is it possible for a bird to sit on a high-voltage wire without being electrocuted?

FIGURE Q18.13 Birds on a high-voltage wire. *(Superstock)*

14. Are the two headlights on a car wired in series or in parallel? How can you tell?

15. Embodied in Kirchhoff's rules are two conservation laws. What are they?

16. A ski resort consists of a few chairlifts and several interconnected downhill runs on the side of a mountain, with a lodge at the bottom. The lifts are analogous to batteries and the runs are analogous to resistors. Describe how two runs can be in series. Describe how three runs can be in parallel. Sketch a junction of one lift and two runs. One of the skiers is carrying an altimeter. State Kirchhoff's junction rule and Kirchhoff's loop rule for ski resorts.

17. Suppose you are flying a kite when it strikes a high-voltage wire (a very dangerous situation). What factors determine how great a shock you will receive?

18. Why is it dangerous to turn on a light when you are in a bathtub?

19. Suppose a parachutist lands on a high-voltage wire and grabs the wire as she prepares to be rescued. Will she be electrocuted? If the wire then breaks, should she continue to hold onto the wire as she falls to the ground?

20. Would a fuse or circuit breaker work successfully if it were placed in parallel with the device it was supposed to protect?

21. A series circuit consists of three identical lamps connected to a battery as in Figure Q18.21. When the switch S is closed, what happens (a) to the intensities of lamps A and B, (b) to the intensity of lamp C, (c) to the current in the circuit, and (d) to the voltage drop across the three lamps? (e) Does the power dissipated in the circuit increase, decrease, or remain the same?

A   B   C

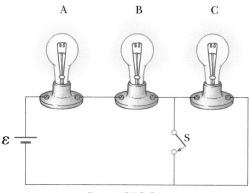

ε S

FIGURE Q18.21

22. Figure Q18.22 shows a series connection of three lamps, all rated at 120 V, with power ratings of 60 W, 75 W, and 200 W. Why do the intensities of the lamps differ? Which lamp has the greatest resistance? How would their intensities differ if they were connected in parallel?

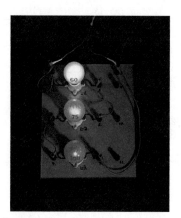

FIGURE Q18.22 *(Courtesy of Henry Leap and Jim Lehman)*

PROBLEMS

1, 2, 3 = straightforward, intermediate, challenging   ☐ = full solution available in Student Solutions Manual/Study Guide

web = solution posted at **http://info.brookscole.com/serway**   = biomedical application

Section 18.1   Sources of emf

Section 18.2   Resistors in Series

Section 18.3   Resistors in Parallel

1. A battery having an emf of 9.00 V delivers 117 mA when connected to a 72.0-Ω load. Determine the internal resistance of the battery.

2. A 4.0-Ω resistor, an 8.0-Ω resistor, and a 12-Ω resistor are connected in series with a 24-V battery. What are (a) the equivalent resistance and (b) the current in each resistor? (c) Repeat for the case in which all three resistors are connected in parallel across the battery.

3. A lightbulb marked "75 W [at] 120 V" is screwed into a socket at one end of a long extension cord in which each of the two conductors has a resistance of 0.800 Ω. The other end of the extension cord is plugged into a 120-V outlet. Draw a circuit diagram, and find the actual power of the bulb in this circuit.

4. A 9.0-Ω resistor and a 6.0-Ω resistor are connected in series with a power supply. (a) The voltage drop across the 6.0-Ω resistor is measured to be 12 V. Find the voltage output of the power supply. (b) The two resistors are connected in parallel across a power supply, and the current through the 9.0-Ω resistor is found to be 0.25 A. Find the voltage setting of the power supply.

5. (a) Find the equivalent resistance between points *a* and *b* in Figure P18.5. (b) Calculate the current in each resistor if a potential difference of 34.0 V is applied between points *a* and *b*.

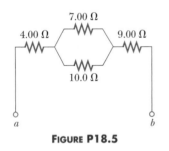

FIGURE P18.5

6. Find the equivalent resistance of the circuit in Figure P18.6.

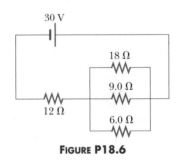

FIGURE P18.6

7. Find the equivalent resistance of the circuit in Figure P18.7.

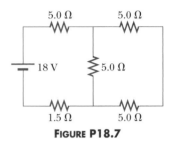

FIGURE P18.7

8. (a) Find the equivalent resistance of the circuit in Figure P18.8. (b) If the total power supplied to the circuit is 4.00 W, find the emf of the battery.

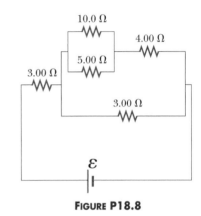

FIGURE P18.8

9. Consider the circuit shown in Figure P18.9. Find (a) the current in the 20.0-Ω resistor and (b) the potential difference between points *a* and *b*.

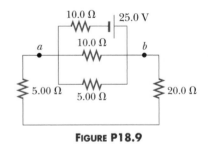

FIGURE P18.9

10. Two resistors, *A* and *B*, are connected in parallel across a 6.0-V battery. The current through *B* is found to be 2.0 A. When the two resistors are connected in series to the 6.0-V battery, a voltmeter connected across resistor *A* measures a voltage of 4.0 V. Find the resistances of *A* and *B*.

11. The resistance between terminals *a* and *b* in Figure P18.11 is 75 Ω. If the resistors labeled *R* have the same value, determine *R*.

12. Three 100-Ω resistors are connected as shown in Figure P18.12. The maximum power that can safely be delivered to any one resistor is 25.0 W. (a) What is the maximum voltage that can be applied to the terminals *a* and *b*?

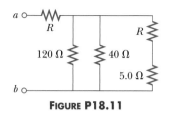

FIGURE P18.11

(b) For the voltage determined in part (a), what is the power delivered to each resistor? What is the total power delivered?

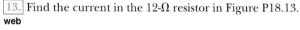

FIGURE P18.12

13. Find the current in the 12-Ω resistor in Figure P18.13.
web

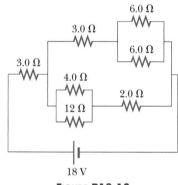

FIGURE P18.13

14. Calculate the power delivered to each resistor in the circuit shown in Figure P18.14.

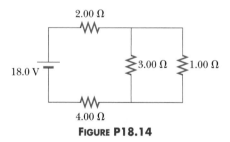

FIGURE P18.14

15. (a) You need a 45-Ω resistor, but the stockroom has only 20-Ω and 50-Ω resistors. How can the desired resistance be achieved under these circumstances? (b) What can you do if you need a 35-Ω resistor?

Section 18.4 Kirchhoff's Rules and Complex DC Circuits

Note: The currents are not necessarily in the direction shown for some circuits.

16. The ammeter shown in Figure P18.16 reads 2.00 A. Find I_1, I_2, and $\mathcal{E}$.

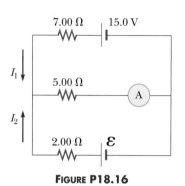

FIGURE P18.16

17. Determine the current in each branch of the circuit shown in Figure P18.17.

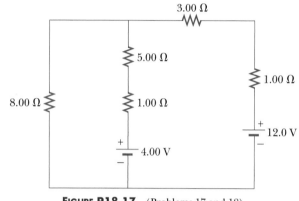

FIGURE P18.17 (Problems 17 and 18)

18. In Figure P18.17, show how to add just enough ammeters to measure every different current in the circuit. Show how to add just enough voltmeters to measure the potential difference across each resistor and across each battery.

19. Figure P18.19 shows a circuit diagram. Determine (a) the current, (b) the potential of wire *A* relative to ground, and (c) the voltage drop across the 1 500-Ω resistor.

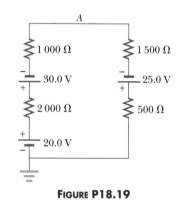

FIGURE P18.19

20. In the circuit of Figure P18.20, the current I_1 is 3.0 A and the values of $\mathcal{E}$ and *R* are unknown. What are the currents I_2 and I_3?

21. What is the emf $\mathcal{E}$ of the battery in the circuit of Figure P18.21?

22. Find the current in each of the three resistors of Figure P18.22 (a) by the rules for resistors in series and parallel and (b) by the use of Kirchhoff's rules.

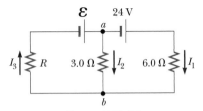

FIGURE P18.20

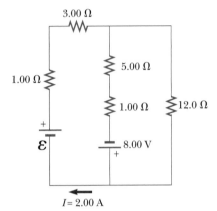

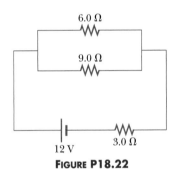

$I = 2.00$ A

FIGURE P18.21

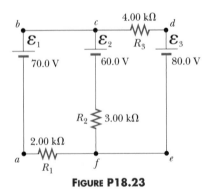

FIGURE P18.22

23. (a) Using Kirchhoff's rules, find the current in each resistor shown in Figure P18.23 and (b) find the potential difference between points *c* and *f*.

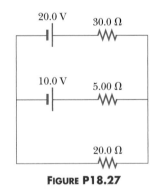

FIGURE P18.23

24. Two 1.50-V batteries—with their positive terminals in the same direction—are inserted in series into the barrel of a flashlight. One battery has an internal resistance of 0.255 Ω, the other an internal resistance of 0.153 Ω. When the switch is closed, a current of 0.600 A passes through the lamp. (a) What is the lamp's resistance? (b) What fraction of the power dissipated is dissipated in the batteries?

25. Calculate each of the unknown currents I_1, I_2, and I_3 for the circuit of Figure P18.25.

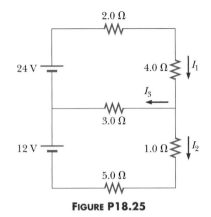

FIGURE P18.25

26. A car with a dead battery is being given a jumpstart by connecting the live battery of another car to the dead battery with jumper cables (Fig. P18.26). Determine the current in the starter and in the dead battery.

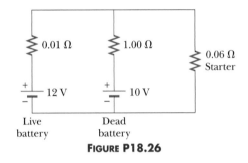

FIGURE P18.26

27. Find the current in each resistor in Figure P18.27.

FIGURE P18.27

28. (a) Determine the potential difference ΔV_{ab} for the circuit in Figure P18.28. Note that each battery has an internal resistance as indicated in the figure. (b) If points *a* and *b* are connected by a 7.0-Ω resistor, what is the current through this resistor?

29. Find the potential difference across each resistor in Figure P18.29.

Section 18.5   *RC Circuits*

30. Show that $\tau = RC$ has units of time.

31. Consider a series *RC* circuit for which $C = 6.0$ μF, $R = 2.0 \times 10^6$ Ω, and $\mathcal{E} = 20$ V. Find (a) the time con-

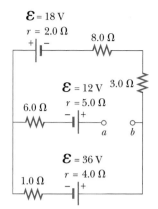

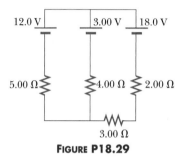

FIGURE P18.28

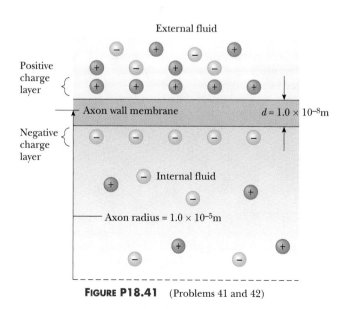

FIGURE P18.29

stant of the circuit and (b) the maximum charge on the capacitor after a switch in the circuit is closed.

32. An uncharged capacitor and a resistor are connected in series to a source of emf. If $\mathcal{E} = 9.00$ V, $C = 20.0$ μF, and $R = 100$ Ω, find (a) the time constant of the circuit, (b) the maximum charge on the capacitor, and (c) the charge on the capacitor after one time constant.

33. Consider a series *RC* circuit for which $R = 1.0$ MΩ, **web** $C = 5.0$ μF, and $\mathcal{E} = 30$ V. The capacitor is initially uncharged when the switch is open. Find the charge on the capacitor 10 s after the switch is closed.

34. A series combination of a 12-kΩ resistor and an unknown capacitor is connected to a 12-V battery. One second after the circuit is completed, the voltage across the capacitor is 10 V. Determine the capacitance.

35. A capacitor in an *RC* circuit is charged to 60.0% of its maximum value in 0.900 s. What is the time constant of the circuit?

36. A series *RC* circuit has a time constant of 0.960 s. The battery has an emf of 48.0 V, and the maximum current in the circuit is 0.500 mA. What are (a) the value of the capacitance and (b) the charge stored in the capacitor 1.92 s after the switch is closed?

Section 18.6 Household Circuits

37. An electric heater is rated at 1 300 W, a toaster is rated at 1 000 W, and an electric grill is rated at 1 500 W. The three appliances are connected in parallel to a common 120-V circuit. (a) How much current does each appliance draw? (b) Is a 30.0-A circuit breaker sufficient in this situation? Explain.

38. A lamp ($R = 150$ Ω), an electric heater ($R = 25$ Ω), and a fan ($R = 50$ Ω) are connected in parallel across a 120-V line. (a) What total current is supplied to the circuit?

(b) What is the voltage across the fan? (c) What is the current in the lamp? (d) What power is expended in the heater?

39. A heating element in a stove is designed to dissipate 3 000 W when connected to 240 V. (a) Assuming that the resistance is constant, calculate the current in this element if it is connected to 120 V. (b) Calculate the power it dissipates at this voltage.

40. Your toaster oven and coffeemaker each dissipate 1 200 W of power. Can you operate them together if the 120-V line that feeds them has a circuit breaker rated at 15 A? Explain.

Section 18.8 Conduction of Electrical Signals by Neurons

41. Assume that a length of axon membrane of about 10 cm is excited by an action potential. (Length excited = nerve speed × pulse duration = 50 m/s × 2.0 ms = 10 cm.) In the resting state, the outer surface of the axon wall is charged positively with K^+ ions and the inner wall has an equal and opposite charge of negative organic ions as shown in Figure P18.41. Model the axon as a parallel plate capacitor and use $C = \kappa \epsilon_0 A/d$ and $Q = C\Delta V$ to investigate the charge as follows. Use typical values for a cylindrical axon of cell wall thickness $d = 1.0 \times 10^{-8}$ m, axon radius $r = 10$ μm, and cell wall dielectric constant $\kappa = 3.0$. (a) Calculate the positive charge on the outside of a 10-cm piece of axon when it is not conducting an electric pulse. How many K^+ ions are on the outside of the axon? Is this a large charge per unit area? [*Hint:* Calculate the charge per unit area in terms of the number of square angstroms (Å^2) per electronic charge. An atom has a cross section of about 1 Å^2 (1 Å = 10^{-10} m).] (b) How much positive charge must flow through the cell membrane to reach the excited state of $+30$ mV from the resting state of -70 mV? How many sodium ions is this? (c) If it takes 2.0 ms for the Na^+ ions to enter the axon, what is the average current in the axon wall in this process? (d) How much energy does it take to raise the potential of the inner axon wall to $+30$ mV starting from the resting potential of -70 mV?

External fluid

Positive charge layer

Axon wall membrane $d = 1.0 \times 10^{-8}$ m

Negative charge layer

Internal fluid

Axon radius = 1.0×10^{-5} m

FIGURE P18.41 (Problems 41 and 42)

42. Continuing with the model of the axon as a capacitor from Problem 41 and Figure P18.41, (a) how much energy does it take to restore the inner wall of the axon to − 70 mV starting from +30 mV? (b) Find the average current in the axon wall during this process, if it takes 3.0 ms.

43. Using Figure 18.27b and the results of Problem 18.41d and Problem 18.42a, find the average power supplied by the axon during firing and recovery.

ADDITIONAL PROBLEMS

44. Consider an *RC* circuit in which the capacitor is being charged by a battery connected in the circuit. After a time equal to two time constants, what percent of the *final* charge is present on the capacitor?

45. Find the equivalent resistance between points *a* and *b* in Figure P18.45.

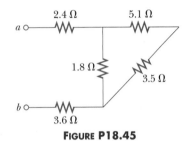

FIGURE P18.45

46. For the circuit in Figure P18.46, calculate (a) the equivalent resistance of the circuit and (b) the power dissipated by the entire circuit. (c) Find the current in the 5.0-Ω resistor.

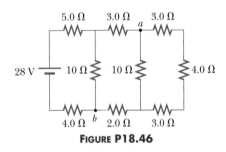

FIGURE P18.46

47. Find (a) the equivalent resistance of the circuit in Figure P18.47, (b) each current in the circuit, (c) the potential difference across each resistor, and (d) the power dissipated by each resistor.

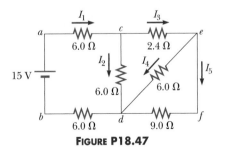

FIGURE P18.47

48. Three 60.0-W, 120-V lightbulbs are connected across a 120-V power source, as shown in Figure P18.48. Find

(a) the total power delivered to the three bulbs and (b) the potential difference across each. Assume that the resistance of each bulb is constant (even though in reality the resistance increases markedly with current).

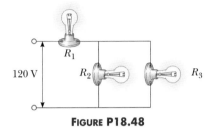

FIGURE P18.48

49. An automobile battery has an emf of 12.6 V and an internal resistance of 0.080 Ω. The headlights have total resistance of 5.00 Ω (assumed constant). What is the potential difference across the headlight bulbs (a) when they are the only load on the battery and (b) when the starter motor is operated, taking an additional 35.0 A from the battery?

50. In Figure P18.50, suppose that the switch has been closed for a length of time sufficiently long for the capacitor to become fully charged. (a) Find the steady-state current in each resistor and (b) find the charge on the capacitor.

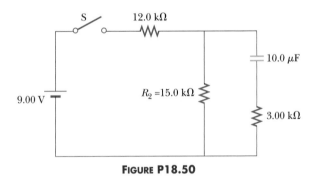

FIGURE P18.50

51. Find the values of I_1, I_2, and I_3 for the circuit in Figure P18.51.

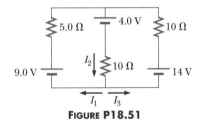

FIGURE P18.51

52. The resistance between points *a* and *b* in Figure P18.52 drops to one half its original value when switch S is closed. Determine the value of *R*.

53. A generator has a terminal voltage of 110 V when it delivers 10.0 A, and 106 V when it delivers 30.0 A. Calculate the emf and the internal resistance of the generator.

54. An emf of 10 V is connected to a series *RC* circuit consisting of a resistor of 2.0×10^6 Ω and a capacitor of 3.0 μF. Find the time required for the charge on the capacitor to reach 90% of its final value.

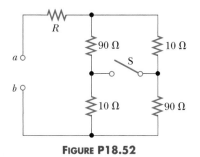

FIGURE P18.52

55. The student engineer of a campus radio station wishes to verify the effectiveness of the lightning rod on the antenna mast (Fig. P18.55). The unknown resistance R_x is between points C and E. Point E is a "true ground" but is inaccessible for direct measurement since this stratum is several meters below the Earth's surface. Two identical rods are driven into the ground at A and B, introducing an unknown resistance R_y. The procedure is as follows: measure resistance R_1 between points A and B, then connect A and B with a heavy conducting wire and measure resistance R_2 between points A and C. (a) Derive a formula for R_x in terms of the observable resistances R_1 and R_2. (b) A satisfactory ground resistance would be $R_x < 2.0 \ \Omega$. Is the grounding of the station adequate if measurements give $R_1 = 13 \ \Omega$ and $R_2 = 6.0 \ \Omega$?

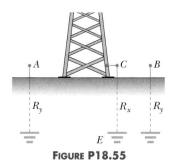

FIGURE P18.55

56. The resistor R in Figure P18.56 dissipates 20 W of power. Determine the value of R.

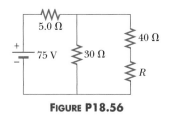

FIGURE P18.56

57. A voltage ΔV is applied to a series configuration of n resistors, each of value R. The circuit components are recon-

nected in a parallel configuration, and voltage ΔV is again applied. Show that the power consumed by the series configuration is $1/n^2$ times the power consumed by the parallel configuration.

58. For the network in Figure P18.58, show that the resistance between points a and b is $R_{ab} = \frac{27}{17} \ \Omega$. (*Hint:* Connect a battery with emf $\mathcal{E}$ across points a and b and determine $\mathcal{E}/I$, where I is the current in the battery.)

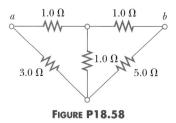

FIGURE P18.58

59. A battery with an internal resistance of 10.0 Ω produces an open-circuit voltage of 12.0 V. A variable load resistance with a range of 0 to 30.0 Ω is connected across the battery. (*Note:* A battery has a resistance that depends on the condition of its chemicals and increases as the battery ages. This internal resistance can be represented in a simple circuit diagram as a resistor in series with the battery.) (a) Graph the power dissipated in the load resistor as a function of the load resistance. (b) With your graph, demonstrate the following important theorem: *the power delivered to a load is a maximum if the load resistance equals the internal resistance of the source.*

60. The circuit in Figure P18.60 contains two resistors, $R_1 = 2.0$ kΩ and $R_2 = 3.0$ kΩ, and two capacitors, $C_1 = 2.0 \ \mu$F and $C_2 = 3.0 \ \mu$F, connected to a battery with emf $\mathcal{E} = 120$ V. If there are no charges on the capacitors before switch S is closed, determine as functions of time the charges q_1 and q_2 on capacitors C_1 and C_2, respectively, after the switch is closed. (*Hint:* First reconstruct the circuit so that it becomes a simple RC circuit containing a single resistor and single capacitor in series, connected to the battery, and then determine the total charge q stored in the circuit.)

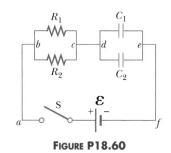

FIGURE P18.60

GROUP ACTIVITIES

G.1 Connect one terminal of a D-cell battery to the base of a flashlight bulb using insulated wire, tape a second wire to the other battery terminal, and tape a third wire to the center conductor of the bulb as shown in the Figure

GA18.1a. Make sure to remove about 1 cm of insulation from the ends of all wires before making connections. Connect the two open wires together to complete the circuit, and note the illumination of the bulb. Now add a second D-cell battery to the circuit as in Figure GA18.1b to give a total voltage of 3.0 V, connect the two open wires together to complete the circuit, and note the illumination of the bulb. Why does the bulb grow brighter in this case?

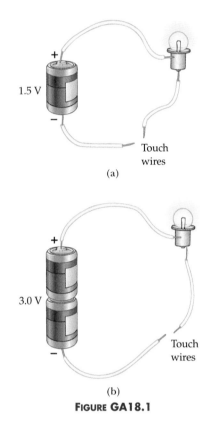

(a)

(b)

FIGURE GA18.1

G.2 Use the basic equipment of activity 1 plus a few more items to test some additional features of circuits. First, note the brightness of a single bulb connected to the battery. Now connect two bulbs in series with each other and the battery. Predict whether the bulbs will be dimmer or brighter than when operated separately. Try it and see. Continue for three bulbs in series.

G.3 Repeat activity 2, but in this activity you must connect the bulbs in parallel with each other. Predict how the brightness will change as you add more bulbs in parallel. Why?

G.4 Continue your experimentation with circuits by connecting the battery to one bulb followed in the same circuit by two bulbs in parallel then back to the battery. Predict the brightness of each bulb in this situation before you connect the circuit. Explain the results you obtain.

G.5 (a) A group of resistors connected in parallel have the same (i) current in them, (ii) potential difference across them, or (iii) neither of the above. (b) A group of resistors connected in series have the same (i) current in them, (ii) potential difference across them, or (iii) neither of the above. (c) The equivalent resistance for a group of resistors connected in parallel is (i) greater than any of the resistors in the group, (ii) less than any of the resistors in the group, or (iii) neither of the above. (d) The equivalent resistance of a group of resistors connected in series is (i) greater than any of the resistors in the group, (ii) less than any of the resistors in

the group, or (iii) neither of the above. (e) Defend your answers by examining a circuit consisting of a 3.0-Ω resistor and a 5.0-Ω resistor connected across a 12-V battery. Be sure that the answers you find are consistent with your answers above.

G.6 (a) Two resistors are connected in series. The power converted by each resistor is (i) the same, or (ii) not necessarily the same. (b) Two resistors are connected in parallel. The power converted by each resistor is (i) the same, or (ii) not necessarily the same. (c) Suppose two resistors of 100 Ω and 200 Ω are connected to a 10.0-V power supply. Find the power converted by the two resistors in the configurations of parts (a) and (b). Are your answers consistent with your selections in parts (a) and (b)?

G.7 In this activity, select the correct answer from within the parentheses. (a) You traverse a 20-Ω resistor in a circuit such that you are moving in the same direction as a current of 2.0 A. In Kirchhoff's loop rule, the potential difference is (− 40 V, 40 V). (b) You traverse the same resistor moving against the current. The potential difference encountered is (− 40 V, 40 V). (c) You traverse a 12-V battery such that you are moving from the negative terminal to the positive terminal. In Kirchhoff's loop rule the potential difference is considered to be (12 V, − 12 V). (d) You traverse the same battery moving from the positive terminal to the negative terminal. The potential difference encountered is (12 V, − 12 V). (e) When working a Kirchhoff's rules problem, if you make an incorrect guess for the direction of the current in a branch, the final answer found for that current will be (positive, negative, meaningless).

G.8 Consider the two arrangements of batteries and bulbs shown in Figure GA18.8. All four bulbs are identical and have resistance R, and the two batteries are identical with output voltage ΔV. Answer the following questions and explain your answer. (a) In case 1, which bulb is brighter? (b) In case 1, what is the current in each bulb and the power supplied to each bulb? (c) In case 2, which bulb is brighter? (d) In case 2, what is the current in each bulb, the voltage drop across each bulb, and the power delivered to each bulb? (e) Which bulbs are brighter, those in case 1 or those in case 2? (f) If one bulb in each case fails, will the other go out as well? If the other bulb doesn't fail, will it get brighter or stay the same?

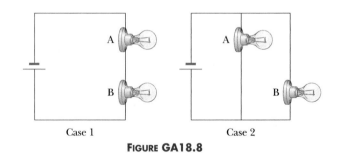

Case 1   Case 2

FIGURE GA18.8

G.9 In Figure GA18.9, some bulbs and batteries are connected together in different ways. The bulbs are all identical. For each case, give the letter that tells which of the bulbs will be brighter. If two or more will be equal in

is moving parallel to the field, all that we can say for sure is that the magnetic field is directed either northward or southward.

EXERCISE Calculate the gravitational force on the proton and compare it with the magnetic force. Note that the mass of the proton is 1.67×10^{-27} kg.

ANSWER 1.6×10^{-26} N; $F_{grav}/F_{max} = 1.9 \times 10^{-8}$

Example 19.2 A Proton Moving in a Magnetic Field

A proton moves at 8.0×10^6 m/s along the x axis. It enters a region in which there is a magnetic field of magnitude 2.5 T, directed at an angle of 60° with the x axis and lying in the xy plane (Fig. 19.8). Calculate the initial force on and acceleration of the proton.

Solution From Equation 19.1 we get

$$F = qvB \sin \theta = (1.6 \times 10^{-19} \text{ C}) (8.0 \times 10^6 \text{ m/s}) (2.5 \text{ T}) (\sin 60°)$$
$$= 2.8 \times 10^{-12} \text{ N}$$

Use right-hand rule #1, noting that the charge is positive, to see that the force is in the positive z direction. Verify that the units of F in the calculation reduce to newtons.

Because the mass of the proton is 1.67×10^{-27} kg, its initial acceleration is

$$a = \frac{F}{m} = \frac{2.8 \times 10^{-12} \text{ N}}{1.67 \times 10^{-27} \text{ kg}} = 1.7 \times 10^{15} \text{ m/s}^2$$

in the positive z direction.

EXERCISE Calculate the acceleration of an electron that moves through the same magnetic field at the same speed as the proton. The mass of an electron is 9.11×10^{-31} kg.

ANSWER 3.0×10^{18} m/s² in the negative z direction

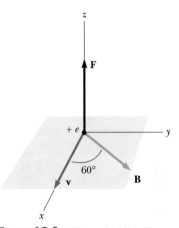

FIGURE 19.8 (Example 19.2) The magnetic force **F** on a proton is in the positive z direction when **v** and **B** lie in the xy plane.

19.4 MAGNETIC FORCE ON A CURRENT-CARRYING CONDUCTOR

If a force is exerted on a single charged particle when it moves through a magnetic field, it should be no surprise that a current-carrying wire also experiences a force when placed in a magnetic field (see Figure 19.9). This follows from the fact that the current is a collection of many charged particles in motion; hence, the resultant force on the wire is due to the sum of the individual forces on the charged particles. The force on the particles is transmitted to the "bulk" of the wire through collisions with the atoms making up the wire.

Before we continue, some explanation is in order concerning notation in many of the figures. To indicate the direction of **B**, we use the following convention.

> If **B** is directed *into* the page, as in Figure 19.10, we use a series of blue crosses, representing the tails of arrows. If **B** is directed *out* of the page, we use a series of blue dots, representing the tips of arrows. If **B** lies in the plane of the page, we use a series of blue field lines with arrowheads.

The force on a current-carrying conductor can be demonstrated by hanging a wire between the poles of a magnet, as in Figure 19.10. In this figure, the magnetic

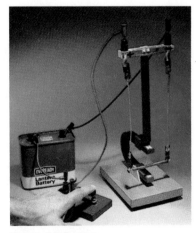

FIGURE 19.9 This apparatus demonstrates the force on a current-carrying conductor in an external magnetic field. Why does the bar swing *away* from the magnet after the switch is closed? (*Courtesy of Henry Leap and Jim Lehman*)

FIGURE 19.10 A segment of a flexible vertical wire partially stretched between the poles of a magnet, with the field (blue crosses) directed into the page. (a) When there is no current in the wire, it remains vertical. (b) When the current is upward, the wire deflects to the left. (c) When the current is downward, the wire deflects to the right.

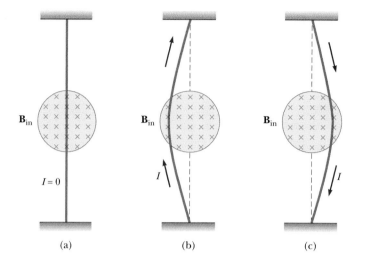

(a)　　　　(b)　　　　(c)

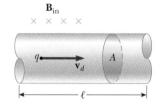

FIGURE 19.11 A section of a wire containing moving charges in an external magnetic field **B**.

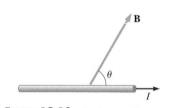

Tip 19.2 THE ORIGIN OF THE MAGNETIC FORCE ON A WIRE

When a magnetic field is applied at some angle to a wire carrying a current, a magnetic force is exerted on each moving charge in the wire. The total magnetic force on the wire is the sum of all the magnetic forces on the individual charges producing the current.

FIGURE 19.12 A wire carrying a current *I* in the presence of an external magnetic field **B** that makes an angle θ with the wire.

APPLICATION ⟩

LOUDSPEAKER OPERATION

field is directed into the page and covers the region within the shaded circle. The wire deflects to the right or left when a current is passed through it.

Let us quantify this discussion by considering a straight segment of wire of length ℓ and cross-sectional area A carrying current I in a uniform external magnetic field **B**, as in Figure 19.11. We assume that the magnetic field is perpendicular to the wire and is directed into the page. Each charge carrier in the wire experiences a force of magnitude $F_{max} = qv_d B$, where v_d is the drift velocity of the charge. To find the total force on the wire, we multiply the force on one charge carrier by the number of carriers in the segment. Because the volume of the segment is $A\ell$, the number of carriers is $nA\ell$, where n is the number of carriers per unit volume. Hence, the magnitude of the total magnetic force on the wire of length ℓ is

Total force = force on each charge carrier × total number of carriers

$$F_{max} = (qv_d B)(nA\ell)$$

From Chapter 17, however, we know that the current in the wire is given by $I = nqv_d A$. Therefore, F_{max} can be expressed as

$$F_{max} = BI\ell \qquad [19.5]$$

This equation can be used *only* when the current and the magnetic field are at right angles to each other.

If the wire is not perpendicular to the field but is at some arbitrary angle, as in Figure 19.12, the magnitude of the magnetic force on the wire is

$$F = BI\ell \sin \theta \qquad [19.6]$$

where θ is the angle between **B** and the direction of the current. The direction of this force can be obtained by using right-hand rule #1. However, in this case you must place your thumb in the direction of the current rather than in the direction of **v**. In Figure 19.12, the direction of the magnetic force on the wire is out of the page.

Finally, when the current is either in the direction of the field or opposite the direction of the field, the magnetic force on the wire is zero.

The fact that a magnetic force acts on a current-carrying wire in a magnetic field is the operating principle of most speakers in sound systems. One speaker design, shown in Figure 19.13, consists of a coil of wire (called the voice coil), a flexible paper cone that acts as the speaker, and a permanent magnet. The coil of wire surrounding the north pole of the magnet is shaped so that the magnetic field lines are directed radially outward from the coil's axis. When an electrical signal is sent to the coil, producing a current in the coil as in Figure 19.13, a magnetic force to the left acts on the coil. (This can be seen by applying right-hand rule #1 to each turn of wire.) When the current reverses direction, as it would for a sinusoidally varying current, the magnetic force on the coil also reverses direction, and the cone, which is attached to the coil, accelerates to the right. An alternating current through the coil causes an alternating force on the coil, which results in vibrations of the cone. The vibrating cone creates sound waves as it pushes and pulls on the air in front of it. In this way a 1-kHz electrical signal is converted to a 1-kHz sound wave.

An unusual application of the force on a current-carrying conductor is illustrated by the electromagnetic pump shown in Figure 19.14. Artificial hearts require a pump to keep the blood flowing, and kidney dialysis machines also require a pump to assist the heart in pumping blood that is to be cleansed. Ordinary mechanical pumps create difficulties because they damage the blood cells as cells move through the pump. The mechanism shown in Figure 19.14 has demonstrated some promise in such applications. A magnetic field is established across a segment of the tube containing the blood, flowing in the direction shown by the velocity **v**. An electric current passing through the fluid perpendicular to the blood velocity as shown has a magnetic force acting on it in the direction of **v**, as application of right-hand rule #1 shows. (Place your right thumb in the direction of the *current*!) This force of magnetic origin helps to keep the blood in motion.

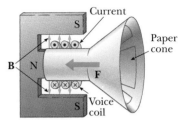

FIGURE 19.13 Diagram of a loudspeaker.

APPLICATION

ELECTROMAGNETIC PUMPS FOR ARTIFICIAL HEARTS AND KIDNEYS

APPLYING PHYSICS 19.2

In a lightning strike, there is a rapid movement of negative charge from a cloud to the ground. In what direction is a lightning strike deflected by Earth's magnetic field?

Explanation The downward flow of negative charge in a lightning stroke is equivalent to an upward-moving current. Thus, we have an upward-moving current in a northward-directed magnetic field. According to right-hand rule #1, the lightning strike deflects toward the west.

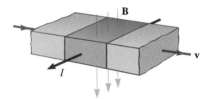

FIGURE 19.14 A simple electromagnetic pump has no moving parts to damage a conducting fluid, such as blood, passing through. Application of right-hand rule #1 (fingers along **B**, thumb along *I* in this case) shows that the force on the current-carrying segment of the fluid is in the direction of the velocity.

Example 19.3 A Current-Carrying Wire in Earth's Magnetic Field

A wire carries a current of 22 A from east to west. Assume that at this location Earth's magnetic field is horizontal and directed from south to north, and that it has a magnitude of 0.50×10^{-4} T. Find the magnetic force on a 36-m length of wire. How does the force change if the current runs west to east?

Solution Because the directions of the current and magnetic field are at right angles, we can use Equation 19.5. The magnitude of the magnetic force is

$$F_{\text{max}} = BI\ell = (0.50 \times 10^{-4}\,\text{T})(22\,\text{A})(36\,\text{m}) = \boxed{4.0 \times 10^{-2}\,\text{N}}$$

Right-hand rule #1 shows that the force on the wire is directed toward Earth.

If the current is directed from west to east, the force has the same magnitude but its direction is upward, away from Earth.

EXERCISE If the current is directed north to south, what is the magnetic force on the wire?

ANSWER Zero

19.5 TORQUE ON A CURRENT LOOP AND ELECTRIC MOTORS

In the preceding section we showed how a force is exerted on a current-carrying conductor when the conductor is placed in an external magnetic field. With this as a starting point, we now show that a torque is exerted on a current loop placed in

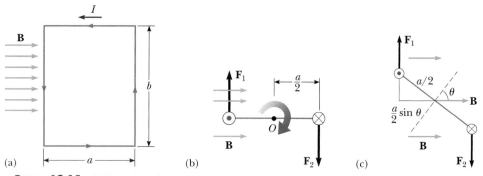

FIGURE 19.15   (a) Top view of a rectangular loop in a uniform magnetic field **B**. No magnetic forces act on the sides of length *a* parallel to **B**, but forces do act on the sides of length *b*. (b) Side view of the rectangular loop shows that the forces **F**$_1$ and **F**$_2$ on the sides of length *b* create a torque that tends to twist the loop clockwise. (c) If **B** is at an angle θ with a line perpendicular to the plane of the loop, the torque is given by *BIA* sin θ.

a magnetic field. The results of this analysis will be of great practical value when we discuss generators and motors, here and in Chapter 20.

Consider a rectangular loop carrying current *I* in the presence of an external uniform magnetic field in the plane of the loop, as shown in Figure 19.15a. The forces on the sides of length *a* are zero because these wires are parallel to the field. The magnitude of the magnetic forces on the sides of length *b*, however, is

$$F_1 = F_2 = BIb$$

The direction of **F**$_1$, the force on the left side of the loop, is out of the page, and that of **F**$_2$, the force on the right side of the loop, is into the page. If we view the loop from the side, as in Figure 19.15b, the forces are directed as shown. If we assume that the loop is pivoted so that it can rotate about point *O*, we see that these two forces produce a torque about *O* that rotates the loop clockwise. The magnitude $\tau_{\max}$ of this torque is

$$\tau_{\max} = F_1 \frac{a}{2} + F_2 \frac{a}{2} = (BIb) \frac{a}{2} + (BIb) \frac{a}{2} = BIab$$

where the moment arm about *O* is *a*/2 for both forces. Because the area of the loop is $A = ab$, the torque can be expressed as

$$\tau_{\max} = BIA \qquad \text{[19.7]}$$

Note that this result is valid only when the magnetic field is parallel to the plane of the loop, as in the view in Figure 19.15b. If the field makes an angle θ with a line perpendicular to the plane of the loop, as in Figure 19.15c, the moment arm for each force is given by $(a/2) \sin \theta$. An analysis such as the one just used produces, for the magnitude of the torque,

$$\tau = BIA \sin \theta \qquad \text{[19.8]}$$

This result shows that the torque has the *maximum* value *BIA* when the field is parallel to the plane of the loop ($\theta = 90°$) and is *zero* when the field is perpendicular to the plane of the loop ($\theta = 0$). As seen in Figure 19.15c, the loop tends to rotate to smaller values of θ (so that the normal to the plane of the loop rotates toward the direction of the magnetic field).

Although this analysis is for a rectangular loop, a more general derivation would indicate that Equation 19.8 applies regardless of the shape of the loop. Furthermore, the torque on a coil with *N* turns is

$$\tau = NBIA \sin \theta \qquad \text{[19.9]}$$

Example 19.4 The Torque on a Circular Loop in a Magnetic Field

A circular wire loop of radius 50.0 cm is placed in a magnetic field of magnitude 0.50 T. The normal to the plane of the loop makes an angle of 30.0° with the magnetic field as shown in an edge view in Figure 19.16. The current in the loop is 2.0 A in the direction shown. Find the magnitude of the torque at this instant.

Solution   Regardless of the shape of the loop, Equation 19.8 is valid:

$$\tau = BIA \sin \theta = (0.50 \text{ T})(2.0 \text{ A})[\pi(0.50 \text{ m})^2](\sin 30.0°) = \boxed{0.39 \text{ N} \cdot \text{m}}$$

Exercise   Find the torque on the loop if it has three turns rather than one.

Answer   The torque is 3 times that on the one-turn loop, or 1.2 N · m.

FIGURE 19.16   (Example 19.4) An edge view of a circular current loop in an external magnetic field **B**.

ELECTRIC MOTORS

It is difficult to imagine life in the 21st century without electric motors. Among the numerous appliances that contain motors are computer disk drives, CD players, VCR and DVD players, food processors and blenders, car starters, furnaces, and air conditioners. The motors convert electrical energy to kinetic energy of rotation, and consist basically of a rigid current-carrying loop that rotates when placed in the field of a magnet.

As we have just seen (Figure 19.15), the torque on such a loop rotates the loop to smaller values of θ until the torque becomes zero when the magnetic field is perpendicular to the plane of the loop and $\theta = 0$. If the loop turns past this angle, and the current remains in the direction shown in Figure 19.15, the torque reverses direction and turns the loop in the opposite direction, that is, counterclockwise. To overcome this difficulty and provide continuous rotation in one direction, the current in the loop must periodically reverse direction. In alternating current (ac) motors, this occurs naturally as the current reverses direction 120 times each second. In direct current (dc) motors, the current reversal is accomplished mechanically with split-ring contacts (commutators) and brushes as shown in Figure 19.17.

Although actual motors contain many current loops and commutators, for simplicity Figure 19.17 shows only a single loop and a single set of split-ring

APPLICATION

ELECTRIC MOTORS

FIGURE 19.17   Simplified sketch of a dc electric motor.

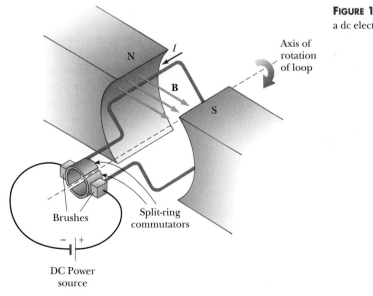

FIGURE 19.18   The Honda Insight combines a three-cylinder gasoline automobile engine with a thin electric motor for improved efficiency and added power when needed. The electric motor (circled) also acts as a generator during braking or coasting downhill to recharge the batteries, with the result that they never need to be recharged by the owner. *(Courtesy of America Honda Motor Co., Inc.)*

The bending of an electron beam in an external magnetic field. The tube contains gas at very low pressure, and the beam is made visible as the electrons collide with the gas atoms, which in turn emit visible light. The beam is projected to the right and deflects downward in the presence of a magnetic field produced by a pair of current-carrying coils. What is the direction of the magnetic field? *(Courtesy of Central Scientific Company)*

contacts rigidly attached to and rotating with the loop. A set of electrical stationary contacts called *brushes* are maintained in electrical contact with the rotating split ring. These brushes are usually made of graphite because graphite is a good electrical conductor as well as a good lubricant. Just as the loop becomes perpendicular to the magnetic field and the torque becomes zero, inertia carries the loop forward in the clockwise direction and the brushes cross the gaps in the ring, causing the loop current to reverse its direction. This provides another pulse of torque in the clockwise direction for another 180°, the current reverses, and the process repeats itself. Figure 19.18 shows a modern motor used to provide added power to a hybrid gasoline-electric car.

19.6   MOTION OF A CHARGED PARTICLE IN A MAGNETIC FIELD

Consider the case of a positively charged particle moving in a uniform magnetic field so that the direction of the particle's velocity is *perpendicular to the field,* as in Figure 19.19. The label $\mathbf{B}_{\text{in}}$ and the blue crosses indicate that $\mathbf{B}$ is directed into the page. Application of right-hand rule #1 at point P shows that the direction of the magnetic force $\mathbf{F}$ at this location is upward. This causes the particle to alter its direction of travel and to follow a curved path. Application of right-hand rule #1 at any point shows that **the magnetic force is always *toward* the center of the circular path;** therefore, the magnetic force causes a centripetal acceleration, which changes only the direction of $\mathbf{v}$ and not its magnitude. Because $\mathbf{F}$ produces the centripetal acceleration, we can equate its magnitude, qvB in this case, to the mass of the particle multiplied by the centripetal acceleration v^2/r. From Newton's second law, we find that

$$F = qvB = \frac{mv^2}{r}$$

which gives

$$r = \frac{mv}{qB} \qquad \text{[19.10]}$$

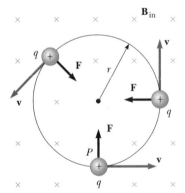

FIGURE 19.19   When the velocity of a charged particle is perpendicular to a uniform magnetic field, the particle moves in a circle whose plane is perpendicular to $\mathbf{B}$, which is directed into the page. (The crosses represent the tails of the magnetic field vectors.) The magnetic force $\mathbf{F}$ on the charge is always directed toward the center of the circle.

This says that the radius of the path is proportional to the momentum mv of the particle and is inversely proportional to the magnetic field.

If the initial direction of the velocity of the charged particle is not perpendicular to the magnetic field but instead is directed at an angle to the field, as shown in Figure 19.20, the path followed by the particle is a spiral (called a helix) along the magnetic field lines.

Webnote 19.2

See an animation of positive and negative charges being deflected by a magnetic field at *http://www.lightlink.com/sergey/java/ java/partmagn/index.html*

APPLYING **PHYSICS** 19.3

Suppose a uniform magnetic field exists in a finite region of space. Can you inject a charged particle into this region from the outside and have it stay trapped in the region by the magnetic force?

Explanation Let us consider separately the components of the particle velocity parallel and perpendicular to the field lines in the region. For the component parallel to the field lines, there is no force on the particle—it continues to move with the parallel component of the velocity. Now consider the component of velocity perpendicular to the field lines. This component results in a magnetic force that is perpendicular to both the field lines and the velocity component. The path of a particle for which the force is always perpendicular to the velocity is a circle. Thus, the particle follows a circular arc and exits the field on the other side of the circle, as shown in Figure 19.21 for a particle with constant kinetic energy. On the other hand, a particle can become trapped if it loses some kinetic energy in a collision after entering the field, as in Figure 19.20.

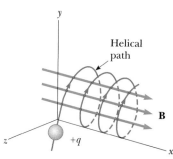

FIGURE 19.20 A charged particle having a velocity directed at an angle with a uniform magnetic field moves in a helical path.

Quick Quiz 19.3	As a charged particle moves freely in a circular path in the presence of a constant magnetic field applied perpendicular to the particle's velocity, its kinetic energy (a) remains constant, (b) increases, or (c) decreases.
Quick Quiz 19.4	Two charged particles are projected into a region in which a magnetic field is perpendicular to their velocities. After they enter the magnetic field, you can conclude that (a) the charges are deflected in opposite directions, (b) the charges continue to move in a straight line, (c) the charges move in circular paths, or (d) the charges move in circular paths but in opposite directions.

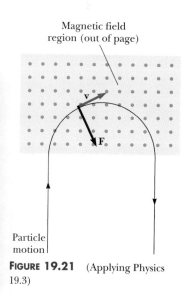

FIGURE 19.21 (Applying Physics 19.3)

Example 19.5 A Proton Moving Perpendicularly to a Uniform Magnetic Field

A proton is moving in a circular orbit of radius 14 cm in a uniform magnetic field of magnitude 0.35 T, directed perpendicularly to the velocity of the proton. Find the orbital speed of the proton.

Solution From Equation 19.10, we get

$$v = \frac{qBr}{m} = \frac{(1.6 \times 10^{-19}\,\text{C})(0.35\,\text{T})(14 \times 10^{-2}\,\text{m})}{1.67 \times 10^{-27}\,\text{kg}} = 4.7 \times 10^{6}\,\text{m/s}$$

EXERCISE If an electron moves perpendicularly to the same magnetic field with this speed, what is the radius of its circular orbit?

ANSWER $7.6 \times 10^{-5}\,\text{m}$

Webnote 19.3
To see how mass spectrometry is used to screen newborns for genetic disorders, visit
http://www.duke.edu/~mdfeezor/NSHome/index.html

APPLICATION

MASS SPECTROMETERS

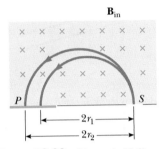

Bin

P   *S*

$2r_1$

$2r_2$

FIGURE 19.22 (Example 19.6) Two isotopes leave the slit at point *S* and travel in different circular paths before striking a photographic plate at *P*.

Example 19.6 The Mass Spectrometer

Two singly ionized atoms move out of a slit at point *S* in Figure 19.22 and into a magnetic field of 0.10 T. Each has a speed of 1.0×10^6 m/s. The nucleus of the first atom contains one proton and has a mass of 1.67×10^{-27} kg, and the nucleus of the second atom contains a proton and a neutron and has a mass of 3.34×10^{-27} kg. Atoms with the same number of protons in the nucleus but different masses are called isotopes. The two isotopes here are hydrogen and deuterium. Find their distance of separation when they strike a photographic plate at *P*.

Solution The radius of the circular path followed by the lighter isotope, hydrogen, is

$$r_1 = \frac{m_1 v}{qB} = \frac{(1.67 \times 10^{-27} \text{ kg})(1.0 \times 10^6 \text{ m/s})}{(1.6 \times 10^{-19} \text{ C})(0.10 \text{ T})} = 0.10 \text{ m}$$

The radius of the path of the heavier isotope, deuterium, is

$$r_2 = \frac{m_2 v}{qB} = \frac{(3.34 \times 10^{-27} \text{ kg})(1.0 \times 10^6 \text{ m/s})}{(1.6 \times 10^{-19} \text{ C})(0.10 \text{ T})} = 0.21 \text{ m}$$

The distance of separation is

$$x = 2r_2 - 2r_1 = \boxed{0.21 \text{ m}}$$

The concepts used in this example underlie the operation of a device called a **mass spectrometer,** which is sometimes used to separate isotopes according to their mass-to-charge ratios, but more often is used to measure masses.

19.7 MAGNETIC FIELD OF A LONG, STRAIGHT WIRE AND AMPÈRE'S LAW

During a lecture demonstration in 1819, the Danish scientist Hans Oersted (1777–1851) found that an electric current in a wire deflected a nearby compass needle. This momentous discovery, linking a magnetic field with an electric current for the first time, was the beginning of our understanding of the origin of magnetism.

A simple experiment first carried out by Oersted in 1820 clearly demonstrates that a current-carrying conductor produces a magnetic field. In this experiment, several compass needles are placed in a horizontal plane near a long, vertical wire, as in Figure 19.23a. When there is no current in the wire, all needles point in the same direction (that of Earth's field), as we would expect. However, when the wire

HANS CHRISTIAN OERSTED (1777–1851), DANISH PHYSICIST AND CHEMIST.

Oersted is best known for observing that a compass needle deflects when placed near a wire carrying a current. This important discovery was the first evidence of the connection between electric and magnetic phenomena. Oersted was also the first to prepare pure aluminum. (*North Wind Picture Archives*)

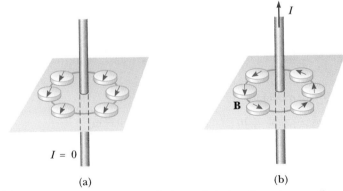

(a)   (b)

FIGURE 19.23 (a) When there is no current in the vertical wire, all compass needles point in the same direction. (b) When the wire carries a strong current, the compass needles deflect in directions tangent to the circle, pointing in the direction of **B** due to the current.

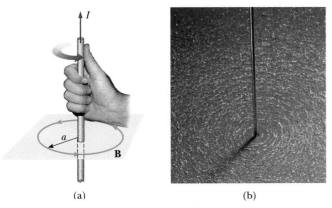

(a)   (b)

FIGURE 19.24   (a) Right-hand rule #2 for determining the direction of the magnetic field due to a long, straight wire carrying a current. Note that the magnetic field lines form circles around the wire. (b) Circular magnetic field lines surrounding a current-carrying wire, displayed by iron filings. *(© Richard Megna, Fundamental Photographs)*

ANDRÉ-MARIE AMPÈRE (1775–1836)

Ampère, a Frenchman, is credited with the discovery of electromagnetism—the relationship between electric currents and magnetic fields. Ampère's genius, particularly in mathematics, became evident by the age of 12, but his personal life was filled with tragedy. His father, a wealthy city official, was guillotined during the French Revolution, and his wife died young, in 1803. Ampère died at the age of 61 of pneumonia. His judgment of his life is clear from the epitaph he chose for his gravestone: *Tandem felix* (Happy at last). *(Leonard de Selva/CORBIS)*

carries a strong, steady current, the needles all deflect in directions tangent to the circle, as in Figure 19.23b. These observations show that the direction of **B** is consistent with the following convenient rule, **right-hand rule #2:**

> **If the wire is grasped in the right hand with the thumb in the direction of the current, as in Figure 19.24a, the fingers will curl in the direction of B.**

When the current is reversed, the filings in Figure 19.24b also reverse.

Because the filings point in the direction of **B**, we conclude that the lines of **B** form circles about the wire. By symmetry, the magnitude of **B** is the same everywhere on a circular path centered on the wire and lying in a plane perpendicular to the wire. By varying the current and distance from the wire, we find that **B** is proportional to the current and inversely proportional to the distance from the wire.

Shortly after Oersted's discovery, scientists arrived at an expression for the strength of the magnetic field due to the current in a long, straight wire. The magnetic field strength at distance r from a wire carrying current I is

$$B = \frac{\mu_0 I}{2\pi r} \qquad \text{[19.11]}$$

◀ Magnetic field due to a long, straight wire

This result shows that the magnitude of the magnetic field is proportional to the current and decreases as the distance from the wire increases, as we might intuitively expect. The proportionality constant μ_0, called the **permeability of free space,** has the value

$$\mu_0 \equiv 4\pi \times 10^{-7}\ \text{T} \cdot \text{m/A} \qquad \text{[19.12]}$$

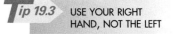

Tip 19.3   USE YOUR RIGHT HAND, NOT THE LEFT

Note that we have introduced two right-hand rules in this chapter. Be sure to use *only* your right hand when applying these rules.

AMPÈRE'S LAW AND A LONG, STRAIGHT WIRE

Equation 19.11 enables us to calculate the magnetic field due to a long, straight wire carrying a current. A general procedure for deriving such equations was proposed by the French scientist André-Marie Ampère (1775–1836); it provides a relation between the current in an arbitrarily shaped wire and the magnetic field produced by the wire.

Consider an arbitrary closed path surrounding a current as in Figure 19.25. The path consists of many short segments, each of length $\Delta\ell$. Let us now multiply one of these lengths by the component of the magnetic field parallel to that segment, where the product is labeled $B_\parallel\,\Delta\ell$. According to Ampère, the sum of all such products over the closed path is equal to μ_0 times the net current I that passes through the surface bounded by the closed path. This statement, known as **Ampère's circuital law,** can be written

Ampère's circuital law ▶

$$\sum B_{\parallel} \Delta \ell = \mu_0 I \qquad [19.13]$$

where $B_{\parallel}$ is the component of **B** parallel to the segment of length $\Delta \ell$ and $\sum B_{\parallel} \Delta \ell$ means that we take the sum over all the products $B_{\parallel} \Delta \ell$ around the closed path. Ampère's law is the fundamental law describing how electric currents create magnetic fields in the surrounding empty space.

We can use Ampère's circuital law to derive the magnetic field due to a long, straight wire carrying a current I. As discussed earlier, the magnetic field lines of this configuration form circles with the wire at their centers. The magnetic field is tangent to this circle at every point and has the same value B over the entire circumference of a circle of radius r, so that $B_{\parallel} = B$, as shown in Figure 19.26. We now calculate the sum $\sum B_{\parallel} \Delta \ell$ over the circular path and note that $B_{\parallel}$ can be removed from the sum (because it has the same value B for each element on the circle). Equation 19.13 then gives

$$\sum B_{\parallel} \Delta \ell = B_{\parallel} \sum \Delta \ell = B(2\pi r) = \mu_0 I$$

Dividing both sides by the circumference $2\pi r$, we obtain

$$B = \frac{\mu_0 I}{2\pi r}$$

This is identical to Equation 19.11, which is the magnetic field of a long, straight current.

Ampère's circuital law provides an elegant and simple method for calculating the magnetic fields of highly symmetric current configurations. However, it cannot be easily used to calculate magnetic fields for complex current configurations that lack symmetry. Furthermore, Ampère's circuital law is valid only when the currents and fields do not change with time.

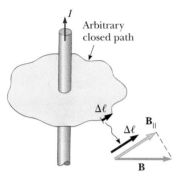

FIGURE 19.25 An arbitrary closed path around a current is used to calculate the magnetic field due to the current by use of Ampère's rule.

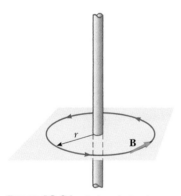

FIGURE 19.26 A closed circular path of radius r around a long, straight current-carrying wire is used to calculate the magnetic field set up by the wire.

APPLYING **PHYSICS** 19.4

Consider a plastic ring encircling a long, straight wire, which is coming out of the page in Figure 19.27. On the plastic ring are fastened two bar magnets as shown. If a current flows in the wire in a direction out of the page, is there a net torque on the ring-magnet combination? If so, in which direction?

Explanation   Each of the poles on the bar magnets experiences a magnetic force due to the circular magnetic field of the wire. The magnetic field falls off inversely with distance from the wire. Thus, the force on the north poles is smaller than the force on the south poles. The torque, however, increases linearly with distance from the wire. Thus, the effects of the decrease in field strength and the increase in torque with distance cancel, and there is no net torque on the ring-magnet combination.

Example 19.7   The Magnetic Field of a Long Wire

A long, straight wire carries a current of 5.00 A. At one instant, a proton, 4.00 mm from the wire, travels at 1.50×10^3 m/s parallel to the wire and in the same direction as the current (Fig. 19.28). Find the magnitude and direction of the magnetic force that is acting on the proton because of the magnetic field produced by the wire.

Solution   From Equation 19.11, the magnitude of the magnetic field produced by the current at a point 4.00 mm from the wire is

$$B = \frac{\mu_0 I}{2\pi r} = \frac{(4\pi \times 10^{-7} \, \text{T} \cdot \text{m/A})(5.00 \, \text{A})}{2\pi(4.00 \times 10^{-3} \, \text{m})} = 2.50 \times 10^{-4} \, \text{T}$$

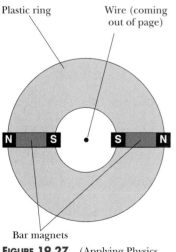

FIGURE 19.27 (Applying Physics 19.4)

This field is directed into the page at the location of the proton, as shown by right-hand rule #2 for a long, straight wire (see Fig. 19.24a).

The magnitude of the magnetic force on the proton is

$$F = qvB = (1.60 \times 10^{-19} \text{ C})(1.50 \times 10^3 \text{ m/s})(2.50 \times 10^{-4} \text{ T})$$
$$= 6.00 \times 10^{-20} \text{ N}$$

The force is directed toward the wire, as shown by right-hand rule #1 for the force on a moving charge (see Fig. 19.7).

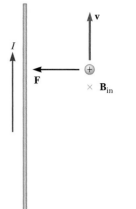

FIGURE 19.28 (Example 19.7) The magnetic field due to the current is into the page at the location of the proton, and the magnetic force on the proton is to the left.

19.8 MAGNETIC FORCE BETWEEN TWO PARALLEL CONDUCTORS

As we have seen, a magnetic force acts on a current-carrying conductor when the conductor is placed in an external magnetic field. Because a conductor carrying a current creates a magnetic field around itself, it is easy to understand that two current-carrying wires placed close together exert magnetic forces on each other. Consider two long, straight, parallel wires separated by the distance d and carrying currents I_1 and I_2 in the same direction, as shown in Figure 19.29. Let us determine the magnetic force on one wire due to a magnetic field set up by the other wire.

Wire 2, which carries current I_2, sets up magnetic field $\mathbf{B}_2$ at wire 1. The direction of $\mathbf{B}_2$ is perpendicular to the wire, as shown in the figure. Using Equation 19.11, we see that the magnitude of this magnetic field is

$$B_2 = \frac{\mu_0 I_2}{2\pi d}$$

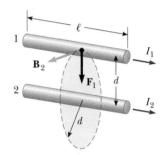

FIGURE 19.29 Two parallel wires, each carrying a steady current, exert forces on each other. The field $\mathbf{B}_2$ at wire 1 due to wire 2 produces a force on wire 1 given by $F_1 = B_2 I_1 \ell$. The force is attractive if the currents have the same direction, as shown, and repulsive if the two currents have opposite directions.

According to Equation 19.5, the magnitude of the magnetic force on wire 1 in the presence of field $\mathbf{B}_2$ due to I_2 is

$$F_1 = B_2 I_1 \ell = \left(\frac{\mu_0 I_2}{2\pi d}\right) I_1 \ell = \frac{\mu_0 I_1 I_2 \ell}{2\pi d}$$

We can rewrite this in terms of the force per unit length:

$$\frac{F_1}{\ell} = \frac{\mu_0 I_1 I_2}{2\pi d} \qquad\qquad \textbf{[19.14]}$$

The direction of $\mathbf{F}_1$ is downward, toward wire 2, as indicated by right-hand rule #1. If we consider the field set up at wire 2 due to wire 1, the force $\mathbf{F}_2$ on wire 2 is found to be equal in magnitude and opposite in direction to $\mathbf{F}_1$. This is what we would expect from Newton's third law of action-reaction.

We have shown that parallel conductors carrying currents in the same direction *attract* each other. You should use the approach indicated by Figure 19.29 and the steps leading to Equation 19.14 to show that parallel conductors carrying currents in opposite directions *repel* each other.

The force between two parallel wires carrying a current is used to define the SI unit of current, the **ampere** (A), as follows:

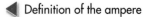 Definition of the ampere

> **If two long, parallel wires 1 m apart carry the same current, and the magnetic force per unit length on each wire is 2×10^{-7} N/m, then the current is defined to be 1 A.**

The SI unit of charge, the **coulomb** (C), can now be defined in terms of the ampere as follows:

> **If a conductor carries a steady current of 1 A, then the quantity of charge that flows through any cross section in 1 s is 1 C.**

Quick Quiz 19.5

If $I_1 = 2$ A and $I_2 = 6$ A in Figure 19.29, which of the following is true: (a) $F_1 = 3F_2$, (b) $F_1 = F_2$, or (c) $F_1 = F_2/3$?

Example 19.8 Levitating a Wire

Two wires, each having a weight per unit length of 1.0×10^{-4} N/m, are strung parallel to one another above Earth's surface, one directly above the other. The wires are aligned in a north-south direction so that Earth's magnetic field will not affect them. When their distance of separation is 0.10 m, what must be the current in each in order for the lower wire to levitate the upper wire? Assume that the wires carry the same currents, traveling in opposite directions.

Solution If the upper wire is to float, it must be in equilibrium under the action of two forces: the force of gravity and magnetic repulsion. The weight per unit length—here 1.0×10^{-4} N/m—must be equal and opposite the magnetic force per unit length given in Equation 19.14. Because the currents are the same, we have

$$\frac{F_1}{\ell} = \frac{mg}{\ell} = \frac{\mu_0 I^2}{2\pi d}$$

$$1.0 \times 10^{-4}\,\text{N/m} = \frac{(4\pi \times 10^{-7}\,\text{T} \cdot \text{m/A})(I^2)}{(2\pi)(0.10\,\text{m})}$$

We solve for the current to find

$$I = \boxed{7.1\,\text{A}}$$

EXERCISE If the current in each wire is doubled, what is the equilibrium separation of the two wires?

ANSWER 0.40 m

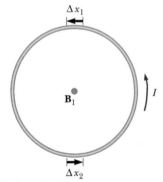

FIGURE 19.30 All segments of the current loop produce a magnetic field at the center of the loop, directed *out of the page.*

19.9 MAGNETIC FIELD OF A CURRENT LOOP

The strength of the magnetic field set up by a piece of wire carrying a current can be enhanced at a specific location if the wire is formed into a loop. You can understand this by considering the effect of several small segments of the current loop, as in Figure 19.30. The small segment at the top of the loop, labeled Δx_1, produces at the loop's center a magnetic field of magnitude B_1, directed out of the page. The direction of **B** can be verified using right-hand rule #2 for a long, straight wire. Imagine holding the wire with your right hand, with your thumb pointing in the direction of the current. Your fingers curl around in the direction of **B**.

A segment at the bottom of the loop of length Δx_2 also contributes to the field at the center, thus increasing its strength. The field produced at the center of the

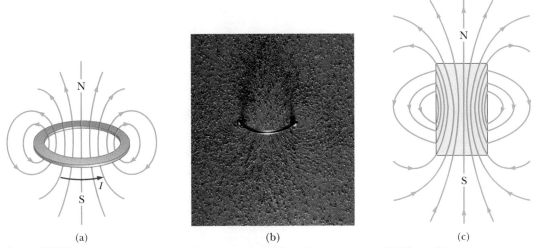

FIGURE 19.31 (a) Magnetic field lines for a current loop. Note that the magnetic field lines of the current loop resemble those of a bar magnet. (b) Field lines of a current loop, displayed by iron filings. (*© Richard Megna, Fundamental Photographs*) (c) The magnetic field of a bar magnet is similar to that of a current loop.

current loop by the segment Δx_2 has the same magnitude as B_1 and is also directed out of the page. Similarly, all other such segments of the current loop contribute to the field. The net effect is a magnetic field for the current loop as pictured in Figure 19.31a.

Notice in Figure 19.31a that the magnetic field lines enter at the bottom of the current loop and exit at the top. Thus, one side of the loop acts as though it were the north pole of a magnet, and the other acts as a south pole. The fact that the field set up by such a current loop bears a striking resemblance to the field of a bar magnet (Fig. 19.31c) will be of interest to us in a future section.

APPLYING PHYSICS 19.5

In electrical circuits, it is often the case that wires carrying currents in opposite directions are twisted together. What is the advantage of doing this?

Explanation If the wires are not twisted together, the combination of the two wires forms a current loop, which produces a relatively strong magnetic field. In fact, the magnetic field generated by the loop could be strong enough to affect adjacent circuits or components.

19.10 MAGNETIC FIELD OF A SOLENOID

If a long, straight wire is bent into a coil of several closely spaced loops, the resulting device is a **solenoid,** often called an **electromagnet.** This device is important in many applications because it acts as a magnet only when it carries a current. As we shall see, the magnetic field inside a solenoid increases with the current and is proportional to the number of coils per unit length.

Figure 19.32 shows the magnetic field lines of a loosely wound solenoid of length ℓ and total number of turns N. Note that the field lines inside the solenoid are nearly parallel, uniformly spaced, and close together. This indicates that the field inside the solenoid is nearly uniform and strong. The exterior field at the sides of the solenoid is nonuniform, is much weaker than the interior field, and is in the *opposite direction* to the field inside the solenoid.

If the turns are closely spaced, the field lines are as shown in Figure 19.33a, entering at one end of the solenoid and emerging at the other. This means that

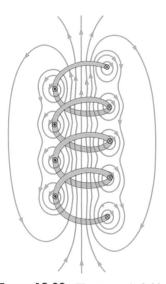

FIGURE 19.32 The magnetic field lines for a loosely wound solenoid.

FIGURE 19.33 (a) Magnetic field lines for a tightly wound solenoid of finite length carrying a steady current. The field inside the solenoid is nearly uniform and strong. Note that the field lines resemble those of a bar magnet, so the solenoid effectively has north and south poles. (b) The magnetic field pattern of a bar magnet, displayed by small iron filings on a sheet of paper. Compare with Figure 19.31c. *(Courtesy of Henry Leap and Jim Lehman)*

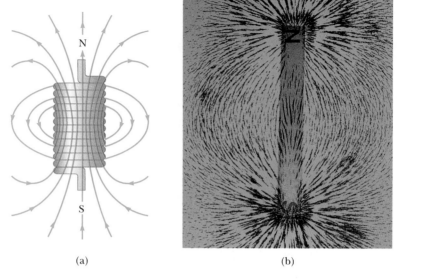

(a)   (b)

one end of the solenoid acts as a north pole and the other end acts as a south pole. If the length of the solenoid is much greater than its radius, the lines that leave the north end of the solenoid spread out over a wide region before returning to enter the south end. Hence, as you can see in Figure 19.33a, the magnetic field lines outside are widely separated, indicative of a weak field. This is in contrast to a much stronger field inside the solenoid, where the lines are close together. Also, the field inside the solenoid has a constant magnitude at all points far from its ends. The expression for the magnetic field inside the solenoid is

The magnetic field inside a solenoid ▶

$$B = \mu_0 n I \qquad \text{[19.15]}$$

where $n = N/\ell$ is the number of turns per unit length of the solenoid.

Example 19.9 The Magnetic Field Inside a Solenoid

A certain solenoid consists of 100 turns of wire and has a length of 10.0 cm.

A Find the magnetic field inside the solenoid when it carries a current of 0.500 A.

Solution The number of turns per unit length is

$$n = \frac{N}{\ell} = \frac{100 \text{ turns}}{0.10 \text{ m}} = 1\,000 \text{ turns/m}$$

so

$$B = \mu_0 n I = (4\pi \times 10^{-7} \text{ T} \cdot \text{m/A})(1\,000 \text{ turns/m})(0.500 \text{ A})$$
$$= 6.28 \times 10^{-4} \text{ T}$$

By comparison, this is about 10 times stronger than Earth's magnetic field.

B Assume that the field has one half of this value just outside the solenoid, at the point labeled N in Figure 19.33a. Find the magnitude and direction of the magnetic force acting on an electron that is moving from right to left in the figure, through point N, at 375 m/s.

Solution The magnitude of the magnetic force on the electron is

$$F = qvB = (1.60 \times 10^{-19} \text{ C})(375 \text{ m/s})(3.14 \times 10^{-4} \text{ T}) = 1.88 \times 10^{-20} \text{ N}$$

By use of right-hand rule #1 as in Figure 19.7, the direction of this force is found to be out of the page. (Remember to change the direction of the magnetic force for the negatively charged electron.) This force will deflect the electron from its original direction of motion.

EXERCISE How many turns should the solenoid have (assuming it carries the same current) if the field inside is to be 5 times as great?

ANSWER 500 turns

So-called steering magnets placed along the neck of the picture tube in a television set, as in Figure 19.34, are used to make the electron beam move to the desired locations on the screen, thus tracing out the images of your favorite program.

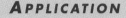

APPLICATION

CONTROLLING THE ELECTRON BEAM IN A TELEVISION SET

AMPÈRE'S LAW APPLIED TO A SOLENOID

We can use Ampère's law to obtain the expression for the magnetic field inside a solenoid carrying a current I. A cross section taken along the length of part of our solenoid is shown in Fig. 19.35. **B** inside the solenoid is uniform and parallel to the axis, and **B** outside is approximately zero. Consider a rectangular path of length L and width w as shown in Figure 19.35. We can apply Ampère's law to this path by evaluating the sum of $B_\parallel \, \Delta \ell$ over each side of the rectangle. The contribution along side 3 is clearly zero, because **B** = 0 in this region. The contributions from sides 2 and 4 are both zero, because **B** is perpendicular to $\Delta \ell$ along these paths. Side 1 of length L gives a contribution of BL to the sum, because **B** along this path is uniform and parallel to $\Delta \ell$. Therefore, the sum over the closed rectangular path has the value

$$\sum B_\parallel \, \Delta \ell = BL$$

The right side of Ampère's law involves the total current that passes through the area bounded by the path chosen. In our case, the total current through the rectangular path equals the current through each turn of the solenoid multiplied by the number of turns. If N is the number of turns in the length L, then the total current through the rectangular path equals NI. Therefore, Ampère's law applied to this path gives

$$\sum B_\parallel \, \Delta \ell = BL = \mu_0 NI$$

$$B = \mu_0 \frac{N}{L} I = \mu_0 nI$$

where $n = N/L$ is the number of turns per unit length.

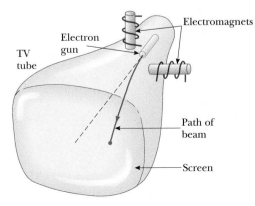

FIGURE 19.34 Electromagnets are used to deflect electrons to desired positions on the screen of a television tube.

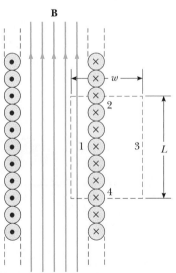

FIGURE 19.35 A cross-sectional view of a tightly wound solenoid. If the solenoid is long relative to its radius, we can assume that the magnetic field inside is uniform and the field outside is zero. Ampère's law applied to the red dashed rectangular path can then be used to calculate the field inside the solenoid.

19.11 MAGNETIC DOMAINS

The magnetic field produced by a current in a coil of wire gives us a hint as to what might cause certain materials to exhibit strong magnetic properties. A single coil like that in Figure 19.31a has a north pole and a south pole, but if this is true for a coil of wire, it should also be true for any current confined to a circular path. *In particular, an individual atom should act as a magnet because of the motion of the electrons about the nucleus.* Each electron, with its charge of 1.6×10^{-19} C, circles the atom once in about 10^{-16} s. If we divide the electronic charge by this time interval, we see that the orbiting electron is equivalent to a current of 1.6×10^{-3} A. Such a current produces a magnetic field on the order of 20 T at the center of the circular path. From this we see that a very strong magnetic field would be produced if several of these atomic magnets could be aligned inside a material. This does not occur, however, because the simple model we have described is not the complete story. A thorough analysis of atomic structure shows that the magnetic field produced by one electron in an atom is often canceled by an oppositely revolving electron in the same atom. The net result is that **the magnetic effect produced by the electrons orbiting the nucleus is either zero or very small for most materials.**

The magnetic properties of many materials are explained by the fact that an electron not only circles in an orbit but also spins on its axis like a top (Fig. 19.36). (This classical description should not be taken too literally. The property of *spin*

FIGURE 19.36 Classical model of a spinning electron.

PHYSICS *IN ACTION*

The Motion of Charged Particles in Magnetic Fields

The white arc in the photograph on the left indicates the circular path followed by an electron beam in a magnetic field. The vessel contains gas at very low pressure, and the beam is made visible as the electrons collide with the gas atoms, which emit visible light. The magnetic field is produced by two coils (not shown). The apparatus can be used to measure the ratio of e/m for the electron.

In the photograph on the right, oxygen, a paramagnetic substance, is attracted to a magnetic field. The liquid oxygen in this photograph is suspended between the poles of a permanent magnet. Paramagnetic substances contain atoms (or ions) that behave as tiny bar magnets. These magnetic atoms interact weakly with each other and are randomly oriented in the absence of an external magnetic field. When the substance is placed in an external magnetic field, the tiny bar magnets associated with the atoms of the substance tend to line up with the field.

(Courtesy of Central Scientific Company)   *(Courtesy of Leon Lewandowski)*

PHYSICS *IN ACTION*

Demonstrations of Electromagnetic Induction

On the left is a simple demonstration involving a permanent magnet, a conducting coil, and a *galvanometer*, which is a device used in the construction of analog ammeters and voltmeters. The galvanometer needle can deflect to either side of the zero mark, indicating the direction of a current. When a strong magnet is moved toward or away from the coil attached to the galvanometer, a current is induced in the coil, indicated by the momentary deflection of the galvanometer during the movement of the magnet. What is the cause of this induced current? Would the galvanometer needle deflect if the coil were moved toward a stationary magnet?

On the right, to demonstrate electromagnetic induction, an ac voltage is applied to the lower coil in the apparatus. A voltage is induced in the upper coil, as indicated by the illuminated flashlight bulb connected to this coil. What do you think happens to the bulb's intensity as the upper coil is moved over the vertical tube? To answer this question, note that the magnetic field associated with the lower coil varies along the axis of the tube.

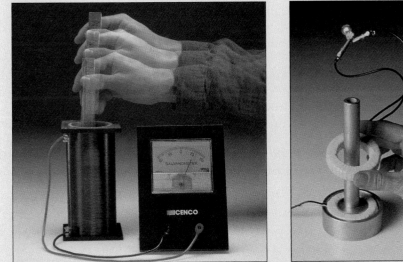

(Richard Megna, Fundamental Photographs)

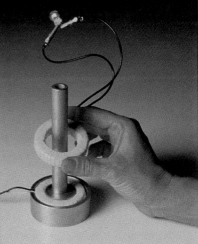

(Courtesy of Central Scientific Company)

Example 20.1 Application of Faraday's Law (Change of *B* with time)

A coil with 25 turns of wire is wrapped on a square frame 1.80 cm on a side. Each turn has the same area, equal to that of the frame, and the total resistance of the coil is 0.35 Ω. A uniform magnetic field is applied perpendicularly to the plane of the coil. If the field changes uniformly from 0 to 0.500 T in 0.800 s, find the magnitude of the induced emf in the coil while the field is changing.

Reasoning   The magnitude of the induced emf can be found from Faraday's law of induction, $\mathcal{E} = -N(\Delta\Phi_B/\Delta t)$. In this equation, $\Delta\Phi_B$ is the difference between the final and initial fluxes, where $\Phi_B = BA\cos\theta = BA\cos 0° = BA$.

Solution   The area of the coil is $(0.018\ 0\ \text{m})^2 = 3.24 \times 10^{-4}\ \text{m}^2$. The magnetic flux through the coil at $t = 0$ is zero because $B = 0$. At $t = 0.800$ s, the magnetic flux through the coil is

$$\Phi_{B,f} = BA = (0.500\ \text{T})(3.24 \times 10^{-4}\ \text{m}^2) = 1.62 \times 10^{-4}\ \text{T} \cdot \text{m}^2$$

Therefore, the *change* in flux through the coil during the 0.800-s interval is

$$\Delta\Phi_B = \Phi_{B,f} - \Phi_{B,i} = 1.62 \times 10^{-4}\ \text{T} \cdot \text{m}^2$$

Faraday's law of induction enables us to find the magnitude of the induced emf:

$$|\mathcal{E}| = N\frac{\Delta\Phi_B}{\Delta t} = (25\ \text{turns})\left(\frac{1.62 \times 10^{-4}\ \text{T} \cdot \text{m}^2}{0.800\ \text{s}}\right) = \boxed{5.1\ \text{mV}}$$

Note that $1\ \text{T} \cdot \text{m}^2/\text{s} = 1(\text{N} \cdot \text{m/C} \cdot \text{s})(\text{m}^2/\text{s}) = 1\ \text{N} \cdot \text{m/C} = 1\ \text{J/C} = 1\ \text{V}$.

EXERCISE   Find the magnitude of the induced current in the coil while the field is changing.

ANSWER   14 mA

20.3   MOTIONAL emf

In Section 20.2, we considered a situation in which an emf is induced in a circuit when the magnetic field changes with time. In this section we describe a particular application of Faraday's law in which a so-called **motional emf** is produced. This is the emf induced in a conductor moving through a magnetic field.

First consider a straight conductor of length ℓ moving with constant velocity through a uniform magnetic field directed into the paper, as in Figure 20.10. For simplicity, we assume that the conductor is moving perpendicularly to the field. The electrons in the conductor experience a force of magnitude $F = qvB$ directed downward along the conductor. Because of this magnetic force, the free electrons move to the lower end and accumulate there, leaving a net positive charge at the upper end. As a result of this charge separation, an electric field is produced in the conductor. The charge at the ends builds up until the downward magnetic force qvB is balanced by the upward electric force qE. At this point, charge stops flowing and the condition for equilibrium requires that

$$qE = qvB \qquad \text{or} \qquad E = vB$$

Because the electric field is uniform, the field produced in the conductor is related to the potential difference across the ends by $\Delta V = E\ell$. Thus,

$$\boxed{\Delta V = E\ell = B\ell v} \qquad \text{[20.3]}$$

Because there is an excess of positive charge at the upper end of the conductor and an excess of negative charge at the lower end, the upper end is at a higher potential than the lower end. Thus,

A potential difference is maintained across the conductor as long as there is motion through the field. If the motion is reversed, the polarity of the potential difference is also reversed.

A more interesting situation occurs if the moving conductor is part of a closed conducting path. This situation is particularly useful for illustrating how a changing loop area induces a current in a closed circuit described by Faraday's law. Consider a circuit consisting of a conducting bar of length ℓ, sliding along two fixed parallel conducting rails, as in Figure 20.11a. For simplicity, we assume that the moving bar has zero resistance and that the stationary part of the circuit has constant resistance R. A uniform and constant magnetic field **B** is applied perpendicularly to the plane of the circuit. As the bar is pulled to the right with velocity **v** under the influence of an applied force $\mathbf{F}_{\text{app}}$, the free charges in the bar experience a magnetic force along the length of the bar. This force in turn sets up an induced current because the charges are free to move in a closed conducting path. In this

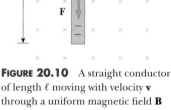

FIGURE 20.10 A straight conductor of length ℓ moving with velocity **v** through a uniform magnetic field **B** directed perpendicularly to **v**. The vector **F** is the force on an electron in the conductor. An emf of $B\ell v$ is induced between the ends of the bar.

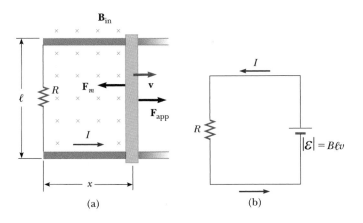

FIGURE 20.11 (a) A conducting bar sliding with velocity **v** along two conducting rails under the action of an applied force $\mathbf{F}_{app}$. The magnetic force $\mathbf{F}_m$ opposes the motion, and a counterclockwise current is induced in the loop. (b) The equivalent circuit of (a).

case, the changing magnetic flux through the loop and the corresponding induced emf across the moving bar arise from the *change in area of the loop* as the bar moves through the magnetic field.

Let us assume that the bar moves a distance Δx in time Δt, as shown in Figure 20.12. The increase in flux $\Delta\Phi_B$ through the loop in that time is the amount of flux that now passes through the portion of the circuit that has area $\ell\,\Delta x$:

$$\Delta\Phi_B = BA = B\ell\,\Delta x$$

Using Faraday's law and noting that there is one loop ($N = 1$), we find that the magnitude of the induced emf is

$$\left|\mathcal{E}\right| = \frac{\Delta\Phi_B}{\Delta t} = B\ell\,\frac{\Delta x}{\Delta t} = B\ell v \qquad \text{[20.4]}$$

This induced emf is often called a **motional emf** because it arises from the motion of a conductor through a magnetic field.

Furthermore, if the resistance of the circuit is R, the magnitude of the induced current in the circuit is

$$I = \frac{\left|\mathcal{E}\right|}{R} = \frac{B\ell v}{R} \qquad \text{[20.5]}$$

Figure 20.11b is the equivalent circuit diagram for this example.

Webnote 20.1

Change the shape of a coil in a magnetic field and watch as a current is induced in the coil. Visit *http://www.lightlink.com/sergey/java/java/indcur/index.html*

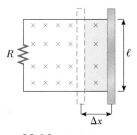

FIGURE 20.12 As the bar moves to the right, the area of the loop increases by the amount $\ell\,\Delta x$ and the magnetic flux through the loop increases by $B\ell\,\Delta x$.

APPLYING PHYSICS 20.2

We have discussed applying a force on the bar, which results in an induced emf in the circuit shown in Figure 20.11. Suppose we remove the external magnetic field in the diagram and replace the resistor with a high-voltage source and a switch, as in Figure 20.13. What happens when the switch is closed? Will the bar move, and does it matter which way we connect the high-voltage source?

Explanation Suppose the source is capable of establishing high current. The two horizontal conducting rods will create a strong magnetic field in the area between them, directed into the page. (The movable bar also creates a magnetic field, but this field cannot exert force on the bar itself.) As the current passes downward through the movable bar, it experiences a magnetic force to the right. Hence, it accelerates along the rails away from the power supply. If the polarity of the power were reversed, the magnetic field would be out of the page, the current in the bar would be upward, and the force on the bar would still be to the right. (This is the essence of a railgun.) The $BI\ell$ force exerted by a magnetic field per Equation 19.6 causes the bar to accelerate away from the voltage source. Studies have shown that it is possible to launch payloads into space with this technology. Very large accelerations can be obtained with currently available technology, with payloads being accelerated to a speed of several kilometers per second in a fraction of a second. This is a larger acceleration than humans can withstand.

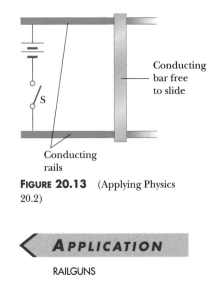

Conducting bar free to slide

Conducting rails

FIGURE 20.13 (Applying Physics 20.2)

◀ **APPLICATION**

RAILGUNS

Quick Quiz 20.2

As an airplane flies due north from Los Angeles to Seattle, it cuts through Earth's magnetic field. As a result, an emf is developed between the wing tips. Which wing tip is positively charged?

Quick Quiz 20.3

You wish to move a rectangular loop of wire into a region of uniform magnetic field at a given speed so as to induce an emf in the loop. The plane of the loop must remain perpendicular to the magnetic field lines. In which orientation should you hold the loop while you move it into the region of magnetic field in order to generate the largest emf? (a) With the long dimension of the loop parallel to the velocity vector. (b) With the short dimension of the loop parallel to the velocity vector. (c) Either way—the emf is the same regardless of orientation.

Example 20.2 The Electrified Airplane Wing

An airplane with a wing span of 30.0 m flies parallel to Earth's surface at a location at which the downward component of Earth's magnetic field is 0.60×10^{-4} T. Find the difference in potential between the wing tips when the speed of the plane is 250 m/s.

Solution Because the plane is flying horizontally, we do not have to concern ourselves with the horizontal component of Earth's magnetic field. Thus, we find that

$$\mathcal{E} = B\ell v = (0.60 \times 10^{-4}\text{ T})(30.0\text{ m})(250\text{ m/s}) = \boxed{0.45\text{ V}}$$

Example 20.3 Where Is the Energy Source?

A The sliding bar in Figure 20.11a has a length of 0.50 m and moves at 2.0 m/s in a magnetic field of magnitude 0.25 T. Find the induced voltage in the moving rod.

Solution We use Equation 20.3 and find that

$$\mathcal{E} = B\ell v = (0.25\text{ T})(0.50\text{ m})(2.0\text{ m/s}) = \boxed{0.25\text{ V}}$$

B If the resistance in the circuit is 0.50 Ω, find the current in the circuit.

Solution The current is found from Ohm's law to be

$$I = \frac{\mathcal{E}}{R} = \frac{0.25\text{ V}}{0.50\text{ }\Omega} = \boxed{0.50\text{ A}}$$

C Find the amount of energy delivered to the 0.50-Ω resistor in 1 s.

Solution The power delivered to the resistor is

$$\mathcal{P} = I\,\Delta V = (0.50\text{ A})(0.25\text{ V}) = 0.13\text{ W}$$

Because power is defined as the rate at which energy is converted in a device, the energy *W* delivered to the resistor in 1 s is

$$W = \mathcal{P}t = (0.125\text{ W})(1.0\text{ s}) = \boxed{0.13\text{ J}}$$

D The source of the energy calculated in part C is some external agent that keeps the bar moving at a constant speed of 2.0 m/s by exerting an applied force F_{app}. Find the value of F_{app}.

Solution   From part C, we know that the work done by the applied force in 1 s is 0.13 J. In 1 s, the bar moves a distance of

$$d = vt = (2.0 \text{ m/s})(1.0 \text{ s}) = 2.0 \text{ m}$$

Thus, from the definition of work, we find that $W = F_{app}d$, or

$$F_{app} = \frac{W}{d} = \frac{0.13 \text{ J}}{2.0 \text{ m}} = \boxed{0.063 \text{ N}}$$

EXERCISE   If the rod is to move at constant speed, the applied force must be equal in magnitude to the retarding magnetic force, $BI\ell$. Show that this approach also gives $F_{app} = 0.063$ N, as found in part (d).

20.4   LENZ'S LAW REVISITED (THE MINUS SIGN IN FARADAY'S LAW)

To attain a better understanding of Lenz's law, let us return to the example of a bar moving to the right on two parallel rails in the presence of a uniform magnetic field directed into the paper (Fig. 20.14a). As the bar moves to the right, the magnetic flux through the circuit increases with time because the area of the loop increases. Lenz's law says that the induced current must be in a direction such that the flux *it* produces opposes the change in the external magnetic flux. Because the flux due to the external field is increasing *into* the paper, the induced current, to oppose the change, must produce a flux *out* of the paper. Hence, the induced current must be counterclockwise when the bar moves to the right. (Use right-hand rule #2 from Chapter 19 to verify this direction.) On the other hand, if the bar is moving to the left, as in Figure 20.14b, the magnetic flux through the loop decreases with time. Because the flux is into the paper, the induced current has to be clockwise to produce its own flux into the paper (which opposes the decrease in the external flux). In either case, the induced current tends to maintain the original flux through the circuit.

Now let us examine this situation from the viewpoint of energy conservation. Suppose that the bar is given a slight push to the right. In the preceding analysis, we found that this motion led to a counterclockwise current in the loop. What would happen if we assume that the current is clockwise, that is, opposite the direction required by Lenz's law? For a clockwise current I, the direction of the magnetic force $BI\ell$ on the sliding bar would be to the right. This force would accelerate the rod and increase its velocity. This, in turn, would cause the area of the loop to increase more rapidly, thereby increasing the induced current, which would increase the force, which would increase the current, which would. . . . In effect, the system would acquire energy with zero input energy. This is clearly inconsistent both with our experience and with the law of conservation of energy. We are forced to conclude that the current must be counterclockwise.

Consider another situation. A bar magnet is moved to the right toward a stationary loop of wire, as in Figure 20.15a. As the magnet moves, the magnetic flux through the loop increases with time. To counteract this, the induced current produces a flux to the left, as in Figure 20.15b; hence, the induced current is in the direction shown. Note that the magnetic field lines associated with the induced current oppose the motion of the magnet. Therefore, the left face of the current loop is a north pole and the right face is a south pole.

On the other hand, if the magnet moves to the left, as in Figure 20.15c, its flux through the loop, which is toward the right, decreases in time. Under these circumstances, the induced current in the loop is in a direction to set up a field directed from left to right through the loop, in an effort to maintain a constant

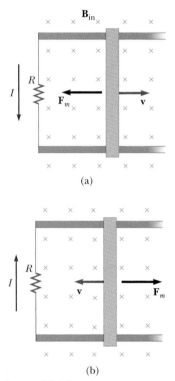

(a)

(b)

FIGURE 20.14   (a) As the conducting bar slides on the two fixed conducting rails, the magnetic flux through the loop increases in time. By Lenz's law, the induced current must be *counterclockwise* so as to produce a counteracting flux *out of the paper.* (b) When the bar moves to the left, the induced current must be *clockwise.* Why?

THERE ARE TWO MAGNETIC FIELDS TO CONSIDER

When applying Lenz's law, note that there are *two* magnetic fields to consider. The first is the external changing magnetic field that induces the current in a conducting loop. The second is the magnetic field produced *by* the induced current in the loop.

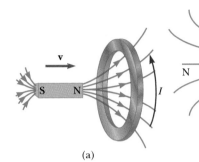

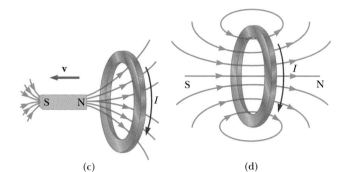

(a)　　　　　　　　　　(b)　　　　　　　　　　(c)　　　　　　　　　　(d)

FIGURE 20.15 (a) When the magnet is moved toward the stationary conducting loop, a current is induced in the direction shown. (b) This induced current produces its own flux to the left to counteract the increasing external flux to the right. (c) When the magnet is moved away from the stationary conducting loop, a current is induced in the direction shown. (d) This induced current produces its own flux to the right to counteract the decreasing external flux to the right.

number of flux lines. Hence, the induced current in the loop is as shown in Figure 20.15d. In this case, the left face of the loop is a south pole and the right face is a north pole.

Quick Quiz 20.4 A bar magnet is falling through a loop of wire with constant velocity with the north pole entering first. Viewed from the same side of the loop as the magnet, as the north pole approaches the loop, the induced current will be in what direction? (a) clockwise (b) zero (c) counterclockwise (d) along the length of the magnet

Example 20.4   Application of Lenz's Law

A coil of wire is placed near an electromagnet, as in Figure 20.16a. Find the direction of the induced current in the coil (a) at the instant the switch is closed, (b) after the switch has been closed for several seconds, and (c) when the switch is opened.

Reasoning and Solution

A When the switch is closed, the situation changes from a condition in which no lines of flux pass through the coil to one in which lines of flux pass through in the direction

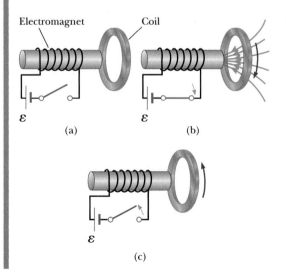

FIGURE 20.16 (Example 20.4)

shown in Figure 20.16b. To counteract this change in the number of lines, the coil must set up a field from left to right in the figure. This requires a current directed as shown in Figure 20.16b.

B After the switch has been closed for several seconds, there is no change in the number of lines through the loop; hence, the induced current is zero.

C Opening the switch causes the magnetic field to change from a condition in which flux lines thread through the coil from right to left to a condition of zero flux. The induced current must then be as shown in Figure 20.16c, so as to set up its own field from right to left.

TAPE RECORDERS

One common practical use of induced currents and emfs is in the tape recorder. Many different types of tape recorders are made, but the basic principles are the same for all. A magnetic tape moves past a recording and playback head, as in Figure 20.17a. The tape is a plastic ribbon coated with iron oxide or chromium oxide.

The recording process uses the fact that a current in an electromagnet produces a magnetic field. Figure 20.17b illustrates the steps in the process. A sound wave sent into a microphone is transformed into an electric current, amplified, and allowed to pass through a wire coiled around a doughnut-shaped piece of iron, which functions as the recording head. The iron ring and the wire constitute an electromagnet, in which the lines of the magnetic field are contained completely inside the iron except at the point where a slot is cut in the ring. Here the magnetic field fringes out of the iron and magnetizes the small pieces of iron oxide embedded in the tape. Thus, as the tape moves past the slot, it becomes magnetized in a pattern that reproduces both the frequency and the intensity of the sound signal entering the microphone.

To reconstruct the sound signal, the previously magnetized tape is allowed to pass through a recorder head operating in the playback mode. A second wire-wound doughnut-shaped piece of iron with a slot in it passes close to the tape, so that the varying magnetic fields on the tape produce changing field lines through the wire coil. The changing flux induces a current in the coil that corresponds to the current in the recording head that originally produced the tape. This changing electric current can be amplified and used to drive a speaker. Playback is thus an example of induction of a current by a moving magnet.

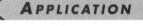

APPLICATION

MAGNETIC TAPE RECORDERS

Webnote 20.2

Take a look at audiotapes, tape decks, and tape recorders by visiting
http://www.howstuffworks.com/cassette.htm

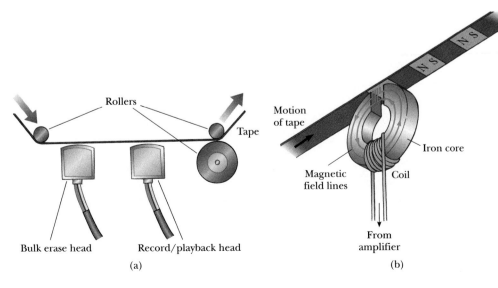

FIGURE 20.17 (a) Major parts of a magnetic tape recorder. If a new recording is to be made, the bulk erase head wipes the tape clean of signals before recording. (b) The fringing magnetic field magnetizes the tape during recording.

APPLICATION

ALTERNATING CURRENT
GENERATORS

20.5 GENERATORS

Generators and motors are important practical devices that operate on the principle of electromagnetic induction. First, let us consider the **alternating current (AC) generator,** a device that converts mechanical energy to electrical energy. In its simplest form, the AC generator consists of a wire loop rotated in a magnetic field by some external means (Fig. 20.18a). In commercial power plants, the energy required to rotate the loop can be derived from a variety of sources. For example, in a hydroelectric plant, falling water directed against the blades of a turbine produces the rotary motion; in a coal-fired plant, heat produced by burning coal is used to convert water to steam, and this steam is directed against the turbine blades. As the loop rotates, the magnetic flux through it changes with time, inducing an emf and a current in an external circuit. The ends of the loop are connected to slip rings that rotate with the loop. Connections to the external circuit are made by stationary brushes in contact with the slip rings.

We can derive an expression for the emf generated in the rotating loop by making use of the expression for motional emf, $\mathcal{E} = B\ell v$. Figure 20.19a shows a loop of wire rotating clockwise in a uniform magnetic field directed to the right. The magnetic force (qvB) on the charges in wires AB and CD is not along the lengths of the wires. (The force on the electrons in these wires is perpendicular to the wires.) Hence, an emf is generated only in wires BC and AD. At any instant, wire BC has velocity **v** at an angle θ with the magnetic field, as shown in Figure 20.19b. (Note that the component of velocity parallel to the field has no effect on the charges in the wire, whereas the component of velocity perpendicular to the field produces a magnetic force on the charges that moves electrons from C to B.) The emf generated in wire BC equals $B\ell v_\perp$, where ℓ is the length of the wire and $v_\perp$ is the component of velocity perpendicular to the field. An emf of $B\ell v_\perp$ is also generated in wire DA, and the sense of this emf is the same as in wire BC. Because $v_\perp = v \sin \theta$, the total induced emf is

$$\mathcal{E} = 2B\ell v_\perp = 2B\ell v \sin \theta \qquad [20.6]$$

If the loop rotates with a constant angular speed ω, we can use the relation $\theta = \omega t$ in Equation 20.6. Furthermore, because every point on wires BC and DA rotates in a circle about the axis of rotation with the same angular speed ω, we have $v = r\omega = (a/2)\omega$, where a is the length of sides AB and CD. Therefore, Equation 20.6 reduces to

$$\mathcal{E} = 2B\ell \left(\frac{a}{2}\right) \omega \sin \omega t = B\ell a \omega \sin \omega t$$

If a coil has N turns, the emf is N times as large because each loop has the same

FIGURE 20.18 (a) Diagram of an AC generator. An emf is induced in a coil, which rotates by some external means in a magnetic field. (b) Plot of the alternating emf induced in the loop versus time.

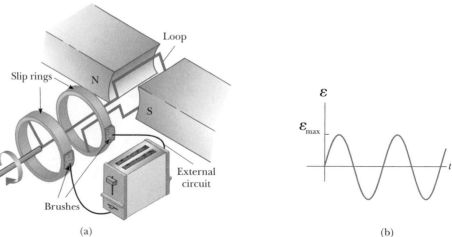

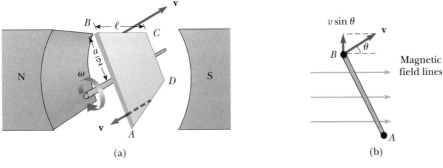

emf induced in it. Furthermore, because the area of the loop is $A = \ell a$, the total emf is

$$\mathcal{E} = NBA\omega \sin \omega t \qquad [20.7]$$

This result shows that the emf varies sinusoidally with time, as plotted in Figure 20.18b. Note that the maximum emf has the value

$$\mathcal{E}_{max} = NBA\omega \qquad [20.8]$$

which occurs when $\omega t = 90°$ or $270°$. In other words, $\mathcal{E} = \mathcal{E}_{max}$ when the plane of the loop is parallel to the magnetic field. Furthermore, the emf is zero when $\omega t = 0$ or $180°$, that is, when the magnetic field is perpendicular to the plane of the loop. In the United States and Canada, the frequency of rotation for commercial generators is 60 Hz, whereas in some European countries 50 Hz is used. (Recall that $\omega = 2\pi f$, where f is the frequency in hertz.)

The **direct current (DC) generator** is illustrated in Figure 20.20a. The components are essentially the same as those of the AC generator, except that the contacts to the rotating loop are made by a split ring, or commutator. In this design, the output voltage always has the same polarity and the current is a pulsating direct current, as in Figure 20.20b. This can be understood by noting that the contacts to the split ring reverse their roles every half cycle. At the same time, the polarity of the induced emf reverses. Hence, the polarity of the split ring remains the same.

A pulsating DC current is not suitable for most applications. To produce a steady DC current, commercial DC generators use many loops and commutators distributed around the axis of rotation so that the sinusoidal pulses from the loops overlap in phase. When these pulses are superimposed, the DC output is almost free of fluctuations.

Turbines turn electric generators at a hydroelectric power plant. (*Luis Castaneda/The IMAGE Bank*)

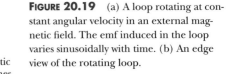

APPLICATION

DIRECT CURRENT GENERATORS

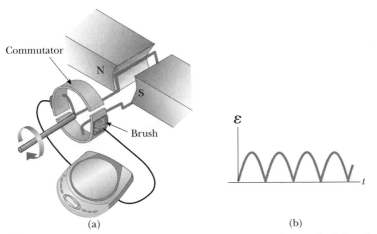

Commutator

Brush

FIGURE 20.20 (a) Diagram of a DC generator. (b) The emf fluctuates in magnitude but always has the same polarity.

Example 20.5 Emf Induced in an AC Generator

An AC generator consists of eight turns of wire of area $A = 0.090\ 0\ m^2$ with a total resistance of 12.0 Ω. The loop rotates in a magnetic field of 0.500 T at a constant frequency of 60.0 Hz.

A Find the maximum induced emf.

Solution First note that $\omega = 2\pi f = 2\pi(60.0\ Hz) = 377\ rad/s$. When we substitute the appropriate numerical values into Equation 20.8, we obtain

$$\mathcal{E}_{max} = NAB\omega = 8(0.90\ 0\ m^2)(0.500\ T)(377\ rad/s) = \boxed{136\ V}$$

B What is the maximum induced current?

Solution From Ohm's law and the result of part A, we find that

$$I_{max} = \frac{\mathcal{E}_{max}}{R} = \frac{136\ V}{12.0\ \Omega} = \boxed{11.3\ A}$$

C Determine the induced emf as a function of time.

Solution We can use Equation 20.7 to obtain the time variation of $\mathcal{E}$:

$$\mathcal{E} = \mathcal{E}_{max} \sin \omega t = \boxed{(136\ V) \sin 377t}$$

where t is in seconds.

EXERCISE Determine the time variation of the induced current.

ANSWER $I = (11.3\ A) \sin 377t$

APPLICATION

MOTORS

MOTORS AND BACK emf

Motors are devices that convert electrical energy to mechanical energy. Essentially, **a motor is a generator run in reverse.** Instead of a current being generated by a rotating loop, a current is supplied to the loop by a source of emf, and the magnetic torque on the current-carrying loop causes it to rotate.

A motor can perform useful mechanical work when a shaft connected to its rotating coil is attached to some external device. As the coil in the motor rotates, however, the changing magnetic flux through it induces an emf, which acts to reduce the current in the coil. If it increased the current, Lenz's law would be violated. The phrase **back emf** is used for an emf that tends to reduce the applied current. The back emf increases in magnitude as the rotational speed of the coil increases. We can picture this state of affairs as the equivalent circuit in Figure 20.21. For illustrative purposes, assume that the external power source attempting to supply current in the coil of the motor has a voltage of 120 V, that the coil has a resistance of 10 Ω, and that the back emf induced in the coil at this instant is 70 V. Thus, the voltage available to supply current equals the difference between the applied voltage and the back emf, 50 V in this case. It is clear that the current is reduced by the back emf.

When a motor is turned on, there is no back emf initially, and the current is very large because it is limited only by the resistance of the coil. As the coil begins to rotate, the induced back emf opposes the applied voltage and the current in the coil is reduced. If the mechanical load increases, the motor slows down, which decreases the back emf. This reduction in the back emf increases the current in the coil and therefore also increases the power needed from the external voltage source. As a consequence, the power requirements for starting a motor and for running it under heavy loads are greater than those for running the motor under average loads. If the motor is allowed to run under no mechanical load, the back

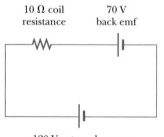

10 Ω coil resistance   70 V back emf

120 V external source

FIGURE 20.21 A motor can be represented as a resistance plus a back emf.

CONCEPTUAL QUESTIONS

1. A circular loop is located in a uniform and constant magnetic field. Describe how an emf can be induced in the loop in this situation.
2. Does dropping a magnet down a copper tube produce a current in the tube? Explain.
3. A spacecraft orbiting Earth has a coil of wire in it. An astronaut measures a small current in the coil, although no battery is connected to it and there are no magnets in the spacecraft. What is causing the current?
4. A loop of wire is placed in a uniform magnetic field. For what orientation of the loop is the magnetic flux a maximum? For what orientation is the flux zero?
5. As the conducting bar in Figure Q20.5 moves to the right, an electric field is set up directed downward. If the bar were moving to the left, explain why the electric field would be upward.

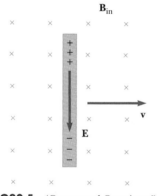

FIGURE Q20.5 (Conceptual Questions 5 and 6)

6. As the bar in Figure Q20.5 moves perpendicular to the field, is an external force required to keep it moving with constant speed?
7. Wearing a metal bracelet in a region of strong magnetic field can be hazardous. Discuss.
8. How is electrical energy produced in dams (that is, how is the energy of motion of the water converted to ac electricity)?
9. Eddy currents are induced currents set up in a piece of metal when it moves through a nonuniform magnetic field. For example, consider the flat metal plate swinging at the end of a bar as a pendulum as shown in Figure Q20.9. At position 1, the pendulum is moving from a region where there is no magnetic field into a region where the field **B**_{in} is directed into the paper. Show that at position 1, the direction of the eddy current is counterclockwise. Also, at position 2

the pendulum is moving out of the field into a region of zero field. Show that the direction of the eddy current is clockwise in this case. Use right-hand rule #2 to show that these eddy currents lead to a magnetic force on the plate directed as shown in the figure. Because the induced eddy current always produces a retarding force when the plate enters or leaves the field, the swinging plate quickly comes to rest.

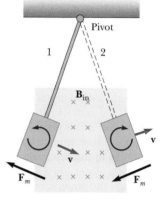

FIGURE Q20.9

10. Suppose you would like to steal power for your home from the electric company by placing a loop of wire near a transmission cable in order to induce an emf in the loop (an illegal procedure). Should you locate the loop so that the transmission cable passes through your loop or simply place your loop near the transmission cable? Does the orientation of the loop matter?
11. A piece of aluminum is dropped vertically downward between the poles of an electromagnet. Does the magnetic field affect the velocity of the aluminum? (*Hint:* See Conceptual Question 9.)
12. A bar magnet is dropped toward a conducting ring lying on the floor. As the magnet falls toward the ring, does it move as a freely falling object?
13. If the current in an inductor is doubled, by what factor does the stored energy change?
14. Is it possible to induce a constant emf for an infinite amount of time?
15. Why is the induced emf that appears in an inductor called a back (counter) emf?
16. A magneto is used to cause the spark in a spark plug in many lawn mowers today. A magneto consists of a permanent magnet mounted on the flywheel so that it spins past a fixed coil. Explain how this arrangement generates a large enough potential difference to cause the spark.

PROBLEMS

1, 2, 3 = straightforward, intermediate, challenging   ☐ = full solution available in Student Solutions Manual/Study Guide
web = solution posted at **http://info.brookscole.com/serway**   = biomedical application

Section 20.1   Induced emf and Magnetic Flux

1. A magnetic field of strength 0.30 T is directed perpendicular to a plane circular loop of wire of radius 25 cm.

Find the magnetic flux through the area enclosed by this loop.
2. A circular loop with a radius of 0.200 m is placed in a uniform magnetic field of magnitude 0.850 T. The normal to

the loop makes an angle of 30.0° with respect to the direction of **B**. If the field increases to 0.950 T, what is the increase in magnetic flux through the loop?

3. Consider a place where Earth's magnetic field has strength of 0.520×10^{-4} T and makes an angle of 62.0° with the horizontal. At this location a magnetic compass points toward true north. What is the magnetic flux in a rectangular loop of wire that is 15.0 by 25.0 cm and is lying on a table? What is the magnetic flux in this loop if it is mounted vertically on a north wall? On an east wall?

4. A long, straight wire carrying a current of 2.00 A is placed along the axis of a cylinder of radius 0.500 m and a length of 3.00 m. Determine the total magnetic flux through the cylinder.

5. A long, straight wire lies in the plane of a circular coil with a radius of 0.010 m. The wire carries a current of 2.0 A and is placed along a diameter of the coil. (a) What is the net flux through the coil? (b) If the wire passes through the center of the coil and is perpendicular to the plane of the coil, find the net flux through the coil.

6. A solenoid 4.00 cm in diameter and 20.0 cm long has 250 turns and carries a current of 15.0 A. Calculate the magnetic flux through the circular cross-sectional area of the solenoid.

7. A cube of edge length $\ell = 2.5$ cm is positioned as shown in Figure P20.7. There is a uniform magnetic field throughout the region with components of $B_x = +5.0$ T, $B_y = +4.0$ T, and $B_z = +3.0$ T. (a) Calculate the flux through the shaded face of the cube. (b) What is the total flux emerging from the volume enclosed by the cube (that is, total flux through all six faces)?

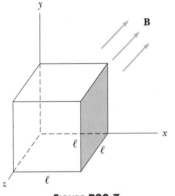

FIGURE P20.7

Section 20.2 Faraday's Law of Induction

8. A circular coil of radius 20 cm is placed in an external magnetic field of strength 0.20 T so that the plane of the coil is perpendicular to the field. The coil is pulled out of the field in 0.30 s. Find the average induced emf during this interval.

9. A 25-turn circular coil of wire has a diameter of 1.00 m. It is placed with its axis along the direction of Earth's magnetic field of 50.0 μT, and then in 0.200 s it is flipped 180°. An average emf of what magnitude is generated in the coil?

10. The flexible loop in Figure P20.10 has a radius of 12 cm and is in a magnetic field of strength 0.15 T. The loop is grasped at points A and B and stretched until its area is

nearly zero. If it takes 0.20 s to close the loop, find the magnitude of the average induced emf in it during this time.

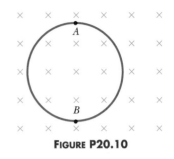

FIGURE P20.10

11. A strong electromagnet produces a uniform field of 1.60 T over a cross-sectional area of 0.200 m². We place a coil having 200 turns and a total resistance of 20.0 Ω around the electromagnet. We then smoothly decrease the current in the electromagnet until it reaches zero in 20.0 ms. What is the current induced in the coil?

12. A 500-turn circular-loop coil 15.0 cm in diameter is initially aligned so that its axis is parallel to Earth's magnetic field. In 2.77 ms, the coil is flipped so that its axis is perpendicular to Earth's magnetic field. If an average voltage of 0.166 V is thereby induced in the coil, what is the value of Earth's magnetic field at that location?

13. The plane of a rectangular coil, 5.0 cm by 8.0 cm, is perpendicular to the direction of a magnetic field **B**. If the coil has 75 turns and a total resistance of 8.0 Ω, at what rate must the magnitude of **B** change to induce a current of 0.10 A in the windings of the coil?
web

14. A square, single-turn wire loop 1.00 cm on a side is placed inside a solenoid that has a circular cross section of radius 3.00 cm, as shown in Figure P20.14. The solenoid is 20.0 cm long and wound with 100 turns of wire. (a) If the current in the solenoid is 3.00 A, find the flux through the loop. (b) If the current in the solenoid is reduced to zero in 3.00 s, find the magnitude of the average induced emf in the loop.

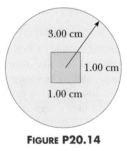

FIGURE P20.14

15. A 300-turn solenoid with a length of 20 cm and a radius of 1.5 cm carries a current of 2.0 A. A second coil of four turns is wrapped tightly about this solenoid so that it can be considered to have the same radius as the solenoid. Find (a) the change in the magnetic flux through the coil and (b) the magnitude of the average induced emf in the coil when the current in the solenoid increases to 5.0 A in a period of 0.90 s.

16. A circular coil, enclosing an area of 100 cm², is made of 200 turns of copper wire. The wire making up the coil has resistance of 5.0 Ω, and the ends of the wire are connected

to form a closed loop. Initially, a 1.1-T uniform magnetic field points perpendicularly upward through the plane of the coil. The direction of the field then reverses so that the final magnetic field has a magnitude of 1.1 T and points downward through the coil. If the time required for the field to reverse directions is 0.10 s, what average current flows through the coil during this time?

17. The person in Figure P20.17 is girded about the chest with a breathing monitor, which is a 100-turn coil. As a breath is inhaled, the area of the coil varies from 0.120 m² to 0.124 m². Earth's magnetic field is 50.0 μT and makes an angle of 22.5° with respect to the normal to the loop. If the patient inhales a breath in 1.59 s, what is the average voltage induced in the coil during the inhalation?

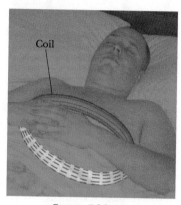

FIGURE P20.17

Section 20.3 Motional emf

18. Consider the arrangement shown in Figure P20.18. Assume that $R = 6.00 \ \Omega$ and $\ell = 1.20$ m, and that a uniform 2.50-T magnetic field is directed *into* the page. At what speed should the bar be moved to produce a current of 0.500 A in the resistor?

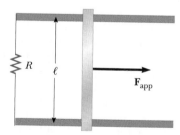

FIGURE P20.18 (Problems 18 and 57)

19. A Boeing-747 jet with a wing span of 60.0 m is flying horizontally at a speed of 300 m/s over Phoenix, Arizona, at a location where Earth's magnetic field is 50.0 μT at 58.0° below the horizontal. What voltage is generated between the wingtips?

20. Over a region where the *vertical* component of Earth's magnetic field is 40.0 μT directed downward, a 5.00-m length of wire is held in an east-west direction and moved horizontally to the north with a speed of 10.0 m/s. Calculate the potential difference between the ends of the wire, and determine which end is positive.

21. An automobile has a vertical radio antenna 1.20 m long. The automobile travels at 65.0 km/h on a horizontal road where Earth's magnetic field is 50.0 μT directed toward

the north and downward at an angle of 65.0° below the horizontal. (a) Specify the direction that the automobile should move in order to generate the maximum motional emf in the antenna, with the top of the antenna positive relative to the bottom. (b) Calculate the magnitude of this induced emf.

22. A helicopter has blades of length 3.0 m, rotating at 2.0 rev/s about a central hub. If the vertical component of Earth's magnetic field is 5.0×10^{-5} T, what is the emf induced between the blade tip and the central hub?

Section 20.4 Lenz's Law Revisited (The Minus Sign in Faraday's Law)

23. A bar magnet is positioned near a coil of wire as shown in Figure P20.23. What is the direction of the current through the resistor when the magnet is moved (a) to the left? (b) to the right?

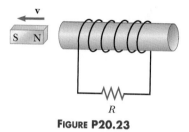

FIGURE P20.23

24. A bar magnet is held above the center of a wire loop in a horizontal plane, as shown in Figure P20.24. The south end of the magnet is toward the loop. The magnet is dropped. Find the direction of the current through the resistor (a) while the magnet is falling toward the loop and (b) after the magnet has passed through the loop and moves away from it.

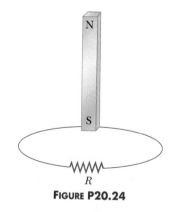

FIGURE P20.24

25. What is the direction of the current induced in the resistor when the current in the long, straight wire in Figure P20.25 decreases rapidly to zero?

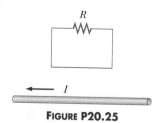

FIGURE P20.25

26. In Figure P20.26, what is the direction of the current induced in the resistor at the instant the switch is closed?

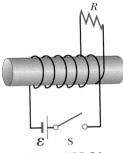

FIGURE P20.26

27. A copper bar is moved to the right while its axis is maintained in a direction perpendicular to a magnetic field, as shown in Figure P20.27. If the top of the bar becomes positive relative to the bottom, what is the direction of the magnetic field?

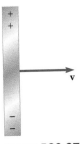

FIGURE P20.27

28. Find the direction of the current through the resistor in Figure P20.28, (a) at the instant the switch is closed, (b) after the switch has been closed for several minutes, and (c) at the instant the switch is opened.

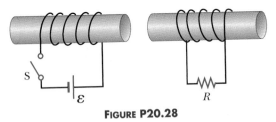

FIGURE P20.28

29. Find the direction of the current in resistor *R* in Figure P20.29 after each of the following steps (taken in the order given). (a) The switch is closed. (b) The variable resistance in series with the battery is decreased. (c) The circuit containing resistor *R* is moved to the left. (d) The switch is opened.

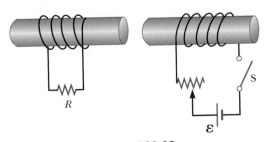

FIGURE P20.29

Section 20.5 Generators

30. A 100-turn square wire coil of area 0.040 m^2 rotates about a vertical axis at 1 500 rev/min, as indicated in Figure P20.30. The horizontal component of Earth's magnetic field at the location of the loop is 2.0×10^{-5} T. Calculate the maximum emf induced in the coil by Earth's field.

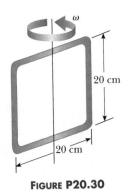

FIGURE P20.30

31. An automobile generator produces 12.0 V when turning at 500 rev/min. What potential difference will it produce at 1 200 rev/min?

32. A motor has coils with a resistance of 30 Ω and operates from a voltage of 240 V. When the motor is operating at its maximum speed, the back emf is 145 V. Find the current in the coils (a) when the motor is first turned on and (b) when the motor has reached maximum speed. (c) If the current in the motor is 6.0 A at some instant, what is the back emf at that time?

33. When the coil of a motor is rotating at maximum speed, the current in the windings is 4.0 A. When the motor is first turned on, the current in the windings is 11 A. If the motor is operated at 120 V, find (a) the resistance of the windings and (b) the back emf in the coil at maximum speed.

34. A loop of area 0.10 m^2 is rotating at 60 rev/s with its axis of rotation perpendicular to a 0.20-T magnetic field. (a) If there are 1 000 turns on the loop, what is the maximum voltage induced in the loop? (b) When the maximum induced voltage occurs, what is the orientation of the loop with respect to the magnetic field?

35. In a model ac generator, a 500-turn rectangular coil, 8.0 cm by 20 cm, rotates at 120 rev/min in a uniform magnetic field of 0.60 T. (a) What is the maximum emf induced in the coil? (b) What is the instantaneous value of the emf in the coil at $t = (\pi/32)$ s? Assume that the emf is zero at $t = 0$. (c) What is the smallest value of t for which the emf will have its maximum value?

Section 20.6 Self-Inductance

36. A coiled telephone cord forms a spiral with 70.0 turns, a diameter of 1.30 cm, and an unstretched length of 60.0 cm. Determine the self-inductance of one conductor in the unstretched cord.

37. A 2.00-H inductor carries a steady current of 0.500 A. When the switch in the circuit is thrown open, the current is effectively zero after 10.0 ms. What is the average induced emf in the inductor during this time?

38. Show that the two expressions for inductance given by

$$L = \frac{N\Phi_B}{I} \quad \text{and} \quad L = \frac{-\mathcal{E}}{\Delta I/\Delta t}$$

have the same units.

39. A solenoid of radius 2.5 cm has 400 turns and a length of 20 cm. Find (a) its inductance and (b) the rate at which current must change through it to produce an emf of 75 mV.

40. An emf of 24.0 mV is induced in a 500-turn coil when the current is changing at a rate of 10.0 A/s. What is the magnetic flux through each turn of the coil at an instant when the current is 4.00 A?

Section 20.7 RL Circuits

41. Show that the SI units for the inductive time constant $\tau = L/R$ are seconds.

42. An RL circuit with $L = 3.00$ H and an RC circuit with $C = 3.00$ μF have the same time constant. If the two circuits have the same resistance R, (a) what is the value of R, and (b) what is this common time constant?

43. A 6.0-V battery is connected in series with a resistor and an inductor. The series circuit has a time constant of 600 μs, and the maximum current is 300 mA. What is the value of the inductance?

44. A 25-mH inductor, an 8.0-Ω resistor, and a 6.0-V battery are connected in series. The switch is closed at $t = 0$. Find the voltage drop across the resistor (a) at $t = 0$ and (b) after one time constant has passed. Also, find the voltage drop across the inductor (c) at $t = 0$ and (d) after one time constant has elapsed.

45. Calculate the resistance in an RL circuit in which $L = 2.50$ H and the current increases to 90.0% of its final value in 3.00 s.

46. Consider the circuit in Figure P20.46, taking $\mathcal{E} = 6.00$ V, $L = 8.00$ mH, and $R = 4.00$ Ω. (a) What is the inductive time constant of the circuit? (b) Calculate the current in the circuit 250 μs after the switch is closed. (c) What is the value of the final steady-state current? (d) How long does it take the current to reach 80.0% of its maximum value?

FIGURE P20.46

20.8 Energy Stored in a Magnetic Field

47. How much energy is stored in a 70.0-mH inductor at an instant when the current is 2.00 A?

48. An air-core solenoid with 68 turns is 8.00 cm long and has a diameter of 1.20 cm. How much energy is stored in its magnetic field when it carries a current of 0.770 A?

49. A 24-V battery is connected in series with a resistor and an inductor, where $R = 8.0$ Ω and $L = 4.0$ H. Find the energy stored in the inductor (a) when the current reaches its maximum value and (b) one time constant after the switch is closed.

ADDITIONAL PROBLEMS

50. What is the time constant for (a) the circuit shown in Figure P20.50a? (b) the circuit shown in Figure P20.50b?

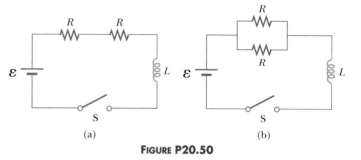

FIGURE P20.50

51. In Figure P20.51, the bar magnet is being moved toward the loop. Is $(V_a - V_b)$ positive, negative, or zero during this motion? Explain.

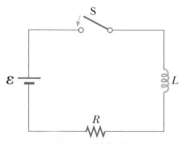

FIGURE P20.51

52. Your physics teacher asks you to help her set up a demonstration of Faraday's law for the class. The apparatus consists of a strong permanent magnet that has a field of 0.10 T, a small 10-turn coil of radius 2.0 cm cemented on a wood frame with a handle, some flexible connecting wires, and an ammeter as in Figure P20.52. The idea is to pull the coil out of the center of the magnetic field as quickly as possible and read the average current registered on the meter. The combined resistance of the coil, leads, and meter is 2.0 Ω and you must flip the coil out of the field in about 0.20 s. The ammeter you must use has a full-scale sensitivity of 1 000 μA. Will this meter be sensitive enough to show clearly the induced current?

53. An 820-turn wire coil of resistance 24.0 Ω is placed on top of a 12 500-turn, 7.00-cm-long solenoid, as in Figure P20.53. Both coil and solenoid have cross-sectional areas of 1.00×10^{-4} m^2. (a) How long does it take the solenoid current to reach 0.632 times its maximum value? (b) Determine the average back emf caused by the self-inductance of the solenoid during this interval. The magnetic field produced by the solenoid at the location of the coil is one-half as strong as the field at the center of the

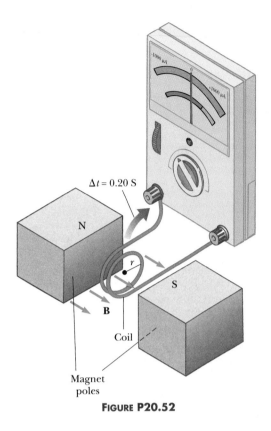

$\Delta t = 0.20$ S

FIGURE P20.52

solenoid. (c) Determine the average rate of change in magnetic flux through each turn of the coil during this interval. (d) Find the magnitude of the average induced current in the coil.

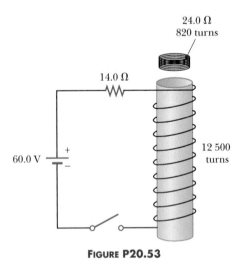

24.0 Ω
820 turns

14.0 Ω

60.0 V

12 500 turns

FIGURE P20.53

54. Figure P20.54 is a graph of induced emf versus time for a coil of N turns rotating with angular speed ω in a uniform magnetic field directed perpendicularly to the axis of rotation of the coil. Copy this sketch (increasing the scale), and on the same set of axes show the graph of emf versus t when (a) the number of turns in the coil is doubled, (b) the angular speed is doubled, and (c) the angular speed is doubled while the number of turns in the coil is halved.

55. The plane of a square loop of wire with edge length $a = 0.200$ m is perpendicular to Earth's magnetic field at a

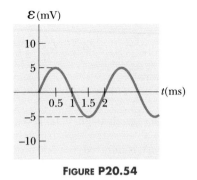

FIGURE P20.54

point where $B = 15.0$ μT, as in Figure P20.55. The total resistance of the loop and the wires connecting it to the ammeter is 0.500 Ω. If the loop is suddenly collapsed by horizontal forces as shown, what total charge passes through the ammeter?

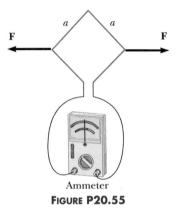

FIGURE P20.55

56. A novel method of storing electrical energy has been proposed. A huge underground superconducting coil, 1.00 km in diameter, would be fabricated. It would carry a maximum current of 50.0 kA through each winding of a 150-turn Nb_3Sn solenoid. (a) If the inductance of this huge coil is 50.0 H, what is the total energy stored? (b) What is the compressive force per meter length acting between two adjacent windings 0.250 m apart? (*Hint:* Because the radius of the coil is so large, the magnetic field created by one winding and acting on an adjacent turn can be considered to be that of a long, straight wire.)

57. A conducting rod of length ℓ moves on two horizontal frictionless rails, as in Figure P20.18. A constant force of magnitude 1.00 N moves the bar at a uniform speed of 2.00 m/s through a magnetic field **B** that is into the page. (a) What is the current in an 8.00-Ω resistor R? (b) What is the rate of energy dissipation in the resistor? (c) What is the mechanical power delivered by the constant force?

58. The square loop in Figure P20.58 is made of wires with total series resistance 10.0 Ω. It is placed in a uniform 0.100-T magnetic field directed perpendicular into the plane of the paper. The loop, which is hinged at each corner, is pulled as shown until the separation between points A and B is 3.00 m. If this process takes 0.100 s, what is the average current generated in the loop? What is the direction of the current?

59. The bolt of lightning depicted in Figure P20.59 passes 200 m from a 100-turn coil oriented as shown. If the cur-

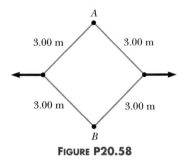

FIGURE P20.58

rent in the lightning bolt falls from 6.02×10^6 A to zero in 10.5 μs, what is the average voltage induced in the coil? Assume that the distance to the center of the coil determines the average magnetic field at the coil's position. Treat the lightning bolt as a long, vertical wire.

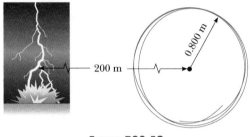

FIGURE P20.59

60. The wire shown in Figure P20.60 is bent in the shape of a "tent" with $\theta = 60°$ and $L = 1.5$ m, and is placed in a uniform magnetic field of 0.30 T perpendicular to the tabletop. The wire is "hinged" at points a and b. If the tent is flattened out on the table in 0.10 s, what is the average induced emf in the wire during this time?

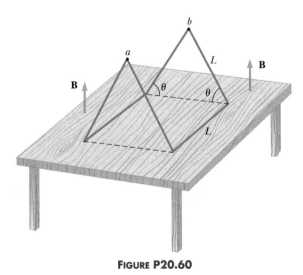

FIGURE P20.60

61. The magnetic field shown in Figure P20.61 has a uniform magnitude of 25.0 mT directed into the paper. The initial diameter of the kink is 2.00 cm. (a) The wire is quickly pulled taut, and the kink shrinks to a diameter of zero in 50.0 ms. Determine the average voltage induced between endpoints A and B. Include the polarity. (b) Suppose the kink is undisturbed, but the magnetic field increases to

100 mT in 4.00×10^{-3} s. Determine the average voltage across terminals A and B, including polarity, during this period.

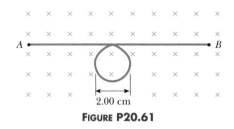

FIGURE P20.61

62. An aluminum ring of radius 5.00 cm and resistance 3.00×10^{-4} Ω is placed around the top of a long air-core solenoid with 1 000 turns per meter and smaller radius 3.00 cm as in Figure P20.62. If the current in the solenoid is increasing at a constant rate of 270 A/s, what is the induced current in the ring? Assume that the magnetic field produced by the solenoid over the area at the end of the solenoid is one-half as strong as the field at the center of the solenoid. Assume that the solenoid produces a negligible field outside its cross-sectional area.

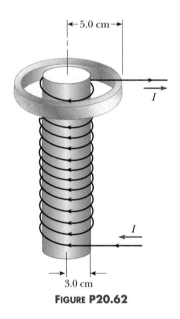

FIGURE P20.62

63. In Figure P20.63, the rolling axle, 1.50 m long, is pushed along horizontal rails at a constant speed $v = 3.00$ m/s. A resistor $R = 0.400$ Ω is connected to the rails at points a and b, directly opposite each other. (The wheels make good electrical contact with the rails, and so the axle, rails, and R form a closed-loop circuit. The only significant resistance in the circuit is R.) There is a uniform magnetic field $B = 0.800$ T vertically downward. (a) Find the induced current I in the resistor. (b) What horizontal force **F** is required to keep the axle rolling at constant speed? (c) Which end of the resistor, a or b, is at the higher electric potential? (d) After the axle rolls past the resistor, does the current in R reverse direction?

64. In 1832 Faraday proposed that the apparatus shown in Figure P20.64 could be used to generate electric current

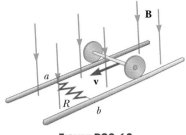

FIGURE P20.63

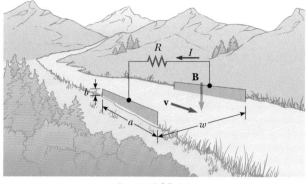

FIGURE P20.64

from the flowing water in the Thames River.[2] Two conducting plates of lengths a and widths b are placed facing one another on opposite sides of the river, a distance w apart and immersed entirely. The flow velocity of the river is **v**, and the vertical component of Earth's magnetic field is B. Show that the current in the load resistor R is

$$I = \frac{abvB}{\rho + abR/w}$$

where ρ is the resistivity of the water. (b) Calculate the short-circuit current ($R = 0$) if $a = 100$ m, $b = 5.00$ m, $v = 3.00$ m/s, $B = 50.0$ μT, and $\rho = 100$ Ω m.

GROUP ACTIVITIES

G.1 Experimenting with induced currents is not easy. For small magnets and small coils of wire, the resulting induced currents are so small that they are difficult to detect. Thus, you may have to try this exercise several times before you are satisfied with your results. Wind a coil of wire on a cardboard mailing tube. Use insulated wire with as small a diameter as possible because you need as many turns as possible on the coil. Connect the coil to a flashlight bulb and see if you can get it to light by moving a bar magnet into and out of the coil in rapid succession. Why does the speed of movement make a difference? If you are unsuccessful, place two magnets side by side and repeat.

    After you have experimented with the bulb, ask your instructor to let you use a galvanometer as a current detector. These devices are capable of measuring very small currents and they have the added advantage of detecting the direction of the current in the circuit.

    Use your equipment to observe or test the following. (a) Does the magnitude of the induced current depend on the speed of movement of the magnet? (b) Can you induce a current by holding the magnet still and moving the coil over it? (c) Does the direction of the current depend on whether the magnet is pushed in or pulled out of the coil? (d) Does the direction of the current depend on whether the inserted pole of the magnet is north or south? (e) Can you predict the direction of the current by using Lenz's law? (f) Replace your bar magnet with the electromagnet you constructed in the last chapter and repeat the observations above.

G.2 As explained in the text, a cassette tape is made up of tiny particles of metal oxide attached to a long plastic strip. Pull a tape out of a cassette that you do not mind destroy-

ing and see if it is repelled or attracted by a refrigerator magnet. Also, try this with an expendable floppy computer disk.

G.3 This experiment takes steady hands, a dime, and a strong magnet. After verifying that a dime is not attracted to the magnet, carefully balance the coin on its edge. (This will not work with other coins because they require too much force to topple them.) Hold one pole of the magnet within a millimeter of the face of the dime, but do not make contact with it. Now very rapidly pull the magnet straight back away from the coin. Which way does the dime tip? Does the coin fall the same way most of the time? Explain what is going on in terms of Lenz's law.

G.4 A ramp runs from the bed of a truck down to the level ground. The ramp holds two parallel conducting rails connected at its base. A metal bar slides on the rails without friction. A magnet supplies an external magnetic field

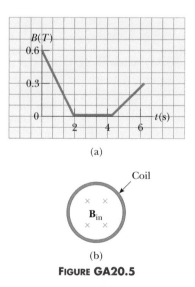

(a)

(b)

FIGURE GA20.5

[2] The idea for this problem and Figure P20.64 is from Oleg D. Jefimenko, *Electricity and Magnetism: An Introduction to the Theory of Electric and Magnetic Fields*, Star City, WV, Electret Scientific Co., 1989.

electric and magnetic fields can move through space as waves. The theory he developed is based on the following four pieces of information:

1. Electric field lines originate on positive charges and terminate on negative charges.
2. Magnetic field lines always form closed loops—that is, they do not begin or end anywhere.
3. A varying magnetic field induces an emf and hence an electric field. This is a statement of Faraday's law (Chapter 20).
4. Magnetic fields are generated by moving charges (or currents), as summarized in Ampère's law (Chapter 19).

Let us examine these statements further in order to understand their significance and Maxwell's great contribution to the theory of electromagnetism. The first statement is a consequence of the nature of the electrostatic force between charged particles, given by Coulomb's law. It embodies the fact that **free charges (electric monopoles) exist in nature.**

The second statement—that magnetic fields form continuous loops—is exemplified by the magnetic field lines around a long, straight wire, which are closed circles, and the magnetic field lines of a bar magnet, which form closed loops.

The third statement is equivalent to Faraday's law of induction, and the fourth statement is equivalent to Ampère's law.

In one of the greatest theoretical developments of the 19th century, Maxwell used these four statements within a corresponding mathematical framework to prove that electric and magnetic fields play symmetric roles in nature. It was already known from experiments that a changing magnetic field produced an electric field according to Faraday's law. Maxwell believed that nature was symmetric, and he therefore hypothesized that a changing electric field should produce a magnetic field. This hypothesis could not be proven experimentally at the time it was developed, because the magnetic fields generated by changing electric fields are generally very weak and therefore difficult to detect.

To justify his hypothesis, Maxwell searched for other phenomena that might be explained by it. He turned his attention to the motion of rapidly oscillating (accelerating) charges, such as those in a conducting rod connected to an alternating voltage. Such charges experience accelerations and, according to Maxwell's predictions, generate changing electric and magnetic fields. The changing fields cause electromagnetic disturbances that travel through space as waves, similar to the spreading water waves created by a pebble thrown into a pool. The waves sent out by the oscillating charges are fluctuating electric and magnetic fields, and so they are called *electromagnetic waves*. From Faraday's law and from his generalization of Ampère's law, Maxwell calculated their speed to be equal to the speed of light, $c = 3 \times 10^8$ m/s. He concluded that visible light and other electromagnetic waves consist of fluctuating electric and magnetic fields traveling through empty space, with each varying field inducing the other! This was truly one of the greatest discoveries of science on a par with Newton's discovery of the laws of motion. Like Newton's laws, it had a profound influence on later scientific developments.

JAMES CLERK MAXWELL, SCOTTISH THEORETICAL PHYSICIST (1831–1879)

Maxwell developed the electromagnetic theory of light and the kinetic theory of gases and explained the nature of Saturn's rings and color vision. Maxwell's successful interpretation of the electromagnetic field resulted in the equations that bear his name. Formidable mathematical ability combined with great insight enabled him to lead the way in the study of electromagnetism and kinetic theory. Although it is difficult to know what influences and characteristics shape a great scientist, Maxwell already showed remarkable curiosity at the age of three. He was described in a letter by his mother as follows: "He is a very happy man, and has improved much since the weather got moderate; he has great work with doors, locks, keys, etc., and 'Show me how it doos' is never out of his mouth. He also investigates the hidden course of streams and bellwires, the way the water gets from the pond through the wall . . . and down a drain into Water Orr (River Urr) . . . As to the bells, they will not rust; he stands sentry in the kitchen . . . or he rings, and sends Bessy to see and shout to let him know, and he drags papa all over to show him the holes where the wires go through." (From *Faraday, Maxwell, and Kelvin*, D. K. C. MacDonald, Doubleday-Anchor Books, 1964.) (*North Wind Picture Archives*)

21.9 HERTZ'S CONFIRMATION OF MAXWELL'S PREDICTIONS

In 1887, after Maxwell's death, Heinrich Hertz (1857–1894) was the first to generate and detect electromagnetic waves in a laboratory setting. To appreciate the details of his experiment, let us re-examine the properties of an *LC* circuit. In such a circuit, a charged capacitor is connected to an inductor, as in Figure 21.16. When the switch is closed, oscillations occur in the current in the circuit and in the charge on the capacitor. If the resistance of the circuit is neglected, no energy is dissipated, and the oscillations continue.

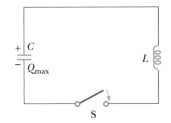

FIGURE 21.16 A simple *LC* circuit. The capacitor has an initial charge of Q_{max} and the switch is closed at $t = 0$.

HEINRICH RUDOLF HERTZ, GERMAN PHYSICIST (1857–1894)

Hertz made his most important discovery of radio waves in 1887. After finding that the speed of a radio wave was the same as that of light, Hertz showed that radio waves, like light waves, could be reflected, refracted, and diffracted. Hertz died of blood poisoning at the age of 36. During his short life, he made many contributions to science. The hertz, equal to one complete vibration or cycle per second, is named after him. *(Hulton-Deutsch Collection/CORBIS)*

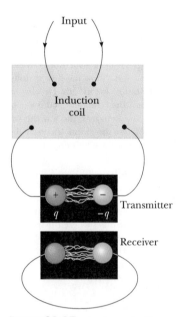

FIGURE 21.17 A schematic diagram of Hertz's apparatus for generating and detecting electromagnetic waves. The transmitter consists of two spherical electrodes connected to an induction coil, which provides short voltage surges to the spheres, setting up oscillations in the discharge. The receiver is a nearby single loop of wire containing a second spark gap.

In the following analysis, we shall neglect the resistance in the circuit. Let us assume that the capacitor has an initial charge of Q_{max} and that the switch is closed at $t = 0$. It is convenient to describe what ensues from an energy viewpoint. When the capacitor is fully charged, the total energy in the circuit is stored in the electric field of the capacitor and is equal to $Q_{max}^2/2C$. At this time, the current is zero and so no energy is stored in the inductor. As the capacitor begins to discharge, the energy stored in its electric field decreases. At the same time, the current increases and energy equal to $LI^2/2$ is now stored in the magnetic field of the inductor. Thus, energy is transferred from the electric field of the capacitor to the magnetic field of the inductor. When the capacitor is fully discharged, it stores no energy. At this time, the current reaches its maximum value, and all of the energy is stored in the inductor. The process then repeats in the reverse direction. The energy continues to transfer between the inductor and the capacitor, corresponding to oscillations in the current and charge.

As we saw in Section 21.6, the frequency of oscillation of an *LC* circuit is called the *resonance frequency* of the circuit and is given by

$$f_0 = \frac{1}{2\pi\sqrt{LC}}$$

The circuit Hertz used in his investigations of electromagnetic waves is similar to that just discussed and is shown schematically in Figure 21.17. An induction coil (a large coil of wire) is connected to two metal spheres with a narrow gap between them to form a capacitor. Oscillations are initiated in the circuit by short voltage pulses sent via the coil to the spheres, charging one positive, the other negative. Because *L* and *C* are quite small in this circuit, the frequency of oscillation is quite high, $f \approx 100$ MHz. This circuit is called a *transmitter* because it produces electromagnetic waves.

Several meters from the transmitter circuit, Hertz placed a second circuit, the receiver, which consisted of a single loop of wire connected to two spheres. It had its own effective inductance, capacitance, and natural frequency of oscillation. Hertz found that energy was being sent from the transmitter to the receiver when the resonance frequency of the receiver was adjusted to match that of the transmitter. The energy transfer was detected when the voltage across the spheres in the receiver circuit became high enough to produce ionization in the air, which caused sparks to appear in the air gap separating the spheres. Hertz's experiment is analogous to the mechanical phenomenon in which a tuning fork picks up the vibrations from another, identical tuning fork.

Hertz hypothesized that the energy transferred from the transmitter to the receiver is carried in the form of waves, which are now known to be electromagnetic waves. In a series of experiments, he also showed that the radiation generated by the transmitter exhibits wave properties: interference, diffraction, reflection, refraction, and polarization. As you will see shortly, all of these properties are exhibited by light. Thus, it became evident that these electromagnetic waves had the same known properties of light waves (already known at the time) and differed only in frequency and wavelength. Hertz effectively confirmed Maxwell's theory by showing that Maxwell's mysterious electromagnetic waves existed and had all the properties of light waves.

Perhaps the most convincing experiment Hertz performed was the measurement of the speed of waves from the transmitter, accomplished as follows. Waves of known frequency from the transmitter were reflected from a metal sheet so that an interference pattern was set up, much like the standing wave pattern on a stretched string. As we saw in our discussion of standing waves, the distance between nodes is $\lambda/2$, so Hertz was able to determine the wavelength λ. Using the relationship $v = \lambda f$, he found that v was close to 3×10^8 m/s, the known speed of visible light. Hertz's experiments thus provided the first evidence in support of Maxwell's theory.

21.10 PRODUCTION OF ELECTROMAGNETIC WAVES BY AN ANTENNA

In the previous section, we found that the energy stored in an *LC* circuit is continually transferred between the electric field of the capacitor and the magnetic field of the inductor. However, this energy transfer continues for prolonged periods of time only when the changes occur slowly. If the current alternates rapidly, the circuit loses some of its energy in the form of electromagnetic waves. In fact, electromagnetic waves are radiated by *any* circuit carrying an alternating current. The fundamental mechanism responsible for this radiation is the acceleration of a charged particle. Whenever a charged particle undergoes an acceleration, it must radiate energy.

An alternating voltage applied to the wires of an antenna forces an electric charge in the antenna to oscillate. This is a common technique for accelerating charged particles and is the source of the radio waves emitted by the broadcast antenna of a radio station.

Figure 21.18 illustrates the production of an electromagnetic wave by oscillating electric charges in an antenna. Two metal rods are connected to an AC source, which causes charges to oscillate between the two rods. The output voltage of the generator is sinusoidal. At $t = 0$, the upper rod is given a maximum positive charge and the bottom rod an equal negative charge, as in Figure 21.18a. The electric field near the antenna at this instant is also shown in Figure 21.18a. As the charges oscillate, the rods become less charged, the field near the rods decreases in strength, and the downward-directed maximum electric field produced at $t = 0$ moves away from the rod. When the charges are neutralized, as in Figure 21.18b, the electric field has dropped to zero. This occurs after an interval equal to one quarter of the period of oscillation. Continuing in this fashion, the upper rod soon obtains a maximum negative charge and the lower rod becomes positive, as in Figure 21.19c, resulting in an electric field directed upward. This occurs after an interval equal to one-half the period of oscillation. The oscillations continue as indicated in Figure 21.18d. Note that the electric field near the antenna oscillates in phase with the charge distribution. That is, the field points down when the upper rod is positive and up when the upper rod is negative. Furthermore, the magnitude of the field at any instant depends on the amount of charge on the rods at that instant.

As the charges continue to oscillate (and accelerate) between the rods, the electric field set up by the charges moves away from the antenna in all directions at the speed of light. Figure 21.18 shows the electric field pattern on one side of the antenna at certain times during the oscillation cycle. As you can see, one cycle of charge oscillation produces one full wavelength in the electric field pattern.

Because the oscillating charges create a current in the rods, a magnetic field is also generated when the current in the rods is upward, as shown in Figure 21.19. The magnetic field lines circle the antenna (right-hand rule #2) and are perpendicular to the electric field at all points. As the current changes with time, the

▶ An accelerating charge radiates energy.

◀ **APPLICATION**

RADIO-WAVE TRANSMISSION

Tip 21.1   ACCELERATED CHARGES PRODUCE EM WAVES

Stationary charges produce only electric fields, while charges in uniform motion (i.e., constant velocity) produce electric and magnetic fields, but no electromagnetic waves. In contrast, accelerated charges produce electromagnetic waves as well as electric and magnetic fields.

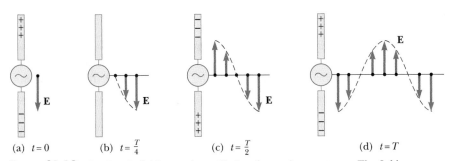

(a) $t = 0$ (b) $t = \frac{T}{4}$ (c) $t = \frac{T}{2}$ (d) $t = T$

FIGURE 21.18   An electric field set up by oscillating charges in an antenna. The field moves away from the antenna at the speed of light.

FIGURE 21.19   Magnetic field lines around an antenna carrying a changing current.

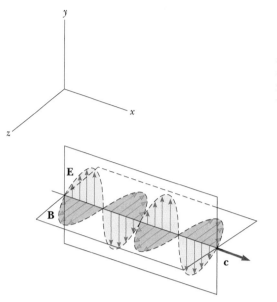

FIGURE 21.20 An electromagnetic wave sent out by oscillating charges in an antenna, represented at one instant of time and far from the antenna. Note that the electric field is perpendicular to the magnetic field, and both are perpendicular to the direction of wave propagation.

Webnote 21.2

Watch the propagation of an electromagnetic wave through the real-time Java applet at *http://www.phy.ntnu.edu.tw/~hwang/ emWave/emWave.html*

magnetic field lines spread out from the antenna. At great distances from the antenna, the strengths of the electric and magnetic fields become very weak. However, at these distances it is necessary to take into account the facts that (1) a changing magnetic field produces an electric field and (2) a changing electric field produces a magnetic field, as predicted by Maxwell. These induced electric and magnetic fields are in phase: At any point, the two fields reach their maximum values at the same instant. This is illustrated at one instant of time in Figure 21.20. Note that (1) **E** and **B** fields are perpendicular to each other, and (2) both fields are perpendicular to the direction of motion of the wave. This second property is characteristic of transverse waves. Hence, we see that **an electromagnetic wave is a transverse wave.**

21.11   PROPERTIES OF ELECTROMAGNETIC WAVES

We have seen that Maxwell's detailed analysis predicted the existence and properties of electromagnetic waves. We have already examined some of those properties. In this section we summarize what we know about electromagnetic waves thus far and consider some additional properties. In our discussion here and in future sections, we shall often make reference to a type of wave called a **plane wave.** A plane electromagnetic wave is a wave traveling from a very distant source. Figure 21.20 pictures such a wave at a given instant of time. In this case, the oscillations of the electric and magnetic fields take place in planes perpendicular to the x axis and thus to the direction of travel for the wave. Because the electric and magnetic fields are perpendicular to the direction of travel of the wave, electromagnetic waves are transverse waves. In Figure 21.20, the electric field **E** is in the y direction and the magnetic field **B** is in the z direction.

Electromagnetic waves travel with the speed of light. In fact, it can be shown that the speed of an electromagnetic wave is related to the permeability and permittivity of the medium through which it travels. Maxwell found this relationship for free space to be

Speed of light ▶

$$c = \frac{1}{\sqrt{\mu_0 \epsilon_0}}$$

[21.24]

where c is the speed of light, $\mu_0 = 4\pi \times 10^{-7} \ \mathrm{N \cdot s^2/C^2}$ is the permeability constant of vacuum, and $\epsilon_0 = 8.854\ 19 \times 10^{-12}\ \text{C}^2/\text{N}\cdot\text{m}^2$ is the permittivity of free space. Substituting these values into Equation 21.24, we find that

$$c = 2.997\ 92 \times 10^8\ \text{m/s} \qquad \text{[21.25]}$$

Because electromagnetic waves travel at a speed that is precisely the same as the speed of light in vacuum, one is led to believe (correctly) that **light is an electromagnetic wave.**

Maxwell also proved that the ratio of the electric to the magnetic field in an electromagnetic wave equals the speed of light. That is,

$$\frac{E}{B} = c \qquad \text{[21.26]}$$

Electromagnetic waves carry energy as they travel through space, and this energy can be transferred to objects placed in their paths. The average rate at which energy passes through an area perpendicular to the direction of travel of a wave, or the average power per unit area, is given by

$$\text{Average power per unit area} = \frac{E_{\text{max}}B_{\text{max}}}{2\mu_0} \qquad \text{[21.27]}$$

where E_{max} and B_{max} are the *maximum* values of E and B. As in Chapter 14, we call this quantity the *intensity* of the wave. From Equation 21.26, we see that $E_{\text{max}} = cB_{\text{max}} = B_{\text{max}}/\sqrt{\mu_0\epsilon_0}$. Therefore, Equation 21.27 can also be expressed as

$$\text{Average power per unit area} = \frac{E_{\text{max}}^2}{2\mu_0 c} = \frac{c}{2\mu_0}B_{\text{max}}^2 \qquad \text{[21.28]}$$

Note that in these expressions we use the *average* power per unit area. A detailed analysis would show that the energy carried by an electromagnetic wave is shared equally by the electric and magnetic fields.

Electromagnetic waves transport linear momentum as well as energy. Hence it follows that pressure is exerted on a surface when an electromagnetic wave impinges on it. In what follows, we assume that the electromagnetic wave transports a total energy U to a surface in a time t. If the surface absorbs all the incident energy U in this time, Maxwell showed that the total momentum **p** delivered to this surface has a magnitude

$$p = \frac{U}{c} \qquad \text{(complete absorption)} \qquad \text{[21.29]}$$

▶ Light is an electromagnetic wave and transports energy and momentum.

If the surface is a perfect reflector, then the momentum delivered in a time t for normal incidence is twice that given by Equation 21.29. That is, a momentum U/c is delivered first by the incident wave and then again by the reflected wave, in analogy with a ball colliding elastically with a wall. Therefore,

$$p = \frac{2U}{c} \qquad \text{(complete reflection)} \qquad \text{[21.30]}$$

Although radiation pressures are very small (about $5 \times 10^{-6}\ \text{N/m}^2$ for direct sunlight), they have been measured with a device such as the one shown in Figure 21.21. Light is allowed to strike a mirror and a black disk that are connected to each other by a horizontal bar suspended from a fine fiber. Light striking the black disk is completely absorbed, so *all* of the momentum of the light is transferred to the disk. Light striking the mirror head on is totally reflected; hence, the momentum transfer to the mirror is twice that transmitted to the disk. As a result, the horizontal bar supporting the disks twists counterclockwise as seen from above. The bar comes to equilibrium at some angle under the action of the torques caused by radiation pressure and the twisting of the fiber. The radiation pressure can be determined by measuring the angle at which equilibrium occurs. The apparatus

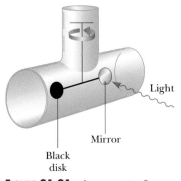

FIGURE 21.21 An apparatus for measuring the radiation pressure of light. In practice, the system is contained in a high vacuum.

must be placed in a high vacuum to eliminate the effects of air currents. It is interesting to note that similar experiments demonstrate that electromagnetic waves carry angular momentum as well.

In summary, electromagnetic waves traveling through free space have the following properties:

1. Electromagnetic waves travel at the speed of light.
2. Electromagnetic waves are transverse waves, because the electric and magnetic fields are perpendicular to the direction of propagation of the wave and to each other.
3. The ratio of the electric field to the magnetic field in an electromagnetic wave equals the speed of light.
4. Electromagnetic waves carry both energy and momentum, which can be delivered to a surface.

APPLYING PHYSICS 21.2

In the interplanetary space in the Solar System, there is a large amount of dust. Although interplanetary dust can in theory have a variety of sizes—from molecular size upward—there are very few dust particles smaller than about 0.2 μm in our Solar System. Why? (*Hint:* The Solar System originally contained dust particles of all sizes.)

Explanation Dust particles in the solar system are subject to two forces—the gravitational force toward the Sun and the force from radiation pressure, which is away from the Sun. The gravitational force is proportional to the cube of the radius of a spherical dust particle, because it is proportional to the mass (ρV) of the particle. The radiation pressure is proportional to the square of the radius, because it depends on the cross-sectional area of the particle. For large particles, the gravitational force is larger than the force of radiation pressure and the weak attraction to the Sun causes such particles to move slowly toward the Sun. For small particles, less than about 0.2 μm, the larger force from radiation pressure sweeps these particles out of the Solar System.

Quick Quiz 21.4

In an apparatus such as that in Figure 21.21, suppose the black disk is replaced by one with half the radius. Which of the following are different after the disk is replaced: (a) radiation pressure on the disk, (b) radiation force on the disk, (c) radiation momentum delivered to the disk in a given time interval?

Example 21.8 Solar Energy

Assume that the Sun delivers an average power per unit area of about 1 000 W/m^2 to Earth's surface. Calculate the total power incident on a roof 8.00 m by 20.0 m. Assume that the radiation is incident *normal* to the roof (the Sun is directly overhead).

Solution The power per unit area, or light intensity, is 1 000 W/m^2. For normal incidence, the power supplied to the roof is

$$\mathcal{P} = (1\ 000 \text{ W/m}^2)(8.00 \times 20.0 \text{ m}^2) = \boxed{1.60 \times 10^5 \text{ W}}$$

Note that if this power could *all* be converted to electric power, it would be more than enough for the average home. Unfortunately, solar energy is not easily harnessed, and the prospects for large-scale conversion are not as bright as they may appear from this

(Example 21.8) A solar home in Oregon. (*John Neal/Photo Researchers, Inc.*)

APPLICATION

SOLAR-POWERED HOMES

simple calculation. For example, the conversion efficiency from solar to electrical energy is far less than 100%; 10% is typical for photovoltaic cells. Roof systems for using solar energy to raise the temperature of water with efficiencies of around 50% have been built. However, other practical problems must be considered, such as overcast days, geographic location, and energy storage.

EXERCISE How much solar energy (in joules) is incident on the roof in 1.00 h?

ANSWER 5.76×10^8 J

Example 21.9 A High-Intensity Laser Beam

The intensity of light ranges from about 1 W/m^2 for a candle to about 30 MW/m^2 for a modest-size laser. (A laser is a high-intensity light source that concentrates several watts of power into a very narrow beam producing MW/m^2 intensity. See Chapter 28.) A particular laser produces a power of 4.0 W in a beam 0.40 mm in diameter. (a) What is its average intensity? (b) Find the peak electric field of the laser light. (c) Find the peak magnetic field of the laser light. (d) If the laser beam is aimed upward to levitate a 20-μm-diameter sphere (this has actually been done), what is the maximum mass and density of this sphere? Assume the sphere is perfectly reflecting.

Solution

A Intensity is power per unit area carried by the beam:

$$\text{Power per unit area} = \frac{\mathcal{P}}{\pi r^2} = \frac{4.0 \text{ W}}{(3.14)(0.20 \times 10^{-3} \text{ m})^2} = \boxed{32 \text{ MW/m}^2}$$

B Since the average power per unit area $= E_{max}^2/2\mu_0 c$,

$$E_{max} = \sqrt{(2\mu_0 c)(\text{average power per unit area})}$$
$$= \sqrt{(8\pi \times 10^{-7} \text{ Ns}^2/\text{C}^2)(3.0 \times 10^8 \text{ m/s})(32 \times 10^6 \text{ W/m}^2)}$$
$$= \boxed{1.6 \times 10^5 \text{ V/m}}$$

C E_{max} and B_{max} are not independent in an electromagnetic wave, but are related by $B_{max} = E_{max}/c$:

$$B_{max} = \frac{E_{max}}{c} = \frac{1.6 \times 10^5 \text{ V/m}}{3.0 \times 10^8 \text{ m/s}} = \boxed{5.3 \times 10^{-4} \text{ T}}$$

D To suspend a 20-μm-diameter sphere, the upward electromagnetic force F_{EM} must balance the sphere's weight.

$$F_{EM} = mg \quad \text{or} \quad m = \frac{F_{EM}}{g}$$

Because $F_{EM} = \Delta p / \Delta t$, and Equation 21.30 states that the momentum p delivered to a perfectly reflecting object is given by $p = 2U/c$ where U is the electromagnetic energy incident on the object, $\Delta p / \Delta t = 2U/c\Delta t$. Finally, we find for the mass

$$m = \frac{2U/\Delta t}{gc}$$

$U/\Delta t$ is the energy per unit time incident on the sphere and is given by the product of the laser intensity and the cross-sectional area of the sphere ($A = \pi r^2$).

$$\frac{U}{\Delta t} = \text{intensity} \cdot \pi r^2 = (3.2 \times 10^7 \text{ W/m}^2)(3.14)(1.0 \times 10^{-10} \text{ m}^2) = 0.010 \text{ W}$$

$$= 0.010 \text{ J/s}$$

The maximum mass that can be supported by this laser beam is

$$m = \frac{2U/\Delta t}{gc} = \frac{0.020 \text{ J/s}}{(9.8 \text{ m/s}^2)(3.0 \times 10^8 \text{ m/s})} = \boxed{6.8 \times 10^{-12} \text{ kg}}$$

The maximum density is given by

$$\rho_{\text{max}} = \frac{\text{mass}}{\text{volume}} = \frac{m}{(4/3)\pi r^3} = \frac{6.8 \times 10^{-12} \text{ kg}}{(4/3)(3.14)(1.0 \times 10^{-5} \text{ m})^3} = \boxed{1.6 \times 10^3 \text{ kg/m}^3}$$

which is about the density of bone.

21.12   THE SPECTRUM OF ELECTROMAGNETIC WAVES

We have seen that all electromagnetic waves travel in vacuum with the speed of light, c. These waves transport energy and momentum from some source to a receiver. In 1887, Hertz successfully generated and detected the radio-frequency electromagnetic waves predicted by Maxwell. Maxwell himself had recognized as electromagnetic waves both visible light and the infrared radiation discovered in 1800 by William Herschel. It is now known that other forms of electromagnetic waves exist that are distinguished by their frequencies and wavelengths.

Because all electromagnetic waves travel through vacuum with a speed c, their frequency f and wavelength λ are related by the important expression

$$c = f\lambda \qquad\qquad \textbf{[21.31]}$$

The types of electromagnetic waves are presented in Figure 21.22. Note the wide and overlapping range of frequencies and wavelengths. For instance, an AM radio wave with a frequency of 5.00 MHz (a typical value) has a wavelength of

$$\lambda = \frac{c}{f} = \frac{3.00 \times 10^8 \text{ m/s}}{5.00 \times 10^6 \text{ s}^{-1}} = 60.0 \text{ m}$$

The following abbreviations are often used to designate short wavelengths and distances:

$$1 \text{ micrometer } (\mu\text{m}) = 10^{-6} \text{ m}$$

$$1 \text{ nanometer } (\text{nm}) = 10^{-9} \text{ m}$$

$$1 \text{ angstrom } (\text{Å}) = 10^{-10} \text{ m}$$

The wavelengths of visible light, for example, range from 0.4 μm to 0.7 μm, or 400 nm to 700 nm, or 4 000 Å to 7 000 Å.

Brief descriptions of these wave types follow, in order of decreasing wavelength. There is no sharp division between one kind of electromagnetic wave and the next. Note that all forms of radiation are produced by accelerating charges.

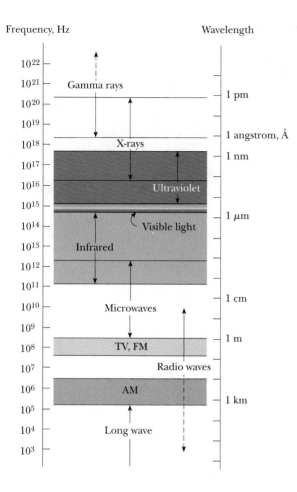

Frequency, Hz   Wavelength

FIGURE 21.22 The electromagnetic spectrum. Note the overlap between one type of wave and the next. There is no sharp division between the types of waves.

Radio waves, which were discussed in Section 21.10, are the result of charges accelerating through conducting wires. They are, of course, used in radio and television communication systems.

Microwaves (short-wavelength radio waves) have wavelengths ranging between about 1 mm and 30 cm and are generated by electronic devices. Their short wavelengths make them well suited for the radar systems used in aircraft navigation and for the study of atomic and molecular properties of matter. Microwave ovens are an interesting domestic application of these waves. It has been suggested that solar energy might be harnessed by beaming microwaves to Earth from a solar collector in space.

Infrared waves (sometimes and incorrectly called "heat waves"), produced by hot objects and molecules, have wavelengths ranging from about 1 mm to the longest wavelength of visible light, 7×10^{-7} m. They are readily absorbed by most materials. The infrared energy absorbed by a substance causes it to get warmer. This is because the energy agitates the atoms of the object, increasing their vibrational or translational motion, and the result is a temperature rise. Infrared radiation has many practical and scientific applications, including physical therapy, infrared photography, and the study of the vibrations of atoms.

Visible light, the most familiar form of electromagnetic waves, may be defined as the part of the spectrum that is detected by the human eye. Light is produced by the rearrangement of electrons in atoms and molecules. The wavelengths of visible light are classified as colors ranging from violet ($\lambda \approx 4 \times 10^{-7}$ m) to red ($\lambda \approx 7 \times 10^{-7}$ m). The eye's sensitivity is a function of wavelength and is greatest at a wavelength of about 5.6×10^{-7} m (yellow-green).

Ultraviolet (UV) light covers wavelengths ranging from about 4×10^{-7} m (400 nm) down to 6×10^{-10} m (0.6 nm). The Sun is an important source of ultraviolet light (which is the main cause of suntans). Most of the ultraviolet light from

Webnote 21.3

For much more detail on the radio spectrum and the electromagnetic spectrum in general, go to
http://www.howstuffworks.com/ radio-spectrum.htm

Webnote 21.4

Microwave ovens produce microwaves at a frequency of 2.45 GHz. How can they heat food from the inside out? Find out at
http://www.howstuffworks.com/ microwave.htm

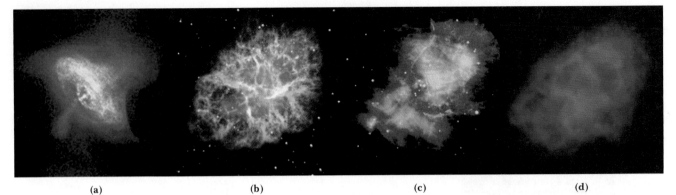

FIGURE 21.23 Observations in different parts of the electromagnetic spectrum show different features of the Crab Nebula. (a) X-ray image *(NASA/CXC/SAO)* (b) Optical image *(Palomar Observatory)* (c) Infrared image *(WM Keck Observatory)* (d) Radio image *(VLA/NRAO).*

the Sun is absorbed by atoms in the upper atmosphere, or stratosphere. This is fortunate, because UV light in large quantities has harmful effects on humans. One important constituent of the stratosphere is ozone (O_3) from reactions of oxygen with ultraviolet radiation. This ozone shield causes lethal high-energy ultraviolet radiation to warm the stratosphere.

X-rays are electromagnetic waves with wavelengths from about 10^{-8} m (10 nm) down to 10^{-13} m (10^{-4} nm). The most common source of x-rays is the acceleration of high-energy electrons bombarding a metal target. X-rays are used as a diagnostic tool in medicine and as a treatment for certain forms of cancer. Because x-rays easily penetrate and damage or destroy living tissues and organisms, care must be taken to avoid unnecessary exposure and overexposure.

Gamma rays, electromagnetic waves emitted by radioactive nuclei, have wavelengths ranging from about 10^{-10} m to less than 10^{-14} m. They are highly penetrating and cause serious damage when absorbed by living tissues. As a consequence, those working near such radiation must be protected by garments containing heavily absorbing materials, such as layers of lead.

When astronomers observe the same celestial object using detectors sensitive to different regions of the electromagnetic spectrum, striking variations in the object's features can be seen. Figure 21.23 shows images of the Crab Nebula made in four different wavelength ranges. The Crab Nebula is the remnant of a supernova explosion that was seen from Earth in 1054 A.D. (Compare Fig. 8.28).

APPLYING PHYSICS 21.3

The center of sensitivity of our eyes coincides with the center of the wavelength distribution of the Sun. Is this an amazing coincidence?

Explanation This is not a coincidence—it is the result of biological evolution. Humans have evolved with vision most sensitive to wavelengths that are strongest from the Sun. It is an interesting conjecture to imagine aliens from another planet, with a Sun of a different temperature, arriving at Earth. Their eyes would have a center of sensitivity at different wavelengths than ours. How would their vision of the Earth compare to ours?

21.13 THE DOPPLER EFFECT FOR ELECTROMAGNETIC WAVES

As we saw in Section 14.6, sound waves exhibit the Doppler effect when the observer, the source, or both are moving relative to the medium of propagation. Recall that in the Doppler effect, the observed frequency of the wave is larger or smaller than the frequency emitted by the source of the wave.

A Doppler effect also occurs for electromagnetic waves, but it differs from the Doppler effect for sound waves in two ways. First, in the Doppler effect for sound waves, motion relative to the medium is most important because sound waves require a medium in which to propagate. In contrast, the medium of propagation plays no role in the Doppler effect for electromagnetic waves because the waves require no medium in which to propagate. Second, the speed of sound that appears in the equation for the Doppler effect for sound depends on the reference frame in which it is measured. In contrast, as we shall see in Chapter 26, the speed of electromagnetic waves has the same value in all coordinate systems that are either at rest or moving at constant velocity with respect to one another.

The single equation that describes the Doppler effect for electromagnetic waves is given by the approximate expression

$$f' = f\left(1 \pm \frac{u}{c}\right) \qquad \text{if } u \ll c \qquad \text{[21.32]}$$

where f' is the observed frequency, f is the frequency emitted by the source, c is the speed of light in vacuum, and u is the *relative* speed of the observer and source. Note that Equation 21.32 is valid only if u is much smaller than c. Furthermore, this result can also be used for sound as long as the relative velocity of the source and observer is much less than the velocity of sound. The positive sign in Equation 21.32 must be used when the source and observer are moving toward one another, and the negative sign must be used when they are moving away from each other. Thus, we anticipate an increase in the observed frequency if the source and observer are approaching each other, and a decrease in observed frequency if the source and observer recede from each other.

Astronomers have made important discoveries using Doppler observations on light reaching Earth from distant stars and galaxies. Such measurements have shown that most distant galaxies are moving away from the Earth. Thus, the Universe is expanding. This is called a *red shift* because the observed wavelengths are shifted toward the red portion (longest wavelength) of the visible spectrum. Furthermore, measurements show that the speed of a galaxy increases with increasing distance from the Earth. More recent Doppler effect measurements made with the Hubble Space Telescope have shown that a galaxy labeled M87 is rotating — one edge moving toward us and the other moving away. Its measured speed of rotation was used to identify a supermassive black hole located at the center of this galaxy.

SUMMARY

If an AC circuit consists of a generator and a resistor, the current in the circuit is in phase with the voltage. That is, the current and voltage reach their maximum values at the same time.

In discussions of voltages and currents in AC circuits, **rms values** of voltages are usually used. One reason is that AC ammeters and voltmeters are designed to read rms values. The rms values of currents and voltage (I_{rms} and ΔV_{rms}) are related to the maximum values of these quantities (I_{max} and ΔV_{max}) as follows:

$$I_{\text{rms}} = \frac{I_{\text{max}}}{\sqrt{2}} \qquad \Delta V_{\text{rms}} = \frac{\Delta V_{\text{max}}}{\sqrt{2}} \qquad \text{[21.2, 21.3]}$$

The rms voltage across a resistor is related to the rms current in the resistor by **Ohm's law:**

$$\Delta V_{R,\text{rms}} = I_{\text{rms}}R \qquad \text{[21.4]}$$

If an AC circuit consists of a generator and a capacitor, the voltage lags behind the current by 90°. That is, the voltage reaches its maximum value one quarter of a period after the current reaches its maximum value.

The impeding effect of a capacitor on current in an AC circuit is given by the **capacitive reactance** X_C, defined as

$$X_C \equiv \frac{1}{2\pi f C} \tag{21.5}$$

where f is the frequency of the AC voltage source.

The rms voltage across and the rms current in a capacitor are related by

$$\Delta V_{C,\text{rms}} = I_{\text{rms}} X_C \tag{21.6}$$

If an AC circuit consists of a generator and an inductor, the voltage leads the current by 90°. That is, the voltage reaches its maximum value one quarter of a period before the current reaches its maximum value.

The effective impedance of a coil in an AC circuit is measured by a quantity called the **inductive reactance** X_L, defined as

$$X_L \equiv 2\pi f L \tag{21.8}$$

The rms voltage across a coil is related to the rms current in the coil by

$$\Delta V_{L,\text{rms}} = I_{\text{rms}} X_L \tag{21.9}$$

In an *RLC* series AC circuit, the maximum applied voltage ΔV is related to the maximum voltages across the resistor (ΔV_R), capacitor (ΔV_C), and inductor (ΔV_L) by

$$\Delta V_{\text{max}} = \sqrt{\Delta V_R^2 + (\Delta V_L - \Delta V_C)^2} \tag{21.10}$$

If an AC circuit contains a resistor, an inductor, and a capacitor connected in series, the limit they place on the current is given by the **impedance Z** of the circuit, defined as

$$Z \equiv \sqrt{R^2 + (X_L - X_C)^2} \tag{21.13}$$

The relationship between the maximum voltage supplied to an *RLC* circuit and the maximum current in the circuit that is the same in every element is

$$\Delta V_{\text{max}} = I_{\text{max}} Z \tag{21.14}$$

In an *RLC* series AC circuit, the applied rms voltage and current are out of phase. The **phase angle** ϕ between the current and voltage is given by

$$\tan \phi = \frac{X_L - X_C}{R} \tag{21.15}$$

The **average power** delivered by the voltage source in an *RLC* series AC circuit is

$$\mathcal{P}_{\text{av}} = I_{\text{rms}} \Delta V_{\text{rms}} \cos \phi \tag{21.17}$$

where the constant $\cos \phi$ is called the **power factor.**

Electromagnetic waves were predicted by James Clerk Maxwell and later experimentally confirmed by Heinrich Hertz. These waves have the following properties:

1. Electromagnetic waves are transverse waves, because the electric and magnetic fields are perpendicular to the direction of travel.
2. Electromagnetic waves travel with the speed of light.
3. The ratio of the electric field to the magnetic field at a given point in an electromagnetic wave equals the speed of light. That is,

$$\frac{E}{B} = c \tag{21.26}$$

4. Electromagnetic waves carry energy as they travel through space. The average power per unit area is

$$\frac{E_{\text{max}} B_{\text{max}}}{2\mu_0} = \frac{E_{\text{max}}^2}{2\mu_0 c} = \frac{c}{2\mu_0} B_{\text{max}}^2 \tag{21.27, 21.28}$$

where E_{max} and B_{max} are the maximum values of the electric and magnetic fields.

5. Electromagnetic waves transport linear and angular momentum as well as energy. The speed c, frequency f, and wavelength λ of an electromagnetic wave are related by

$$c = f\lambda \qquad\qquad [21.31]$$

The **electromagnetic spectrum** includes waves covering a broad range of frequencies and wavelengths. These waves have a variety of applications and characteristics, depending on their frequencies or wavelengths.

CONCEPTUAL QUESTIONS

1. Before the advent of cable television and satellite dishes, homeowners either mounted a television antenna on the roof or used "rabbit ears" atop their sets (see Fig. Q21.1). Certain orientations of the receiving antenna on a television set gave better reception than others. Furthermore, the best orientation varied from station to station. Explain.

FIGURE Q21.1 *(George Semple)*

2. What is the impedance of an *RLC* circuit at the resonance frequency?
3. When a DC voltage is applied to a transformer, the primary coil sometimes will overheat and burn. Why?
4. Why are the primary and secondary coils of a transformer wrapped on an iron core that passes through both coils?
5. Receiving radio antennas can be in the form of conducting lines or loops. What should the orientation of each of these antennas be relative to a broadcasting antenna that is vertical?
6. If the fundamental source of a sound wave is a vibrating object, what is the fundamental source of an electromagnetic wave?
7. In radio transmission, a radio wave serves as a carrier wave and the sound signal is superimposed on the carrier wave. In amplitude modulation (AM) radio, the amplitude of the carrier wave varies according to the sound wave. The Navy sometimes uses flashing lights to send Morse code between neighboring ships, a process that has similarities to radio broadcasting. Is this AM or FM? What is the carrier frequency? What is the signal frequency? What is the broadcasting antenna? What is the receiving antenna?
8. When light (or other electromagnetic radiation) travels across a given region, what is it that moves?
9. In space sailing, which is a proposed alternative method to reach the planets, a spacecraft carries a very large sail that experiences a force due to radiation pressure from the Sun. Should the sail be absorptive or reflective to be most effective?
10. How can the average value of an alternating current be zero yet the square root of the average squared value not be zero?
11. Suppose a creature from another planet had eyes that were sensitive to infrared radiation. Describe what it would see if it looked around the room that you are now in. That is, what would be bright and what would be dim?
12. Why should an infrared photograph of a person look different from a photograph taken with visible light?
13. Radio stations often advertise "instant news." If what they mean is that you hear the news at the instant they speak it, is their claim true? About how long would it take for a message to travel across this country by radio waves, assuming that these waves could travel this great distance and still be detected?
14. Would an inductor and a capacitor used together in an AC circuit dissipate any energy?
15. Does a wire connected to a battery emit an electromagnetic wave?
16. Do all current-carrying conductors emit electromagnetic waves? Explain.
17. If the resistance in an *RLC* circuit is doubled but the capacitance and inductance are unchanged, how will the resonance frequency change?
18. If the resistance in an *RLC* circuit remains the same but the capacitance and inductance are each doubled, how will the resonance frequency change?

PROBLEMS

1, 2, 3 = straightforward, intermediate, challenging   ☐ = full solution available in Student Solutions Manual/Study Guide

web = solution posted at **http://info.brookscole.com/serway**   = biomedical application

Section 21.1   Resistors in an AC Circuit

1. An rms voltage of 100 V is applied to a purely resistive load of 5.00 Ω. Find (a) the maximum voltage applied, (b) the rms current supplied, (c) the maximum current supplied, and (d) the power dissipated.

2. (a) What is the resistance of a lightbulb that uses an average power of 75.0 W when connected to a 60-Hz power source with a peak voltage of 170 V? (b) What is the resistance of a 100-W bulb?

3. An AC power supply produces a maximum voltage of $\Delta V_{max} = 100$ V. This power supply is connected to a 24.0-Ω resistor, and the current and resistor voltage are measured with an ideal AC ammeter and an ideal AC voltmeter, as shown in Figure P21.3. What does each meter read? Recall that an ideal ammeter has zero resistance and an ideal voltmeter has infinite resistance.

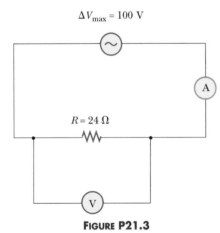

$\Delta V_{max} = 100$ V

$R = 24$ Ω

FIGURE P21.3

4. Figure P21.4 shows three lamps connected to a 120-V AC (rms) household supply voltage. Lamps 1 and 2 have 150-W bulbs; lamp 3 has a 100-W bulb. Find the rms current and resistance of each bulb.

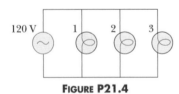

120 V   1   2   3

FIGURE P21.4

5. An audio amplifier, represented by the AC source and the resistor R in Figure P21.5, delivers alternating voltages at audio frequencies to the speaker. If the source puts out an alternating voltage of 15.0 V (rms), resistance R is 8.20 Ω, and the speaker is equivalent to a resistance of 10.4 Ω, what is the time-averaged power delivered to the speaker?

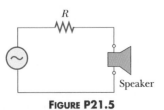

R

Speaker

FIGURE P21.5

6. An AC voltage source has an output of $\Delta v = 150 \sin 377t$. Find (a) the rms voltage output, (b) the frequency of the source, and (c) the voltage at $t = 1/120$ s. (d) Find the maximum current in the circuit when the generator is connected to a 50.0-Ω resistor.

Section 21.2   Capacitors in an AC Circuit

7. Show that the SI unit of capacitive reactance, X_C, is the ohm.

8. What maximum current is delivered by a 2.20-μF capacitor when connected across (a) a North American outlet having $\Delta V_{rms} = 120$ V, $f = 60.0$ Hz; and (b) a European outlet having $\Delta V_{rms} = 240$ V, $f = 50.0$ Hz?

9. When a 4.0-μF capacitor is connected to a generator whose rms output is 30 V, the current in the circuit is observed to be 0.30 A. What is the frequency of the source?

10. What maximum current is delivered by an AC generator with $\Delta V_{max} = 48.0$ V and $f = 90.0$ Hz when connected across a 3.70-μF capacitor?

11. What value of capacitor must be inserted in a 60-Hz circuit **web** in series with a generator of 170 V maximum output voltage to produce an rms current output of 0.75 A?

12. The generator in a purely capacitive AC circuit has an angular frequency of 120π rad/s. If $\Delta V_{max} = 140$ V and $C = 6.00$ μF, what is the rms current in the circuit?

Section 21.3   Inductors in an AC Circuit

13. Show that the inductive reactance X_L has SI units of ohms.

14. In a purely inductive AC circuit, as in Figure 21.6, $\Delta V_{max} = 100$ V. (a) If the maximum current is 7.50 A at 50.0 Hz, calculate the inductance L. (b) At what angular frequency ω is the maximum current 2.50 A?

15. An inductor has a 54.0-Ω reactance at 60.0 Hz. What will be the *maximum* current if this inductor is connected to a 50.0-Hz source that produces a 100-V rms voltage?

16. A 2.40-μF capacitor is connected across an alternating voltage with an rms value of 9.00 V. The rms current in the circuit is 25.0 mA. (a) What is the source frequency? (b) If the capacitor is replaced by an ideal coil with an inductance of 0.160 H, what is the rms current in the coil?

17. Determine the maximum magnetic flux through an inductor connected to a standard outlet ($\Delta V_{rms} = 120$ V, $f = 60.0$ Hz).

Section 21.4   The *RLC* Series Circuit

18. An inductor ($L = 400$ mH), a capacitor ($C = 4.43$ μF), and a resistor ($R = 500$ Ω) are connected in series. A 50.0-Hz AC generator connected in series to these elements produces a maximum current of 250 mA in the circuit. (a) Calculate the required maximum voltage ΔV_{max}. (b) Determine the phase angle by which the current leads or lags the applied voltage.

19. A 10.0-μF capacitor and a 2.00-H inductor are connected in series with a 60.0-Hz source whose rms output is 100 V. Find (a) the rms current in the circuit, (b) the voltage drop across the inductor, (c) the voltage drop across the capacitor, and (d) the phase angle for the circuit. (e) Sketch the phasor diagram for this circuit.

20. A series AC circuit contains the following components: $R = 150 \ \Omega$, $L = 250$ mH, $C = 2.00 \ \mu$F, and a generator with $\Delta V_{max} = 210$ V operating at 50.0 Hz. Calculate the (a) inductive reactance, (b) capacitive reactance, (c) impedance, (d) maximum current, and (e) phase angle between the current and the generator voltage.

21. A resistor ($R = 900 \ \Omega$), a capacitor ($C = 0.25 \ \mu$F), and an inductor ($L = 2.5$ H) are connected in series across a 240-Hz AC source for which $\Delta V_{max} = 140$ V. Calculate the (a) impedance of the circuit, (b) maximum current delivered by the source, and (c) phase angle between the current and voltage. (d) Is the current leading or lagging behind the voltage?

22. A sinusoidal voltage $\Delta v(t) = (40.0 \text{ V}) \sin(100t)$ is applied to a series *RLC* circuit with $L = 160$ mH, $C = 99.0 \ \mu$F, and $R = 68.0 \ \Omega$. (a) What is the impedance of the circuit? (b) What is the maximum current?

23. A 60.0-Ω resistor, a 3.00-μF capacitor, and a 0.400-H inductor are connected in series to a 90.0-V, 60.0-Hz source. Find (a) the voltage drop across the *LC* combination and (b) the voltage drop across the *RC* combination.

24. A 50.0-Ω resistor is connected in series with a 15.0-μF capacitor and a 60.0-Hz, 120-V source. (a) Find the current in the circuit. (b) What is the value of the inductor that must be inserted in the circuit to reduce the current to one-half that found in (a)?

25. A person is working near the secondary of a transformer, as shown in Figure P21.25. The primary voltage is 120 V (rms) at 60.0 Hz. The capacitance C_s, which is the stray capacitance between the hand and the secondary winding, is 20.0 pF. Assuming the person has a body resistance to ground of $R_b = 50.0$ kΩ, determine the rms voltage across the body. (*Hint:* Redraw the circuit with the secondary of the transformer as a simple AC source.)

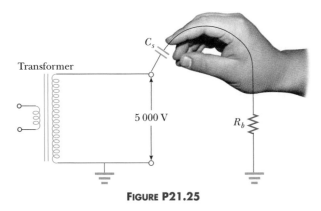

Transformer
C_s
5 000 V
R_b

FIGURE P21.25

26. A coil of resistance 35.0 Ω and inductance 20.5 H is in series with a capacitor and a 200-V (rms), 100-Hz source. The rms current in the circuit is 4.00 A. (a) Calculate the capacitance in the circuit. (b) What is ΔV_{rms} across the coil?

27. An AC source with a maximum voltage of 150 V and **web** $f = 50.0$ Hz is connected between points *a* and *d* in Figure P21.27. Calculate the rms voltages between points (a) *a* and *b*, (b) *b* and *c*, (c) *c* and *d*, (d) *b* and *d*.

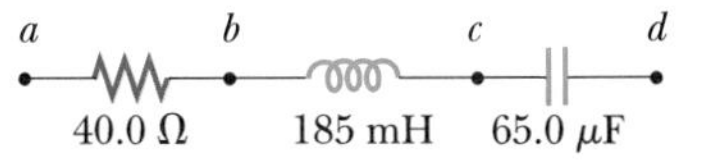

a — 40.0 Ω — *b* — 185 mH — *c* — 65.0 μF — *d*

FIGURE P21.27

Section 21.5 Power in an AC Circuit

28. A 50.0-Ω resistor is connected to a 30.0-μF capacitor and to a 60.0-Hz, 100-V (rms) source. (a) Find the power factor and the average power delivered to the circuit. (b) Repeat part (a) when the capacitor is replaced with a 0.300-H inductor.

29. A multimeter in an *RL* circuit records an rms current of 0.500 A and a 60.0-Hz rms generator voltage of 104 V. A wattmeter shows that the average power delivered to the resistor is 10.0 W. Determine (a) the impedance in the circuit, (b) the resistance R, and (c) the inductance L.

30. In a certain *RLC* circuit, the rms current is 6.0 A, the rms voltage is 240 V, and the current leads the voltage by 53°. (a) What is the total resistance of the circuit? (b) Calculate the total reactance $X_L - X_C$. (c) Find the average power delivered to the circuit.

31. An inductor and a resistor are connected in series. When connected to a 60-Hz, 90-V source, the voltage drop across the resistor is found to be 50 V and the power delivered to the circuit is 14 W. Find (a) the value of the resistance and (b) the value of the inductance.

32. An AC voltage of the form $\Delta v = (100 \text{ V}) \sin(1\ 000t)$ is applied to a series *RLC* circuit. If $R = 400 \ \Omega$, $C = 5.00 \ \mu$F, and $L = 0.500$ H, find the average power delivered to the circuit.

Section 21.6 Resonance in a Series *RLC* Circuit

33. An *RLC* circuit is used to tune a radio to an FM station broadcasting at 88.9 MHz. The resistance in the circuit is 12.0 Ω and the capacitance is 1.40 pF. What inductance should be present in the circuit?

34. A resonant circuit in a radio receiver is tuned to a certain station when the inductor has a value of 0.200 mH and the capacitor has a value of 30.0 pF. Find the frequency of the radio station and the wavelength sent out by the station.

35. The AM band extends from approximately 500 kHz to 1 600 kHz. If a 2.0-μH inductor is used in a tuning circuit for a radio, what are the extremes that a capacitor must reach in order to cover the complete band of frequencies?

36. A series circuit contains a 3.00-H inductor, a 3.00-μF capacitor, and a 30.0-Ω resistor connected to a 120-V (rms) source of variable frequency. Find the power delivered to the circuit when the frequency of the source is (a) the resonance frequency, (b) one-half the resonance frequency, (c) one-fourth the resonance frequency, (d) two times the resonance frequency, (e) four times the resonance frequency. From your calculations, can you draw a conclusion about the frequency at which the maximum power is delivered to the circuit?

37. A 10.0-Ω resistor, 10.0-mH inductor, and a 100-μF capacitor are connected in series to a 50.0-V (rms) source having variable frequency. Find the energy delivered to the circuit during one period if the operating frequency is twice the resonance frequency.

Section 21.7 The Transformer

38. A step-down transformer is used for recharging the batteries of portable devices such as tape players. The turns ratio inside the transformer is 13:1 and it is used with 120 V (rms) household service. If a particular ideal transformer draws 0.350 A from the house outlet, what are (a) the voltage and (b) the current supplied to a tape player from the transformer? (c) How much power is delivered?

39. A step-up transformer is designed to have an output voltage of 2 200 V (rms) when the primary is connected across a 110-V (rms) source. (a) If there are 80 turns on the primary winding, how many turns are required on the secondary? (b) If a load resistor across the secondary draws a current of 1.5 A, what is the current in the primary, assuming ideal conditions?

40. A transformer has $N_1 = 350$ turns and $N_2 = 2\,000$ turns. If the input voltage is $\Delta v(t) = (170\text{ V})\cos \omega t$, what rms voltage is developed across the secondary coil?

41. A transformer on a pole near a factory steps the voltage down from 3 600 V to 120 V. The transformer is to deliver 1 000 kW to the factory at 90% efficiency. Find (a) the power delivered to the primary, (b) the current in the primary, and (c) the current in the secondary.

42. A transmission line that has a resistance per unit length of $4.50 \times 10^{-4}\ \Omega/\text{m}$ is to be used to transmit 5.00 MW over 400 miles (6.44×10^5 m). The output voltage of the generator is 4.50 kV. (a) What is the line loss if a transformer is used to step up the voltage to 500 kV? (b) What fraction of the input power is lost to the line under these circumstances? (c) What difficulties would be encountered on attempting to transmit the 5.00 MW at the generator voltage of 4.50 kV?

Section 21.10 Production of Electromagnetic Waves by an Antenna

Section 21.11 Properties of Electromagnetic Waves

43. The U.S. Navy has long proposed the construction of extremely low-frequency (ELF) communications systems; such waves could penetrate the oceans to reach distant submarines. Calculate the length of a quarter-wavelength antenna for a transmitter generating ELF waves of frequency 75 Hz. How practical is this?

44. Experimenters at the National Institute of Standards and Technology have made precise measurements of the speed of light using the fact that, in vacuum, the speed of electromagnetic waves is $c = 1/\sqrt{\mu_0 \epsilon_0}$, where the constants $\mu_0 = 4\pi \times 10^{-7}\ \text{N} \cdot \text{s}^2/\text{C}^2$ and $\epsilon_0 = 8.854 \times 10^{-12}\ \text{C}^2/\text{N} \cdot \text{m}^2$. What value (to four significant figures) does this give for the speed of light in vacuum?

45. An electromagnetic wave in vacuum has an electric field amplitude of 220 V/m. Calculate the amplitude of the corresponding magnetic field.

46. A particular electromagnetic wave traveling in vacuum has a magnetic field amplitude of 1.5×10^{-7} T. Find (a) the electric field amplitude and (b) the average power per unit area associated with the wave.

47. A microwave oven is powered by an electron tube called a magnetron that generates electromagnetic waves of frequency 2.45 GHz. The microwaves enter the oven and are reflected by the walls. The standing wave pattern produced in the oven can cook food unevenly, with hot spots in the food at antinodes and cool spots at nodes, so a turntable is often used to rotate the food and distribute the energy. If a microwave oven is used with a cooking dish in a fixed position, the antinodes can appear as burn marks on foods such as carrot strips or cheese. The separation distance between the burns is measured to be 6.00 cm. Calculate the speed of the microwaves from these data.

48. Assume that the solar radiation incident on Earth is 1 340 W/m^2 (at the top of Earth's atmosphere). Calculate the total power radiated by the Sun, taking the average separation between Earth and the Sun to be 1.49×10^{11} m.

49. The Sun delivers an average power of 1 340 W/m^2 to the top of Earth's atmosphere. Find the magnitudes of $\mathbf{E}_{\text{max}}$ and $\mathbf{B}_{\text{max}}$ for the electromagnetic waves at the top of the atmosphere.

Section 21.12 The Spectrum of Electromagnetic Waves

50. The eye is most sensitive to light of wavelength 5.50×10^{-7} m, which is in the green-yellow region of the visible electromagnetic spectrum. What is the frequency of this light?

51. What are the wavelength ranges in the (a) AM radio band (540–1 600 kHz) and (b) the FM radio band (88–108 MHz)?

52. A diathermy machine, used in physiotherapy, generates electromagnetic radiation that gives the effect of "deep heat" when absorbed in tissue. One assigned frequency for diathermy is 27.33 MHz. What is the wavelength of this radiation?

53. A radar pulse returns to the receiver after a total travel time of 4.00×10^{-4} s. How far away is the object that reflected the wave?

54. An important news announcement is transmitted by radio waves to people who are 100 km away, sitting next to their radios, and by sound waves to people sitting across the newsroom, 3.0 m from the newscaster. Who receives the news first? Explain. Take the speed of sound in air to be 343 m/s.

ADDITIONAL PROBLEMS

55. An AC adapter for a telephone answering unit uses a transformer to reduce the line voltage of 120 V (rms) to a voltage of 9.0 V. The rms current delivered to the answering system is 400 mA. (a) If the primary (input) coil in the transformer in the adapter has 240 turns, how many turns are there on the secondary (output) coil? (b) What is the rms power delivered to the transformer? Assume an ideal transformer.

56. The primary coil of a certain transformer has an inductance of 2.50 H and a resistance of 80.0 Ω. (a) If the primary coil is connected to an AC source with a frequency of 60.0 Hz and a voltage of 110 V (rms), what is the rms current in the primary? (b) If the primary is connected to 110 V DC, what is the current in the primary? (c) In each case, compare the power delivered to the coil.

57. As a way of determining the inductance of a coil used in a research project, a student first connects the coil to a 12.0-V battery and measures a current of 0.630 A. The student then connects the coil to a 24.0-V (rms), 60.0-Hz generator and measures an rms current of 0.570 A. What is the inductance?

58. The intensity of solar radiation at the top of Earth's atmosphere is 1 340 W/m^3. Assuming that 60% of the incoming solar energy reaches Earth's surface and assuming that you absorb 50% of the incident energy, make an order-of-magnitude estimate of the amount of solar energy you absorb in a 60-minute sunbath.

59. A 200-Ω resistor is connected in series with a 5.0-μF capacitor and a 60-Hz, 120-V rms line. If electrical energy costs $0.080/kWh, how much does it cost to leave this circuit connected for 24 h?

60. A series *RLC* circuit has a resonance frequency of $2\,000/\pi$ Hz. When it is operating at a frequency of

$\omega > \omega_0$, $X_L = 12\ \Omega$ and $X_C = 8.0\ \Omega$. Calculate the values of L and C for the circuit.

61. Two connections allow contact with two circuit elements in
web series inside a box, but it is not known whether the circuit elements are R, L, or C. In an attempt to find what is inside the box, you make some measurements, with the following results. When a 3.0-V DC power supply is connected across the terminals, there is a direct current of 300 mA in the circuit. When a 3.0-V, 60-Hz source is connected, the current becomes 200 mA. (a) What are the two elements in the box? (b) What are their values of R, L, or C?

62. (a) What capacitance will resonate with a one-turn loop of inductance 400 pH to give a radar wave of wavelength 3.0 cm? (b) If the capacitor has square parallel plates separated by 1.0 mm of air, what should the edge length of the plates be? (c) What is the common reactance of the loop and capacitor at resonance?

63. A dish antenna with a diameter of 20.0 m receives (at normal incidence) a radio signal from a distant source, as shown in Figure P21.63. The radio signal is a continuous sinusoidal wave with amplitude $E_{max} = 0.20\ \mu V/m$. Assume the antenna absorbs all the radiation that falls on the dish. (a) What is the amplitude of the magnetic field in this wave? (b) What is the intensity of the radiation received by this antenna? (c) What is the power received by the antenna?

64. A particular inductor has appreciable resistance. When the inductor is connected to a 12-V battery, the current in the inductor is 3.0 A. When it is connected to an AC source with an rms output of 12 V and a frequency of 60 Hz, the current drops to 2.0 A. What are (a) the impedance at 60 Hz and (b) the inductance of the inductor?

FIGURE P21.63

65. A possible means of space flight is to place a perfectly reflecting aluminized sheet into Earth's orbit and to use the light from the Sun to push this solar sail. Suppose such a sail, of area $6.00 \times 10^4\ m^2$ and mass 6 000 kg, is placed in orbit facing the Sun. (a) What force is exerted on the sail? (b) What is the sail's acceleration? (c) How long does it take for this sail to reach the Moon, 3.84×10^8 m away? Ignore all gravitational effects, and assume a solar intensity of 1 340 W/m². [*Hint:* The radiation pressure by a reflected wave is given by 2(average power per unit area)/c.]

66. Suppose you wish to use a transformer as an impedance-matching device between an audio amplifier that has an output impedance of 8.0 kΩ and a speaker that has an input impedance of 8.0 Ω. What should be the ratio of primary to secondary turns on the transformer?

67. Compute the average energy content of a liter of sunlight as it reaches the top of Earth's atmosphere, where its intensity is 1 340 W/m².

GROUP ACTIVITIES

G.1 For this observation, you will need some items that can be found at many electronics stores. You will need a bicolored LED (light-emitting diode), a resistor of about 100 Ω, 2 m of flexible wire, and a step-down transformer with an output of 3 to 6 V. Use the wire to connect the LED and the resistor in series with the transformer. A bicolored LED is designed such that it emits a red color when the current in the LED is in one direction and green when the current reverses. When connected to an AC source, the LED is yellow. Why?

Hold the wires and whirl the LED in a circular path. In a darkened room, you will see red and green bars at equally spaced intervals along the path of the LED. Why?

As you continue to whirl the LED in a circular path, have your partner count the number of green bars in the circle then measure the time it takes for the LED to travel ten times around the circular path. Based on this information, determine the time it takes for the color of the LED to change from green to red to green. You should obtain an answer of (1/60) s. Why?

G.2 Rotate a portable radio (with a telescoping antenna) about a horizontal axis while it is tuned to a weak station. Such an antenna detects the varying electric field produced by the station. What can you determine about the direction of the electric field produced by the transmitter?

Now turn on your radio to a nearby station and experiment with shielding the radio from incoming waves. Is the reception affected by surrounding the radio by aluminum foil? By plastic wrap? Use any other material you have available. What kinds of material block the signal? Why?

G.3 An *RLC* circuit is brought to resonance. (a) What is the impedance of the circuit at resonance? (b) When will the current be greatest—at resonance, at 10% below the resonant frequency, or at 10% above the resonant frequency? (c) Check your guesses by examining a circuit with $\Delta V = 12$ V, $R = 15\ \Omega$, $C = 75\ \mu F$, and $L = 200$ mH.

G.4 A transformer is to be used to provide power for a computer disk drive that needs 6.0 V instead of the 120 V from the wall outlet. (a) Should the power supply have more turns in the primary as compared to the secondary, or the reverse? (b) Will the larger current be in the primary or the secondary? (c) Assume the number of turns in the primary is 400 and that it delivers 4.0 A at the 6.0-V output. Do the calculations to confirm your answers.

G.5 Consider an *RLC* circuit with $R = 25\ \Omega$, $L = 6.0$ mH, and $C = 25\ \mu F$. This circuit is connected to a 10-V, 600-Hz AC source. (a) Is the sum of the voltage drops across R, L, and C equal to 10 V? (b) Which is greatest, the power delivered to the resistor, the capacitor, or the inductor? (c) Do the calculations to verify your predictions.

PART Five

Light and Optics

Scientists have long been intrigued by the nature of light, and philosophers have argued endlessly concerning the proper definition and perception of light. It is important to understand the nature of this basic ingredient of life on Earth. Plants convert light energy from the Sun to chemical energy through photosynthesis. Light is the means by which we transmit and receive information from objects around us and throughout the Universe.

The ancient Greeks believed that vision worked by means of "fire" emanating from the eye, which then interacted with objects and returned to the eye carrying object properties with it. Around 1 000 A.D., it was proposed that light consisted of tiny particles that were emitted by a light source. These particles stimulated the perception of vision on striking the observer's eye. Newton used this particle model to explain the reflection and refraction of light. In 1670 one of Newton's contemporaries, the Dutch scientist Christian Huygens, succeeded in explaining many properties of light by proposing that light was wave-like. In 1801 Thomas Young gave strong support to the wave theory by showing that light beams can interfere with one another. In 1865 Maxwell developed a brilliant theory that electromagnetic waves travel with the speed of light (Chapter 21). By that time, the wave theory of light seemed to be on firm ground.

However, at the beginning of the 20th century, Max Planck introduced the notion that vibrating electrons in the wall of a glowing solid had quantized (discrete) energies in order to explain the light spectrum emitted by hot objects. Albert Einstein extended this idea to a full-fledged particle theory of light to explain the photoelectric effect (electrons emitted by a metal exposed to light). We shall discuss those and other modern topics in the last part of this book.

Today, scientists view light as having a dual nature. Experiments can be devised that will display either its particle-like or its wave-like nature. All we can say about light is that whenever we design an experiment to determine its wave-like nature, we will find it, and whenever we look for its particle-like nature, we will find that. Nature prevents us from testing both qualities at the same time.

This striking photograph shows the Grand Tetons in Wyoming reflected in a lake which acts like a mirror. Note that the image of the mountains has the same size as the mountains.
(David Muench/CORBIS)

22

Reflection and Refraction of Light

Chapter Outline

The prism in this photograph demonstrates both refraction and reflection. A beam of white light at the top left is refracted and strikes the right face of the prism. At this face, some of the light is refracted once more, exiting the prism as a spectrum of light with different colors. The spectrum arises from the fact that the incident white light contains different wavelengths that refract at different angles. Part of the light is reflected at the right face and exits the prism at the bottom left as white light. *(David Parker/Science Photo Library/Photo Researchers, Inc.)*

*I*n this part of the book, we concentrate on the aspects of light that are best understood through the wave model. First, we discuss the reflection of light at the boundary between two media and the refraction (bending) of light as it travels from one medium into another. We use these ideas to study the refraction of light as it passes through lenses and the reflection of light from mirrored surfaces. Finally, we describe how lenses and mirrors can be used to view objects with telescopes and microscopes and how lenses are used in photography.

22.1 THE NATURE OF LIGHT

Until the beginning of the 19th century, light was considered to be a stream of particles, emitted by a light source, that stimulated the sense of sight on entering the eye. The chief architect of the particle theory of light was Newton. With this theory he provided simple explanations of some known experimental facts concerning the nature of light—namely, the laws of reflection and refraction.

Most scientists accepted Newton's particle theory of light. However, during Newton's lifetime another theory was proposed. In 1678 a Dutch physicist and astronomer, Christian Huygens (1629–1695), showed that a wave theory of light could also explain the laws of reflection and refraction. The wave theory did not receive immediate acceptance for several reasons. All the waves known at the time (sound, water, and so on) traveled through some sort of medium, but light from the Sun could travel to the Earth through empty space. Furthermore, it was argued that if light were some form of wave, it would bend around obstacles; hence, we should be able to see around corners. It is now known that light does indeed bend around the edges of objects. This phenomenon, known as *diffraction,* is not easy to observe because light waves have such short wavelengths. Even though experimental evidence for the diffraction of light was discovered by Francesco Grimaldi (1618–1663) around 1660, for more than a century most scientists rejected the wave theory and adhered to Newton's particle theory. This was also due to Newton's great reputation as a scientist.

The first clear demonstration of the wave nature of light was provided in 1801 by Thomas Young (1773–1829), who showed that under appropriate conditions, light exhibits interference behavior. That is, light waves emitted by a single source and traveling along two different paths can arrive at some point, combine, and cancel each other by destructive interference. Such behavior could not be explained at that time by a particle model, because scientists could not imagine how two or more particles could come together and cancel one another.

The most important development concerning the theory of light was the work of Maxwell, who in 1865 predicted that light was a form of high-frequency electromagnetic wave (Chapter 21). His theory predicted that these waves should have a speed of 3×10^8 m/s, in agreement with the measured value.

Although the classical theory of electricity and magnetism explained most known properties of light, some subsequent experiments could not be explained by the assumption that light was a wave. The most striking of these was the *photoelectric effect* (which we shall examine more closely in Chapter 27), discovered by Hertz. Hertz found that clean metal surfaces emit charges when exposed to ultraviolet light.

In 1905 Einstein published a paper that formulated the theory of light quanta ("particles") and explained the photoelectric effect. He reached the conclusion that light is composed of corpuscles, or discontinuous quanta of energy. These corpuscles or quanta are now called *photons* to emphasize their particle-like nature. According to Einstein's theory, the energy of a photon is proportional to the frequency of the electromagnetic wave:

$$E = hf \qquad [22.1]$$

◀ Energy of a photon

CHRISTIAN HUYGENS, DUTCH PHYSICIST AND ASTRONOMER (1629–1695)

Huygens is best known for his contributions to the fields of optics and dynamics. To Huygens, light was a vibratory motion in the ether, spreading out and producing the sensation of light when impinging on the eye. On the basis of this theory, he deduced the laws of reflection and refraction and explained the phenomenon of double refraction. *(Courtesy of Rijksmuseum voor de Geschiedenis der Natuurwetenschappen. Courtesy AIP Niels Bohr Library)*

where $h = 6.63 \times 10^{-34}$ J·s is *Planck's constant*. This theory retains some features of both the wave and particle theories of light. As we shall discuss later, the photo-electric effect is the result of energy transfer from a single photon to an electron in the metal. That is, the electron interacts with one photon of light as if the electron had been struck by a particle. Yet the photon has wave-like characteristics, as implied by the fact that a frequency is used in its definition.

In view of these developments, light must be regarded as having a *dual nature*. That is, **in some cases light acts as a wave and in others it acts as a particle.** Classical electromagnetic wave theory provides adequate explanations of light propagation and of the effects of interference, whereas the photoelectric effect and other experiments involving the interaction of light with matter are best explained by assuming that light is a particle. However, the question "Is light a wave or a particle?" is inappropriate; sometimes it acts as one, sometimes as the other. Fortunately, it never acts as both in the same experiment.

22.2 THE RAY APPROXIMATION IN GEOMETRIC OPTICS

In studying geometric optics here and in future chapters, we shall make use of an important property of light that can be understood based on common experience. **Light travels in a straight-line path in a homogeneous medium until it encounters a boundary between two different materials.** As we shall see, when light strikes a boundary it either is reflected from that boundary or passes into the material on the other side of the boundary or partially does both.

Based on this observation, we will use what is called the **ray approximation** to represent beams of light. As shown in Figure 22.1, a ray of light is an imaginary line drawn along the direction of travel of the light beam. For example, a beam of sunlight passing through a darkened room traces out the path of a light ray. We will also have occasion to refer to wave fronts of light. A **wave front** is a surface passing through the points of a wave that have the same phase and amplitude. For instance, the wave fronts in Figure 22.1 could be surfaces passing through the crests of waves. You should note that the rays, corresponding to the direction of wave motion, are straight lines perpendicular to the wave fronts. Also, note that when light rays travel in parallel paths, the wave fronts are planes perpendicular to the rays.

22.3 REFLECTION AND REFRACTION

REFLECTION OF LIGHT

When a light ray traveling in a transparent medium encounters a boundary leading into a second medium, part of the incident ray is reflected back into the first medium. Figure 22.2a shows several rays of a beam of light incident on a smooth, mirror-like, reflecting surface. The reflected rays are parallel to each other, as indicated in the figure. Reflection of light from such a smooth surface is called **specular reflection.** On the other hand, if the reflecting surface is rough, as in Figure 22.2b, the surface reflects the rays in a variety of directions. Reflection from any rough surface is known as **diffuse reflection.** A surface behaves as a smooth surface as long as the surface variations are small compared with the wavelength of the incident light. Figures 22.2c and 22.2d are photographs of specular and diffuse reflection of laser light.

For instance, consider the two types of reflection from a road surface that one sees while driving at night. When the road is dry, light from oncoming vehicles is scattered off the road in different directions (diffuse reflection) and the road is quite visible. On a rainy night, when the road is wet, the road irregularities are filled with water. Because the wet surface is quite smooth, the light undergoes

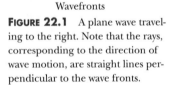

Rays

Wavefronts

FIGURE 22.1 A plane wave traveling to the right. Note that the rays, corresponding to the direction of wave motion, are straight lines perpendicular to the wave fronts.

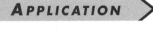

APPLICATION

SEEING THE ROAD ON A RAINY NIGHT

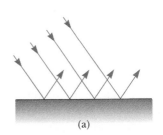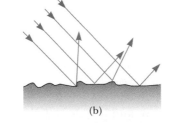

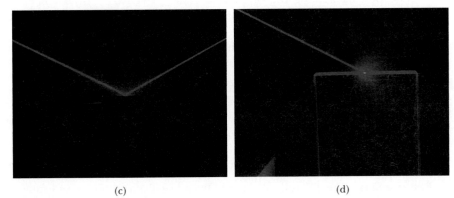

FIGURE 22.2 A schematic representation of (a) specular reflection, where the reflected rays are all parallel to each other, and (b) diffuse reflection, where the reflected rays travel in random directions. (c, d) Photographs of specular and diffuse reflection, using laser light. *(Photographs courtesy of Henry Leap and Jim Lehman)*

FIGURE 22.3 (Quick Quiz 22.1) *(Photos by Charles D. Winters)*

specular reflection. This means that the light is reflected straight ahead, and the driver of a car sees only what is directly in front. Light from the side never reaches the driver's eye. In this book we concern ourselves only with specular reflection, and we use the term *reflection* to mean specular reflection.

Quick Quiz 22.1 Which part of Figure 22.3, (a) or (b), shows specular reflection of light from the roadway?

Consider a light ray traveling in air and incident at some angle on a flat, smooth surface, as in Figure 22.4. The incident and reflected rays make angles θ_1 and θ_1', respectively, **with a line perpendicular to the surface** at the point where the incident ray strikes the surface. We call this line the *normal* to the surface. Experiments

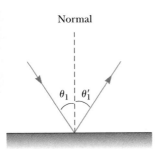

FIGURE 22.4 According to the law of reflection, $\theta_1 = \theta_1'$.

show that the angle of reflection equals *the angle of incidence;* that is,

$$\theta_1' = \theta_1 \qquad\qquad [22.2]$$

A P P L I C A T I O N

RED EYES IN FLASH
PHOTOGRAPHS

You may have noticed a common occurrence in photographs of individuals—their eyes appear to be glowing red. This occurs when a photographic flash device is used and the flash unit is very close to the camera lens. Light from the flash unit enters the eye and is reflected back along its original path from the retina. This type of reflection back along the original direction is called *retroreflection.* If the flash unit and lens are close together, this retroreflected light can enter the lens. Most of the light reflected from the retina is red, due to the blood vessels at the back of the eye, giving the red-eye effect in the photograph.

APPLYING **PHYSICS** 22.1

An observer on the west-facing beach of a large lake is watching the beginning of a sunset. The water is very smooth except for some areas with small ripples. The observer notices that some areas of the water are blue and some are pink. Why does the water appear to be different colors in different areas?

A P P L I C A T I O N

THE COLORS OF WATER RIP-
PLES AT SUNSET

Explanation The different colors arise from specular and diffuse reflection. The smooth areas of the water will specularly reflect the light from the west, which is the pink light from the sunset. The areas with small ripples will reflect the light diffusely. Thus, light from all parts of the sky will be reflected into the observer's eyes. Because most of the sky is still blue at the beginning of the sunset, these areas will appear to be blue.

APPLYING **PHYSICS** 22.2

When looking outdoors through a glass window at night, you sometimes see a double image of yourself. Why?

Explanation Reflection occurs whenever there is an interface between two different media. For the glass in the window, there are two such surfaces. The first is the inner surface of the glass, and the second is the outer surface. Each of these interfaces results in an image.

Example 22.1 The Double-Reflecting Light Ray

Two mirrors make an angle of 120° with each other, as in Figure 22.5. A ray is incident on mirror M_1 at an angle of 65° to the normal. Find the angle the ray makes with the normal to M_2 after it is reflected from both mirrors.

Reasoning and Solution From the law of reflection, we see that the first reflected ray also makes an angle of 65° with the normal to M_1. It follows that this same ray makes an angle of $90° - 65°$, or 25°, with the horizontal. From the triangle made by the first reflected ray and the two mirrors, we see that the first reflected ray makes an angle of 35° with M_2 (because the sum of the interior angles of any triangle is 180°). This means that this ray makes an angle of 55° with the normal to M_2. From the law of reflection, it follows that the second reflected ray makes an angle of 55° with the normal to M_2.

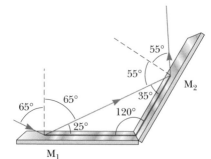

FIGURE 22.5 (Example 22.1) Mirrors M_1 and M_2 make an angle of 120° with each other.

REFRACTION OF LIGHT

When a ray of light traveling through a transparent medium encounters a boundary leading into another transparent medium, as in Figure 22.6a, part of the ray is reflected and part enters the second medium. The ray that enters the second

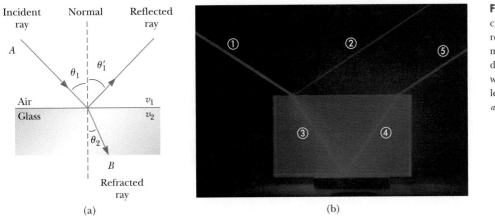

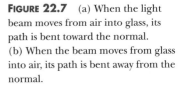

FIGURE 22.6 (a) A ray obliquely incident on an air-glass interface. The refracted ray is bent toward the normal because $v_2 < v_1$. (b) Light incident on the Lucite block bends both when it enters the block and when it leaves the block. *(Courtesy of Henry Leap and Jim Lehman)*

medium is bent at the boundary and is said to be *refracted*. The incident ray, the reflected ray, the refracted ray, and the normal at the point of incidence all lie in the same plane. The **angle of refraction,** θ_2, in Figure 22.6a depends on the properties of the two media and on the angle of incidence, through the relationship

$$\frac{\sin \theta_2}{\sin \theta_1} = \frac{v_2}{v_1} = \text{constant} \qquad \text{[22.3]}$$

where v_1 is the speed of light in medium 1 and v_2 is the speed of light in medium 2. Note that the angle of refraction is also measured with respect to the normal. In Section 22.7 we shall derive the laws of reflection and refraction using Huygens's principle.

Experiment shows that **the path of a light ray through a refracting surface is reversible.** For example, the ray in Figure 22.6a travels from point *A* to point *B*. If the ray originated at *B*, it would follow the same path to reach point *A*, but the reflected ray would be in the glass.

Quick Quiz 22.2	If beam 1 is the incoming beam in Figure 22.6b, which of the other four beams are reflected and which are refracted?

When light moves from a material in which its speed is high to a material in which its speed is lower, the angle of refraction θ_2 is less than the angle of incidence. The refracted ray therefore bends toward the normal, as shown in Figure 22.7a. If the ray moves from a material in which it travels slowly to a material in

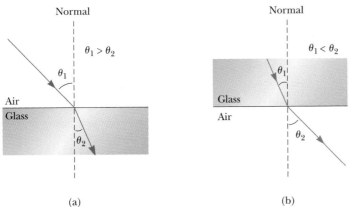

FIGURE 22.7 (a) When the light beam moves from air into glass, its path is bent toward the normal. (b) When the beam moves from glass into air, its path is bent away from the normal.

which it travels more rapidly, θ_2 is greater than θ_1, and so the ray bends away from the normal, as shown in Figure 22.7b.

22.4 THE LAW OF REFRACTION

When light passes from one transparent medium to another, it is refracted because the speed of light is different in the two media.[1] It is convenient to define the **index of refraction,** n, of a medium as the ratio

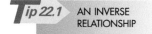

Index of refraction ▶

$$n \equiv \frac{\text{speed of light in vacuum}}{\text{speed of light in a medium}} = \frac{c}{v} \qquad [22.4]$$

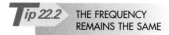

Tip 22.1 AN INVERSE RELATIONSHIP

The index of refraction is *inversely* proportional to the wave speed. Therefore, as the wave speed v decreases, the index of refraction n *increases*.

Tip 22.2 THE FREQUENCY REMAINS THE SAME

Note that the *frequency* of a wave does *not* change as the wave passes from one medium to another. Both the wave speed and wavelength *do* change, but the frequency remains constant.

From this definition, we see that the index of refraction is a dimensionless number that is greater than or equal to unity because v is always less than c. Furthermore, n is equal to unity for vacuum. Table 22.1 lists the indices of refraction for various substances.

As light travels from one medium to another, its frequency does not change. To see why, consider Figure 22.8. Wave fronts pass an observer at point A in medium 1 with a certain frequency and are incident on the boundary between medium 1 and medium 2. The frequency at which the wave fronts pass an observer at point B in medium 2 must equal the frequency at which they arrive at point A. If this were not the case, the wave fronts would either pile up at the boundary or be destroyed or created at the boundary. Because this does not occur, the frequency must remain the same as a light ray passes from one medium into another.

Therefore, because the relation $v = f\lambda$ must be valid in both media and because $f_1 = f_2 = f$, we see that

$$v_1 = f\lambda_1 \qquad \text{and} \qquad v_2 = f\lambda_2$$

Because $v_1 \neq v_2$, it follows that $\lambda_1 \neq \lambda_2$. A relationship between index of refraction and wavelength can be obtained by dividing these two equations and making use of the definition of index of refraction given by Equation 22.4:

$$\frac{\lambda_1}{\lambda_2} = \frac{v_1}{v_2} = \frac{c/n_1}{c/n_2} = \frac{n_2}{n_1} \qquad [22.5]$$

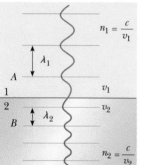

FIGURE 22.8 As the wave moves from medium 1 to medium 2, its wavelength changes but its frequency remains constant.

TABLE 22.1	Indices of Refraction for Various Substances, Measured with Light of Vacuum Wavelength $\lambda_0 = 589$ nm		
Substance	**Index of Refraction**	**Substance**	**Index of Refraction**
Solids at 20°C		**Liquids at 20°C**	
Diamond (C)	2.419	Benzene	1.501
Fluorite (CaF$_2$)	1.434	Carbon disulfide	1.628
Fused quartz (SiO$_2$)	1.458	Carbon tetrachloride	1.461
Glass, crown	1.52	Ethyl alcohol	1.361
Glass, flint	1.66	Glycerine	1.473
Ice (H$_2$O) (at 0°C)	1.309	Water	1.333
Polystyrene	1.49		
Sodium chloride (NaCl)	1.544	**Gases at 0°C, 1 atm**	
Zircon	1.923	Air	1.000 293
		Carbon dioxide	1.000 45

[1] The speed of light varies between media because the time lags caused by absorption and reemission of light as it travels from atom to atom depends on the particular electronic structure of the atoms constituting each material.

which gives

$$\lambda_1 n_1 = \lambda_2 n_2 \qquad \text{[22.6]}$$

Let medium 1 be the vacuum so that $n_1 = 1$. It follows from Equation 22.6 that the index of refraction of any medium can be expressed as the ratio

$$n = \frac{\lambda_0}{\lambda_n} \qquad \text{[22.7]}$$

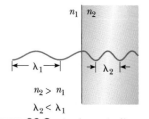

where λ_0 is the wavelength of light in vacuum and λ_n is the wavelength in the medium whose index of refraction is n. Figure 22.9 is a schematic representation of this reduction in wavelength when light passes from vacuum into a transparent medium.

We are now in a position to express Equation 22.3 in an alternative form. If we substitute Equation 22.5 into Equation 22.3, we get

$$n_1 \sin \theta_1 = n_2 \sin \theta_2 \qquad \text{[22.8]}$$

FIGURE 22.9 A schematic diagram of the *reduction* in wavelength when light travels from a medium with a low index of refraction to one with a higher index of refraction.

◀ Snell's law of refraction

The experimental discovery of this relationship is usually credited to Willebord Snell (1591–1627) and is therefore known as **Snell's law of refraction.**

Quick Quiz 22.3
A material has an index of refraction that increases continuously from top to bottom. Of the three paths shown in Figure 22.10, which path will a light ray follow as it passes through the material.

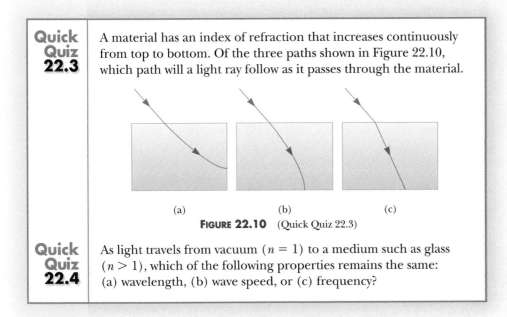

(a)   (b)   (c)

FIGURE 22.10 (Quick Quiz 22.3)

Quick Quiz 22.4
As light travels from vacuum ($n = 1$) to a medium such as glass ($n > 1$), which of the following properties remains the same: (a) wavelength, (b) wave speed, or (c) frequency?

Example 22.2   Angle of Refraction for Glass

A light ray of wavelength 589 nm (produced by a sodium lamp) traveling through air is incident on a smooth, flat slab of crown glass at an angle of 30.0° to the normal, as sketched in Figure 22.11. Find the angle of refraction, θ_2.

Solution   Snell's law (Eq. 22.8) can be rearranged as

$$\sin \theta_2 = \frac{n_1}{n_2} \sin \theta_1$$

From Table 22.1, we find that $n_1 = 1.00$ for air and $n_2 = 1.52$ for crown glass. Therefore, the unknown refraction angle is determined by

$$\sin \theta_2 = \left(\frac{1.00}{1.52}\right)(\sin 30.0°) = 0.329$$

$$\theta_2 = \sin^{-1}(0.329) = \boxed{19.2°}$$

We see that the ray is bent *toward* the normal, as expected.

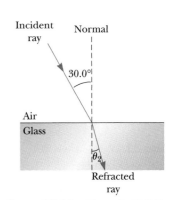

FIGURE 22.11 (Example 22.2) Refraction of light by glass.

EXERCISE   If the light ray moves from inside the glass toward the glass-air interface at an angle of 30.0° to the normal, determine the angle of refraction.

ANSWER   49.5° *away* from the normal

Example 22.3   The Speed of Light in Fused Quartz

Light of wavelength 589 nm in vacuum passes through a piece of fused quartz of index of refraction $n = 1.458$.

A Find the speed of light in fused quartz.

Solution   The speed of light in fused quartz can be obtained from Equation 22.4:

$$v = \frac{c}{n} = \frac{3.00 \times 10^8 \text{ m/s}}{1.458} = \boxed{2.06 \times 10^8 \text{ m/s}}$$

It is interesting to note that the speed of light in vacuum, 3.00×10^8 m/s, is an upper limit for the speed of material objects. In our treatment of relativity in Chapter 26, we shall find that this upper limit is consistent with experimental observations. However, it is possible for a particle moving in a medium to have a speed that exceeds the speed of light in that medium. For example, it is theoretically possible for a particle to travel through fused quartz at a speed greater than 2.06×10^8 m/s, but it must have a speed less than 3.00×10^8 m/s.

B What is the wavelength of this light in fused quartz?

Solution   We can use $\lambda_n = \lambda_0/n$ (Eq. 22.7) to calculate the wavelength in fused quartz, noting that we are given $\lambda_0 = 589$ nm $= 589 \times 10^{-9}$ m:

$$\lambda_n = \frac{\lambda_0}{n} = \frac{589 \text{ nm}}{1.458} = \boxed{404 \text{ nm}}$$

EXERCISE   Find the frequency of the light passing through the fused quartz.

ANSWER 5.09×10^{14} Hz

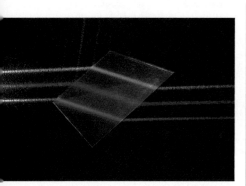

A transparent block has refracted three incident light beams, which exit parallel to each other but are off-set from their original direction of travel. Note the partial reflection at both interfaces. (© *Leonard Lessin/Peter Arnold, Inc.*)

Webnote 22.1

Scientists have been able to slow down the speed of light to as low as 38 mph! See how this was achieved at
http://ajanta.sci.ccny.cuny.edu/~jupiter/ pub/sciinfo/slowlight.html

Example 22.4   Light Passing Through a Slab

A light beam traveling through a transparent medium of index of refraction n_1 passes through a thick transparent slab with parallel faces and index of refraction n_2 (Fig. 22.12). Show that the emerging beam is parallel to the incident beam.

Reasoning   To solve this problem, it is necessary to apply Snell's law twice, once at the upper surface and once at the lower surface. The two equations will be related because the angle of refraction at the upper surface equals the angle of incidence at the lower surface. The ray passing through the slab makes equal angles with the normal at the entry and exit points. This procedure will enable us to compare angles θ_1 and θ_3.

Solution   First, let us apply Snell's law to the upper surface:

$$\sin \theta_2 = \frac{n_1}{n_2} \sin \theta_1 \tag{1}$$

Applying Snell's law to the lower surface gives

$$\sin \theta_3 = \frac{n_2}{n_1} \sin \theta_2 \tag{2}$$

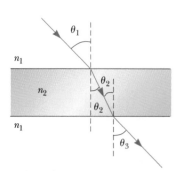

FIGURE 22.12   (Example 22.4) When light passes through a flat slab of material, the emerging beam is parallel to the incident beam, and therefore $\theta_1 = \theta_3$.

Substituting (1) into (2) gives

$$\sin \theta_3 = \frac{n_2}{n_1} \left(\frac{n_1}{n_2} \sin \theta_1 \right) = \sin \theta_1$$

Thus, $\theta_3 = \theta_1$, and so the slab does not alter the direction of the beam as shown by the dashed line in Figure 22.12. It does, however, produce a lateral displacement of the beam.

Example 22.5　Refraction of Laser Light in a DVD (Digital Video Disk)

A DVD is a video recording consisting of a spiral track of digital information about 1.0 μm wide (see Fig. 22.13a). The digital information consists of a series of pits that are "read" by a laser beam sharply focused on a track in the information layer. If the width of the laser beam a at the information layer must equal 1.0 μm to distinguish individual tracks, and the width of the beam w as it enters the plastic is 0.70 mm, find the angle θ_1 at which the conical beam should enter the plastic (see Fig. 22.13b). Assume the plastic has a thickness $t = 1.2$ mm and an index of refraction $n = 1.55$. Note that this system is relatively immune to small dust particles degrading the video quality, because particles would have to be as large as 0.70 mm to obscure the beam at the point where it enters the plastic.

APPLICATION

READING DATA ON A DVD

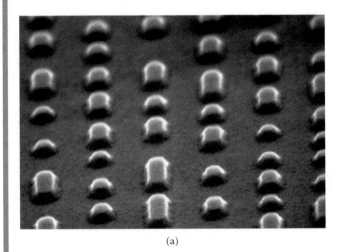

(a)

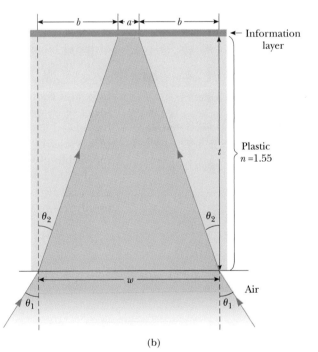

(b)

FIGURE 22.13 (a) A micrograph of a DVD surface showing tracks and pits along each track. *(Courtesy of Sony Disc Manufacturing)* (b) Cross section of a cone-shaped laser beam used to read a DVD.

Solution　From Figure 22.13b we see that

$$w = 2b + a$$

$$b = \frac{w - a}{2} = \frac{699 \ \mu m}{2} = 349.5 \ \mu m$$

Also,

$$\tan \theta_2 = \frac{b}{t} = \frac{349.5 \ \mu m}{1200 \ \mu m}; \qquad \theta_2 = 16.2°$$

Finally we can use Snell's law to find θ_1.

$$n_1 \sin \theta_1 = n_2 \sin \theta_2$$

$$\sin \theta_1 = \frac{n_2 \sin \theta_2}{n_1} = \frac{1.55 \sin 16.2°}{1.00} = 0.433$$

$$\theta_1 = \boxed{26°}$$

22.5 DISPERSION AND PRISMS

In Table 22.1, we presented the index of refraction values for various materials. If we make careful measurements, however, we find that the index of refraction in anything but vacuum depends on the wavelength of light. The dependence of the index of refraction on wavelength is called **dispersion.** Figure 22.14 is a graphical representation of this variation in index of refraction with wavelength. Because n is a function of wavelength, Snell's law indicates that **the angle of refraction when light enters a material depends on the wavelength of the light.** As seen in Figure 22.14, the index of refraction for a material usually decreases with increasing wavelength. This means that violet light ($\lambda \cong 400$ nm) refracts more than red light ($\lambda \cong 650$ nm) when passing from air into a material.

To understand the effects of dispersion on light, consider what happens when light strikes a prism, as in Figure 22.15a. A ray of light of a single wavelength that is incident on the prism from the left emerges bent away from its original direction of travel by an angle δ, called the **angle of deviation.** Now suppose a beam of white light (a combination of all visible wavelengths) is incident on a prism. Because of dispersion, the different colors refract through different angles of deviation, and the rays that emerge from the second face of the prism spread out in a series of colors known as a visible **spectrum,** as shown in Figure 22.16. These colors, in order of decreasing wavelength, are red, orange, yellow, green, blue, and violet. Violet light deviates the most, red light the least, and the remaining colors in the visible spectrum fall between these extremes.

Prisms are often used in an instrument known as a **prism spectrometer,** the essential elements of which are shown in Figure 22.17a. This instrument is commonly used to study the wavelengths emitted by a light source, such as a sodium

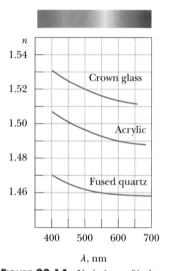

FIGURE 22.14 Variations of index of refraction in the visible spectrum with respect to vacuum wavelength for three materials.

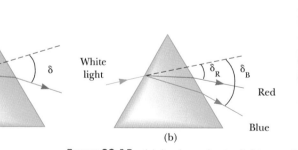

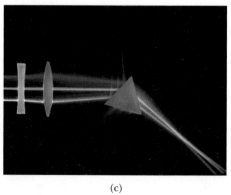

FIGURE 22.15 (a) A prism refracts a light ray and deviates the light through the angle δ. (b) When light is incident on a prism, the blue light is bent more than the red. (c) Light of different colors passes through a prism and two lenses. Note that as the light passes through the prism, different wavelengths are refracted at different angles. *(David Parker/SPL/Photo Researchers)*

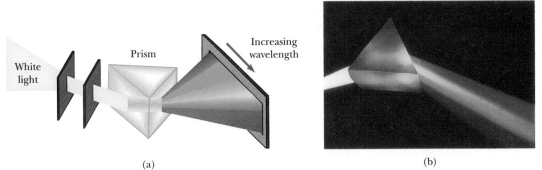

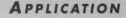

(a)

(b)

FIGURE 22.16 (a) Dispersion of white light by a prism. Since *n* varies with wavelength, the prism disperses the white light into its various spectral components. (b) Different colors of light that pass through a prism are refracted at different angles because the index of refraction of the glass depends on wavelength. The violet light bends the most; red light bends the least. *(b, courtesy of Bausch & Lomb)*

vapor lamp. Light from the source is sent through a narrow, adjustable slit and lens to produce a parallel, or collimated, beam. The light then passes through the prism and is dispersed into a spectrum. The refracted light is observed through a telescope. The experimenter sees different colored images of the slit through the eyepiece of the telescope. The telescope can be moved or the prism can be rotated in order to view the various wavelengths, which have different angles of deviation. Figure 22.17b shows one type of prism spectrometer used in undergraduate laboratories.

All hot, low-pressure gases emit their own characteristic spectra. Thus, one use of a prism spectrometer is to identify gases. For example, sodium emits only two wavelengths in the visible spectrum: two closely spaced yellow lines. (The bright line-like images of the slit seen in a spectroscope are called *spectral lines*.) A gas emitting these and only these colors can thus be identified as sodium. Likewise, mercury vapor has its own characteristic spectrum, consisting of four prominent wavelengths—orange, green, blue, and violet lines—along with some wavelengths of lower intensity. The particular wavelengths emitted by a gas serve as "fingerprints" of that gas. Spectral analysis, which is the measurement of the wavelengths emitted or absorbed by a substance, has proven to be a powerful general tool in many areas: Chemists and biologists use infrared spectroscopy to identify molecules, astronomers use visible light spectroscopy to identify elements on distant stars, and geologists use spectral analysis to identify minerals.

APPLICATION

IDENTIFYING GASES USING A SPECTROMETER

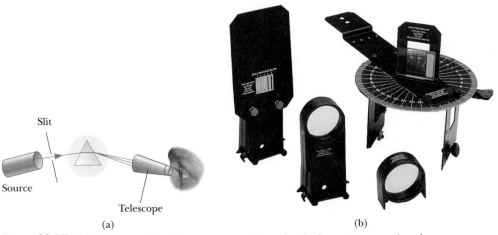

FIGURE 22.17 (a) A diagram of a prism spectrometer. The colors in the spectrum are viewed through a telescope. (b) A prism spectrometer with interchangeable components. *(Courtesy of PASCO Scientific)*

APPLYING **PHYSICS** 22.3

When a beam of light enters a glass prism, which has nonparallel sides, the rainbow of color exiting the prism is a testimonial to the dispersion occurring in the glass. Suppose a beam of light enters a slab of material with parallel sides. When the beam exits the other side, traveling in the same direction as the original beam, is there any evidence of dispersion?

Explanation   Due to dispersion, light at the violet end of the spectrum will exhibit a larger angle of refraction on entering the glass than light at the red end. All colors of light will return to the original direction of propagation on refracting back out into the air. Thus, the outgoing beam will be white. But the net shift in the position of the violet light along the edge of the slab will be larger than that of the red light. Thus, one edge of the outgoing beam will have a bluish tinge (it will appear blue rather than violet, because the eye is not very sensitive to violet light), whereas the other edge will have a reddish tinge. This effect is indicated in Figure 22.18. The colored edges of the outgoing beam of white light are evidence of dispersion.

FIGURE 22.18   (Applying Physics 22.3)

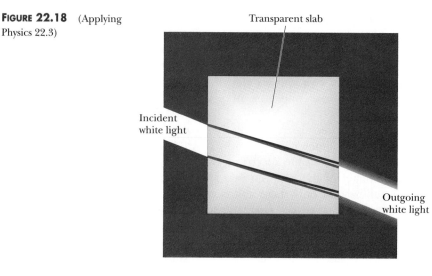

Transparent slab

Incident white light

Outgoing white light

22.6 THE RAINBOW

The dispersion of light into a spectrum is demonstrated most vividly in nature through the formation of a rainbow, often seen by an observer positioned between the Sun and a rain shower. To understand how a rainbow is formed, consider Figure 22.19. A ray of light passing overhead strikes a drop of water in the atmosphere and is refracted and reflected as follows. It is first refracted at the front surface of the drop, with the violet light deviating the most and the red light the least. At the back surface of the drop, the light is reflected and returns to the front surface, where it again undergoes refraction as it moves from water into air. The rays leave the drop so that the angle between the incident white light and the returning violet ray is 40°, and the angle between the white light and the returning red ray is 42° (Fig. 22.19). This small angular difference between the returning rays causes us to see the bow as explained in the next paragraph.

Now consider an observer viewing a rainbow, as in Figure 22.20a. If a raindrop high in the sky is being observed, the red light returning from the drop can reach the observer because it is deviated the most, but the violet light passes over the observer because it is deviated the least. Hence, the observer sees this drop as being red. Similarly, a drop lower in the sky would direct violet light toward the observer and appear to be violet. (The red light from this drop would strike the ground

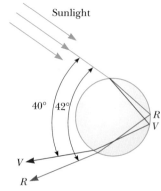

FIGURE 22.19   Refraction of sunlight by a spherical raindrop.

Webnote 22.2

See what makes the bow in a rainbow. Go to
http://www.unidata.ucar.edu/staff/blynds/rnbw.html

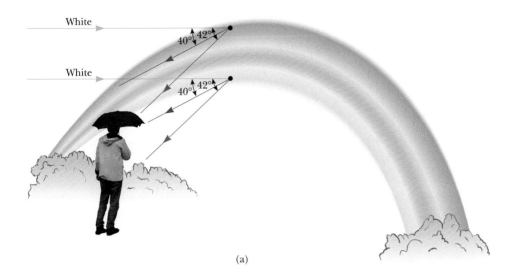

(b)

FIGURE 22.20 (a) The formation of a rainbow. (b) This photograph of a rainbow shows a distinct secondary rainbow with the colors reversed. *(Mark D. Phillips/Photo Researchers, Inc.)*

and not be seen.) The remaining colors of the spectrum would reach the observer from raindrops lying between these two extreme positions. Figure 22.20b shows a beautiful rainbow.

22.7 HUYGENS'S PRINCIPLE

The laws of reflection and refraction can be deduced using a geometric method proposed by Huygens in 1678. Huygens assumed that light is a form of wave motion rather than a stream of particles. He had no knowledge of the nature of light or of its electromagnetic character. Nevertheless, his simplified wave model is adequate for understanding many practical aspects of the propagation of light.

Huygens's principle is a geometric construction for determining at some instant the position of a new wave front from the knowledge of the wave front that preceded it. (A wave front is a surface passing through those points of a wave that have the same phase and amplitude. For instance, a wave front could be a surface passing through the crests of waves.) In Huygens's construction, **all points on a**

FIGURE 22.21 Huygens's constructions for (a) a plane wave propagating to the right and (b) a spherical wave.

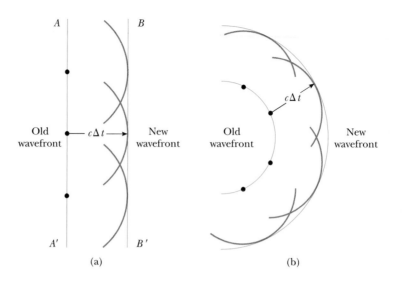

(a)

(b)

Huygens's principle ▶ **given wave front are taken as point sources for the production of spherical secondary waves, called wavelets, which propagate in the forward direction with speeds characteristic of waves in that medium. After some time has elapsed, the new position of the wave front is the surface tangent to the wavelets.**

Figure 22.21 illustrates two simple examples of Huygens's construction. First, consider a plane wave moving through free space, as in Figure 22.21a. At $t = 0$, the wave front is indicated by the plane labeled AA'. In Huygens's construction, each point on this wave front is considered a point source. For clarity, only a few points on AA' are shown. With these points as sources for the wavelets, we draw circles each of radius $c \Delta t$, where c is the speed of light in vacuum and Δt is the period of propagation from one wave front to the next. The surface drawn tangent to these wavelets is the plane BB', which is parallel to AA'. In a similar manner, Figure 22.21b shows Huygens's construction for an outgoing spherical wave.

Figure 22.22 shows a convincing demonstration of Huygens's principle. Plane waves coming from far off shore emerge from the openings between the barriers as two-dimensional circular waves propagating outward.

FIGURE 22.22 This photograph of the beach at Tel Aviv, Israel, shows Huygens wavelets radiating from each opening between breakwalls. Note how the beach has been shaped by the wave action. *(Courtesy of Sabina Zigman/Benjamin Cardozo High School, and by permission of PHYSICS TEACHER, vol. 37, January 1999, p. 55)*

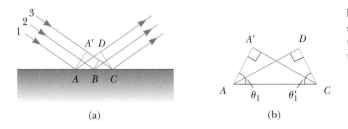

FIGURE 22.23 (a) Huygens's construction for proving the law of reflection. (b) Triangle ADC is identical to triangle $AA'C$.

HUYGENS'S PRINCIPLE APPLIED TO REFLECTION AND REFRACTION

The laws of reflection and refraction were stated earlier in this chapter without proof. We shall now derive these laws using Huygens's principle. For the law of reflection, refer to Figure 22.23a. The line AA' represents a wave front of the incident light. As ray 3 travels from A' to C, ray 1 reflects from A and produces a spherical wavelet of radius AD. (Recall that the radius of a Huygens wavelet is $v\Delta t$.) Because the two wavelets having radii $A'C$ and AD are in the same medium, they have the same speed, v, and thus $AD = A'C$. Meanwhile, the spherical wavelet centered at B has spread only half as far as the one centered at A, because ray 2 strikes the surface later than ray 1.

From Huygens's principle, we find that the reflected wave front is CD, a line tangent to all the outgoing spherical wavelets. The remainder of our analysis depends on geometry, as summarized in Figure 22.23b. Note that the right triangles ADC and $AA'C$ are congruent because they have the same hypotenuse, AC, and because $AD = A'C$. From Figure 22.23b we have

$$\sin \theta_1 = \frac{A'C}{AC} \qquad \text{and} \qquad \sin \theta_1' = \frac{AD}{AC}$$

Thus,

$$\sin \theta = \sin \theta_1'$$

$$\theta_1 = \theta_1'$$

which is the law of reflection.

Now let us use Huygens's principle and Figure 22.24a to derive Snell's law of refraction. Note that in the time interval Δt, ray 1 moves from A to B and ray 2 moves from A' to C. The radius of the outgoing spherical wavelet centered at A is equal to $v_2 \Delta t$. The distance $A'C$ is equal to $v_1 \Delta t$. Geometric considerations show

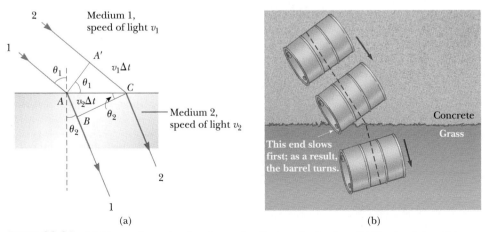

FIGURE 22.24 (a) Huygens's construction for proving the law of refraction. (b) Overhead view of a barrel rolling from concrete onto grass.

that angle $A'AC$ equals θ_1 and angle ACB equals θ_2. From triangles $AA'C$ and ACB we find that

$$\sin \theta_1 = \frac{v_1 \, \Delta t}{AC} \quad \text{and} \quad \sin \theta_2 = \frac{v_2 \, \Delta t}{AC}$$

If we divide these two equations, we get

$$\frac{\sin \theta_1}{\sin \theta_2} = \frac{v_1}{v_2}$$

But from Equation 22.4 we know that $v_1 = c/n_1$ and $v_2 = c/n_2$. Therefore,

$$\frac{\sin \theta_1}{\sin \theta_2} = \frac{c/n_1}{c/n_2} = \frac{n_2}{n_1}$$

$$n_1 \sin \theta_1 = n_2 \sin \theta_2$$

which is the law of refraction.

A mechanical analog of refraction is shown in Figure 22.24b. When the left end of the rolling barrel reaches the grass, it slows down, while the right end remains on the concrete and moves at its original speed. This difference in speeds causes the barrel to pivot, and this changes the direction of travel.

This photograph shows nonparallel light rays entering a glass prism. The bottom two rays undergo total internal reflection at the longest side of the prism. The top three rays are refracted at the longest side as they leave the prism. *(Courtesy of Henry Leap and Jim Lehman)*

22.8   TOTAL INTERNAL REFLECTION

An interesting effect called *total internal reflection* can occur when light attempts to move from a medium with a *high* index of refraction to one with a *lower* index of refraction. Consider a light beam traveling in medium 1 and meeting the boundary between medium 1 and medium 2, where n_1 is greater than n_2 (Fig. 22.25). Possible directions of the beam are indicated by rays 1 through 5. Note that the refracted rays are bent away from the normal because n_1 is greater than n_2. At some particular angle of incidence θ_c, called the **critical angle,** the refracted light ray moves parallel to the boundary so that $\theta_2 = 90°$ (Fig. 22.25b). *For angles of incidence greater than θ_c, the beam is entirely reflected at the boundary, as is ray 5 in Figure 22.25a.* This ray is reflected at the boundary as though it had struck a perfectly reflecting surface. It and all rays like it obey the law of reflection; that is, the angle of incidence equals the angle of reflection.

We can use Snell's law to find the critical angle. When $\theta_1 = \theta_c$, $\theta_2 = 90°$ and Snell's law (Eq. 22.8) gives

$$n_1 \sin \theta_c = n_2 \sin 90° = n_2$$

$$\sin \theta_c = \frac{n_2}{n_1} \quad \text{for } n_1 > n_2 \qquad \text{[22.9]}$$

FIGURE 22.25   (a) Rays from a medium with index of refraction n_2 travel to a medium with index of refraction n_2, where $n_1 > n_2$. As the angle of incidence increases, the angle of refraction θ_2 increases until θ_2 is 90° (ray 4). For even larger angles of incidence, total internal reflection occurs (ray 5). (b) The angle of incidence producing a 90° angle of refraction is often called the *critical angle, θ_c.*

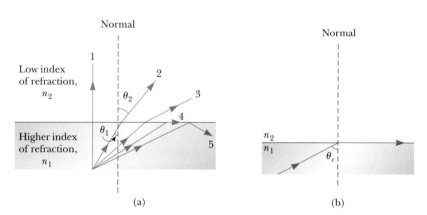

(a)   (b)

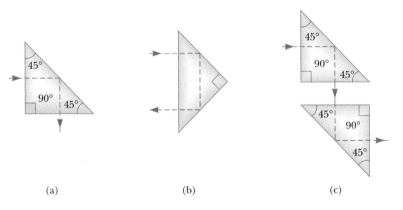

FIGURE 22.26 Internal reflection in a prism. (a) The ray is deviated by 90°. (b) The direction of the ray is reversed. (c) Two prisms used as a periscope.

(a)   (b)   (c)

Note that this equation can be used only when n_1 is greater than n_2. That is, **total internal reflection occurs only when light attempts to move from a medium of high index of refraction to a medium of lower index of refraction.** If n_1 were less than n_2, Equation 22.9 would give sin $\theta_c > 1$, which is an absurd result because the sine of an angle can never be greater than unity.

When medium 2 is air, the critical angle is small for substances with large indices of refraction, such as diamond, where $n = 2.42$ and $\theta_c = 24.0°$. By comparison, for crown glass, $n = 1.52$ and $\theta_c = 41.0°$. This property, combined with proper faceting, causes a diamond to sparkle brilliantly.

One can use a prism and the phenomenon of total internal reflection to alter the direction of travel of a light beam. Figure 22.26 illustrates two such possibilities. In one case the light beam is deflected by 90.0° (Fig. 22.26a), and in the second case the path of the beam is reversed (Fig. 22.26b). A common application of total internal reflection is a submarine periscope. In this device, two prisms are arranged, as in Figure 22.26c so that an incident beam of light follows the path shown and the user can "see around corners."

◀ **APPLICATION**

SUBMARINE PERISCOPES

APPLYING **PHYSICS** **22.4**

A beam of white light is incident on the curved edge of a semicircular piece of glass, as shown in Figure 22.27. The light enters the curved surface along the normal, so it shows no refraction. It encounters the straight side at the center of curvature of the curved side and refracts into the air. The incoming beam is moved clockwise (so that the angle θ increases) such that the beam always enters along the normal to the curved side and encounters the straight side at the center of curvature of the curved side. As the refracted beam approaches a direction parallel to the straight side, it becomes redder. Why?

Explanation When the outgoing beam approaches the direction parallel to the straight side, the incident angle is approaching the critical angle for total internal reflection. Dispersion occurs as the light passes out of the glass. The index of refraction for light at the violet end of the visible spectrum is larger than at the red end. Thus, as the outgoing beam approaches the straight side, the violet light experiences total internal reflection first, followed by the other colors. The red light is the last to experience total internal reflection, so just before the outgoing light disappears, it is composed of light from the red end of the visible spectrum.

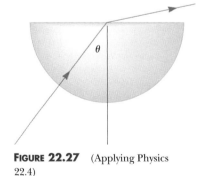

FIGURE 22.27 (Applying Physics 22.4)

Example 22.6   A View from the Fish's Eye

A Find the critical angle for a water-air boundary if the index of refraction of water is 1.33.

Webnote 22.3
See what total internal reflection would look like beneath the surface of a pond. Go to
http://www.lightlink.com/sergey/java/ java/totintrefl/index.html

This finger tip is being illuminated by a bundle of optical fibers. The light source is not shown in the photograph. (© *Novastock/Photo Researchers, Inc.*)

FIBER OPTICS IN MEDICAL DIAGNOSIS AND SURGERY

FIBER OPTICS IN TELECOMMUNICATIONS

FIGURE 22.28 Light travels in a curved transparent rod by multiple internal reflections.

Solution Applying Equation 22.9, we find the critical angle to be

$$\sin \theta_c = \frac{n_2}{n_1} = \frac{1.00}{1.33} = 0.752$$

$$\theta_c = 48.8°$$

B Use the results of (part A) to predict what a fish will see if it looks up toward the water surface at angles of 40.0°, 48.8°, and 60.0°.

Reasoning and Solution Because the path of a light ray is reversible, the fish can see out of the water if it looks toward the surface at an angle less than the critical angle. Thus, at 40.0°, the fish can see into the air above the water. At an angle of 48.8°, the critical angle for water, the light that reaches the fish has to skim along the water surface before being refracted to the fish's eye. At angles greater than the critical angle, the light reaching the fish comes via internal reflection at the surface. Thus, beyond 60.0°, the fish sees a reflection of some object on the bottom of the pool.

FIBER OPTICS

Another interesting application of total internal reflection is the use of solid glass or transparent plastic rods to "pipe" light from one place to another. As indicated in Figure 22.28, light is confined to traveling within the rods, even around gentle curves, as a result of successive internal reflections. Such a light pipe can be quite flexible if thin fibers are used rather than thick rods. If a bundle of parallel fibers is used to construct an optical transmission line, images can be transferred from one point to another.

Very little light intensity is lost in these fibers as a result of reflections on the sides. Any loss of intensity is due essentially to reflections from the two ends and absorption by the fiber material. Fiber-optic devices are particularly useful for viewing images produced at inaccessible locations. Physicians often use fiber-optic cables to aid in the diagnosis and correction of certain medical problems without the intrusion of major surgery. For example, a fiber-optic cable can be threaded through the esophagus and into the stomach to look for ulcers. In this application, the cable actually consists of two fiber-optic lines, one to transmit a beam of light into the stomach for illumination and the other to allow this light to be transmitted out of the stomach. The resulting image can, in some cases, be viewed directly by the physician but most often is displayed on a television monitor or captured on film. In a similar fashion, the cables can be used to examine the colon or to do repair work without the need for large incisions.

Most important, the field of fiber optics has revolutionized the entire communications industry. Billions of kilometers of optical fiber have been installed in the United States to carry high-speed Internet traffic, radio and television signals, and telephone calls. The fibers can carry much higher volumes of telephone calls and other forms of communication than electrical wires because of the higher frequency of the infrared light used to carry the information on optical fibers. Optical fibers are also preferable to copper wires because they are insulators and do not pick up stray electric and magnetic fields or electronic "noise."

APPLYING **PHYSICS** 22.5

An optical fiber consists of a transparent core, surrounded by cladding, which is a material with a lower index of refraction than the core. There is a cone of angles, called the *acceptance cone*, at the entrance to the fiber. Incoming light at angles within this cone will be transmitted through the fiber, whereas light entering the core from angles outside the cone will not be transmitted. Figure 22.29 shows a light ray entering the fiber just within the acceptance cone and experiencing total internal reflection at the interface between the core and the cladding. If it is technologically difficult to produce light entering the fiber from a small range of angles, how could you adjust the indices of refraction of the core and cladding to increase the size of the acceptance cone—would you design them to be farther apart or closer together?

APPLICATION

DESIGN OF AN OPTICAL FIBER

Explanation   The acceptance cone would become larger if the critical angle (θ_c in the diagram) could be made smaller. This can be done by making the index of refraction of the cladding material smaller, so that the indices of refraction of the core and cladding material would be farther apart.

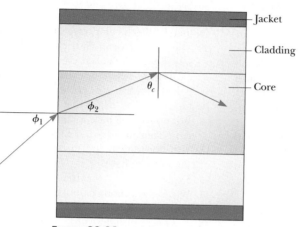

FIGURE 22.29   (Applying Physics 22.5)

SUMMARY

The **index of refraction** of a material, n, is defined as

$$n \equiv \frac{c}{v} \qquad \text{[22.4]}$$

where c is the speed of light in a vacuum and v is the speed of light in the material. The index of refraction of a material is also

$$n = \frac{\lambda_0}{\lambda_n} \qquad \text{[22.7]}$$

where λ_0 is the wavelength of the light in vacuum and λ_n is its wavelength in the material.

The **law of reflection** states that a wave reflects from a surface so that the *angle of reflection* θ_1' equals the *angle of incidence* θ_1.

The **law of refraction,** or **Snell's law,** states that

$$n_1 \sin \theta_1 = n_2 \sin \theta_2 \qquad \text{[22.8]}$$

where n_1 and n_2 are the indices of refraction in the two media. The incident ray, the reflected ray, the refracted ray, and the normal to the surface all lie in the same plane.

Huygens's principle states that all points on a wave front are point sources for the production of spherical secondary waves called wavelets. These wavelets propagate forward at a speed characteristic of waves in a particular medium. After some time has elapsed, the new position of the wave front is the surface tangent to the wavelets.

Total internal reflection can occur when light attempts to move from a material with a high index of refraction to one with a lower index of refraction. The *maximum angle of incidence* θ_c for which light can move from a medium with index n_1 into a medium with index n_2, where n_1 is greater than n_2, is called the **critical angle** and is given by

$$\sin \theta_c = \frac{n_2}{n_1} \qquad \text{for } n_1 > n_2 \qquad \text{[22.9]}$$

CONCEPTUAL QUESTIONS

1. Under certain circumstances, sound can be heard from extremely far away. This frequently happens over a body of water, where the air near the water surface is cooler than the air at higher altitudes. Explain how the refraction of sound waves could increase the distance over which sound can be heard.

2. What are some reasons that most ceilings are made of white textured material?

3. The color of an object is said to depend on wavelength. So, if you view colored objects under water, in which the wavelength of the light will be different, does the color change?

4. How is it possible that a complete circle of a rainbow can sometimes be seen from an airplane?

5. Pass white light through a prism to produce a multicolored spectrum. Let that spectrum fall on a piece of cardboard that has a slit in it such that only green light passes through. Now let this green light fall on a second prism. Describe the light that comes out of the second prism.

6. Why does the arc of a rainbow appear with red on top and violet on the bottom?

7. A scientific supply catalog advertises a material having an index of refraction of 0.85. Is this a good product to buy? Why or why not?

8. Under what conditions is a mirage formed? On a hot day, what are we seeing when we observe a mirage water puddle on the road?

9. In dispersive materials, the angle of refraction for a light ray depends on the wavelength of the light. Does the angle of reflection from the surface of the material depend on the wavelength? Why or why not?

10. A type of mirage called a *pingo* is observed often in Alaska. These occur when the light from a small hill passes to an observer by a path that takes the light over a body of water warmer than the air. What is seen is the hill and an inverted image directly below it. Explain how these mirages are formed.

11. Explain why a diamond loses most of its sparkle when submerged in carbon disulfide.

12. Suppose you are told that only two colors of light (X and Y) are sent through a glass prism and that X is bent more than Y. Which color travels more slowly in the prism?

13. The level of water in a clear, colorless glass is easily observed with the naked eye. The level of liquid helium in a clear glass vessel is extremely difficult to see with the naked eye. Explain. (*Hint:* The index of refraction of liquid helium is close to that of air.)

14. Is it possible to have total internal reflection for light incident from air on water? Explain.

15. Why does a diamond show flashes of color when observed under white light?

16. Explain why an oar partially in water appears to be bent.

17. Why do astronomers looking at distant galaxies talk about looking backward in time?

18. If a beam of light with a given cross section enters a new medium, the cross section of the refracted beam is different from the incident beam. Is it larger or smaller, or is there no definite direction to the change?

PROBLEMS

1, 2, 3 = straightforward, intermediate, challenging ☐ = full solution available in Student Solutions Manual/Study Guide

web = solution posted at **http://info.brookscole.com/serway** 🧬 = biomedical application

Section 22.1 The Nature of Light

1. During the Apollo XI Moon landing, a retroreflecting panel was erected on the Moon's surface. The speed of light may be found by measuring the time it takes a laser beam to travel from Earth, reflect from the panel, and return to Earth. If this interval is measured to be 2.51 s, what is the measured speed of light? Take the center-to-center distance from Earth to the Moon to be 3.84×10^8 m. Assume the Moon is directly overhead and do not neglect the sizes of Earth and the Moon.

2. Figure P22.2 shows the apparatus used by Armand H. L. Fizeau (1819–1896) to measure the speed of light. The basic idea is to measure the total time it takes light to travel from some point to a distant mirror and back. If d is the distance between the light source and the mirror and if the transit time for one round trip is t, then the speed of light is $c = 2d/t$. To measure the transit time, Fizeau used a rotating toothed wheel, which converts an otherwise continuous beam of light to a series of light pulses. The rotation of the wheel controls what an observer at the light source sees. For example, assume that the toothed wheel of the Fizeau experiment has 360 teeth and is rotating at a speed of 27.5 rev/s when the light from the source is extinguished—that is, when a burst of light passing through opening A in Figure P22.2 is blocked by tooth B on return. If the distance to the mirror is 7 500 m, find the speed of light.

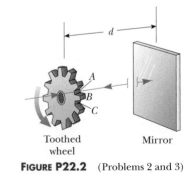

Toothed wheel   Mirror

FIGURE P22.2 (Problems 2 and 3)

3. In an experiment to measure the speed of light using the apparatus of Fizeau described in the preceding problem, the distance between light source and mirror was 11.45 km and the wheel had 720 notches. The experimen-

tally determined value of c was 2.998×10^8 m/s. Calculate the minimum angular speed of the wheel for this experiment.

4. Albert A. Michelson very carefully measured the speed of light using an alternative version of the technique developed by Fizeau. (See Problem 2.) Figure P22.4 shows the approach he used. Light was reflected from one face of a rotating eight-sided mirror toward a stationary mirror 35.0 km away. At certain rates of rotation, the returning beam of light was directed toward the eye of an observer as shown. (a) What minimum angular speed must the rotating mirror have in order that side A will have rotated to position B, causing the light to be reflected to the eye? (b) What is the next-higher angular velocity that will enable the source of light to be seen?

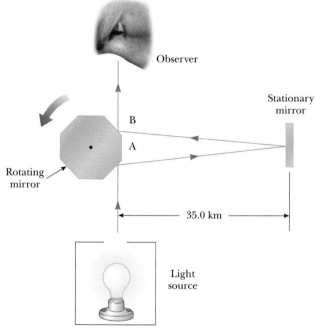

FIGURE P22.4

5. Figure P22.5 shows an apparatus used to measure the speed distribution of gas molecules. It consists of two slotted rotating disks separated by a distance d, with the slots displaced by the angle θ. Suppose the speed of light is measured by sending a light beam toward the left disk of this apparatus. (a) Show that a light beam will be seen in the detector (that is, will make it through both slots) only

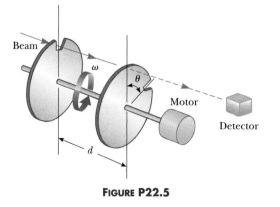

FIGURE P22.5

if its speed is given by $c = \omega d/\theta$, where ω is the angular speed of the disks and θ is measured in radians. (b) What is the measured speed of light if the distance between the two slotted rotating disks is 2.500 m, the slot in the second disk is displaced 1/60 of one degree from the slot in the first disk, and the disks are rotating at 5 555 rev/s?

Section 22.3 Reflection and Refraction

Section 22.4 The Law of Refraction

6. The two mirrors in Figure P22.6 meet at a right angle. The beam of light in the vertical plane P strikes mirror 1 as shown. (a) Determine the distance the reflected light beam travels before striking mirror 2. (b) In what direction does the light beam travel after being reflected from mirror 2?

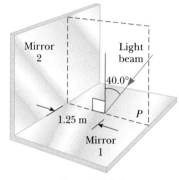

FIGURE P22.6

7. An underwater scuba diver sees the Sun at an apparent angle of 45.0° from the vertical. What is the actual direction of the Sun?

8. Light is incident normally on a 1.00-cm layer of water that lies on top of a flat Lucite plate with a thickness of 0.500 cm. How much more time is required for light to pass through this double layer than is required to traverse the same distance in air ($n_{\text{Lucite}} = 1.59$)?

9. A laser beam is incident at an angle of 30.0° to the vertical onto a solution of corn syrup in water. If the beam is refracted to 19.24° to the vertical, (a) what is the index of refraction of the syrup solution? Suppose the light is red, with vacuum wavelength 632.8 nm. Find its (b) wavelength, (c) frequency, and (d) speed in the solution.

10. Find the speeds of light in (a) flint glass, (b) water, and (c) zircon.

11. Light of wavelength λ_0 in vacuum has a wavelength of 438 nm in water and a wavelength of 390 nm in benzene. (a) What is the wavelength λ_0 of this light in vacuum? (b) Using only the given wavelengths, determine the ratio of the index of refraction of benzene to that of water.

12. Light of wavelength 436 nm in air enters a fishbowl filled with water, then exits through the crown-glass wall of the container. Find the wavelengths of the light (a) in the water and (b) in the glass.

13. A ray of light is incident on the surface of a block of clear ice at an angle of 40.0° with the normal. Part of the light is reflected and part is refracted. Find the angle between the reflected and refracted light.

14. A narrow beam of sodium yellow light ($\lambda_0 = 589$ nm) is incident from air on a smooth surface of water at an angle

of $\theta_1 = 35.0°$. Determine the angle of refraction θ_2 and the wavelength of the light in water.

15. A beam of light, traveling in air, strikes the surface of mineral oil at an angle of 23.1° with the normal to the surface. If the light travels at 2.17×10^8 m/s through the oil, what is the angle of refraction?

16. A flashlight on the bottom of a 4.00-m-deep swimming pool sends a ray upward and at an angle so that the ray strikes the surface of the water 2.00 m from the point directly above the flashlight. What angle (in air) does the emerging ray make with the water's surface?

17. How many times will the incident beam shown in Figure P22.17 be reflected by each of the parallel mirrors?

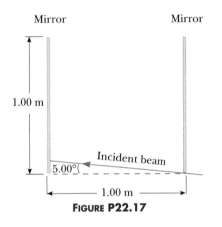

FIGURE P22.17

18. A ray of light strikes a flat, 2.00-cm-thick block of glass ($n = 1.50$) at an angle of 30.0° with the normal (Fig. P22.18). Trace the light beam through the glass and find the angles of incidence and refraction at each surface.

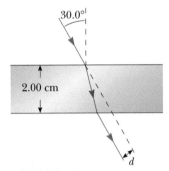

FIGURE P22.18 (Problems 18, 19, and 20)

19. When the light ray in Problem 18 passes through the glass **web** block, it is shifted laterally by a distance d (Fig. P22.18). Find the value of d.

20. Find the time required for the light to pass through the glass block described in Problem 19.

21. The light beam shown in Figure P22.21 makes an angle of 20.0° with the normal line NN' in the linseed oil. Determine the angles θ and θ'. (The refractive index for linseed oil is 1.48.)

22. A submarine is 300 m horizontally out from the shore and 100 m beneath the surface of the water. A laser beam is sent from the sub so that it strikes the surface of the water at a point 210 m from the shore. If the beam just strikes the top of a building standing directly at the water's edge, find the height of the building.

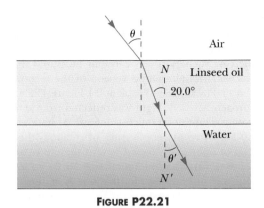

FIGURE P22.21

23. Two light pulses are emitted simultaneously from a source. The pulses take parallel paths to a detector 6.20 m away, but one moves through air and the other through a block of ice. Determine the difference in the pulses' times of arrival at the detector.

24. A narrow beam of ultrasonic waves reflects off the liver tumor in Figure P22.24. If the speed of the wave is 10.0% less in the liver than in the surrounding medium, determine the depth of the tumor.

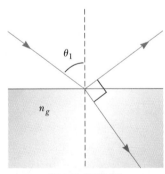

FIGURE P22.24

25. A beam of light both reflects and refracts at the surface between air and glass as shown in Figure P22.25. If the index of refraction of the glass is n_g, find the angle of incidence θ_1 in the air that would result in the reflected ray and the refracted ray being perpendicular to each other. [*Hint:* Remember the identity $\sin(90° - \theta) = \cos\theta$.]

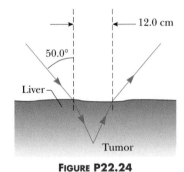

FIGURE P22.25

26. Three sheets of plastic have unknown indices of refraction. Sheet 1 is placed on top of sheet 2, and a laser beam is directed onto the sheets from above so that it strikes the interface at an angle of 26.5° with the normal. The refracted beam in sheet 2 makes an angle of 31.7° with the

normal. The experiment is repeated with sheet 3 on top of sheet 2 and, with the same angle of incidence, the refracted beam makes an angle of 36.7° with the normal. If the experiment is repeated again with sheet 1 on top of sheet 3, what is the expected angle of refraction in sheet 3? Assume the same angle of incidence.

27. An opaque cylindrical tank with an open top has a diameter of 3.00 m and is completely filled with water. When the afternoon Sun reaches an angle of 28.0° above the horizon, sunlight ceases to illuminate the bottom of the tank. How deep is the tank?

28. A cylindrical cistern, constructed below ground level, is 3.0 m in diameter and 2.0 m deep and is filled to the brim with a liquid whose index of refraction is 1.5. A small object rests on the bottom of the cistern at its center. How far from the edge of the cistern can a girl whose eyes are 1.2 m from the ground stand and still see the object?

Section 22.5 Dispersion and Prisms

29. The index of refraction for red light in water is 1.331, and that for blue light is 1.340. If a ray of white light enters the water at an angle of incidence of 83.00°, what are the underwater angles of refraction for the blue and red components of the light?

30. A certain kind of glass has an index of refraction of 1.650 for blue light of wavelength 430 nm and an index of 1.615 for red light of wavelength 680 nm. If a beam containing these two colors is incident at an angle of 30.00° on a piece of this glass, what is the angle between the two beams inside the glass?

31. A ray of light strikes the midpoint of one face of an equiangular (60°-60°-60°) glass prism ($n = 1.5$) at an angle of incidence of 30°. (a) Trace the path of the light ray through the glass, and find the angles of incidence and refraction at each surface. (b) If a small fraction of light is also reflected at each surface, find the angles of reflection at these surfaces.

32. The index of refraction for violet light in silica flint glass is 1.66 and that for red light is 1.62. What is the angular dispersion of visible light passing through an equilateral prism of apex angle 60.0° if the angle of incidence is 50.0°? (See Fig. P22.32.)

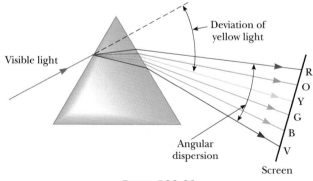

FIGURE P22.32

Section 22.8 Total Internal Reflection

33. Calculate the critical angles for the following materials when surrounded by air: (a) zircon, (b) fluorite, (c) ice. Assume that $\lambda = 589$ nm.

34. For 589-nm light, calculate the critical angle for the following materials surrounded by air: (a) diamond and (b) flint glass.

35. Repeat Problem 34 when the materials are surrounded by water.

36. A beam of light is incident from air on the surface of a liquid. If the angle of incidence is 30.0° and the angle of refraction is 22.0°, find the critical angle for the liquid when surrounded by air.

37. A light pipe consists of a central strand of material surrounded by an outer coating. The interior portion of the pipe has an index of refraction of 1.60. If all rays striking the interior walls of the pipe with incident angles greater than 59.5° are subject to total internal reflection, what is the index of refraction of the coating?

38. Determine the maximum angle θ for which the light rays incident on the end of the light pipe in Figure P22.38 are subject to total internal reflection along the walls of the pipe. Assume that the light pipe has an index of refraction of 1.36 and that the outside medium is air.

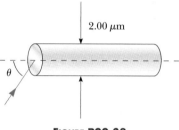

FIGURE P22.38

39. Consider a common mirage formed by super-heated air just above a roadway. A truck driver whose eyes are 2.00 m above the road, where $n = 1.000\ 3$, looks forward. She has the illusion of seeing a patch of water ahead on the road, where her line of sight makes an angle of 1.20° below the horizontal. Find the index of refraction of the air just above the road surface. (*Hint:* Treat this as a problem in total internal reflection.)

40. A jewel thief hides a diamond by placing it on the bottom of a public swimming pool. He places a circular raft on the surface of the water directly above and centered on the diamond, as shown in Figure P22.40. If the surface of the water is calm and the pool is 2.00 m deep, find the minimum diameter of the raft that would prevent the diamond from being seen.

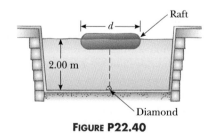

FIGURE P22.40

41. A room contains air in which the speed of sound is 343 m/s. The walls of the room are made of concrete, in which the speed of sound is 1 850 m/s. (a) Find the critical angle for total internal reflection of sound at the

concrete-air boundary. (b) In which medium must the sound be traveling in order to undergo total internal reflection? (c) "A bare concrete wall is a highly efficient mirror for sound." Give evidence for or against this statement.

42. A light ray is incident normally to the long face (the hypotenuse) of a 45°-45°-90° prism surrounded by air, as shown in Figure 22.26b. Calculate the minimum index of refraction of the prism for which the ray will follow the path shown.

43. The light beam in Figure P22.43 strikes surface 2 at the critical angle. Determine the angle of incidence θ_i.

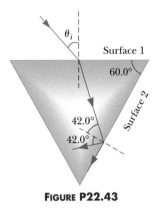

FIGURE P22.43

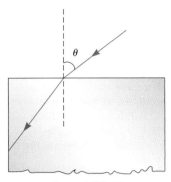

FIGURE P22.46

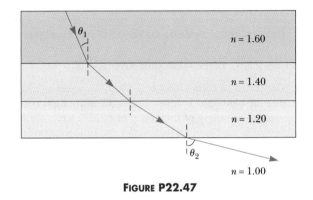

FIGURE P22.47

ADDITIONAL PROBLEMS

44. (a) Consider a horizontal interface between air above and glass with an index of 1.55 below. Draw a light ray incident from the air at an angle of incidence of 30.0°. Determine the angles of the reflected and refracted rays and show them on the diagram. (b) Suppose instead that the light ray is incident from the glass at an angle of incidence of 30.0°. Determine the angles of the reflected and refracted rays and show all three rays on a new diagram. (c) For rays incident from the air onto the air-glass surface, determine and tabulate the angles of reflection and refraction for all the angles of incidence at 10.0° intervals from 0° to 90.0°. (d) Do the same for light rays traveling up to the interface through the glass.

45. A layer of ice, having parallel sides, floats on water. If light **web** is incident on the upper surface of the ice at an angle of incidence of 30.0°, what is the angle of refraction in the water?

46. A light ray of wavelength 589 nm is incident at an angle θ on the top surface of a block of polystyrene surrounded by air, as shown in Figure P22.46. (a) Find the maximum value of θ for which the refracted ray will undergo total internal reflection at the left vertical face of the block. (b) Repeat the calculation for the case in which the polystyrene block is immersed in water. (c) What happens if the block is immersed in carbon disulfide?

47. Figure P22.47 shows the path of a beam of light through several layers of different indices of refraction. (a) If $\theta_1 = 30.0°$, what is the angle θ_2 of the emerging beam? (b) What must the incident angle θ_1 be in order to have total internal reflection at the surface between the $n = 1.20$ medium and the $n = 1.00$ medium?

48. The walls of a prison cell are perpendicular to the four cardinal compass directions. On the first day of spring, light from the rising Sun enters a rectangular window in the eastern wall. The light traverses 2.37 m horizontally to shine perpendicularly on the wall opposite the window. A prisoner observes the patch of light moving across this western wall and for the first time forms his own understanding of the rotation of the Earth. (a) With what speed does the illuminated rectangle move? (b) The prisoner holds a small square mirror flat against the wall at one corner of the rectangle of light. The mirror reflects light back to a spot on the eastern wall close beside the window. How fast does the smaller square of light move across that wall? (c) Seen from a latitude of 40.0° north, the rising Sun moves through the sky along a line making a 50.0° angle with the southeastern horizon. In what direction does the rectangular patch of light on the western wall of the prisoner's cell move? (d) In what direction does the smaller square of light on the eastern wall move?

49. As shown in Figure P22.49, a light ray is incident normally on one face of a 30°-60°-90° block of dense flint glass (a prism) that is immersed in water. (a) Determine the exit angle θ_4 of the ray. (b) A substance is dissolved in the water to increase the index of refraction. At what value of n_2 does total internal reflection cease at point P?

50. A narrow beam of light is incident from air onto a glass surface with index of refraction 1.56. Find the angle of incidence for which the corresponding angle of refraction is one half the angle of incidence. (*Hint:* You might want to use the trigonometric identity $\sin 2\theta = 2 \sin \theta \cos \theta$.)

51. One technique to measure the angle of a prism is shown in Figure P22.51. A parallel beam of light is directed on

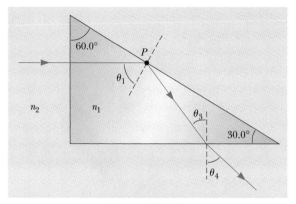

FIGURE P22.49

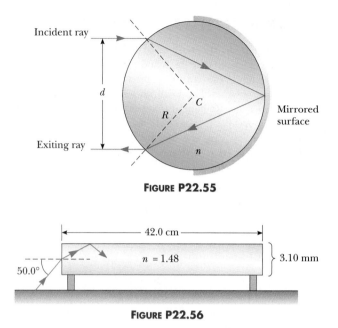

FIGURE P22.55

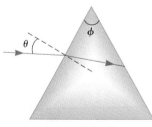

FIGURE P22.56

the apex of the prism so that the beam reflects from opposite faces of the prism. Show that the angular separation of the two reflected beams is given by $B = 2A$.

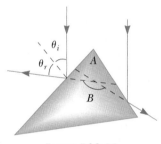

FIGURE P22.51

52. A 4.00-m-long pole stands vertically in a lake having a depth of 2.00 m. When the Sun is 40.0° above the horizontal, determine the length of the pole's shadow on the bottom of the lake. Take the index of refraction for water to be 1.33.

53. A piece of wire is bent through an angle θ. The bent wire is partially submerged in benzene (index of refraction = 1.50) so that to a person looking along the dry part, the wire appears to be straight and makes an angle of 30.0° with the horizontal. Determine the value of θ.

54. When you look through a window, by how much time is the light you see delayed by having to go through glass instead of air? Make an order-of-magnitude estimate on the basis of data you specify. By how many wavelengths is it delayed?

55. A transparent cylinder of radius $R = 2.00$ m has a mirrored surface on its right half, as shown in Figure P22.55. A light ray traveling in air is incident on the left side of the cylinder. The incident light ray and the exiting light ray are parallel and $d = 2.00$ m. Determine the index of refraction of the material.

56. A laser beam strikes one end of a slab of material, as in Figure P22.56. The index of refraction of the slab is 1.48. Determine the number of internal reflections of the beam before it emerges from the opposite end of the slab.

57. For this problem, refer to Figure 22.15. For various angles of incidence, it can be shown that the deviation angle δ is a minimum when the ray passes through the glass so that the interior ray is parallel to the base of the prism. A measurement of this minimum angle of deviation enables one to find the index of refraction of the prism material. Show that n is given by the expression

$$n = \frac{\sin[\frac{1}{2}(A + \delta_{min})]}{\sin\left(\frac{A}{2}\right)}$$

where A is the apex angle of the prism.

58. A hiker stands on a mountain peak near sunset and observes a rainbow caused by water droplets in the air about 8.00 km away. The valley is 2.00 km below the mountain peak and entirely flat. What fraction of the complete circular arc of the rainbow is visible to the hiker?

59. A light ray is incident on a prism and refracted at the first surface, as shown in Figure P22.59. Let ϕ represent the apex angle of the prism and n its index of refraction. Find, in terms of n and ϕ, the smallest allowed value of the angle of incidence at the first surface for which the refracted ray will not undergo internal reflection at the second surface.

FIGURE P22.59

GROUP ACTIVITIES

G.1 Tape a coin to the bottom of a large opaque bowl, as shown in Figure GA22.1a. Stand over the bowl so that you are looking at the coin and then move backwards away from the bowl until you can no longer see the coin over the bowl's rim. Remain at that position and have your co-investigator fill the bowl with water, as shown in Figure GA22.1b. You can now see the coin again because the light is refracted at the water-air interface.

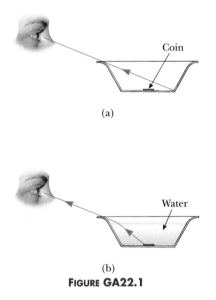

(a)

(b)

FIGURE GA22.1

G.2 Tape a piece of black paper to the end of a flashlight and cut a narrow slit in the middle of the paper, as shown in Figure GA22.2. Lean a flat mirror against one end of a tray partially filled with water. Shine your flashlight on that part of the mirror that is under water, and hold a sheet of white paper such that the reflected light shines on the paper. You should observe a spectrum of colors on the paper as the light is dispersed when it travels from air into water, and then from water into air. According to your observations, which color is bent the most? Which is bent the least?

G.3 Create an artificial rainbow by standing with your back to the Sun and spraying water into the air with a hose. You should cover the end of the hose slightly with a finger or use a nozzle so that the water is broken up into tiny

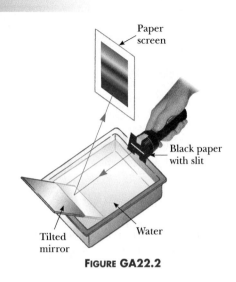

Paper screen

Black paper with slit

Tilted mirror

Water

FIGURE GA22.2

droplets. You should be able to form a rainbow in this way. Which color is on the outside of the arc? The inside?

If the droplets are close to you in the experiment above, you may be able to see two rainbows, one for each eye. Close one eye and only one bow is seen.

With the Sun high in the sky, stand on a ladder and spray the hose toward the ground. In this case, you should be able to form a complete circle rainbow.

G.4 You can demonstrate that light rays travel in straight-line paths by observing the shadows formed by large and small light sources. Figure GA22.4 shows the shadows formed on a screen by a baseball when the light from a lightbulb (a large source) falls on it. Use the figure to explain why the shadow is dark at location A and less dark at B. Replace the light source with a small source such as a high-intensity lamp with a small clear bulb. (A substitute for a high-intensity lamp is an ordinary lightbulb with a piece of cardboard close in front of it with a hole about the size of a penny punched in the cardboard. The hole serves as the small source of light.) What kind of shadow is formed in this case? Why?

G.5 A ray of light is moving from a material having a high index of refraction into a material with a lower index of refraction. (a) Is the ray bent toward the normal or away from it? (b) If the wavelength is 600 nm in the high-index-of-refraction material, is it greater, smaller, or the

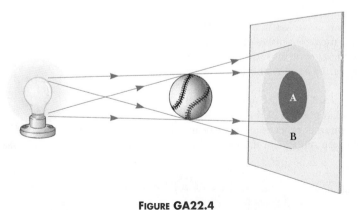

FIGURE GA22.4

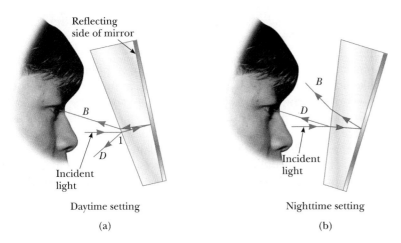

FIGURE 23.5 Cross-sectional views of a rearview mirror. (a) With the day setting, the silvered back surface of the mirror reflects a bright ray *B* into the driver's eyes. (b) With the night setting, the glass of the unsilvered front surface of the mirror reflects a dim ray *D* into the driver's eyes.

behind the car strikes the mirror at point 1. Most of the light enters the wedge, is refracted, and reflects from the back of the mirror to return to the front surface, where it is refracted again as it re-enters the air as ray *B* (for *bright*). In addition, a small portion of the light is reflected at the front surface, as indicated by ray *D* (for *dim*). This dim reflected light is responsible for the image observed when the mirror is in the night setting, as in Figure 23.5b. In this case, the wedge is rotated so that the path followed by the bright light (ray *B*) does not lead to the eye. Instead, the dim light reflected from the front surface travels to the eye, and the brightness of trailing headlights does not become a hazard.

APPLYING PHYSICS 23.1

The professor in the box shown in Figure 23.6 appears to be balancing himself on a few fingers with both of his feet elevated from the floor. The professor can maintain this position for a long time, and he appears to defy gravity. How do you suppose this illusion was created?

Explanation This is one example of an optical illusion, used by magicians, that makes use of a mirror. The box that the professor is standing in is a cubical frame that contains a flat vertical mirror through a diagonal plane. The professor straddles the mirror so that the foot you see is in front of the mirror and the other foot is behind the mirror where you cannot see it. When he raises the foot that you see in front of the mirror, the reflection of this foot also rises, so he appears to float in air.

FIGURE 23.6 (Applying Physics 23.1) *(Courtesy of Henry Leap and Jim Lehman)*

23.2 IMAGES FORMED BY SPHERICAL MIRRORS

CONCAVE MIRRORS

A **spherical mirror,** as its name implies, has the shape of a segment of a sphere. Figure 23.7 shows a spherical mirror with light reflecting from its silvered inner, concave surface; this is called a **concave mirror.** The mirror has radius of curvature *R*, and its center of curvature is at point *C*. Point *V* is the center of the spherical segment, and a line drawn from *C* to *V* is called the **principal axis** of the mirror.

Now consider a point source of light placed at point *O* in Figure 23.7, on the principal axis and outside point *C*. Several diverging rays originating at *O* are shown. After reflecting from the mirror, these rays converge to meet at *I*, called the **image point.** The rays then continue and diverge from *I* as if there were an

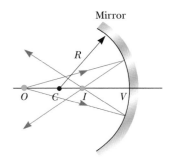

FIGURE 23.7 A point object placed at *O*, outside the center of curvature of a concave spherical mirror, forms a real image at *I* as shown. If the rays diverge from *O* at small angles, they all reflect through the same image point.

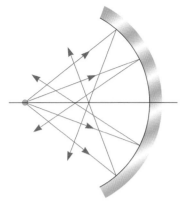

FIGURE 23.8   Rays at large angles from the horizontal axis reflect from a spherical concave mirror to intersect the principal axis at different points, resulting in a blurred image. This is called *spherical aberration.*

object there. As a result, a real image is formed. **Whenever reflected light actually passes through a point, the image formed there is real.**

We often assume that all rays that diverge from the object make small angles with the principal axis. All such rays reflect through the image point, as in Figure 23.7. Rays that make a large angle with the principal axis, as in Figure 23.8, converge to other points on the principal axis, producing a blurred image. This effect, called **spherical aberration,** is present to some extent with any spherical mirror and will be discussed in Section 23.7.

We can use the geometry shown in Figure 23.9 to calculate the image distance q from the object distance p and radius of curvature R. By convention, these distances are measured from point V. Figure 23.9 shows two rays of light leaving the tip of the object. One ray passes through the center of curvature C of the mirror, hitting the mirror head on (perpendicularly to the mirror surface) and reflecting back on itself. The second ray strikes the mirror at point V and reflects as shown, obeying the law of reflection. The image of the tip of the arrow is at the point at which the two rays intersect. From the largest triangle in Figure 23.9, we see that $\tan \theta = h/p$; the light blue triangle gives $\tan \theta = -h'/q$. The negative sign has been introduced to satisfy our convention that h' is negative when the image is inverted with respect to the object, as it is here. Thus, from Equation 23.1 and these results, we find that the magnification of the mirror is

$$M = \frac{h'}{h} = -\frac{q}{p}$$ [23.2]

We also note, from two other triangles in the figure, that

$$\tan \alpha = \frac{h}{p - R} \quad \text{and} \quad \tan \alpha = -\frac{h'}{R - q}$$

from which we find that

$$\frac{h'}{h} = -\frac{R - q}{p - R}$$ [23.3]

If we compare Equation 23.2 to Equation 23.3, we see that

$$\frac{R - q}{p - R} = \frac{q}{p}$$

Simple algebra reduces this to

Mirror equation ▶

$$\frac{1}{p} + \frac{1}{q} = \frac{2}{R}$$ [23.4]

This expression is called the **mirror equation.**

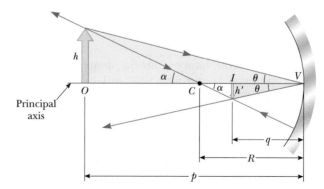

FIGURE 23.9   The image formed by a spherical concave mirror, where the object at O lies outside the center of curvature, C.

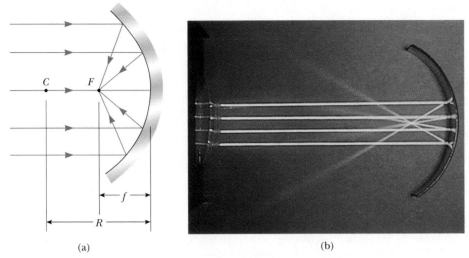

(a)   (b)

FIGURE 23.10   (a) Light rays from a distant object ($p = \infty$) reflect from a concave mirror through the focal point F. In this case, the image distance $q = R/2 = f$, where f is the focal length of the mirror. (b) A photograph of the reflection of parallel rays from a concave mirror. *(Courtesy of Jim Lehman, James Madison University)*

If the object is very far from the mirror—that is, if the object distance p is great enough compared with R that p can be said to approach infinity—then $1/p \approx 0$, and we see from Equation 23.4 that $q \approx R/2$. In other words, when the object is very far from the mirror, **the image point is halfway between the center of curvature and the center of the mirror,** as in Figure 23.10a. The incoming rays are essentially parallel in this figure because the source is assumed to be very far from the mirror. In this special case we call the image point the **focal point,** F, and the image distance the **focal length,** f, where

$$ f = \frac{R}{2} \qquad\qquad \text{[23.5]} $$

◀ Focal length

The mirror equation can therefore be expressed in terms of the focal length:

$$ \frac{1}{p} + \frac{1}{q} = \frac{1}{f} \qquad\qquad \text{[23.6]} $$

Note that rays from objects at infinity are always focused at the focal point.

Tip 23.2   FOCAL POINT $\neq$ FOCUS POINT

The focal point *is not* the point at which light rays focus to form an image. The focal point of a mirror is determined *solely* by its curvature—it does not depend on the location of any object.

23.3 CONVEX MIRRORS AND SIGN CONVENTIONS

Figure 23.11 shows the formation of an image by a **convex mirror,** which is silvered so that light is reflected from the outer, convex surface. This is sometimes called a **diverging mirror** because the rays from any point on the object diverge after reflection as though they were coming from some point behind the mirror. The image in Figure 23.11 is virtual rather than real because it lies behind the mirror at the point at which the reflected rays appear to originate. In general, as shown in the figure, the image formed by a convex mirror is upright, virtual, and smaller than the object.

We shall not derive any equations for convex spherical mirrors. If we did, we would find that the equations developed for concave mirrors can be used with convex mirrors if particular sign conventions are used. Let us call the region in which light rays move the *front side* of the mirror, and the other side, where virtual images are formed, the *back side.* For example, in Figures 23.9 and 23.11, the side to the left of the mirror is the front side, and the side to the right is the back side.

FIGURE 23.11 Formation of an image by a spherical convex mirror. Note that the image is virtual and upright.

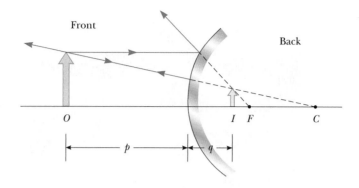

Front, or real, side | Back, or virtual, side
p and q positive | p and q negative

Incident light →

← Reflected light | No light

Convex or concave mirror

FIGURE 23.12 A diagram describing the signs of p and q for convex and concave mirrors.

Figure 23.12 is helpful for understanding the rules for object and image distances, and Table 23.1 summarizes the sign conventions for all the necessary quantities.

RAY DIAGRAMS FOR MIRRORS

We can determine conveniently the positions and sizes of images formed by mirrors by constructing *ray diagrams* similar to the ones we have been using. This kind of graphical construction tells us the overall nature of the image and can be used to check parameters calculated from the mirror and magnification equations. To make a ray diagram, one needs to know the position of the object and the location of the center of curvature. To locate the image, three rays are constructed (rather than just the two we have been constructing so far), as shown by the examples in Figure 23.13. All three rays start from the same object point; for these examples the tip of the arrow was chosen. For the concave mirrors in Figure 23.13a and b, the rays are drawn as follows:

1. **Ray 1 is drawn parallel to the principal axis and is reflected back through the focal point F.**
2. **Ray 2 is drawn through the focal point. Thus, it is reflected parallel to the principal axis.**
3. **Ray 3 is drawn through the center of curvature C and is reflected back on itself.**

Note that rays actually go in all directions from the object; we choose to follow those moving in a direction that simplifies our drawing.

The intersection of any *two* of these rays at a point locates the image. The third ray serves as a check of construction. The image point obtained in this fashion must always agree with the value of q calculated from the mirror formula.

In the case of a concave mirror, note what happens as the object is moved closer to the mirror. The real, inverted image in Figure 23.13a moves to the left as the object approaches the focal point. When the object is at the focal point, the

TABLE 23.1	Sign Conventions for Mirrors	
Quantity	**Positive When**	**Negative When**
Object location (p)	Object is in front of mirror	Object is behind mirror
Image location (q)	Image is in front of mirror	Image is behind mirror
Image height (h')	Image is upright	Image is inverted
Focal length (f) and radius (R)	Mirror is concave	Mirror is convex
Magnification (M)	Image is upright	Image is inverted

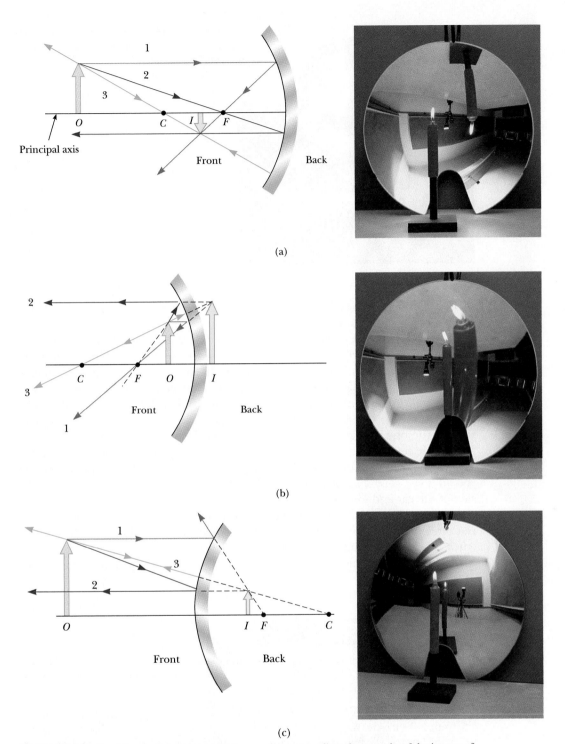

FIGURE 23.13 Ray diagrams for spherical mirrors, and corresponding photographs of the images of candles. (a) When an object is outside the center of curvature of a concave mirror, the image is real, inverted, and reduced in size. (b) When an object is between a concave mirror and the focal point, the image is virtual, upright, and magnified. (c) When an object is in front of a convex mirror, the image is virtual, upright, and reduced in size. *(Photos courtesy of David Rogers)*

image is infinitely far to the left. However, when the object lies between the focal point and the mirror surface, as in Figure 23.13b, the image is virtual and upright.

With the convex mirror shown in Figure 23.13c, the image of a real object is always virtual and upright. As the object distance increases, the virtual image

FIGURE 23.14   A convex sideview mirror on a vehicle produces an upright image that is smaller than the object. *(© Junebug Clark 1988/Photo Researchers, Inc.)*

shrinks and approaches the focal point as p approaches infinity. You should construct a ray diagram to verify this.

The image-forming characteristics of curved mirrors obviously determine their uses. For example, suppose you want to design a mirror that will help people shave or apply cosmetics, such as the one in Figure 23.13b. That is, you need a concave mirror that puts the user inside the focal point. In such a situation, the image is upright and greatly enlarged. In contrast, suppose that the primary purpose of a mirror is to observe a large field of view, in which case you need a convex mirror such as the one in Figure 23.13c. The diminished size of the image means that a fairly large field of view is seen in the mirror. Mirrors such as this are often placed in stores to help employees watch for shoplifting. A second use is as a sideview mirror on a car (Fig. 23.14). This kind of mirror is usually placed on the passenger side of the car and carries the warning "Objects are closer than they appear." Without this warning, a driver might think she is looking into a flat mirror, which does not alter the size of the image. Thus, she could be fooled into believing that a truck is far away because it looks small, when it is actually a large semi very close behind her but diminished in size because of the image formation characteristics of the convex mirror.

APPLYING **PHYSICS** *23.2*

For a concave mirror, a virtual image can be anywhere behind the mirror. For a convex mirror, however, there is a maximum distance at which the image can exist behind the mirror. Why?

Explanation   Let us consider the concave mirror first and imagine two different light rays leaving a tiny object and striking the mirror. If the object is at the focal point, the light rays reflecting from the mirror will be parallel to the mirror axis. They can be interpreted as forming a virtual image infinitely far away behind the mirror. As the object is brought closer to the mirror, the reflected rays will diverge through larger and larger angles, resulting in their extensions converging closer and closer to the back of the mirror. When the object is brought right up to the mirror, the image is right behind the mirror. When the object is much closer to the mirror than the focal length, the mirror acts like a flat mirror, and the image is just as far behind the mirror as the object is in front of it. Thus, the image can be anywhere from infinitely far away to right at the surface of the mirror. For the convex mirror, an object at infinity produces a virtual image at the focal point. As the object is brought closer, the reflected rays diverge more sharply and the image moves closer to the mirror. Thus, the virtual image is restricted to the region between the mirror and the focal point.

APPLYING **PHYSICS** 23.3

Large trucks often have a sign on the back saying, "If you can't see my mirror, I can't see you." Explain this sign.

Explanation   The trucking companies are making use of the principle of reversibility of light rays. In order for an image of you to be formed in the driver's mirror, there must be a pathway for rays of light to reach the mirror, allowing the driver to see your image. If you can't see the mirror, obviously there is no such pathway.

Example 23.2   Images Formed by a Concave Mirror

Assume that a certain concave spherical mirror has a focal length of 10.0 cm. Locate the images for object distances of (a) 25.0 cm, (b) 10.0 cm, and (c) 5.00 cm. Describe the image in each case.

Solution

A For an object distance of 25.0 cm, we find the image distance using the mirror equation:

$$\frac{1}{p} + \frac{1}{q} = \frac{1}{f}$$

$$\frac{1}{25.0 \text{ cm}} + \frac{1}{q} = \frac{1}{10.0 \text{ cm}}$$

$$q = \boxed{16.7 \text{ cm}}$$

The magnification is given by Equation 23.2:

$$M = -\frac{q}{p} = -\frac{16.7 \text{ cm}}{25.0 \text{ cm}} = -0.667$$

Thus, the image is smaller than the object. Furthermore, the image is inverted because M is negative. Finally, because q is positive, the image is on the front side of the mirror and is real. This situation is pictured in Figure 23.13a.

B When the object distance is 10.0 cm, the object is at the focal point. Substituting the values $p = 10.0$ cm and $f = 10.0$ cm into the mirror equation, we find that

$$\frac{1}{10.0 \text{ cm}} + \frac{1}{q} = \frac{1}{10.0 \text{ cm}}$$

$$q = \boxed{\infty}$$

Thus, we see that rays of light originating from an object at the focal point of a concave mirror are reflected so that the image is formed an infinite distance from the mirror—that is, the rays travel parallel to one another after reflection. Furthermore, because the image distance is infinite, the magnification M is infinite. In other words, light diverging from a point of zero height forms an image of height $2R$ and so M is infinite.

C When the object is at 5.00 cm, inside the focal point of the mirror, the mirror equation gives

$$\frac{1}{5.00 \text{ cm}} + \frac{1}{q} = \frac{1}{10.0 \text{ cm}}$$

$$q = \boxed{-10.0 \text{ cm}}$$

That is, the image is virtual because it is behind the mirror. The magnification is

$$M = -\frac{q}{p} = -\left(\frac{-10.0 \text{ cm}}{5.00 \text{ cm}}\right) = 2.00$$

We see that the image height is magnified by a factor of 2, and the positive sign indicates that the image is upright (Fig. 23.13b).

Note the characteristics of an image formed by a concave spherical mirror. When the object is outside the focal point, the image is inverted and real; at the focal point, the image is formed at infinity; inside the focal point, the image is upright and virtual.

EXERCISE If the object distance is 20.0 cm, find the image distance and the magnification of the mirror.

ANSWER $q = 20.0$ cm, $M = -1.00$

Example 23.3 Images Formed by a Convex Mirror

An object 3.00 cm high is placed 20.0 cm from a convex mirror with a focal length of 8.00 cm. Find (a) the position of the final image and (b) the magnification of the mirror.

Solution

A Because the mirror is convex, its focal length is negative. To find the image position, we use the mirror equation:

$$\frac{1}{p} + \frac{1}{q} = \frac{1}{f}$$

$$\frac{1}{20.0 \text{ cm}} + \frac{1}{q} = \frac{1}{-8.00 \text{ cm}}$$

$$q = \boxed{-5.71 \text{ cm}}$$

The negative value of q indicates that the image is virtual, or behind the mirror, as in Figure 23.13c.

B The magnification of the mirror is

$$M = -\frac{q}{p} = -\left(\frac{-5.71 \text{ cm}}{20.0 \text{ cm}}\right) = \boxed{0.286}$$

The image is upright because M is positive.

EXERCISE Find the height of the image.

ANSWER 0.857 cm

Webnote 23.1

To see how objects are formed into virtual images for a convex (diverging) mirror, go to
http://www.lightlink.com/sergey/java/ java/dmirr/index.html

Example 23.4 An Enlarged Image

When a woman stands with her face 40.0 cm from a cosmetic mirror, the upright image is twice as tall as her face. What is the focal length of the mirror?

Reasoning Most of the problems we have encountered so far have been simple applications of the mirror equation. However, to find f in this example, we must first find q, the image distance. Because the problem states that the image is upright, the magnification must be positive (in this case, $M = +2$), and because $M = -q/p$, we can determine q.

Solution The magnification equation gives us a relationship between the object and image distances:

$$M = -\frac{q}{p} = 2$$

$$q = -2p = -2(40.0 \text{ cm}) = -80.0 \text{ cm}$$

First, note that a virtual image is formed because the woman is able to see her upright image in the mirror. This explains why the image distance is negative. Substitute $q = -80.0$ cm into the mirror equation to obtain

$$\frac{1}{40.0 \text{ cm}} - \frac{1}{80.0 \text{ cm}} = \frac{1}{f}$$

$$f = \boxed{80.0 \text{ cm}}$$

The positive sign for the focal length indicates that the mirror is concave, a fact that we already knew because the mirror magnified the object. (A convex mirror would have produced a diminished image.)

23.4 IMAGES FORMED BY REFRACTION

In this section we describe how images are formed by refraction at a spherical surface. Consider two transparent media with indices of refraction n_1 and n_2, where the boundary between the two media is a spherical surface of radius R (Fig. 23.15). Let us assume that the medium to the right has a higher index of refraction than the one to the left; that is, $n_2 > n_1$. This would be the case for light entering a curved piece of glass from air or for light entering the water in a fishbowl from air. The rays originating at the object location O are refracted at the spherical surface and then converge to the image point I. We can begin with Snell's law of refraction and use simple geometric techniques to show that the object distance, image distance, and radius of curvature are related by the equation

$$\frac{n_1}{p} + \frac{n_2}{q} = \frac{n_2 - n_1}{R} \qquad \text{[23.7]}$$

Furthermore, the magnification of a refracting surface is

$$M = \frac{h'}{h} = -\frac{n_1 q}{n_2 p} \qquad \text{[23.8]}$$

As with mirrors, we must use a sign convention if we are to apply these equations to a variety of circumstances. First note that real images are formed on the side of the surface *opposite* the side from which the light comes. This is in contrast with mirrors, where real images are formed on the *same* side of the reflecting surface. We define the side of the surface in which light rays originate as the front side. The other side is called the back side. Because of the difference in location of real images, the refraction sign conventions for q and R are the opposite of those for reflection. For example, p, q, and R are all positive in Figure 23.15. The sign convention for spherical refracting surfaces is summarized in Table 23.2.

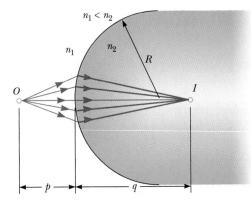

FIGURE 23.15   An image formed by refraction at a spherical surface. Rays making small angles with the principal axis diverge from a point object at O and pass through the image point I.

TABLE 23.2	Sign Conventions for Refracting Surfaces	
Quantity	Positive When	Negative When
Object location (p)	Object is in front of surface	Object is in back of surface
Image location (q)	Image is in back of surface	Image is in front of surface
Image height (h')	Image is upright	Image is inverted
Radius (R)	Center of curvature is in back of surface	Center of curvature is in front of surface

APPLYING **PHYSICS** 23.4

APPLICATION

OPENING YOUR EYES
UNDERWATER

Why does a person with normal vision see a blurry image if the eyes are opened underwater without goggles or a diving mask?

Explanation   The eye presents a spherical refraction surface. The eye normally functions so that light entering from the air is refracted to form an image in the retina located at the back of the eyeball. The difference in index of refraction between water and the eye is smaller than the difference in index of refraction between air and the eye. Thus, light entering the eye from the water does not experience as much refraction as does light entering from the air, and the image is formed behind the retina. A diving mask or swimming goggles have no optical action of their own; they are simply flat pieces of glass or plastic in a rubber mount. However, they provide a region of air adjacent to the eyes, so that the correct refraction relationship is established, and images will be in focus.

FLAT REFRACTING SURFACES

If the refracting surface is flat, then R approaches infinity and Equation 23.7 reduces to

$$\frac{n_1}{p} = -\frac{n_2}{q}$$

$$q = -\frac{n_2}{n_1}p \qquad [23.9]$$

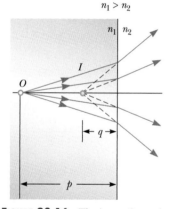

$n_1 > n_2$

FIGURE 23.16   The image formed by a flat refracting surface is virtual; that is, it forms to the left of the refracting surface. Note that if the light rays are reversed in direction, we have the situation described in Example 22.6.

From Equation 23.9 we see that the sign of q is opposite that of p. Thus, **the image formed by a flat refracting surface is on the same side of the surface as the object.** This is illustrated in Figure 23.16 for the situation in which n_1 is greater than n_2, where a virtual image is formed between the object and the surface. Note that the refracted ray bends *away* from the normal in this case, because $n_1 > n_2$.

Quick Quiz 23.2   A person spear fishing from a boat sees a fish located 3 m from the boat at an apparent depth of 1 m. To spear the fish, should the person aim (a) at, (b) above, or (c) below the image of the fish?

Quick Quiz 23.3   True or false? (a) The image of an object placed in front of a concave mirror is always upright. (b) The height of the image of an object placed in front of a concave mirror must be smaller than or equal to the height of the object. (c) The image of an object placed in front of a convex mirror is always upright and smaller than the object.

Example 23.5 Gaze into the Crystal Ball

A coin 2.00 cm in diameter is embedded in a solid glass ball of radius 30.0 cm (Fig. 23.17). The index of refraction of the ball is 1.50, and the coin is 20.0 cm from the surface. Find the position and height of the image of the coin.

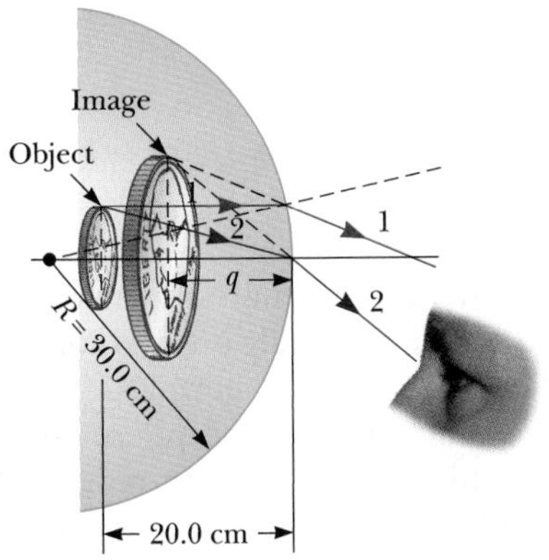

FIGURE 23.17 (Example 23.5) A coin embedded in a glass ball forms a virtual image between the coin and the glass surface.

Image
Object
$R = 30.0$ cm
q
1
2
2
20.0 cm

Solution Because they are moving from a medium of high index of refraction to a medium of lower index of refraction, the rays originating at the coin are refracted away from the normal at the surface and diverge outward. The image is formed in the glass and is virtual. Applying Equation 23.7 and taking $n_1 = 1.50$, $n_2 = 1.00$, $p = 20.0$ cm, and $R = -30.0$ cm, we get

$$\frac{n_1}{p} + \frac{n_2}{q} = \frac{n_2 - n_1}{R}$$

$$\frac{1.50}{20.0 \text{ cm}} + \frac{1.00}{q} = \frac{1.00 - 1.50}{-30.0 \text{ cm}}$$

$$q = \boxed{-17.1 \text{ cm}}$$

The negative sign indicates that the image is in the same medium as the object (the side of incident light), in agreement with our ray diagram, and therefore must be virtual.

To find the image height, we first use Equation 23.8 for the magnification:

$$M = -\frac{n_1 q}{n_2 p} = -\frac{1.50(-17.1 \text{ cm})}{1.00(20.0 \text{ cm})} = \frac{h'}{h}$$

Therefore,

$$h' = 1.28h = (1.28)(2.00 \text{ cm}) = \boxed{2.56 \text{ cm}}$$

The positive value for M indicates an upright image.

Example 23.6 The One That Got Away

A small fish is swimming at a depth of d below the surface of a pond (Fig. 23.18). What is the *apparent depth* of the fish as viewed from directly overhead?

Reasoning In this example, the refracting surface is flat, and so R is infinite. Hence, we can use Equation 23.9 to determine the location of the image, which is the apparent location of the fish.

Solution The facts that $n_1 = 1.33$ for water and $p = d$ give us

$$q = -\frac{n_2}{n_1} p = -\frac{1}{1.33} d = \boxed{-0.752d}$$

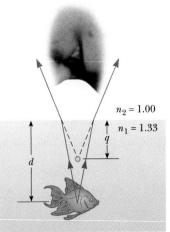

FIGURE 23.18 (Example 23.6) The apparent depth q of the fish is less than the true depth d.

Again, because q is negative, the image is virtual, as indicated in Figure 23.18. The apparent depth is three-fourths the actual depth. For instance, if $d = 4.0$ m, $q = -3.0$ m.

EXERCISE   If the fish is 12 cm long, how long is its image?

ANSWER   12 cm

23.5   ATMOSPHERIC REFRACTION

Images formed by refraction in our atmosphere lead to some interesting results. In this section we look at two examples. A situation that occurs daily is the visibility of the Sun at dusk even though it has passed below the horizon. Figure 23.19 shows why this occurs. Rays of light from the Sun strike the Earth's atmosphere (represented by the shaded area around the Earth) and are bent as they pass into a medium that has an index of refraction different from that of the almost empty space in which they have been traveling. The bending in this situation differs somewhat from the bending we have considered previously in that it is gradual and continuous as the light moves through the atmosphere toward an observer at point O. This is because the light moves through layers of air that have a continuously changing index of refraction. When the rays reach the observer, the eye follows them back along the direction from which they appear to have come (indicated by the dashed path in the figure). The end result is that the Sun is seen to be above the horizon even after it has fallen below it.

The **mirage** is another phenomenon of nature produced by refraction in the atmosphere. A mirage can be observed when the ground is so hot that the air directly above it is warmer than the air at higher elevations. The desert is a region in which such circumstances prevail, but mirages are also seen on heated roadways during the summer. The layers of air at different heights above the Earth have different densities and different refractive indices. The effect this can have is pictured in Figure 23.20a. In this situation the observer sees the sky and a tree in two different ways. One group of light rays reaches the observer by the straight-line path A, and the eye traces these rays back to see the tree in the normal fashion. In addition, a second group of rays travels along the curved path B. These rays are directed toward the ground and are then bent as a result of refraction. As a consequence, the observer also sees an inverted image of the tree and sky background as he traces these rays back to the point at which they appear to have originated. Because an upright image and an inverted image are seen when the image of a tree is observed in a reflecting pool of water, the observer unconsciously calls on this past experience and concludes that the sky background is reflected by a pool of water in front of the tree.

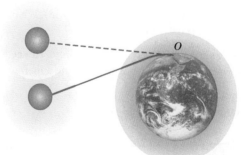

FIGURE 23.19   Because light is refracted by the Earth's atmosphere, an observer at O sees the Sun even though it has fallen below the horizon.

(a)

(b)

FIGURE 23.20   (a) A mirage is produced by the bending of light rays in the atmosphere when there are large temperature differences between the ground and the air. (b) *(John M. Dunay IV, Fundamental Photographs, NYC)*

23.6   THIN LENSES

A typical **thin lens** consists of a piece of glass or plastic, ground so that each of its two refracting surfaces is a segment of either a sphere or a plane. Lenses are commonly used to form images by refraction in optical instruments, such as cameras, telescopes, and microscopes. The equation that relates object and image distances for a lens is virtually identical to the mirror equation derived earlier, and the method used to derive it is also similar.

Figure 23.21 shows some representative shapes of lenses. Notice that we have placed these lenses in two groups. Those in Figure 23.21a are thicker at the center

Biconvex   Convex–
concave

Plano–
convex

(a)

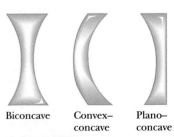

Biconcave   Convex–
concave

Plano–
concave

FIGURE 23.21   Lens shapes. (a) Converging lenses have positive focal lengths and are thickest at the middle. (b) Diverging lenses have negative focal lengths and are thickest at the edges.

than at the rim, and those in Figure 23.21b are thinner at the center than at the rim. The lenses in the first group are examples of **converging lenses,** and those in the second group are **diverging lenses.** The reason for these names will become apparent shortly.

As we did for mirrors, it is convenient to define a point called the **focal point** for a lens. For example, in Figure 23.22a, a group of rays parallel to the axis passes through the focal point *F* after being converged by the lens. The distance from the focal point to the lens is called the **focal length,** *f.* **The focal length is the image distance that corresponds to an infinite object distance.** Recall that we are considering the lens to be very thin. As a result, it makes no difference whether we take the focal length to be the distance from the focal point to the surface of the lens or the distance from the focal point to the center of the lens, because the difference between these two lengths is negligible. A thin lens has *two* focal points, as illustrated in Figure 23.22, corresponding to parallel rays traveling from the left and from the right.

Rays parallel to the axis diverge after passing through a lens of biconcave shape in Figure 23.22b. In this case, the focal point is defined to be the point at which the diverged rays appear to originate, labeled *F* in the figure. Figures 23.22a and 23.22b indicate why the names *converging* and *diverging* are applied to these lenses.

Consider a ray of light passing through the center of a lens, labeled ray 1 in Figure 23.23. For a thin lens, a ray passing through the center is undeviated. Ray 2 in Figure 23.23 is parallel to the principal axis of the lens (the horizontal axis passing through *O*), and as a result it passes through the focal point *F* after refraction. Rays 1 and 2 intersect at the point which is the tip of the image arrow.

We first note that the tangent of the angle α can be found by using the blue and gold shaded triangles in Figure 23.23:

$$\tan \alpha = \frac{h}{p} \qquad \text{or} \qquad \tan \alpha = -\frac{h'}{q}$$

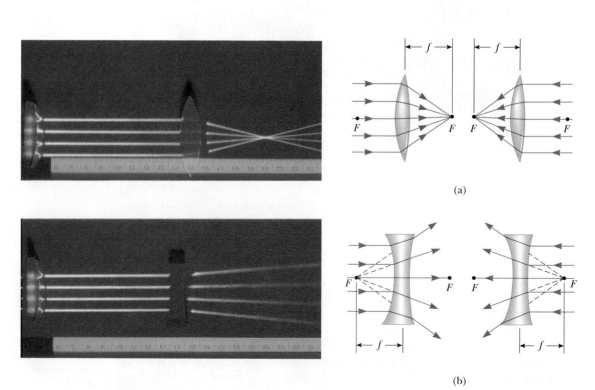

(a)

(b)

FIGURE 23.22 (*Left*) Photographs of the effects of converging and diverging lenses on parallel rays. (*Courtesy of Jim Lehman, James Madison University*) (*Right*) The focal points of (a) the biconvex lens and (b) the biconcave lens.

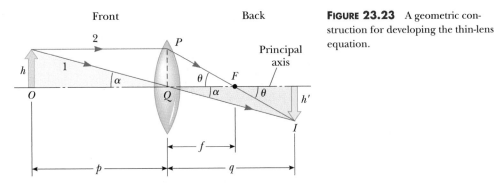

FIGURE 23.23 A geometric construction for developing the thin-lens equation.

From this we find that

$$M = \frac{h'}{h} = -\frac{q}{p} \qquad [23.10]$$

Thus, the equation for magnification by a lens is the same as the equation for magnification by a mirror. We also note from Figure 23.23 that the tangent of θ is

$$\tan \theta = \frac{PQ}{f} \qquad \text{or} \qquad \tan \theta = -\frac{h'}{q - f}$$

However, the height PQ used in the first of these equations is the same as h, the height of the object. Therefore,

$$\frac{h}{f} = -\frac{h'}{q - f}$$

$$\frac{h'}{h} = -\frac{q - f}{f}$$

Using this in combination with Equation 23.10 gives

$$\frac{q}{p} = \frac{q - f}{f}$$

which reduces to

$$\frac{1}{p} + \frac{1}{q} = \frac{1}{f} \qquad [23.11]$$

◀ Thin-lens equation

This equation, called the **thin-lens equation,** can be used with both converging and diverging lenses if we adhere to a set of sign conventions. Figure 23.24 is useful for obtaining the signs of p and q, and Table 23.3 gives the complete sign conventions for lenses. Note that **a converging lens has a positive focal length** under this convention, and **a diverging lens has a negative focal length.** Hence the names *positive* and *negative* are often given to these lenses.

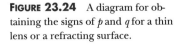

FIGURE 23.24 A diagram for obtaining the signs of p and q for a thin lens or a refracting surface.

TABLE 23.3	Sign Conventions for Thin Lenses	
Quantity	**Positive When**	**Negative When**
Object location (p)	Object is in front of lens	Object is in back of lens
Image location (q)	Image is in back of lens	Image is in front of lens
Image height (h')	Image is upright	Image is inverted
R_1 and R_2	Center of curvature is in back of lens	Center of curvature is in front of lens
Focal length (f)	Converging lens	Diverging lens

The focal length for a lens in air is related to the curvatures of its front and back surfaces and to the index of refraction n of the lens material by

Lens maker's equation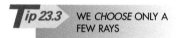

$$\frac{1}{f} = (n - 1)\left(\frac{1}{R_1} - \frac{1}{R_2}\right)$$ [23.12]

where R_1 is the radius of curvature of the front surface of the lens and R_2 is the radius of curvature of the back surface. (As with mirrors, we arbitrarily call the side from which the light approaches the *front* of the lens.) Table 23.3 gives the sign conventions for R_1 and R_2. Equation 23.12 enables us to calculate the focal length from the known properties of the lens. It is called the **lens maker's equation.**

RAY DIAGRAMS FOR THIN LENSES

Ray diagrams are essential for understanding the overall image formation by a thin lens or a system of lenses. They should also help clarify the sign conventions we have already discussed. Figure 23.25 illustrates this method for three single-lens situations. To locate the image formed by a converging lens (Fig. 23.25a and b), the following three rays are drawn from the top of the object:

1. The first ray is drawn parallel to the principal axis. After being refracted by the lens, this ray passes through (or appears to come from) one of the focal points.
2. The second ray is drawn through the center of the lens. This ray continues in a straight line.
3. The third ray is drawn through the other focal point and emerges from the lens parallel to the principal axis.

A similar construction is used to locate the image formed by a diverging lens, as shown in Figure 23.25c. The point of intersection of *any two* of the rays in these diagrams can be used to locate the image. The third ray serves as a check on construction.

For the converging lens in Figure 23.25a, where the object is *outside* the front focal point ($p > f$), the ray diagram shows the image is real and inverted. When the real object is *inside* the front focal point ($p < f$), as in Figure 23.25b, the image is virtual and upright. For the diverging lens of Figure 23.25c, the image is virtual and upright.

Tip 23.3   WE CHOOSE ONLY A FEW RAYS

Note that although our ray diagrams in Figure 23.25 only show three rays leaving an object, there are an infinite number of rays that can be drawn between the object and its image.

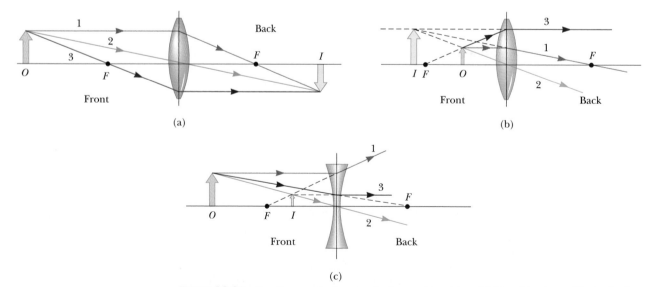

FIGURE 23.25   Ray diagrams for locating the image of an object. (a) The object is outside the focal point of a converging lens. (b) The object is inside the focal point of a converging lens. (c) The object is outside the focal point of a diverging lens.

Quick Quiz 23.4	A plastic sandwich bag filled with water can act as a crude converging lens in air. If the bag is filled with air and placed under water, is the effective lens (a) converging or (b) diverging?
Quick Quiz 23.5	In Figure 23.25a, the blue object arrow is replaced by one that is much taller than the lens. How many rays from the object will strike the lens?
Quick Quiz 23.6	An object is placed to the left of a converging lens. Which of the following statements are true and which are false? (a) The image is always to the right of the lens. (b) The image can be upright or inverted. (c) The image is always smaller or the same size as the object. Justify your answers with ray diagrams.

PROBLEM-SOLVING *STRATEGY* Lenses and Mirrors

Your success or failure in working lens and mirror problems will be determined largely by whether or not you make sign errors when substituting into the lens and mirror equations. The only way to ensure that you don't make sign errors is to become adept at using the sign conventions. The best way to do this is to work a multitude of problems on your own and to make confirming ray diagrams. Watching an instructor or reading the example problems is no substitute for practice.

APPLYING PHYSICS 23.5

Diving masks often have a lens built into the glass for divers who do not have perfect vision. This allows the individual to dive without the necessity of glasses, because the lenses in the faceplate perform the necessary refraction to produce clear vision. Normal glasses have lenses that are curved on both the front and rear surfaces. The lenses in a diving mask faceplate often only have curved surfaces on the inside of the glass. Why is this design desirable?

Explanation The main reason for curving only the inner surface of the lenses in the diving mask faceplate is so that the diver can see clearly while underwater and in the air. If there were curved surfaces on both the front and the back of the diving lens, there would be two refractions. The lens could be designed so that these two refractions would give clear vision while the diver is in air. When the diver goes underwater, however, the refraction between the water and the glass at the first interface is now different, because the index of refraction of water is different from that of air. Thus, the vision will not be clear underwater.

APPLYING PHYSICS 23.6

Consider a glass plano-convex lens, one that is flat on one side and convex on the other. You project three laser beams through it, as shown in Figure 23.26a, and measure the focal length f—that is, the distance from the lens to the point at which the three beams cross. Now you hold the flat side of the lens against the glass of an aquarium filled with water, as in Figure 23.26b. When you shine the

◀ **APPLICATION**

VISION AND DIVING MASKS

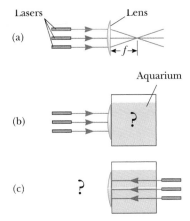

FIGURE 23.26 (Applying Physics 23.6)

laser beams through the lens from the outside of the aquarium, will the beams cross at a point closer to the lens than before, farther away, or at the same distance? What if you direct the laser beams through the lens from the other side of the aquarium, as in Figure 23.26c? What happens in this case?

Explanation The laser beam going through the center of the lens is undeviated by the lens in all three cases. We need to look at what happens to the outer beams. In Figure 23.26a, these two laser beams will refract toward the normal as they pass from the air into the glass of the lens. Thus, they will be deviated *toward* the central beam. As they reach the other side, and pass from the glass back into the air, they will deviate away from the normal, which causes additional deviation toward the central beam. As a result, all three beams cross at the focal point. When the flat side of the lens is held against the glass side of the aquarium as in Figure 23.26b, the outgoing laser beams pass through the glass of the aquarium without additional refraction, but then enter *water*. The change in index of refraction in going from glass to water is much smaller than in the previous case, and the deviation toward the central beam is less. As a result, the crossing point for the three beams is farther from the lens. When the laser beams are directed through the water and then through the lens, as in Figure 23.26c, we must consider two factors. First, the focal length of a lens is independent of which way the light passes through it. Second, in the situation shown, the laser beams experience no refraction on entering the flat side of the lens, because they strike the glass at normal incidence. Thus, all of the refraction occurs at the interface of the air with the curved side of the lens. This is exactly the same situation as if the laser beams had struck the flat side of the lens while it was in air. Thus, there is no effect of the water in the aquarium. The laser beams cross at the same distance from the lens as they did in the first diagram. This third situation is very similar to the curvature of a lens on the inside of a diving mask, discussed previously.

Example 23.7 Images Formed by a Converging Lens

A converging lens of focal length 10.0 cm forms images of objects placed (a) 30.0 cm, (b) 10.0 cm, and (c) 5.00 cm from the lens. In each case, construct a ray diagram, find the image distance, and describe the image.

Solution

A First we construct a ray diagram as shown in Figure 23.27a. The diagram shows that a real, inverted, smaller image is formed on the far side of the lens. The thin-lens equation, Equation 23.11, can be used to find the image distance:

$$\frac{1}{p} + \frac{1}{q} = \frac{1}{f}$$

$$\frac{1}{30.0 \text{ cm}} + \frac{1}{q} = \frac{1}{10.0 \text{ cm}}$$

$$q = \boxed{+15.0 \text{ cm}}$$

The positive sign for the image distance tells us that the image is real and on the back side of the lens. The magnification of the lens is

$$M = -\frac{q}{p} = -\frac{15.0 \text{ cm}}{30.0 \text{ cm}} = \boxed{-0.500}$$

Thus, the image is reduced in height by one half, and the negative sign for *M* tells us that the image is inverted.

B No calculation is necessary for this case because we know that when the object is placed at the focal point, the image is formed at infinity. This is readily verified by substituting $p = 10.0$ cm into the lens equation.

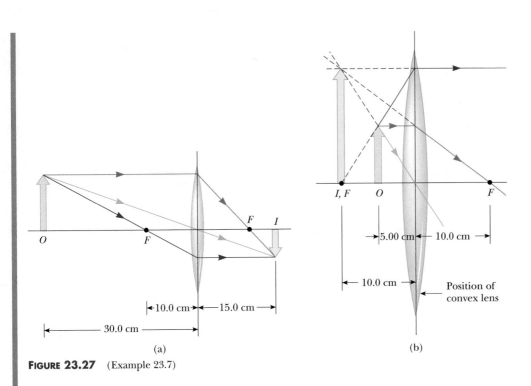

FIGURE 23.27 (Example 23.7)

C We now move inside the focal point. The ray diagram in Figure 23.27b immediately shows that in this case the lens is being used as a magnifying glass; that is, the image is magnified, erect, on the same side as the object, and virtual. Because the object distance is 5.00 cm, the lens equation gives

$$\frac{1}{5.00 \text{ cm}} + \frac{1}{q} = \frac{1}{10.0 \text{ cm}}$$

$$q = -10.0 \text{ cm}$$

$$M = -\frac{q}{p} = -\left(\frac{-10.0 \text{ cm}}{5.00 \text{ cm}}\right) = +2.00$$

The negative image distance tells us that the image is virtual and formed on the side of the lens from which the light is incident, the front side. The image is enlarged, and the positive sign for *M* tells us that the image is upright.

Example 23.8 The Case of a Diverging Lens

Repeat the problem of Example 23.7 for a *diverging* lens of focal length 10.0 cm.

Solution

A We begin by constructing a ray diagram as in Figure 23.28a taking the object distance to be 30.0 cm. The diagram shows that the image is virtual, smaller than the object, and upright. Let us now apply the lens equation with $p = 30.0$ cm:

$$\frac{1}{p} + \frac{1}{q} = \frac{1}{f}$$

$$\frac{1}{30.0 \text{ cm}} + \frac{1}{q} = -\frac{1}{10.0 \text{ cm}}$$

$$q = -7.50 \text{ cm}$$

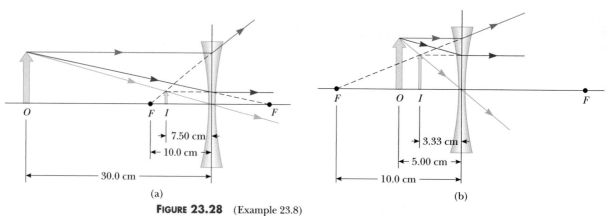

FIGURE 23.28   (Example 23.8)

The magnification is

$$M = -\frac{q}{p} = -\left(\frac{-7.50 \text{ cm}}{30.0 \text{ cm}}\right) = \boxed{+0.250}$$

This result confirms that the image is virtual, smaller than the object, and upright.

B When the object is at the focal point, $p = 10.0$ cm, we have

$$\frac{1}{10.0 \text{ cm}} + \frac{1}{q} = -\frac{1}{10.0 \text{ cm}}$$

$$q = \boxed{-5.00 \text{ cm}}$$

$$M = -\frac{q}{p} = -\left(\frac{-5.00 \text{ cm}}{10.0 \text{ cm}}\right) = \boxed{+0.500}$$

C When the object is inside the focal point, at $p = 5.00$ cm, the ray diagram in Figure 23.28b shows that we have a virtual image that is smaller than the object and upright. In this case, the lens equation gives

$$\frac{1}{5.00 \text{ cm}} + \frac{1}{q} = -\frac{1}{10.0 \text{ cm}}$$

$$q = \boxed{-3.33 \text{ cm}}$$

$$M = -\left(\frac{-3.33 \text{ cm}}{5.00 \text{ cm}}\right) = \boxed{+0.667}$$

This confirms that the image is virtual, smaller than the object, and upright.

COMBINATIONS OF THIN LENSES

If two thin lenses are used to form an image, the system can be treated in the following manner. First, the image produced by the first lens is calculated as though the second lens were not present. The light then approaches the second lens *as if* it had come from the image formed by the first lens. Hence, **the image formed by the first lens is treated as the object for the second lens.** The image formed by the second lens is the final image of the system. If the image formed by the first lens lies on the back side of the second lens, then the image is treated as a virtual object for the second lens (that is, p is negative). The same procedure can be extended to a system of three or more lenses. The overall magnification of a system of thin lenses is the *product* of the magnifications of the separate lenses.

Webnote 23.2

For a more advanced version of lens and mirror combinations, go to
*http://webphysics.ph.msstate.edu/
javamirror/ntnujava/Lens/lens_e.html*

Example 23.9   Two Lenses in a Row

Two converging lenses are placed 20.0 cm apart, as shown in Figure 23.29a. If the first lens has a focal length of 10.0 cm and the second has a focal length of 20.0 cm, locate the final image formed of an object 30.0 cm in front of the first lens. Find the magnification of the system.

Reasoning   We apply the thin-lens equation to both lenses. The image formed by the first lens is treated as the object for the second lens. Also, we use the fact that the total magnification of the system is the product of the magnifications produced by the separate lenses.

Solution   First we make ray diagrams roughly to scale to see where the image from the first lens falls and how it acts as the object for the second lens. (See Fig. 23.29b.) The location of the image formed by the first lens is found via the thin-lens equation:

$$\frac{1}{30.0 \text{ cm}} + \frac{1}{q} = \frac{1}{10.0 \text{ cm}}$$

$$q = \boxed{+15.0 \text{ cm}}$$

The magnification of this lens is

$$M_1 = -\frac{q}{p} = -\frac{15.0 \text{ cm}}{30.0 \text{ cm}} = \boxed{-0.500}$$

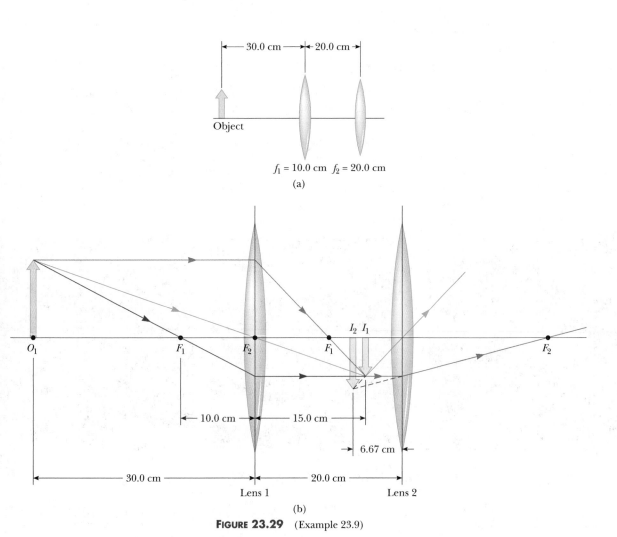

FIGURE 23.29   (Example 23.9)

The image formed by this lens becomes the object for the second lens. Thus, the object distance for the second lens is 20.0 cm − 15.0 cm = 5.00 cm. We again apply the thin-lens equation to find the location of the final image.

$$\frac{1}{5.00 \text{ cm}} + \frac{1}{q} = \frac{1}{20.0 \text{ cm}}$$

$$q = -6.67 \text{ cm}$$

The magnification of the second lens is

$$M_2 = -\frac{q}{p} = -\frac{(-6.67 \text{ cm})}{5.00 \text{ cm}} = +1.33$$

Thus, the final image is 6.67 cm to the left of the second lens, and the overall magnification of the system is

$$M = M_1 M_2 = (-0.500)(1.33) = -0.667$$

The negative sign indicates that the final image is inverted with respect to the initial object.

EXERCISE   If the two lenses in Figure 23.29 are separated by 10.0 cm, locate the final image and find the magnification of the system.

ANSWER   4.00 cm behind the second lens; $M = -0.400$

23.7   LENS AND MIRROR ABERRATIONS

One of the basic problems of systems containing mirrors and lenses is the imperfect quality of the images, which is largely the result of defects in shape and form. The simple theory of mirrors and lenses assumes that rays make small angles with the principal axis and that all rays reaching the lens or mirror from a point source are focused at a single point, producing a sharp image. Clearly, this is not always true in the real world. Where the approximations used in this theory do not hold, imperfect images are formed.

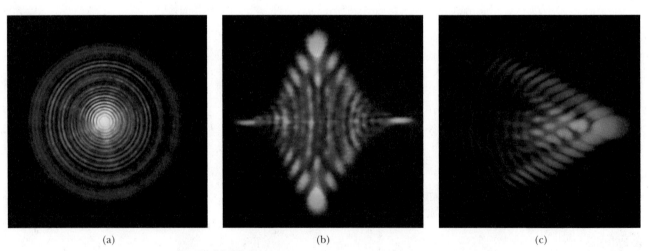

(a)   (b)   (c)

FIGURE 23.30   Lenses can produce varied forms of aberrations, as shown by these blurred photographic images of a point source. (a) Spherical aberration occurs when light passing through the lens at different distances from the principal axis is focused at different points. (b) Astigmatism is an aberration that occurs when the object is not on the principal axis of the lens. (c) Coma. This aberration occurs when light passing through the lens far from the principal axis focuses at a different part of the focal plane from light passing near the center of the lens. *(Photos by Norman Goldberg)*

If one wishes to analyze image formation precisely, it is necessary to trace each ray, using Snell's law, at each refracting surface. This procedure shows that there is no single-point image; instead, the image is blurred. The departures of real (imperfect) images from the ideal predicted by the simple theory are called **aberrations.** Two common types of aberrations are spherical aberration and chromatic aberration. Photographs of three forms of lens aberrations are shown in Figure 23.30.

SPHERICAL ABERRATION

Spherical aberration results from the fact that the focal points of light rays far from the principal axis of a spherical lens (or mirror) are different from the focal points of rays with the same wavelength passing near the axis. Figure 23.31 illustrates spherical aberration for parallel rays passing through a converging lens. Rays near the middle of the lens are imaged farther from the lens than rays at the edges. Hence, there is no single focal length for a spherical lens.

Most cameras are equipped with an adjustable aperture to control the light intensity and, when possible, reduce spherical aberration. (An aperture is an opening that controls the amount of light transmitted through the lens.) As the aperture size is reduced, sharper images are produced, because only the central portion of the lens is exposed to the incident light when the aperture is very small. At the same time, however, progressively less light is imaged. To compensate for this loss, a longer exposure time is used. An example of the results obtained with small apertures is the sharp image produced by a "pinhole" camera, with an aperture size of approximately 0.1 mm.

In the case of mirrors used for very distant objects, one can eliminate, or at least minimize, spherical aberration by employing a parabolic rather than spherical surface. Parabolic surfaces are not used in many applications, however, because they are very expensive to make with high-quality optics. Parallel light rays incident on such a surface focus at a common point. Parabolic reflecting surfaces are used in many astronomical telescopes to enhance the image quality. They are also used in flashlights, in which a nearly parallel light beam is produced from a small lamp placed at the focus of the reflecting surface.

CHROMATIC ABERRATION

The fact that different wavelengths of light refracted by a lens focus at different points gives rise to chromatic aberration. In Chapter 22 we described how the index of refraction of a material varies with wavelength. When white light passes through a lens, one finds, for example, that violet light rays are refracted more than red light rays (Fig. 23.32); thus, the focal length for red light is greater than that for violet light. Other wavelengths (not shown in Fig. 23.32) would have intermediate focal points. The chromatic aberration for a diverging lens is opposite that for a converging lens. Chromatic aberration can be greatly reduced by the use of a combination of converging and diverging lenses.

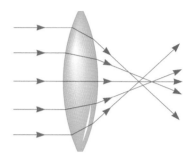

FIGURE 23.31 Spherical aberration produced by a converging lens. Does a diverging lens produce spherical aberration? (Angles are greatly exaggerated for clarity.)

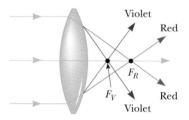

FIGURE 23.32 Chromatic aberration produced by a converging lens. Rays of different wavelengths focus at different points. (Angles are greatly exaggerated for clarity.)

Webnote 23.3

To learn more about optical aberrations, visit
http://nikon.topica.ne.jp/bi_e/encyclo/ad.htm

SUMMARY

Images are formed where rays of light intersect or where they appear to originate. A **real image** is one in which light intersects, or passes through, an image point. A **virtual image** is one in which the light does not pass through the image point but appears to diverge from that point.

The image formed by a flat mirror has the following properties:

1. The image is as far behind the mirror as the object is in front.
2. The image is unmagnified, virtual, and upright.

The **magnification** M of a mirror is defined as the ratio of **image height** h' to **object height** h, which is the negative of the ratio of image distance q to object distance p:

$$M = \frac{h'}{h} = -\frac{q}{p} \qquad [23.2]$$

The **object distance** and **image distance** for a spherical mirror of radius R are related by the **mirror equation:**

$$\frac{1}{p} + \frac{1}{q} = \frac{1}{f} \qquad [23.6]$$

where $f = R/2$ is the **focal length** of the mirror.

An image can be formed by refraction at a spherical surface of radius R. The object and image distances for refraction from such a surface are related by

$$\frac{n_1}{p} + \frac{n_2}{q} = \frac{n_2 - n_1}{R} \qquad [23.7]$$

The **magnification of a refracting surface** is

$$M = \frac{h'}{h} = -\frac{n_1 q}{n_2 p} \qquad [23.8]$$

where the object is located in the medium with index of refraction n_1 and the image is formed in the medium with index of refraction n_2.

The **magnification for a thin lens** is

$$M = \frac{h'}{h} = -\frac{q}{p} \qquad [23.10]$$

and the object and image distances are related by the **thin-lens equation:**

$$\frac{1}{p} + \frac{1}{q} = \frac{1}{f} \qquad [23.11]$$

Aberrations are responsible for the formation of imperfect images by lenses and mirrors. **Spherical aberration** results from the fact that the focal points of light rays far from the principal axis of a spherical lens or mirror are different from those of rays passing through the center. **Chromatic aberration** arises from the fact that light rays of different wavelengths focus at different points when refracted by a lens.

CONCEPTUAL QUESTIONS

1. Tape a picture of yourself on a bathroom mirror. Stand several centimeters away from the mirror. Can you focus your eyes on *both* the picture taped to the mirror *and* your image in the mirror *at the same time?* So, where is the image of yourself?

2. One method for determining the position of an image, either real or virtual, is by means of *parallax*. If a finger or another object is placed at the position of the image, as shown in Figure Q23.2, and the finger and the image are viewed simultaneously (the image is viewed through the lens if it is virtual), the finger and image have the same parallax; that is, if the image is viewed from different positions, it will appear to move along with the finger. Use this method to locate the image formed by a lens. Explain why this method works.

3. A flat mirror creates a virtual image of your face. Suppose the flat mirror is combined with another optical element. Can a flat mirror form a real image in such a combination?

4. Explain why a mirror cannot give rise to chromatic aberration.

5. You are taking a picture of yourself with a camera that uses an ultrasonic range finder to measure the distance to the object. When you take a picture of yourself in a mirror with this camera, your image is out of focus. Why?

6. A solar furnace can be constructed by using a concave mirror to reflect and focus sunlight into a furnace enclosure. What factors in the design of the reflecting mirror will guarantee that very high temperatures can be achieved?

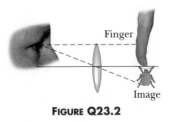

FIGURE Q23.2

7. A virtual image is often described as one through which the light rays do not actually travel, as they do for a real image. Can a virtual image be photographed?

8. What is wrong with the caption of the cartoon shown in Figure Q23.8?

FIGURE Q23.8 "Most mirrors reverse left and right. This one reverses top and bottom."

9. Suppose you want to use a converging lens to project the image of two trees onto a screen. One tree is distance x from the lens; the other is at $2x$, as in Figure Q23.9. You adjust the screen so that the near tree is in focus. If you now want the far tree to be in focus, do you move the screen toward or away from the lens?

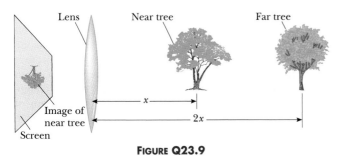

FIGURE Q23.9

10. Why does a clear stream always appear to be shallower than it actually is?

11. Can a converging lens be made to diverge light if placed in a liquid? How about a converging mirror?

12. A common mirage is formed when the air gets gradually cooler as the height above the ground increases. What might happen if the air grows gradually warmer as the height is increased? This often happens over bodies of water or snow-covered ground; the effect is called *looming*.

13. In a Jules Verne novel, a piece of ice is shaped into a magnifying lens to focus sunlight to start a fire. Is this possible?

14. Lenses used in eyeglasses, whether converging or diverging, are always designed such that the middle of the lens curves away from the eye. Why?

15. Why does the focal length of a mirror not depend on the mirror material whereas the focal length of a lens does depend on the lens material?

16. If a cylinder of solid glass or clear plastic is placed above the words LEAD OXIDE and viewed from the side, as shown in Figure Q23.16, the word LEAD appears inverted but the word OXIDE does not. Explain.

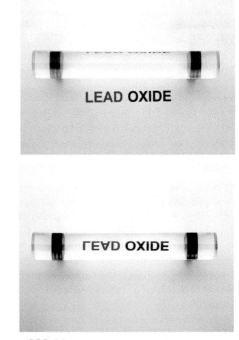

FIGURE Q23.16 *(Richard Megna/Fundamental Photographs, NYC)*

17. What is the focal length of a flat mirror? Is that value consistent with the mirror equation?

PROBLEMS

1, **2**, 3 = straightforward, intermediate, challenging   ☐ = full solution available in Student Solutions Manual/Study Guide

web = solution posted at **http://info.brookscole.com/serway**   [icon] = biomedical application

Section 23.1 Flat Mirrors

1. Does your bathroom mirror show you older or younger than your actual age? Compute an order-of-magnitude estimate for the age difference, based on data that you specify.

2. Use Figure 23.2 to give a geometric proof that the virtual image formed by a plane mirror is the same distance behind the mirror as the object is in front of it.

3. A person walks into a room that has, on opposite walls, two plane mirrors producing multiple images. When the person is 5.00 ft from the mirror on the left wall and 10.0 ft

from the mirror on the right wall, find the distances from the person to the first three images seen in the left-hand mirror.

4. In a church choir loft, two parallel walls are 5.30 m apart. The singers stand against the north wall. The organist faces the south wall, sitting 0.800 m away from it. So that she can see the choir, a flat mirror 0.600 m wide is mounted on the south wall, straight in front of the organist. What width of the north wall can she see? (*Hint:* Draw a top-view diagram to justify your answer.)

Section 23.2 Images Formed by Spherical Mirrors

Section 23.3 Convex Mirrors and Sign Conventions

In the following problems, algebraic signs are not given. We leave it to you to determine the correct sign to use with each quantity, based on an analysis of the problem and the sign conventions in Table 23.1.

5. At an intersection of hospital hallways, a convex mirror is mounted high on a wall to help people avoid collisions. The mirror has a radius of curvature of 0.550 m. Locate and describe the image of a patient 10.0 m from the mirror. Determine the magnification.

6. A spherical Christmas tree ornament is 6.00 cm in diameter. What is the magnification of an object placed 10.0 cm away from the ornament?

7. A concave spherical mirror has a radius of curvature of 20.0 cm. Locate the images for object distances of (a) 40.0 cm, (b) 20.0 cm, and (c) 10.0 cm. In each case, state whether the image is real or virtual and upright or inverted, and find the magnification.

8. A dentist uses a mirror to examine a tooth. The tooth is 1.00 cm in front of the mirror, and the image is formed 10.0 cm behind the mirror. Determine (a) the mirror's radius of curvature and (b) the magnification of the image.

9. A large church has a niche in one wall. On the floor plan it appears as a semicircular indentation of radius 2.50 m. A worshiper stands on the center line of the niche, 2.00 m out from its deepest point, and whispers a prayer. Where is the sound concentrated after reflection from the back wall of the niche?

10. A convex mirror has a focal length of 20.0 cm. Determine the object location for which the image will be one-half as tall as the object.

11. A 2.00-cm-high object is placed 3.00 cm in front of a concave mirror. If the image is 5.00 cm high and virtual, what is the focal length of the mirror?

12. A 2.00-cm-high object is placed 10.0 cm in front of a mirror. What type of mirror and what radius of curvature are needed to create an upright image that is 4.00 cm high?

13. A concave makeup mirror is designed so that a person 25 cm in front of it sees an upright image magnified by a factor of 2. What is the radius of curvature of the mirror?

14. A concave mirror has a focal length of 40.0 cm. Determine the object position for which the resulting image is upright and four times the size of the object.

15. A man standing 1.52 m in front of a shaving mirror produces an inverted image 18.0 cm in front of it. How close to the mirror should he stand if he wants to form an upright image of his chin that is twice the chin's actual size?

16. A convex spherical mirror with a radius of curvature of 10.0 cm produces a virtual image one-third the size of the real object. Where is the object?

17. A child holds a candy bar 10.0 cm in front of a convex mirror and notices that the image is only one-half the size of the candy bar. What is the radius of curvature of the mirror?

18. It is observed that the size of a *real* image formed by a concave mirror is four times the size of the object when the object is 30.0 cm in front of the mirror. What is the radius of curvature of this mirror?

19. A spherical mirror is to be used to form an image, five times as tall as an object, on a screen positioned 5.0 m from the mirror. (a) Describe the type of mirror required. (b) Where should the mirror be positioned relative to the object?

20. A ball is dropped from rest 3.00 m directly above the vertex of a concave mirror having a radius of 1.00 m and lying in a horizontal plane. (a) Describe the motion of the ball's image in the mirror. (b) At what time do the ball and its image coincide?

Section 23.4 Images Formed by Refraction

21. A cubical block of ice 50.0 cm on an edge is placed on a level floor over a speck of dust. Locate the image of the speck, when viewed from directly above, if the index of refraction of ice is 1.309.

22. The top of a swimming pool is at ground level. If the pool is 2 m deep, how far below ground level does the bottom of the pool appear to be located when (a) the pool is completely filled with water? (b) the pool is filled halfway with water?

23. A paperweight is made of a solid glass hemisphere of index of refraction 1.50. The radius of the circular cross section is 4.0 cm. The hemisphere is placed on its flat surface with the center directly over a 2.5-mm-long line drawn on a sheet of paper. What length of line is seen by someone looking vertically down on the hemisphere?

24. A flint glass plate ($n = 1.66$) rests on the bottom of an aquarium tank. The plate is 8.00 cm thick (vertical dimension) and covered with water ($n = 1.33$) to a depth of 12.0 cm. Calculate the apparent thickness of the plate as viewed from above the water. (Assume nearly normal incidence.)

25. One end of a long glass rod ($n = 1.50$) is formed into the shape of a convex surface of radius 8.00 cm. An object is positioned in air along the axis of the rod. Find the image position that corresponds to each of the following object positions: (a) 20.0 cm, (b) 8.00 cm, (c) 4.00 cm, (d) 2.00 cm.

26. A goldfish is swimming at 2.00 cm/s toward the front wall of a rectangular aquarium. What is the apparent speed of the fish measured by an observer looking in from outside the front wall of the tank? The index of refraction of water is 1.333.

Section 23.6 Thin Lenses

27. A contact lens is made of plastic with an index of refraction of 1.50. The lens has an outer radius of curvature of $+2.00$ cm and an inner radius of curvature of $+2.50$ cm. What is the focal length of the lens?

28. The left face of a biconvex lens has a radius of curvature of 12.0 cm, and the right face has a radius of curvature of 18.0 cm. The index of refraction of the glass is 1.44. (a) Calculate the focal length of the lens. (b) Calculate the focal length if the radii of curvature of the two faces are interchanged.

29. A converging lens has a focal length of 20.0 cm. Locate the images for object distances of (a) 40.0 cm, (b) 20.0 cm, and (c) 10.0 cm. For each case, state whether the image is real or virtual and upright or inverted, and find the magnification.

30. Where must an object be placed to have unit magnification ($|M| = 1.00$) (a) for a converging lens of focal length 12.0 cm? (b) for a diverging lens of focal length 12.0 cm?

31. A diverging lens has a focal length of 20.0 cm. Locate the images for object distances of (a) 40.0 cm, (b) 20.0 cm, and (c) 10.0 cm. For each case, state whether the image is real or virtual and upright or inverted, and find the magnification.

32. A convex lens of focal length 15.0 cm is used as a magnifying glass. At what distance from a postage stamp should you hold this lens to get a magnification of $+2.00$?

33. A transparent photographic slide is placed in front of a converging lens with a focal length of 2.44 cm. The lens forms an image of the slide 12.9 cm from the slide. How far is the lens from the slide if the image is (a) real? (b) virtual?

34. The nickel's image in Figure P23.34 has twice the diameter of the nickel when the lens is 2.84 cm from the nickel. Determine the focal length of the lens.

FIGURE P23.34

35. A certain LCD projector contains a single thin lens. An object 24.0 mm high is to be projected so that its image fills a screen 1.80 m high. The object-to-screen distance is 3.00 m. (a) Determine the focal length of the projection lens. (b) How far from the object should the lens of the projector be placed in order to form the image on the screen?

36. An object's distance from a converging lens is ten times the focal length. How far is the image from the focal point? Express the answer as a fraction of the focal length.

37. A diverging lens is to be used to produce a virtual image
web one-third as tall as the object. Where should the object be placed?

38. An object is 5.00 m to the left of a flat screen. A converging lens for which the focal length is $f = 0.800$ m is placed between object and screen. (a) Show that there are two lens positions that form an image on the screen, and determine how far these positions are from the object. (b) How do the two images differ from each other?

39. A converging lens is placed 30.0 cm to the right of a diverging lens of focal length 10.0 cm. A beam of parallel light enters the diverging lens from the left, and the beam is again parallel when it emerges from the converging lens. Calculate the focal length of the converging lens.

40. An object is placed 20.0 cm to the left of a converging lens of focal length 25.0 cm. A diverging lens of focal length 10.0 cm is 25.0 cm to the right of the converging lens. Find the position and magnification of the final image.

41. Two converging lenses, each of focal length 15.0 cm, are placed 40.0 cm apart, and an object is placed 30.0 cm in front of the first. Where is the final image formed, and what is the magnification of the system?

42. Object O_1 is 15.0 cm to the left of a converging lens of 10.0-cm focal length. A second lens is positioned 10.0 cm to the right of the first lens and is observed to form a final image at the position of the original object O_1. (a) What is the focal length of the second lens? (b) What is the overall magnification of this system? (c) What is the nature (i.e., real or virtual, upright or inverted) of the final image?

43. A 1.00-cm-high object is placed 4.00 cm to the left of a converging lens of focal length 8.00 cm. A diverging lens of focal length -16.00 cm is 6.00 cm to the right of the converging lens. Find the position and height of the final image. Is the image inverted or upright? Real or virtual?

44. Two converging lenses having focal lengths of 10.0 cm and 20.0 cm are placed 50.0 cm apart, as shown in Figure P23.44. The final image is to be located between the lenses, at the position indicated. (a) How far to the left of the first lens should the object be positioned? (b) What is the overall magnification? (c) Is the final image upright or inverted?

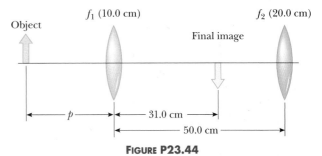

FIGURE P23.44

45. Lens L_1 in Figure P23.45 has a focal length of 15.0 cm and is located a fixed distance in front of the film plane of a camera. Lens L_2 has a focal length of 13.0 cm, and its distance d from the film plane can be varied from 5.00 cm to 10.0 cm. Determine the range of distances for which objects can be focused on the film.

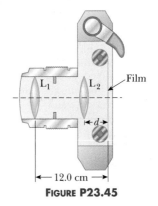

FIGURE P23.45

46. Consider two thin lenses, one of focal length f_1 and the other of focal length f_2, placed in contact with each other as shown in Figure P23.46. Apply the thin-lens equation to each of these lenses and combine these results to show that this combination of lenses behaves like a thin lens having a focal length f given by $1/f = 1/f_1 + 1/f_2$. Assume that the thicknesses of the lenses can be ignored in comparison to the other distances involved.

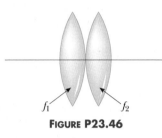

FIGURE P23.46

ADDITIONAL PROBLEMS

47. An object placed 10.0 cm from a concave spherical mirror produces a real image 8.00 cm from the mirror. If the object is moved to a new position 20.0 cm from the mirror, what is the position of the image? Is the final image real or virtual?

48. An object is placed 12 cm to the left of a diverging lens of focal length -6.0 cm. A converging lens of focal length 12 cm is placed a distance of d to the right of the diverging lens. Find the distance d that places the final image at infinity.

49. The distance between an object and its upright image is 20.0 cm. If the magnification is 0.500, what is the focal length of the lens being used to form the image?

50. The object in Figure P23.50 is midway between the lens and the mirror. The mirror's radius of curvature is 20.0 cm, and the lens has a focal length of -16.7 cm. Considering only the light that leaves the object and travels first toward the mirror, locate the final image formed by this system. Is this image real or virtual? Is it upright or inverted? What is the overall magnification?

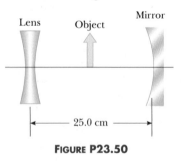

FIGURE P23.50

51. The lens and mirror in Figure P23.51 are separated by 1.00 m and have focal lengths of $+80.0$ cm and -50.0 cm, respectively. If an object is placed 1.00 m to the left of the lens, locate the final image. State whether the image is upright or inverted, and determine the overall magnification.

52. A diverging lens ($n = 1.50$) is shaped like that in Figure 23.25c. The radius of the first surface is 15.0 cm and that of the second surface is 10.0 cm. (a) Find the focal length of the lens. Determine the positions of the images for object distances of (b) infinity, (c) $3|f|$, (d) $|f|$, and (e) $|f|/2$.

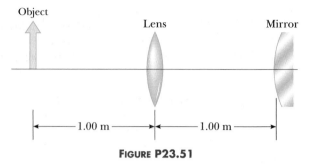

FIGURE P23.51

53. A parallel beam of light enters a glass hemisphere perpendicular to the flat face, as shown in Figure P23.53. The radius is $R = 6.00$ cm, and the index of refraction is $n = 1.560$. Determine the point at which the beam is focused. (Assume paraxial rays; that is, all rays are located close to the principal axis.)

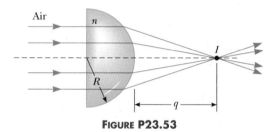

FIGURE P23.53

54. A converging lens of focal length 20.0 cm is separated by 50.0 cm from a converging lens of focal length 5.00 cm. (a) Find the position of the final image of an object placed 40.0 cm in front of the first lens. (b) If the height of the object is 2.00 cm, what is the height of the final image? Is the image real or virtual? (c) If the two lenses are now placed in contact with each other and the object is 5.00 cm in front of this combination, where will the image be located? (See Problem 46.)

55. To work this problem, use the fact that the image formed by the first surface becomes the object for the second surface. Figure P23.55 shows a piece of glass with index of refraction of 1.50. The ends are hemispheres with radii 2.00 cm and 4.00 cm, and the centers of the hemispherical ends are separated by a distance of 8.00 cm. A point object is in air, 1.00 cm from the left end of the glass. Locate the image of the object due to refraction at the two spherical surfaces.

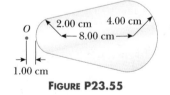

FIGURE P23.55

56. In a darkened room, a burning candle is placed 1.50 m from a white wall. A lens is placed between candle and wall at a location that causes a larger, inverted image to form on the wall. When the lens is moved 90.0 cm toward the wall, another image of the candle is formed. Find (a) the two object distances that produce the images and (b) the focal length of the lens. (c) Characterize the second image.

57. An object 2.00 cm high is placed 40.0 cm to the left of a converging lens having a focal length of 30.0 cm. A diverging lens having a focal length of -20.0 cm is placed 110 cm to the right of the converging lens. (a) Determine the final position and magnification of the final image. (b) Is the image upright or inverted? (c) Repeat parts (a) and (b) for the case where the second lens is a converging lens having a focal length of $+20.0$ cm.

58. A "floating strawberry" illusion consists of two parabolic mirrors, each with a focal length of 7.5 cm, facing each other so that their centers are 7.5 cm apart (Fig. P23.58). If a strawberry is placed on the lower mirror, an image of the strawberry forms at the small opening at the center of the top mirror. Show that the final image forms at that location, and describe its characteristics. (*Note:* A flashlight beam shone on these *images* has a very startling effect. Even at a glancing angle, the incoming light beam is seemingly reflected off the *images* of the strawberry! Do you understand why?)

Small hole

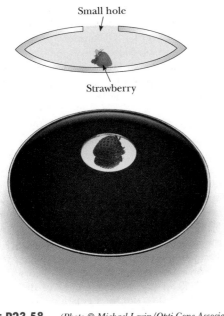

Strawberry

FIGURE P23.58   *(Photo © Michael Levin/Opti-Gone Associates)*

59. Figure P23.59 shows a converging lens with radii $R_1 = 9.00$ cm and $R_2 = -11.00$ cm in front of a concave spherical mirror of radius $R = 8.00$ cm. The focal points (F_1 and F_2) for the thin lens and the center of curvature (C) of the mirror are also shown. (a) If the focal points F_1 and F_2 are 5.00 cm from the vertex of the thin lens, determine the index of refraction for the lens. (b) If the lens and mirror are 20.0 cm apart, and an object is placed 8.00 cm to the left of the lens, determine the position of the final image and its magnification as seen by the eye in the figure. (c) Is the final image inverted or upright? Explain.

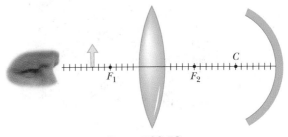

FIGURE P23.59

60. Find the object distances (in terms of f) for a thin converging lens of focal length f if (a) the image is real and the image distance is four times the focal length; (b) the image is virtual and the image distance is three times the focal length. (c) Calculate the magnification of the lens for case (a) and (b).

61. The lens maker's equation for a lens with index n_1 immersed in a medium with index n_2 takes the form

$$\frac{1}{f} = \left(\frac{n_1}{n_2} - 1\right)\left(\frac{1}{R_1} - \frac{1}{R_2}\right)$$

A thin diverging glass lens (index = 1.50) with $R_1 = -3.00$ m and $R_2 = -6.00$ m is surrounded by air. An arrow is placed 10.0 m to the left of the lens. (a) Determine the position of the image. Repeat part (a) with the arrow and lens immersed in (b) water (index = 1.33); (c) a medium with an index of refraction of 2.00. (d) How can a lens that is diverging in air be changed into a converging lens?

GROUP ACTIVITIES

G.1 This experiment will enable you to examine some of the properties of images formed by flat mirrors. As shown in Figure GA23.1, place a clear plastic sheet between two small candles of the same height as in the figure. Light one candle placed about 6 inches from the sheet and observe the reflected flame from the front side. Move the unlit candle until it also appears to be lit. The candles should be approximately equidistant from the sheet at this point. Why? When you place your finger on the wick of the unlit candle, you will get the illusion that your finger is burning. Explain your observation.

A similar experiment can be performed with a flat mirror and two pencils. Hold one of the pencils in front of the mirror at a distance of about 10 inches from the

mirror and at a position such that about half of the image appears in the mirror. Move the second pencil back and forth behind the mirror until it appears to align perfectly with the image in the mirror. When alignment is achieved, measure the distance from the mirror to the location of the second pencil. This pencil is located at the apparent position of the virtual image. Why? The two pencils should be at equal distances from the mirror. Why?

G.2 Fill a clear glass tumbler with water and place a pencil or straw into the tumbler. Now observe the pencil from the side at an angle of about 45° to the surface, and note that the line of the portion of the pencil under water is not parallel with the line of the portion in air. That is, the

FIGURE GA23.1

pencil appears to be bent at the point where it enters the water. Use the techniques of Section 23.4 to explain this observation.

FIGURE GA23.2

G.3 View yourself in a full-length mirror. Standing close to the mirror, place one piece of tape at the top of the image of your head and another piece of tape at the very bottom of the image of your feet. Now step back a few meters and observe your image. How big is it relative to the original size? How does the distance between the pieces of tape compare with your actual height?

Move to a position in front of the mirror such that you can see a full image of yourself with the top of your head just level with the top of the mirror. Have your co-worker gradually block off the lower portion of the mirror with a sheet or newspaper page until you can see your complete image but no more. Measure the length of the mirror and compare this measurement to your height. How do the two compare?

G.4 Compare the images formed of your face when you look first at the front side and then at the back side of a shiny soupspoon. For each side observe the change in the image of your face as you move closer to and farther away from the spoon.

G.5 An object is placed 25 cm in front of a concave makeup mirror that has a focal length of 40 cm. (a) The sign of the focal length is _____ (plus or minus). (b) The sign of the object distance is _____. (c) The sign of the image distance will be _____. (d) Use the mirror equation to find the image distance. (e) Repeat for the case in which the object is 5.0 m away.

G.6 Describe in words what happens to the following rays of light when they strike a converging lens: (a) a ray parallel to the optic axis, (b) a ray passing through the focal point before striking the lens, (c) a ray passing through the center of curvature of the lens. (d) Repeat for the case of a diverging lens.

G.7 An object is placed 25 cm in front of a converging lens that has a focal length of 10 cm. (a) The sign of the focal length to be used in the lens equation is _____. (b) The sign of the object distance is _____. (c) The sign of the image distance will be _____. (d) Verify your answer to (c) by a calculation. (d) Repeat for the case when the lens is diverging with a focal length of 15 cm.

G.8 Draw a sketch of you and your image in a plane mirror when you are walking away from it with a velocity **v**. Indicate on your diagram the velocity vectors for you and your image. Keep the relative lengths of the two velocities as you believe they will be. Explain your answer. (b) Repeat for when you are walking toward the mirror. Explain why you have the vectors drawn as you do.

Wave Optics

Peacock feathers appear to be brilliant in color, especially in the blues and greens. However, these colors are not caused by pigments in the feathers. This chapter describes diffraction, the optical process responsible for this bird's magnificent display. *(Terry Qing/FPG International, Inc.)*

Thus far our discussion of light has been concerned with what happens when light passes through a lens or reflects from a mirror. Because explanations of such phenomena rely on a geometric analysis of light rays, that part of optics is often called geometric optics. We now expand our study of light into an area called *wave optics*. The three primary topics we examine in this chapter are interference, diffraction, and polarization. These phenomena cannot be adequately explained with ray optics, but the wave theory leads us to satisfying descriptions.

24.1   CONDITIONS FOR INTERFERENCE

When you blew bubbles as a child, did you notice the brilliant colors playing on the surface of the bubbles? Have you ever wondered what causes the vivid rainbow of colors reflected from the oil films floating in parking lot puddles on rainy days? Or, have you noticed that good-quality lenses used in cameras have a violet hue? All of these effects are caused by interference in thin films, a subject we shall discuss in this chapter. However, before we can examine interference, we must pause to consider the conditions necessary for it to occur.

In our discussion of interference of mechanical waves in Chapter 13, we found that two waves can add together either constructively or destructively. In constructive interference, the amplitude of the resultant wave is greater than that of either of the individual waves, whereas in destructive interference, the resultant amplitude is less than that of either individual wave. Light waves also interfere with each other. Fundamentally, all interference associated with light waves arises when the electromagnetic fields that constitute the individual waves combine.

Interference effects in light waves are not easy to observe because of the short wavelengths involved (about 4×10^{-7} m to about 7×10^{-7} m). For sustained interference between two sources of light to be observed, the following conditions must be met:

Conditions for interference ▶

1. The sources must be **coherent**—that is, they must maintain a constant phase with respect to each other.
2. The waves must have identical wavelengths.

Let us examine the characteristics of coherent sources. Two sources (producing two traveling waves) are needed to create interference. To produce a stable interference pattern, the individual waves must maintain a constant phase with one another. When this situation prevails, the sources are said to be coherent. The sound waves emitted by two side-by-side loudspeakers driven by a single amplifier can produce interference because the two speakers respond to the amplifier in the same way at the same time—that is, they are in phase.

If two light sources are placed side by side, however, no interference effects are observed, because the light waves from one source are emitted independently of the waves from the other source. Hence, the emissions from the two sources do not maintain a constant phase relationship with each other during the time of observation. An ordinary light source undergoes random changes about once every 10^{-8} s. Therefore, the conditions for constructive interference, destructive interference, and intermediate states have durations on the order of 10^{-8} s. The result is that no interference effects are observed, because the eye cannot follow such short-term changes. Such light sources are said to be **incoherent.**

An older method for producing two coherent light sources is to pass light from a single wavelength (monochromatic) source through a narrow slit and then allow this light to fall on a screen containing two narrow slits. The first slit is needed to insure that light comes from a tiny region of the light source in which excited atoms emit coherently. The light emerging from the two slits is coherent because a single source produces the original light beam and the slits serve only to separate the original beam into two parts (which is exactly what was done to the sound signal just mentioned). Any random change in the light emitted by the source will occur in the two separate beams at the same time, and interference effects can be observed.

Currently it is much more common to use a laser as a coherent source to demonstrate interference. A laser produces an intense, coherent, monochromatic beam over a width of several millimeters. This means that the laser can be used to illuminate multiple slits directly, and that interference effects can be observed easily in a fully lighted room. The principles of operation of lasers are explained in Chapter 28.

24.2 YOUNG'S DOUBLE-SLIT INTERFERENCE

Thomas Young first demonstrated interference in light waves from two sources in 1801. Figure 24.1a is a diagram of the apparatus used in this experiment. (Young used pinholes rather than slits in his original experiments.) Light is incident on a screen in which there is a narrow slit S_0. The light waves emerging from this slit arrive at a second screen that contains two narrow, parallel slits S_1 and S_2. These slits serve as a pair of coherent light sources because waves emerging from them originate from the same wave front and therefore are always in phase. The light from the two slits produces a visible pattern on screen C consisting of a series of bright and dark parallel bands called **fringes** (Fig. 24.1b). When the light from slits S_1 and S_2 arrives at a point on the screen so that constructive interference occurs at that location, a bright fringe appears. When the light from the two slits combines destructively at any location on the screen, a dark fringe results.

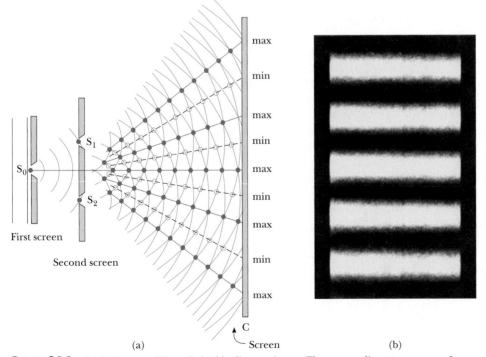

(a)   (b)

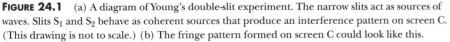

FIGURE 24.1 (a) A diagram of Young's double-slit experiment. The narrow slits act as sources of waves. Slits S_1 and S_2 behave as coherent sources that produce an interference pattern on screen C. (This drawing is not to scale.) (b) The fringe pattern formed on screen C could look like this.

FIGURE 24.2 An interference pattern involving water waves is produced by two vibrating sources at the water's surface. The pattern is analogous to that observed in Young's double-slit experiment. Note the regions of constructive and destructive interference. *(Richard Megna, Fundamental Photographs)*

Figure 24.2 is a photograph of an interference pattern produced by two coherent vibrating sources in a water tank.

Figure 24.3 is a diagram of some of the ways in which the two waves can combine at screen C. In Figure 24.3a, the two waves, which leave the two slits in phase, strike the screen at the central point P. Because these waves travel equal distances, they arrive in phase at P, and as a result constructive interference occurs there and a bright fringe is observed. In Figure 24.3b, the two light waves again start in phase, but the upper wave has to travel one wavelength farther to reach point Q on the screen. Because the upper wave falls behind the lower one by exactly one wavelength, the two waves still arrive in phase at Q, and so a second bright fringe appears at that location. Now consider point R, midway between P and Q in Figure 24.3c. At that point, the upper wave has fallen half a wavelength behind the lower wave. This means that the trough of the bottom wave overlaps the crest of the upper wave, giving rise to destructive interference at R. As a consequence, a dark fringe can be observed at R.

We can describe Young's experiment quantitatively with the help of Figure 24.4. Consider point P on the viewing screen; the screen is positioned a perpendicular distance L from the screen containing slits S_1 and S_2, which are separated by distance d, and r_1 and r_2 are the distances the secondary waves travel from slit to screen. Let us assume the waves emerging from S_1 and S_2 have the same constant frequency, the same amplitude, and start out in phase. The light intensity on the screen at P is the resultant of the light from both slits. Note that a wave from the lower slit travels farther than a wave from the upper slit by the amount $d \sin \theta$. This distance is called the **path difference** δ (lowercase Greek delta), where

▶ Path difference

$$\delta = r_2 - r_1 = d \sin \theta \qquad [24.1]$$

This equation assumes that the two waves travel in parallel lines, which is approximately true, because L is much greater than d. As noted earlier, the value of this path difference determines whether the two waves are in phase when they arrive at P. If the path difference is either zero or some integral multiple of the wavelength, the two waves are in phase at P and constructive interference results. Therefore, the condition for bright fringes, or **constructive interference,** at P is

▶ Condition for constructive interference (two slits)

$$\delta = d \sin \theta_{bright} = m\lambda \qquad m = 0, \pm 1, \pm 2, \ldots \qquad [24.2]$$

The number m is called the **order number.** The central bright fringe at $\theta_{bright} = 0$ ($m = 0$) is called the *zeroth-order maximum.* The first maximum on either side, where $m = \pm 1$, is called the *first-order maximum,* and so forth.

When δ is an odd multiple of $\lambda/2$, the two waves arriving at P are 180° out of phase and give rise to **destructive interference.** Therefore, the condition for

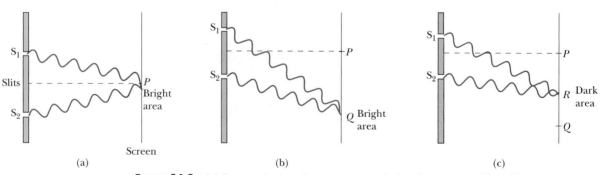

(a)   (b)   (c)

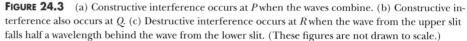

FIGURE 24.3 (a) Constructive interference occurs at P when the waves combine. (b) Constructive interference also occurs at Q. (c) Destructive interference occurs at R when the wave from the upper slit falls half a wavelength behind the wave from the lower slit. (These figures are not drawn to scale.)

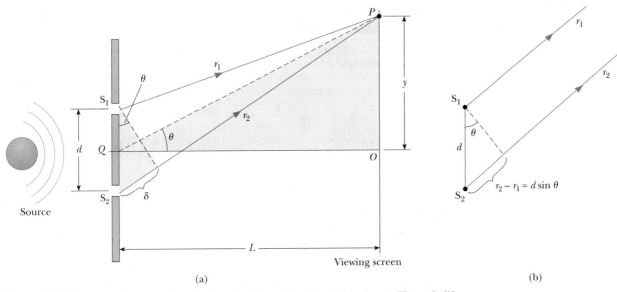

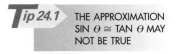

FIGURE 24.4 A geometric construction that describes Young's double-slit experiment. The path difference between the two rays is $r_2 - r_1 = d \sin \theta$. (This figure is not drawn to scale.)

dark fringes at P is

$$\delta = d \sin \theta_{\text{dark}} = (m + \tfrac{1}{2})\lambda \qquad m = 0, \pm 1, \pm 2, \ldots \qquad \textbf{[24.3]}$$

◀ Condition for destructive interference (two slits)

If $m = 0$ in this equation, the path difference is $\delta = \lambda/2$, which is the condition for the location of the first dark fringe on either side of the central (bright) maximum. Likewise, if $m = 1$, $\delta = 3\lambda/2$, which is the condition for the second dark fringe on each side, and so forth.

It is useful to obtain expressions for the positions of the bright and dark fringes measured vertically from O to P. In addition to our assumption that $L \gg d$, we assume that $d \gg \lambda$. These can be valid assumptions because in practice, L is often of the order of 1 m, d a fraction of a millimeter, and λ a fraction of a micrometer for visible light. Under these conditions, θ is small; we can thus use the approximation $\sin \theta \cong \tan \theta$. Then, from triangle OPQ in Figure 24.4, we see that

$$y = L \tan \theta \approx L \sin \theta \qquad \textbf{[24.4]}$$

Solving Equation 24.2 for $\sin \theta$ and substituting the result into Equation 24.4, we see that the positions of the *bright fringes*, measured from O are

$$y_{\text{bright}} = \frac{\lambda L}{d} m \qquad m = 0, \pm 1, \pm 2, \ldots \qquad \textbf{[24.5]}$$

Using Equations 24.3 and 24.4, we find that the *dark fringes* are located at

$$y_{\text{dark}} = \frac{\lambda L}{d}(m + \tfrac{1}{2}) \qquad m = 0, \pm 1, \pm 2, \ldots \qquad \textbf{[24.6]}$$

As we shall demonstrate in Example 24.1, Young's double-slit experiment provides a method for measuring the wavelength of light. In fact, Young used this technique to do just that. In addition, his experiment gave the wave model of light a great deal of credibility. It was inconceivable that particles of light coming through the slits could cancel each other in a way that would explain the dark fringes.

Tip 24.1 THE APPROXIMATION SIN $\theta \cong$ TAN θ MAY NOT BE TRUE

The small-angle approximation $\sin \theta \cong \tan \theta$ is true to three-digit precision only for angles less than about 4°.

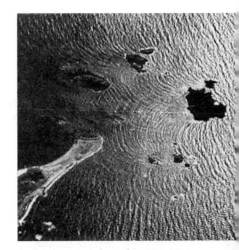

As the waves pass through a narrow gap, they spread out (diffract), and the interference of two waveforms is manifested in cross-patterned areas. *(John S. Shelton)*

Webnote 24.1

See the Young's two-slit diffraction pattern by visiting
http://vsg.tripod.com/interfer.htm

APPLYING PHYSICS 24.1

Consider a double-slit experiment in which a laser beam is passed through a pair of very closely spaced slits, and a clear interference pattern is displayed on a distant screen. Now suppose you place smoke particles between the double slit and the screen. With the presence of the smoke particles, will you see the effects of interference in the space between the slits and the screen, or will you only see the effects on the screen?

Explanation You will see the effects in the area filled with smoke. There will be bright lines directed toward the bright areas on the screen and dark lines directed toward the dark areas on the screen.

APPLYING PHYSICS 24.2

Suppose you are watching television by means of an antenna rather than a cable system. If an airplane flies near your location, you will sometimes notice wavering ghost images in the television picture. What might cause this?

Explanation Your television antenna receives two signals—the direct signal from the transmitting antenna and a signal reflected from the surface of the airplane. As the airplane changes position, there are times when these two signals are in phase and other times when they are out of phase. As a result, the intensity of the combined signal received at your antenna varies. The wavering of the ghost images of the picture is evidence for this variation.

APPLICATION

TELEVISION SIGNAL
INTERFERENCE

Example 24.1 Measuring the Wavelength of a Light Source

A screen is separated from a double-slit source by 1.2 m. The distance between the two slits is 0.030 mm. The second-order bright fringe ($m = 2$) is measured to be 4.5 cm from the centerline. Determine (a) the wavelength of the light and (b) the distance between adjacent bright fringes.

Reasoning and Solution

A We can use Equation 24.5 with $m = 2$, $y_2 = 4.5 \times 10^{-2}$ m, $L = 1.2$ m, and $d = 3.0 \times 10^{-5}$ m:

$$\lambda = \frac{y_2 d}{mL} = \frac{(4.5 \times 10^{-2} \text{ m})(3.0 \times 10^{-5} \text{ m})}{2(1.2 \text{ m})} = 5.6 \times 10^{-7} \text{ m} = \boxed{560 \text{ nm}}$$

B Because the positions of the bright fringes are given by Equation 24.5, we see that the distance between *any* adjacent bright fringes (say, those characterized by m and $m + 1$) is

$$\Delta y = y_{m+1} - y_m = \frac{\lambda L}{d}(m + 1) - \frac{\lambda L}{d}m = \frac{\lambda L}{d}$$

$$= \frac{(5.6 \times 10^{-7} \text{ m})(1.2 \text{ m})}{3.0 \times 10^{-5} \text{ m}} = \boxed{2.2 \text{ cm}}$$

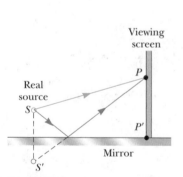

FIGURE 24.5 Lloyd's mirror. An interference pattern is produced on a screen at *P* as a result of the combination of the direct ray (blue) and the reflected ray (brown). The reflected ray undergoes a phase change of 180°.

24.3 CHANGE OF PHASE DUE TO REFLECTION

Young's method of producing two coherent light sources involves illuminating a pair of slits with a single source. Another simple, although ingenious, arrangement for producing an interference pattern with a single light source is known as *Lloyd's mirror*. A light source is placed at point *S*, close to a mirror, as illustrated in Figure 24.5. Waves can reach the viewing point *P* either by the direct path *SP* or by the

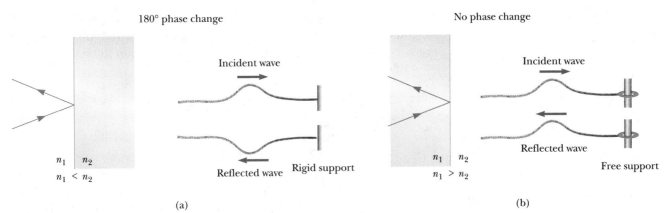

FIGURE 24.6 (a) A ray reflecting from a medium of higher refractive index undergoes a 180° phase change. The right side shows the analogy with a reflected pulse on a string. (b) A ray reflecting from a medium of lower refractive index undergoes no phase change.

path involving reflection from the mirror. The reflected ray can be treated as a ray originating at the source S' behind the mirror. Source S', which is the image of S, can be considered a virtual source.

At points far from the source, we would expect an interference pattern due to waves from S and S', just as is observed for two real coherent sources. An interference pattern is indeed observed. However, the positions of the dark and bright fringes are *reversed* relative to the pattern of two real coherent sources (Young's experiment). This is because the coherent sources S and S' differ in phase by 180°. This 180° phase change is produced by reflection.

To illustrate this further, consider point P', at which the mirror meets the screen. This point is equidistant from S and S'. If path difference alone were responsible for the phase difference, we would expect to see a bright fringe at P' (because the path difference is zero for this point), corresponding to the central fringe of the two-slit interference pattern. Instead, we observe a *dark* fringe at P' because of the 180° phase change produced by reflection. In general, an electromagnetic wave undergoes a phase change of 180° upon reflection from a medium of higher index of refraction than the one in which it was traveling.

It is useful to draw an analogy between reflected light waves and the reflections of a transverse wave on a stretched string when the wave meets a boundary, as in Figure 24.6. The reflected pulse on a string undergoes a phase change of 180° when it is reflected from the boundary of a denser string, or from a rigid barrier, and no phase change when it is reflected from the boundary of a less dense string. Similarly, an electromagnetic wave undergoes a 180° phase change when reflected from the boundary of a medium of higher index of refraction than the one in which it has been traveling. There is no phase change when the wave is reflected from a boundary leading to a medium of lower index of refraction. The transmitted wave that crosses the boundary also undergoes no phase change.

24.4 INTERFERENCE IN THIN FILMS

Interference effects are commonly observed in thin films, such as soap bubbles and thin layers of oil on water. The varied colors observed with incoherent white light result from the interference of waves reflected from the opposite surfaces of the film.

Consider a film of uniform thickness t and index of refraction n, as in Figure 24.7. Let us assume that the light rays traveling in air are nearly normal to the two surfaces of the film. To determine whether the reflected rays interfere constructively or destructively, we must first note the following facts:

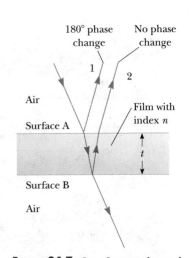

FIGURE 24.7 Interference observed in light reflected from a thin film is due to a combination of rays reflected from the upper and lower surfaces.

1. An electromagnetic wave traveling from a medium of index of refraction n_1 toward a medium of index of refraction n_2 undergoes a 180° phase change on reflection when $n_2 > n_1$. There is no phase change in the reflected wave if $n_2 < n_1$.

2. The wavelength of light λ_n in a medium with index of refraction n is

$$\lambda_n = \frac{\lambda}{n} \qquad [24.7]$$

where λ is the wavelength of light in vacuum.

We apply these rules to the film of Figure 24.7. According to the first rule, ray 1, which is reflected from the upper surface A, undergoes a phase change of 180° with respect to the incident wave. Ray 2, which is reflected from the lower surface B, undergoes no phase change with respect to the incident wave. Therefore, ray 1 is 180° out of phase with respect to ray 2, a situation that is equivalent to a path difference of $\lambda_n/2$. However, we must also consider the fact that ray 2 travels an extra distance of $2t$ before the waves recombine. For example, if $2t = \lambda_n/2$, rays 1 and 2 recombine in phase and constructive interference results. In general, the condition for *constructive interference* is

$$2t = (m + \tfrac{1}{2})\lambda_n \qquad m = 0, 1, 2, \ldots \qquad [24.8]$$

This condition takes into account two factors: (a) the difference in optical path length for the two rays (the term $m\lambda_n$) and (b) the 180° phase change upon reflec-

PHYSICS *IN ACTION*

Interference

The image on the left shows a layer of soap-film bubbles on water. The colors, produced just before the bubbles burst, are due to interference between light rays reflected from the front and back of the thin film of soap making up the bubble. The color depends on the thickness of the film, ranging from black where the film is at its thinnest to magenta where it is thickest.

On the right, a thin film of oil on water displays interference, evidenced by the pattern of colors when white light is incident on the film. Variations in the film's thickness in the vicinity of the blade produces the intersecting color pattern.

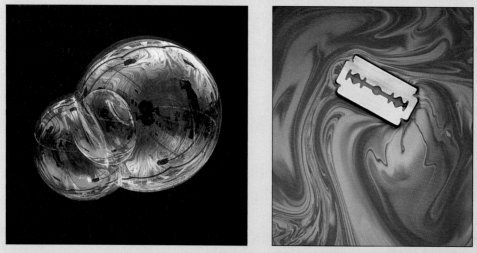

(Dr. Jeremy Burgess/Science Photo Library)

(Peter Aprahamian/Science Photo Library/Photo Researchers, Inc.)

tion (the term $\lambda_n/2$). Because $\lambda_n = \lambda/n$, we can write Equation 24.8 in the form

$$2nt = (m + \tfrac{1}{2})\lambda \qquad m = 0, 1, 2, \ldots \qquad \text{[24.9]}$$

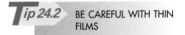 Condition for constructive interference (thin film)

If the extra distance $2t$ traveled by ray 2 is a multiple of λ_n, the two waves combine out of phase and destructive interference results. The general equation for *destructive interference* is

$$2nt = m\lambda \qquad m = 0, 1, 2, \ldots \qquad \text{[24.10]}$$

Condition for destructive interference (thin film)

It is important to realize that two factors influence interference: (1) possible phase reversals on reflection and (2) differences in travel distance. The foregoing conditions for constructive and destructive interference are valid when the medium above the top surface of the film is the same as the medium below the bottom surface. The surrounding medium may have a refractive index less than or greater than that of the film. In either case, the rays reflected from the two surfaces will be out of phase by 180°. If the film is placed between two *different* media, one of lower refractive index than the film and one of higher refractive index, the conditions for constructive and destructive interference are reversed. In this case, either there is a phase change of 180° for both ray 1 reflecting from surface A and ray 2 reflecting from surface B, or there is no phase change for either ray; hence, the net change in relative phase due to the reflections is *zero*.

Tip 24.2 BE CAREFUL WITH THIN FILMS

Be sure to include *both* effects—path length and phase change—when analyzing an interference pattern from a thin film.

NEWTON'S RINGS

Another method for observing interference of light waves is to place a planoconvex lens on top of a flat glass surface, as in Figure 24.8a. With this arrangement, the air film between the glass surfaces varies in thickness from zero at the point of contact to some value t at P. If the radius of curvature R of the lens is very large compared with the distance r, and if the system is viewed from above using light of wavelength λ, a pattern of light and dark rings is observed (Fig. 24.8b). These circular fringes are called **Newton's rings** after their discoverer. Newton's particle model of light could not explain the origin of the rings.

The interference is due to the combination of ray 1, reflected from the plate, with ray 2, reflected from the lower surface of the lens. Ray 1 undergoes a phase change of 180° on reflection, because it is reflected from a boundary leading into a medium of higher refractive index, whereas ray 2 undergoes no phase change. Hence, the conditions for constructive and destructive interference are given by

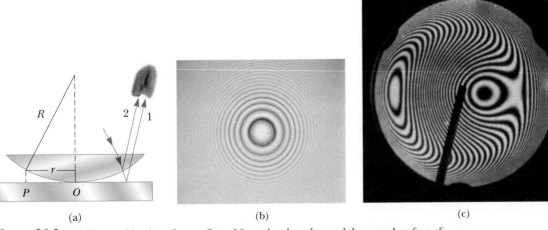

(a)   (b)   (c)

FIGURE 24.8 (a) The combination of rays reflected from the glass plate and the curved surface of the lens gives rise to an interference pattern known as Newton's rings. (b) A photograph of Newton's rings. *(Richard Megna, Fundamental Photographs, NYC)* (c) This asymmetric interference pattern indicates imperfections in the lens. *(From Physical Science Study Committee, College Physics, Lexington, Mass., D. C. Heath and Co., 1968)*

Equations 24.9 and 24.10, respectively, with $n = 1$, because the "film" is air. Here again, we might guess that the contact point O would be bright, corresponding to constructive interference. Instead, it is dark, as seen in Figure 24.8b, because ray 1, reflected from the plate, undergoes a 180° phase change with respect to ray 2. Using the geometry shown in Figure 24.8a, we can obtain expressions for the radii of the bright and dark bands in terms of the radius of curvature R, and vacuum wavelength λ. For example, the dark rings have radii of $r \approx \sqrt{m\lambda R/n}$. In Problem 64 at the end of the chapter, you will be asked to supply the details.

One of the important uses of Newton's rings is in the testing of optical lenses. A circular pattern like that in Figure 24.8b is achieved only when the lens is ground to a perfectly spherical curvature. Variations from such symmetry might produce a pattern like that in Figure 24.8c. These variations give an indication of how the lens must be ground and polished to remove the imperfections.

APPLICATION

CHECKING FOR IMPERFECTIONS IN OPTICAL LENSES

PROBLEM-SOLVING *STRATEGY*   **Thin-Film Interference**

The following features should be kept in mind when you work thin-film interference problems:

1. Identify the thin film causing the interference.
2. The type of interference—constructive or destructive—that occurs is determined by the phase relationship between the portion of the wave reflected at the upper surface of the film and the portion reflected at the lower surface.
3. Phase differences between the two portions of the wave have two causes: (a) differences in the distances traveled by the two portions and (b) phase changes occurring on reflection. *Both* causes must be considered when you are determining whether constructive or destructive interference occurs.
4. The interference is constructive if the path difference between the two waves is an integral multiple of λ, and destructive if the equivalent path difference is $\lambda/2$, $3\lambda/2$, $5\lambda/2$, and so forth. However, the conditions for constructive and destructive interference are reversed if one of the waves undergoes a phase change on reflection.

Example 24.2   Interference in a Soap Film

Calculate the minimum thickness of a soap-bubble film ($n = 1.33$) that will result in constructive interference in the reflected light if the film is illuminated by light with a wavelength in free space of 602 nm.

Reasoning   The minimum film thickness for constructive interference corresponds to $m = 0$ in Equation 24.9. This gives $2nt = \lambda/2$.

Solution   Because $2nt = \lambda/2$, we have

$$ t = \frac{\lambda}{4n} = \frac{602 \text{ nm}}{(4)(1.33)} = \boxed{113 \text{ nm}} $$

EXERCISE   What other film thicknesses will produce constructive interference?

ANSWER   338 nm, 564 nm, 789 nm, and so on

Example 24.3 **Nonreflective Coatings for Solar Cells**

Semiconductors such as silicon are used to fabricate solar cells—devices that generate electric energy when exposed to sunlight. Solar cells are often coated with a transparent thin film, such as silicon monoxide (SiO; $n = 1.45$) to minimize reflective losses (Fig. 24.9). A silicon solar cell ($n = 3.50$) is coated with a thin film of silicon monoxide for this purpose. Assuming normal incidence, determine the minimum thickness of the film that will produce the least reflection at a wavelength of 552 nm.

Reasoning Reflection is least when rays 1 and 2 in Figure 24.9 meet the condition of destructive interference. Note that both rays undergo 180° phase changes on reflection. Hence, the net change in phase due to reflection is zero, and the condition for a reflection *minimum* is a path difference of $\lambda_n/2$; therefore, $2t = \lambda_n/2 = \lambda/2n$.

Solution Because $2t = \lambda/2n$, the required thickness is

$$t = \frac{\lambda}{4n} = \frac{552 \text{ nm}}{(4)(1.45)} = \boxed{95.2 \text{ nm}}$$

Typically, such coatings reduce the reflective loss from 30% (with no coating) to 10% (with coating), thereby increasing the cell's efficiency, because more light is available to create charge carriers in the cell. In reality, the coating is never perfectly nonreflecting, because the required thickness is wavelength-dependent and the incident light covers a wide range of wavelengths.

Glass lenses used in cameras and other optical instruments are usually coated with a transparent thin film, such as magnesium fluoride (MgF_2), to reduce or eliminate unwanted reflection. More important, such coatings enhance the transmission of light through the lenses.

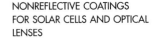**APPLICATION**

NONREFLECTIVE COATINGS FOR SOLAR CELLS AND OPTICAL LENSES

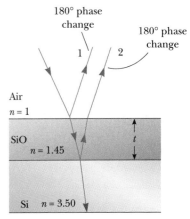

FIGURE 24.9 (Example 24.3) Reflective losses from a silicon solar cell are minimized by coating it with a thin film of silicon monoxide, SiO.

Example 24.4 **Interference in a Wedge-Shaped Film**

A thin, wedge-shaped film of refractive index n is illuminated with monochromatic light of wavelength λ, as illustrated in Figure 24.10a. Describe the interference pattern observed in this case.

Reasoning and Solution The interference pattern is that of a thin film of variable thickness consisting of air. Hence, the pattern is a series of alternating bright and dark

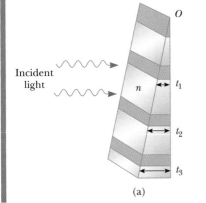

FIGURE 24.10 (Example 24.4) (a) Interference bands in reflected light can be observed by illuminating a wedge-shaped film with monochromatic light. The dark areas correspond to positions of destructive interference. (b) Interference in a vertical film of variable thickness. The top of the film appears darkest where the film is thinnest. *(Richard Megna, Fundamental Photographs)*

parallel bands. A dark band corresponding to destructive interference appears at point *O*, the apex, because the upper reflected ray undergoes a 180° phase change and the lower one does not. According to Equation 24.10, other dark bands appear when $2nt = m\lambda$, so that $t_1 = \lambda/2n$, $t_2 = \lambda/n$, $t_3 = 3\lambda/2n$, and so on. Similarly, bright bands are observed when the thickness satisfies the condition $2nt = (m + \frac{1}{2})\lambda$, corresponding to thicknesses of $\lambda/4n$, $3\lambda/4n$, $5\lambda/4n$, and so on. If white light is used, bands of different colors are observed at different points, corresponding to the different wavelengths of light present. This is shown in the soap film in Figure 24.10b.

24.5 USING INTERFERENCE TO READ CDs AND DVDs

APPLICATION

THE PHYSICS OF CDS AND DVDS

Compact disks (CDs) and digital video disks (DVDs) have revolutionized the computer and entertainment industries by providing fast access; high-density storage of text, graphics, and movies; and high-quality sound recordings. The data on these disks are stored digitally—that is, as a series of zeros and ones—and these zeros and ones are read by laser light reflected from the disk. Strong reflections (constructive interference) from the disk are chosen to represent zeros and weak reflections (destructive interference) represent ones.

To see in more detail how thin-film interference plays a crucial role in reading CDs, consider Figure 24.11. This shows a photomicrograph of several CD tracks, which consist of a sequence of pits (when viewed from the top or label side of the disk) of varying length formed in a reflecting-metal information layer. A cross-sectional view of a CD as shown in Figure 24.12 reveals that the pits appear as bumps to the laser beam, which shines on the metallic layer through a clear plastic coating from below. As the disk rotates, the laser reflects off the sequence of bumps and lower areas into a photodetector, which converts the fluctuating reflected light intensity into an electrical string of zeros and ones. To make the light fluctuations more pronounced and easier to detect, the pit depth *t* is made equal to one quarter of a wavelength of the laser light in the plastic. When the beam hits a rising or falling bump edge, part of the beam reflects from the top of the bump and part from the lower adjacent area, ensuring destructive interference and very low intensity when the reflected beams combine at the detector. Thus bump edges are read as ones, and flat bump tops and intervening flat plains are read as zeros.

Example 24.5 determines the pit depth for a standard CD using an infrared laser of wavelength 780 nm. DVDs use shorter-wavelength lasers of 635 nm, and the track separation, pit depth, and minimum pit length are all smaller. This allows a DVD to store about 30 times more information than a CD.

FIGURE 24.11 A photomicrograph of adjacent tracks on a compact disc (CD). The information encoded in these pits and smooth areas is read by a laser beam. *(Courtesy of Sony Disc Manufacturing)*

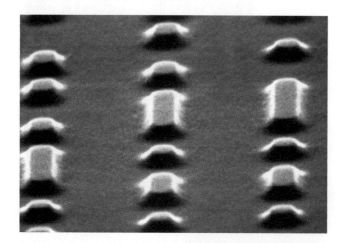

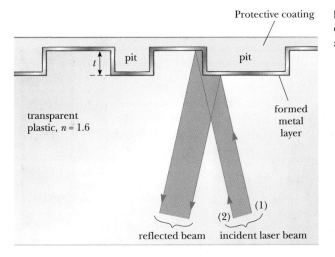

Protective coating

pit   pit

t

transparent
plastic, *n* = 1.6

formed
metal
layer

(1)
(2)
reflected beam   incident laser beam

FIGURE 24.12 Cross section of a
CD showing metallic pits of depth *t*
and a laser beam detecting a pit edge.

Example 24.5   Pit Depth in a CD

Find the pit depth in a CD, which has a plastic transparent layer with index of refraction
of 1.6 and is designed for use in a CD player, using a laser that has a wavelength of
780 nm in air.

Reasoning and Solution   Rays 1 and 2 undergo a 180° phase change on reflection
from the metal layer, which acts like a mirror. (See Fig. 24.12.) Thus, there is no phase
difference between rays 1 and 2 because of reflection, but there is a phase difference
caused by the extra distance $2t$ traveled by ray 2. The wavelength of light in a substance
with index of refraction n is λ/n, where λ is the wavelength in air. Hence, for destructive
interference of rays 1 and 2, we know that $2t = \lambda/2n$, or

$$t = \frac{\lambda}{4n} = \frac{780 \text{ nm}}{(4)(1.60)} = 1.2 \times 10^2 \text{ nm}$$

24.6   DIFFRACTION

Suppose a light beam is incident on two slits, as in Young's double-slit experiment.
If the light truly traveled in straight-line paths after passing through the slits, as in
Figure 24.13a, the waves would not overlap and no interference pattern would be
seen. Instead, Huygens's principle requires that the waves spread out from the
slits, as shown in Figure 24.13b. In other words, the light deviates from a straight-
line path and enters the region that would otherwise be shadowed. This spreading
out of light from its initial line of travel is called **diffraction.**

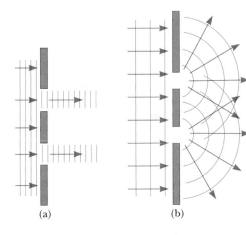

(a)   (b)

FIGURE 24.13   (a) If light did not
spread out after passing through the
slits, no interference would occur.
(b) The light from the two slits overlaps as it spreads out, filling the expected shadowed regions with light
and producing interference fringes.

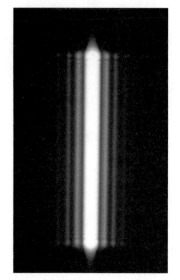

FIGURE 24.14 The diffraction pattern that appears on a screen when light passes through a narrow, vertical slit. The pattern consists of a broad central band and a series of less intense and narrower side bands.

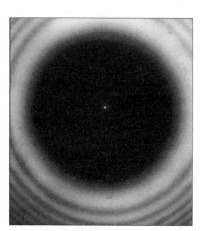

FIGURE 24.15 The diffraction pattern of a penny placed midway between the screen and the source. *(Courtesy of P. M. Rinard, from* Am. J. Phys., *44:70, 1976)*

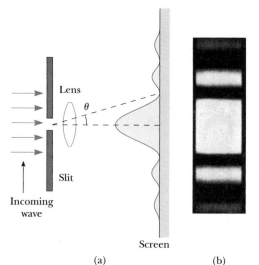

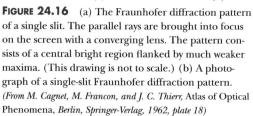

(a)   (b)

FIGURE 24.16 (a) The Fraunhofer diffraction pattern of a single slit. The parallel rays are brought into focus on the screen with a converging lens. The pattern consists of a central bright region flanked by much weaker maxima. (This drawing is not to scale.) (b) A photograph of a single-slit Fraunhofer diffraction pattern. *(From M. Cagnet, M. Francon, and J. C. Thierr,* Atlas of Optical Phenomena, *Berlin, Springer-Verlag, 1962, plate 18)*

In general, diffraction occurs when waves pass through small openings, around obstacles, or by sharp edges. For example, when a single narrow slit is placed between a distant light source (or a laser beam) and a screen, the light produces a diffraction pattern like that in Figure 24.14. The pattern consists of a broad, intense central band flanked by a series of narrower, less intense secondary bands (called **secondary maxima**) and a series of dark bands, or **minima.** This cannot be explained within the framework of geometric optics, which says that light rays traveling in straight lines should cast a sharp image of the slit on the screen.

Figure 24.15 shows the diffraction pattern and shadow of a penny. The pattern consists of the shadow, a bright spot at its center, and a series of bright and dark circular bands of light near the edge of the shadow. The bright spot at the center (called the *Fresnel bright spot*) is explained by Augustin Fresnel's wave theory of light, which predicts constructive interference at this point for certain locations of the penny. In contrast, from the viewpoint of geometric optics, the center of the pattern would be completely screened by the penny, and so we would never observe a central bright spot.

One type of diffraction, called **Fraunhofer diffraction,** occurs when the rays leave the diffracting object in parallel directions. This can be achieved experimentally either by placing the observing screen far from the slit or by using a converging lens to focus the parallel rays on a nearby screen, as in Figure 24.16a. A bright fringe is observed along the axis at $\theta = 0$, with alternating dark and bright fringes on each side of the central bright fringe. Figure 24.16b is a photograph of a single-slit Fraunhofer diffraction pattern.

24.7 SINGLE-SLIT DIFFRACTION

Until now we have assumed that slits are line sources of light. In this section we determine how their finite widths are the basis for understanding the nature of the Fraunhofer diffraction pattern produced by a single slit.

We can deduce some important features of this problem by examining waves coming from various portions of the slit, as shown in Figure 24.17. According to Huygens's principle, **each portion of the slit acts as a source of waves. Hence, light from one portion of the slit can interfere with light from another portion,** and the resultant intensity on the screen depends on the direction θ.

To analyze the diffraction pattern, it is convenient to divide the slit into halves, as in Figure 24.17. All the waves that originate at the slit are in phase. Consider waves 1 and 3, which originate at the bottom and center of the slit, respectively. Wave 1 travels farther than wave 3 by an amount equal to the path difference $(a/2)\sin\theta$, where a is the width of the slit. Similarly, the path difference between waves 3 and 5 is also $(a/2)\sin\theta$. If this path difference is exactly half of a wavelength (corresponding to a phase difference of 180°), the two waves cancel each other and destructive interference results. This is true, in fact, for any two waves that originate at points separated by half the slit width, because the phase difference between two such points is 180°. Therefore, waves from the upper half of the slit interfere *destructively* with waves from the lower half of the slit when

$$\frac{a}{2}\sin\theta = \frac{\lambda}{2} \qquad \text{or when} \qquad \sin\theta = \frac{\lambda}{a}$$

If we divide the slit into four parts rather than two, and use similar reasoning, we find that the screen is also dark when

$$\sin\theta = \frac{2\lambda}{a}$$

Likewise, we can divide the slit into six parts and show that darkness occurs on the screen when

$$\sin\theta = \frac{3\lambda}{a}$$

Therefore, the general condition for **destructive interference** for a single slit of width a is

$$\sin\theta_{\text{dark}} = m\frac{\lambda}{a} \qquad m = \pm1, \pm2, \pm3, \ldots \qquad \text{[24.11]}$$

Equation 24.11 gives the values of θ for which the diffraction pattern has zero intensity—that is, a dark fringe is formed. However, this equation tells us nothing about the variation in intensity along the screen. The general features of the intensity distribution along the screen are shown in Figure 24.18. A broad central bright fringe is observed, flanked by much weaker bright fringes alternating with dark fringes. The various dark fringes (points of zero intensity) occur at the values of θ that satisfy Equation 24.11. The points of constructive interference lie approximately halfway between the dark fringes. Note that the central bright fringe is twice as wide as the weaker maxima having $m > 1$.

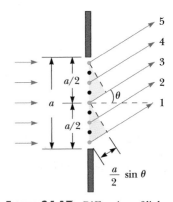

FIGURE 24.17 Diffraction of light by a narrow slit of width a. Each portion of the slit acts as a point source of waves. The path difference between rays 1 and 3 or between rays 2 and 4 is equal to $(a/2)\sin\theta$. (This drawing is not to scale, and the rays are assumed to converge at a distant point.)

Webnote 24.2

To see the diffraction pattern of a single slit as a function of the incident wavelength, explore
http://www.lightlink.com/sergey/java/java/slitdiffr/index.html

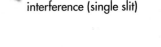 Condition for destructive interference (single slit)

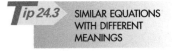 *Tip 24.3* SIMILAR EQUATIONS WITH DIFFERENT MEANINGS

Although Equations 24.2 and 24.11 have the same form, they have different meanings. Equation 24.2 describes the *bright* regions in a two-slit interference pattern, whereas Equation 24.11 describes the *dark* regions in a single-slit interference pattern.

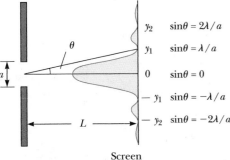

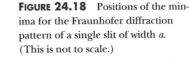

FIGURE 24.18 Positions of the minima for the Fraunhofer diffraction pattern of a single slit of width a. (This is not to scale.)

$y_2 \qquad \sin\theta = 2\lambda/a$
$y_1 \qquad \sin\theta = \lambda/a$
$0 \qquad \sin\theta = 0$
$-y_1 \qquad \sin\theta = -\lambda/a$
$-y_2 \qquad \sin\theta = -2\lambda/a$

Screen

In a single-slit diffraction experiment, as the width of the slit is made smaller, the width of the central maximum of the diffraction pattern becomes (a) smaller, (b) larger, or (c) remains the same.

APPLYING PHYSICS 24.3

If a classroom door is open even just a small amount, you can hear sounds coming from the hallway. Yet you cannot see what is going on in the hallway. Why is there this difference?

Explanation The space between the slightly open door and the wall is acting as a single slit for waves. Sound waves have wavelengths larger than the slit width, so sound is effectively diffracted by the opening and the central maximum spreads throughout the room. Light wavelengths are much smaller than the slit width, so there is virtually no diffraction for the light. You must have a direct line of sight to detect the light waves.

Example 24.6 Where Are the Dark Fringes?

Light of wavelength 580 nm is incident on a slit of width 0.30 mm. The observing screen is placed 2.0 m from the slit. Find the positions of the first dark fringes and the width of the central bright fringe.

Solution The first dark fringes that flank the central bright fringe correspond to $m = \pm 1$ in Equation 24.11:

$$\sin \theta = \pm \frac{\lambda}{a} = \pm \frac{5.8 \times 10^{-7} \, \text{m}}{0.30 \times 10^{-3} \, \text{m}} = \pm 1.9 \times 10^{-3}$$

From the triangle in Figure 24.16, we see that $\tan \theta = y_1 / L$. Because θ is very small, we can use the approximation $\sin \theta \approx \tan \theta$, so that $\sin \theta \approx y_1 / L$. Therefore, the positions of the first minima, measured from central axis, are

$$y_1 \approx L \sin \theta = \pm L \frac{\lambda}{a} = \boxed{\pm 3.9 \times 10^{-3} \, \text{m}}$$

The positive and negative signs correspond to the first dark fringes on either side of the central bright fringe. Hence, the width of the central bright fringe is given by $2|y_1| = 7.8 \times 10^{-3} \, \text{m} = \boxed{7.8 \, \text{mm}}$. Note that this value is much greater than the width of the slit. However, as the width of the slit is *increased*, the diffraction pattern *narrows*, corresponding to smaller values of θ. In fact, for large values of a, the maxima and minima are so closely spaced that the only observable pattern is a large central bright area resembling the geometric image of the slit. Because the width of the geometric image increases as the slit width increases, the narrowest image occurs when the geometric and diffraction widths are equal.

EXERCISE Determine the width of the first-order bright fringe.

ANSWER 3.9 mm

24.8 THE DIFFRACTION GRATING

The diffraction grating, a very useful device for analyzing light sources, consists of many equally spaced parallel slits. A grating can be made by scratching parallel lines on a glass plate with a precision machining technique. The scratches act to scatter the light and so behave as slits. A typical grating contains several thousand lines per centimeter. For example, a grating ruled with 5 000 lines/cm has a slit spacing d equal to the inverse of this number; hence, $d = (1/5\ 000)$ cm $= 2 \times 10^{-4}$ cm.

Figure 24.19 is a diagram of a section of a plane diffraction grating. A plane wave is incident from the left, normal to the plane of the grating. The intensity of the pattern on the screen is the result of the combined effects of interference and diffraction. Each slit produces diffraction, and the diffracted beams in turn interfere with one another to produce the pattern. Moreover, each slit acts as a source of waves, and all waves start at the slits in phase. However, for some arbitrary direction θ measured from the horizontal, the waves must travel *different* path lengths before reaching a particular point P on the screen. From Figure 24.19, note that the path difference between waves from any two adjacent slits is $d \sin \theta$. If this path difference equals one wavelength or some integral multiple of a wavelength, waves from all slits will be in phase at P and a bright line will be observed at this point. Therefore, the condition for **maxima** in the interference pattern at the angle θ is

$$d \sin \theta_{\text{bright}} = m\lambda \qquad m = 0, 1, 2, \ldots \qquad \text{[24.12]}$$

◀ Condition for maxima in the interference pattern of a diffraction grating

Light emerging from a slit at an angle other than that for a maximum interferes with nearly complete destruction with light from some other slit on the grating. All such pairs will result in little or no transmission in that direction, as illustrated in Figure 24.20.

Equation 24.12 can be used to calculate the wavelength from the grating spacing and the angle of deviation, θ. The integer m is the **order number** of the diffraction pattern. If the incident radiation contains several wavelengths, each wavelength deviates through a specific angle, which can be found from Equation 24.12. All wavelengths are focused at $\theta = 0$, corresponding to $m = 0$. This is called the *zeroth-order maximum*. The *first-order maximum*, corresponding to $m = 1$, is observed at an angle that satisfies the relationship $\sin \theta = \lambda / d$; the *second-order maximum*, corresponding to $m = 2$, is observed at a larger angle θ, and so on. Figure 24.20 is a sketch of the intensity distribution for some of the orders produced by a diffrac-

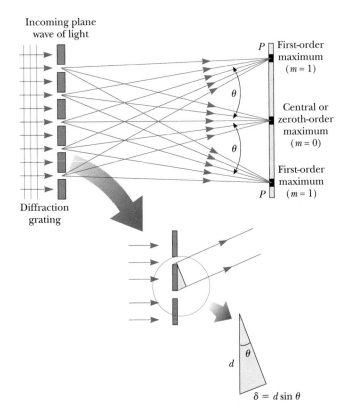

Incoming plane wave of light

P — First-order maximum ($m = 1$)

θ

Central or zeroth-order maximum ($m = 0$)

θ

First-order maximum ($m = 1$)

P

Diffraction grating

d

θ

$\delta = d \sin \theta$

FIGURE 24.19 A side view of a diffraction grating. The slit separation is d, and the path difference between adjacent slits is $d \sin \theta$.

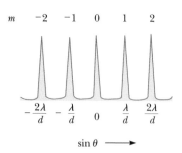

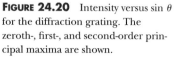

$$\sin \theta \longrightarrow$$

FIGURE 24.20 Intensity versus $\sin \theta$ for the diffraction grating. The zeroth-, first-, and second-order principal maxima are shown.

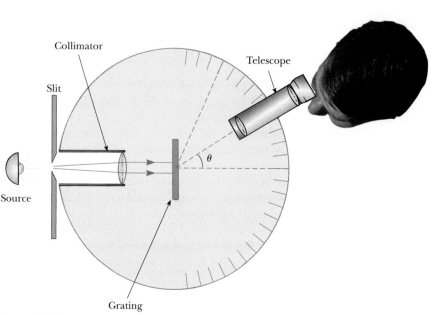

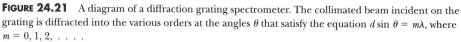

FIGURE 24.21 A diagram of a diffraction grating spectrometer. The collimated beam incident on the grating is diffracted into the various orders at the angles θ that satisfy the equation $d \sin \theta = m\lambda$, where $m = 0, 1, 2, \ldots$.

tion grating. Note the sharpness of the principal maxima and the broad range of the dark areas. This is in contrast to the broad, bright fringes characteristic of the two-slit interference pattern.

A simple arrangement that can be used to measure the angles in a diffraction pattern is shown in Figure 24.21. This is a form of diffraction-grating spectrometer. The light to be analyzed passes through a slit and is formed into a parallel beam by a lens. The light then strikes the grating at a 90° angle. The diffracted light leaves the grating at angles that satisfy Equation 24.12. A telescope is used to view the image of the slit. The wavelength can be determined by measuring the angles at which the images of the slit appear for the various orders.

Quick Quiz 24.2 If laser light is reflected from a phonograph record or a compact disc, a diffraction pattern appears. This occurs because both devices contain parallel tracks of information that act as a reflection diffraction grating. Which device, record or compact disc, results in diffraction maxima that are farther apart?

APPLYING PHYSICS 24.4

White light enters through an opening in an opaque box, exits through an opening on the other side of the box, and a spectrum of colors appears on the wall. How would you determine whether the box *contains* a prism or a diffraction grating?

Explanation The determination could be made by noticing the order of the colors in the spectrum relative to the direction of the original beam of white light. For a prism, in which the separation of light is a result of dispersion, the violet light will be refracted more than the red light. Hence, the order of the spectrum from a prism will be from red, closest to the original direction, to violet. For a diffraction grating, the angle of diffraction increases with wavelength. Thus, the spectrum from the diffraction grating will have colors in the order of violet, closest to the original direction, to red. Furthermore, the diffraction grating will produce *two* first-order spectra on either side of the grating, whereas the prism will only produce a single spectrum.

APPLYING **PHYSICS** 24.5

White light reflected from the surface of a compact disc has a multicolored appearance, as shown in Figure 24.22. Furthermore, the observation depends on the orientation of the disc relative to the eye and the position of the light source. Explain how this works.

Explanation The surface of a compact disc has a spiral-shaped track (with a spacing of approximately 1 μm) that acts as a reflection grating. The light scattered by these closely spaced parallel tracks interferes constructively in certain directions that depend on the wavelength and on the direction of the incident light. Any one section of the disc serves as a diffraction grating for white light, sending beams of constructive interference for different colors in different directions. The different colors you see when viewing one section of the disc change as the light source, the disc, or you move to change the angles of incidence or diffraction.

Figure 24.22 (Applying Physics 24.5) Compact discs act as diffraction gratings when observed under white light. (*© Kristen Brochmann/Fundamental Photographs*)

USE OF DIFFRACTION GRATING IN CD TRACKING

If a CD player is to reproduce sound faithfully, the laser beam must follow the spiral track of information perfectly. However, sometimes the laser beam can drift off track, and without a feedback procedure to let the player know this is happening, the fidelity of the music would be greatly reduced.

Figure 24.23 shows how a diffraction grating is used in a three-beam method to keep the beam on track. The central maximum of the diffraction pattern is

APPLICATION

TRACKING INFORMATION ON A CD

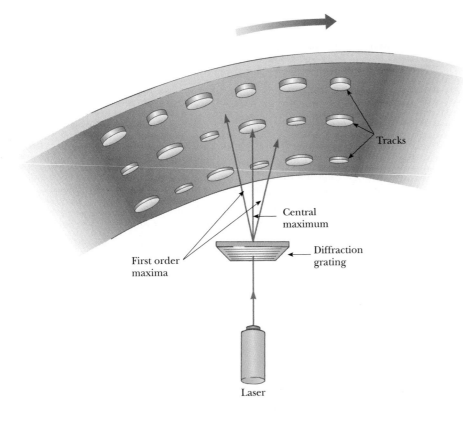

Figure 24.23 The laser beam in a CD player is able to follow the spiral track by using three beams produced with a diffraction grating.

Tracks

Central maximum

Diffraction grating

First order maxima

Laser

used to read the information on the CD track, and the two first-order maxima are used for steering. The grating is designed so that the first-order maxima fall on the smooth surfaces on either side of the information track. Both of these reflected beams have their own detectors, and because both beams are reflected from smooth surfaces, they should both have the same strong intensity when they are detected. If the central beam wanders off the track however, one of the steering beams will begin to strike bumps on the information track and the amount of light reflected will decrease. This information is then used by electronic circuits to drive the main beam back to its desired location.

Example 24.7 The Orders of a Diffraction Grating

Monochromatic light from a helium-neon laser ($\lambda = 632.8$ nm) is incident normally on a diffraction grating containing 6 000 lines/cm. Find the angles at which we would observe the first-order maximum, the second-order maximum, and so forth.

Solution First we must calculate the slit separation, which is the inverse of the number of lines per centimeter:

$$d = \frac{1}{6\,000} \text{ cm} = 1.667 \times 10^{-4} \text{ cm} = 1\,667 \text{ nm}$$

For the first-order maximum ($m = 1$), we get

$$\sin \theta_1 = \frac{\lambda}{d} = \frac{632.8 \text{ nm}}{1\,667 \text{ nm}} = 0.379\,6$$

$$\theta_1 = \boxed{22.31°}$$

For $m = 2$, we find that

$$\sin \theta_2 = \frac{2\lambda}{d} = \frac{2(632.8 \text{ nm})}{1\,667 \text{ nm}} = 0.759\,2$$

$$\theta_2 = \boxed{49.39°}$$

However, for $m = 3$, we find that $\sin \theta_3 = 1.139$. Because $\sin \theta$ cannot exceed unity, this is not a realistic solution. Hence, only zeroth-, first-, and second-order maxima would be observed in this situation.

24.9 POLARIZATION OF LIGHT WAVES

In Chapter 21 we described the transverse nature of electromagnetic waves. Figure 24.24 shows that the electric and magnetic field vectors associated with an electromagnetic wave are at right angles to each other and also to the direction of wave propagation. The phenomenon of polarization, described in this section, is firm evidence of the transverse nature of electromagnetic waves.

An ordinary beam of light consists of a large number of electromagnetic waves emitted by the atoms or molecules of the light source. The vibrating charges asso-

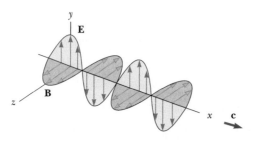

FIGURE 24.24 A diagram of a polarized electromagnetic wave propagating in the *x* direction. The electric field vector **E** vibrates in the *xy* plane, and the magnetic field vector **B** vibrates in the *xz* plane.

ciated with the atoms act as tiny antenna. Each atom produces a wave with its own orientation of **E**, as in Figure 24.24, corresponding to the direction of atomic vibration. However, because all directions of vibration are possible, the resultant electromagnetic wave is a superposition of waves produced by the individual atomic sources. The result is an **unpolarized** light wave, represented schematically in Figure 24.25a. The direction of wave propagation in this figure is perpendicular to the page. Note that *all* directions of the electric field vector are equally probable and lie in a plane (such as the plane of this page) perpendicular to the direction of propagation.

A wave is said to be **linearly polarized** if the resultant electric field **E** vibrates in the same direction *at all times* at a particular point, as in Figure 24.25b. (Sometimes such a wave is described as *plane-polarized* or simply *polarized*.) The wave in Figure 24.24 is an example of a wave linearly polarized in the y direction. As the wave propagates in the x direction, **E** is always in the y direction. The plane formed by **E** and the direction of propagation is called the *plane of polarization* of the wave. In Figure 24.24, the plane of polarization is the xy plane.

It is possible to obtain a linearly polarized beam from an unpolarized beam by removing all waves from the beam except those whose electric field vectors oscillate in a single plane. We now discuss three processes for doing this: (1) selective absorption, (2) reflection, and (3) scattering.

POLARIZATION BY SELECTIVE ABSORPTION

The most common technique for polarizing light is to use a material that transmits waves whose electric field vectors vibrate in a plane parallel to a certain direction and absorbs those waves whose electric field vectors vibrate in directions perpendicular to that direction.

In 1932, E. H. Land discovered a material, which he called **polaroid,** that polarizes light through selective absorption by oriented molecules. This material is fabricated in thin sheets of long-chain hydrocarbons, which are stretched during manufacture so that the molecules align. After a sheet is dipped into a solution containing iodine, the molecules become good electrical conductors. However, the conduction takes place primarily along the hydrocarbon chains, because the valence electrons of the molecules can move easily only along the chains. (Recall that valence electrons are "free" electrons that can readily move through the conductor.) As a result, the molecules readily *absorb* light whose electric field vector is parallel to their lengths, and *transmit* light whose electric field vector is perpendicular to their lengths. It is common to refer to the direction perpendicular to the molecular chains as the **transmission axis.** In an ideal polarizer, all light with **E** parallel to the transmission axis is transmitted, and all light with **E** perpendicular to the transmission axis is absorbed.

Let us now describe the intensity of light that passes through a polarizing material. In Figure 24.26, an unpolarized light beam is incident on the first polarizing sheet, called the **polarizer,** where the transmission axis is as indicated. The light that is passing through this sheet is polarized vertically, and the transmitted elec-

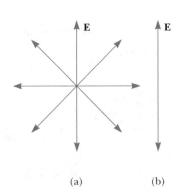

FIGURE 24.25 (a) An unpolarized light beam viewed along the direction of propagation (perpendicular to the page). The transverse electric field vector can vibrate in any direction with equal probability. (b) A linearly polarized light beam with the electric field vector vibrating in the vertical direction.

Webnote 24.3

Polarized light has helped dermatologic surgeons estimate the size (area) of skin cancers. Explore *http://omlc.ogi.edu/news/feb98/polarization/index.html*

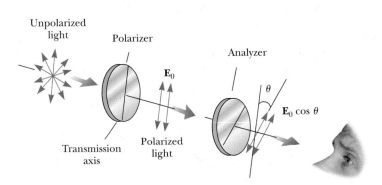

FIGURE 24.26 Two polarizing sheets whose transmission axes make an angle θ with each other. Only a fraction of the polarized light incident on the analyzer is transmitted.

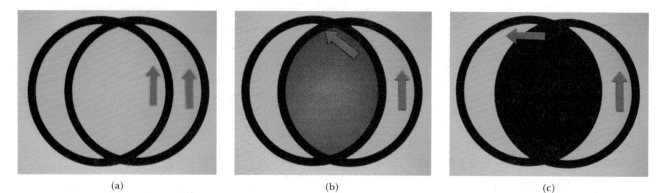

(a) (b) (c)

FIGURE 24.27 The intensity of light transmitted through two polarizers depends on the relative orientations of their transmission axes. (a) The transmitted light has *maximum* intensity when the transmission axes are *aligned* with each other. (b) The transmitted light intensity diminishes when the transmission axes are at an angle of 45° with each other. (c) The transmitted light intensity is a *minimum* when the transmission axes are at *right angles* to each other. *(Photos courtesy of Henry Leap)*

tric field vector is $\mathbf{E}_0$. A second polarizing sheet, called the **analyzer,** intercepts this beam with its transmission axis at an angle of θ to the axis of the polarizer. The component of $\mathbf{E}_0$ that is perpendicular to the axis of the analyzer is completely absorbed. The component of $\mathbf{E}_0$ parallel to the analyzer axis, $E_0 \cos \theta$, is allowed to pass through the analyzer. Because the intensity of the transmitted beam varies as the *square* of its amplitude E, we conclude that the intensity of the (polarized) beam transmitted through the analyzer varies as

$$I = I_0 \cos^2 \theta \qquad\qquad \text{[24.13]}$$

where I_0 is the intensity of the polarized wave incident on the analyzer. This expression, known as **Malus's law,** applies to any two polarizing materials whose transmission axes are at an angle of θ to each other. From this expression, note that the transmitted intensity is a maximum when the transmission axes are parallel ($\theta = 0$ or 180°) and zero (complete absorption by the analyzer) when the transmission axes are perpendicular to each other. This variation in transmitted intensity through a pair of polarizing sheets is illustrated in Figure 24.27.

When unpolarized light of intensity I_0 is sent through a single ideal polarizer, the transmitted linearly polarized light has intensity $I_0/2$. This fact follows from Malus's law because the average value of $\cos^2 \theta$ is one half.

APPLYING **PHYSICS** 24.6

APPLICATION

POLARIZING MICROWAVES

A polarizer for microwaves can be made as a grid of parallel metal wires about a centimeter apart. Is the electric field vector for microwaves transmitted through this polarizer parallel to, or perpendicular to, the metal wires?

Explanation Electric field vectors parallel to the metal wires cause electrons in the metal to oscillate parallel to the wires. Thus, the energy from the waves with these electric field vectors is transferred to the metal by accelerating these electrons and is eventually transformed to internal energy through the metal's resistance. Waves with electric field vectors perpendicular to the metal wires are not able to accelerate electrons and therefore pass through. Thus, the electric field polarization is perpendicular to the metal wires.

POLARIZATION BY REFLECTION

When an unpolarized light beam is reflected from a surface, the reflected light is completely polarized, partially polarized, or unpolarized, depending on the angle of incidence. If the angle of incidence is either 0° or 90° (a normal or grazing an-

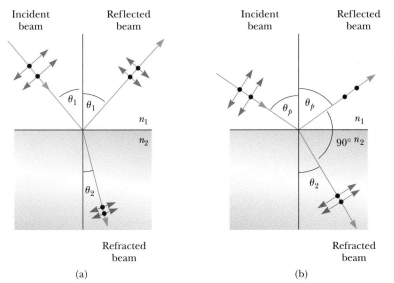

FIGURE 24.28 (a) When unpolarized light is incident on a reflecting surface, the reflected and refracted beams are partially polarized. (b) The reflected beam is completely polarized when the angle of incidence equals the polarizing angle θ_p, satisfying the equation $n = \tan \theta_p$.

gle), the reflected beam is unpolarized. However, for angles of incidence between 0° and 90°, the reflected light is polarized to some extent. For one particular angle of incidence, the reflected beam is completely polarized. Let us now investigate that special angle.

Suppose an unpolarized light beam is incident on a surface, as in Figure 24.28a. The beam can be described by two electric field components, one parallel to the surface (represented by dots) and the other perpendicular to the first component and to the direction of propagation (represented by brown arrows). It is found that the parallel component reflects more strongly than the other components, and this results in a partially polarized beam. Furthermore, the refracted beam is also partially polarized.

Now suppose that the angle of incidence θ_1 is varied until the angle between the reflected and refracted beams is 90° (Fig. 24.28b). At this particular angle of incidence, the reflected beam is completely polarized, with its electric field vector parallel to the surface, whereas the refracted beam is partially polarized. The angle of incidence at which this occurs is called the **polarizing angle** θ_p.

An expression relating the polarizing angle to the index of refraction of the reflecting surface can be obtained by use of Figure 24.28b. From this figure we see that, at the polarizing angle, $\theta_p + 90° + \theta_2 = 180°$, so that $\theta_2 = 90° - \theta_p$. Using Snell's law and taking $n_1 = n_{air} = 1.00$ and $n_2 = n$,

$$n = \frac{\sin \theta_1}{\sin \theta_2} = \frac{\sin \theta_p}{\sin \theta_2}$$

Because $\sin \theta_2 = \sin(90° - \theta_p) = \cos \theta_p$, the expression for n can be written

$$n = \frac{\sin \theta_p}{\cos \theta_p} = \tan \theta_p \qquad [24.14]$$

◀ Brewster's law

This expression is called **Brewster's law,** and the polarizing angle θ_p is sometimes called **Brewster's angle** after its discoverer, Sir David Brewster (1781–1868). For example, Brewster's angle for crown glass ($n = 1.52$) has the value $\theta_p = \tan^{-1}(1.52) = 56.7°$. Because n varies with wavelength for a given substance, Brewster's angle is also a function of wavelength.

Polarization by reflection is a common phenomenon. Sunlight reflected from water, glass, or snow is partially polarized. If the surface is horizontal, the electric

APPLICATION

POLAROID SUNGLASSES

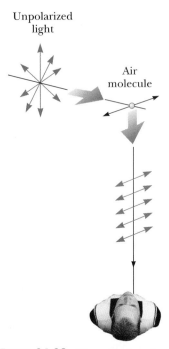

FIGURE 24.29 The scattering of unpolarized sunlight by air molecules. The light observed at right angles is linearly polarized because the vibrating molecule has a horizontal component of vibration.

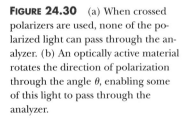

field vector of the reflected light has a strong horizontal component. Sunglasses made of polarizing material reduce the glare, which *is* the reflected light. The transmission axes of the lenses are oriented vertically to absorb the strong horizontal component of the reflected light. Because the reflected light is mostly polarized, most of the glare can be eliminated without removing most of the normal light.

POLARIZATION BY SCATTERING

When light is incident on a system of particles, such as a gas, the electrons in the medium can absorb and reradiate part of the light. The absorption and reradiation of light by the medium, called **scattering**, is what causes sunlight reaching an observer on Earth from straight overhead to be polarized. You can observe this effect by looking directly up through a pair of sunglasses made of polarizing glass. Less light passes through at certain orientations of the lenses than at others.

Figure 24.29 illustrates how the sunlight becomes polarized. The left side of the figure shows an incident unpolarized beam of sunlight on the verge of striking an air molecule. When the beam strikes the air molecule, it sets the electrons of the molecule into vibration. These vibrating charges act like those in an antenna except that they vibrate in a complicated pattern. The horizontal part of the electric field vector in the incident wave causes the charges to vibrate horizontally, and the vertical part of the vector simultaneously causes them to vibrate vertically. A horizontally polarized wave is emitted by the electrons as a result of their horizontal motion, and a vertically polarized wave is emitted parallel to Earth as a result of their vertical motion.

Scientists have found that bees and homing pigeons use the polarization of sunlight as a navigational aid.

OPTICAL ACTIVITY

Many important practical applications of polarized light involve the use of certain materials that display the property of **optical activity.** A substance is said to be optically active if it rotates the plane of polarization of transmitted light. Suppose unpolarized light is incident on a polarizer from the left, as in Figure 24.30a. The transmitted light is polarized vertically, as shown. If this light is then incident on an analyzer with its axis perpendicular to that of the polarizer, no light emerges from it. If an optically active material is placed between the polarizer and analyzer, as in Figure 24.30b, the material causes the direction of the polarized beam to ro-

FIGURE 24.30 (a) When crossed polarizers are used, none of the polarized light can pass through the analyzer. (b) An optically active material rotates the direction of polarization through the angle θ, enabling some of this light to pass through the analyzer.

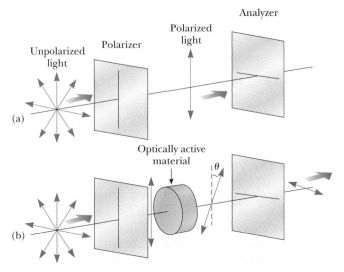

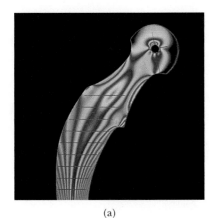
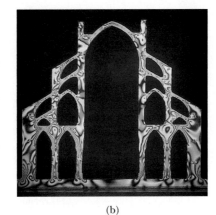

(a) (b)

FIGURE 24.31 (a) Strain distribution in a plastic model of a hip replacement used in a medical research laboratory. The pattern is produced when the plastic model is placed between a polarizer and analyzer oriented perpendicular to each other. *(Sepp Seitz 1981)* (b) A plastic model of an arch structure under load conditions observed between perpendicular polarizers. Such patterns are useful in the optimum design of architectural components. *(Peter Aprahamian/Science Photo Library)*

tate through the angle θ. As a result, some light is able to pass through the analyzer. The angle through which the light is rotated by the material can be found by rotating the polarizer until the light is again extinguished. It is found that the angle of rotation depends on the length of the sample and, if the substance is in solution, on the concentration. One optically active material is a solution of common sugar, dextrose. A standard method for determining the concentration of sugar solutions is to measure the rotation produced by a fixed length of the solution.

Optical activity occurs in a material because of an asymmetry in the shape of its constituent molecules. For example, some proteins are optically active because of their spiral shapes. Other materials, such as glass and plastic, become optically active when placed under stress. If polarized light is passed through an unstressed piece of plastic and then through an analyzer with an axis perpendicular to that of the polarizer, none of the polarized light is transmitted. However, if the plastic is placed under stress, the regions of greatest stress produce the largest angles of rotation of polarized light. Hence, we observe a series of light and dark bands in the transmitted light. Engineers often use this procedure in the design of structures ranging from bridges to small tools. A plastic model is built and analyzed under different load conditions to determine positions of potential weakness and failure under stress. If the design is poor, patterns of light and dark bands will indicate the points of greatest weakness, and the design can be corrected at an early stage. Figure 24.31 shows examples of stress patterns in plastic.

APPLICATION

FINDING THE CONCENTRATIONS OF SOLUTIONS USING THEIR OPTICAL ACTIVITY

LIQUID CRYSTALS

An effect similar to rotation of the plane of polarization is used to create the familiar displays on pocket calculators, wristwatches, laptop computers, and so forth. The properties of a unique substance called a liquid crystal make these displays (called LCDs for *liquid crystal displays*) possible. As its name implies, a **liquid crystal** is a substance with properties intermediate between those of a crystalline solid and those of a liquid—that is, the molecules of the substance are more orderly than those in a liquid but less than those in a pure crystalline solid. The forces that hold the molecules together in such a state are just barely strong enough to enable the substance to maintain a definite shape, so it is reasonable to call it a solid. However, small inputs of mechanical or electrical energy can disrupt these weak bonds and make the substance flow, rotate, or twist.

To see how liquid crystals are used to create a display, consider Figure 24.32a. The liquid crystal is placed between two glass plates in the pattern shown, and electrical contacts, indicated by the thin lines, are made to the liquid crystal. When a voltage is applied across any segment in the display, that segment turns dark. In this fashion, any number between 0 and 9 can be formed by the pattern, depending on the voltages applied to the seven segments.

APPLICATION

LIQUID CRYSTAL DISPLAYS (LCDS)

FIGURE 24.32 (a) The light-segment pattern of a liquid crystal display. (b) Rotation of a polarized light beam by a liquid crystal when the applied voltage is zero. (c) Molecules of the liquid crystal align with the electric field when a voltage is applied.

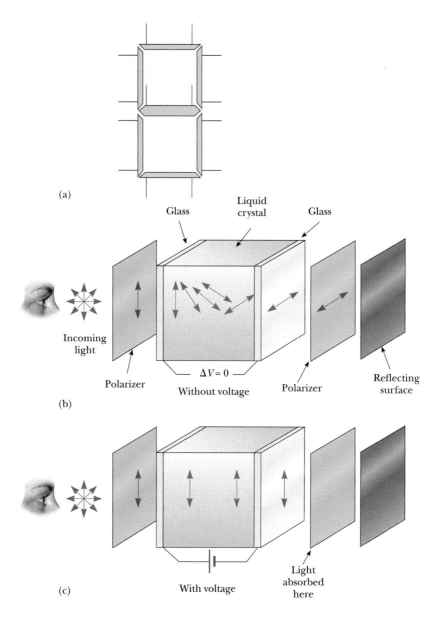

To see why a segment can be changed from dark to light by the application of a voltage, consider Figure 24.32b, which shows the basic construction of a portion of the display. The liquid crystal is placed between two glass substrates that are packaged between two pieces of Polaroid material with their transmission axes perpendicular. A reflecting surface is placed behind one of the pieces of Polaroid. First consider what happens when light falls on this package and no voltages are applied to the liquid crystal, as shown in Figure 24.32b. Incoming light is polarized by the polarizer on the left and then falls on the liquid crystal. As the light passes through the crystal, its plane of polarization is rotated by 90°, allowing it to pass through the polarizer on the right. It reflects from the reflecting surface and retraces its path through the crystal. Thus, an observer to the left of the crystal sees the segment as being bright. When a voltage is applied as in Figure 24.32c, the molecules of the liquid crystal do not rotate the plane of polarization of the light. In this case, the light is absorbed by the polarizer on the right and none is reflected back to the observer to the left of the crystal. Thus, the observer sees this segment as black. Changing the applied voltage to the crystal in a precise pattern and at precise times can make the pattern tick off the seconds on a watch, display a letter on a computer display, and so forth.

of a distant object (Formed by refraction at the cornea) falls 2.53 cm behind the retina. The lens is designed to put the image of the distant object on the retina. What is the power of the implanted lens? (*Hint:* Consider the image formed by the cornea as a virtual object.)

16. A person is to be fitted with bifocals. She can see clearly when the object is between 30 cm and 1.5 m from the eye. (a) The upper portions of the bifocals (Fig. P25.16) should be designed to enable her to see distant objects clearly. What power should they have? (b) The lower portions of the bifocals should enable her to see objects comfortably at 25 cm. What power should they have?

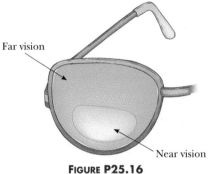

Far vision

Near vision

FIGURE P25.16

Section 25.3   The Simple Magnifier

17. A philatelist examines the printing detail on a stamp using a biconvex lens of focal length 10.0 cm as a simple magnifier. The lens is held close to the eye, and the lens-to-object distance is adjusted so that the virtual image is formed at the normal near point (25.0 cm). Calculate the magnification.

18. A lens having a focal length of 25 cm is used as a simple magnifier. (a) What is the angular magnification obtained when the image is formed at the normal near point ($q = -25$ cm)? (b) What is the angular magnification produced by this lens when the eye is relaxed?

19. A biology student uses a simple magnifier to examine the structural features of the wing of an insect. The wing is held 3.50 cm in front of the lens, and the image is formed 25.0 cm from the eye. (a) What is the focal length of the lens? (b) What angular magnification is achieved?

20. A lens that has a focal length of 5.00 cm is used as a magnifying glass. (a) To obtain maximum magnification, where should the object be placed? (b) What is the magnification?

21. A leaf of length h is positioned 71.0 cm in front of a converging lens with a focal length of 39.0 cm. An observer views the image of the leaf from a position 1.26 m behind the lens, as shown in Figure P25.21. (a) What is the magnitude of the lateral magnification (ratio of image size to object size) produced by the lens? (b) What angular magnifi-

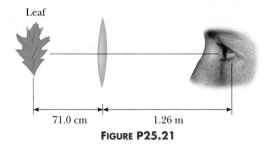

Leaf

71.0 cm   1.26 m

FIGURE P25.21

cation is achieved by viewing the image of the leaf rather than viewing the leaf directly?

Section 25.4   The Compound Microscope

Section 25.5   The Telescope

22. The distance between eyepiece and objective lens in a certain compound microscope is 23.0 cm. The focal length of the eyepiece is 2.50 cm, and that of the objective is 0.400 cm. What is the overall magnification of the microscope?

23. The desired overall magnification of a compound microscope is 140×. The objective alone produces a lateral magnification of 12×. Determine the required focal length of the eyepiece.

24. A microscope has an objective lens with a focal length of 16.22 mm and an eyepiece with a focal length of 9.50 mm. With the length of the barrel set at 29.0 cm, the diameter of a red blood cell's image subtends an angle of 1.43 mrad with the eye. If the final image distance is 29.0 cm from the eyepiece, what is the actual diameter of the red blood cell?

25. The length of a microscope tube is 15.0 cm. The focal length of the objective is 1.00 cm, and the focal length of the eyepiece is 2.50 cm. What is the magnification of the microscope, assuming it is adjusted so that the eye is relaxed?

26. The Yerkes refracting telescope has a 1.00-m-diameter objective lens of focal length 20.0 m. Assume it is used with an eyepiece of focal length 2.50 cm. (a) Determine the angular magnification of the planet Mars as seen through this telescope. (b) Are the Martian polar caps right side up or upside down?

27. The lenses of an astronomical telescope are 92 cm apart when adjusted for viewing a distant object with minimum eyestrain. The angular magnification produced by the telescope is 45. Compute the focal length of each lens.

28. An elderly sailor is shipwrecked on a desert island but manages to save his eyeglasses. The lens for one eye has a power of + 1.20 diopters, and the other lens has a power of + 9.00 diopters. (a) What is the magnifying power of the telescope he can construct with these lenses? (b) How far apart are the lenses when the telescope is adjusted for minimum eyestrain?

29. Astronomers often take photographs with the objective lens or mirror of a telescope alone, without an eyepiece. (a) Show that the image size h' for this telescope is given by $h' = fh/(f - p)$ where h is the object size, f the objective focal length, and p the object distance. (b) Simplify the expression in part (a) if the object distance is much greater than objective focal length. (c) The "wingspan" of the International Space Station is 108.6 m, the overall width of its solar panel configuration. When it is orbiting at an altitude of 407 km, find the width of the image formed by a telescope objective of focal length 4.00 m.

30. Galileo devised a simple terrestrial telescope that produces an upright image. It consists of a converging objective lens and a diverging eyepiece at opposite ends of the telescope tube. For distant objects, the tube length is the objective focal length less the absolute value of the eyepiece focal length. (a) Does the user of the telescope see a real or virtual image? (b) Where is the final image? (c) If a telescope is to be constructed with a tube of length 10.0 cm and a magnification of 3.00, what are the focal lengths of the objective and eyepiece?

31. A person decides to use an old pair of eyeglasses to make some optical instruments. He knows that the near point in his left eye is 50.0 cm and the near point in his right eye is 100 cm. (a) What is the maximum angular magnification he can produce in a telescope? (b) If he places the lenses 10.0 cm apart, what is the maximum overall magnification he can produce in a microscope? (Go back to basics and use the thin-lens equation to solve part (b).)

Section 25.6 Resolution of Single-Slit and Circular Apertures

32. If the distance from Earth to the Moon is 3.8×10^8 m, what diameter would be required for a telescope objective to resolve a Moon crater 300 m in diameter? Assume a wavelength of 500 nm.

33. A converging lens with a diameter of 30.0 cm forms an image of a satellite passing overhead. The satellite has two green lights (wavelength 500 nm) spaced 1.00 m apart. If the lights can just be resolved according to the Rayleigh criterion, what is the altitude of the satellite?

34. The pupil of a cat's eye narrows to a vertical slit of width 0.500 mm in daylight. What is the angular resolution for a pair of horizontally separated mice? (Use 500-nm light in your calculation.)

35. To increase the resolving power of a microscope, the object and the objective are immersed in oil ($n = 1.5$). If the limiting angle of resolution without the oil is 0.60 μrad, what is the limiting angle of resolution with the oil? (*Hint:* The oil changes the wavelength of the light.)

36. Two motorcycles, separated laterally by 2.00 m, are approaching an observer holding an infrared detector that is sensitive to radiation of wavelength 885 nm. What aperture diameter is required in the detector if the two headlights are to be resolved at a distance of 10.0 km?

37. A helium-neon laser emits light that has a wavelength of 632.8 nm. The circular aperture through which the beam emerges has a diameter of 0.500 cm. Estimate the diameter of the beam 10.0 km from the laser.

38. A spy satellite circles Earth at an altitude of 200 km and carries out surveillance with a special high-resolution telescopic camera having a lens diameter of 35 cm. If the angular resolution of this camera is limited by diffraction, estimate the separation of two small objects on Earth's surface that are just resolved in yellow-green light ($\lambda = 550$ nm).

39. Suppose a 5.00-m-diameter telescope is constructed on the Moon, where the absence of atmospheric distortion permits excellent viewing. If observations are made using 500-nm light, what minimum separation between two objects could just be resolved on Mars at closest approach (when Mars is 8.0×10^7 km from the Moon)?

Section 25.7 The Michelson Interferometer

40. Light of wavelength 550 nm is used to calibrate a Michelson interferometer. By use of a micrometer screw, the platform on which one mirror is mounted is moved 0.180 mm. How many fringe shifts are counted?

41. An interferometer is used to measure the length of a bacterium. The wavelength of the light used is 650 nm. As one arm of the interferometer is moved from one end of the cell to the other, 310 fringe shifts are counted. How long is the bacterium?

42. Mirror M_1 in Figure 25.14 is displaced a distance ΔL. During this displacement, 250 fringe shifts are counted. The

light being used has a wavelength of 632.8 nm. Calculate the displacement ΔL.

43. Monochromatic light is beamed into a Michelson interferometer. The movable mirror is displaced 0.382 mm, causing the central spot in the interferometer pattern to change from bright to dark and back to bright 1 700 times. Determine the wavelength of the light. What color is it?

44. The Michelson interferometer can be used to measure the index of refraction of a gas by placing an evacuated transparent tube in the light path along one arm of the device. Fringe shifts occur as the gas is slowly added to the tube. Assume that 600-nm light is used, the tube is 5.00 cm long, and that 160 fringe shifts occur as the pressure of the gas in the tube increases to atmospheric pressure. What is the index of refraction of the gas? (*Hint:* The fringe shifts occur because the wavelength of the light changes inside the gas-filled tube.)

45. The light path in one arm of a Michelson interferometer includes a transparent cell that is 5.00 cm long. How many fringe shifts would be observed if all the air were evacuated from the cell? The wavelength of the light source is 590 nm and the refractive index of air is 1.000 29. (See the Hint in Problem 44.)

46. The H_α line in hydrogen has a wavelength of 656.20 nm. This line differs in wavelength from the corresponding spectral line in deuterium (the heavy stable isotope of hydrogen) by 0.18 nm. (a) Determine the minimum number of lines a grating must have to resolve these two wavelengths in the first order. (b) Repeat part (a) for the second order.

47. A 15.0-cm-long grating has 6 000 slits per centimeter. Can two lines of wavelengths 600.000 nm and 600.003 nm be separated using this grating? Explain.

ADDITIONAL PROBLEMS

48. A person with a nearsighted eye has near and far points of 16 cm and 25 cm, respectively. (a) Assuming a lens is placed 2.0 cm from the eye, what power must the lens have to correct this condition? (b) Contact lenses placed directly on the cornea are used to correct the eye in this example. What is the power of the lens required in this case, and what is the new near point? (*Hint:* The contact lens and the eyeglass lens require slightly different powers because they are at different distances from the eye.)

49. The near point of an eye is 75.0 cm. (a) What should be the power of a corrective lens prescribed to enable the eye to see an object clearly at 25.0 cm? (b) If, using the corrective lens, the user can see an object clearly at 26.0 cm but not 25.0 cm, by how many diopters did the lens grinder miss the prescription?

50. A compound microscope has an objective of focal length 0.300 cm and an eyepiece of focal length 2.50 cm. If an object is 3.40 mm from the objective, what is the magnification? (*Hint:* Use the lens equation for the objective.)

51. A cataract-impaired lens in an eye can be surgically removed and replaced by a manufactured lens. The focal length required for the new lens is determined by the lens-to-retina distance, which is measured by a sonar-like device, and by the requirement that the implant provide for correct distant vision. (a) If the distance from lens to retina is 22.4 mm, calculate the power of the implanted lens in diopters. (b) Since there is no accommodation and the implant allows for correct distant vision, a corrective lens for close work or reading must be used. Assume a

reading distance of 33.0 cm and calculate the power of the lens in the reading glasses.

52. Estimate the minimum angle subtended at the eye of a hawk flying at an altitude of 50 m to recognize a mouse on the ground.

53. The wavelengths of the sodium spectrum are $\lambda_1 = 589.00$ nm and $\lambda_2 = 589.59$ nm. Determine the minimum number of lines in a grating that will allow resolution of the sodium spectrum in (a) the first order and (b) the third order.

54. The text discusses the astronomical telescope. Another type is the Galilean telescope, in which an objective lens gathers light (Fig. P25.54), and tends to form an image at point *A*. An eyepiece, consisting of a diverging lens, intercepts the light before it comes to a focus and forms a virtual image at point *B*. When adjusted for minimum eyestrain, point *B* is an infinite distance in front of the lens and parallel rays emerge from the lens, as in Figure P25.54b. An opera glass, which is a Galilean telescope, is used to view a 30.0-cm-tall singer's head that is 40.0 m from the objective lens. The focal length of the objective is +8.00 cm, and that of the eyepiece is −2.00 cm. The telescope is adjusted so parallel rays enter the eye. Compute (a) the size of the real image that would have been formed by the objective, (b) the virtual object distance for the diverging lens, (c) the distance between the lenses, and (d) the overall angular magnification.

55. A laboratory (astronomical) telescope is used to view a scale that is 300 cm from the objective, which has a focal length of 20.0 cm; the eyepiece has a focal length of

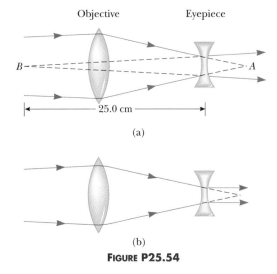

FIGURE P25.54

2.00 cm. Calculate the angular magnification when the telescope is adjusted for minimum eyestrain. (*Note:* The object is not at infinity, and so the simple expression $m = f_o/f_e$ is not sufficiently accurate for this problem. Also, assume small angles so that $\tan \theta \approx \theta$.)

56. If the aqueous humor of the eye has an index of refraction of 1.34 and the distance from the vertex of the cornea to the retina is 2.00 cm, what is the radius of curvature of the cornea for which distant objects will be focused on the retina? (For simplicity, assume that all refraction occurs in the aqueous humor.)

GROUP ACTIVITIES

G.1 (a) Move this book toward your face until the letters just begin to blur. The distance from the book to your eye is your near point. (b) On a sheet of paper make a dot near the center and about 3 inches to the left and right place an **x**. With one eye shut and while looking at the dot, slowly move the paper toward your eye. You will notice at some distance from your eye, one of the x's will disappear. This is the location of the blind spot of your eye, the point where the optic nerve enters the eye. (c) Stand before a mirror in a darkened room for a few minutes. Then, turn on a light in the room and observe your pupils in the mirror as they change size. Such dark adaptation also takes place at the rods and cones as they chemically adjust their sensitivity. This adjustment takes 15–30 minutes as you have noted when you enter a darkened movie theater. Iris aperture control takes less than a second and helps protect the retina from overload.

G.2 On a sunny day, hold a magnifying glass above a nonflammable surface, such as a sidewalk, so that the image of the Sun forms a round spot of light on the surface. Note where the spot formed by the lens is most distinct, or smallest. Use a ruler to measure the distance between the glass and the image. The distance is equal to the focal length of the lens.

G.3 Hold a pair of prescription glasses about 12 cm from your eye, and look at different objects through the lenses. Try this with different types of glasses, such as those for farsightedness and nearsightedness, and describe what effects the differences have on the image you see. If you

have bifocals, how do the images produced by the top and bottom portions of the bifocal lens compare?

G.4 If you have never experimented with a 35-mm camera with adjustable *f*-numbers and shutter speeds, use up a couple of rolls of film to see what happens. Take several shots of the same object with different settings for these two variables. (You should record your *f*-numbers and shutter speeds for each photograph.) Explain any differences you see in the final images in terms of the settings used.

G.5 (a) A patient has a near point of 1.25 m. Is she nearsighted or farsighted? (b) Should the corrective lens be converging or diverging? (c) For a comfortable reading distance of 25.0 cm from her eye, what diopter lens should be prescribed? (d) Does the power of the lens change if you assume that the lens is always placed 2.00 cm from the eye?

G.6 A lens with a certain power is used as a simple magnifier. (a) If the power of the lens is doubled, does the angular magnification increase or decrease? (b) What assumptions are made when you use the equation $m = 25$ cm$/f$? (c) What is the angular magnification of a lens having a focal length of 5.0 cm?

G.7 A telescope uses an objective lens of focal length 40 cm and an eyepiece of 2.0 cm. (a) Without doing a calculation, how will the view through the telescope change if the eyepiece is changed to one having a focal length of 4.0 cm? (b) Do the calculations to verify your prediction. (c) Does your answer depend on whether you are using a refracting or a reflecting telescope?

Six

Modern Physics

At the end of the 19th century, scientists believed that they had learned most of what there was to know about physics. Newton's laws of motion and his universal theory of gravitation, Maxwell's theoretical work in unifying electricity and magnetism, and the laws of thermodynamics and kinetic theory were highly successful in explaining a wide variety of phenomena.

However, at the turn of the 20th century, a major revolution shook the world of physics. In 1900 Planck provided the basic ideas that led to the formulation of the quantum theory, and in 1905 Einstein formulated his brilliant special theory of relativity. The excitement of the times is captured in Einstein's own words: "It was a marvelous time to be alive." Both ideas were to have a profound effect on our understanding of nature. Within a few decades, these two theories inspired new developments in the fields of atomic physics, nuclear physics, and condensed-matter physics.

Our discussion of modern physics will begin with a treatment of the special theory of relativity in Chapter 26. The theory provides us with a new and deeper view of physical laws. In Chapter 27 we shall discuss various developments in quantum theory, which provides us with a successful model for understanding electrons, atoms, and molecules. The last three chapters of the text are concerned with applications of quantum theory. Chapter 28 discusses the structure and properties of atoms using concepts from quantum mechanics. Chapter 29 is concerned with the structure and properties of the atomic nucleus. Chapter 30 discusses practical applications of nuclear physics and concludes with a discussion of elementary particles.

Keep in mind that although modern physics was developed during the 20th century and has led to a multitude of important technological achievements, the story is still incomplete as we begin this 21st century. Surprising discoveries will continue to be made during our lifetimes, and many of these discoveries will deepen or refine our understanding of nature and the world around us. It is still "a marvelous time to be alive."

astronaut's pulse are radioed to Earth when the vehicle is moving perpendicularly to a line that connects the vehicle with an Earth observer. What pulse rate does Earth observer measure? What would be the pulse rate if the speed of the space vehicle were increased to $0.990c$?

9. A muon formed high in Earth's atmosphere travels at a
web speed $v = 0.99c$ for a distance of 4.6 km before it decays into an electron, a neutrino, and an antineutrino $(\mu^- \rightarrow e^- + \nu + \bar{\nu})$. (a) How long does the muon live, as measured in its reference frame? (b) How far does the muon travel, as measured in its frame?

10. A box is cubical with sides of proper lengths $L_1 = L_2 = L_3$, as shown in Figure P26.10, when viewed in its own rest frame. If this block moves parallel to one of its edges with a speed of $0.80c$ past an observer, (a) what shape does it appear to have to this observer, and (b) what is the length of each side as measured by this observer?

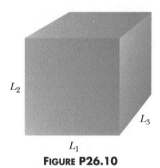

L_2 L_3 L_1

FIGURE P26.10

11. The proper length of one spaceship is three times that of another. The two spaceships are traveling in the same direction and, while both are passing overhead, an Earth observer measures the two spaceships to be the same length. If the slower spaceship is moving with a speed of $0.350c$, determine the speed of the faster spaceship.

12. An atomic clock moves at 1 000 km/h for one hour as measured by an identical clock on Earth. How many nanoseconds slow will the moving clock be at the end of the one-hour interval? (*Hint:* The following approximation will prove helpful: $\sqrt{1 - x} \approx 1 - (x/2)$ for $x \ll 1$.)

13. A supertrain of proper length 100 m travels at a speed of $0.95c$ as it passes through a tunnel having proper length 50 m. As seen by a trackside observer, is the train ever completely within the tunnel? If so, by how much?

14. An observer moving at a speed of $0.995c$ relative to a rod (Fig. P26.14) measures its length to be 2.00 m and sees its length to be oriented at $30.0°$ with respect to the direction of motion. What is the proper length of the rod? (b) What is the orientation angle in a reference frame moving with the rod?

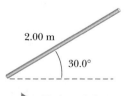

2.00 m

$30.0°$

Motion of observer

FIGURE P26.14 View of rod as seen by an observer moving to the right.

15. In 1963 when Mercury astronaut Gordon Cooper orbited Earth 22 times, the press stated that for each orbit he aged 2 millionths of a second less than if he had remained on Earth. (a) Assuming that he was 160 km above Earth in a circular orbit, determine the time difference between someone on Earth and the orbiting astronaut for the 22 orbits. You will need to use the approximation

$$\frac{1}{\sqrt{1 - x}} \approx 1 + \frac{x}{2} \qquad \text{for } x \ll 1$$

(b) Did the press report accurate information? Explain.

16. An interstellar space probe is launched from Earth. After a brief period of acceleration, it moves with a constant velocity, 70.0% of the speed of light. Its nuclear-powered batteries supply the energy to keep its data transmitter active continuously. The batteries have a lifetime of 15.0 years as measured in a rest frame. (a) How long do the batteries on the space probe last as measured by mission control on Earth? (b) How far is the probe from Earth when its batteries fail, as measured by mission control? (c) How far is the probe from Earth, as measured by its built-in trip odometer, when its batteries fail? (d) For what total time after launch is data received from the probe by mission control? Note that radio waves travel at the speed of light and fill the space between the probe and Earth at the time of battery failure.

Section 26.7 Relativistic Momentum

17. An electron has a speed of $v = 0.90c$. At what speed will a proton have a momentum equal to that of the electron?

18. Calculate the momentum of an electron moving with a speed of (a) $0.010c$, (b) $0.50c$, (c) $0.90c$.

19. An unstable particle at rest breaks up into two fragments of *unequal mass*. The mass of the lighter fragment is 2.50×10^{-28} kg, and that of the heavier fragment is 1.67×10^{-27} kg. If the lighter fragment has a speed of $0.893c$ after the breakup, what is the speed of the heavier fragment?

20. The nonrelativistic expression for the momentum of a particle, $p = mv$, can be used if $v \ll c$. For what speed does the use of this formula give an error in the momentum of (a) 1.00% and (b) 10.0%?

Section 26.8 Relativistic Addition of Velocities

21. An electron moves to the right with a speed of $0.90c$ relative to the laboratory frame. A proton moves to the left with a speed of $0.70c$ relative to the electron. Find the speed of the proton relative to the laboratory frame.

22. Spaceship R is moving to the right at a speed of $0.70c$ with respect to Earth. A second spaceship, L, moves to the left at the same speed with respect to Earth. What is the speed of L with respect to R?

23. A Klingon space ship moves away from Earth at a speed of $0.800c$ (Fig. P26.23). The Starship *Enterprise* pursues at a speed of $0.900c$ relative to Earth. Observers on Earth see the *Enterprise* overtaking the Klingon ship at a relative speed of $0.100c$. With what speed is the *Enterprise* overtaking the Klingon ship as seen by the crew of the *Enterprise*?

24. A spaceship travels at $0.750c$ relative to Earth. If the spaceship fires a small rocket in the forward direction, how fast

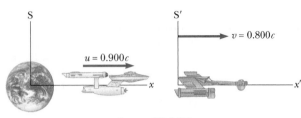

FIGURE P26.23

(relative to the ship) must it be fired for it to travel at 0.950c relative to Earth?

25. A rocket moves with a velocity of 0.92c to the right with re-
web spect to a stationary observer A. An observer B moving relative to observer A finds that the rocket is moving with a
velocity of 0.95c to the left. What is the velocity of observer
B relative to observer A? (*Hint:* Consider observer B's velocity in the frame of reference of the rocket.)

26. A pulsar is a stellar object that emits light in short bursts.
Suppose a pulsar with a speed of 0.950c approaches Earth,
and a rocket with a speed of 0.995c heads toward the pulsar (both speeds measured in Earth's frame of reference).
If the pulsar emits 10.0 pulses per second in its own frame
of reference, at what rate are the pulses emitted in the
rocket's frame of reference?

27. Spaceship I, which contains students taking a physics
exam, approaches Earth with a speed of 0.60c, while spaceship II, which contains instructors proctoring the exam,
moves away from Earth at 0.28c, as in Figure P26.27. If the
instructors in spaceship II stop the exam after 50 min have
passed on *their clock*, how long does the exam last as measured by (a) the students? (b) an observer on Earth?

FIGURE P26.27

Section 26.9   Relativistic Energy and the Equivalence of Mass and Energy

28. A proton moves with a speed of 0.950c. Calculate its
(a) rest energy, (b) total energy, and (c) kinetic energy.

29. What is the speed of a particle whose kinetic energy is
equal to its own rest energy?

30. A proton in a high-energy accelerator is given a kinetic energy of 50.0 GeV. Determine (a) the momentum and
(b) the speed of the proton.

31. In a color television tube, electrons are accelerated
through a potential difference of 20 000 V. With what
speed do the electrons strike the screen?

32. Determine the energy required to accelerate an electron
from (a) 0.500c to 0.900c and (b) 0.900c to 0.990c.

33. A cube of steel has a volume of 1.00 cm³ and a mass of
8.00 g when at rest on Earth. If this cube is now given a
speed of $v = 0.900c$, what is its density as measured by a
stationary observer? Note that relativistic density is E_R/c^2V.

34. An unstable particle with a mass equal to 3.34×10^{-27} kg
is initially at rest. The particle decays into two fragments
that fly off with velocities of 0.987c and $-0.868c$. Find the

masses of the fragments. (*Hint:* Conserve both mass-
energy and momentum.)

Section 26.10   Pair Production and Annihilation

35. How much total kinetic energy will an electron-positron
pair have if produced by a photon of energy 3.00 MeV?

36. If an electron-positron pair with a total kinetic energy of
2.50 MeV is produced, find (a) the energy of the photon
that produced the pair and (b) its frequency.

37. Two photons are produced when a proton and an antipro-
web ton annihilate each other. What is the minimum frequency and corresponding wavelength of each photon?

38. An electron moving at a speed of 0.60c collides head on
with a positron also moving at 0.60c. Determine the energy
and momentum of each photon produced in the process.

ADDITIONAL PROBLEMS

39. What is the speed of a proton that has been accelerated
from rest through a difference of potential of (a) 500 V
and (b) 5.00×10^8 V?

40. An electron has a total energy equal to 5 times its rest energy. (a) What is its momentum? (b) Repeat for a proton.

41. What is the momentum (in units of MeV/c) of an electron
with a kinetic energy of 1.00 MeV?

42. An astronomer on Earth observes a meteoroid in the
southern sky approaching at a speed of 0.800c. At the time
of its discovery, the meteoroid is 20.0 lightyears from
Earth. Calculate (a) the time required for the meteoroid
to reach Earth as measured by the Earthbound astronomer, (b) this time as measured by a tourist on the
meteoroid, and (c) the distance to Earth as measured by
the tourist.

43. Ted and Mary are playing a game of catch in frame S',
which is moving with a speed of 0.60c; Jim in frame S is
watching (Fig. P26.43). Ted throws the ball to Mary with a
speed of 0.80c (according to Ted) and their separation
(measured in S') is 1.8×10^{12} m. (a) According to Mary,
how fast is the ball moving? (b) According to Mary, how
long will it take the ball to reach her? (c) According to
Jim, how far apart are Ted and Mary, and how fast is the
ball moving?

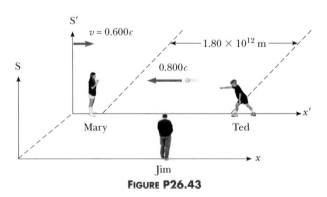

FIGURE P26.43

44. An alarm clock is set to sound in 10 h. At $t = 0$, the clock
is placed in a spaceship moving with a speed of 0.75c (relative to Earth). What distance, as determined by an Earth
observer, does the spaceship travel before the alarm clock
sounds?

45. A radioactive nucleus moves with a speed of v relative to a laboratory observer. The nucleus emits an electron in the positive x direction with a speed of $0.70c$ relative to the decaying nucleus, and a speed of $0.85c$ in the $+x$ direction relative to the laboratory observer. What is the value of v?

46. A certain quasar recedes from Earth at $v = 0.870c$. A jet of material ejected from the quasar back toward Earth moves at $0.550c$ relative to the quasar. Find the speed of the ejected material relative to Earth.

47. An astronaut wishes to visit the Andromeda galaxy, making a one-way trip that will take 30.0 y in the spaceship's frame of reference. Assume that the galaxy is 2.00 million lightyears away and that the speed is constant. (a) How fast must the astronaut travel relative to Earth? (b) What will be the kinetic energy of the spacecraft, whose mass is 1.00×10^6 kg? (c) What is the cost of this energy if it is purchased at a typical consumer price for electric energy, 13.0 cents per kWh? The following approximation will prove useful:

$$\frac{1}{\sqrt{1 + x}} \approx 1 - \frac{x}{2} \quad \text{for } x \ll 1$$

48. The cosmic rays of highest energy are protons that have kinetic energy on the order of 10^{13} MeV. (a) How long would it take a proton of this energy to travel across the Milky Way galaxy, having a diameter $\sim 10^5$ lightyears, as measured in the proton's frame? (b) From the point of view of the proton, how many kilometers across is the galaxy?

49. A spaceship of proper length 300 m takes 0.75 μs to pass an Earth observer. Determine the speed of this spaceship as measured by Earth observer.

50. Find the kinetic energy of a 78.0-kg spacecraft launched out of the solar system with a speed of 106 km/s by using (a) the classical equation $KE = \frac{1}{2}mv^2$ and (b) the relativistic equation. You will need to use the approximation

$$\frac{1}{\sqrt{1 - x}} \approx 1 + \frac{x}{2} \quad \text{for } x \ll 1$$

51. An alien spaceship traveling at $0.600c$ toward Earth
web launches a landing craft with an advance guard of purchasing agents. The lander travels in the same direction with a velocity $0.800c$ relative to the spaceship. As observed on Earth, the spaceship is 0.200 lightyears from Earth when the lander is launched. (a) With what velocity is the lander observed to be approaching by observers on Earth? (b) What is the distance to Earth at the time of lander launch, as observed by the aliens on the mother ship? (c) How long does it take the lander to reach Earth as observed by the aliens on the mother ship? (d) If the lander has a mass of 4.00×10^5 kg, what is its kinetic energy as observed in Earth's reference frame?

52. (a) Show that a potential difference of 1.02×10^6 V would be sufficient to give an electron a speed equal to twice the speed of light if Newtonian mechanics remained valid at high speeds. (b) What speed would an electron actually acquire in falling through a potential difference of 1.02×10^6 V?

53. The muon is an unstable particle that spontaneously decays into an electron and two neutrinos. In a reference frame in which the muons are stationary, if the number of muons at $t = 0$ is N_0, the number at time t is given by $N = N_0 e^{-t/\tau}$, where τ is the mean lifetime, equal to 2.2 μs. Suppose that the muons move at a speed of $0.95c$ and that there are 5.0×10^4 muons at $t = 0$. (a) What is the observed lifetime of the muons? (b) How many muons remain after traveling a distance of 3.0 km?

54. An observer in a rocket moves toward a mirror at speed v relative to the reference frame labeled by S in Figure P26.54. The mirror is stationary with respect to S. A light pulse emitted by the rocket travels toward the mirror and is reflected back to the rocket. The front of the rocket is a distance d from the mirror (as measured by observers in S) at the moment the light pulse leaves the rocket. What is the total travel time of the pulse as measured by observers in (a) the S frame and (b) the front of the rocket?

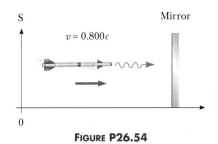

FIGURE P26.54

55. A physics professor on Earth gives an exam to her students who are on a rocket ship traveling at a speed of v with respect to Earth. The moment the ship passes the professor, she signals the start of the exam. If she wishes her students to have T_0 (rocket time) to complete the exam, show that she should wait a time of

$$T = T_0 \sqrt{\frac{1 - v/c}{1 + v/c}}$$

(Earth time) before sending a light signal telling them to stop. (*Hint:* Remember that it takes some time for the second light signal to travel from the professor to the students.)

56. Imagine that the entire Sun collapses to a sphere of radius R_g such that the work required to remove a small mass m from the surface would be equal to its rest energy, mc^2. This radius is called the *gravitational radius* for the Sun. Find R_g. (It is believed that the ultimate fate of very massive stars is to collapse beyond their gravitational radii into black holes.)

57. A rod of length L_0 moves with a speed of v along the horizontal direction. The rod makes an angle of θ_0 with respect to the axis of a coordinate system moving with the rod. (a) Show that the length of the rod as measured by a stationary observer is given by

$$L = L_0 \left[1 - \left(\frac{v^2}{c^2} \right) \cos^2 \theta_0 \right]^{1/2}$$

(b) Show that the angle the rod makes with the axis as seen by the stationary observer is given by the expression $\tan \theta = \gamma \tan \theta_0$. These results show that the rod is both contracted and rotated. (Take the lower end of the rod to be at the origin of the moving coordinate system.)

GROUP ACTIVITIES

G.1 How would your life change if the speed of light were 55 mi/h? As you and your co-worker go about your daily activities, make notes about how events or occurrences would be changed. See which of the two of you can find the largest number of ways in which your day-to-day life would be affected.

G.2 Consider two travelers on the surface of Earth walking directly toward the North Pole but from different starting locations. How is the fact that they are getting closer together as they go north similar to the curvature of space discussed in the general theory of relativity?

G.3 Find a comfortable location on campus to observe life going on about you. Take note of all the motion going on about you and describe how this motion would look in a system moving at a constant velocity with respect to you.

G.4 Two rockets are of length L and each carries an antenna also of length L that is supported in a direction perpendicular to the length of the rocket. Two pilots, A and B, are in the rockets, which are initially at rest with respect to one another. Rocket A takes off and travels at a speed of $0.400c$ away from B. Answer the following questions for observers A and B either with greater than L, less than L, or equal to L.
(a) The length of A as seen by pilot A is _____.
(b) The length of A as seen by pilot B is _____.
(c) The length of B as seen by pilot B is _____.

(d) The length of B as seen by pilot A is _____.
(e) The length of the antenna of A as seen by pilot B is _____.
(f) The length of the antenna of B as seen by pilot A is _____.
(g) Take L to be 100 m to find precise values for all of the above answers.

G.5 Two rockets carry identical clocks. When they are at rest with respect to each other, pilot A on ship A and pilot B on ship B agree that the second hand on each other's watch takes precisely time ΔT to go around the dial of the watch once. Rocket A takes off and travels at a speed of $0.500c$ away from B. Answer the following questions for observers A and B either with greater than ΔT, less than ΔT, or equal to ΔT.
(a) The interval on A as seen by pilot A is _____.
(b) The interval on A as seen by pilot B is _____.
(c) The interval on B as seen by pilot B is _____.
(d) The interval on B as seen by pilot A is _____.
(e) Find the time intervals when ΔT is equal to 1 min.

G.6 An alien spacecraft fires torpedoes forward with a speed of $0.75c$ relative to the spacecraft. The spacecraft has a speed of $0.30c$ relative to Earth. In such a case, will the torpedoes travel at $1.05c$ with respect to Earth? If you do not agree with the speed of $1.05c$, find the speed at which the torpedoes do travel with respect to Earth.

Quantum Physics

27

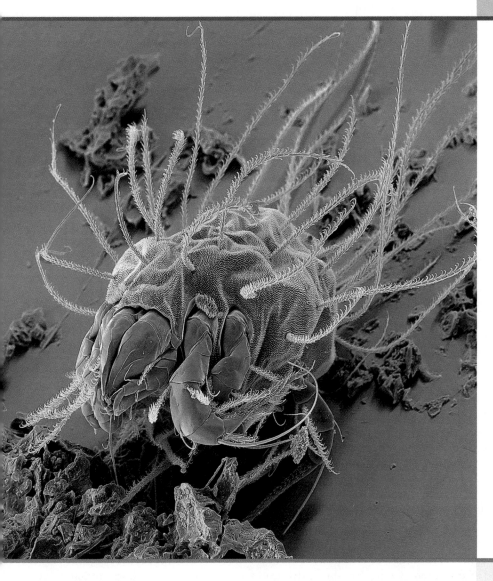

Color-enhanced scanning electron micrograph of the storage mite *Lepidoglyphus destructor*. These common mites grow up to 0.75 mm and feed on molds, flour and rice. They thrive at 25°C and high humidity, and can trigger allergies. *(© Eye of Science/Science Source/Photo Researchers, Inc.)*

*A*lthough many problems were resolved by the theory of relativity in the early part of the 20th century, many other problems remained unsolved. Attempts to explain the behavior of matter on the atomic level with the laws of classical physics were consistently unsuccessful. Various phenomena, such as the electromagnetic radiation emitted by a heated object (blackbody radiation), the emission of electrons by illuminated metals (photoelectric effect), and the emission of sharp spectral lines by gas atoms in an electric discharge tube, could not be understood within the framework of classical physics. Between 1900 and 1930, however, a modern version of mechanics called *quantum mechanics* or *wave mechanics* was highly successful in explaining the behavior of atoms, molecules, and nuclei. Moreover, quantum mechanics reduces to classical physics when applied to macroscopic systems. As with relativity, the quantum theory requires a modification of our everyday ideas concerning the physical world.

The earliest and most basic ideas of quantum theory were introduced by Planck, and most of the subsequent mathematical developments, interpretations, and improvements were made by a number of distinguished physicists, including Einstein, Bohr, Schrödinger, de Broglie, Heisenberg, Born, and Dirac. An extensive study of quantum theory is certainly beyond the scope of this book. This chapter is simply an introduction to the underlying ideas of quantum theory and the wave-particle nature of matter. We also discuss some simple applications of quantum theory, including the photoelectric effect, the Compton effect, and x-rays.

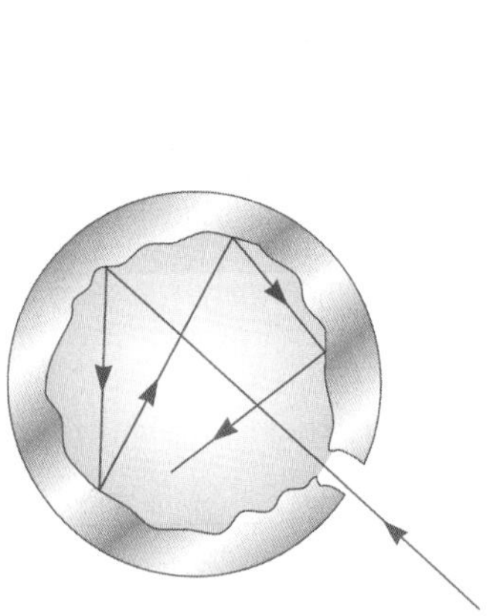

27.1 BLACKBODY RADIATION AND PLANCK'S HYPOTHESIS

An object at any temperature is known to emit electromagnetic radiation that is sometimes referred to as **thermal radiation.** Stefan's law, which we discussed in Section 11.7, describes the total power radiated. The spectrum of the radiation depends on the temperature and properties of the object. At low temperatures, the wavelengths of the thermal radiation are mainly in the infrared region and hence not observable by the eye. As the temperature of an object increases, the object eventually begins to glow red. At sufficiently high temperatures, it appears to be white, as in the glow of the hot tungsten filament of a lightbulb. A careful study of thermal radiation shows that it consists of a continuous distribution of wavelengths from the infrared, visible, and ultraviolet portions of the spectrum.

From a classical viewpoint, thermal radiation originates from accelerated charged particles near the surface of an object; such charges emit radiation much as small antennas do. The thermally agitated charges can have a distribution of frequencies, which accounts for the continuous spectrum of radiation emitted by the object. By the end of the 19th century, it had become apparent that the classical theory of thermal radiation was inadequate. The basic problem was in understanding the observed distribution of energy as a function of wavelength in the radiation emitted by a blackbody. By definition, a blackbody is an ideal system that absorbs *all* radiation incident on it. A good approximation of a blackbody is a small hole leading to the inside of a hollow object, as shown in Figure 27.1. The nature of the radiation emitted through the small hole leading to the cavity depends *only on the temperature* of the cavity walls, and not at all on the material composition of the object, its shape, or other factors.

Experimental data for the distribution of energy in blackbody radiation at three temperatures are shown in Figure 27.2. The radiated energy varies with wavelength and temperature. As the temperature of the blackbody increases, the total amount of energy (area under the curve) it emits increases. Also, with increasing temperature, the peak of the distribution shifts to shorter wavelengths. This shift was found to obey the following relationship, called **Wien's displacement law:**

$$\lambda_{\max} T = 0.2898 \times 10^{-2} \text{ m} \cdot \text{K} \qquad [27.1]$$

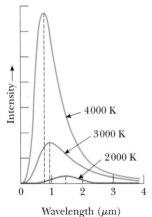

FIGURE 27.1 The opening in the cavity of a body is a good approximation of a blackbody. As light enters the cavity through the small opening, part is reflected and part is absorbed on each reflection from the interior walls. After many reflections, essentially all of the incident energy is absorbed.

FIGURE 27.2 Intensity of blackbody radiation versus wavelength at three different temperatures. Note that the total radiation emitted (the area under the curve) increases with increasing temperature.

where $\lambda_{\max}$ is the wavelength at which the curve peaks and T is the absolute temperature of the object emitting the radiation.

APPLYING **PHYSICS** *27.1*

If you look carefully at stars in the night sky, you can distinguish three main colors—red, white, and blue. What is the reason for these colors?

Explanation These colors are a result of the thermal radiation from the surfaces of the stars. A relatively cool star, with a surface temperature of 3 000 K, has a radiation curve like the middle curve in Figure 27.2. Notice that the peak in this curve is above the visible wavelengths, 0.4 μm–0.7 μm. Thus, significantly more radiation is emitted within the visible range at the red end than the blue end. As a result, the star appears to be reddish in color, similar to the red glow from the burner of an electric stove.

A hotter star has a radiation curve more like the upper curve in Figure 27.2. In this case, the star emits significant radiation throughout the visible range and the combination of all colors causes the star to look white. This is the case with our own Sun, with a surface temperature of 5 800 K. For very hot stars, the peak can be shifted so far below the visible range that significantly more blue radiation is emitted than red, so that the star appears bluish in color.

APPLICATION

COLOR OF THE STARS

Attempts to use classical ideas to explain the shape of the curves shown in Figure 27.2 failed. Figure 27.3 shows an experimental plot of the blackbody radiation spectrum (red curve) together with the theoretical picture of what this curve should look like based on classical theories (blue curve). At long wavelengths, classical theory is in good agreement with the experimental data. At short wavelengths, however, major disagreement exists between classical theory and experiment. As λ approaches zero, classical theory predicts that the amount of energy being radiated should increase. In fact, theory predicts that the intensity should be infinite. This is contrary to the experimental data, which show that as λ approaches zero, the amount of energy carried by short-wavelength radiation also approaches zero. This contradiction is called the **ultraviolet catastrophe** because theory and experiment disagree strongly in the short-wavelength ultraviolet region of the spectrum.

In 1900 Planck developed a formula for blackbody radiation that was in complete agreement with experiments at all wavelengths, yielding an intensity versus wavelength distribution that matches the red line in Figure 27.3. Planck's original theoretical approach is rather abstract in that it involves arguments based on entropy and thermodynamics. We shall present arguments that are easier to visualize physically, while attempting to convey the spirit and revolutionary impact of Planck's original work.

Planck hypothesized that blackbody radiation was produced by submicroscopic charged oscillators, which he called *resonators*. He assumed that the walls of a glowing cavity were composed of literally billions of these resonators whose exact nature was unknown. In his theory, Planck made a bold and controversial assumption concerning the nature of the oscillating molecular charges in the blackbody:

The resonators were allowed to have only certain discrete energies, E_n, given by

$$E_n = nhf \qquad \text{[27.2]}$$

where n is a positive integer called a **quantum number,** f is the frequency of vibra-

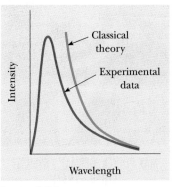

FIGURE 27.3 Comparison of experimental data with the classical theory of blackbody radiation. Planck's theory perfectly matches the experimental data.

Webnote 27.1

Physics 2000 from the University of Colorado at Boulder is an interactive journey through modern physics using applets, animations, and cartoons. Go to *http://www.colorado.edu/physics/ 2000/index.pl*

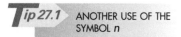

ip 27.1   ANOTHER USE OF THE SYMBOL *n*

The symbol *n* in this chapter represents an integer quantum number that is used to identify a particular state of a system. Do not confuse this with the same symbol *n* used for the index of refraction (which was not an integer) in our study of optics.

tion of the resonator, and *h* is a constant, known as **Planck's constant,** which has the value

$$h = 6.626 \times 10^{-34} \, \text{J} \cdot \text{s} \qquad [27.3]$$

Because the energy of each resonator can have only discrete values given by Equation 27.2, we say the energy is *quantized*. Each discrete energy value represents a different *quantum state,* with each value of *n* representing a specific quantum state. (When the resonator is in the *n* = 1 quantum state, its energy is *hf*; when it is in the *n* = 2 quantum state, its energy is 2*hf*; and so on.)

The key point in Planck's theory is the radical assumption of quantized energy states. This development marked the birth of the quantum theory. When Planck presented his theory, most scientists (including Planck!) did not consider the quantum concept to be realistic since it violated several centuries of belief that energy was a continuous quantity. Hence, Planck and others continued to search for a classical explanation of blackbody radiation. However, subsequent developments showed that a theory based on the quantum concept (rather than on classical concepts) had to be used to explain a number of other phenomena at the atomic level.

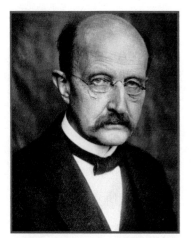

MAX PLANCK, GERMAN PHYSICIST (1858–1947)

Planck was born in Germany. His first son was killed in World War I, and his second son was executed by the Nazis for conspiring to assassinate Hitler. During an air raid on Berlin, he lost his home and valuable library. Planck introduced the concept of a "quantum of action" (Planck's constant, *h*) in an attempt to explain the spectral distribution of blackbody radiation, which laid the foundations for quantum theory. In 1918 he was awarded the Nobel prize for this discovery of the quantized nature of energy. The work leading to the "lucky" blackbody radiation formula was described by Planck in his Nobel prize acceptance speech (1920): "But even if the radiation formula proved to be perfectly correct, it would after all have been only an interpolation formula found by lucky guesswork and, thus, would have left us rather unsatisfied." (© *Bettmann/CORBIS)*

Example 27.1   Thermal Radiation from the Human Body

The temperature of the skin is approximately 35°C. At what wavelength does the radiation emitted from the skin reach its peak?

Solution   From Wien's displacement law, Equation 27.1, we have

$$\lambda_{\text{max}} T = 0.2898 \times 10^{-2} \, \text{m} \cdot \text{K}$$

Solving for λ_{max}, noting that 35°C corresponds to an absolute temperature of 308 K, we find

$$\lambda_{\text{max}} = \frac{0.2898 \times 10^{-2} \, \text{m} \cdot \text{K}}{308 \, \text{K}} = \boxed{940 \, \mu\text{m}}$$

This radiation is in the infrared region of the spectrum.

EXERCISE   (a) Find the wavelength corresponding to the peak of the radiation curve for the heating element of an electric oven at a temperature of 1.20×10^3 K. (Note that although this radiation peak lies in the infrared, there is enough visible radiation at this temperature to give the element a red glow.) (b) Calculate the wavelength corresponding to the peak of the radiation curve for an object whose temperature is 5.00×10^3 K, an approximate temperature for the surface of the Sun.

ANSWER   (a) 2.42 μm; (b) 580 nm in the visible region

Example 27.2   The Quantized Macroscopic Oscillator

A 2.0-kg object is attached to a massless spring of force constant *k* = 25 N/m. The spring is stretched 0.40 m from its equilibrium position and released.

A Find the total energy and frequency of oscillation according to classical calculations.

Solution   The total energy of a simple harmonic oscillator having an amplitude *A* is $kA^2/2$. Therefore,

$$E = \tfrac{1}{2}kA^2 = \tfrac{1}{2}(25 \, \text{N/m})(0.40 \, \text{m})^2 = \boxed{2.0 \, \text{J}}$$

The frequency of oscillation is

$$f = \frac{1}{2\pi}\sqrt{\frac{k}{m}} = \frac{1}{2\pi}\sqrt{\frac{25 \text{ N/m}}{2.0 \text{ kg}}} = \boxed{0.56 \text{ Hz}}$$

B Assume that Planck's law of energy quantization applies to any oscillator (atomic or large-scale) and find the quantum number n for this system.

Solution If the energy is quantized, we have $E_n = nhf$, and from the result of Part A, we have

$$E_n = nhf = n(6.63 \times 10^{-34} \text{ J} \cdot \text{s})(0.56 \text{ Hz}) = 2.0 \text{ J}$$

Therefore,

$$n = \boxed{5.4 \times 10^{33}}$$

and we see that macroscopic systems have huge quantum numbers.

C How much energy would be carried away in a one-quantum change?

Solution The energy carried away in a one-quantum change of energy is $\Delta E = E_{n+1} - E_n = hf$, so

$$\Delta E = hf = (6.63 \times 10^{-34} \text{ J} \cdot \text{s})(0.56 \text{ Hz}) = \boxed{3.7 \times 10^{-34} \text{ J}}$$

The energy carried away by a one-quantum change is such a small fraction of the total energy of the oscillator that we could not expect to measure it. Thus, the energy of an object-spring system decreases by such small quantum transitions that the decrease in energy of this macroscopic oscillator appears to be continuous.

27.2 THE PHOTOELECTRIC EFFECT AND THE PARTICLE THEORY OF LIGHT

In the latter part of the 19th century, experiments showed that when light is incident on certain metallic surfaces, electrons are emitted from the surfaces. This phenomenon is known as the **photoelectric effect,** and the emitted electrons are called **photoelectrons.** The first discovery of this phenomenon was made by Hertz, who was also the first to produce the electromagnetic waves predicted by Maxwell.

Figure 27.4 is a schematic diagram of a photoelectric effect apparatus. An evacuated glass tube known as a photocell contains a metal plate E connected to the negative terminal of a variable power supply. Another metal plate C is maintained at a positive potential by the power supply. When the tube is kept in the dark, the ammeter reads zero, indicating that there is no current in the circuit. However, when plate E is illuminated by light having a wavelength shorter than some particular wavelength that depends on the material used to make plate E, a current is detected by the ammeter, indicating a flow of charges across the gap between E and C. This current arises from photoelectrons emitted from the negative plate (the emitter) and collected at the positive plate (the collector).

Figure 27.5 is a plot of the photoelectric current versus the potential difference (voltage) ΔV between E and C for two light intensities. At large values of ΔV, the current reaches a maximum value. In addition, the current increases as the incident light intensity increases, as you might expect. Finally, when ΔV is negative—that is, when the power supply in the circuit is reversed to make E positive and C negative—the current drops to a low value because most of the emitted photoelectrons are repelled by the now negative plate C. In this situation, only those electrons having a kinetic energy greater than the magnitude of $e\,\Delta V$ reach C, where e is the charge on the electron.

When ΔV is equal to or more negative than $-\Delta V_s$, the **stopping potential,** no electrons reach C and the current is zero. The stopping potential is *independent* of the radiation intensity. The maximum kinetic energy of the photoelectrons is

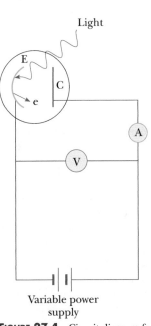

FIGURE 27.4 Circuit diagram for observing the photoelectric effect. When light strikes plate E, photoelectrons are ejected from the plate. Electrons collected at C and passing through the ammeter constitute a current in the circuit.

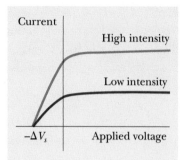

FIGURE 27.5 Photoelectric current versus applied voltage for two light intensities. The current increases with intensity but reaches a saturation level for large values of ΔV. At voltages equal to or less than $-\Delta V_s$, the current is zero.

related to the stopping potential through the relationship

$$KE_{max} = e\,\Delta V_s \qquad\qquad [27.4]$$

Several features of the photoelectric effect cannot be explained with classical physics or with the wave theory of light:

- No electrons are emitted if the incident light frequency falls below some **cutoff frequency,** f_c, that is characteristic of the material being illuminated. This is inconsistent with the wave theory, which predicts that the photoelectric effect should occur at any frequency, provided the light intensity is sufficiently high.
- The maximum kinetic energy of the photoelectrons is independent of light intensity. According to wave theory, light of higher intensity should carry more energy into the metal per unit time and therefore eject photoelectrons having higher kinetic energies.
- The maximum kinetic energy of the photoelectrons increases with increasing light frequency. The wave theory predicts no relationship between photoelectron energy and incident light frequency.
- Electrons are emitted from the surface almost instantaneously (less than 10^{-9} s after the surface is illuminated), even at low light intensities. Classically, we expect the photoelectrons to require some time to absorb the incident radiation before they acquire enough kinetic energy to escape from the metal.

A successful explanation of the photoelectric effect was given by Einstein in 1905, the same year he published his special theory of relativity. As part of a general paper on electromagnetic radiation, for which he received the Nobel prize in 1921, Einstein extended Planck's concept of quantization to electromagnetic waves. He suggested that a tiny packet of light energy or **photon** would be emitted when a quantized oscillator made a jump from an energy state $E_n = nhf$ to the next lower state $E_{n-1} = (n - 1)hf$. Conservation of energy would require the decrease in oscillator energy, hf, to be equal to the photon's energy, E, so that

Energy of a photon ▶

$$E = hf \qquad\qquad [27.5]$$

h is Planck's constant and f is the frequency of the light which is equal to the frequency of the Planckian oscillator.

The key point here is that the light energy lost by the emitter, hf, stays sharply localized in a tiny packet or particle called a photon. In Einstein's model, a photon is so localized that it can give *all* its energy hf to a single electron in the metal. According to Einstein, the maximum kinetic energy for these liberated photoelectrons is

Photoelectric effect equation ▶

$$KE_{max} = hf - \phi \qquad\qquad [27.6]$$

where ϕ is called the **work function** of the metal. The work function represents the minimum energy with which an electron is bound in the metal and is on the order of a few electron volts. Table 27.1 lists work functions for various metals.

With the photon theory of light, we can explain the previously mentioned features of the photoelectric effect that cannot be understood using concepts of classical physics:

- That the effect is not observed below a certain cutoff frequency follows from the fact that the photon energy must be greater than or equal to ϕ. If the energy of the incoming photon does not satisfy this condition, the electrons are never ejected from the surface, regardless of light intensity.
- That KE_{max} is independent of the light intensity can be understood with the following argument. If the light intensity is doubled, the number of photons is doubled, which doubles the number of photoelectrons emitted. However, their maximum kinetic energy, which equals $hf - \phi$, depends only on the light frequency and the work function, not on the light intensity.
- That KE_{max} increases with increasing frequency is easily understood with Equation 27.6.

TABLE 27.1

Work Functions of Selected Metals

Metal	ϕ (eV)
Na	2.46
Al	4.08
Cu	4.70
Zn	4.31
Ag	4.73
Pt	6.35
Pb	4.14
Fe	4.50

- That the electrons are emitted almost instantaneously is consistent with the particle theory of light, in which the incident energy arrives at the surface in small spatial packets and there is a one-to-one interaction between photons and photoelectrons. In this interaction, the photon's energy is imparted to an electron that then has enough energy to leave the metal. This is in contrast to the wave theory, in which the incident energy is distributed uniformly over a large area of the metal surface.

Experimental observation of a linear relationship between f and KE_{max} would be a final confirmation of Einstein's theory. Indeed, such a linear relationship is observed, as sketched in Figure 27.6. The intercept on the horizontal axis gives the cutoff frequency below which no photoelectrons are emitted, regardless of light intensity. The cutoff frequency is related to the work function through the relationship $f_c = \phi/h$. This cutoff frequency corresponds to a **cutoff wavelength** of

$$\lambda_c = \frac{c}{f_c} = \frac{c}{\phi/h} = \frac{hc}{\phi} \qquad [27.7]$$

where c is the speed of light. Wavelengths *greater* than λ_c incident on a material with a work function of ϕ do not result in the emission of photoelectrons.

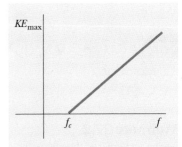

FIGURE 27.6 A sketch of KE_{max} versus frequency of incident light for photoelectrons in a typical photoelectric effect experiment. Photons with frequency less than f_c do not have sufficient energy to eject an electron from the metal.

Example 27.3   The Energy of a "Yellow" Photon

Yellow light with a frequency of approximately 6.0×10^{14} Hz is the predominant frequency in sunlight. What is the energy carried by a photon (a quantum of light) having this frequency?

Solution   The energy carried by a photon having this frequency is given by Equation 27.5:

$$E = hf = (6.63 \times 10^{-34} \, \text{J} \cdot \text{s})(6.0 \times 10^{14} \, \text{Hz}) = 4.0 \times 10^{-19} \, \text{J} = \boxed{2.5 \, \text{eV}}$$

Example 27.4   The Photoelectric Effect for Sodium

A sodium surface is illuminated with light of wavelength 0.300 μm. The work function for sodium is 2.46 eV. Find (a) the maximum kinetic energy of the ejected photoelectrons and (b) the cutoff wavelength for sodium.

Solution

A The energy of each photon of the illuminating light beam is

$$E = hf = \frac{hc}{\lambda} = \frac{(6.63 \times 10^{-34} \, \text{J} \cdot \text{s})(3.00 \times 10^8 \, \text{m/s})}{3.00 \times 10^{-7} \, \text{m}}$$

$$= 6.63 \times 10^{-19} \, \text{J} = \frac{6.63 \times 10^{-19} \, \text{J}}{1.60 \times 10^{-19} \, \text{J/eV}} = 4.14 \, \text{eV}$$

where we have used the conversion 1 eV = 1.6×10^{-19} J. Using Equation 27.6 gives

$$KE_{max} = hf - \phi = 4.14 \, \text{eV} - 2.46 \, \text{eV} = \boxed{1.68 \, \text{eV}}$$

B The cutoff wavelength can be calculated from Equation 27.7 after we convert ϕ from electron volts to joules:

$$\phi = 2.46 \, \text{eV} = (2.46 \, \text{eV})(1.6 \times 10^{-19} \, \text{J/eV}) = 3.94 \times 10^{-19} \, \text{J}$$

$$\lambda_c = \frac{hc}{\phi} = \frac{(6.63 \times 10^{-34} \, \text{J} \cdot \text{s})(3.00 \times 10^8 \, \text{m/s})}{3.94 \times 10^{-19} \, \text{J}}$$

$$= 5.05 \times 10^{-7} \, \text{m} = \boxed{505 \, \text{nm}}$$

This wavelength is in the yellow-green region of the visible spectrum.

Webnote 27.2

To see an applet of the photoelectric effect, go to
http://home.a-city.de/walter.fendt/physengl/photoeffect.htm

FIGURE 27.7 X-ray diffraction pattern of NaCl.

PHOTOCELLS

The photoelectric effect has many interesting applications using a device called the *photocell*. The photocell shown in Figure 27.4 acts much like a switch in an electric circuit with associated active elements in that it produces a current in the circuit when light of sufficiently high frequency falls on the cell, but it does not allow a current in the dark. Many practical devices in our everyday lives make use of photocells. For example, a use familiar to everyone is that of turning streetlights on at night and off in the morning. A photoelectric control unit in the base of the light activates a switch to turn off the streetlight when ambient light strikes it. Many garage-door systems and elevators use a light beam and a photocell as a safety feature in their design. When the light beam strikes the photocell, the electric current generated is sufficiently large to maintain a closed circuit. When an object or person blocks the light beam, the current is interrupted, which signals the door to open.

27.3 X-RAYS

In 1895 at the University of Wurzburg, Wilhelm Roentgen (1845–1923) was studying electrical discharges in low-pressure gases when he noted that a fluorescent screen glowed even when placed several meters from the gas discharge tube and even when black cardboard was placed between the tube and the screen. He concluded that the effect was caused by a mysterious type of radiation, which he called **x-rays** because of their unknown nature. Subsequent study showed that these rays traveled at or near the speed of light and that they could not be deflected by either electric or magnetic fields. This last fact indicated that x-rays did not consist of beams of charged particles, although the possibility that they were beams of uncharged particles remained.

In 1912 Max von Laue (1879–1960) suggested that if x-rays were electromagnetic waves with very short wavelengths, one should be able to diffract x-rays by using the regular atomic spacings of a crystal lattice as a diffraction grating, just as visible light is diffracted by a ruled grating. Shortly thereafter, researchers demonstrated that such a diffraction pattern could be observed, similar to that shown in Figure 27.7 for NaCl. The wavelengths of the x-rays were then determined from the diffraction data and the known values of the spacing between atoms in the crystal. X-ray diffraction has proved to be an invaluable technique for understanding the structure of matter. We shall discuss this subject in more detail in the next section.

Typical x-ray wavelengths are about 0.1 nm, which is on the order of the atomic spacing in a solid. We now know that x-rays are a part of the electromagnetic spectrum, characterized by frequencies higher than those of ultraviolet radiation and having the ability to penetrate most materials with relative ease.

X-rays are produced when high-speed electrons are suddenly slowed down, for example, when a metal target is struck by electrons that have been accelerated through a potential difference of several thousand volts. Figure 27.8a shows a schematic diagram of an x-ray tube. A current in the filament causes electrons to be emitted, and these freed electrons are accelerated toward a dense metal target, such as tungsten, which is held at a higher potential than the filament.

Figure 27.9 represents a plot of x-ray intensity versus wavelength for the spectrum of radiation emitted by an x-ray tube. Note that the spectrum has two distinct components. One component is a continuous broad spectrum that depends on the voltage applied to the tube. Superimposed on this component is a series of sharp, intense lines that depend on the nature of the target material. The accelerating voltage must exceed a certain value, called the **threshold voltage,** in order to observe these sharp lines, which represent radiation emitted by the target atoms as their electrons undergo rearrangements. We shall discuss this further in Chap-

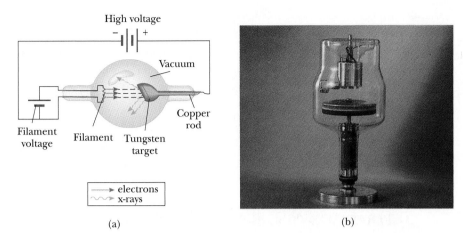

(a)

FIGURE 27.8 (a) Diagram of an x-ray tube. (b) Photograph of an x-ray tube. *(Courtesy of GE Medical Systems)*

(b)

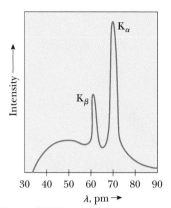

FIGURE 27.9 The x-ray spectrum of a metal target consists of a broad continuous spectrum plus a number of sharp lines, which are due to *characteristic x-rays*. The data shown were obtained when 35-keV electrons bombarded a molybdenum target. Note that 1 pm = 10^{-12} m = 10^{-3} nm.

ter 28. The continuous radiation is sometimes called **bremsstrahlung,** a German word meaning "braking radiation." The term arises from the nature of the mechanism responsible for the radiation. That is, electrons emit radiation when they undergo an acceleration inside the target.

Figure 27.10 illustrates how x-rays are produced when an electron passes near a charged target nucleus. As an electron passes close to a positively charged nucleus contained in the target material, it is deflected from its path because of its electrical attraction to the nucleus, and hence experiences an acceleration. An analysis from classical physics shows that any charged particle will emit electromagnetic radiation when it is accelerated. (An example of this is the production of electromagnetic waves by accelerated charges in a radio antenna, as described in Chapter 21.) According to quantum theory, this radiation must appear in the form of photons. Because the radiated photon shown in Figure 27.10 carries energy, the electron must lose kinetic energy because of its encounter with the target nucleus. Let us consider an extreme example in which the electron loses all of its energy in a single collision. In this case, the initial energy of the electron ($e \Delta V$) is transformed completely into the energy of the photon (hf_{max}). In equation form, we have

$$e \Delta V = hf_{max} = \frac{hc}{\lambda_{min}} \qquad [27.8]$$

where $e \Delta V$ is the energy of the electron after it has been accelerated through a potential difference of ΔV volts, and e is the charge on the electron. This says that the shortest-wavelength radiation that can be produced is

$$\lambda_{min} = \frac{hc}{e \Delta V} \qquad [27.9]$$

The reason that all the radiation produced does not have this particular wavelength is because many of the electrons are not stopped in a single collision. This results in the production of the continuous spectrum of wavelengths.

Interesting insights into the process of painting and revising a masterpiece are being revealed by x-rays. Long-wavelength x-rays are absorbed in varying degrees by some paints, such as those having lead, cadmium, chromium, or cobalt as a base. The x-ray interactions with the paints give contrast because the different elements in the paints have different electron densities. Also, thicker layers will absorb more than thin layers. To examine a painting by an old master, a film is placed behind it while it is x-rayed from the front. Ghost outlines of earlier paintings and earlier forms of the final masterpiece are sometimes revealed when the film is developed.

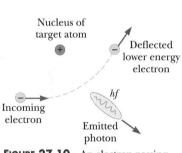

FIGURE 27.10 An electron passing near a charged target atom experiences an acceleration, and a photon is emitted in the process.

APPLICATION

USING X-RAYS TO STUDY THE WORK OF MASTER PAINTERS

Example 27.5   The Minimum X-Ray Wavelength

Calculate the minimum wavelength produced when electrons are accelerated through a potential difference of 100 000 V, a not-uncommon voltage for an x-ray tube.

Solution   From Equation 27.9, we have

$$\lambda_{min} = \frac{(6.63 \times 10^{-34}\,\text{J} \cdot \text{s})(3.00 \times 10^8\,\text{m/s})}{(1.60 \times 10^{-19}\,\text{C})(10^5\,\text{V})} = \boxed{1.24 \times 10^{-11}\,\text{m}}$$

27.4   DIFFRACTION OF X-RAYS BY CRYSTALS

In Chapter 24 we described how a diffraction grating can be used to measure the wavelength of light. In principle, the wavelength of *any* electromagnetic wave can be measured if a grating having a suitable line spacing can be found. The spacing between lines must be approximately equal to the wavelength of the radiation to be measured. X-rays are electromagnetic waves with wavelengths on the order of 0.1 nm. It would be impossible to construct a grating with such a small spacing. However, as noted in the previous section, Max von Laue suggested that the regular array of atoms in a crystal could act as a three-dimensional grating for observing the diffraction of x-rays.

One experimental arrangement for observing x-ray diffraction is shown in Figure 27.11. A narrow beam of x-rays with a continuous wavelength range is incident on a crystal such as sodium chloride. The diffracted radiation is very intense in certain directions, corresponding to constructive interference from waves reflected from layers of atoms in the crystal. The diffracted radiation is detected by a photographic film and forms an array of spots known as a Laue pattern. The crystal structure is determined by analyzing the positions and intensities of the various spots in the pattern.

The arrangement of atoms in a crystal of NaCl is shown in Figure 27.12. The smaller red spheres represent Na^+ ions and the larger blue spheres represent Cl^- ions. The spacing between successive Na^+ (or Cl^-) ions in this cubic structure, denoted by the symbol a in Figure 27.12, is approximately 0.563 nm.

A careful examination of the NaCl structure shows that the ions lie in various planes. The shaded areas in Figure 27.12 represent one example in which the atoms lie in equally spaced planes. Now suppose an x-ray beam is incident at grazing angle θ on one of the planes, as in Figure 27.13. The beam can be reflected from both the upper and lower plane of atoms. However, the geometric construction in Figure 27.13 shows that the beam reflected from the lower surface travels farther than the beam reflected from the upper surface by a distance of $2d \sin \theta$. The two portions of the reflected beam will combine to produce constructive in-

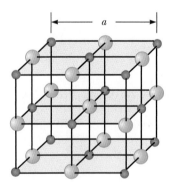

FIGURE 27.11   Schematic diagram of the technique used to observe the diffraction of x-rays by a single crystal. The array of spots formed on the film by the diffracted beams is called a Laue pattern. (See Figure 27.7.)

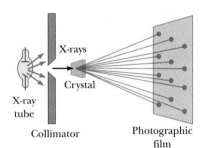

FIGURE 27.12   A model of the cubic crystalline structure of sodium chloride. The blue spheres represent the Cl^- ions, and the red spheres represent the Na^+ ions. The length of the cube edge is $a = 0.563$ nm.

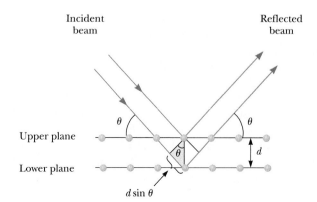

FIGURE 27.13   A two-dimensional description of the reflection of an x-ray beam from two parallel crystalline planes separated by a distance d. The beam reflected from the lower plane travels farther than the one reflected from the upper plane by an amount equal to $2d \sin \theta$.

terference when this path difference equals some integral multiple of the wavelength λ. The condition for constructive interference is given by

$$2d \sin \theta = m\lambda \qquad (m = 1, 2, 3, \ldots)$$ [27.10] ◀ Bragg's law

This condition is known as **Bragg's law** after W. L. Bragg (1890–1971), who first derived the relationship. If the wavelength and diffraction angle are measured, Equation 27.10 can be used to calculate the spacing between atomic planes.

The method of x-ray diffraction to determine crystalline structures was thoroughly developed in England by W. H. Bragg and his son W. L. Bragg, who shared a Nobel prize in 1915 for their work. Since then, thousands of crystalline structures have been investigated. Most important, the technique of x-ray diffraction has been and is being used to determine the atomic arrangement of complex organic molecules such as proteins. Proteins are large molecules containing thousands of atoms that help to regulate and speed up chemical life processes in cells. Some proteins are amazing catalysts, speeding up the slow room-temperature reactions in cells by 17 orders of magnitude. In order to understand this incredible biochemical reactivity, it is important to determine the structure of these intricate molecules.

The main technique used to determine the molecular structure of proteins, DNA, and RNA is x-ray diffraction using x-rays of wavelength of about 1.0 A. This allows the experimenter to "see" individual atoms that are separated by about this distance in molecules. Because the biochemical x-ray diffraction sample is prepared in crystal form, the *geometry* (position of the bright spots in space) of the diffraction pattern is determined by the regular three-dimensional crystal lattice arrangement of molecules in the sample. The *intensities* of the bright diffraction spots are determined by the atoms and their electronic distributions in the fundamental building block of the crystal, the unit cell. Using complicated computational techniques, investigators can essentially deduce the molecular structure by matching the observed intensities of diffracted beams with a series of assumed atomic positions, resulting in the atomic structure and electron density of the molecule. Figure 27.14 shows a classic x-ray diffraction image of DNA made by Rosalind Franklin in 1952. This and similar x-ray diffraction photos played an important role in the determination of the double-helix structure of DNA by F. H. C. Crick and J. D. Watson in 1953. A model of the famous DNA double helix is shown in Figure 27.15.

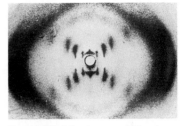

FIGURE 27.14 An x-ray diffraction photograph of DNA taken by Rosalind Franklin. The cross pattern of spots was a clue that DNA has a helical structure. *(Science Source/Photo Researchers, Inc.)*

|◀—— 2 nm ——▶|

FIGURE 27.15 The double-helix structure of DNA.

Example 27.6 X-Ray Diffraction from Calcite

If the spacing between certain planes in a crystal of calcite ($CaCO_3$) is 0.314 nm, find the grazing angles at which first- and third-order interference will occur for x-rays of wavelength 0.070 nm.

Solution For first-order interference, the value of *m* in Equation 27.10 is 1. Thus, the grazing angle corresponding to this order of interference is found as follows:

$$\sin \theta = \frac{m\lambda}{2d} = \frac{(0.0700 \text{ nm})}{2(0.314 \text{ nm})} = 0.111$$

$$\theta = \boxed{6.37°}$$

In third-order interference, $m = 3$, and we find

$$\sin \theta = \frac{m\lambda}{2d} = \frac{3(0.0700 \text{ nm})}{2(0.314 \text{ nm})} = 0.334$$

$$\theta = \boxed{19.5°}$$

Webnote 27.3

To learn much more about x-ray crystallography, its techniques, and the equipment used for measurement, explore *http://www-structure.llnl.gov/xray/101index.html*

27.5 THE COMPTON EFFECT

Further justification for the photon nature of light came from an experiment conducted by Arthur H. Compton in 1923. In his experiment, Compton directed an x-ray beam of wavelength λ_0 toward a block of graphite. He found that the scattered x-rays had a slightly longer wavelength λ than the incident x-rays, and hence the energies of the scattered rays were lower. The amount of energy reduction depended on the angle at which the x-rays were scattered. The change in wavelength, $\Delta\lambda$, between a scattered x-ray and an incident x-ray is called the **Compton shift.**

In order to explain this effect, Compton assumed that if a photon behaves like a particle, its collision with other particles is similar to that between two billiard balls. Hence, the x-ray photon carries both measurable *energy and momentum* and these two quantities must be conserved in a collision. If the incident photon collides with an electron initially at rest, as in Figure 27.16, the photon transfers some of its energy and momentum to the electron. As a consequence, the energy and frequency of the scattered photon are lowered and its wavelength increases. Applying relativistic energy and momentum conservation to the collision described in Figure 27.16, the shift in wavelength of the scattered photon is given by

◄ The Compton shift formula

$$\Delta\lambda = \lambda - \lambda_0 = \frac{h}{m_e c}(1 - \cos\theta) \qquad [27.11]$$

where m_e is the mass of the electron and θ is the angle between the directions of the scattered and incident photons. The quantity $h/m_e c$ is called the **Compton wavelength** and has a value of $h/m_e c = 0.002\,43$ nm. Note that the Compton wavelength is very small relative to the wavelengths of visible light, and hence the shift in wavelength would be difficult to detect if visible light were used. Furthermore, note that the Compton shift depends on the scattering angle θ and not on the wavelength. Experimental results for x-rays scattered from various targets obey Equation 27.11 and strongly support the photon concept.

Quick Quiz 27.1	An x-ray photon is scattered by an electron. The frequency of the scattered photon relative to that of the incident photon (a) increases, (b) decreases, or (c) remains the same.

ARTHUR HOLLY COMPTON, AMERICAN PHYSICIST (1892–1962)

Compton was born in Wooster, Ohio, and attended Wooster College and Princeton University. He became director of the laboratory at the University of Chicago where experimental work concerned with sustained chain reactions was conducted. This work was of central importance to the construction of the first atomic bomb. His discovery of the Compton effect and his work with cosmic rays led to his sharing the 1927 Nobel prize in physics with Charles Wilson. *(Courtesy of AIP Niels Bohr Library)*

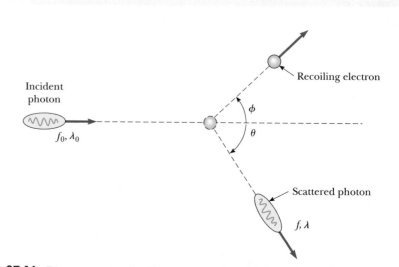

FIGURE 27.16 Diagram representing Compton scattering of a photon by an electron. The scattered photon has less energy (or longer wavelength) than the incident photon.

Quick Quiz 27.2

A photon of energy E_0 strikes a free electron, with the scattered photon of energy E moving in the direction opposite that of the incident photon. In this Compton effect interaction, the resulting kinetic energy of the electron is (a) E_0, (b) E, (c) $E_0 - E$, (d) $E_0 + E$, (e) none of the above.

Quick Quiz 27.3

A photon of energy E_0 strikes a free electron with the scattered photon of energy E moving in the direction opposite that of the incident photon. In this Compton effect interaction, the resulting momentum of the electron is (a) E_0/c, (b) $< E_0/c$, (c) $> E_0/c$, (d) $(E_0 - E)/c$, (e) $(E - E_0)/c$.

APPLYING **PHYSICS** 27.2

The Compton effect involves a change in wavelength as photons are scattered through different angles. Suppose we illuminate a piece of material with a beam of light and then view the material from different angles relative to the beam of light. Will we see a color change corresponding to the change in wavelength of the scattered light?

Explanation There will be a wavelength change for visible light scattered by the material, but the change will be far too small to detect as a color change. The largest possible wavelength change, at $180°$ scattering, will be twice the Compton wavelength, about 0.005 nm. This represents a change of less than 0.001% of the wavelength of red light. The Compton effect is only detectable for wavelengths that are very short to begin with, so that the Compton wavelength is an appreciable fraction of the incident wavelength. As a result, the usual radiation for observing the Compton effect is in the x-ray range of the electromagnetic spectrum.

Example 27.7 Compton Scattering at 45°

X-rays of wavelength $\lambda_0 = 0.200\,000$ nm are scattered from a block of material. The scattered x-rays are observed at an angle of $45.0°$ to the incident beam. Calculate the wavelength of the x-rays scattered at this angle.

Solution The shift in wavelength of the scattered x-rays is given by Equation 27.11. Taking $\theta = 45.0°$, we find that

$$\Delta\lambda = \frac{h}{m_e c}(1 - \cos\theta)$$

$$= \frac{6.626 \times 10^{-34}\,\text{J}\cdot\text{s}}{(9.11 \times 10^{-31}\,\text{kg})(3.00 \times 10^8\,\text{m/s})}(1 - \cos 45.0°)$$

$$= 7.10 \times 10^{-13}\,\text{m} = 0.000\,710\,\text{nm}$$

Hence, the wavelength of the scattered x-ray at this angle is

$$\lambda = \Delta\lambda + \lambda_0 = \boxed{0.200\,710\,\text{nm}}$$

EXERCISE Find the fraction of energy lost by the photon in this collision.

ANSWER Fraction $= \Delta E/E = 0.003\,54$

27.6   PHOTONS AND ELECTROMAGNETIC WAVES

Phenomena such as the photoelectric effect and the Compton effect offer iron-clad evidence that when light (or other forms of electromagnetic radiation) and matter interact, the light behaves as if it were composed of particles having energy hf and momentum h/λ. An obvious question that arises at this point is, "How can light be considered a photon (in other words, a particle) when we know it is a wave?" On the one hand, we describe light in terms of photons having energy and momentum. On the other hand, we must also recognize that light and other electromagnetic waves exhibit interference and diffraction effects that are consistent only with a wave interpretation.

Which model is correct? Is light a wave or a particle? The answer depends on the phenomenon being observed. Some experiments can be better explained with the photon concept, whereas others are best described with a wave model. The end result is that we must use both models and admit that the true nature of light is not describable in terms of a single classical picture. We can say that

Light has a dual nature ▶ | **Light has a dual nature. It exhibits both wave and particle characteristics.**

To understand why photons are compatible with electromagnetic waves, consider 2.5-MHz radio waves as an example. The energy of a photon having this frequency is only about 10^{-8} eV, too small to allow the photon to be detected. A sensitive radio receiver might require as many as 10^{10} of these photons to produce a detectable signal. Such a large number of photons would appear, on the average, as a continuous wave. With so many photons reaching the detector every second, it is unlikely that any graininess would appear in the detected signal. That is, with 2.5-MHz waves, we would not be able to detect the individual photons striking the antenna.

Now consider what happens as we go to higher frequencies. In the visible region, it is possible to observe both the particle characteristics and the wave characteristics of light. As we mentioned earlier, a light beam shows interference phenomena (thus, it is a wave) and at the same time can produce photoelectrons (thus, it is a particle). At even higher frequencies, the momentum and energy of the photons increase. Consequently, the particle nature of light becomes more evident than its wave nature. For example, absorption of an x-ray photon is easily detected as a single event, but wave effects are difficult to observe.

**LOUIS DE BROGLIE, FRENCH PHYSICIST
(1892–1987)**

De Broglie was born in Dieppe, France. At the Sorbonne in Paris, he studied history in preparation for what he hoped to be a career in the diplomatic service. The world of science is lucky that he changed his career path to become a theoretical physicist. De Broglie was awarded the Nobel prize in 1929 for his discovery of the wave nature of electrons. "It would seem that the basic idea of quantum theory is the impossibility of imaging an isolated quantity of energy without associating with it a certain frequency." (*AIP Niels Bohr Library*)

27.7   THE WAVE PROPERTIES OF PARTICLES

The "magic" duality of light mentioned in the last section is not the only duality we have encountered in this course. The mass-energy duality mentioned in Chapter 26 is probably about as easy or difficult to accept as the duality of light. All these dualities should make it easy to accept that *matter* has a dual nature as well!

In 1924, in his doctoral dissertation, Louis de Broglie postulated that **because photons have wave and particle characteristics, perhaps all forms of matter have both properties.** This was a highly revolutionary idea with no experimental confirmation at that time. According to de Broglie, electrons, just like light, have a dual particle-wave nature. Let us follow de Broglie's argument for the wavelength of a material particle.

In Chapter 26 we found that the relationship between energy and momentum for a photon, which has a rest energy of zero, is $p = E/c$. We also know from Equation 27.5 that the energy of a photon is

Energy of a photon ▶ | $$E = hf = \frac{hc}{\lambda}$$ [27.12]

SUMMARY

The characteristics of **blackbody radiation** cannot be explained using classical concepts. The peak of a blackbody radiation curve is given by **Wien's displacement law:**

$$\lambda_{\max} T = 0.2898 \times 10^{-2}\,\text{m} \cdot \text{K} \qquad \text{[27.1]}$$

where $\lambda_{\max}$ is the wavelength at which the curve peaks and T is the absolute temperature of the object emitting the radiation.

Planck first introduced the quantum concept when he assumed that the subatomic oscillators responsible for blackbody radiation could have only discrete amounts of energy given by

$$E_n = nhf \qquad \text{[27.2]}$$

where n is a positive integer called a **quantum number** and f is the frequency of vibration of the resonator.

The **photoelectric effect** is a process whereby electrons are ejected from a metal surface when light is incident on that surface. Einstein provided a successful explanation of this effect by extending Planck's quantum hypothesis to electromagnetic waves. In this model, light is viewed as a stream of particles called photons, each with energy $E = hf$, where f is the light frequency and h is **Planck's constant.** The maximum kinetic energy of the ejected photoelectrons is

$$KE_{\max} = hf - \phi \qquad \text{[27.6]}$$

where ϕ is the **work function** of the metal.

X-rays are produced when high-speed electrons are suddenly decelerated. When electrons have been accelerated through a voltage V, the shortest-wavelength radiation that can be produced is

$$\lambda_{\min} = \frac{hc}{e\,\Delta V} \qquad \text{[27.9]}$$

The regular array of atoms in a crystal can act as a diffraction grating for x-rays and for electrons. The condition for constructive interference of the diffracted rays is given by **Bragg's law:**

$$2d\sin\theta = m\lambda \qquad (m = 1, 2, 3, \ldots) \qquad \text{[27.10]}$$

X-rays from an incident beam are scattered at various angles by electrons in a target such as carbon. In such a scattering event, a shift in wavelength is observed for the scattered x-rays. This phenomenon is known as the **Compton shift.** Conservation of momentum and energy applied to a photon-electron collision yields the following expression for the shift in wavelength of the scattered x-rays:

$$\Delta\lambda = \lambda - \lambda_0 = \frac{h}{m_e c}(1 - \cos\theta) \qquad \text{[27.11]}$$

where m_e is the mass of the electron, c is the speed of light, and θ is the scattering angle.

De Broglie proposed that all matter has both a particle and a wave nature. The **de Broglie wavelength** of any particle of mass m and speed v is

$$\lambda = \frac{h}{p} = \frac{h}{mv} \qquad \text{[27.14]}$$

De Broglie also proposed that the frequencies of the waves associated with particles obey the Einstein relationship, $E = hf$.

In the theory of **quantum mechanics,** each particle is described by a quantity Ψ called the **wave function.** The probability of finding the particle at a particular point at some instant is proportional to Ψ^2. Quantum mechanics is very successful in describing the behavior of atomic and molecular systems.

According to Heisenberg's **uncertainty principle,** it is impossible to measure simultaneously the exact position and exact momentum of a particle. If Δx is the uncertainty in the measured position and Δp_x the uncertainty in the momentum, the product $\Delta x \, \Delta p_x$ is given by

$$\Delta x \, \Delta p_x \geq \frac{h}{4\pi} \qquad \text{[27.16]}$$

Also,

$$\Delta E \, \Delta t \geq \frac{h}{4\pi} \qquad \text{[27.17]}$$

where ΔE is the uncertainty in the energy of the particle and Δt is the uncertainty in the time it takes to measure the energy.

CONCEPTUAL QUESTIONS

1. If you observe objects inside a very hot kiln, it is difficult to discern the shapes of the objects. Why?
2. Why is an electron microscope more suitable than an optical microscope for "seeing" objects of an atomic size?
3. Are blackbodies black?
4. Why is it impossible to simultaneously measure the position and velocity of a particle with infinite accuracy?
5. All objects radiate energy. Why, then, are we not able to see all objects in a dark room?
6. Is light a wave or a particle? Support your answer by citing specific experimental evidence.
7. A student claims that he is going to eject electrons from a piece of metal by placing a radio transmitter antenna adjacent to the metal and sending a strong AM radio signal into the antenna. The work function of a metal is typically a few electron volts. Will this work?
8. In view of Heisenberg's uncertainty principle, how can we predict such things as maximum height and range when describing something like a baseball?
9. In the photoelectric effect, explain why the stopping potential depends on the frequency of the light but not on the intensity.
10. Which has more energy, a photon of ultraviolet radiation or a photon of yellow light?
11. Why does the existence of a cutoff frequency in the photoelectric effect favor a particle theory of light rather than a wave theory?
12. What effect, if any, would you expect the temperature of a material to have on the ease with which electrons can be ejected from it in the photoelectric effect?
13. If photons that cause photoelectrons to be released all have the same energy, how can the photoelectrons have different energies?
14. The brightest star in the constellation Lyra is the bluish star Vega, whereas the brightest star in Bootes is the reddish star Arcturus. How do you account for the difference in color of the two stars?
15. If the photoelectric effect is observed for one metal, can you conclude that the effect will also be observed for another metal under the same conditions? Explain.
16. Discuss whether the behavior of an electron is mainly wave-like or particle-like in each of the following situations: (a) traversing a circular orbit in a magnetic field; (b) absorbing a photon and being photoelectrically ejected from the surface of a metal; (c) forming an interference pattern.
17. If a photon is deflected via the Compton effect, can its wavelength ever become shorter?
18. In an adult, 13% of the bone marrow is in the head. Discuss this in relation to dental x-rays.

PROBLEMS

1, 2, 3 = straightforward, intermediate, challenging   ☐ = full solution available in Student Solutions Manual/Study Guide

web = solution posted at **http://info.brookscole.com/serway**   🔬 = biomedical application

Section 27.1   Blackbody Radiation and Planck's Hypothesis

Section 27.2   The Photoelectric Effect and the Particle Theory of Light

1. (a) What is the surface temperature of Betelgeuse, a red giant star in the constellation of Orion, which radiates with a peak wavelength of about 970 nm? (b) Rigel, a bluish-white star in Orion, radiates with a peak wavelength of 145 nm. Find the temperature of Rigel's surface.
2. (a) Lightning produces a maximum air temperature on the order of 10^4 K, while (b) a nuclear explosion produces a temperature on the order of 10^7 K. Use Wien's displacement law to find the order of magnitude of the wavelength of the thermally produced photons radiated with greatest intensity by each of these sources. Name the part of the electromagnetic spectrum where you would expect each to radiate most strongly.
3. (a) Assuming that the tungsten filament of a lightbulb is a blackbody, determine its peak wavelength if its temperature is 2 900 K. (b) Why does your answer to part (a) suggest that more energy from a lightbulb goes into heat than into light?
4. Calculate the energy, in electron volts, of a photon whose frequency is (a) 620 THz, (b) 3.10 GHz, (c) 46.0 MHz. (d) Determine the corresponding wavelengths for these

photons and state the classification of each on the electromagnetic spectrum.

5. Calculate the energy in electron volts of a photon having a wavelength in (a) the microwave range, 5.00 cm; (b) the visible light range, 500 nm; and (c) the x-ray range, 5.00 nm.

6. A sodium-vapor lamp has a power output of 1 000 W. Using 589.3 nm as the average wavelength of this source, calculate the number of photons emitted per second.

7. An FM radio transmitter has a power output of 150 kW and operates at a frequency of 99.7 MHz. How many photons per second does the transmitter emit?

8. The threshold of dark-adapted (scotopic) vision is 4.0×10^{-11} W/m^2 at a central wavelength of 500 nm. If light with this intensity and wavelength enters the eye when the pupil is open to its maximum diameter of 8.5 mm, how many photons per second enter the eye?

9. A 1.5-kg mass vibrates at an amplitude of 3.0 cm on the end of a spring of spring constant 20 N/m. (a) If the energy of the spring is quantized, find its quantum number. (b) If n changes by 1, find the fractional change in energy of the spring.

10. A 0.50-kg mass falls from a height of 3.0 m. If all of the energy of this mass could be converted to visible light of wavelength 5.0×10^{-7} m, how many photons would be produced?

11. When light of wavelength 350 nm falls on a potassium surface, electrons are emitted that have a maximum kinetic energy of 1.31 eV. Find (a) the work function of potassium, (b) the cutoff wavelength, and (c) the frequency corresponding to the cutoff wavelength.

12. Electrons are ejected from a metallic surface with speeds ranging up to 4.6×10^5 m/s when light with a wavelength of $\lambda = 625$ nm is used. (a) What is the work function of the surface? (b) What is the cutoff frequency for this surface?

13. Molybdenum has a work function of 4.20 eV. (a) Find the cutoff wavelength and threshold frequency for the photoelectric effect. (b) Calculate the stopping potential if the incident light has a wavelength of 180 nm.

14. Lithium, beryllium, and mercury have work functions of 2.30 eV, 3.90 eV, and 4.50 eV, respectively. If 400-nm light is incident on each of these metals, determine (a) which metals exhibit the photoelectric effect and (b) the maximum kinetic energy for the photoelectrons in each case.

15. From the scattering of sunlight, Thomson calculated that the classical radius of the electron has a value of 2.82×10^{-15} m. If sunlight having an intensity of 500 W/m^2 falls on a disk with this radius, find the time required to accumulate 1.00 eV of energy. Assume that light is a classical wave and that the light striking the disk is completely absorbed. How does your value compare with the observation that photoelectrons are promptly (within 10^{-9} s) emitted?

16. An isolated copper sphere of radius 5.00 cm, initially uncharged, is illuminated by ultraviolet light of wavelength 200 nm. What charge will the photoelectric effect induce on the sphere? The work function for copper is 4.70 eV.

17. When light of wavelength 254 nm falls on cesium, the required stopping potential is 3.00 V. If light of wavelength 436 nm is used, the stopping potential is 0.900 V. Use this information to plot a graph like that shown in Figure 27.6, and from the graph determine the cutoff frequency for cesium and its work function.

18. Ultraviolet light is incident normally on the surface of a certain substance. The binding energy of the electrons in this substance is 3.44 eV. The incident light has an intensity of 0.055 W/m^2. The electrons are photoelectrically emitted with a maximum speed of 4.2×10^5 m/s. How many electrons are emitted from a square centimeter of the surface each second? Assume that the absorption of every photon ejects an electron.

Section 27.3 X-Rays

19. The extremes of the x-ray portion of the electromagnetic spectrum range from approximately 1.0×10^{-8} m to 1.0×10^{-13} m. Find the minimum accelerating voltages required to produce wavelengths at these two extremes.

20. Calculate the minimum wavelength x-ray that can be produced when a target is struck by an electron that has been accelerated through a potential difference of (a) 15.0 kV, (b) 100 kV.

21. What minimum accelerating voltage would be required to produce an x-ray with a wavelength of 0.0300 nm?

Section 27.4 Diffraction of X-Rays by Crystals

22. A monochromatic x-ray beam is incident on a NaCl crystal surface where $d = 0.353$ nm. The second-order maximum in the reflected beam is found when the angle between the incident beam and the surface is 20.5°. Determine the wavelength of the x-rays.

23. Potassium iodide has an interplanar spacing of $d = 0.296$ nm. A monochromatic x-ray beam shows a first-order diffraction maximum when the grazing angle is 7.6°. Calculate the x-ray wavelength.

24. The spacing between certain planes in a crystal is known to be 0.30 nm. Find the smallest grazing angle at which constructive interference will occur for wavelength 0.070 nm.

25. X-rays of wavelength 0.140 nm are reflected from a certain crystal, and the first-order maximum occurs at an angle of 14.4°. What value does this give for the interplanar spacing of this crystal?

Section 27.5 The Compton Effect

26. X-rays are scattered from electrons in a carbon target. The measured wavelength shift is 1.50×10^{-3} nm. Calculate the scattering angle.

27. Calculate the energy and momentum of a photon of wavelength 700 nm.

28. A beam of 0.68-nm photons undergoes Compton scattering from free electrons. What are the energy and momentum of the photons that emerge at a 45° angle with respect to the incident beam?

29. A 0.001 6-nm photon scatters from a free electron. For what (photon) scattering angle will the recoiling electron and scattered photon have the same kinetic energy?

30. X-rays with an energy of 300 keV undergo Compton scattering from a target. If the scattered rays are deflected at 37.0° relative to the direction of the incident rays, find (a) the Compton shift at this angle, (b) the energy of the scattered x-ray, and (c) the kinetic energy of the recoiling electron.

31. A 0.110-nm photon collides with a stationary electron. After the collision, the electron moves forward and the photon recoils backward. Find the momentum and kinetic energy of the electron.

32. After a 0.800 nm x-ray photon scatters from a free electron, the electron recoils with a speed equal to 1.40×10^6 m/s. (a) What was the Compton shift in the photon's wavelength? (b) Through what angle was the photon scattered?

33. A 0.45-nm x-ray photon is deflected through a 23° angle after scattering from a free electron. (a) What is the kinetic energy of the recoiling electron? (b) What is its speed?

Section 27.7 The Wave Properties of Particles

34. Calculate the de Broglie wavelength for an electron that has kinetic energy (a) 50.0 eV and (b) 50.0 keV (ignore relativistic effects).

35. (a) If the wavelength of an electron is equal to 5.00×10^{-7} m, how fast is it moving? (b) If the electron has a speed of 1.00×10^7 m/s, what is its wavelength?

36. Through what potential difference would an electron have to be accelerated from rest to give it a de Broglie wavelength of 1.0×10^{-10} m?

37. The nucleus of an atom is on the order of 10^{-14} m in diameter. For an electron to be confined to a nucleus, its de Broglie wavelength would have to be of this order of magnitude or smaller. (a) What would be the kinetic energy of an electron confined to this region? (b) On the basis of this result, would you expect to find an electron in a nucleus? Explain.

38. After learning about de Broglie's hypothesis that particles of momentum p have wave characteristics with wavelength $\lambda = h/p$, an 80-kg student has grown concerned about being diffracted when passing through a 75-cm-wide doorway. Assume that significant diffraction occurs when the width of the diffraction aperture is less than 10 times the wavelength of the wave being diffracted. (a) Determine the order of magnitude of the maximum speed at which the student can pass through the doorway in order to be significantly diffracted. (b) With that speed, how long will it take the student to pass through a doorway in a wall 15 cm thick? Compare your order-of-magnitude result to the currently accepted age of the Universe, which is 4×10^{17} s. (c) Should this student worry about being diffracted?

39. De Broglie postulated that the relationship $\lambda = h/p$ is
web valid for relativistic particles. What is the de Broglie wavelength for a (relativistic) electron whose kinetic energy is 3.00 MeV?

40. At what speed must an electron move so that its de Broglie wavelength equals its Compton wavelength? (*Hint:* This electron is relativistic.)

41. The resolving power of a microscope is proportional to the wavelength used. A resolution of approximately 1.0×10^{-11} m (0.010 nm) would be required in order to "see" an atom. (a) If electrons were used (electron microscope), what minimum kinetic energy would be required for the electrons? (b) If photons were used, what minimum photon energy would be needed to obtain 1.0×10^{-11} m resolution?

Section 27.9 The Uncertainty Principle

42. A 50.0-g ball moves at 30.0 m/s. If its speed is measured to an accuracy of 0.10%, what is the minimum uncertainty in its position?

43. A 0.50-kg block rests on the icy surface of a frozen pond, which we can assume to be frictionless. If the location of the block is measured to a precision of 0.50 cm, what speed must the block acquire because of the measurement process?

44. Suppose Fuzzy, a quantum-mechanical duck, lives in a world in which $h = 2\pi \, \text{J} \cdot \text{s}$. Fuzzy has a mass of 2.00 kg and is initially known to be within a pond 1.00 m wide. (a) What is the minimum uncertainty in his speed? (b) Assuming this uncertainty in speed to prevail for 5.00 s, determine the uncertainty in position after this time.

45. Suppose optical radiation ($\lambda = 5.00 \times 10^{-7}$ m) is used to determine the position of an electron to within the wavelength of the light. What will be the resulting uncertainty in the electron's velocity?

46. (a) Show that the kinetic energy of a nonrelativistic particle can be written in terms of its momentum as $KE = p^2/2m$. (b) Use the results of (a) to find the minimum kinetic energy of a proton confined within a nucleus having a diameter of 1.0×10^{-15} m.

ADDITIONAL PROBLEMS

47. Figure P27.47 shows the spectrum of light emitted by a firefly. Determine the temperature of a blackbody that would emit radiation peaked at the same frequency. Based on your result, would you say firefly radiation is blackbody radiation?

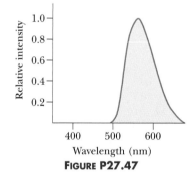

FIGURE P27.47

48. An x-ray tube is operated at 50 000 V. (a) Find the minimum wavelength of the radiation emitted by this tube. (b) If this radiation is directed at a crystal, the first-order maximum in the reflected radiation occurs when the grazing angle is 2.50°. What is the spacing between reflecting planes in the crystal?

49. The spacing between planes of nickel atoms in a nickel crystal is 0.352 nm. At what angle does a second-order Bragg reflection occur in nickel for 11.3-keV x-rays?

50. Johnny Jumper's favorite trick is to step out of his 16th story window and fall 50.0 m into a pool. A news reporter takes a picture of 75.0-kg Johnny just before he makes a splash, using an exposure time of 5.00 ms. Find (a) Johnny's de Broglie wavelength at this moment, (b) the uncertainty of his kinetic energy measurement during such a period of time, and (c) the percent error caused by such an uncertainty.

51. Photons of wavelength 450 nm are incident on a metal. The most energetic electrons ejected from the metal are bent into a circular arc of radius 20.0 cm by a magnetic field with a magnitude of 2.00×10^{-5} T. What is the work function of the metal?

52. A 200-MeV photon is scattered at 40.0° by a free proton initially at rest. Find the energy (in MeV) of the scattered photon.

53. A light source of wavelength λ illuminates a metal and ejects photoelectrons with a maximum kinetic energy of 1.00 eV. A second light source of wavelength $\lambda/2$ ejects photoelectrons with a maximum kinetic energy of 4.00 eV. What is the work function of the metal?

web

54. Red light of wavelength 670 nm produces photoelectrons from a certain photoemissive material. Green light of wavelength 520 nm produces photoelectrons from the same material with 1.50 times the maximum kinetic energy. What is the material's work function?

55. How fast must an electron be moving if all its kinetic energy is lost to a single x-ray photon (a) at the long-wavelength end of the x-ray electromagnetic spectrum with a wavelength of 1.00×10^{-8} m; (b) at the short-wavelength end of the x-ray electromagnetic spectrum with a wavelength of 1.00×10^{-13} m?

56. Show that if an electron were confined inside an atomic nucleus of diameter 2.0×10^{-15} m, it would have to be moving relativistically, while a proton confined to the same nucleus can be moving at less than one-tenth the speed of light.

57. A photon strikes a metal with a work function of ϕ and produces a photoelectron with a de Broglie wavelength equal to the wavelength of the original photon. (a) Show that the energy of this photon must have been given by

$$E = \frac{\phi(m_e c^2 - \phi/2)}{m_e c^2 - \phi}$$

where m_e is the mass of the electron. (*Hint:* Begin with conservation of energy, $E + m_e c^2 = \phi + \sqrt{(pc)^2 + (m_e c^2)^2}$.) (b) If one of these photons strikes platinum ($\phi = 6.35$ eV), determine the resulting maximum speed of the photoelectron.

58. In a Compton scattering event, the scattered photon has an energy of 120.0 keV and the recoiling electron has a kinetic energy of 40.0 keV. Find (a) the wavelength of the incident photon, (b) the angle θ at which the photon is scattered, and (c) the recoil angle of the electron. (*Hint:* Conserve both mass-energy and relativistic momentum.)

59. A woman on a ladder drops small pellets toward a spot on the floor. (a) Show that according to the uncertainty principle, the average miss distance must be at least

$$\Delta x = \left(\frac{h}{2\pi m}\right)^{1/2}\left(\frac{H}{2g}\right)^{1/4}$$

where H is the initial height of each pellet above the floor and m is the mass of each pellet. (b) If $H = 2.00$ m and $m = 0.500$ g, what is Δx?

60. Show that the speed of a particle having de Broglie wavelength λ and Compton wavelength $\lambda_C = h/(mc)$ is

$$v = \frac{c}{\sqrt{1 + (\lambda/\lambda_C)^2}}$$

GROUP ACTIVITIES

G.1 Use a black marker or pieces of dark electrical tape to make a very dark area on the outside of a shoebox. Poke a hole in the center of the dark area with a pencil. Now put a lid on the box and compare the blackness of the hole with the blackness of the surrounding dark area. Based on your observation explain why the radiation emitted from the hole is like that emitted from a blackbody.

G.2 On a clear night, go outdoors far from city lights and find the constellation Orion. Your instructor should be able to furnish you with a star chart to assist you in locating this grouping of stars. Look very carefully at the color of the two stars Betelgeuse and Rigel. Can you tell which star is hotter? Orion is visible from November through April in the evening sky. Thus, if Orion is not visible, compare two of the brightest stars you can see, such as Vega in the constellation Lyra and Arcturus in Bootes.

G.3 (a) If a pulse of blue light and a pulse of red light both carry energy E, which pulse contains more photons? (b) If the wavelength of the red light is 690 nm, the wavelength of the blue light is 420 nm, and the energy E is 1 000 eV, find the number of photons in each beam to verify your prediction of part (a).

G.4 (a) The threshold frequency for the photoelectric effect for a material is f_0. Are electrons emitted from the material when light incident on the material is of frequency (i) greater than f_0? (ii) less than f_0? (iii) equal to f_0? (b) The work function for zinc is 4.31 eV. What is the lowest frequency of light that releases photoelectrons when incident on the material?

FIGURE GA27.2 (*John Chumack/Photo Researchers, Inc.*)

G.5 Light sometimes acts like a wave and sometimes like a particle. For the following situations, which best describes the behavior of light? Defend your answers. (a) The photoelectric effect. (b) The Compton effect. (c) Young's double-slit experiment.

28

Atomic Physics

Chapter Outline

"Neon lights", commonly used in advertising signs, consist of thin glass tubes filled with various gases such as neon and helium. The gas atoms are excited to higher energy levels by electric discharge through the tube. When the electrons in these excited levels return to lower energy levels, the atoms emit light having a wavelength (color) that depends on the type of gas in the tube. For example, a tube filled with neon produces a red-orange color, while helium produces pink.

(Terry Gleason, Visuals Unlimited)

A large portion of this chapter is concerned with the study of the hydrogen atom. Although the hydrogen atom is the simplest atomic system, it is especially important for several reasons:

- The quantum numbers used to characterize the allowed states of hydrogen can also be used to describe (approximately) the allowed states of more complex atoms. This enables us to understand the periodic table of the elements, one of the great triumphs of quantum mechanics.
- The hydrogen atom is an ideal system for performing precise comparisons of theory and experiment and for improving our overall understanding of atomic structure.
- Much of what we know about the hydrogen atom with its single electron can be extended to such single-electron ions as He^+ and Li^{2+}.

In this chapter we first discuss the Bohr model of hydrogen, which helps us understand many features of hydrogen but fails to explain the finer details of atomic structure. Next we examine the hydrogen atom from the viewpoint of quantum mechanics and the quantum numbers used to characterize various atomic states. In addition, we examine the physical significance of the quantum numbers and the effect of a magnetic field on certain quantum states. The Pauli exclusion principle is also presented. This physical principle is extremely important in understanding the properties of complex atoms and the arrangement of elements in the periodic table. Finally, we apply our knowledge of atomic structure to describe the mechanisms involved in the production of x-rays, the operation of lasers, and the behavior of solid-state devices such as diodes and transistors.

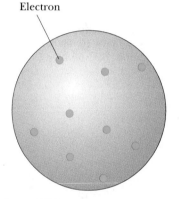

FIGURE 28.1 Thomson's model of the atom, with the electrons embedded inside the positive charge like seeds in a watermelon.

28.1 EARLY MODELS OF THE ATOM

The model of the atom in the days of Newton was a tiny, hard, indestructible sphere. Although this model was a good basis for the kinetic theory of gases, new models had to be devised when later experiments revealed the electronic nature of atoms. J. J. Thomson (1856–1940) suggested a model of the atom as a volume of positive charge with electrons embedded throughout the volume, much like the seeds in a watermelon (Fig. 28.1).

In 1911 Ernest Rutherford (1871–1937) and his students Hans Geiger and Ernest Marsden performed a critical experiment showing that Thomson's model could not be correct. In this experiment, a beam of positively charged **alpha particles** was projected against a thin metal foil, as in Figure 28.2a. The results of the experiment astounded scientists. Most of the alpha particles passed through the foil as if it were empty space, but a few particles deflected from their original direc-

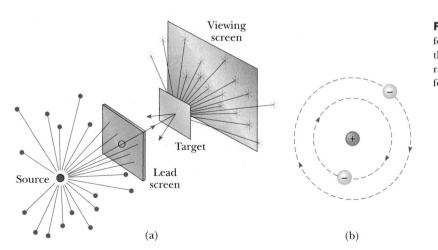

(a)   (b)

FIGURE 28.2 (a) Geiger and Marsden's technique for observing the scattering of alpha particles from a thin foil target. The source is a naturally occurring radioactive substance, such as radium. (b) Rutherford's planetary model of the atom.

SIR JOSEPH JOHN THOMSON, ENGLISH PHYSICIST (1856–1940)

Thomson, usually considered the discoverer of the electron, opened up the field of subatomic particle physics with his extensive work on the deflection of cathode rays (electrons) in an electric field. Thomson received the 1906 Nobel prize for his discovery of the electron. *(Stock Montage, Inc.)*

tion of travel were scattered through large angles. Some particles even deflected backward, reversing their direction of travel. When Geiger informed Rutherford of these results, Rutherford wrote, "It was quite the most incredible event that has ever happened to me in my life. It was almost as incredible as if you fired a 15-inch [artillery] shell at a piece of tissue paper and it came back and hit you."

Such large deflections were not expected on the basis of Thomson's model. According to this model, a positively charged alpha particle would never come close enough to a large positive charge to cause large-angle deflections. Rutherford explained these astounding results by assuming that the positive charge in an atom was concentrated in a region that was small relative to the size of the atom. He called this concentration of positive charge the **nucleus** of the atom. Any electrons belonging to the atom were assumed to be in the relatively large volume outside the nucleus. In order to explain why electrons in this outer region of the atom were not pulled into the nucleus, Rutherford viewed them as moving in orbits about the positively charged nucleus in the same manner as the planets orbit the Sun, as shown in Figure 28.2b. Alpha particles themselves were later identified as the nuclei of helium atoms.

There are two basic difficulties with Rutherford's planetary model. As we shall see in the next section, an atom emits certain discrete characteristic frequencies of electromagnetic radiation and no others; the Rutherford model is unable to explain this phenomenon. A second difficulty is that Rutherford's electrons are undergoing a centripetal acceleration. According to Maxwell's theory of electromagnetism, centripetally accelerated charges revolving with frequency f should radiate electromagnetic waves of the same frequency. Unfortunately, this classical model falls apart when applied to the atom. As the electron radiates energy, the radius of its orbit steadily decreases and its frequency of revolution increases. This leads to an ever-increasing frequency of emitted radiation and a rapid collapse of the atom as the electron spirals into the nucleus.

28.2   ATOMIC SPECTRA

As you may have already learned in chemistry, the hydrogen atom is the simplest atomic system, and an especially important one to understand. Much of what we know about the hydrogen atom (which consists of one proton and one electron) can be extended directly to other single-electron ions such as He^+ and Li^{2+}. Furthermore, a thorough understanding of the physics underlying the hydrogen atom can then be used to describe more complex atoms and the periodic table of the elements.

Suppose an evacuated glass tube is filled with hydrogen (or some other gas) at low pressure. If a voltage applied between metal electrodes in the tube is great enough to produce an electric current in the gas, the tube emits light whose color is characteristic of the gas in the tube. (This is how a neon sign works.) When the emitted light is analyzed with a spectrometer, a series of discrete bright lines is observed, each line having a different wavelength, or color. Such a series of spectral lines is commonly referred to as an **emission spectrum.** The wavelengths contained in a given line spectrum are characteristic of the element emitting the light (Fig. 28.3a). Because no two elements emit the same line spectrum, this phenomenon represents a marvelous and reliable technique for identifying elements in a gaseous substance.

The emission spectrum of hydrogen shown in Figure 28.4 includes four prominent lines that occur at wavelengths of 656.3 nm, 486.1 nm, 434.1 nm, and 410.2 nm. In 1885 Johann Balmer (1825–1898) found that the wavelengths of these and less prominent lines can be described by the simple empirical equation:

Balmer series ▶

$$\frac{1}{\lambda} = R_H \left(\frac{1}{2^2} - \frac{1}{n^2} \right)$$

[28.1]

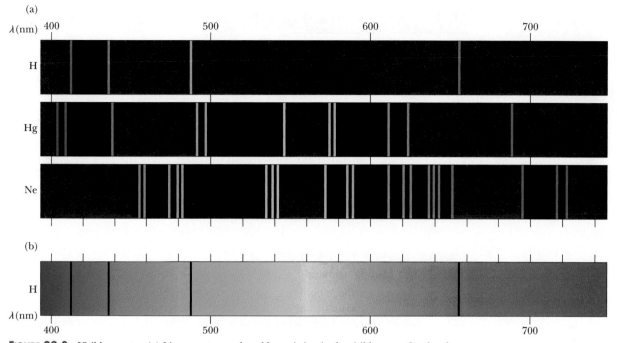

FIGURE 28.3   Visible spectra. (a) Line spectra produced by emission in the visible range for the elements hydrogen, mercury, and neon. (b) The absorption spectrum for hydrogen. The dark absorption lines occur at the same wavelengths as the emission lines for hydrogen shown in (a).

where n has integral values of 3, 4, 5, . . . , and R_H is a constant, called the **Rydberg constant.** If the wavelength is in meters, R_H has the value

$$R_H = 1.097\ 373\ 2 \times 10^7\ \text{m}^{-1} \qquad \text{[28.2]}$$

◀ Rydberg constant

The first line in the Balmer series, at 656.3 nm, corresponds to $n = 3$ in Equation 28.1; the line at 486.1 nm corresponds to $n = 4$; and so on. In addition to this Balmer series of spectral lines, a Lyman series was subsequently discovered in the far ultraviolet, with the radiated wavelengths described by a similar equation.

In addition to emitting light at specific wavelengths, an element can also absorb light at specific wavelengths. The spectral lines corresponding to this process form what is known as an **absorption spectrum.** An absorption spectrum can be obtained by passing a continuous radiation spectrum (one containing all wavelengths) through a vapor of the element being analyzed. The absorption spectrum consists of a series of dark lines superimposed on the otherwise bright continuous spectrum. Each line in the absorption spectrum of a given element coincides with a line in the emission spectrum of the element. That is, if hydrogen is the absorbing vapor, dark lines will appear at the visible wavelengths 656.3 nm, 486.1 nm, 434.1 nm, and 410.2 nm, as shown in Figures 28.3b and 28.4.

The absorption spectrum of an element has many practical applications. For example, the continuous spectrum of radiation emitted by the Sun must pass through the cooler gases of the solar atmosphere before reaching Earth. The various absorption lines observed in the solar spectrum have been used to identify elements in the solar atmosphere. It is interesting to note that when the solar spectrum was first studied, some lines were found that did not correspond to any known element. A new element had been discovered! Because the Greek word for Sun is *helios,* the new element was named *helium.* It was later identified in underground gases on Earth. Scientists are able to examine the light from stars other than our Sun in this fashion, but elements other than those present on Earth have never been detected.

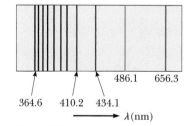

FIGURE 28.4   A series of spectral lines for atomic hydrogen. The prominent labeled lines are part of the Balmer series.

APPLICATION

DISCOVERY OF HELIUM

APPLYING **PHYSICS** *28.1*

You are observing a yellow candle flame, and your laboratory partner claims that the light from the flame originates from excited sodium atoms in the flame. You disagree, claiming that the candle flame is hot, so the radiation must be thermal in origin. Before your disagreement leads to fisticuffs, how can you determine who is correct?

Explanation   A simple determination can be made by observing the light from the candle flame through a spectrometer, which is a slit and diffraction grating combination discussed in Chapter 25. If the spectrum of the light is continuous, then it is most likely thermal in origin. If the spectrum shows discrete lines, it is atomic in origin. The results of the experiment show that the light is indeed thermal in origin and originates from random molecular motion in the candle flame.

APPLYING **PHYSICS** *28.2*

At extreme northern latitudes, the aurora borealis provides a beautiful and colorful display in the nighttime sky. A similar display occurs near the southern polar region and is called the aurora australis. What is the origin of the various colors of radiation seen in the auroras?

Explanation   The auroras are due to high-speed particles interacting with Earth's magnetic field and entering the atmosphere, as described on the back cover of this book. When these particles collide with molecules in the atmosphere, they excite the molecules in a way similar to the voltage in the spectrum tubes discussed earlier in this section. In response, the molecules emit colors of light according to the characteristic spectrum of their atomic constituents. For our atmosphere, the primary constituents are nitrogen and oxygen, which provide the red, blue, and green colors of the auroras.

APPLICATION

THE NORTHERN AND SOUTH-
ERN LIGHTS

28.3   THE BOHR THEORY OF HYDROGEN

At the beginning of the 20th century, scientists were perplexed by the failure of classical physics to explain the characteristics of spectra. Why did atoms of a given element emit only certain lines? Furthermore, why did the atoms absorb only those wavelengths that they emitted? In 1913 Bohr provided an explanation of atomic spectra that includes some features of the currently accepted theory. Using the simplest atom, hydrogen, Bohr developed a model of what he thought must be the atom's structure in an attempt to explain why the atom was stable. His model of the hydrogen atom contains some classical features as well as some revolutionary postulates that could not be justified within the framework of classical physics. The basic assumptions of the Bohr theory as it applies to the hydrogen atom are as follows:

1. The electron moves in circular orbits about the proton under the influence of the Coulomb force of attraction, as in Figure 28.5. In this case, the Coulomb force produces the electron's centripetal acceleration.
2. Only certain electron orbits are stable. These are orbits in which the hydrogen atom does not emit energy in the form of electromagnetic radiation. Hence, the total energy of the atom remains constant, and classical mechanics can be used to describe the electron's motion.
3. Radiation is emitted by the hydrogen atom when the electron "jumps" from a more energetic initial state to a lower state. The "jump" cannot be visualized or treated classically. In particular, the frequency, f, of the radiation emitted in the jump is related to the change in the atom's energy and is

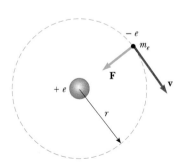

FIGURE 28.5   Bohr's model of the hydrogen atom. In this model, the orbiting electron is allowed only in specific orbits of discrete radius.

generally not the same as the frequency of the electron's orbital motion. The frequency of the emitted radiation is given by

$$E_i - E_f = hf \qquad [28.3]$$

where E_i is the energy of the initial state, E_f is the energy of the final state, h is Planck's constant, and $E_i > E_f$.

4. The size of the allowed electron orbits is determined by a condition imposed on the electron's orbital angular momentum: The allowed orbits are those for which the electron's orbital angular momentum about the nucleus is an integral multiple of $\hbar$ (pronounced "h bar"), where $\hbar = h/2\pi$.

$$m_e vr = n\hbar \qquad n = 1, 2, 3, \ldots \qquad [28.4]$$

With these assumptions, we can calculate the allowed energies and emission wavelengths of the hydrogen atom. We shall use the model pictured in Figure 28.5, in which the electron travels in a circular orbit of radius r with an orbit speed v.

The electrical potential energy of the atom is

$$PE = k_e \frac{q_1 q_2}{r} = k_e \frac{(-e)(e)}{r} = -k_e \frac{e^2}{r}$$

where k_e is the Coulomb constant. Assuming the nucleus is at rest the total energy E of the atom is the sum of the kinetic and potential energy:

$$E = KE + PE = \tfrac{1}{2}m_e v^2 - k_e \frac{e^2}{r} \qquad [28.5]$$

Let us apply Newton's second law to the electron. We know that the electric force of attraction on the electron, $k_e e^2/r^2$, must equal $m_e a_r$, where $a_r = v^2/r$ is the centripetal acceleration of the electron. Thus,

$$k_e \frac{e^2}{r^2} = m_e \frac{v^2}{r} \qquad [28.6]$$

From this equation, we see that the kinetic energy of the electron is

$$\tfrac{1}{2}mv^2 = \frac{k_e e^2}{2r} \qquad [28.7]$$

We can combine this result with Equation 28.5 and express the energy of the atom as

$$E = -\frac{k_e e^2}{2r} \qquad [28.8]$$

◀ Energy of the hydrogen atom

where the negative value of the energy indicates that the electron is bound to the proton.

An expression for r is obtained by solving Equations 28.4 and 28.6 for v and equating the results:

$$v^2 = \frac{n^2\hbar^2}{m_e^2 r^2} = \frac{k_e e^2}{m_e r}$$

$$r_n = \frac{n^2\hbar^2}{m_e k_e e^2} \qquad n = 1, 2, 3, \ldots \qquad [28.9]$$

◀ The radii of the Bohr orbits are quantized

This equation is based on the assumption that the **electron can exist only in certain allowed orbits determined by the integer n.**

The orbit with the smallest radius, called the **Bohr radius,** a_0, corresponds to $n = 1$ and has the value

$$a_0 = \frac{\hbar^2}{m_e k_e e^2} = 0.052\ 9 \text{ nm} \qquad [28.10]$$

NIELS BOHR, DANISH PHYSICIST (1885–1962)
Bohr was an active participant in the early development of quantum mechanics and provided much of its philosophical framework. During the 1920s and 1930s, Bohr headed the Institute for Advanced Studies in Copenhagen. The institute was a magnet for many of the world's best physicists and provided a forum for the exchange of ideas. When Bohr visited the United States in 1939 to attend a scientific conference, he brought news that the fission of uranium had been observed by Hahn and Strassman in Berlin. The results were the foundations of the atomic bomb developed in the United States during World War II. Bohr was awarded the 1922 Nobel prize for his investigation of the structure of atoms and of the radiation emanating from them. *(Princeton University/Courtesy of AIP Emilio Segre Visual Archives)*

A general expression for the radius of any orbit in the hydrogen atom is obtained by substituting Equation 28.10 into Equation 28.9:

$$r_n = n^2 a_0 = n^2(0.052\ 9\ \text{nm}) \qquad \text{[28.11]}$$

The first three Bohr orbits for hydrogen are shown in Figure 28.6.

Equation 28.9 may be substituted into Equation 28.8 to give the following expression for the energies of the quantum states:

$$E_n = -\frac{m_e k_e^2 e^4}{2\hbar^2}\left(\frac{1}{n^2}\right) \qquad n = 1, 2, 3, \ldots \qquad \text{[28.12]}$$

If we insert numerical values into Equation 28.12, we find

Allowed energies of the ▶
hydrogen atom

$$E_n = -\frac{13.6}{n^2}\ \text{eV} \qquad \text{[28.13]}$$

The lowest energy state, or **ground state,** corresponds to $n = 1$ and has an energy $E_1 = -m_e k_e^2 e^4/2\hbar^2 = -13.6$ eV. The next state, corresponding to $n = 2$, has an energy $E_2 = E_1/4 = -3.40$ eV, and so on. An energy level diagram showing the energies (horizontal lines) of these stationary states and the corresponding quantum numbers is shown in Figure 28.7. The uppermost level shown, corresponding to $E = 0$ and $n \rightarrow \infty$, represents the state for which the electron is completely removed from the atom. In this case, both the electron's *KE* and *PE* are zero, which means that the electron is at rest infinitely far away from the proton. The minimum energy required to ionize the atom, that is, to completely remove the electron, is called the **ionization energy.** The ionization energy for hydrogen is 13.6 eV.

Equations 28.3 and 28.12 and the third Bohr postulate show that if the electron jumps from one orbit, whose quantum number is n_i, to a second orbit, whose quantum number is n_f, it emits a photon of frequency f, given by

$$f = \frac{E_i - E_f}{h} = \frac{m_e k_e^2 e^4}{4\pi\hbar^3}\left(\frac{1}{n_f^2} - \frac{1}{n_i^2}\right) \qquad \text{[28.14]}$$

where $n_f < n_i$.

Finally, to compare this result with the empirical formulas for the various spectral series, we use the fact that for light, $\lambda f = c$ and Equation 28.14 to get

$$\frac{1}{\lambda} = \frac{f}{c} = \frac{m_e k_e^2 e^4}{4\pi c\hbar^3}\left(\frac{1}{n_f^2} - \frac{1}{n_i^2}\right) \qquad \text{[28.15]}$$

A comparison of this result with Equation 28.1 gives the following expression for the Rydberg constant:

$$R_{\text{H}} = \frac{m_e k_e^2 e^4}{4\pi c\hbar^3} \qquad \text{[28.16]}$$

If we insert the known values of $m_e, k_e, e, c,$ and $\hbar$ into this expression, the resulting theoretical value for R_{H} is found to be in excellent agreement with the value determined experimentally for the Rydberg constant. When Bohr demonstrated this agreement, it was recognized as a major accomplishment of his theory.

In order to compare Equation 28.15 with spectroscopic data, it is convenient to express it in the form

$$\frac{1}{\lambda} = R_{\text{H}}\left(\frac{1}{n_f^2} - \frac{1}{n_i^2}\right) \qquad \text{[28.17]}$$

We can use this expression to evaluate the wavelengths for the various series in the hydrogen spectrum. For example, in the Balmer series, $n_f = 2$ and $n_i = 3, 4, 5, \ldots$

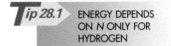

Tip 28.1   ENERGY DEPENDS ON *N* ONLY FOR HYDROGEN

According to Equation 28.13, the energy depends only on the quantum number *n*. Note that this is only true for the hydrogen atom. For more complicated atoms, the energy levels depend primarily on *n*, but also on other quantum numbers.

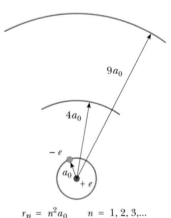

$$r_n = n^2 a_0 \qquad n = 1, 2, 3, \ldots$$

FIGURE 28.6   The first three circular orbits predicted by the Bohr model of the hydrogen atom.

(Eq. 28.1). For the Lyman series, we take $n_f = 1$ and $n_i = 2, 3, 4, \ldots$. The energy level diagram for hydrogen, shown in Figure 28.7, indicates the origin of the spectral lines described previously. The transitions between levels are represented by vertical arrows. Note that whenever a transition occurs between a state designated by n_i to one designated by n_f (where $n_i > n_f$), a photon with a frequency of $(E_i - E_f)/h$ is emitted. This can be interpreted as follows. The lines in the visible part of the hydrogen spectrum arise when the electron jumps from the third, fourth, or even higher orbit to the second orbit. Likewise, the lines of the Lyman series (in the ultraviolet) arise when the electron jumps from the second, third, or even higher orbit to the innermost ($n_f = 1$) orbit. Hence, the Bohr theory successfully predicts the wavelengths of all observed spectral lines of hydrogen.

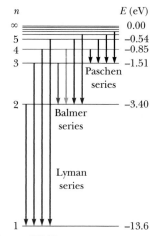

FIGURE 28.7 An energy level diagram for hydrogen. In such diagrams the discrete allowed energies are plotted on the vertical axis. Nothing is plotted on the horizontal axis, but the horizontal extent of the diagram is made large enough to show allowed transitions. Note that the quantum numbers are given on the left and the energies (in eV) are on the right.

APPLYING PHYSICS 28.3

According to the Bohr model of the hydrogen atom, the electron in the ground state moves in a circular orbit of radius 0.529×10^{-10} m, and the speed of the electron in this state is 2.2×10^6 m/s. How could Bohr's model be so successful initially when its specific orbital radii seem to contradict the "fuzziness" demanded by the Heisenberg uncertainty principle?

Explanation We have no answer. The Bohr theory works much better than it should. The theory pictures an electron as having definite position and momentum, ignoring quantum uncertainty. The theory proceeds from Newton's second law, but Schrödinger's equation describes the wave motion of the electron. Scattering experiments show that the electron does not lie on a flat circle, but fills a sphere around the nucleus.

Example 28.1 An Electronic Transition in Hydrogen

The electron in the hydrogen atom makes a transition from the $n = 2$ energy state to the ground state (corresponding to $n = 1$). Find the wavelength and frequency of the emitted photon.

Solution We can use Equation 28.17 directly to obtain λ, with $n_i = 2$ and $n_f = 1$:

$$\frac{1}{\lambda} = R_H \left(\frac{1}{n_f^2} - \frac{1}{n_i^2} \right)$$

$$\frac{1}{\lambda} = R_H \left(\frac{1}{1^2} - \frac{1}{2^2} \right) = \frac{3R_H}{4}$$

$$\lambda = \frac{4}{3R_H} = \frac{4}{3(1.097 \times 10^7 \text{ m}^{-1})} = 1.215 \times 10^{-7} \text{ m} = \boxed{121.5 \text{ nm}}$$

This wavelength lies in the ultraviolet region.

Because $c = f\lambda$, the frequency of the photon is

$$f = \frac{c}{\lambda} = \frac{3.00 \times 10^8 \text{ m/s}}{1.215 \times 10^{-7} \text{ m}} = \boxed{2.47 \times 10^{15} \text{ Hz}}$$

EXERCISE What is the wavelength of the photon emitted by hydrogen when the electron makes a transition from the $n = 3$ state to the $n = 1$ state?

ANSWER $\dfrac{9}{8R_H} = 102.6$ nm

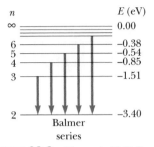

FIGURE 28.8   (Example 28.2) Transitions responsible for the Balmer series for the hydrogen atom. All transitions terminate at the $n = 2$ level.

Example 28.2   The Balmer Series for Hydrogen

The Balmer series for the hydrogen atom corresponds to electronic transitions that terminate in the state of quantum number $n = 2$, as shown in Figure 28.8.

A Find the longest-wavelength photon emitted and determine its energy.

Solution   The longest-wavelength photon in the Balmer series results from the transition from $n = 3$ to $n = 2$. Using Equation 28.17 gives

$$\frac{1}{\lambda} = R_H \left(\frac{1}{n_f^2} - \frac{1}{n_i^2} \right)$$

$$\frac{1}{\lambda_{max}} = R_H \left(\frac{1}{2^2} - \frac{1}{3^2} \right) = \frac{5}{36} R_H$$

$$\lambda_{max} = \frac{36}{5 R_H} = \frac{36}{5(1.097 \times 10^7 \ \text{m}^{-1})} = \boxed{656.3 \ \text{nm}}$$

This wavelength is in the red region of the visible spectrum.

The energy of this photon is

$$E_{photon} = hf = \frac{hc}{\lambda_{max}}$$

$$= \frac{(6.626 \times 10^{-34} \ \text{J} \cdot \text{s})(3.00 \times 10^8 \ \text{m/s})}{656.3 \times 10^{-9} \ \text{m}}$$

$$= 3.03 \times 10^{-19} \ \text{J} = \boxed{1.89 \ \text{eV}}$$

We could also obtain the energy of the photon by using Equation 28.3 in the form $hf = E_3 - E_2$, where E_2 and E_3 are the energy levels of the hydrogen atom, calculated from Equation 28.13. Note that this is the lowest-energy photon in this series, because it involves the smallest energy change.

B Find the shortest-wavelength photon emitted in the Balmer series.

Solution   The shortest-wavelength photon in the Balmer series is emitted when the electron makes a transition from $n = \infty$ to $n = 2$. Therefore,

$$\frac{1}{\lambda_{min}} = R_H \left(\frac{1}{2^2} - \frac{1}{\infty} \right) = \frac{R_H}{4}$$

$$\lambda_{min} = \frac{4}{R_H} = \frac{4}{1.097 \times 10^7 \ \text{m}^{-1}} = \boxed{364.6 \ \text{nm}}$$

This wavelength is in the ultraviolet region and corresponds to the series limit.

BOHR'S CORRESPONDENCE PRINCIPLE

In our study of relativity in Chapter 26, we found that Newtonian mechanics cannot be used to describe phenomena that occur at speeds approaching the speed of light. Newtonian mechanics is a special case of relativistic mechanics and is usable only when v is much smaller than c. Similarly, **quantum mechanics is in agreement with classical physics when the energy differences between quantized levels are very small.** This principle, first set forth by Bohr, is called the **correspondence principle.**

For example, consider the hydrogen atom with $n > 10\ 000$. For such large values of n, the energy differences between adjacent levels approach zero and the levels are nearly continuous as Equation 28.13 shows. As a consequence, the classical model is reasonably accurate in describing the system for large values of n. According to the classical model, the frequency of the light emitted by the atom is equal to the frequency of revolution of the electron in its orbit about the nucleus. Calculations show that for $n > 10\,000$, the frequency of the emitted light is the same as the electron's frequency of revolution to within 0.015%.

28.4 MODIFICATION OF THE BOHR THEORY

The Bohr theory of the hydrogen atom was a tremendous success in certain areas because it explains several features of the hydrogen spectrum that previously defied explanation. It accounts for the Balmer series and other series; it predicts a value for the Rydberg constant that is in excellent agreement with the experimental value; it gives an expression for the radius of the atom; and it predicts the energy levels of hydrogen. Although these successes are important to scientists, it is perhaps even more significant that the Bohr theory gives us a model of what the atom looks like and how it behaves. Once a basic model is constructed, refinements and modifications can be made to enlarge on the concept and to explain finer details.

The analysis used in the Bohr theory is also successful when applied to *hydrogen-like* atoms. An atom is said to be hydrogen-like when it contains only one electron. Examples are singly ionized helium, doubly ionized lithium, triply ionized beryllium, and so forth. The results of the Bohr theory for hydrogen can be extended to hydrogen-like atoms by substituting Ze^2 for e^2 in the hydrogen equations, where Z is the atomic number of the element. For example, Equations 28.12 and 28.15 become

$$E_n = -\frac{m_e k_e^2 Z^2 e^4}{2\hbar^2}\left(\frac{1}{n^2}\right) \qquad n = 1, 2, 3, \ldots \qquad \textbf{[28.18]}$$

and

$$\frac{1}{\lambda} = \frac{m_e k_e^2 Z^2 e^4}{4\pi c \hbar^3}\left(\frac{1}{n_f^2} - \frac{1}{n_i^2}\right) \qquad \textbf{[28.19]}$$

Although many attempts have been made to extend the Bohr theory to more complex multi-electron atoms, the results have thus far been unsuccessful. Even today, only approximate methods are available for treating multi-electron atoms.

Within a few months following the publication of Bohr's theory, Arnold Sommerfeld (1868–1951) extended the results to include elliptical orbits. We shall examine his model briefly because much of the nomenclature used in this treatment is still in use today. Bohr's concept of quantization of angular momentum led to the **principal quantum number** n, which determines the energy of the allowed states of hydrogen. Sommerfeld's theory retained n, but also introduced a new quantum number ℓ called the **orbital quantum number,** where the value of ℓ ranges from 0 to $n - 1$ in integer steps. According to this model, an electron in any one of the allowed energy states of a hydrogen atom may move in any one of a number of orbits corresponding to different ℓ values. For each value of n there are n possible orbits corresponding to different ℓ values. Because $n = 1$ and $\ell = 0$ for the first energy level (ground state), there is only one possible orbit for this state. The second energy level, with $n = 2$, has two possible orbits corresponding to $\ell = 0$ and $\ell = 1$. The third energy level, with $n = 3$, has three possible orbits corresponding to $\ell = 0$, $\ell = 1$, and $\ell = 2$.

For historical reasons, **all states with the same principal quantum number n are said to form a shell.** Shells are identified by the letters K, L, M, . . . , which designate the states for which $n = 1, 2, 3, \ldots$. Likewise, **the states with given values of n and ℓ are said to form a subshell.** The letters s, p, d, f, g, . . . are used to designate the states for which $\ell = 0, 1, 2, 3, 4, \ldots$. These notations are summarized in Table 28.1.

States that violate the rules given in Table 28.1 cannot exist. For instance, the $2d$ state, which would have $n = 2$ and $\ell = 2$, cannot exist because the highest al-

TABLE 28.1	Shell and Subshell Notations		
n	Shell Symbol	ℓ	Subshell Symbol
1	K	0	s
2	L	1	p
3	M	2	d
4	N	3	f
5	O	4	g
6	P	5	h
⋮		⋮	

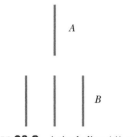

FIGURE 28.9 A single line (*A*) can split into three separate lines (*B*) in a magnetic field.

lowed value of ℓ is $n - 1$, or 1 in this case. Thus, for $n = 2$, $2s$ and $2p$ are allowed subshells but $2d$, $2f$, . . . are not. For $n = 3$, the allowed states are $3s$, $3p$, and $3d$. The maximum number of electrons allowed in any given subshell is $2(2\ell + 1)$. For example, the p subshell ($\ell = 1$) is filled when it contains six electrons. This fact will be important to us later when we discuss the *Pauli exclusion principle*.

Another modification of the Bohr theory arose when it was discovered that the spectral lines of a gas are split into several closely spaced lines when the gas is placed in a strong magnetic field. (This is called the *Zeeman effect*, after its discoverer.) Figure 28.9 shows a single spectral line being split into three closely spaced lines. This observation indicates that the energy of an electron is slightly modified when the atom is immersed in a magnetic field. In order to explain this observation, a new quantum number m_ℓ called the **orbital magnetic quantum number,** was introduced. The theory is in accord with experimental results when m_ℓ is restricted to values ranging from $-\ell$ to $+\ell$, in integer steps.

Finally, very high resolution spectrometers revealed that spectral lines of gases are in fact two very closely spaced lines even in the absence of an external magnetic field. This splitting was referred to as **fine structure.** In 1925 Samuel Goudsmit and George Uhlenbeck introduced the idea of an electron spinning about its own axis to explain the origin of fine structure. The results of their work introduced yet another quantum number m_s called the **spin magnetic quantum number.** We shall save further discussion of this quantum number for a later section. It is interesting to note that each of the new concepts introduced into the original Bohr theory added new quantum numbers and improved the original model. However, the most profound step forward in our understanding of atomic structure came with the development of quantum mechanics.

Example 28.3   Singly Ionized Helium

Singly ionized helium, He$^+$, a hydrogen-like system, has one electron in the $1s$ orbit when the atom is in its ground state.

A Find the energy of the system in the ground state.

Solution   From Equation 28.18, the energy of a hydrogen-like system whose principal quantum number is n is given by

$$E_n = -\frac{m_e k_e^2 Z^2 e^4}{2\hbar^2}\left(\frac{1}{n^2}\right)$$

This can be expressed in eV units as

$$E_n = -\frac{Z^2(13.6)}{n^2} \text{ eV}$$

Because $Z = 2$ for helium, and $n = 1$ in the ground state, we have

$$E_1 = -4(13.6) \text{ eV} = \boxed{-54.4 \text{ eV}}$$

B Find the radius of the ground-state orbit.

Solution   The radius of the ground-state orbit can be found with the help of Equation 28.9. This equation must be modified in the case of a hydrogen-like atom by substituting Ze^2 for e^2 to obtain

$$r_n = \frac{n^2 \hbar^2}{m_e k_e Z e^2} = \frac{n^2}{Z}(0.052\ 9 \text{ nm})$$

For our case, $n = 1$ and $Z = 2$, and the result is

$$r_1 = \boxed{0.026\ 5 \text{ nm}}$$

28.5 DE BROGLIE WAVES AND THE HYDROGEN ATOM

One of the postulates made by Bohr in his theory of the hydrogen atom was that angular momentum of the electron is quantized in units of $\hbar$ or

$$m_e v r = n\hbar$$

For more than a decade following Bohr's publication, no one was able to explain why the angular momentum of the electron was restricted to these discrete values. Finally, de Broglie gave a direct physical way of interpreting this condition. He assumed that an electron orbit would be stable (allowed) only if it contained an integral number of electron wavelengths. Figure 28.10a demonstrates this point when three complete wavelengths are contained in one circumference of the orbit. Similar patterns can be drawn for orbits containing one wavelength, two wavelengths, four wavelengths, five wavelengths, and so forth. This situation is analogous to that of standing waves on a string, discussed in Chapter 14. There we found that strings have preferred (resonant) frequencies of vibration. Figure 28.10b shows a standing-wave pattern containing three wavelengths for a string fixed at each end. Now imagine that the vibrating string is removed from its supports at A and B and bent into a circular shape that brings points A and B together. The end result is a pattern such as the one shown in Figure 28.10a.

In general, the condition for a de Broglie standing wave in an electron orbit is that the circumference must contain an integral number of electron wavelengths. We can express this condition as

$$2\pi r = n\lambda \qquad n = 1, 2, 3, \ldots$$

Because the de Broglie wavelength of an electron is $\lambda = h/m_e v$, we can write the preceding equation as $2\pi r = nh/m_e v$, or

$$m_e v r = n\hbar$$

This is precisely the quantization of angular momentum condition imposed by Bohr in his original theory of hydrogen.

The electron orbit shown in Figure 28.10a contains three complete wavelengths and corresponds to the case in which the principal quantum number $n = 3$. The orbit with one complete wavelength in its circumference corresponds to the first Bohr orbit, $n = 1$; the orbit with two complete wavelengths corresponds to the second Bohr orbit, $n = 2$, and so forth.

By applying the wave theory of matter to electrons in atoms, de Broglie was able to explain the appearance of integers in the Bohr theory as a natural consequence of standing-wave patterns. This was the first convincing argument that the wave nature of matter was at the heart of the behavior of atomic systems. Although the analysis provided by de Broglie was a promising first step, gigantic strides were made subsequently with the development of Schrödinger's wave equation and its application to atomic systems.

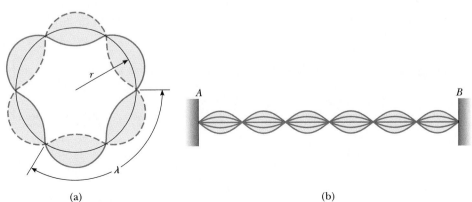

(a)   (b)

FIGURE 28.10 (a) Standing-wave pattern for an electron wave in a stable orbit of hydrogen. There are three full wavelengths in this orbit. (b) Standing-wave pattern for a vibrating stretched string fixed at its ends. This pattern has three full wavelengths.

<table>
<tr><td>Quick
Quiz
28.1</td><td>In an analysis relating Bohr's theory to the de Broglie wavelength of electrons, when an electron moves from the $n = 1$ level to the $n = 3$ level, the circumference of its orbit becomes 9 times greater. This occurs because (a) there are 3 times as many wavelengths in the new orbit, (b) there are 3 times as many wavelengths and each wavelength is 3 times as long, (c) the wavelength of the electron becomes 9 times as long, or (d) the electron is moving 9 times as fast.</td></tr>
</table>

28.6   QUANTUM MECHANICS AND THE HYDROGEN ATOM

One of the first great achievements of quantum mechanics was the solution of the wave equation for the hydrogen atom. We shall not attempt to carry out this solution. Rather, we will simply describe its properties and some of its implications with regard to atomic structure.

According to quantum mechanics, the energies of the allowed states are in exact agreement with the values obtained by the Bohr theory (Eq. 28.12), when the allowed energies depend only on the principal quantum number n.

In addition to the principal quantum number, two other quantum numbers emerge from the solution of the wave equation, ℓ and m_ℓ. The quantum number ℓ is called the **orbital quantum number,** and m_ℓ is called the **orbital magnetic quantum number.** As pointed out in Section 28.4, these quantum numbers had already appeared in modifications made to the Bohr theory. The significance of quantum mechanics is that these quantum numbers and the restrictions placed on their values arise directly from mathematics and not from any ad hoc assumptions to make the theory consistent with experimental observation. Because we shall need to make use of the various quantum numbers in the next sections, we repeat the allowed ranges of their values here.

> The values of n can range from 1 to ∞ in integer steps. The values of ℓ can range from 0 to $n - 1$ in integer steps. The values of m_ℓ can range from $-\ell$ to ℓ in integer steps.

For example, if $n = 1$, only $\ell = 0$ and $m_\ell = 0$ are permitted. If $n = 2$, the value of ℓ may be 0 or 1; if $\ell = 0$, then $m_\ell = 0$, but if $\ell = 1$, then m_ℓ may be 1, 0, or -1. Table 28.2 summarizes the rules for determining the allowed values of ℓ and m_ℓ for a given value of n.

States that violate the rules given in Table 28.2 cannot exist. For instance, one state that cannot exist is the $2d$ state, which would have $n = 2$ and $\ell = 2$. This state

TABLE 28.2	Three Quantum Numbers for the Hydrogen Atom		
Quantum Number	Name	Allowed Values	Number of Allowed States
n	Principal quantum number	1, 2, 3, . . .	Any number
ℓ	Orbital quantum number	0, 1, 2, . . ., $n - 1$	n
m_ℓ	Orbital magnetic quantum number	$-\ell, -\ell + 1, \ldots,$ $0, \ldots, \ell - 1, \ell$	$2\ell + 1$

is not allowed because the highest allowed value of ℓ is $n - 1$, or 1 in this case. Thus, for $n = 2$, 2s and 2p are allowed states but 2d, 2f, . . . are not. For $n = 3$, the allowed states are 3s, 3p, and 3d.

Example 28.4   The *n* = 2 Level of Hydrogen

Determine the number of states in the hydrogen atom for $n = 2$ and calculate the energies of these states.

Solution   For $n = 2$, ℓ can have the values 0 and 1. For $\ell = 0$, m_ℓ can only be 0; for $\ell = 1$, m_ℓ can be -1, 0, or 1. Hence we have one state designated as the 2s state associated with the quantum numbers $n = 2$, $\ell = 0$, and $m_\ell = 0$, and three states designated as 2p states, for which the quantum numbers are $n = 2$, $\ell = 1$, $m_\ell = -1$; $n = 2$, $\ell = 1$, $m_\ell = 0$; and $n = 2$, $\ell = 1$, $m_\ell = 1$.

Because all these states have the same principal quantum number, $n = 2$, they also have the same energy, which can be calculated using Equation 28.13, $E_n = -(13.6/n^2)\text{eV}$. For $n = 2$, this gives

$$E_2 = -\frac{13.6}{2^2}\text{ eV} = -3.40\text{ eV}$$

EXERCISE   How many possible states are there for the $n = 3$ level of hydrogen? For the $n = 4$ level?

ANSWER   9 states with different values of ℓ or m_ℓ for $n = 3$, and 16 states for $n = 4$

Quick Quiz 28.2	How many possible orbital states are there for (a) the $n = 3$ level of hydrogen? (b) the $n = 4$ level?
Quick Quiz 28.3	When the principal quantum number is $n = 5$, how many different values of (a) ℓ and (b) m_ℓ are possible?

28.7   THE SPIN MAGNETIC QUANTUM NUMBER

Example 28.4 was presented to give you some practice in manipulating quantum numbers, but as we shall see in this section, there actually are *eight* states corresponding to $n = 2$ for hydrogen, not four. This happens because another quantum number, m_s, the **spin magnetic quantum number,** has to be introduced to explain the splitting of each level into two.

As pointed out in Section 28.4, the need for this new quantum number first came about because of an unusual feature in the spectra of certain gases, such as sodium vapor. Close examination of one of the prominent lines of sodium shows that it is, in fact, two very closely spaced lines. The wavelengths of these lines occur in the yellow region at 589.0 nm and 589.6 nm. In 1925, when this doublet was first noticed, atomic theory could not explain it. To resolve the dilemma, Samuel Goudsmit and George Uhlenbeck, following a suggestion by the Austrian physicist Wolfgang Pauli, proposed that a fourth quantum number, called the *spin quantum number,* be introduced to describe any atomic level.

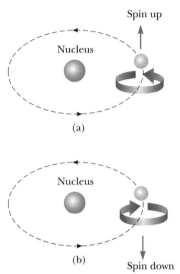

Spin up

Nucleus

(a)

Nucleus

(b)

Spin down

FIGURE 28.11 As an electron moves in its orbit about the nucleus, its spin can be either (a) up or (b) down.

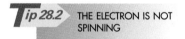

Tip 28.2 THE ELECTRON IS NOT SPINNING

The electron is *not* physically spinning. Electron spin is a purely quantum effect that gives the electron an angular momentum as if it were physically spinning.

In order to describe the spin quantum number, it is convenient (but technically incorrect) to think of the electron as spinning on its axis as it orbits the nucleus, just as Earth spins on its axis as it orbits the Sun. Strangely, there are only two ways in which the electron can spin as it orbits the nucleus, as shown in Figure 28.11. If the direction of spin is as shown in Figure 28.11a, the electron is said to have "spin up." If the direction of spin is reversed, as in Figure 28.11b, the electron is said to have "spin down." The energy of the electron is slightly different for the two spin directions, and this energy difference accounts for the sodium doublet. The quantum numbers associated with electron spin are $m_s = \frac{1}{2}$ for the spin-up state and $m_s = -\frac{1}{2}$ for the spin-down state. As we shall see in Example 28.5, this new quantum number doubles the number of allowed states specified by the quantum numbers n, ℓ, and m_ℓ.

Any classical description of electron spin is technically incorrect because quantum mechanics tells us that, because the electron cannot be precisely located in space, it cannot be considered to be a spinning solid object, as pictured in Figure 28.11. In spite of this conceptual difficulty, all experimental evidence supports the fact that an electron does have some intrinsic property that can be described by the spin magnetic quantum number.

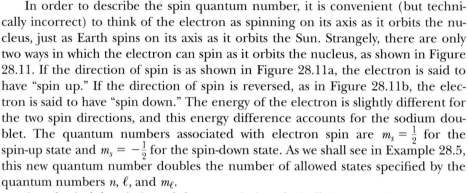

Example 28.5 The Quantum Numbers for the 2p Subshell

List the quantum numbers for electrons in the $2p$ subshell.

Solution For this subshell, $n = 2$ and $\ell = 1$. The magnetic quantum number can have the values -1, 0, 1, and the spin quantum number is always $+\frac{1}{2}$ or $-\frac{1}{2}$. Thus, the six possibilities are as follows:

n	ℓ	m_ℓ	m_s
2	1	-1	$-\frac{1}{2}$
2	1	-1	$\frac{1}{2}$
2	1	0	$-\frac{1}{2}$
2	1	0	$\frac{1}{2}$
2	1	1	$-\frac{1}{2}$
2	1	1	$\frac{1}{2}$

28.8 ELECTRON CLOUDS

The solution of the wave equation, discussed in Section 27.10, yields a wave function Ψ that depends on the quantum numbers n, ℓ, and m_ℓ. Let us assume that we have found a wave function Ψ and see what it may tell us about the hydrogen atom. We choose a value of $n = 1$ for the principal quantum number, which corresponds to the lowest energy state for hydrogen. For $n = 1$, the restrictions placed on the remaining quantum numbers are that $\ell = 0$ and $m_\ell = 0$.

The quantity Ψ^2 has great physical significance because it is proportional to the probability of finding the electron at a given position. Figure 28.12 gives the probability of finding the electron at various radial distances from the nucleus in the $1s$ state of hydrogen. Some useful and surprising information can be extracted from this curve. First, the curve peaks at a value of $r = 0.052\ 9$ nm, the Bohr value of the radius of the first electron orbit in hydrogen. This means that there is a maximum probability of finding the electron in a small interval centered at this distance from the nucleus. However, as the curve indicates, there is also a probabil-

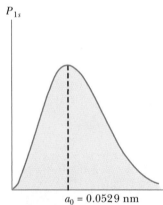

P_{1s}

$a_0 = 0.0529$ nm

r

FIGURE 28.12 The probability of finding the electron versus distance from the nucleus for the hydrogen atom in the $1s$ (ground) state. Note that the probability has its maximum value when r equals the first Bohr radius, a_0.

ity of finding the electron in a small interval centered at any other distance from the nucleus. In other words, the electron is not confined to a particular orbital distance from the nucleus, as assumed in the Bohr model. The electron may be found at various distances from the nucleus, but **the probability of finding it at a distance corresponding to the first Bohr orbit is a maximum.** Quantum mechanics also predicts that the wave function for the hydrogen atom in the ground state is spherically symmetric; hence, the electron can be found in a spherical region surrounding the nucleus. This is in contrast to the Bohr theory, which confines the position of the electron to points in a plane. The quantum mechanical result is often interpreted by viewing the electron as a cloud surrounding the nucleus. An attempt at picturing this cloud-like behavior is shown in Figure 28.13. The densest regions of the cloud represent those locations where the electron is most likely to be found.

If a similar analysis is carried out for the $n = 2$, $\ell = 0$ state of hydrogen, a peak of the probability curve is found at $4a_0$. Likewise, for the $n = 3$, $\ell = 0$ state, the curve peaks at $9a_0$. Thus, quantum mechanics predicts a most-probable electron distance to the nucleus that is in agreement with the location predicted by the Bohr theory.

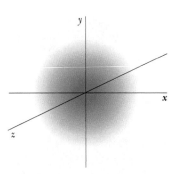

FIGURE 28.13 The spherical electron cloud for the hydrogen atom in its $1s$ state.

28.9 THE EXCLUSION PRINCIPLE AND THE PERIODIC TABLE

Earlier we found that the state of an electron in an atom is specified by four quantum numbers: n, ℓ, m_ℓ, and m_s. For example, an electron in the ground state of hydrogen could have quantum numbers of $n = 1$, $\ell = 0$, $m_\ell = 0$, $m_s = \frac{1}{2}$. As it turns out, the state of an electron in any other atom may also be specified by this same set of quantum numbers. In fact, these four quantum numbers can be used to describe all the electronic states of an atom regardless of the number of electrons in its structure.

An obvious question that arises here is, How many electrons in an atom can have a particular set of quantum numbers? This important question was answered by Pauli in 1925 in a powerful statement known as the **Pauli exclusion principle:**

> **No two electrons in an atom can ever be in the same quantum state; that is, no two electrons in the same atom can have exactly the same values for the set of quantum numbers n, ℓ, m_ℓ, and m_s.**

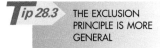

Tip 28.3 THE EXCLUSION PRINCIPLE IS MORE GENERAL

The exclusion principle stated here is a limited form of the more general exclusion principle, which states that no two *fermions* (particles with spin $\frac{1}{2}$, $\frac{3}{2}$, . . .) can be in the same quantum state.

 The Pauli exclusion principle

The Pauli exclusion principle explains the electronic structure of complex atoms as a succession of filled levels with different quantum numbers increasing in energy, where the outermost electrons are primarily responsible for the chemical properties of the element. It is interesting to note that if this principle were not valid, every electron would end up in the lowest energy state of the atom and the chemical behavior of the elements would be grossly different. Nature as we know it would not exist—and *we* would not exist to wonder about it!

As a general rule, the order that electrons fill an atom's subshell is as follows. Once one subshell is filled, the next electron goes into the vacant subshell that is lowest in energy. We can understand this principle by recognizing that if the atom were not in the lowest energy state available to it, it would radiate energy until it reached this state. A subshell is filled when it contains $2(2\ell + 1)$ electrons. This rule is based on the analysis of quantum numbers to be described later. Following this rule, shells and subshells can contain a number of electrons according to the pattern given in Table 28.3.

The exclusion principle can be illustrated by examining the electronic arrangement in a few of the lighter atoms.

Hydrogen has only one electron, which, in its ground state, can be described by either of two sets of quantum numbers: $1, 0, 0, \frac{1}{2}$ or $1, 0, 0, -\frac{1}{2}$. The electronic

Wolfgang Pauli and Niels Bohr watch a spinning top. *(Courtesy of AIP Niels Bohr Library, Margarethe Bohr Collection)*

WOLFGANG PAULI (1900–1958)

An extremely talented Austrian theoretical physicist who made important contributions in many areas of modern physics. Pauli gained public recognition at the age of 21 with a masterful review article on relativity, which is still considered one of the finest and most comprehensive introductions to the subject. Other major contributions were the discovery of the exclusion principle, the explanation of the connection between particle spin and statistics, and theories of relativistic quantum electrodynamics, the neutrino hypothesis, and the hypothesis of nuclear spin. *(CERN/Courtesy of AIP Emilio Segre Visual Archives)*

TABLE 28.3	Number of Electrons in Filled Subshells and Shells		
Shell	**Subshell**	**Number of Electrons in Filled Subshell**	**Number of Electrons in Filled Shell**
K ($n = 1$)	$s(\ell = 0)$	2	2
L ($n = 2$)	$s(\ell = 0)$	2	8
	$p(\ell = 1)$	6	
M ($n = 3$)	$s(\ell = 0)$	2	18
	$p(\ell = 1)$	6	
	$d(\ell = 2)$	10	
N ($n = 4$)	$s(\ell = 0)$	2	32
	$p(\ell = 1)$	6	
	$d(\ell = 2)$	10	
	$f(\ell = 3)$	14	

configuration of this atom is often designated as $1s^1$. The notation $1s$ refers to a state for which $n = 1$ and $\ell = 0$, and the superscript indicates that one electron is present in this level.

Neutral *helium* has two electrons. In the ground state, the quantum numbers for these two electrons are 1, 0, 0, $\frac{1}{2}$ and 1, 0, 0, $-\frac{1}{2}$. No other possible combinations of quantum numbers exist for this level, and we say that the K shell is filled. The helium electronic configuration is designated as $1s^2$.

Neutral *lithium* has three electrons. In the ground state, two of these are in the $1s$ subshell and the third is in the $2s$ subshell because it is lower in energy than the $2p$ subshell. Hence, the electronic configuration for lithium is $1s^2 2s^1$.

A list of electronic ground-state configurations for a number of atoms is provided in Table 28.4. In 1871 Dmitri Mendeleev (1834–1907), a Russian chemist, arranged the elements known at that time in a table according to their atomic masses and chemical similarities. The first table Mendeleev proposed contained many blank spaces, and he boldly stated that the gaps were there only because those elements had not yet been discovered. By noting the column in which these missing elements should be located, he was able to make rough predictions about their chemical properties. Within 20 years of this announcement, these elements were indeed discovered.

The elements in our current version of the periodic table are still arranged so that all those in a vertical column have similar chemical properties. For example, consider the elements in the last column: He (helium), Ne (neon), Ar (argon), Kr (krypton), Xe (xenon), and Rn (radon). The outstanding characteristic of these elements is that they do not normally take part in chemical reactions; that is, they do not join with other atoms to form molecules, and are therefore classified as inert. Because of this aloofness, they are referred to as the *noble gases*. We can partially understand their behavior by looking at the electronic configurations shown in Table 28.4. The element helium has the electronic configuration $1s^2$. In other words, one shell is filled. The electrons in this filled shell are considerably separated in energy from the next available level, the $2s$ level.

The electronic configuration for neon is $1s^2 2s^2 2p^6$. Again, the outer shell is filled and the difference in energy between the $2p$ level and the $3s$ level is large. Argon has the configuration $1s^2 2s^2 2p^6 3s^2 3p^6$. Here, the $3p$ subshell is filled and a wide gap in energy exists between the $3p$ subshell and the $3d$ subshell. Through all the noble gases, the pattern remains the same. A noble gas is formed when either a shell or a subshell is filled and there is a large gap in energy before the next possible level is encountered.

Example 29.6 The Beta Decay of Carbon-14

Find the energy liberated in the beta decay of $^{14}_{6}C$ to $^{14}_{7}N$ as represented by Equation 29.13. Equation 29.13 refers to nuclei, whereas Appendix B shows masses of neutral atoms. Adding six electrons to both sides of Equation 29.13 gives

$$^{14}_{6}C \text{ atom} \longrightarrow {}^{14}_{7}N \text{ atom}$$

Solution We find from Appendix B that $^{14}_{6}C$ has a mass of 14.003 242 u and $^{14}_{7}N$ has a mass of 14.003 074 u. Here, the mass difference between the initial and final states is

$$\Delta m = 14.003\ 242\ u - 14.003\ 074\ u = 0.000\ 168\ u$$

This corresponds to an energy release of

$$E = (0.000\ 168\ u)(931.494\ \text{MeV/u}) = \boxed{0.156\ \text{MeV}}$$

From Example 29.6, we see that the energy released in the beta decay of ^{14}C is approximately 0.16 MeV. As with alpha decay, we expect the electron to carry away virtually all of this as kinetic energy because apparently it is the lightest particle produced in the decay. However, as Figure 29.8 shows, only a small number of electrons have this maximum kinetic energy, represented as KE_{max} on the graph; most of the electrons emitted have kinetic energies lower than this predicted value. If the daughter nucleus and the electron are not carrying away this liberated energy, then the energy conservation requirement leads to the question, What accounts for the missing energy? As an additional complication, further analysis of beta decay shows that the principles of conservation of both angular momentum and linear momentum appear to be violated!

In 1930 Pauli proposed that a third particle must be present to carry away the "missing" energy and to conserve momentum. Enrico Fermi later developed a complete theory of beta decay and named this particle the **neutrino** ("little neutral one") because it had to be electrically neutral and have little or no mass. Although it eluded detection for many years, the neutrino (ν) was finally detected experimentally in 1956. The neutrino has the following properties:

- Zero electric charge
- A mass much smaller than that of the electron, but probably not zero (Recent experiments suggest that the neutrino definitely has mass, but the value is uncertain—perhaps less than 1 eV/c^2.)
- A spin of $\frac{1}{2}$
- Very weak interaction with matter, making it quite difficult to detect

Thus, with the introduction of the neutrino, we are now able to represent the beta decay process of Equation 29.13 in its correct form:

$$^{14}_{6}C \longrightarrow {}^{14}_{7}N + e^- + \bar{\nu} \qquad [29.15]$$

where the bar in the symbol $\bar{\nu}$ indicates an **antineutrino.** To explain what an antineutrino is, let us first consider the following decay:

$$^{12}_{7}N \longrightarrow {}^{12}_{6}C + e^+ + \nu \qquad [29.16]$$

Here we see that when ^{12}N decays into ^{12}C, a particle is produced that is identical to the electron except that it has a positive charge of $+e$. This particle is called a **positron.** Because it is like the electron in all respects except charge, the positron is said to be the **antiparticle** to the electron. We shall discuss antiparticles further in Chapter 30. For now, suffice it to say that **in beta decay, an electron and an antineutrino are emitted or a positron and a neutrino are emitted.**

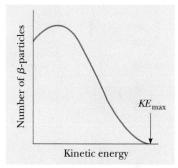

FIGURE 29.8 A typical beta-decay spectrum.

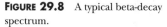 Properties of the neutrino

Webnote 29.3

Visit a neutrino measurement facility in Japan. Check out *http://neutrino.kek.jp*

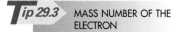 **MASS NUMBER OF THE ELECTRON**

Another notation that is sometimes used for an electron is $_{-1}^{0}e$. This notation does not imply that the electron has zero rest energy. The mass of the electron is much smaller than that of the lightest nucleon, so we can approximate it as zero when we study nuclear decays and reactions.

GAMMA DECAY

Very often a nucleus that undergoes radioactive decay is left in an excited energy state. The nucleus can then undergo a second decay to a lower energy state, perhaps to the ground state, by emitting one or more high-energy photons. The process is very similar to the emission of light by an atom. An atom emits radiation to release some extra energy when an electron "jumps" from a state of high energy to a state of lower energy. Likewise, the nucleus uses essentially the same method to release any extra energy it may have following a decay or some other nuclear event. In nuclear de-excitation, the "jumps" that release energy are made by protons or neutrons in the nucleus as they move from a higher energy level to a lower level. The photons emitted in such a de-excitation process are called **gamma rays,** which have very high energy relative to the energy of visible light.

A nucleus may reach an excited state as the result of a violent collision with another particle. However, it is more common for a nucleus to be in an excited state as a result of alpha or beta decay. The following sequence of events represents a typical situation in which gamma decay occurs:

$$^{12}_{5}\text{B} \longrightarrow {}^{12}_{6}\text{C*} + \text{e}^- + \bar{\nu} \qquad \text{[29.17]}$$

$$^{12}_{6}\text{C*} \longrightarrow {}^{12}_{6}\text{C} + \gamma \qquad \text{[29.18]}$$

Equation 29.17 represents a beta decay in which ^{12}B decays to ^{12}C* where the asterisk indicates that the carbon nucleus is left in an excited state following the decay. The excited carbon nucleus then decays to the ground state by emitting a gamma ray, as indicated by Equation 29.18. Note that gamma emission does not result in any change in either Z or A.

PRACTICAL USES OF RADIOACTIVITY

Carbon Dating

The beta decay of ^{14}C given by Equation 29.15 is commonly used to date organic samples. Cosmic rays (high-energy particles from outer space) in the upper atmosphere cause nuclear reactions that create ^{14}C from ^{14}N. In fact, the ratio of ^{14}C to ^{12}C (by numbers of nuclei) in the carbon dioxide molecules of our atmosphere has a constant value of about 1.3×10^{-12} as determined by measuring carbon ratios in tree rings. All living organisms have the same ratio of ^{14}C to ^{12}C because they continuously exchange carbon dioxide with their surroundings. When an organism dies, however, it no longer absorbs ^{14}C from the atmosphere, and so the ratio of ^{14}C to ^{12}C decreases as the result of the beta decay of ^{14}C. It is therefore possible to determine the age of a material by measuring its activity per unit mass as a result of the decay of ^{14}C. Using carbon dating, samples of wood, charcoal, bone, and shell have been identified as having lived from 1 000 to 25 000 years ago. This knowledge has helped researchers to reconstruct the history of living organisms—including humans—during this time span.

A particularly interesting example is the dating of the Dead Sea Scrolls. This group of manuscripts was first discovered by a young Bedouin boy in a cave at Qumran near the Dead Sea in 1947. Translation showed them to be religious documents, including most of the books of the Old Testament. Because of their historical and religious significance, scholars wanted to know their age. Carbon dating applied to fragments of the scrolls and to the material in which they were wrapped established that they were about 1 950 years old. The scrolls are now stored at the Israel Museum in Jerusalem.

Smoke Detectors Smoke detectors are frequently used in homes and industry for fire protection. Most of the common ones are the ionization type that use radioactive materials (see Fig. 29.9). A smoke detector consists of an ionization chamber, a sensitive current detector, and an alarm. A weak radioactive source ionizes the

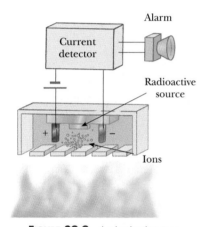

FIGURE 29.9 An ionization-type smoke detector. Smoke entering the chamber reduces the detected current, causing the alarm to sound.

Webnote 29.4

Learn a lot more about ^{14}C dating by exploring
http://abcnews.go.com/sections/ science/DailyNews/carbon0220.html

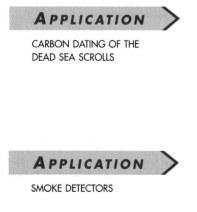

> **APPLICATION**

CARBON DATING OF THE DEAD SEA SCROLLS

> **APPLICATION**

SMOKE DETECTORS

air in the chamber of the detector, which creates charged particles. A voltage is maintained between the plates inside the chamber, setting up a small but detectable current in the external circuit. As long as the current is maintained, the alarm is deactivated. However, if smoke drifts into the chamber, the ions become attached to the smoke particles. These heavier particles do not drift as readily as do the lighter ions, which causes a decrease in the detector current. The external circuit senses this decrease in current and sets off the alarm.

Radon Detecting   Radioactivity can also affect our daily lives in harmful ways. Soon after the discovery of radium by the Curies, it was found that the air in contact with radium compounds becomes radioactive. It was shown that this radioactivity came from the radium itself, and the product was therefore called "radium emanation." Rutherford and Soddy succeeded in condensing this "emanation," confirming that it is a real substance—the inert, gaseous element now called **radon,** Rn. We now know that the air in uranium mines is radioactive because of the presence of radon gas. The mines must therefore be well ventilated to help protect miners. The fear of radon pollution has now moved from uranium mines into our own homes (see Example 29.4). Because certain types of rock, soil, brick, and concrete contain small quantities of radium, some of the resulting radon gas finds its way into our homes and other buildings. The most serious problems arise from leakage of radon from the ground into the structure. One practical remedy is to exhaust the air through a pipe just above the underlying soil or gravel directly to the outdoors by means of a small fan or blower.

APPLICATION

RADON POLLUTION

Webnote 29.5

Read more on the risks associated with radon gas. Go to *http://www.howstuffworks.com/radon.htm*

APPLYING *PHYSICS* 29.4

In 1991 a German tourist discovered the well-preserved remains of the Iceman trapped in a glacier in the Italian Alps (see Fig. 29.10). Radioactive dating of a sample of Iceman revealed an age of 5 300 years. Why did scientists date the sample using the isotope ^{14}C, rather than ^{11}C, a beta emitter with a half-life of 20.4 min?

Explanation   Carbon-14 has a long half-life of 5 730 years, so the fraction of ^{14}C nuclei remaining after one half-life is high enough to measure accurate changes in the sample's activity. The ^{11}C isotope, which has a very short half-life, is not useful because its activity decreases to a vanishingly small value over the age of the sample, making it impossible to detect.

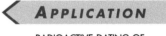

APPLICATION

RADIOACTIVE DATING OF ICEMAN

FIGURE 29.10   (Applying Physics 29.4) The body of an ancient man (dubbed the Iceman) was exposed by a melting glacier in the Alps. *(Hanny Paul/Gamma Liaison)*

If a sample to be dated is not very old, say about 50 years, then you should select the isotope of some other element whose half-life is comparable with the age of the sample. For example, if the sample contained hydrogen, you could measure the activity of ^{3}H (tritium), a beta emitter of half-life 12.3 years. As a general rule, the expected age of the sample should be long enough to measure a change in activity, but not so long that its activity cannot be detected.

APPLYING PHYSICS 29.5

A wooden coffin is found that contains a skeleton holding a gold statue. Which of the three objects (coffin, skeleton, and gold statue) can be carbon dated to find out how old it is?

Explanation   Only the coffin and the skeleton can be carbon dated. These two items exchanged air with the environment during their lifetimes, resulting in a fixed ratio of ^{14}C to ^{12}C. Once the human and the tree died, the amount of ^{14}C began to decrease due to radioactive decay. The gold in the statue was never alive and did not have an uptake of carbon; therefore, there is no ^{14}C to detect.

Example 29.7   Should We Report This to Homicide?

A 50.0-g sample of carbon is taken from the pelvis bone of a skeleton and is found to have a ^{14}C decay rate of 200.0 decays/min. It is known that carbon from a living organism has a decay rate of 15.0 decays/min · g and that ^{14}C has a half-life of 5 730 yr $= 3.01 \times 10^9$ min. Find the age of the skeleton.

Solution   Let us start with Equation 29.4

$$N = N_0 e^{-\lambda t}$$

and multiply both sides by λ to get

$$\lambda N = \lambda N_0 e^{-\lambda t}$$

But from Equation 29.3 we see that this is equivalent to

$$R = R_0 e^{-\lambda t}$$

or

$$\frac{R}{R_0} = e^{-\lambda t}$$

where R is the present activity and R_0 was the activity when the skeleton was a part of a living organism. We can solve for the time by taking the natural log of both sides of this equation.

$$\ln\left(\frac{R}{R_0}\right) = \ln(e^{-\lambda t}) = -\lambda t$$

$$t = -\frac{\ln\left(\dfrac{R}{R_0}\right)}{\lambda}$$

Because we are given the decay rate and mass of the sample, we can find R_0 as

$$R_0 = \left(15.0 \, \frac{\text{decays}}{\text{min} \cdot \text{g}}\right)(50.0 \text{ g}) = 750 \, \frac{\text{decays}}{\text{min}}$$

The decay constant is found from Equation 29.5 as

$$\lambda = \frac{0.693}{T_{1/2}} = \frac{0.693}{3.01 \times 10^9 \, \text{min}} = 2.30 \times 10^{-10} \, \text{min}^{-1}$$

Thus, we make the following substitutions:

$$t = -\frac{\ln\left(\dfrac{R}{R_0}\right)}{\lambda} = -\frac{\ln\left(\dfrac{200.0 \text{ decays/min}}{750 \text{ decays/min}}\right)}{2.30 \times 10^{-10} \text{ min}^{-1}} = \frac{1.32}{2.30 \times 10^{-1} \text{min}^{-1}}$$

$$= 5.74 \times 10^9 \text{ min} = \boxed{10\,900 \text{ yr}}$$

CARBON-14 AND THE SHROUD OF TURIN

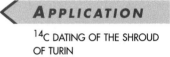

APPLICATION

^{14}C DATING OF THE SHROUD OF TURIN

Since the Middle Ages, many people have marveled at a 14-foot-long, yellowing piece of linen found in Turin, Italy, purported to be the burial shroud of Jesus Christ (Fig. 29.11). The cloth bears a remarkable, full-size likeness of a crucified body, with wounds on the head that could have been caused by a crown of thorns, and another in the side that could have been the cause of death. Skepticism over the authenticity of the shroud has existed since its first public showing in 1354; in fact, a French bishop declared it to be a fraud at the time. Because of its controversial nature, religious bodies have taken a neutral stance on its authenticity.

In 1978 the bishop of Turin allowed the cloth to be subjected to scientific analysis, but notably missing from these tests was a ^{14}C dating. The reason for this omission was that, at the time, carbon-dating techniques required a piece of cloth about the size of a handkerchief. By 1988 the process had been refined to the point that pieces as small as 1 in^2 were sufficient, and at that time permission was granted to allow the dating to proceed. Three labs were selected for the testing, and each was given four pieces of material. One of these was a piece of the shroud, and the other three pieces were control pieces similar in appearance to the shroud.

The testing procedure consisted of burning the cloth to produce carbon dioxide, which was then converted chemically to graphite. The graphite sample was subjected to ^{14}C analysis, and in the end all three labs agreed amazingly well on the age of the shroud. The average of their results gave a date for the cloth of A.D. 1320 ± 60 years, with an assurance that the cloth could not be older than A.D. 1200. Carbon-14 dating has thus unraveled the most important mystery concerning the shroud, but others remain. For example, investigators have not yet been able to explain how the image was imprinted.

FIGURE 29.11 The Shroud of Turin as it appears in a photographic negative image. *(Santi Visali/The IMAGE Bank)*

29.5 NATURAL RADIOACTIVITY

Radioactive nuclei are generally classified into two groups: (1) unstable nuclei found in nature, which give rise to what is called **natural radioactivity,** and (2) nuclei produced in the laboratory through nuclear reactions, which exhibit **artificial radioactivity.**

Three series of naturally occurring radioactive nuclei exist (Table 29.2). Each series starts with a specific long-lived radioactive isotope whose half-life exceeds

TABLE 29.2	The Four Radioactive Series		
Series	**Starting Isotope**	**Half-life (yr)**	**Stable End Product**
Uranium	$^{238}_{92}\text{U}$	4.47×10^9	$^{206}_{82}\text{Pb}$
Actinium	$^{235}_{92}\text{U}$	7.04×10^8	$^{207}_{82}\text{Pb}$
Thorium	$^{232}_{90}\text{Th}$	1.41×10^{10}	$^{208}_{82}\text{Pb}$
Neptunium	$^{237}_{93}\text{Np}$	2.14×10^6	$^{209}_{83}\text{Bi}$

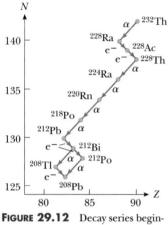

FIGURE 29.12 Decay series beginning with ^{232}Th.

that of any of its descendants. The fourth series in Table 29.2 begins with ^{237}Np, a transuranic element (one having an atomic number greater than that of uranium) not found in nature. This element has a half-life of "only" 2.14×10^6 yr.

The two uranium series are somewhat more complex than the ^{232}Th series (Fig. 29.12). Also, several other naturally occurring radioactive isotopes, such as ^{14}C and ^{40}K, are not part of either decay series.

Natural radioactivity constantly supplies our environment with radioactive elements that would otherwise have disappeared long ago. For example, because the Solar System is about 5×10^9 yr old, the supply of ^{226}Ra (whose half-life is only 1 600 yr) would have been depleted by radioactive decay long ago if it were not for the decay series that starts with ^{238}U, with a half-life of 4.47×10^9 yr.

29.6   NUCLEAR REACTIONS

It is possible to change the structure of nuclei by bombarding them with energetic particles. Such changes are called **nuclear reactions.** Rutherford was the first to observe nuclear reactions, using naturally occurring radioactive sources for the bombarding particles. He found that protons were released when alpha particles were allowed to collide with nitrogen atoms. The process can be represented symbolically as

$$\textstyle{}^4_2\text{He} + {}^{14}_7\text{N} \longrightarrow X + {}^1_1\text{H} \qquad \text{[29.19]}$$

This equation says that an alpha particle ($^4_2$He) strikes a nitrogen nucleus and produces an unknown product nucleus (X) and a proton ($^1_1$H). Balancing atomic numbers and mass numbers, as we did for radioactive decay, enables us to conclude that the unknown is characterized as $^{17}_8$X. Because the element with atomic number 8 is oxygen, we see that the reaction is

$$\textstyle{}^4_2\text{He} + {}^{14}_7\text{N} \longrightarrow {}^{17}_8\text{O} + {}^1_1\text{H} \qquad \text{[29.20]}$$

This nuclear reaction starts with two stable isotopes, helium and nitrogen, and produces two different stable isotopes, hydrogen and oxygen.

Since the time of Rutherford, thousands of nuclear reactions have been observed, particularly following the development of charged-particle accelerators in the 1930s. With today's advanced technology in particle accelerators and particle accelerators, it is possible to achieve particle energies of at least 1 000 GeV = 1 TeV. These high-energy particles are used to create new particles whose properties are helping us solve the mysteries of the nucleus.

Quick Quiz 29.4

Which of the following are possible reactions?
(a) $^1_0\text{n} + {}^{235}_{92}\text{U} \rightarrow {}^{140}_{54}\text{Xe} + {}^{94}_{38}\text{Sr} + 2\,({}^1_0\text{n})$
(b) $^1_0\text{n} + {}^{235}_{92}\text{U} \rightarrow {}^{132}_{50}\text{Sn} + {}^{101}_{42}\text{Mo} + 3\,({}^1_0\text{n})$
(c) $^1_0\text{n} + {}^{239}_{94}\text{Pu} \rightarrow {}^{127}_{53}\text{I} + {}^{93}_{41}\text{Nb} + 3\,({}^1_0\text{n})$

Example 29.8   **The Discovery of the Neutron**

A nuclear reaction of significant note occurred in 1932 when Chadwick, in England, bombarded a beryllium target with alpha particles. Analysis of the experiment indicated that the following reaction occurred:

$$\textstyle{}^4_2\text{He} + {}^9_4\text{Be} \longrightarrow {}^{12}_6\text{C} + X$$

What is X in this reaction?

Solution Balancing mass numbers and atomic numbers, we see that the unknown particle must be represented as $_0^1X$, that is, a particle having a mass of 1 and zero charge.

Hence, the particle X is the neutron, $_0^1n$. This experiment was the first to provide positive proof of the existence of neutrons.

Example 29.9 Synthetic Elements

A A beam of neutrons is directed at a target of $_{92}^{238}U$. The reaction products are a gamma ray and another isotope. What is the isotope?

Solution
Balancing input with output gives

$$_0^1n + _{92}^{238}U \longrightarrow _{92}^{239}U + \gamma$$

B Isotope ^{239}U is radioactive and undergoes beta decay. Write the equation symbolizing this decay and identify the resulting isotope.

Solution The decay of ^{239}U by beta emission is

$$_{92}^{239}U \longrightarrow _{93}^{239}Np + e^- + \bar{\nu}$$

C The isotope ^{239}Np is also radioactive and decays by beta emission. What is the end product?

Solution The decay of ^{239}Np by beta emission gives

$$_{93}^{239}Np \longrightarrow _{94}^{239}Pu + e^- + \bar{\nu}$$

D What is the significance of these reactions?

Solution The interesting feature of these reactions is the fact that uranium is the element with the greatest number of protons, 92, that exists in nature in any appreciable amount. The reactions in parts A, B, and C do occur occasionally in nature; hence, minute traces of neptunium and plutonium are present. In 1940, however, researchers bombarded uranium with neutrons to produce plutonium and neptunium by the steps given previously. These two elements were thus the first elements made in the laboratory, and by bombarding them with neutrons and other particles, the list of synthetic elements has been extended to include those up to atomic number 112.

Q VALUES

We have just examined some nuclear reactions for which mass numbers and atomic numbers must be balanced in the equations. We shall now consider the energy involved in these reactions, because energy is another important quantity that must be conserved.

Let us illustrate this procedure by analyzing the following nuclear reaction:

$$_1^2H + _7^{14}N \longrightarrow _6^{12}C + _2^4He \qquad \text{[29.21]}$$

The total mass on the left side of the equation is the sum of the mass of $_1^2H$ (2.014 102 u) and the mass of $_7^{14}N$ (14.003 074 u), which equals 16.017 176 u. Similarly, the mass on the right side of the equation is the sum of the mass of $_6^{12}C$ (12.000 000 u) plus the mass of $_2^4He$ (4.002 602 u), for a total of 16.002 602 u. Thus, the total mass before the reaction is greater than the total mass after the reaction. The mass difference in this reaction is equal to 16.017 176 u − 16.002 602 u = 0.014 574 u. This "lost" mass is converted to the kinetic energy of the nuclei present after the reaction. In energy units, 0.014 574 u is equivalent to 13.576 MeV of kinetic energy carried away by the carbon and helium nuclei.

The energy required to balance the equation is called the Q value of the reaction. In Equation 29.21 the Q value is 13.576 MeV. Nuclear reactions in which there is a release of energy—that is, positive Q values—are said to be **exothermic reactions.**

The energy balance sheet is not complete, however. We must also consider the kinetic energy of the incident particle before the collision. As an example, let us assume that the deuteron in Equation 29.21 has a kinetic energy of 5 MeV. Adding this to our Q value, we find that the carbon and helium nuclei have a total kinetic energy of 18.576 MeV following the reaction.

Now consider the reaction

$$\frac{4}{2}\text{He} + \frac{17}{7}\text{N} \longrightarrow \frac{17}{8}\text{O} + \frac{1}{1}\text{H} \qquad \text{[29.22]}$$

Before the reaction, the total mass is the sum of the masses of the alpha particle and the nitrogen nucleus: 4.002 602 u + 14.003 074 u = 18.005 676 u. After the reaction, the total mass is the sum of the masses of the oxygen nucleus and the proton: 16.999 133 u + 1.007 825 u = 18.006 958 u. In this case, the total mass after the reaction is *greater* than the total mass before the reaction. The mass deficit is 0.001 282 u, equivalent to an energy deficit of 1.194 MeV. This deficit is expressed by the negative Q value of the reaction, -1.194 MeV. Reactions with negative Q values are called **endothermic reactions.** Such reactions will not take place unless the incoming particle has at least enough kinetic energy to overcome the energy deficit.

At first it might appear that the reaction in Equation 29.22 could take place if the incoming alpha particle had a kinetic energy of 1.194 MeV. In practice, however, the alpha particle must have more energy than this. If it had an energy of only 1.194 MeV, energy would be conserved but careful analysis would show that momentum was not. This can easily be understood by recognizing that the incoming alpha particle has some momentum before the reaction. However, if its kinetic energy were only 1.194 MeV, the products (oxygen and a proton) would be created with zero kinetic energy and, thus, zero momentum. It can be shown that, in order to conserve both energy and momentum, the incoming particle must have a minimum kinetic energy given by

$$KE_{\text{min}} = \left(1 + \frac{m}{M}\right)|Q| \qquad \text{[29.23]}$$

where m is the mass of the incident particle, M is the mass of the target, and the absolute value of the Q value is used. For the reaction given by Equation 29.22, we find

$$KE_{\text{min}} = \left(1 + \frac{4.002\ 602}{14.003\ 074}\right)|-1.194\ \text{MeV}| = 1.535\ \text{MeV}$$

This minimum value of the kinetic energy of the incoming particle is called the **threshold energy.** The nuclear reaction shown in Equation 29.22 will not occur if the incoming alpha particle has a kinetic energy of less than 1.535 MeV, but can occur if its kinetic energy is equal to or greater than 1.535 MeV.

Webnote 29.6

Take a step-by-step tour of an atom smasher. Start by going to *http://www.howstuffworks.com/atom-smasher.htm*

Quick Quiz 29.5 If the Q value of an endothermic reaction is -2.17 MeV, the minimum kinetic energy needed in the reactant nuclei if the reaction is to occur must be (a) equal to 2.17 MeV, (b) greater than 2.17 MeV, (c) less than 2.17 MeV, or (d) precisely half of 2.17 MeV.

29.7 MEDICAL APPLICATIONS OF RADIATION

RADIATION DAMAGE IN MATTER

Radiation absorbed by matter can cause severe damage. The degree and type of damage depend on several factors, including the type and energy of the radiation and the properties of the absorbing material. Radiation damage in biological organisms is primarily due to ionization effects in cells. The normal function of a cell may be disrupted when highly reactive ions or radicals are formed as the result of ionizing radiation. For example, hydrogen and hydroxyl radicals produced from water molecules can induce chemical reactions that may break bonds in proteins and other vital molecules. Large acute doses of radiation are especially dangerous because damage to a great number of molecules in a cell can cause death of the cell. Also, cells that do survive the radiation may become defective, which can lead to cancer.

In biological systems, it is common to separate radiation damage into two categories: somatic damage and genetic damage. **Somatic damage** is radiation damage to any cells except the reproductive cells. Such damage can lead to cancer at high radiation levels or seriously alter the characteristics of specific organisms. **Genetic damage** affects only reproductive cells. Damage to the genes in reproductive cells can lead to defective offspring. Clearly, we must be concerned about the effect of diagnostic treatments, such as x-rays and other forms of radiation exposure.

Several units are used to quantify radiation exposure and dose. The **roentgen** (R) is defined as **that amount of ionizing radiation that will produce 2.08×10^9 ion pairs in 1 cm^3 of air under standard conditions.** Equivalently, the roentgen is **that amount of radiation that deposits 8.76×10^{-3} J of energy into 1 kg of air.**

For most applications, the roentgen has been replaced by the **rad** (which is an acronym for *radiation absorbed dose*), defined as follows: **1 rad is that amount of radiation that deposits 10^{-2} J of energy into 1 kg of absorbing material.**

Although the rad is a perfectly good physical unit, it is not the best unit for measuring the degree of biological damage produced by radiation. This is because the degree of biological damage depends not only on the dose but also on the type of radiation. For example, a given dose of alpha particles causes about 10 times more biological damage than an equal dose of x-rays. The **RBE** (relative *biological effectiveness*) factor is defined as **the number of rad of x-radiation or gamma radiation that produces the same biological damage as 1 rad of the radiation being used.** The RBE factors for different types of radiation are given in Table 29.3. Note that the values are only approximate because they vary with particle energy and form of damage.

Finally, the **rem** (*roentgen equivalent in man*) is defined as the product of the dose in rad and the RBE factor:

$$\text{Dose in rem} = \text{dose in rad} \times \text{RBE}$$

TABLE 29.3	RBE Factors for Several Types of Radiation
Radiation	**RBE Factor**
x-Rays and gamma rays	1.0
Beta particles	1.0–1.7
Alpha particles	10–20
Slow neutrons	4–5
Fast neutrons and protons	10
Heavy ions	20

According to this definition, 1 rem of any two radiations produces the same amount of biological damage. From Table 29.3, we see that a dose of 1 rad of fast neutrons represents an effective dose of 10 rem and that 1 rad of x-radiation is equivalent to a dose of 1 rem.

Low-level radiation from natural sources, such as cosmic rays and radioactive rocks and soil, delivers to each of us a dose of about 0.13 rem/yr. The upper limit of radiation dose recommended by the U.S. government (apart from background radiation and exposure related to medical procedures) is 0.5 rem/yr. Many occupations involve higher levels of radiation exposure, and for individuals in these occupations an upper limit of 5 rem/yr has been set for whole-body exposure. Higher upper limits are permissible for certain parts of the body, such as the hands and forearms. An acute whole-body dose of 400–500 rem results in a mortality rate of about 50%. The most dangerous form of exposure is ingestion or inhalation of radioactive isotopes, especially those elements the body retains and concentrates, such as ^{90}Sr. In some cases, a dose of 1 000 rem can result from ingesting 1 mCi of radioactive material.

Sterilizing objects by exposing them to radiation has been going on for at least 25 years, but in recent years, the methods used have become safer to use and more economical. Most bacteria, worms, and insects are easily destroyed by exposure to gamma radiation from radioactive cobalt. There is no intake of radioactive nuclei by an organism in such sterilizing processes, as there is in the use of radioactive tracers. The process is very effective in destroying Trichinella worms in pork, salmonella bacteria in chickens, insect eggs in wheat, and surface bacteria on fruit and vegetables that can lead to rapid spoilage. Recently, this procedure has been expanded to include the sterilization of medical equipment while in its protective covering. Surgical gloves, sponges, sutures, and so forth are irradiated while packaged. Also, bone, cartilage, and skin used for grafting is often irradiated to reduce the chance for infection.

TRACING

Radioactive particles can be used to trace chemicals participating in various reactions. One of the most valuable uses of radioactive tracers is in medicine. For example, ^{131}I is an artificially produced isotope of iodine (the natural, nonradioactive isotope is ^{127}I). Iodine, a necessary nutrient for our bodies, is obtained largely through the intake of seafood and iodized salt. The thyroid gland plays a major role in the distribution of iodine throughout the body. In order to evaluate the performance of the thyroid, the patient drinks a small amount of radioactive sodium iodide. Two hours later, the amount of iodine in the thyroid gland is determined by measuring the radiation intensity at the neck area.

A medical application of the use of radioactive tracers occurring in emergency situations is that of locating a hemorrhage inside the body. Often the location of the site cannot easily be determined, but radioactive chromium can identify the location with a high degree of precision. Chromium is taken up by red blood cells and carried uniformly throughout the body. However, the blood will be dumped at a hemorrhage site, and the radioactivity of that region will increase markedly.

The tracer technique is also useful in agricultural research. Suppose the best method of fertilizing a plant is to be determined. A certain material in the fertilizer, such as nitrogen, can be tagged with one of its radioactive isotopes. The fertilizer is then sprayed on one group of plants, sprinkled on the ground for a second group, and raked into the soil for a third. A Geiger counter is then used to track the nitrogen through the three types of plants.

Tracing techniques are as wide-ranging as human ingenuity can devise. Present applications range from checking the absorption of fluorine by teeth to checking contamination of food-processing equipment by cleansers to monitoring deterioration inside an automobile engine. In the latter case, a radioactive material is used in the manufacture of the pistons, and the oil is checked for radioactivity to determine the amount of wear on the pistons.

COMPUTED AXIAL TOMOGRAPHY (CAT SCANS)

The normal x-ray of a human body has two primary disadvantages when used as a source of clinical diagnosis. First, it is difficult to distinguish between various types of tissue in the body because they all have similar x-ray absorption properties. Second, a conventional x-ray absorption picture is indicative of the average amount of absorption along a particular direction in the body, leading to somewhat obscured pictures. To overcome these problems, a device called a CAT scanner was developed in England in 1973; it is capable of producing pictures of much greater clarity and detail than were previously obtainable.

The operation of a CAT scanner can be understood by considering the following hypothetical experiment. Suppose a box consists of four compartments, labeled A, B, C, and D, as in Figure 29.13a. Each compartment has a different amount of absorbing material from any other compartment. What set of experimental procedures will enable us to determine the relative amounts of material in each compartment? The following steps outline one method that will provide this information. First, a beam of x-rays is passed through compartments A and C, as Figure 29.13b. The intensity of the exiting radiation is reduced by absorption by some number that we assign as 8. (The number 8 could mean, for example, that the intensity of the exiting beam is reduced by eight tenths of 1% from its initial value.) Because we do not know which of the compartments, A or C, was responsible for this reduction in intensity, half the loss is assigned to each compartment, as in Figure 29.13c. Next, a beam of x-rays is passed through compartments B and D, as in Figure 29.13b. The reduction in intensity for this beam is 10, and again we assign half the loss to each compartment. We now redirect the x-ray source so that it sends one beam through compartments A and B and another through compartments C and D, as in Figure 29.13d, and again measure the absorption. Suppose the absorption through compartments A and B in this experiment is measured to be 7 units. On the basis of our first experiment, we would have guessed it would be 9 units, 4 by compartment A, and 5 by compartment B. Thus, we have reduced the guessed absorption for each compartment by 1 unit so that the sum is 7 rather than 9, to give the numbers shown in Figure 29.13e. Likewise, when the beam is passed through compartments C and D, as in Figure 29.13d, we may find the total absorption to be 11 as compared to our first experiment of 9. In this case, we add

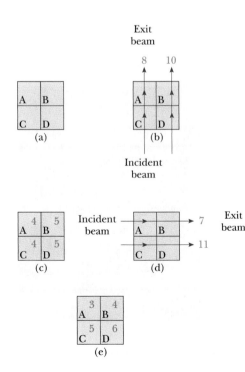

(a)

Exit
beam

8   10

(b)

Incident
beam

(c)

Incident
beam

Exit
beam

7

11

(d)

(e)

FIGURE 29.13 An experimental procedure for determining the relative amounts of x-ray absorption by four different compartments in a box.

1 unit of absorption to each compartment to give a sum of 11 as in Figure 29.13e. This somewhat crude procedure could be improved by measuring the absorption along other paths. However, these simple measurements are sufficient to enable us to conclude that compartment D contains the most absorbing material and A the least. A visual representation of these results can be obtained by assigning to each compartment a shade of gray corresponding to the particular number associated with the absorption. In our example, compartment D would be very dark and compartment A would be very light.

APPLICATION

CAT SCANS

The steps outlined above are representative of how a CAT scanner produces images of the human body. A thin slice of the body is subdivided into perhaps 10 000 compartments, rather than 4 compartments as in our simple example. The function of the CAT scanner is to determine the relative absorption in each of these 10 000 compartments and to display a picture of its calculations in various shades of gray. Note that CAT stands for **computed axial tomography.** The term *axial* is used because the slice of the body to be analyzed corresponds to a plane perpendicular to the head-to-toe axis. *Tomos* is the Greek word for slice and *graph* is the Greek word for picture. In a typical diagnosis, the patient is placed in the position shown in Figure 29.14 and a narrow beam of x-rays is sent through the plane of interest. The emerging x-rays are detected and measured by photomultiplier tubes behind the patient. The x-ray tube is then rotated a few degrees, and the intensity is recorded again. An extensive amount of information is obtained by rotating the beam through 180° at intervals of about 1° per measurement, resulting in a set of numbers assigned to each of the 10 000 "compartments" in the slice. These numbers are then converted by the computer to a photograph in various shades of gray for this segment of the body.

A brain scan of a patient can now be made in about 2 s, and a full-body scan requires about 6 s. The final result is a picture containing much greater quantitative information and clarity than a conventional x-ray photograph. Because CAT scanners use x-rays, which are an ionizing form of radiation, the technique presents a modest health risk to the patient being diagnosed.

MAGNETIC RESONANCE IMAGING (MRI)

APPLICATION

MAGNETIC RESONANCE
IMAGING (MRI)

At the heart of magnetic resonance imaging (MRI) is the fact that when a nucleus having a magnetic moment is placed in an external magnetic field, its moment precesses about the magnetic field with a frequency that is proportional to the field. For example, a proton, whose spin is $\frac{1}{2}$, can occupy one of two energy states when placed in an external magnetic field. The lower-energy state corresponds to

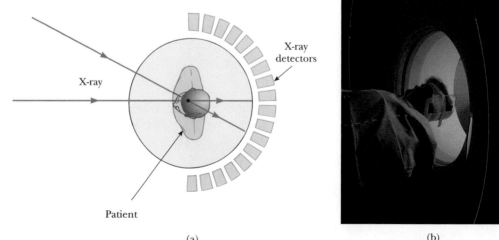

(a) (b)

FIGURE 29.14   (a) CAT scanner detector assembly. (b) Photograph of a patient undergoing a CAT scan in a hospital. (*Jay Freis/The Image Bank*)

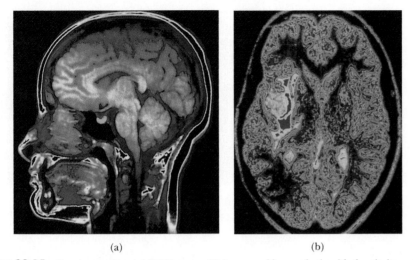

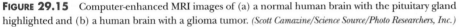

(a)   (b)

FIGURE 29.15 Computer-enhanced MRI images of (a) a normal human brain with the pituitary gland highlighted and (b) a human brain with a glioma tumor. *(Scott Camazine/Science Source/Photo Researchers, Inc.)*

the case in which the spin is aligned with the field, where as the higher-energy state corresponds to the case in which the spin is opposite the field. Transitions between these two states can be observed using a technique known as **nuclear magnetic resonance.** A DC magnetic field is applied to align the magnetic moments, and a second, weak oscillating magnetic field is applied perpendicular to the DC field. When the frequency of the oscillating field is adjusted to match the precessional frequency of the magnetic moments, the nuclei "flip" between the two spin states. These transitions result in a net absorption of energy by the spin system, which can be detected electronically.

In MRI, image reconstruction is obtained using spatially varying magnetic fields and a procedure for encoding each point in the sample being imaged. Some MRI images taken on a human head are shown in Figure 29.15. In practice, a computer-controlled pulse sequencing technique is used to produce signals that are captured by a suitable processing device. This signal is then subjected to appropriate mathematical manipulations to provide data for the final image. The main advantage of MRI over other imaging techniques in medical diagnostics is that it causes minimal damage to cellular structures. Photons associated with the radio frequency signals used in MRI have energies of only about 10^{-7} eV. Because molecular bond strengths are much larger (of the order of 1 eV), the rf photons cause little cellular damage. In comparison, x-rays or gamma-rays have energies ranging from 10^4 to 10^6 eV and can cause considerable cellular damage.

Webnote 29.7

Learn about other medical applications of nuclear physics by going to *http://www.howstuffworks.com/nuclear-medicine.htm*

29.8 RADIATION DETECTORS

Although most medical applications of radiation require instruments to make quantitative measurements of radioactive intensity, we have not yet explained how such instruments operate. Various devices have been developed to detect the energetic particles emitted when a radioactive nucleus decays. The **Geiger counter** (Fig. 29.16) is perhaps the most common form of device used to detect radioactivity. It can be considered the prototype of all counters that use the ionization of a medium as the basic detection process. It consists of a thin wire electrode aligned along the central axis of a cylindrical metallic tube filled with a gas at low pressure. The wire is maintained at a high positive voltage of about 1 000 V relative to the tube. When an energetic charged particle or gamma-ray photon enters the tube through a thin window at one end, some of the gas atoms are ionized. The electrons removed from these atoms are attracted toward the wire electrode, and in the process they ionize other atoms in their path. This sequential ionization results in an *avalanche* of electrons that produces a current pulse. After the pulse has been amplified, it can either be used to trigger an electronic counter or delivered

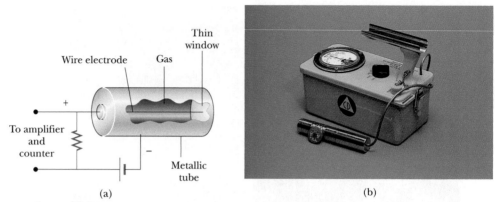

FIGURE 29.16 (a) Diagram of a Geiger counter. The voltage between the wire electrode and the metallic tube is usually about 1 000 V. (b) A Geiger counter. *(David Rogers)*

to a loudspeaker that clicks each time a particle is detected. Although a Geiger counter reliably detects the presence and quantity of radiation, it cannot be used to measure the energy of the detected radiation.

A **semiconductor diode detector** is essentially a reverse-biased *p-n* junction. As an energetic particle passes through the junction, it produces electron-hole pairs that are separated by the internal electric field. This movement of electrons and holes creates a brief pulse of current that is measured with an electronic counter. In a typical device, the duration of the pulse is 10^{-8} s.

A **scintillation counter** typically uses a solid or liquid material whose atoms are easily excited by radiation. The excited atoms then emit visible-light photons when they return to their ground state. Common materials used as scintillators are transparent crystals of sodium iodide and certain plastics. If the scintillator material is attached to one end of a device called a **photomultiplier** (PM) tube as shown in Figure 29.17, the photons emitted by the scintillator can be converted to an electrical signal. The PM tube consists of numerous electrodes, called *dynodes,* whose electric potentials increase in succession along the length of the tube. Between the top of the tube and the scintillator material is a plate called a photocathode. When photons leaving the scintillator hit this plate, electrons are emitted because of the photoelectric effect. As one of these emitted electrons strikes the first

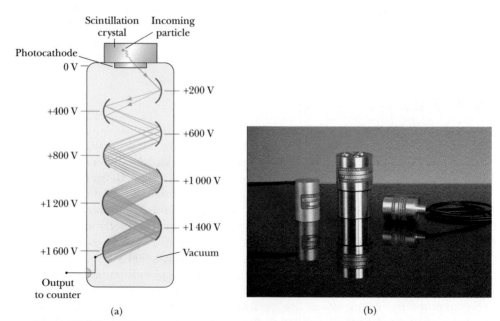

FIGURE 29.17 (a) Diagram of a scintillation counter connected to a photomultiplier tube. (b) The sodium iodide in these scintillation crystals flashes when an energetic particle passes through it, something like the way the atmosphere flashes when a meteor passes through it.

dynode, the electron has sufficient kinetic energy to eject several other electrons from the dynode surface. When these electrons are accelerated to the second dynode, many more electrons are ejected, and thus a multiplication process occurs. The end result is 1 million or more electrons striking the last dynode. Hence, one particle striking the scintillator produces a sizable electrical pulse at the PM output, and this pulse is sent to an electronic counter.

Both the scintillator and the semiconductor diode detector are much more sensitive than a Geiger counter, mainly because of the higher mass density of the detecting medium. Both can also be used to measure particle energy from the height of the pulses produced.

Track detectors are various devices used to view the tracks or paths of charged particles directly. High-energy particles produced in particle accelerators may have energies ranging from 10^9 to 10^{12} eV. Thus, they cannot be stopped and cannot have their energy measured with the small detectors already mentioned. Instead, the energy and momentum of these energetic particles are found from the curvature of their path in a magnetic field of known magnitude and direction.

A **photographic emulsion** is the simplest example of a track detector. A charged particle ionizes the atoms in an emulsion layer. The path of the particle corresponds to a family of points at which chemical changes have occurred in the emulsion. When the emulsion is developed, the particle's track becomes visible.

A **cloud chamber** contains a gas that has been supercooled to just below its usual condensation point. An energetic charged particle passing through ionizes the gas along its path. The ions serve as centers for condensation of the supercooled gas. The track can be seen with the naked eye and can be photographed. A magnetic field can be applied to determine the charges of the radioactive particles, as well as their momentum and energy.

A device called a **bubble chamber,** invented in 1952 by D. Glaser, uses a liquid (usually liquid hydrogen) maintained near its boiling point. Ions produced by incoming charged particles leave bubble tracks, which can be photographed (Fig. 29.18). Because the density of the liquid in a bubble chamber is much higher than the density of the gas in a cloud chamber, the bubble chamber has a much higher sensitivity.

A **wire chamber** consists of thousands of closely spaced parallel wires that collect the electrons created by a passing ionizing particle. A second grid, with wires

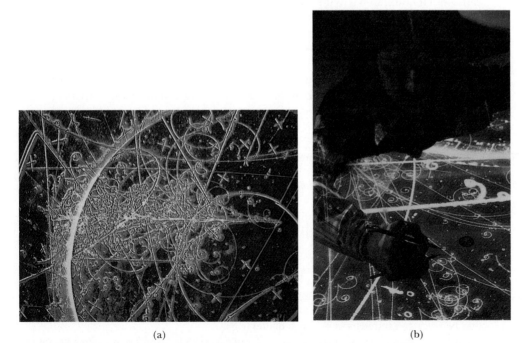

(a)   (b)

FIGURE 29.18 (a) Artificially colored bubble-chamber photograph showing tracks of particles that have passed through the chamber. *(Photo Researchers, Inc./Science Photo Library)* (b) This research scientist is studying a photograph of particle tracks made in a bubble chamber at Fermilab.

Webnote 29.8

Nuclear aircraft? Nuclear automobiles? Nuclear excavation? These were some of the ideas that were proposed following World War II. To learn why one of them failed, go to
http://www.megazone.org/ANP

perpendicular to the first, allows the x,y position of the particle in the plane of the two sets of wires to be determined. Finally several such x,y grids arranged parallel to each other along the z axis can be used to determine the particle's track in three dimensions. Wire chambers form a part of most detectors used at high-energy accelerator labs, and provide electronic readout to a computer for rapid track reconstruction and display.

SUMMARY

Nuclei are represented symbolically as $^A_Z X$, where X represents the chemical symbol for the element. The quantity A is the **mass number,** which equals the total number of nucleons (neutrons plus protons) in the nucleus. The quantity Z is the **atomic number,** which equals the number of protons in the nucleus. Nuclei that contain the same number of protons but different numbers of neutrons are called **isotopes.** In other words, isotopes have the same Z value but different A values.

Most nuclei are approximately spherical, with an average radius given by

$$r = r_0 A^{1/3} \qquad \text{[29.1]}$$

where A is the mass number and r_0 is a constant equal to 1.2×10^{-15} m.

The total mass of a nucleus is always less than the sum of the masses of its individual nucleons. This mass difference, Δm, multiplied by c^2 gives the **binding energy** of the nucleus.

The spontaneous emission of radiation by certain nuclei is called **radioactivity.** There are three processes by which a radioactive substance can decay: alpha (α) decay, in which the emitted particles are $^4_2$He nuclei; beta (β) decay, in which the emitted particles are electrons or positrons; and gamma (γ) decay, in which the emitted particles are high-energy photons.

The **decay rate,** or **activity,** R, of a sample is given by

$$R = \left| \frac{\Delta N}{\Delta t} \right| = \lambda N \qquad \text{[29.3]}$$

where N is the number of radioactive nuclei at some instant and λ is a constant for a given substance called the **decay constant.**

Nuclei in a radioactive substance decay in such a way that the number of nuclei present varies with time according to the expression

$$N = N_0 e^{-\lambda t} \qquad \text{[29.4]}$$

where N is the number of radioactive nuclei present at time t, N_0 is the number at time $t = 0$, and $e = 2.718. \ldots$

The **half-life,** $T_{1/2}$, of a radioactive substance is the time required for half of a given number of radioactive nuclei to decay. The half-life is related to the decay constant as

$$T_{1/2} = \frac{0.693}{\lambda} \qquad \text{[29.5]}$$

If a nucleus decays by alpha emission, it loses two protons and two neutrons. A typical alpha decay is

$$^{238}_{92}\text{U} \longrightarrow {}^{234}_{90}\text{Th} + {}^4_2\text{He} \qquad \text{[29.9]}$$

Note that in this decay, as in all radioactive decay processes, the sum of the Z values on the left equals the sum of the Z values on the right; the same is true for the A values.

A typical beta decay is

$$^{14}_{6}\text{C} \longrightarrow {}^{14}_{7}\text{N} + \text{e}^- + \bar{\nu} \qquad \text{[29.15]}$$

When a nucleus beta decays, an **antineutrino** is emitted along with an electron, or a **neutrino** along with a position. A neutrino has zero electric charge and a small mass (which may be zero) and interacts weakly with matter.

Nuclei are often in an excited state following radioactive decay, and release their extra energy by emitting a high-energy photon called a **gamma ray** (γ). A typical gamma-ray emission is

$$^{12}_{6}\text{C}^* \longrightarrow {}^{12}_{6}\text{C} + \gamma \qquad \text{[29.18]}$$

where the asterisk indicates that the carbon nucleus was in an excited state before gamma emission.

Nuclear reactions can occur when a bombarding particle strikes another nucleus. A typical nuclear reaction is

$$^4_2\text{He} + ^{14}_7\text{N} \longrightarrow ^{17}_8\text{O} + ^1_1\text{H} \qquad \text{[29.20]}$$

In this reaction an alpha particle strikes a nitrogen nucleus, producing an oxygen nucleus and a proton. As in radioactive decay, atomic numbers and mass numbers balance on the two sides of the arrow.

Nuclear reactions in which energy is released are said to be **exothermic reactions** and are characterized by positive Q values. Reactions with negative Q values, called **endothermic reactions,** cannot occur unless the incoming particle has at least enough kinetic energy to overcome the energy deficit. In order to conserve both energy and momentum, the incoming particle must have a minimum kinetic energy, called the **threshold energy,** given by

$$KE_{\text{min}} = \left(1 + \frac{m}{M}\right)|Q| \qquad \text{[29.23]}$$

where m is the mass of the incident particle and M is the mass of the target atom.

CONCEPTUAL QUESTIONS

1. Isotopes of a given element have many different properties, such as mass, but the same chemical properties. Why is this?
2. Physicists have attempted to measure the solar neutrinos associated with nuclear processes in the Sun's interior. These studies involve placing a large volume of detecting material in a deep mine shaft to reduce the background from cosmic rays. Explain why neutrinos are not similarly reduced.
3. A student claims that a heavy form of hydrogen decays by alpha emission. How do you respond?
4. Explain the main differences between alpha, beta, and gamma rays.
5. In beta decay, the energy of the electron or positron emitted from the nucleus lies somewhere in a relatively large range of possibilities. In alpha decay, however, the alpha-particle energy can only have discrete values. Why is there this difference?
6. If film is kept in a box, alpha particles from a radioactive source outside the box cannot expose the film but beta particles can. Explain.
7. In positron decay, a proton in the nucleus becomes a neutron, and the positive charge is carried away by the positron. But a neutron has a larger rest energy than a proton. How is this possible?
8. An alpha particle has twice the charge of a beta particle. Why does the former deflect less than the latter when passing between electrically charged plates, assuming they both have the same speed?
9. Can carbon-14 dating be used to measure the age of a stone?
10. Pick any beta-decay process and show that the neutrino must have zero charge.
11. Why do heavier elements require more neutrons in order to maintain stability?
12. Suppose it could be shown that cosmic-ray intensity was much greater 10 000 years ago. How would this affect the ages we assign to ancient samples of once-living matter?
13. If no more people were to be born, the law of population growth would strongly resemble the radioactive decay law. Discuss this statement.
14. Why is carbon dating unable to provide accurate estimates of very old materials?
15. Two samples of the same radioactive nuclide are prepared. Sample A has twice the intial activity of sample B. How does the half-life of A compare with the half-life of B? After each has passed through five half-lives, what is the ratio of their activities?
16. Which number determines the chemical characteristics of an atom: A, N, or Z?
17. Compare and contrast a photon and a neutrino.
18. Why do you think neutrons are more effective at starting a nuclear reaction than alpha particles or protons?

PROBLEMS

1, 2, 3 = straightforward, intermediate, challenging ☐ = full solution available in Student Solutions Manual/Study Guide

web = solution posted at **http://info.brookscole.com/serway** 🔥 = biomedical application

Note: Table 29.4 will be useful for many of these problems. A more complete list of atomic masses is given in Appendix B.

Section 29.1 Some Properties of Nuclei

1. Compare the nuclear radii of the following nuclides: ^2_1H, $^{60}_{27}\text{Co}$ $^{197}_{79}\text{Au}$ $^{239}_{94}\text{Pu}$

2. What is the order of magnitude of the number of protons in your body? Of the number of neutrons? Of the number of electrons?

3. Using the result of Example 29.1, find the radius of a sphere of nuclear matter that would have a mass equal to that of Earth. (Earth has a mass of 5.98×10^{24} kg and equatorial radius of 6.38×10^6 m.)

TABLE 29.4	**Some Atomic Masses**
Element	**Atomic Mass (u)**
$(_{-1}^{0}e)$	0.000 549
$(_{1}^{0}n)$	1.008 665
$_{1}^{1}H$	1.007 825
$_{1}^{2}H$	2.014 102
$_{2}^{4}He$	4.002 602
$_{3}^{7}Li$	7.016 003
$_{4}^{9}Be$	9.012 174
$_{5}^{10}B$	10.012 936
$_{6}^{12}C$	12.000 000
$_{6}^{13}C$	13.003 355
$_{7}^{14}N$	14.003 074
$_{7}^{15}N$	15.000 108
$_{8}^{15}O$	15.003 065
$_{8}^{17}O$	16.999 131
$_{8}^{18}O$	17.999 160
$_{9}^{18}F$	18.000 937
$_{10}^{20}Ne$	19.992 435
$_{11}^{23}Na$	22.989 770
$_{12}^{23}Mg$	22.994 127
$_{13}^{27}Al$	26.981 538
$_{15}^{30}P$	29.978 310
$_{20}^{40}Ca$	39.962 591
$_{20}^{42}Ca$	41.958 63
$_{20}^{43}Ca$	42.958 770
$_{26}^{56}Fe$	55.934 940
$_{30}^{64}Zn$	63.929 144
$_{29}^{64}Cu$	63.929 599
$_{41}^{93}Nb$	92.906 376 8
$_{79}^{197}Au$	196.966 543
$_{80}^{202}Hg$	201.970 617
$_{84}^{216}Po$	216.001 790
$_{86}^{220}Rn$	220.011 401
$_{90}^{234}Th$	234.043 583
$_{92}^{238}U$	238.050 784

4. Consider the hydrogen atom to be a sphere of radius equal to the Bohr radius, 0.53×10^{-10} m, and calculate the approximate value of the ratio of the nuclear density to the atomic density.

5. An alpha particle ($Z = 2$, mass 6.64×10^{-27} kg) approaches to within 1.00×10^{-14} m of a carbon nucleus ($Z = 6$). What are (a) the maximum Coulomb force on the alpha particle, (b) the acceleration of the alpha particle at this point, and (c) the potential energy of the alpha particle at this point?

6. Singly ionized carbon is accelerated through 1 000 V and passed into a mass spectrometer to determine the isotopes present (see Chapter 19). The magnetic field strength in the spectrometer is 0.200 T. (a) Determine the orbit radii for the ^{12}C and ^{13}C isotopes as they pass through the field. (b) Show that the ratio of radii may be written in the form

$$\frac{r_1}{r_2} = \sqrt{\frac{m_1}{m_2}}$$

and verify that your radii in part (a) agree with this.

7. (a) Find the speed an alpha particle requires to come within 3.2×10^{-14} m of a gold nucleus. (b) Find the energy of the alpha particle in MeV.

8. Find the nucleus that has a radius approximately equal to one half of the radius of uranium $_{92}^{238}U$.

Section 29.2 Binding Energy

9. Calculate the average binding energy per nucleon of $_{41}^{93}Nb$ and $_{79}^{197}Au$.

10. Calculate the binding energy per nucleon for (a) ^{2}H, (b) ^{4}He, (c) ^{56}Fe, and (d) ^{238}U.

11. A pair of nuclei for which $Z_1 = N_2$ and $Z_2 = N_1$ are called *mirror isobars* (the atomic and neutron numbers are interchangeable). Binding energy measurements on such pairs can be used to obtain evidence of the charge independence of nuclear forces. Charge independence means that the proton-proton, proton-neutron, and neutron-neutron nuclear forces are approximately equal. Calculate the difference in binding energy for the two mirror nuclei, $_{8}^{15}O$ and $_{7}^{15}N$.

12. The peak of the stability curve occurs at ^{56}Fe. Elements up to iron are produced in the cores of massive stars by exothermic fusion reactions. This is the fundamental reason that iron and lighter elements are much more common in the Universe than elements with higher mass numbers. Show that ^{56}Fe has a higher binding energy per nucleon than its neighbors ^{55}Mn and ^{59}Co. Compare your results with Figure 29.4.

13. Two nuclei having the same mass number are known as *isobars*. Calculate the difference in binding energy per nucleon for the isobars $_{11}^{23}Na$ and $_{12}^{23}Mg$. How do you account for the difference?

14. Compare the average binding energy per nucleon of $_{12}^{24}Mg$ and $_{37}^{85}Rb$.

Section 29.3 Radioactivity

15. The half-life of an isotope of phosphorus is 14 days. If a sample contains 3.0×10^{16} such nuclei, determine its activity. Express your answer in curies.

16. A drug tagged with $_{43}^{99}Tc$ (half-life = 6.05 h) is prepared for a patient. If the original activity of the sample was 1.1×10^{4} Bq, what is its activity after it has sat on the shelf for 2.0 h?

17. The half-life of ^{131}I is 8.04 days. (a) Calculate the decay constant for this isotope. (b) Find the number of ^{131}I nuclei necessary to produce a sample with an activity of 0.50 μCi.

18. The half-life of ^{131}I is 8.04 days. On a certain day, the activity of an ^{131}I sample is 6.40 mCi. What is its activity 40.2 days later?

19. A radioactive sample contains 3.50 μg of pure ^{11}C, which has a half-life of 20.4 min. (a) How many moles of ^{11}C is present initially? (b) Determine the number of nuclei present initially. What is the activity of the sample (c) initially and (d) after 8.00 h?

20. How much time elapses before 90.0% of the radioactivity of a sample of $^{72}_{33}$As disappears, as measured by its activity? The half-life of $^{72}_{33}$As is 26 h.

21. Many smoke detectors use small quantities of the isotope ^{241}Am in their operation. The half-life of ^{241}Am is 432 yr. How long will it take for the activity of this material to decrease to 1.00×10^{-3} of the original activity?

22. After a plant or animal dies, its ^{14}C content decreases with a half-life of 5 730 yr. If an archaeologist finds an ancient firepit containing partially consumed firewood, and the ^{14}C content of the wood is only 12.5% that of an equal carbon sample from a present-day tree, what is the age of the ancient site?

23. A freshly prepared sample of a certain radioactive isotope has an activity of 10.0 mCi. After 4.00 h, the activity is 8.00 mCi. (a) Find the decay constant and half-life of the isotope. (b) How many atoms of the isotope were contained in the freshly prepared sample? (c) What is the sample's activity 30 h after it is prepared?

24. A building has become accidentally contaminated with radioactivity. The longest-lived material in the building is strontium-90 (the atomic mass of $^{90}_{38}$Sr is 89.907 7). If the building initially contained 5.0 kg of this substance and the safe level is less than 10.0 counts/min, how long will the building be unsafe?

Section 29.4 The Decay Processes

25. Complete the following radioactive decay formulas:

$$^{212}_{83}\text{Bi} \longrightarrow \text{?} + ^{4}_{2}\text{He}$$
$$^{95}_{36}\text{Kr} \longrightarrow \text{?} + e^-$$
$$\text{?} \longrightarrow ^{4}_{2}\text{He} + ^{140}_{58}\text{Ce}$$

26. Complete the following radioactive decay formulas:
    (a) $^{12}_{5}\text{B} \rightarrow \text{?} + e^-$
    (b) $^{234}_{90}\text{Th} \rightarrow ^{230}_{88}\text{Ra} + \text{?}$
    (c) $\text{?} \rightarrow ^{14}_{7}\text{N} + e^-$

27. Complete the following radioactive decay formulas:
    (a) $\text{?} \rightarrow e^+ + ^{40}_{19}\text{K}$
    (b) $^{94}_{44}\text{Ru} \rightarrow ^{4}_{2}\text{He} + \text{?}$
    (c) $^{144}_{60}\text{Nd} \rightarrow \text{?} + ^{140}_{58}\text{Ce}$

28. Figure P29.28 shows the steps by which $^{235}_{92}\text{U}$ decays to $^{207}_{82}\text{Pb}$. Enter the correct isotope symbol in each square.

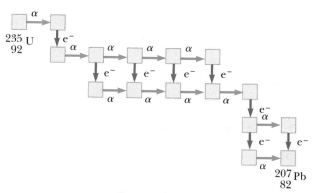

FIGURE P29.28

29. The mass of ^{56}Fe is 55.934 9 u and the mass of ^{56}Co is 55.939 9 u. Which isotope decays into the other, and by what process?

30. Find the energy released in the alpha decay of $^{238}_{92}\text{U}$. The following mass value will be useful: $^{234}_{90}\text{Th}$ has a mass of 234.043 583 u.

31. A nucleus of mass 228 u, initially at rest, undergoes alpha decay. If the alpha particle emitted has a kinetic energy of 4.00 MeV, what is the kinetic energy of the recoiling daughter nucleus?

32. $^{66}_{28}\text{Ni}$ (mass = 65.929 1 u) undergoes beta decay to $^{66}_{29}\text{Cu}$ (mass = 65.928 9 u). (a) Write the complete decay formula for this process. (b) Find the maximum kinetic energy of the emerging electrons.

33. An ^{3}H nucleus beta-decays into ^{3}He by creating an electron and an antineutrino according to the reaction

$$^{3}_{1}\text{H} \longrightarrow ^{3}_{2}\text{He} + e^- + \bar{\nu}$$

Use Appendix B to determine the total energy released in this reaction.

34. A piece of charcoal used for cooking is found at the remains of an ancient campsite. A 1.00-kg sample of carbon from the wood has an activity of 2.00×10^3 decays per minute. Find the age of the charcoal. (*Hint:* Living material has an activity of 15.0 decays/minute per gram of carbon present.)

35. A wooden artifact is found in an ancient tomb. Its ^{14}C activity is measured to be 60.0% of that in a fresh sample of wood from the same region. Assuming the same amount of ^{14}C was initially present in the wood from which the artifact was made, determine the age of the artifact.

36. A living specimen in equilibrium with the atmosphere contains one atom of ^{14}C (half-life = 5 730 yr) for every 7.70×10^{11} stable carbon atoms. An archeological sample of wood (cellulose, $C_{12}H_{22}O_{11}$) contains 21.0 mg of carbon. When the sample is placed inside a shielded beta counter with 88.0% counting efficiency, 837 counts are accumulated in one week. Assuming that the cosmic-ray flux and Earth's atmosphere have not changed appreciably since the sample was formed, find the age of the sample.

Section 29.6 Nuclear Reactions

37. The first known reaction in which the product nucleus was radioactive (achieved in 1934) was one in which $^{27}_{13}\text{Al}$ was bombarded with alpha particles. Produced in the reaction were a neutron and a product nucleus. (a) What was the product nucleus? (b) Find the Q value of the reaction.

38. Complete the following nuclear reactions:
    (a) $\text{?} + ^{14}_{7}\text{N} \rightarrow ^{1}_{1}\text{H} + ^{17}_{8}\text{O}$
    (b) $^{7}_{3}\text{Li} + ^{1}_{1}\text{H} \rightarrow ^{4}_{2}\text{He} + \text{?}$

39. Identify the unknown particles X and X′ in the following nuclear reactions:
    (a) $\text{X} + ^{4}_{2}\text{He} \rightarrow ^{24}_{12}\text{Mg} + ^{1}_{0}\text{n}$
    (b) $^{235}_{92}\text{U} + ^{1}_{0}\text{n} \rightarrow ^{90}_{38}\text{Sr} + \text{X} + 2\,^{1}_{0}\text{n}$
    (c) $2\,^{1}_{1}\text{H} \rightarrow ^{2}_{1}\text{H} + \text{X} + \text{X}′$

40. The first nuclear reaction utilizing particle accelerators was performed by Cockcroft and Walton. Accelerated protons were used to bombard lithium nuclei, producing the following reaction:

$$^{1}_{1}\text{H} + ^{7}_{3}\text{Li} \longrightarrow ^{4}_{2}\text{He} + ^{4}_{2}\text{He}$$

Since the masses of the particles involved in the reaction were well known, these results were used to obtain an early proof of the Einstein mass-energy relation. Calculate the Q value of the reaction.

41. (a) Suppose $^{10}_{5}$B is struck by an alpha particle, releasing a proton and a product nucleus in the reaction. What is the product nucleus? (b) An alpha particle and a product nucleus are produced when $^{13}_{6}$C is struck by a proton. What is the product nucleus?

42. (a) Determine the product of the reaction $^{7}_{3}$Li $+ ^{4}_{2}$He $\rightarrow$? $+$ n. (b) What is the Q value of the reaction?

43. A beam of 6.61-MeV protons is incident on a target of $^{27}_{13}$Al. Those that collide produce the reaction

$$p + ^{27}_{13}\text{Al} \longrightarrow ^{27}_{14}\text{Si} + n$$

($^{27}_{14}$Si has a mass of 26.986 721 u.) Neglect any recoil of the product nucleus and determine the kinetic energy of the emerging neutrons.

44. Find the threshold energy that the incident neutron must have to produce the reaction:

$$^{1}_{0}\text{n} + ^{4}_{2}\text{He} \longrightarrow ^{2}_{1}\text{H} + ^{3}_{1}\text{H}$$

45. When ^{18}O is struck by a proton, ^{18}F and another particle are produced. (a) What is the other particle? (b) This reaction has a Q value of -2.453 MeV, and the atomic mass of ^{18}O is 17.999 160 u. What is the atomic mass of ^{18}F?

Section 29.7 Medical Applications of Radiation

46. In terms of biological damage, how many rad of heavy ions are equivalent to 100 rad of x-rays?

47. A person whose mass is 75.0 kg is exposed to a whole-body dose of 25.0 rad. How many joules of energy are deposited in the person's body?

48. A 200-rad dose of radiation is administered to a patient in an effort to combat a cancerous growth. Assuming all of the energy deposited is absorbed by the growth, (a) calculate the amount of energy delivered per unit mass. (b) Assuming the growth has a mass of 0.25 kg and a specific heat equal to that of water, calculate its temperature rise.

49. A "clever" technician decides to heat some water for his coffee with an x-ray machine. If the machine produces 10 rad/s, how long will it take to raise the temperature of a cup of water by 50°C? Ignore heat losses during this time.

50. An x-ray technician works 5 days per week, 50 weeks per year. Assume that the technician takes an average of eight x-rays per day and receives a dose of 5.0 rem/yr as a result. (a) Estimate the dose in rem per x-ray taken. (b) How does this result compare with the amount of low-level background radiation the technician is exposed to?

51. A patient swallows a radiopharmaceutical tagged with phosphorus-32 ($^{32}_{15}$P), a β^- emitter with a half-life of 14.3 days. The average kinetic energy of the emitted electrons is 700 keV. If the initial activity of the sample is 1.31 MBq, determine (a) the number of electrons emitted in a 10-day period, (b) the total energy deposited in the body during the 10 days, and (c) the absorbed dose if the electrons are completely absorbed in 100 g of tissue.

52. A particular radioactive source produces 100 mrad of 2-MeV gamma rays per hour at a distance of 1.0 m. (a) How long could a person stand at this distance before accumulating an intolerable dose of 1 rem? (b) Assuming the gamma radiation is emitted uniformly in all directions, at what distance would a person receive a dose of 10 mrad/h from this source?

ADDITIONAL PROBLEMS

53. A 200.0-mCi sample of a radioactive isotope is purchased by a medical supply house. If the sample has a half-life of 14.0 days, how long will it keep before its activity is reduced to 20.0 mCi?

54. One method for producing neutrons for experimental use is to bombard $^{7}_{3}$Li with protons. The neutrons are emitted according to the following reaction:

$$^{1}_{1}\text{H} + ^{7}_{3}\text{Li} \longrightarrow ^{7}_{4}\text{Be} + ^{1}_{0}\text{n}$$

What is the minimum kinetic energy the incident proton must have if this reaction is to occur?

55. Deuterons that have been accelerated are used to bombard other deuterium nuclei, resulting in the reaction

$$^{2}_{1}\text{H} + ^{2}_{1}\text{H} \longrightarrow ^{3}_{2}\text{He} + ^{1}_{0}\text{n}$$

Does this reaction require a threshold energy? If so, what is its value?

56. Consider a radioactive sample. Determine the ratio of the number of atoms decaying during the first half of its half-life to the number of atoms decaying during the second half of its half-life.

57. A by-product of some fission reactors is the isotope $^{239}_{94}$Pu, an alpha emitter having a half-life of 24 120 yr:

$$^{239}_{94}\text{Pu} \longrightarrow ^{235}_{92}\text{U} + \alpha$$

Consider a sample of 1.00 kg of pure $^{239}_{94}$Pu at $t = 0$. Calculate (a) the number of $^{239}_{94}$Pu nuclei present at $t = 0$ and (b) the initial activity in the sample. (c) How long does the sample have to be stored if a "safe" activity level is 0.100 Bq?

58. (a) Find the radius of the $^{12}_{6}$C nucleus. (b) Find the force of repulsion between a proton at the surface of a $^{12}_{6}$C nucleus and the remaining five protons. (c) How much work (in MeV) has to be done to overcome this electrostatic repulsion in order to put the last proton into the nucleus? (d) Repeat (a), (b), and (c) for $^{238}_{92}$U.

59. In a piece of rock from the Moon, the ^{87}Rb content is assayed to be 1.82×10^{10} atoms per gram of material, and the ^{87}Sr content is found to be 1.07×10^{9} atoms per gram. (The relevant decay is ^{87}Rb $\rightarrow ^{87}$Sr $+ e^-$. The half-life of the decay is 4.8×10^{10} yr.) (a) Determine the age of the rock. (b) Could the material in the rock actually be much older? What assumption is implicit in using the radioactive dating method?

60. Many radioisotopes have important industrial, medical, and research applications. One of these is ^{60}Co, which has a half-life of 5.2 yr and decays by the emission of a beta particle (energy 0.31 MeV) and two gamma photons (energies 1.17 MeV and 1.33 MeV). A scientist wishes to prepare a ^{60}Co sealed source that will have an activity of at least 10 Ci after 30 months of use. What is the minimum initial mass of ^{60}Co required?

61. A medical laboratory stock solution is prepared with an initial activity due to ^{24}Na of 2.5 mCi/mL, and 10.0 mL of the stock solution is diluted at $t_0 = 0$ to a working solution whose total volume is 250 mL. After 48 h, a 5.0-mL sample of the working solution is monitored with a counter. What is the measured activity? (Note that 1 mL = 1 milliliter.)

62. A theory of nuclear astrophysics is that all the heavy elements like uranium are formed in supernova explosions of massive stars, which immediately release the elements into space. If we assume that at the time of explosion there were equal amounts of ^{235}U and ^{238}U, how long ago were the elements that formed our Earth released, given that the present ^{235}U/^{238}U ratio is 0.007? (The half-lives of ^{235}U and ^{238}U are 0.70×10^9 yr and 4.47×10^9 yr, respectively.)

63. A fission reactor is hit by a nuclear weapon, causing 5.0×10^6 Ci of ^{90}Sr ($T_{1/2} = 28.7$ yr) to evaporate into the air. The ^{90}Sr falls out over an area of 10^4 km^2. How long will it take the activity of the ^{90}Sr to reach the agriculturally "safe" level of 2.0 μCi/m^2?

64. After the sudden release of radioactivity from the Chernobyl nuclear reactor accident in 1986, the radioactivity of milk in Poland rose to 2 000 Bq/L due to iodine-131, with a half-life of 8.04 days. Radioactive iodine is particularly hazardous, because the thyroid gland concentrates iodine. The Chernobyl accident caused a measurable increase in thyroid cancers among children in Belarus. (a) For comparison, find the activity of milk due to potassium. Assume that 1 L of milk contains 2.00 g of potassium, of which 0.011 7% is the isotope ^{40}K which has a half-life of 1.28×10^9 yr. (b) After what time would the activity due to iodine fall below that due to potassium?

65. During the manufacture of a steel engine component, radioactive iron (^{59}Fe) is included in the total mass of 0.20 kg. The component is placed in a test engine when the activity due to the isotope is 20.0 μCi. After a 1 000-h test period, oil is removed from the engine and found to contain enough ^{59}Fe to produce 800 disintegrations/min per liter of oil. The total volume of oil in the engine is 6.5 L. Calculate the total mass worn from the engine component per hour of operation. (The half-life for ^{59}Fe is 45.1 days.)

66. After determining that the Sun has existed for hundreds of millions of years, but before the discovery of nuclear physics, scientists could not explain why the Sun has continued to burn for such a long time. For example, if it were a coal fire, it would have burned up in about 3 000 yr. Assume that the Sun, whose mass is 1.99×10^{30} kg, originally consisted entirely of hydrogen and that its total power output is 3.76×10^{26} W. (a) If the energy-generating mechanism of the Sun is the transforming of hydrogen into helium via the net reaction

$$4\,^1_1\text{H} + 2e^- \longrightarrow\,^4_2\text{He} + 2\nu + \gamma$$

calculate the energy (in joules) given off by this reaction. (b) Determine how many hydrogen atoms constitute the Sun. Take the mass of one hydrogen atom to be 1.67×10^{-27} kg. (c) Assuming that the total power output remains constant, after what time will all the hydrogen be converted into helium, making the Sun die? The actual projected lifetime of the Sun is about 10 billion years, because only the hydrogen in a relatively small core is available as a fuel. Only in the core are temperatures and densities high enough for the fusion reaction to be self-sustaining.

GROUP ACTIVITIES

G.1 This experiment will take a little longer to do than most that we have suggested, but the time spent is worthwhile to help you understand the concept of half-life. Obtain a box of sugar cubes and with a pencil make a mark on one side of each of about 200 cubes. Each of these cubes will represent the nucleus of a radioactive substance. Thus, at $t = 0$, you have 200 undecayed nuclei. Now, put the 200 marked cubes in a box and roll them out on a table, just as you would roll dice. Next, count and remove any cubes that have landed marked-side up. These cubes represent nuclei that emitted radiation during the roll. They are no longer radioactive and thus do not participate in the rest of the action. Record the number of undecayed cubes remaining as the number of undecayed nuclei at $t = 1$ roll.

Continue rolling, counting, and removing until you have completed 12 to 15 rolls. By then, you should have only a few cubes remaining. Plot a graph of undecayed cubes versus the roll number and from this determine the "half-roll" of the cubes.

G.2 Use a nail to punch a hole in the bottom of a large tin can. Hold the can beneath a faucet and adjust the water flow from the faucet to a fine constant stream. Although water flows from the hole at the bottom, you will note that the level of the water in the can rises. As it does so, however, the flow of water leaving the can increases due to increased water pressure caused by the greater depth of water. Unless the flow of water is too great, an equilibrium point will be reached at which the amount of water flowing out of the can each second exactly equals the amount flowing in each second. When this happens, the level of water in the can is constant. As noted in the text, carbon-14 is continually being produced in the atmosphere and is also continually disappearing as it decays into nitrogen. What is the analogy between water entering the can, remaining in the can, and flowing out of the can and the behavior of carbon-14 in the atmosphere?

G.3 (a) Describe what happens to the number of protons and neutrons in a nucleus when the nucleus undergoes alpha decay. (b) Repeat for a beta decay. (c) In alpha decay how does the mass of the parent nuclei compare to the sum of the masses of the daughter plus alpha? Is the parent mass greater, less, equal to, or no relation to the sum of the daughter plus alpha? (d) Use the decay relationship $^{230}_{90}\text{Th} \rightarrow\,^{226}_{98}\text{Ra} +\,^4_2\text{He}$ to check your answer to part (c). The mass of ^{230}Th is 230.033 131 u and the mass of ^{226}Ra is 226.025 406 u.

G.4 (a) Assume the energy released in a beta decay is 0.150 MeV. If the daughter nucleus is found to have an energy of 0.100 MeV, then for energy conservation, is it true that the beta must have an energy of 0.050 MeV? (b) If you find that the beta actually has only an energy of 0.025 MeV, what happens to the remaining energy? (c) Correct the following beta decay reaction to agree with your answers to the above: $^{14}_6\text{C} \rightarrow\,^{14}_7\text{N} + e^-$.

G.5 (a) In a radioactive decay process, which of the following are reduced by one half after one half-life, (i) the number of nuclei, (ii) the activity, or (iii) the decay constant? (b) If the half-life of a given material is 14 days, what is the ratio of the number of particles remaining after 10 days to the original number? What is the ratio of the activity after 10 days to the initial activity?

30

Nuclear Energy and Elementary Particles

Chapter Outline

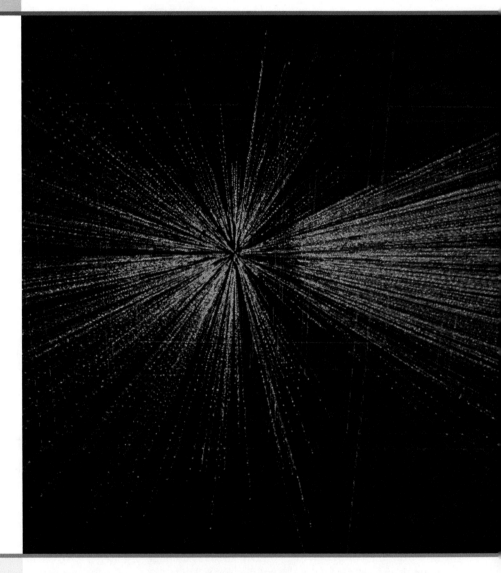

In this image from the NA49 experiment at CERN, hundreds of subatomic particles are created in the collision of high-energy lead nuclei with a lead target. The aim of the experiment was to create extreme densities in matter to break down the strong force that normally locks quarks within protons and neutrons.

(Courtesy of CERN)

*I*n this concluding chapter we discuss the two means by which energy can be derived from nuclear reactions. These two techniques are fission, in which a nucleus of large mass number splits, or fissions, into two smaller nuclei, and fusion, in which two light nuclei fuse to form a heavier nucleus. In either case, large amounts of energy are released that can be used destructively, through bombs, or constructively, through the production of electric power.

We end our study of physics by examining the known subatomic particles and the fundamental interactions that govern their behavior. We also discuss the current theory of elementary particles, which states that all matter in nature is constructed from only two families of particles, quarks and leptons. Finally, we describe how clarifications of such models might help us understand the evolution of the Universe.

30.1 NUCLEAR FISSION

Nuclear fission occurs when a heavy nucleus, such as ^{235}U, splits, or fissions, into two smaller nuclei. In such a reaction, **the total mass of the products is less than the original mass of the heavy nucleus.**

Nuclear fission was first observed in 1939 by Otto Hahn and Fritz Strassman, following some basic studies by Fermi. After bombarding uranium ($Z = 92$) with neutrons, Hahn and Strassman discovered among the reaction products two medium-mass elements, barium and lanthanum. Shortly thereafter, Lisa Meitner and Otto Frisch explained what had happened. The uranium nucleus had split into two nearly equal fragments after absorbing a neutron. Such an occurrence was of considerable interest to physicists attempting to understand the nucleus, but it was to have even more far-reaching consequences. Measurements showed that about 200 MeV of energy is released in each fission event, and this fact was to affect the course of human history.

The fission of ^{235}U by slow (low energy) neutrons can be represented by the equation

$$^{1}_{0}n + ^{235}_{92}U \longrightarrow ^{236}_{92}U^{*} \longrightarrow X + Y + \text{neutrons} \qquad \text{[30.1]}$$

where $^{236}U^{*}$ is an intermediate state that lasts only for about 10^{-12} s before splitting into X and Y. The resulting nuclei, X and Y, are called **fission fragments.** Many combinations of X and Y satisfy the requirements of conservation of energy and charge. In the fission of uranium, there are about 90 different daughter nuclei that can be formed. The process also results in the production of several (typically two or three) neutrons per fission event. On the average, 2.47 neutrons are released per event.

A typical reaction of this type is

$$^{1}_{0}n + ^{235}_{92}U \longrightarrow ^{141}_{56}Ba + ^{92}_{36}Kr + 3\,^{1}_{0}n \qquad \text{[30.2]}$$

The fission fragments, barium and krypton, and the released neutrons have a great deal of kinetic energy following the fission event.

The breakup of the uranium nucleus can be likened to what happens to a drop of water when excess energy is added to it. All of the atoms in the drop have energy, but not enough to break up the drop. However, if enough energy is added to set the drop vibrating, it will undergo elongation and compression until the amplitude of vibration becomes large enough to cause the drop to break apart. In the uranium nucleus, a similar process occurs (Fig. 30.1). The sequence of events is as follows:

1. The ^{235}U nucleus captures a thermal (slow-moving) neutron.
2. This capture results in the formation of $^{236}U^{*}$, and the excess energy of this nucleus causes it to undergo violent oscillations.
3. The $^{236}U^{*}$ nucleus becomes highly elongated, and the force of repulsion

◀ Sequence of events in a nuclear fission process

FIGURE 30.1   The stages involved in a nuclear fission event as described by the liquid-drop model of the nucleus.

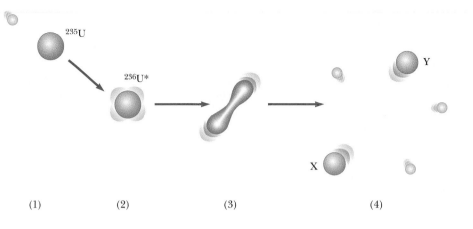

^{235}U

^{236}U*

Y

X

(1)   (2)   (3)   (4)

between protons in the two halves of the dumbbell shape tends to increase the distortion.

4. The nucleus splits into two fragments, emitting several neutrons in the process.

Let us estimate the disintegration energy, Q, released in a typical fission process. From Figure 29.4 we see that the binding energy per nucleon is about 7.2 MeV for heavy nuclei (those having a mass number of approximately 240) and about 8.2 MeV for nuclei of intermediate mass. This means that the nucleons in the fission fragments are more tightly bound and therefore have less mass than the nucleons in the original heavy nucleus. This decrease in mass per nucleon appears as released energy when fission occurs. The amount of energy released is (8.2–7.2) MeV per nucleon. Assuming a total of 240 nucleons, we find that the energy released per fission event is

$$Q = (240 \text{ nucleons})(8.2 \text{ MeV/nucleon} - 7.2 \text{ MeV/nucleon}) = 240 \text{ MeV}$$

This is indeed a very large amount of energy relative to the amount released in chemical processes. For example, the energy released in the combustion of one molecule of the octane used in gasoline engines is about one hundred-millionth the energy released in a single fission event!

APPLYING **PHYSICS** *30.1*

If a heavy nucleus were to fission into just two product nuclei, they would be very unstable. Why is this?

Explanation   According to Figure 29.3, the ratio of the number of neutrons to the number of protons increases with Z. As a result, when a heavy nucleus splits in a fission reaction to two lighter nuclei, the lighter nuclei tend to have too many neutrons. This leads to instability, as the nucleus returns to the curve in Figure 29.3 by decay processes that reduce the number of neutrons.

Example 30.1   **The Fission of Uranium**

Two other possible ways by which ^{235}U can undergo fission when bombarded with a neutron are (1) by the release of ^{140}Xe and ^{94}Sr as fission fragments and (2) by the release of ^{132}Sn and ^{101}Mo as fission fragments. In each case, neutrons are also released. Find the number of neutrons released in each of these events.

Solution   By balancing mass numbers and atomic numbers, we find that these reactions can be written

$$^{1}_{0}n + ^{235}_{92}U \longrightarrow ^{140}_{54}Xe + ^{94}_{38}Sr + 2\,^{1}_{0}n$$

$$^{1}_{0}n + ^{235}_{92}U \longrightarrow ^{132}_{50}Sn + ^{101}_{42}Mo + 3\,^{1}_{0}n$$

Thus, two neutrons are released in the first event and three in the second.

Quick Quiz 30.1   In the first atomic bomb, the energy released was equivalent to about 30 kilotons of TNT, where a ton of TNT releases an energy of 4.0×10^9 J. The amount of mass converted into energy in this event is nearest to (a) 1 μg,   (b) 1 mg,   (c) 1 g,   (d) 1 kg, (e) 20 kilotons

---

Example 30.2   The Energy Released in the Fission of ^{235}U

Calculate the total energy released if 1.00 kg of ^{235}U undergoes fission, taking the disintegration energy per event to be $Q = 208$ MeV (a more accurate value than the estimate given previously).

Solution   We need to know the number of nuclei in 1.00 kg of uranium. Because $A = 235$, the number of nuclei is

$$N = \left(\frac{6.02 \times 10^{23} \text{ nuclei/mol}}{235 \text{ g/mol}} \right) (1.00 \times 10^3 \text{ g}) = 2.56 \times 10^{24} \text{ nuclei}$$

Hence the disintegration energy is

$$E = NQ = (2.56 \times 10^{24} \text{ nuclei}) \left(208 \frac{\text{MeV}}{\text{nucleus}} \right) = 5.32 \times 10^{26} \text{ MeV}$$

Because 1 MeV is equivalent to 4.45×10^{-20} kWh, $E = 2.37 \times 10^7$ kWh. This is enough energy to keep a 100-W lightbulb burning for about 30 000 years.

30.2   NUCLEAR REACTORS

We have seen that neutrons are emitted when ^{235}U undergoes fission. These neutrons can in turn trigger other nuclei to undergo fission, with the possibility of a chain reaction (Fig. 30.2). Calculations show that if the chain reaction is not controlled (that is, if it does not proceed slowly), it can result in a violent explosion, with the release of an enormous amount of energy, even from only 1 g of ^{235}U. If the energy in 1 kg of ^{235}U were released, it would equal that released by the detonation of about 20 000 tons of TNT! This, of course, is the principle behind the first nuclear bomb, an uncontrolled fission reaction.

A nuclear reactor is a system designed to maintain what is called a **self-sustained chain reaction.** This important process was first achieved in 1942 by a group led by Fermi at the University of Chicago, with natural uranium as the fuel. Most reactors in operation today also use uranium as fuel. Natural uranium contains only about 0.7% of the ^{235}U isotope, with the remaining 99.3% being the

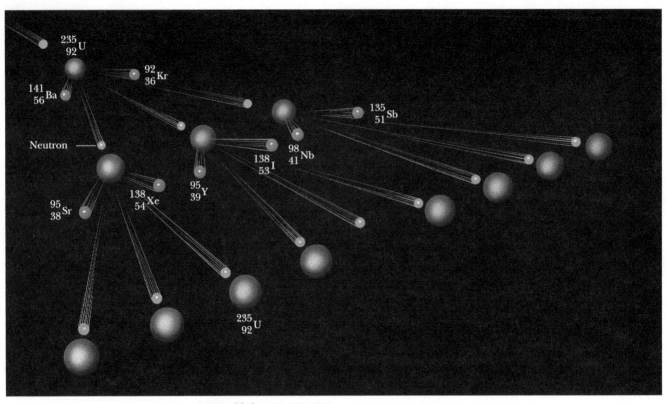

FIGURE 30.2 A nuclear chain reaction initiated by capture of a neutron.

^{238}U isotope. This is important to the operation of a reactor because ^{238}U almost never undergoes fission. Instead, it tends to absorb neutrons, producing neptunium and plutonium. For this reason, reactor fuels must be artificially enriched so that they contain several percent of the ^{235}U isotope.

Earlier, we mentioned that an average of about 2.5 neutrons are emitted in each fission event of ^{235}U. In order to achieve a self-sustained chain reaction, one of these neutrons must be captured by another ^{235}U nucleus and cause it to un-

Painting of the world's first reactor. Because of wartime secrecy, there are no photographs of the completed reactor. The reactor was composed of layers of graphite interspersed with uranium. A self-sustained chain reaction was first achieved on December 2, 1942. Word of the success was telephoned immediately to Washington with this message: "The Italian navigator has landed in the New World and found the natives very friendly." The historic event took place in an improvised laboratory in the racquet court under the west stands of the University of Chicago's Stagg Field; the Italian navigator was Fermi. *(Courtesy of Chicago Historical Society)*

dergo fission. A useful parameter for describing the level of reactor operation is the **reproduction constant *K*, defined as the average number of neutrons from each fission event that will cause another event.** As we have seen, *K* can have a maximum value of 2.5 in the fission of uranium. However, in practice *K* is less than this because of several factors, which we shall discuss in a moment.

A self-sustained chain reaction is achieved when *K* = 1. Under this condition, the reactor is said to be **critical.** When *K* is less than unity, the reactor is subcritical and the reaction dies out. When *K* is greater than unity, the reactor is said to be supercritical, and a runaway reaction occurs. In a nuclear reactor used to furnish power to a utility company, it is necessary to maintain a *K* value close to unity.

The basic design of a nuclear reactor is shown in Figure 30.3. The fuel elements consist of enriched uranium. Now let us look at the function of the remaining parts of the reactor and some aspects of its design.

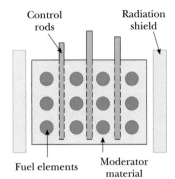

FIGURE 30.3 Cross section of a reactor core surrounded by a radiation shield.

NEUTRON LEAKAGE

In any reactor, a fraction of the neutrons produced in fission will leak out of the core before inducing other fission events. If the fraction leaking out is too large, the reactor will not operate. The percentage lost is large if the reactor is very small because leakage is a function of the ratio of surface area to volume. Therefore, a critical requirement of reactor design is choosing the correct surface-area-to-volume ratio so that a sustained reaction can be achieved.

REGULATING NEUTRON ENERGIES

The neutrons released in fission events are very energetic, with kinetic energies of about 2 MeV. It is found that slow neutrons are far more likely than fast neutrons to produce fission events in ^{235}U. Furthermore, ^{238}U does not absorb slow neutrons. Therefore, in order for the chain reaction to continue, the neutrons must be slowed down. This is accomplished by surrounding the fuel with a **moderator** substance.

To understand how neutrons are slowed down, consider a collision between a light object and a very massive one. In such an event, the light object rebounds from the collision with most of its original kinetic energy. However, if the collision is between objects with nearly the same masses, the incoming projectile transfers a large percentage of its kinetic energy to the target. In the first nuclear reactor ever constructed, Fermi placed bricks of graphite (carbon) between the fuel elements. Carbon nuclei are about 12 times more massive than neutrons, but after about 100 collisions with carbon nuclei, a neutron is slowed sufficiently to increase its likelihood of fission with ^{235}U. In this design the carbon is the moderator; most modern reactors use heavy water (D_2O) as the moderator.

NEUTRON CAPTURE

In the process of being slowed down, the neutrons may be captured by nuclei that do not undergo fission. The most common event of this type is neutron capture by ^{238}U. The probability of neutron capture by ^{238}U is very high when the neutrons have high kinetic energies and very low when they have low kinetic energies. Thus, the slowing down of the neutrons by the moderator serves the dual purpose of making them available for reaction with ^{235}U and decreasing their chances of being captured by ^{238}U.

CONTROL OF POWER LEVEL

It is possible for a reactor to reach the critical stage (*K* = 1) after all neutron losses described previously are minimized. However, a method of control is needed to adjust *K* to a value near unity. If *K* were to rise above this value, the heat produced

APPLICATION

NUCLEAR REACTOR DESIGN

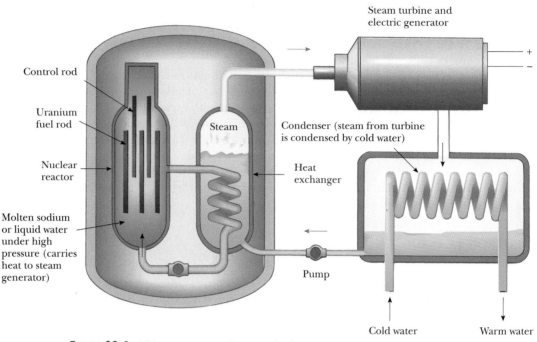

FIGURE 30.4   Main components of a pressurized-water reactor.

in the runaway reaction would melt the reactor. To control the power level, control rods are inserted into the reactor core (see Fig. 30.3). These rods are made of materials such as cadmium that are very efficient in absorbing neutrons. By adjusting the number and position of these control rods in the reactor core, the K value can be varied and any power level within the design range of the reactor can be achieved.

A diagram of a pressurized-water reactor is shown in Figure 30.4. This type of reactor is commonly used in electric power plants in the United States. Fission events in the reactor core supply heat to the water contained in the primary (closed) system, which is maintained at high pressure to keep it from boiling. This water also serves as the moderator. The hot water is pumped through a heat exchanger, and the heat is transferred to the water contained in the secondary system. There the hot water is converted to steam, which drives a turbine-generator to create electric power. Note that the water in the secondary system is isolated from the water in the primary system in order to prevent contamination of the secondary water and steam by radioactive nuclei from the reactor core.

REACTOR SAFETY[1]

The safety aspects of nuclear power reactors are often sensationalized by the media and misunderstood by the public. The 1979 near-disaster of Three Mile Island in Pennsylvania and the accident at the Chernobyl reactor in the Ukraine rightfully focused attention on reactor safety. Yet the safety record in the United States is enviable. The records show no fatalities attributed to commercial nuclear power generation in the history of the United States nuclear industry.

Commercial reactors achieve safety through careful design and rigid operating procedures. Radiation exposure and the potential health risks associated with such exposure are controlled by three layers of containment. The fuel and radioactive fission products are contained inside the reactor vessel. Should this vessel rupture, the reactor building acts as a second containment structure to prevent

[1] The authors are grateful to Professor Gene Skluzacek of the University of Nebraska at Omaha for rewriting this section on reactor safety.

radioactive material from contaminating the environment. Finally, the reactor facilities must be in a remote location to protect the general public from exposure should radiation escape the reactor building.

According to the Oak Ridge National Laboratory Review, "the health risk of living within 8 km (5 miles) of a nuclear reactor for 50 years is no greater than the risk of smoking 1.4 cigarettes, drinking 0.5 liters of wine, traveling 240 km by car, flying 9 600 km by jet, or having one chest x-ray in a hospital. Each of these activities is estimated to increase a person's chances of dying in any given year by one in a million."

Another potential danger in nuclear reactor operations is the possibility that the water flow could be interrupted. Even if the nuclear fission chain reaction were stopped immediately, residual heat could build up in the reactor to the point of melting the fuel elements. The molten reactor core would melt to the bottom of the reactor vessel and conceivably melt its way into the ground below—the so-called "China syndrome." Although it might appear that this deep underground burial site would be an ideal safe haven for a radioactive blob, there would be danger of a steam explosion should the molten mass encounter water. This nonnuclear explosion could spread radioactive material to the areas surrounding the power plant. To prevent such an unlikely chain of events, nuclear reactors are designed with emergency core-cooling systems, requiring no power, that automatically flood the reactor with water in the event of coolant loss. The emergency cooling water moderates heat buildup in the core, which in turn prevents the occurrence of melting.

A continuing concern in nuclear fission reactors is the safe disposal of radioactive material when the reactor core is replaced. This waste material contains long-lived, highly radioactive isotopes and must be stored over long periods of time in such a way that there is no chance of environmental contamination. At present, sealing radioactive wastes in waterproof containers and burying them in deep salt mines seems to be the most promising solution.

Transportation of reactor fuel and reactor wastes poses additional safety risks. However, neither the waste nor the fuel of nuclear power reactors can be used to construct a nuclear bomb.

Accidents during transportation of nuclear fuel could expose the public to harmful levels of radiation. The Department of Energy requires stringent crash tests on all containers used to transport nuclear materials. Container manufacturers must demonstrate that their containers will not rupture even in high-speed collisions.

The safety issues associated with nuclear power reactors are complex and often emotional. All sources of energy have associated risks. In each case, we must weigh the risks against the benefits and the availability of the energy source.

30.3 NUCLEAR FUSION

Figure 29.4 shows that the binding energy for light nuclei (those having a mass number lower than 20) is much smaller than the binding energy for heavier nuclei. This suggests a possible process that is the reverse of fission. **When two light nuclei combine to form a heavier nucleus, the process is called nuclear fusion.** Because the mass of the final nucleus is less than the masses of the original nuclei, there is a loss of mass accompanied by a release of energy. Although fusion power plants have not yet been developed, a worldwide effort is under way to harness the energy from fusion reactions in the laboratory. Later we shall discuss the possibilities and advantages of this process for generating electric power.

FUSION IN THE SUN

All stars generate their energy through fusion processes. About 90% of the stars, including the Sun, fuse hydrogen, whereas some older stars fuse helium or other heavier elements. Stars are born in regions of space containing vast clouds of dust

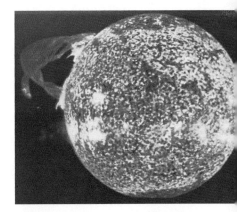

This photograph of the Sun, taken on December 19, 1973, during the third and final manned Skylab mission, shows one of the most spectacular solar flares ever recorded, spanning more than 588 000 km (365 000 mi) across the solar surface. The last picture, taken some 17 hours earlier, showed this feature as a large quiescent prominence on the eastern side of the Sun. The flare gives the distinct impression of a twisted sheet in the process of unwinding itself. In this photograph the solar poles are distinguished by a relative absence of granulation and a much darker tone than the central portions of the disk. Several active regions are seen on the eastern side of the disk. The photograph was taken in the light of ionized helium by the extreme ultraviolet spectroheliograph instrument of the U.S. Naval Research Laboratory. *(NASA)*

and gas. Recent mathematical models of these clouds indicate that star formation is triggered by shock waves passing through a cloud. These shock waves are similar to sonic booms and are produced by events such as the explosion of a nearby star, called a *supernova explosion*. The shock wave compresses certain regions of the cloud, causing these regions to collapse under their own gravity. As the gas falls inward toward the center, the atoms gain speed, which causes the temperature of the gas to rise. Two conditions must be met before fusion reactions in the star can sustain its energy needs: (1) The temperature must be high enough (about 10^7 K for hydrogen) to allow the kinetic energy of the positively charged hydrogen nuclei to overcome their mutual Coulomb repulsion as they collide, and (2) the density of nuclei must be high enough to ensure a high rate of collision.

When fusion reactions occur at the core of a star, the energy liberated eventually becomes sufficient to prevent further collapse of the star under its own gravity. The star then continues to live out the remainder of its life under a balance between the inward force of gravity pulling it toward collapse and the outward force due to thermal effects and radiation pressure.

The **proton-proton cycle** is a series of three nuclear reactions that are believed to be the stages in the liberation of energy in the Sun and other stars rich in hydrogen. An overall view of the proton-proton cycle is that four protons combine to form an alpha particle and two positrons, with the release of 25 MeV of energy in the process.

The specific steps in the proton-proton cycle are

$$\begin{aligned} {}_1^1\text{H} + {}_1^1\text{H} &\longrightarrow {}_1^2\text{H} + e^+ + \nu \\ {}_1^1\text{H} + {}_1^2\text{H} &\longrightarrow {}_2^3\text{He} + \gamma \end{aligned}$$

[30.3]

This second reaction is followed by either hydrogen-helium fusion or helium-helium fusion:

$$ {}_1^1\text{H} + {}_2^3\text{He} \longrightarrow {}_2^4\text{He} + e^+ + \nu $$

or

$$ {}_2^3\text{He} + {}_2^3\text{He} \longrightarrow {}_2^4\text{He} + {}_1^1\text{H} + {}_1^1\text{H} $$

The energy liberated is carried primarily by gamma rays, positrons, and neutrinos, as can be seen from the reactions. The gamma rays are soon absorbed by the dense gas, thus raising its temperature. The positrons combine with electrons to produce gamma rays, which in turn are also absorbed by the gas within a few centimeters. The neutrinos, however, almost never interact with matter; hence they escape from the star, carrying about 2% of the generated energy with them. These energy-liberating fusion reactions are called **thermonuclear fusion reactions.** The hydrogen (fusion) bomb, first exploded in 1952, is an example of an uncontrolled thermonuclear fusion reaction.

FUSION REACTORS

APPLICATION

FUSION REACTORS

The enormous amount of energy released in fusion reactions suggests the possibility of harnessing this energy for useful purposes on Earth. A great deal of effort is focused on developing a sustained and controllable thermonuclear reactor—a fusion power reactor. Controlled fusion is often called the ultimate energy source because of the availability of its fuel source: water. For example, if deuterium were used as the fuel, 0.06 g of it could be extracted from 1 gal of water at a cost of about four cents. Such rates would make the fuel costs of even an inefficient reactor almost insignificant. An additional advantage of fusion reactors is that comparatively few radioactive by-products are formed. As noted in Equation 30.3, the end product of the fusion of hydrogen nuclei is safe, nonradioactive helium. Unfortunately, a thermonuclear reactor that can deliver a net power output over a reasonable time interval is not yet a reality, and many difficulties must be solved before a successful device is constructed.

We have seen that the Sun's energy is based, in part, on a set of reactions in which ordinary hydrogen is converted to helium. Unfortunately, the proton-proton interaction is not suitable for use in a fusion reactor because the event requires very high pressures and densities. The process works in the Sun only because of the extremely high density of protons in the Sun's interior. In fact, even at the densities and temperatures that exist at the center of the Sun, the average proton takes 14 billion years to react!

The fusion reactions that appear most promising in the construction of a fusion power reactor involve deuterium and tritium, which are isotopes of hydrogen. These reactions are

$$^2_1\text{H} + \,^2_1\text{H} \longrightarrow \,^3_2\text{He} + \,^1_0\text{n} \qquad Q = 3.27 \text{ MeV}$$

$$^2_1\text{H} + \,^2_1\text{H} \longrightarrow \,^3_1\text{H} + \,^1_1\text{H} \qquad Q = 4.03 \text{ MeV} \qquad \textbf{[30.4]}$$

$$^2_1\text{H} + \,^3_1\text{H} \longrightarrow \,^4_2\text{He} + \,^1_0\text{n} \qquad Q = 17.59 \text{ MeV}$$

where the Q values refer to the amount of energy released per reaction. As noted earlier, deuterium is available in almost unlimited quantities from our lakes and oceans and is very inexpensive to extract. Tritium, however, is radioactive ($T_{1/2} = 12.3$ yr) and undergoes beta decay to ^3He. For this reason, tritium does not occur naturally to any great extent and must be artificially produced.

One of the major problems in obtaining energy from nuclear fusion is the fact that the Coulomb repulsion force between two charged nuclei must be overcome before they can fuse. The fundamental challenge is to give the two nuclei enough kinetic energy to overcome this repulsive force. This can be accomplished by heating the fuel to extremely high temperatures (about 10^8 K, far greater than the interior temperature of the Sun). As you might expect, such high temperatures are not easy to obtain in a laboratory or a power plant. At these high temperatures the atoms are ionized, and the system consists of a collection of electrons and nuclei, commonly referred to as a *plasma*.

In addition to the high temperature requirements, two other critical factors determine whether or not a thermonuclear reactor will be successful: **plasma ion density,** n, and **plasma confinement time,** τ, the time the interacting ions are maintained at a temperature equal to or greater than that required for the reaction to proceed successfully. The density and confinement time must both be large enough to ensure that more fusion energy will be released than is required to heat the plasma.

Lawson's criterion states that a net power output in a fusion reactor is possible under the following conditions:

$$n\tau \geq 10^{14} \text{ s/cm}^3 \qquad \text{Deuterium-tritium interaction}$$

$$n\tau \geq 10^{16} \text{ s/cm}^3 \qquad \text{Deuterium-deuterium interaction}$$

[30.5]   ◀ Lawson's criterion

The problem of plasma confinement time has yet to be solved. How can a plasma be confined at a temperature of 10^8 K for times on the order of 1 s? The basic plasma-confinement technique under investigation is discussed following Example 30.3.

Example 30.3   The Deuterium-Deuterium Reaction

Find the energy released in the deuterium-deuterium reaction

$$^2_1\text{H} + \,^2_1\text{H} \longrightarrow \,^3_1\text{H} + \,^1_1\text{H}$$

Solution   The mass of the ^2_1H atom is 2.014 102 u. Thus, the total mass before the reaction is 4.028 204 u. After the reaction, the sum of the masses is equal to 3.016 049 u + 1.007 825 u = 4.023 874 u. Thus, the excess mass is 0.004 33 u. In energy units, this is equivalent to   4.03 MeV .

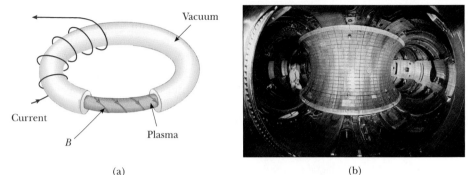

(a)   (b)

(c)

FIGURE 30.5   (a) Diagram of a tokamak used in the magnetic confinement scheme. The plasma is trapped within the spiraling magnetic field lines as shown. (b) Interior view of the Tokamak Fusion Test Reactor (TFTR) vacuum vessel located at the Princeton Plasma Physics Laboratory, Princeton University, New Jersey. *(Courtesy of Princeton Plasma Physics Laboratory)* (c) The National Spherical Torus Experiment (NSTX) that began operation in March 1999. *(Courtesy of Princeton University)*

MAGNETIC FIELD CONFINEMENT

Most fusion experiments use magnetic field confinement to contain a plasma. One device, called a **tokamak,** has a doughnut-shaped geometry (a toroid), as shown in Figure 30.5a. This device, first developed in the former Soviet Union, uses a combination of two magnetic fields to confine the plasma inside the doughnut. A strong magnetic field is produced by the current in the windings, and a weaker magnetic field is produced by the current in the toroid. The resulting magnetic field lines are helical, as in Figure 30.5a. In this configuration, the field lines spiral around the plasma and prevent it from touching the walls of the vacuum chamber.

In order for the plasma to reach ignition temperature, some form of auxiliary heating is necessary. A successful and efficient auxiliary heating technique that has been used recently is the injection of a beam of energetic neutral particles into the plasma.

When it was in operation, the Tokamak Fusion Test Reactor (TFTR) at Princeton reported central ion temperatures of 510 million degrees Celsius, more than 30 times hotter than the center of the Sun. TFTR $n\tau$ values for the D-T reaction were well above 10^{13} s/cm^3 and close to the value required by Lawson's criterion. In 1991, reaction rates of 6×10^{17} D-T fusions per second were reached in the JET tokamak at Abington, England.

One of the new generations of fusion experiments is the National Spherical Torus Experiment (NSTX) shown in Figure 30.5c. Rather than the donut-shaped

Webnote 30.1

Want to operate a tokamak on your own? Do it here with a Java applet.
http://ippex.pppl.gov/tokamak/default.htm

plasma of a tokamak, the NSTX produces a spherical plasma that has a hole through its center. The major advantage of the spherical configuration is its ability to confine the plasma at a higher pressure in a given magnetic field. This approach could lead to the development of smaller and more economical fusion reactors.

An international collaboration involving four major fusion programs is currently working on building a fusion reactor called ITER (International Thermonuclear Experimental Reactor). This facility will address the remaining technological and scientific issues concerning the feasibility of fusion power. The design is completed, and site and construction negotiations are under way. If the planned device works as expected, the Lawson number for ITER will be about six times greater than the current record holder, the JT-60U tokamak in Japan.

30.4 ELEMENTARY PARTICLES

The word "atom" comes from the Greek word *atomos,* meaning "indivisible." At one time atoms were thought to be the indivisible constituents of matter; that is, they were regarded as elementary particles. Discoveries in the early part of the 20th century revealed that the atom is not elementary, but has as its constituents protons, neutrons, and electrons. Until 1932 physicists viewed these three constituent particles as elementary because, with the exception of the free neutron, they are very stable. The theory soon fell apart, however, and beginning in 1937, many new particles were discovered in experiments involving high-energy collisions between known particles. These new particles are characteristically unstable and have very short half-lives, ranging between 10^{-6} s and 10^{-23} s. So far more than 300 of them have been cataloged.

Until the 1960s, physicists were bewildered by the large number and variety of subatomic particles being discovered. They wondered if the particles were like animals in a zoo or if a pattern was emerging that would provide a better understanding of the elaborate structure in the subnuclear world. In the last 30 years, physicists have made tremendous advances in our knowledge of the structure of matter by recognizing that all particles (with the exception of electrons, photons, and a few others) are made of smaller particles called *quarks.* Thus, protons and neutrons, for example, are not truly elementary but are systems of tightly bound quarks. The quark model has reduced the bewildering array of particles to a manageable number and has successfully predicted new quark combinations that were subsequently found in many experiments.

30.5 THE FUNDAMENTAL FORCES IN NATURE

The key to understanding the properties of elementary particles is to be able to describe the forces between them. All particles in nature are subject to four fundamental forces: strong, electromagnetic, weak, and gravitational.

The **strong force** is responsible for the tight binding of quarks to form neutrons and protons and for the nuclear force, a sort of residual strong force, binding neutrons and protons into nuclei. This force represents the "glue" that holds the nucleons together and is the strongest of all the fundamental forces. It is very short-ranged and is negligible for separations greater than about 10^{-15} m (the approximate size of the nucleus). The **electromagnetic force,** which is about 10^{-2} times the strength of the strong force, is responsible for the binding of atoms and molecules. It is a long-range force that decreases in strength as the inverse square of the separation between interacting particles. The **weak force** is a short-range nuclear force that tends to produce instability in certain nuclei. It is responsible for beta decay, and its strength is only about 10^{-6} times that of the strong force. (As we shall discuss later, scientists now believe that the weak and electromagnetic

Webnote 30.2
One means of creating high plasma-ion density is through magnetic field confinement. Another approach is to use laser fusion. Find out how laser fusion can produce electricity at *http://www.llnl.gov/nif/library/ife.pdf* (*Note:* You will need Acrobat Reader to view this file.)

TABLE 30.1	Particle Interactions		
Interaction (Force)	**Relative Strength[a]**	**Range of Force**	**Mediating Field Particle**
Strong	1	Short (~ 1 fm)	Gluon
Electromagnetic	10^{-2}	Long ($\propto 1/r^2$)	Photon
Weak	10^{-6}	Short ($\sim 10^{-3}$ fm)	$W^{\pm}$ and Z^0 bosons
Gravitational	10^{-43}	Long ($\propto 1/r^2$)	Graviton

[a] For two quarks separated by 3×10^{-17} m

forces are two manifestations of a single force called the *electroweak* force). Finally, the **gravitational force** is a long-range force with a strength of only about 10^{-43} times that of the strong force. Although this familiar interaction is the force that holds the planets, stars, and galaxies together, its effect on elementary particles is negligible. Thus, the gravitational force is the weakest of all the fundamental forces.

Modern physics often describes the forces between particles in terms of the actions of field particles or quanta. In the case of the familiar electromagnetic interaction, the field particles are photons. In the language of modern physics, it can be said that the electromagnetic force is *mediated* (carried) by photons, which are the quanta of the electromagnetic field. Likewise, the strong force is mediated by field particles called *gluons,* the weak force is mediated by particles called the W and Z *bosons,* and the gravitational force is mediated by quanta of the gravitational field called *gravitons.* All of these field quanta have been detected except for the graviton, which may never be found directly because of the weakness of the gravitational field. These interactions, their ranges, and their relative strengths are summarized in Table 30.1.

30.6 POSITRONS AND OTHER ANTIPARTICLES

In the 1920s, the theoretical physicist Paul Adrien Maurice Dirac (1902–1984) developed a version of quantum mechanics that incorporated special relativity. Dirac's theory successfully explained the origin of the electron's spin and its magnetic moment. But it had one major problem: Its relativistic wave equation required solutions corresponding to negative energy states even for free electrons. If negative energy states existed, we would expect a normal free electron in a state of positive energy to make a rapid transition to one of these lower states, emitting a photon in the process. Thus, normal electrons would not exist, and we would be left with a universe of photons and electrons locked up in negative energy states.

Dirac circumvented this difficulty by postulating that all negative energy states are normally filled. The electrons that occupy the negative energy states are said to be in the "Dirac sea" and are not directly observable when all negative energy states are filled. However, if one of these negative energy states is vacant, leaving a hole in the sea of filled states, the hole can react to external forces and therefore can be observed as the electron's positive antiparticle. The general and profound implication of Dirac's theory is that **for every particle, there is an antiparticle.** The antiparticle has the same mass as the particle, but the opposite charge. For example, the electron's antiparticle, the *positron,* has a mass of 0.511 MeV/c^2 and a positive charge of 1.6×10^{-19} C. As noted in Chapter 29, we usually designate an antiparticle with a bar over the symbol for the particle. Thus, $\bar{p}$ denotes the antiproton and $\bar{\nu}$ the antineutrino. In this book, the notation e$^+$ is preferred for the positron.

The positron was discovered by Carl Anderson in 1932, and in 1936 he was awarded the Nobel prize for his achievement. Anderson discovered the positron while examining tracks created by electron-like particles of positive charge in a

PAUL ADRIEN MAURICE DIRAC (1902–1984)

Winner of the Nobel prize for physics in 1933. *(Courtesy AIP Emilio Segre Visual Archives)*

Tip 30.1   ANTIPARTICLES

An antiparticle is not identified solely on the basis of opposite charge. For example, even neutral particles have antiparticles.

Webnote 30.3

In 1933, Dirac shared the Nobel Prize in Physics with Erwin Schroedinger. Read about it at *http://www.nobel.se/physics/laureates/1933*

cloud chamber. (These early experiments used cosmic rays—mostly energetic protons passing through interstellar space—to initiate high-energy reactions on the order of several GeV.) In order to discriminate between positive and negative charges, the cloud chamber was placed in a magnetic field, causing moving charges to follow curved paths. Anderson noted that some of the electron-like tracks deflected in a direction corresponding to a positively charged particle.

Since Anderson's initial discovery, the positron has been observed in a number of experiments. Perhaps the most common process for producing positrons is pair production, which was introduced in Chapter 26. In this process, a gamma ray with sufficiently high energy collides with a nucleus, creating an electron-positron pair. Because the total rest energy of the electron-positron pair is $2m_ec^2 = 1.02$ MeV, the gamma ray must have at least this much energy to create an electron-positron pair.

Practically every known elementary particle has a distinct antiparticle. Among the exceptions are the photon and the neutral pion (π^0), which are their own antiparticles. Following the construction of high-energy accelerators in the 1950s, many of these antiparticles were discovered. They included the antiproton $\bar{p}$ discovered by Emilio Segrè and Owen Chamberlain in 1955 and the antineutron $\bar{n}$ discovered shortly thereafter.

The process of electron-positron annihilation is used in the medical diagnostic technique of positron emission tomography (PET). The patient is injected with glucose solution containing a radioactive substance that decays by positron emission. Examples of such substances are oxygen-15, nitrogen-13, carbon-11, and fluorine-18. The radioactive material is carried to the brain. When a decay occurs, the emitted positron annihilates with an electron in the brain tissue, resulting in two gamma-ray photons. With the assistance of a computer, we can display an image of the sites in the brain at which the glucose accumulates.

The images from a PET scan can indicate a wide variety of disorders in the brain, including Alzheimer's disease. In addition, because glucose metabolizes more rapidly in active areas of the brain, the PET scan can indicate which areas of the brain are involved in various processes such as language, music, and vision.

HIDEKI YUKAWA, JAPANESE PHYSICIST (1907–1981)

Yukawa was awarded the Nobel Prize in 1949 for predicting the existence of mesons. This photograph of Yukawa at work was taken in 1950 in his office at Columbia University. *(UPI/Corbis-Bettman)*

APPLICATION

POSITRON EMISSION TOMOGRAPHY (PET SCANNING)

30.7 MESONS AND THE BEGINNING OF PARTICLE PHYSICS

Physicists in the mid-1930s had a fairly simple view of the structure of matter. The building blocks were the proton, the electron, and the neutron. Three other particles were known or postulated at the time: the photon, the neutrino, and the positron. These six particles were considered the fundamental constituents of matter. Although the accepted picture of the world was marvelously simple, no one was able to provide an answer to the following important question: Because the many protons in proximity in any nucleus should strongly repel each other due to their like charges, what is the nature of the force that holds the nucleus together? Scientists recognized that this mysterious nuclear force must be much stronger than anything encountered up to that time.

The first theory to explain the nature of the nuclear force was proposed in 1935 by the Japanese physicist Hideki Yukawa (1907–1981), an effort that later earned him the Nobel prize. In order to understand Yukawa's theory, it is useful to first note that **two atoms can form a covalent chemical bond by the exchange of electrons.** Similarly, in the modern view of electromagnetic interactions, **charged particles interact by exchanging a photon.** Yukawa used this same idea to explain the nuclear force by proposing a new particle that is exchanged by nucleons in the nucleus to produce the strong force. Furthermore, he established that the range of the force is inversely proportional to the mass of this particle, and predicted that the mass would be about 200 times the mass of the electron. Because the new particle would have a mass between that of the electron and the proton, it was called a **meson** (from the Greek *meso*, meaning "middle").

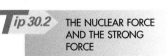

Tip 30.2   THE NUCLEAR FORCE AND THE STRONG FORCE

The nuclear force discussed in Chapter 29 was originally called the *strong force*. Once the quark theory was established, however, the phrase *strong force* was reserved for the force between quarks. We shall follow this convention—the strong force is between quarks and the nuclear force is between nucleons.

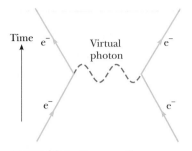

FIGURE 30.6 Feynman diagram representing a photon mediating the electromagnetic force between two electrons.

Webnote 30.4

For a good summary on Feynman diagrams, take a look at *http://www2.slac.stanford.edu/vvc/theory/feynman.html*

RICHARD FEYNMAN, AMERICAN PHYSICIST (1918–1988)

With his son, Carl, in 1965. Feynman, together with Julian S. Schwinger and Shinichiro Tomonaga, won the 1965 Nobel prize for physics for fundamental work in the principles of quantum electrodynamics. His many important contributions to physics include the invention of simple diagrams to represent particle interactions graphically, the theory of the weak interaction of subatomic particles, a reformulation of quantum mechanics, the theory of superfluid helium, and his contribution to physics education through the magnificent three-volume text *The Feynman Lectures on Physics. (UPI Telephotos)*

In an effort to substantiate Yukawa's predictions, physicists began looking for the meson by studying cosmic rays that enter Earth's atmosphere. In 1937 Carl Anderson and his collaborators discovered a particle whose mass was 106 MeV/c^2, about 207 times the mass of the electron. However, subsequent experiments showed that the particle interacted very weakly with matter, and hence could not be the carrier of the nuclear force. This puzzling situation inspired several theoreticians to propose that there are two mesons with slightly different masses, an idea that was confirmed in 1947 with the discovery of the pi meson (π), or simply *pion*, by Cecil Frank Powell (1903–1969) and Guiseppe P. S. Occhialini (1907–1993). The particle discovered earlier by Anderson in 1937, the one thought to be a meson, is not really a meson. Instead, it takes part in weak and electromagnetic interactions only, and it is now called the *muon* (μ).

The pion comes in three varieties, corresponding to three charge states: π^+, π^-, and π^0. The π^+ and π^- particles have masses of 139.6 MeV/c^2, and the π^0 has a mass of 135.0 MeV/c^2. Pions and muons are very unstable particles. For example, the π^-, which has a lifetime of about 2.6×10^{-8} s, decays into a muon and an antineutrino. The muon, with a lifetime of 2.2 μs, then decays into an electron, a neutrino, and an antineutrino. The sequence of decays is

$$\pi^- \longrightarrow \mu^- + \bar{\nu}$$
$$\mu^- \longrightarrow e^- + \nu + \bar{\nu} \qquad \text{[30.6]}$$

The interaction between two particles can be understood in general with a simple diagram called a *Feynman diagram,* developed by Richard P. Feynman (1918–1988). Figure 30.6 is a Feynman diagram for the electromagnetic interaction between two electrons. In this simple case, a photon is the field particle that mediates the electromagnetic force between the electrons. The photon transfers energy and momentum from one electron to the other in this interaction. Such a photon, called a *virtual photon,* can never be detected directly because it is absorbed by the second electron very shortly after being emitted by the first electron. The existence of a virtual photon would violate the law of conservation of energy, but because of the uncertainty principle and its very short lifetime Δt, the photon's excess energy is less than the uncertainty in its energy, given by $\Delta E \approx \hbar/\Delta t$.

Now consider the pion exchange between a proton and a neutron via the nuclear force (Fig. 30.7). We can reason that the energy ΔE needed to create a pion of mass m_π is given by $\Delta E = m_\pi c^2$. Again, the existence of the pion is allowed in spite of conservation of energy if this energy is surrendered in a short enough time Δt, the time it takes the pion to transfer from one nucleon to the other. From the uncertainty principle, $\Delta E \, \Delta t \approx \hbar$, we get

$$\Delta t \approx \frac{\hbar}{\Delta E} = \frac{\hbar}{m_\pi c^2} \qquad \text{[30.7]}$$

Because the pion cannot travel faster than the speed of light, the maximum distance d it can travel in a time Δt is $c \, \Delta t$. Using Equation 30.7 and $d = c \, \Delta t$, we find this maximum distance to be

$$d \approx \frac{\hbar}{m_\pi c} \qquad \text{[30.8]}$$

The measured range of the nuclear force is about 1.5×10^{-15} m. Using this value for d in Equation 30.8, the rest energy of the pion is calculated to be

$$m_\pi c^2 \approx \frac{\hbar c}{d} = \frac{(1.05 \times 10^{-34}\,\text{J} \cdot \text{s})(3.00 \times 10^8\,\text{m/s})}{1.5 \times 10^{-15}\,\text{m}}$$
$$= 2.1 \times 10^{-11}\,\text{J} \cong 130\,\text{MeV}$$

This corresponds to a mass of 130 MeV/c^2 (about 250 times the mass of the electron), which is in good agreement with the observed mass of the pion.

The concept we have just described is quite revolutionary. In effect, it says that a proton can change into a proton plus a pion, as long as it returns to its original state in a very short time. High-energy physicists often say that a nucleon undergoes "fluctuations" as it emits and absorbs pions. As we have seen, these fluctuations are a consequence of a combination of quantum mechanics (through the uncertainty principle) and special relativity (through Einstein's energy-mass relation $E = mc^2$).

This section has dealt with Yukawa's early theory of particles that mediate the nuclear force, namely the pions, and the mediators of the electromagnetic force, the photons. Although Yukawa's model led to the modern view, it has been superseded by the more basic quark-gluon theory as explained in Sections 30.12 and 30.13.

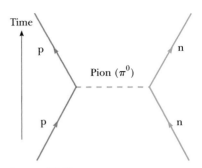

FIGURE 30.7 Feynman diagram representing a proton interacting with a neutron via the strong force. In this case, the pion mediates the nuclear force.

30.8 CLASSIFICATION OF PARTICLES

HADRONS

All particles other than photons can be classified into two broad categories, hadrons and leptons, according to their interactions. Particles that interact through the strong force are called *hadrons*. There are two classes of hadrons, known as *mesons* and *baryons*, distinguished by their masses and spins. All mesons are known to decay finally into electrons, positrons, neutrinos, and photons. The pion is the lightest of known mesons, with a mass of about 140 MeV/c^2 and a spin of 0. Another is the K meson, with a mass of about 500 MeV/c^2 and spin 0.

Baryons have masses equal to or greater than the proton mass (the name *baryon* means "heavy" in Greek), and their spin is always a noninteger value ($\frac{1}{2}$ or $\frac{3}{2}$). Protons and neutrons are baryons, as are many other particles. With the exception of the proton, all baryons decay in such a way that the end products include a proton. For example, the baryon called the Ξ hyperon first decays to a Λ^0 in about 10^{-10} s. The Λ^0 then decays to a proton and a π^- in about 3×10^{-10} s.

Today it is believed that hadrons are composed of quarks. (We shall have more to say about the quark model later.) Some of the important properties of hadrons are listed in Table 30.2.

LEPTONS

Leptons (from the Greek *leptos* meaning "small" or "light") are a group of particles that participate in the weak interaction. All leptons have a spin of $\frac{1}{2}$. Included in this group are electrons, muons, and neutrinos, which are less massive than the lightest hadron. Although hadrons have size and structure, leptons appear to be truly elementary, with no structure (that is, point-like).

Quite unlike hadrons, the number of known leptons is small. Currently, scientists believe there only are six leptons (each having an antiparticle) — the electron, the muon, the tau, and a neutrino associated with each:

$$\begin{pmatrix} e^- \\ \nu_e \end{pmatrix} \qquad \begin{pmatrix} \mu^- \\ \nu_\mu \end{pmatrix} \qquad \begin{pmatrix} \tau^- \\ \nu_\tau \end{pmatrix}$$

The tau lepton, discovered in 1975, has a mass about twice that of the proton.

Although neutrinos have masses of about zero, there is strong indirect evidence that the electron neutrino has a finite mass of about 3 eV/c^2 or a mass 1/180 000 of the electron mass. As we shall see, a firm knowledge of the neutrino's mass could have great significance in cosmological models and the future of the Universe.

Webnote 30.5

Take a look at particle physics in terms of the Big Bang Theory by visiting
http://hepwww.rl.ac.uk/Pub/Phil/ppintro/ppintro.html

Webnote 30.6

Protons showering from outer space can create muons in the atmosphere. Understand more about this effect by visiting
http://www2.slac.stanford.edu/vvc/cosmic_rays.html

| TABLE 30.2 | Some Particles and Their Properties |

Category	Particle Name	Symbol	Anti-particle	Mass (MeV/c^2)	B	L_e	L_μ	L_τ	S	Lifetime(s)	Principal Decay Modes[a]
Leptons	Electron	e^-	e^+	0.511	0	+1	0	0	0	Stable	
	Electron–neutrino	ν_e	$\bar\nu_e$	<7 eV/c^2	0	+1	0	0	0	Stable	
	Muon	μ^-	μ^+	105.7	0	0	+1	0	0	2.20×10^{-6}	$e^- \bar\nu_e \nu_\mu$
	Muon–neutrino	ν_μ	$\bar\nu_\mu$	<0.3	0	0	+1	0	0	Stable	
	Tau	τ^-	τ^+	1 784	0	0	0	+1	0	$<4 \times 10^{-13}$	$\mu^- \bar\nu_\mu \nu_\tau, e^- \bar\nu_e \nu_\tau$
	Tau–neutrino	ν_τ	$\bar\nu_\tau$	<30	0	0	0	+1	0	Stable	
Hadrons											
Mesons	Pion	π^+	π^-	139.6	0	0	0	0	0	2.60×10^{-8}	$\mu^+ \nu_\mu$
		π^0	Self	135.0	0	0	0	0	0	0.83×10^{-16}	2γ
	Kaon	K^+	K^-	493.7	0	0	0	0	+1	1.24×10^{-8}	$\mu^+ \nu_\mu, \pi^+ \pi^0$
		K^0_S	$\bar K^0_S$	497.7	0	0	0	0	+1	0.89×10^{-10}	$\pi^+ \pi^-, 2\pi^0$
		K^0_L	$\bar K^0_L$	497.7	0	0	0	0	+1	5.2×10^{-8}	$\pi^\pm e^\mp \bar\nu_e, 3\pi^0$
											$\pi^\pm \mu^\mp \bar\nu_\mu$
	Eta	η	Self	548.8	0	0	0	0	0	$<10^{-18}$	$2\gamma, 3\pi$
		η'	Self	958	0	0	0	0	0	2.2×10^{-21}	$\eta\pi^+ \pi^-$
Baryons	Proton	p	$\bar p$	938.3	+1	0	0	0	0	Stable	
	Neutron	n	$\bar n$	939.6	+1	0	0	0	0	920	$pe^- \bar\nu_e$
	Lambda	Λ^0	$\bar\Lambda^0$	1 115.6	+1	0	0	0	-1	2.6×10^{-10}	$p\pi^-, n\pi^0$
	Sigma	Σ^+	$\bar\Sigma^-$	1 189.4	+1	0	0	0	-1	0.80×10^{-10}	$p\pi^0, n\pi^+$
		Σ^0	$\bar\Sigma^0$	1 192.5	+1	0	0	0	-1	6×10^{-20}	$\Lambda^0 \gamma$
		Σ^-	$\bar\Sigma^+$	1 197.3	+1	0	0	0	-1	1.5×10^{-10}	$n\pi^-$
	Xi	Ξ^0	$\bar\Xi^0$	1 315	+1	0	0	0	-2	2.9×10^{-10}	$\Lambda^0 \pi^0$
		Ξ^-	Ξ^+	1 321	+1	0	0	0	-2	1.64×10^{-10}	$\Lambda^0 \pi^-$
	Omega	Ω^-	Ω^+	1 672	+1	0	0	0	-3	0.82×10^{-10}	$\Xi^0 \pi^0, \Lambda^0 K^-$

[a] Notations in this column such as $p\pi^-$, $n\pi^0$ mean two possible decay modes. In this case, the two possible decays are $\Lambda^0 \rightarrow p + \pi^-$ and $\Lambda^0 \rightarrow n + \pi^0$.

30.9 CONSERVATION LAWS

A number of conservation laws are important in the study of elementary particles. Although the two described here have no theoretical foundation, they are supported by abundant empirical evidence.

BARYON NUMBER

The law of conservation of baryon number tells us that whenever a baryon is created in a reaction or decay, an antibaryon is also created. This can be quantified by assigning a baryon number: $B = +1$ for all baryons, $B = -1$ for all antibaryons, and $B = 0$ for all other particles. Thus, the **law of conservation of baryon number** states that whenever a nuclear reaction or decay occurs, the sum of the baryon numbers before the process must equal the sum of the baryon numbers after the process.

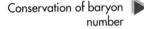

Conservation of baryon number

Note that if baryon number is absolutely conserved, the proton must be absolutely stable. If it were not for the law of conservation of baryon number, the proton could decay into a positron and a neutral pion. However, such a decay has never been observed. At present, we can only say that the proton has a half-life of at least 10^{31} years (the estimated age of the Universe is about 10^{10} years). In one recent version of a so-called grand unified theory or (GUT), physicists have predicted that the proton is actually unstable. According to this theory, the baryon number (sometimes called the *baryonic charge*) is not absolutely conserved, whereas electric charge is always conserved.

Example 30.4 Checking Baryon Numbers

Determine whether or not each of the following reactions can occur based on the law of conservation of baryon number.

$$p + n \longrightarrow p + p + n + \bar{p} \tag{1}$$

$$p + n \longrightarrow p + p + \bar{p} \tag{2}$$

Solution Recall that $B = +1$ for baryons and $B = -1$ for antibaryons. Hence, the left side of (1) gives a total baryon number of $1 + 1 = 2$. The right side gives a total baryon number of $1 + 1 + 1 + (-1) = 2$. Thus, reaction (1) can occur provided the incoming proton has sufficient energy.

The left side of (2) gives a total baryon number of $1 + 1 = 2$. However, the right side gives $1 + 1 + (-1) = 1$. Because baryon number is not conserved, reaction (2) cannot occur.

LEPTON NUMBER

There are three conservation laws involving lepton numbers, one for each variety of lepton. The **law of conservation of electron-lepton number** states that the sum of the electron-lepton numbers before a reaction or decay must equal the sum of the electron-lepton numbers after the reaction or decay. The electron and the electron neutrino are assigned a positive electron-lepton number, $L_e = +1$; the antileptons e^+ and $\bar{\nu}_e$ are assigned the electron-lepton number $L_e = -1$; and all other particles have $L_e = 0$. For example, consider the decay of the neutron

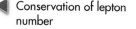 ◀ Conservation of lepton number

$$n \longrightarrow p^+ + e^- + \bar{\nu}_e$$

◀ Neutron decay

Before the decay, the electron-lepton number is $L_e = 0$; after the decay it is $0 + 1 + (-1) = 0$. Thus, the electron-lepton number is conserved. It is important to recognize that the baryon number must also be conserved. This can easily be seen by noting that before the decay $B = +1$, whereas after the decay $B = +1 + 0 + 0 = +1$.

Similarly, when a decay involves muons, the muon-lepton number L_μ is conserved. The μ^- and the ν_μ are assigned $L_\mu = +1$, the antimuons μ^+ and $\bar{\nu}_\mu$ are assigned $L_\mu = -1$, and all other particles have $L_\mu = 0$. Finally, the τ-lepton number L_τ is conserved, and similar assignments can be made for the τ lepton and its neutrino.

Example 30.5 Checking Lepton Numbers

Determine which of the following decay schemes can occur on the basis of conservation of lepton number.

$$\mu^- \longrightarrow e^- + \bar{\nu}_e + \nu_\mu \tag{1}$$

$$\pi^+ \longrightarrow \mu^+ + \nu_\mu + \nu_e \tag{2}$$

Solution Because decay 1 involves both a muon and an electron, L_μ and L_e must both be conserved. Before the decay, $L_\mu = +1$ and $L_e = 0$. After the decay, $L_\mu = 0 + 0 + 1 = +1$, and $L_e = +1 - 1 + 0 = 0$. Thus, both numbers are conserved, and on this basis the decay mode is possible.

Before decay 2 occurs, $L_\mu = 0$ and $L_e = 0$. After the decay, $L_\mu = -1 + 1 + 0 = 0$, but $L_e = +1$. Thus, the decay is not possible because the electron-lepton number is not conserved.

EXERCISE Determine whether the decay $\mu^- \rightarrow e^- + \bar{\nu}_e$ can occur.

ANSWER No. The muon-lepton number is +1 before the decay and 0 after the decay.

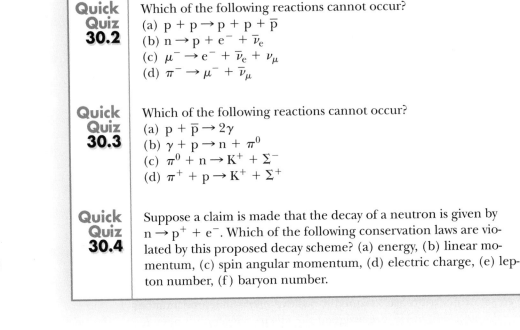

Which of the following reactions cannot occur?

(a) $p + p \rightarrow p + p + \bar{p}$
(b) $n \rightarrow p + e^- + \bar{\nu}_e$
(c) $\mu^- \rightarrow e^- + \bar{\nu}_e + \nu_\mu$
(d) $\pi^- \rightarrow \mu^- + \bar{\nu}_\mu$

Quick Quiz 30.2

Which of the following reactions cannot occur?

(a) $p + \bar{p} \rightarrow 2\gamma$
(b) $\gamma + p \rightarrow n + \pi^0$
(c) $\pi^0 + n \rightarrow K^+ + \Sigma^-$
(d) $\pi^+ + p \rightarrow K^+ + \Sigma^+$

Quick Quiz 30.3

Suppose a claim is made that the decay of a neutron is given by $n \rightarrow p^+ + e^-$. Which of the following conservation laws are violated by this proposed decay scheme? (a) energy, (b) linear momentum, (c) spin angular momentum, (d) electric charge, (e) lepton number, (f) baryon number.

Quick Quiz 30.4

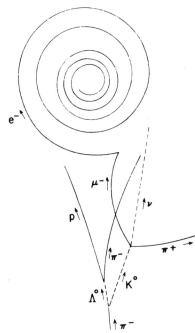

FIGURE 30.8 This drawing represents tracks of many events obtained by analyzing a bubble-chamber photograph. The strange particles Λ^0 and K^0 are formed (at the bottom) as the π^- interacts with a proton according to $\pi^- + p \rightarrow \Lambda^0 + K^0$. (Note that the neutral particles leave no tracks, as indicated by the dashed lines.) The Λ^0 and K^0 then decay according to $\Lambda^0 \rightarrow \pi^- + p$ and $K^0 \rightarrow \pi + \mu^- + \nu_\mu$. *(Courtesy Lawrence Berkeley Laboratory, University of California)*

▶ Conservation of strangeness number

30.10   STRANGE PARTICLES AND STRANGENESS

Many particles discovered in the 1950s were produced by the nuclear interaction of pions with protons and neutrons in the atmosphere. A group of these particles, namely the K, Λ, and Σ, were found to exhibit unusual properties in their production and decay, and hence were called *strange particles*.

One unusual property of strange particles is that they are always produced in pairs. For example, when a pion collides with a proton, two neutral strange particles are produced with high probability (Fig. 30.8) following the reaction

$$\pi^- + p^+ \longrightarrow K^0 + \Lambda^0$$

On the other hand, the reaction $\pi^- + p^+ \rightarrow K^0 + n$ never occurs, even though no known conservation laws are violated and the energy of the pion is sufficient to initiate the reaction.

The second peculiar feature of strange particles is that although they are produced by the strong interaction at a high rate, they do not decay into particles that interact via the strong force at a very high rate. Instead, they decay very slowly, which is characteristic of the weak interaction. Their half-lives are in the range 10^{-10} s to 10^{-8} s; most other particles that interact via the strong force have lifetimes on the order of 10^{-23} s.

To explain these unusual properties of strange particles, physicists introduced a law called *conservation of strangeness*, together with a new quantum number S called **strangeness.** The strangeness numbers for some particles are given in Table 30.2. The production of strange particles in pairs is explained by assigning $S = +1$ to one of the particles and $S = -1$ to the other. All nonstrange particles are assigned strangeness $S = 0$. The **law of conservation of strangeness** states that whenever a nuclear reaction or decay occurs, the sum of the strangeness numbers before the process must equal the sum of the strangeness numbers after the process.

We can explain the slow decay of strange particles by assuming that the strong and electromagnetic interactions obey the law of conservation of strangeness, whereas the weak interaction does not. Because the decay reaction involves the loss of one strange particle, it violates strangeness conservation and hence proceeds slowly via the weak interaction.

Webnote 30.7
Does a 600-ton bubble chamber sound big? Find out more by exploring the Imaging Cosmic and Rare Underground Signal (ICARUS) in Italy. Visit
http://www.aquila.infn.it/icarus

APPLYING **PHYSICS** **30.2**

A student claims to have observed a decay of an electron into two neutrinos, traveling in opposite directions. What conservation laws would be violated by this decay?

Explanation Several conservation laws are violated. Conservation of electric charge is violated because the negative charge of the electron has disappeared. Conservation of electron-lepton number is also violated, because there is one lepton before the decay and two afterward. If both neutrinos were electron-neutrinos, electron-lepton number conservation would be violated in the final state. However, if one of the product neutrinos were other than an electron-neutrino, then another lepton conservation law would be violated, because no other leptons were in the initial state. Other conservation laws are obeyed by this decay. Energy can be conserved—the rest energy of the electron appears as the kinetic energy (and possibly some small rest energy) of the neutrinos. The opposite directions of the velocities of the two neutrinos allows for conservation of momentum. Conservation of baryon number and conservation of other lepton numbers are also upheld in this decay.

Example 30.6 **Is Strangeness Conserved?**

A Determine whether the following reaction occurs on the basis of conservation of strangeness.

$$\pi^0 + n \longrightarrow K^+ + \Sigma^-$$

Solution The initial state has a total strangeness of $S = 0 + 0 = 0$. Because the strangeness of the K^+ is $S = +1$, and the strangeness of the Σ^- is $S = -1$, the total strangeness of the final state is $+1 - 1 = 0$. Thus, strangeness is conserved and the reaction is allowed.

B Show that the following reaction does not conserve strangeness.

$$\pi^- + p \longrightarrow \pi^- + \Sigma^+$$

Solution The initial state has strangeness $S = 0 + 0 = 0$, and the final state has strangeness $S = 0 + (-1) = -1$. Thus, strangeness is not conserved.

EXERCISE Show that the observed reaction $p^+ + \pi^- \rightarrow K^0 + \Lambda^0$ obeys the law of conservation of strangeness.

30.11 THE EIGHTFOLD WAY

As we have seen, quantities such as spin, baryon number, lepton number, and strangeness are labels we associate with particles. Many classification schemes that group particles into families based on such labels have been proposed. First, consider the first eight baryons listed in Table 30.2, all having a spin of $\frac{1}{2}$. The family consists of the proton, the neutron, and six other particles. If we plot their strangeness versus their charge using a sloping coordinate system, as in Figure 30.9a, a fascinating pattern emerges. Six of the baryons form a hexagon, and the remaining two are at the hexagon's center. (Particles with spin quantum number $\frac{1}{2}$ or $\frac{3}{2}$ are called fermions.)

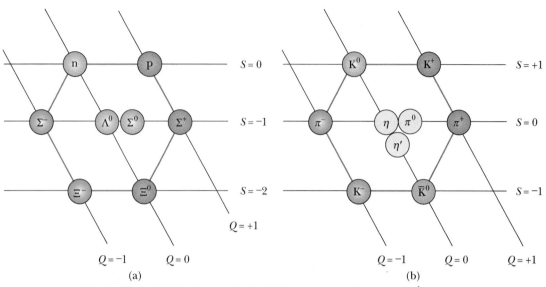

FIGURE 30.9   (a) The hexagonal eightfold-way pattern for the eight spin-$\frac{1}{2}$ baryons. This strangeness versus charge plot uses a horizontal axis for the strangeness values S, but a sloping axis for the charge number Q. (b) The eightfold-way pattern for the nine spin-0 mesons.

Now consider the family of mesons listed in Table 30.2 with spins of 0 (Particles with spin quantum number 0 or 1 are called bosons.) If we count both particles and antiparticles, there are nine such mesons. Figure 30.9b is a plot of strangeness versus charge for this family. Again, a fascinating hexagonal pattern emerges. In this case, the particles on the perimeter of the hexagon lie opposite their antiparticles, and the remaining three (which form their own antiparticles) are at its center. These and related symmetric patterns, called the **eightfold way,** were proposed independently in 1961 by Murray Gell-Mann and Yuval Ne'eman.

The groups of baryons and mesons can be displayed in many other symmetric patterns within the framework of the eightfold way. For example, the family of spin-$\frac{3}{2}$ baryons contains ten particles arranged in a pattern such as the tenpins in a bowling alley. After the pattern was proposed, one of the particles was missing—it had yet to be discovered. Gell-Mann predicted that the missing particle, which he called the *omega minus* (Ω^-), should have a spin of $\frac{3}{2}$, a charge of -1, a strangeness of -3, and a mass of about 1 680 MeV/c^2. Shortly thereafter, in 1964, scientists at the Brookhaven National Laboratory found the missing particle through careful analyses of bubble chamber photographs, and confirmed all its predicted properties.

The patterns of the eightfold way in the field of particle physics have much in common with the periodic table. Whenever a vacancy (a missing particle or element) occurs in the organized patterns, experimentalists have a guide for their investigations.

30.12   QUARKS

As we have noted, leptons appear to be truly elementary particles because they have no measurable size or internal structure, are limited in number, and do not seem to break down into smaller units. Hadrons, on the other hand, are complex particles with size and structure. Furthermore, we know that hadrons decay into other hadrons and are many in number. Table 30.2 lists only those hadrons that are stable against hadronic decay; hundreds of others have been discovered. These facts strongly suggest that hadrons cannot be truly elementary but have some substructure.

MURRAY GELL-MANN, AMERICAN PHYSICIST (B. 1929)

Gell-Mann was awarded the Nobel prize in 1969 for his theoretical studies dealing with subatomic particles. *(Photo courtesy of Michael R. Dressler)*

TABLE 30.3	Properties of Quarks and Antiquarks

Quarks

Name	Symbol	Spin	Charge	Baryon Number	Strangeness	Charm	Bottomness	Topness
Up	u	$\frac{1}{2}$	$+\frac{2}{3}e$	$\frac{1}{3}$	0	0	0	0
Down	d	$\frac{1}{2}$	$-\frac{1}{3}e$	$\frac{1}{3}$	0	0	0	0
Strange	s	$\frac{1}{2}$	$-\frac{1}{3}e$	$\frac{1}{3}$	-1	0	0	0
Charmed	c	$\frac{1}{2}$	$+\frac{2}{3}e$	$\frac{1}{3}$	0	$+1$	0	0
Bottom	b	$\frac{1}{2}$	$-\frac{1}{3}e$	$\frac{1}{3}$	0	0	$+1$	0
Top	t	$\frac{1}{2}$	$+\frac{2}{3}e$	$\frac{1}{3}$	0	0	0	$+1$

Antiquarks

Name	Symbol	Spin	Charge	Baryon Number	Strangeness	Charm	Bottomness	Topness
Anti-up	$\bar{u}$	$\frac{1}{2}$	$-\frac{2}{3}e$	$-\frac{1}{3}$	0	0	0	0
Anti-down	$\bar{d}$	$\frac{1}{2}$	$+\frac{1}{3}e$	$-\frac{1}{3}$	0	0	0	0
Anti-strange	$\bar{s}$	$\frac{1}{2}$	$+\frac{1}{3}e$	$-\frac{1}{3}$	$+1$	0	0	0
Anti-charmed	$\bar{c}$	$\frac{1}{2}$	$-\frac{2}{3}e$	$-\frac{1}{3}$	0	-1	0	0
Anti-bottom	$\bar{b}$	$\frac{1}{2}$	$+\frac{1}{3}e$	$-\frac{1}{3}$	0	0	-1	0
Anti-top	$\bar{t}$	$\frac{1}{2}$	$-\frac{2}{3}e$	$-\frac{1}{3}$	0	0	0	-1

THE ORIGINAL QUARK MODEL

In 1963 Gell-Mann and George Zweig independently proposed that hadrons have a more elementary substructure. According to their model, all hadrons are composite systems of two or three fundamental constitutents called **quarks,** which rhymes with "forks." Gell-Mann borrowed the word *quark* from the passage "Three quarks for Muster Mark" in James Joyce's book *Finnegan's Wake*. In the original model, there were three types of quarks designated by the symbols *u*, *d*, and *s*. These were given the arbitrary names *up, down,* and *sideways* (or, now more commonly, *strange*).

An unusual property of quarks is that they have fractional electronic charges, as shown—along with other properties—in Table 30.3. Associated with each quark is an antiquark of opposite charge, baryon number, and strangeness. The compositions of all hadrons known when Gell-Mann and Zweig presented their models could be completely specified by three simple rules:

1. Mesons consist of one quark and one antiquark, giving them a baryon number of 0, as required.
2. Baryons consist of three quarks.
3. Antibaryons consist of three antiquarks.

Table 30.4 lists the quark compositions of several mesons and baryons. Note that just two of the quarks, u and d, are contained in all hadrons encountered in ordinary matter (protons and neutrons). The third quark, s, is needed only to construct strange particles with a strangeness of either $+1$ or -1. Figure 30.10 shows the quark compositions of several particles.

TABLE 30.4

Quark Composition of Several Hadrons

Particle	Quark Composition
	Mesons
π^+	$u\bar{d}$
π^-	$\bar{u}d$
K^+	$u\bar{s}$
K^-	$\bar{u}s$
K^0	$d\bar{s}$
	Baryons
p	uud
n	udd
Λ^0	uds
Σ^+	uus
Σ^0	uds
Σ^-	dds
Ξ^0	uss
Ξ^-	dss
Ω^-	sss

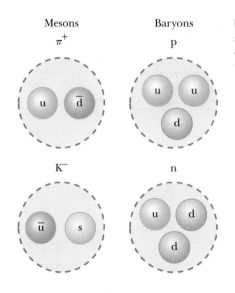

Mesons   Baryons

π^+ p

K^- n

FIGURE 30.10 Quark compositions of two mesons and two baryons. Note that the mesons on the left contain two quarks, and the baryons on the right contain three quarks.

APPLYING **PHYSICS** **30.3**

We have seen a law of conservation of lepton number and a law of conservation of baryon number. Why isn't there a law of conservation of meson number?

Explanation   We can argue this from the point of view of creating particle-antiparticle pairs from available energy. If energy is converted to rest energy of a lepton-antilepton pair, then there is no net change in lepton number, because the lepton has a number of $+1$ and the antilepton has a number of -1. Energy could also be transformed into rest energy of a baryon-antibaryon pair. The baryon has baryon number $+1$, the antibaryon -1, and there is no net change in baryon number.

But now suppose energy is transformed into rest energy of a quark-antiquark pair. By definition in quark theory, a quark antiquark pair is a meson. Thus, we have created a meson from energy—there was no meson before; now there is. Thus, there is no conservation of meson number. With more energy, we can create more mesons, with no restriction from a conservation law other than that of energy.

CHARM AND OTHER RECENT DEVELOPMENTS

Although the original quark model was highly successful in classifying particles into families, there were some discrepancies between predictions of the model and certain experimental decay rates. As a consequence, a fourth quark was proposed by several physicists in 1967. The fourth quark, designated by c, was given a property called **charm.** A charmed quark would have the charge $+2e/3$, but its charm would distinguish it from the other three quarks. The new quark would have a charm of $C = +1$, its antiquark would have a charm of $C = -1$, and all other quarks would have $C = 0$, as indicated in Table 30.3. Charm, like strangeness, would be conserved in strong and electromagnetic interactions but not in weak interactions.

In 1974 a new heavy meson called the J/ψ particle (or simply ψ) was discovered independently by a group led by Burton Richter at the Stanford Linear Accelerator (SLAC) and another group led by Samuel Ting at the Brookhaven National Laboratory. Richter and Ting were awarded the Nobel prize in 1976 for this work. The J/ψ particle does not fit into the three-quark model, but has the properties of a combination of a charmed quark and its antiquark ($c\bar{c}$). It is much heavier than the other known mesons ($\sim 3\ 100\ \text{MeV}/c^2$) and its lifetime is much longer than those of other particles that decay via the strong force. In 1975 researchers at Stan-

TABLE 30.5	The Fundamental Particles and Some of Their Properties

Quarks

Particle	Rest Energy	Charge
u	360 MeV	$+\frac{2}{3}e$
d	360 MeV	$-\frac{1}{3}e$
c	1 500 MeV	$+\frac{2}{3}e$
s	540 MeV	$-\frac{1}{3}e$
t	173 GeV	$+\frac{2}{3}e$
b	5 GeV	$-\frac{1}{3}e$

Leptons

Particle	Rest Energy	Charge
e^-	511 keV	$-e$
μ^-	107 MeV	$-e$
τ^-	1 784 MeV	$-e$
ν_e	<30 eV	0
ν_μ	<0.5 MeV	0
ν_τ	<250 MeV	0

ford University reported strong evidence for the tau (τ) lepton, with a mass of 1 784 MeV/c^2. Such discoveries led to more elaborate quark models and the proposal of two new quarks, named *top* (t) and *bottom* (b). (Some physicists prefer the whimsical names *truth* and *beauty*.) To distinguish these quarks from the old ones, quantum numbers called *topness* and *bottomness* were assigned to these new particles and are included in Table 30.3. In 1977 researchers at the Fermi National Laboratory, under the direction of Leon Lederman, reported the discovery of a very massive new meson Υ whose composition is considered to be b$\bar{\text{b}}$. In March of 1995, researchers at Fermilab announced the discovery of the top quark (supposedly the last of the quarks to be found) having mass 173 GeV/c^2.

You are probably wondering whether or not such discoveries will ever end. How many "building blocks" of matter really exist? At the present, physicists believe that the fundamental particles in nature include six quarks and six leptons (together with their antiparticles). Some of the properties of these particles are given in Table 30.5.

Despite extensive experimental efforts, no isolated quark has ever been observed. Physicists now believe that quarks are permanently confined inside ordinary particles because of an exceptionally strong force that prevents them from escaping. This force, called the color force (which we discuss in Section 30.13), increases with separation distance (similar to the force of a spring). The great strength of the force between quarks has been described by one author as follows:[2]

> **Quarks are slaves of their own color charge, . . . bound like prisoners of a chain gang. . . . Any locksmith can break the chain between two prisoners, but no locksmith is expert enough to break the gluon chains between quarks. Quarks remain slaves forever.**

Webnote 30.8

Further investigate the importance of the search for the top quark by exploring
http://www.duke.edu/vertices/update/fall94/quark.html

[2] Harald Fritzsch, *Quarks, The Stuff of Matter*, London: Allen Lane, 1983.

30.13   COLORED QUARKS

Shortly after the theory of quarks was proposed, scientists recognized that certain particles had quark compositions that were in violation of the Pauli exclusion principle. Because all quarks have spins of $\frac{1}{2}$, they are expected to follow the exclusion principle. One example of a particle that violates the exclusion principle is the Ω^- (sss) baryon that contains three s quarks having parallel spins, giving it a total spin of $\frac{3}{2}$. Other examples of baryons that have identical quarks with parallel spins are the Δ^{++} (uuu) and the Δ^- (ddd). To resolve this problem, Moo-Young Han and Yoichiro Nambu suggested in 1965 that quarks possess a new property called **color or color charge.** This "charge" property is similar in many respects to electric charge except that it occurs in three varieties labeled *red, green,* and *blue!* (The antiquarks are labled *antired, antigreen,* and *antiblue.*) To satisfy the exclusion principle, all three quarks in a baryon must have different colors. Just as a combination of actual colors of light can produce the neutral color white, a combination of three quarks with different colors is also white, or colorless. A meson consists of a quark of one color and an antiquark of the corresponding anticolor. The result is that baryons and mesons are always colorless (or white).

Although the concept of color in the quark model was originally conceived to satisfy the exclusion principle, it also provided a better theory for explaining certain experimental results. For example, the modified theory correctly predicts the lifetime of the π^0 meson. The theory of how quarks interact with each other by means of color charge is called **quantum chromodynamics,** or QCD, to parallel quantum electrodynamics (the theory of interaction between electric charges). In QCD, the quark is said to carry a **color charge,** in analogy to electric charge. The strong force between quarks is often called the **color force.**

The strong force between quarks is carried by massless particles called **gluons** (analogous to photons for the electromagnetic force). According to QCD, there are eight gluons, all with color charge. When a quark emits or absorbs a gluon, its color changes. For example, a blue quark that emits a gluon may become a red quark, and the red quark that absorbs this gluon becomes a blue quark. The color force between quarks is analogous to the electric force between charges; like colors repel and opposite colors attract. Therefore, two red quarks repel each other, but a red quark will be attracted to an antired quark. The attraction between quarks of opposite color to form a meson ($q\bar{q}$) is indicated in Figure 30.11a. Differently colored quarks also attract each other, but with less intensity than opposite colors of quark and antiquark. For example, a cluster of red, blue, and green quarks all attract each other to form baryons as indicated in Figure 30.11b. Thus, every baryon contains three quarks of three different colors.

Tip 30.3   COLOR IS NOT REALLY COLOR

When we use the word color to describe a quark, it has nothing to do with visual sensation from light. It is simply a convenient name for a property analogous to electric charge.

Webnote 30.9

For more introductory remarks on quantum chromodynamics, take a look at
http://www.infoplease.com/ce6/sci/A0840717.html

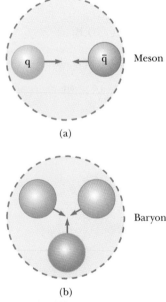

FIGURE 30.11   (a) A red quark is attracted to an antired quark. This forms a meson whose quark structure is ($q\bar{q}$). (b) Three different colored quarks attract each other to form a baryon.

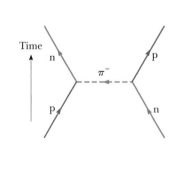

(a) Yukawa's pion model

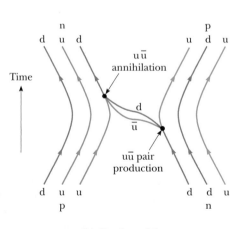

(b) Quark model

FIGURE 30.12   (a) A nuclear interaction between a proton and a neutron explained in terms of Yukawa's pion exchange model. (b) The same interaction explained in terms of quarks and gluons. Note that the exchanged $\bar{u}d$ quark pair makes up a π^- meson.

Although the color force between two color-neutral hadrons (like a proton and a neutron) is negligible at large separations, the strong color force between their constituent quarks does not exactly cancel at small separations of about 1 fm. **This residual strong force is in fact the nuclear force that binds protons and neutrons to form nuclei.** It is similar to the residual electromagnetic force that binds neutral atoms into molecules. According to QCD, a more basic explanation of nuclear force can be given in terms of quarks and gluons, as shown in Figure 30.12, which shows contrasting Feynman diagrams of the same process. Each quark within the neutron and proton is continually emitting and absorbing virtual gluons and creating and annihilating virtual (q$\bar{\text{q}}$) pairs. When the neutron and proton approach within 1 fm of each other, these virtual gluons and quarks can be exchanged between the two nucleons, and such exchanges produce the nuclear force. Figure 30.12b depicts one likely possibility or contribution to the process shown in Figure 30.12a. A down quark emits a virtual gluon (represented by a wavy line), which creates a u$\bar{\text{u}}$ pair. Both the recoiling d quark and the $\bar{\text{u}}$ are transmitted to the proton where the $\bar{\text{u}}$ annihilates a proton u quark (with the creation of a gluon) and the d is captured.

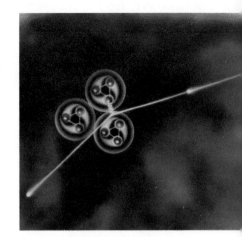

An artist's version of a high-energy particle colliding with a nucleus. The quark structure of the nucleus is indicated by the small colored spheres inside the nucleus. *(Courtesy of Janie Martz/CEBAF)*

30.14 ELECTROWEAK THEORY AND THE STANDARD MODEL

Recall that the weak interaction is an extremely short-range force having an interaction distance of approximately 10^{-18} m (Table 30.1). Such a short-range interaction implies that the quantized particles that carry the weak field (the spin-1 W$^+$, W$^-$, and Z^0 bosons) are quite massive as is indeed the case. These amazing bosons can be thought of as structureless, point-like particles as massive as krypton atoms! The weak interaction is responsible for the decay of the c, s, b, and t quarks into lighter, more stable u and d quarks as well as the decay of the massive μ and τ leptons into (lighter) electrons. Thus, **the weak interaction is very important because it governs the stability of the basic matter particles.**

A mysterious feature of the weak interaction is its lack of symmetry, especially compared to the high degree of symmetry shown by the strong, electromagnetic, and gravitational interactions. For example, the weak interaction, unlike the strong interaction, is not symmetric under mirror reflection or charge exchange. (*Mirror reflection* means that all the quantities in a given particle reaction are exchanged as in a mirror reflection—left for right, an inward motion toward the mirror for an outward motion. *Charge exchange* means that all the electric charges in a particle reaction are converted to their opposites—all positives to negatives, and vice versa.) When we say that the weak interaction is not symmetric, we mean that the reaction with all quantities changed occurs less frequently than the direct reaction. For example, the decay of the K^0, which is governed by the weak interaction, is not symmetric under charge exchange because K$^0 \rightarrow \pi^- + \text{e}^+ + \nu_e$ occurs much more frequently than K$^0 \rightarrow \pi^+ + \text{e}^- + \bar{\nu}_e$.

In 1979, Sheldon Glashow, Abdus Salam, and Steven Weinberg won a Nobel prize for developing a theory called the **electroweak theory** that unified the electromagnetic and weak interactions. This theory postulates that the weak and electromagnetic interactions have the same strength at very high particle energies. Thus, the two interactions are viewed as two different manifestations of a single unifying electroweak interaction. The photon and the three massive bosons (W$^\pm$ and Z^0) play a key role in the electroweak theory. The theory makes many concrete predictions, but perhaps the most spectacular is the prediction of the masses of the W and Z particles at about 82 GeV/c^2 and 93 GeV/c^2, respectively. A 1984 Nobel prize was awarded to Carlo Rubbia and Simon van der Meer for their work leading to the discovery of these particles at just these energies at the CERN Laboratory in Geneva, Switzerland.

The combination of the electroweak theory and QCD for the strong interaction form what is referred to in high-energy physics as the **Standard Model.** Al-

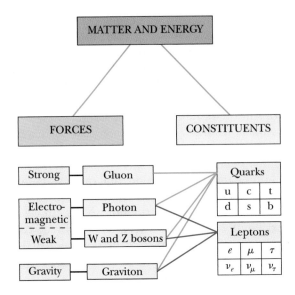

FIGURE 30.13   The Standard Model of particle physics.

Webnote 30.10

Visit the European Laboratory for Particle Physics in Geneva, Switzerland for more on accelerators and the invention of the World Wide Web. Go to the "General Public" link at *http://www.CERN.ch*

Webnote 30.11

Take a look at a full-view chart of the particles in the Standard Model of particle physics. Visit *http://www.particleadventure.org/ particleadventure*

A view from inside the LEP (Large Electron-Positron Collider) tunnel, which is 27 km in circumference. *(Courtesy of CERN)*

though the details of the Standard Model are complex, its essential ingredients can be summarized with the help of Figure 30.13. The strong force, mediated by gluons, holds quarks together to form composite particles such as protons, neutrons, and mesons. Leptons participate only in the electromagnetic and weak interactions. The electromagnetic force is mediated by photons, and the weak force is mediated by W and Z bosons. Note that all fundamental forces are mediated by bosons (particles with spin 1) whose properties are given, to a large extent, by symmetries involved in the theories.

However, the Standard Model does not answer all questions. A major question is why the photon has no mass, whereas the W and Z bosons do. Because of this mass difference, the electromagnetic and weak forces are quite distinct at low energies, but become similar in nature at very high energies, where the rest energies of the W and Z bosons are insignificant fractions of their total energies. This behavior as we go from high to low energies, called **symmetry breaking,** does not answer the question of the origin of particle masses. To resolve this problem, a hypothetical particle called the **Higgs boson** has been proposed that provides a mechanism for breaking the electroweak symmetry and bestowing different particle masses on different particles. The Standard Model, including the Higgs mechanism, provides a logically consistent explanation of the massive nature of the W and Z bosons. Unfortunately, the Higgs boson has not yet been found, but physicists know that its mass should be less than $1 \text{ TeV}/c^2$ (10^{12} eV).

In order to determine whether the Higgs boson exists, two quarks of at least 1 TeV of energy must collide, but calculations show that this requires injecting 40 TeV of energy within the volume of a proton. Scientists are convinced that because of the limited energy available in conventional accelerators using fixed targets, it is necessary to build colliding-beam accelerators called **colliders.** The concept of colliders is straightforward. Particles with equal masses and kinetic energies, traveling in opposite directions in an accelerator ring, collide head-on to produce the required reaction and the formation of new particles. Because the total momentum of the interacting particles is zero, all of their kinetic energy is available for the reaction. The Large Electron-Positron collider (LEP) at CERN near Geneva, Switzerland, and the Stanford Linear Collider in California collide both electrons and positrons. The Super Proton Synchrotron at CERN accelerates protons and antiprotons to energies of 270 GeV, and the world's highest-energy proton acclerator, the Tevatron, at the Fermi National Laboratory in Illinois produces protons at almost 1 000 GeV (or 1 TeV). CERN has started construction of the Large Hadron Collider (LHC), a proton-proton collider that will provide a center of mass energy of 14 TeV and allow an exploration of Higgs-boson physics.

The accelerator is being constructed in the same 27-km circumference tunnel as CERN's Large Electron-Positron collider, and construction is expected to be completed in 2005.

Following the success of the electroweak theory, scientists attempted to combine it with QCD in a **grand unification theory** known as GUT. In this model, the electroweak force was merged with the strong color force to form a grand unified force. One version of the theory considers leptons and quarks as members of the same family that are able to change into each other by exchanging an appropriate particle. Many GUT theories predict that protons are unstable, and will decay with a lifetime of about 10^{31} years. This is far greater than the age of the Universe, and as yet proton decays have not been observed.

APPLYING *PHYSICS* 30.4

Consider a car making a head-on collision with an identical car moving in the opposite direction at the same speed. Compare that collision to one in which one of the cars collides with the second car at rest. In which collision is there the larger transformation of kinetic energy to other forms during the collision? How does this relate to producing exotic particles in collisions?

Explanation In the head-on collision with both cars moving, conservation of momentum causes most, if not all, of the kinetic energy to be transformed to other forms. In the collision between a moving car and a stationary car, the cars are still moving after the collision, in the direction of the moving car, but with reduced speed. Thus, only part of the kinetic energy is transformed to other forms. This suggests the advantage of using colliding beams to produce exotic particles, as opposed to firing a beam into a stationary target. When particles moving in opposite directions collide, all of the kinetic energy is available for transformation into other forms—in this case, the creation of new particles. When a beam is fired into a stationary target, only part of the energy is available for transformation, so higher-mass particles cannot be created.

30.15 THE COSMIC CONNECTION

In this section we describe one of the most fascinating theories in all of science—the Big Bang theory of the creation of the Universe—and the experimental evidence that supports it. This theory of cosmology states that the Universe had a beginning, and that this beginning was so cataclysmic that it is impossible to look back beyond it. According to the theory, the Universe erupted from an infinitely dense singularity about 15–20 billion years ago. The first few minutes after the Big Bang saw such extremes of energy that it is believed that all four interactions of physics were unified and all matter was contained in an undifferentiated "quark soup."

The evolution of the four fundamental forces from the Big Bang to the present is shown in Figure 30.14. During the first 10^{-43} s (the ultra-hot epoch, $T \sim 10^{32}$ K), it is presumed that the strong, electroweak, and gravitational forces were joined to form a completely unified force. In the first 10^{-35} s following the Big Bang (the hot epoch, $T \sim 10^{29}$ K), gravity broke free of this unification and the strong and electroweak forces remained as one, described by a grand unification theory. This was a period when particle energies were so great ($> 10^{16}$ GeV) that very massive particles as well as quarks, leptons, and their antiparticles existed. Then, after 10^{-35} s, the Universe rapidly expanded and cooled (the warm epoch, $T \sim 10^{29}$ to 10^{15} K), the strong and electroweak forces parted company, and the grand unification scheme was broken. As the Universe continued to cool, the electroweak force split into the weak force and the electromagnetic force about 10^{-10} s after the Big Bang.

GEORGE GAMOW (1904–1968)

Gamow and two of his students, Ralph Alpher and Robert Herman, were the first to take the first half-hour of the Universe seriously. In a mostly overlooked paper published in 1948, they made truly remarkable cosmological predictions. They correctly calculated the abundances of hydrogen and helium after the first half-hour (75% H and 25% He) and predicted that radiation from the Big Bang should still be present and have an apparent temperature of about 5 K. *(Courtesy of AIP Emilio Segre Visual Archives)*

FIGURE 30.14 A brief history of the Universe from the Big Bang to the present. The four forces became distinguishable during the first microsecond. Following this, all the quarks combined to form particles that interact via the strong force. The leptons remained separate, however, and exist as individually observable particles to this day.

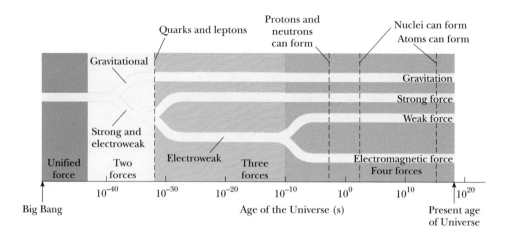

After a few minutes, protons condensed out of the hot soup. For half an hour the Universe underwent thermonuclear detonation, exploding as a hydrogen bomb and producing most of the helium nuclei now present. The Universe continued to expand, and its temperature dropped. Until about 700 000 years after the Big Bang, the Universe was dominated by radiation. Energetic radiation prevented matter from forming single hydrogen atoms because collisions would instantly ionize any atoms that might form. Photons experienced continuous Compton scattering from the vast number of free electrons, resulting in a Universe that was opaque to radiation. By the time the Universe was about 700 000 years old, it had expanded and cooled to about 3 000 K, and protons could bind to electrons to form neutral hydrogen atoms. Because the energies of the atoms were quantized, far more wavelengths of radiation were not absorbed by atoms than were, and the Universe suddenly became transparent to photons. Radiation no longer dominated the Universe, and clumps of neutral matter steadily grew—first atoms, followed by molecules, gas clouds, stars, and finally galaxies.

OBSERVATION OF RADIATION FROM THE PRIMORDIAL FIREBALL

In 1965 Arno A. Penzias (b. 1933) and Robert W. Wilson (b. 1936) of Bell Laboratories made an amazing discovery while testing a sensitive microwave receiver. A pesky signal producing a faint background hiss was interfering with their satellite communications experiments. In spite of their valiant efforts, the signal remained. Ultimately it became clear that they were observing microwave background radiation (at a wavelength of 7.35 cm) representing the leftover "glow" from the Big Bang.

The microwave horn that served as their receiving antenna is shown in Figure 30.15. The intensity of the detected signal remained unchanged as the antenna was pointed in different directions. The fact that the radiation had equal strengths in all directions suggested that the entire Universe was the source of this radiation. Evicting a flock of pigeons from the 20-foot horn and cooling the microwave detector both failed to remove the signal. Through a casual conversation, Penzias and Wilson discovered that a group at Princeton had predicted the residual radiation from the Big Bang and were planning an experiment to confirm the theory. The excitement in the scientific community was high when Penzias and Wilson announced that they had already observed an excess microwave background compatible with a 3-K blackbody source.

Because Penzias and Wilson made their measurements at a single wavelength, they did not completely confirm the radiation as 3-K blackbody radiation. Subsequent experiments by other groups added intensity data at different wavelengths as shown in Figure 30.16. The results confirm that the radiation is that of a black body at 2.9 K. This figure is perhaps the most clear-cut evidence for the Big Bang

FIGURE 30.15 Robert W. Wilson *(left)* and Arno A. Penzias *(right)* with Bell Telephone Laboratories' horn-reflector antenna. *(AT&T Bell Laboratories)*

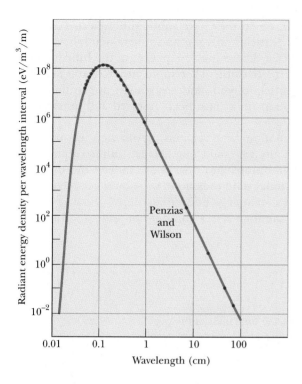

FIGURE 30.16 Theoretical blackbody (red curve) and measured radiation spectra (black points) of the Big Bang. Most of the data were collected from the Cosmic Background Explorer (COBE) satellite. The data of Wilson and Penzias are indicated.

theory. The 1978 Nobel prize in physics was awarded to Penzias and Wilson for this most important discovery.

The discovery of the cosmic background radiation produced a problem, however—the radiation was too uniform. Scientists believed there had to be slight fluctuations in this background in order for such objects as galaxies to form. In 1989 NASA launched a satellite called COBE (KOH-bee), for Cosmic Background Explorer, to study this radiation in greater detail. In 1992 George Smoot (b. 1945) at the Lawrence Berkeley Laboratory, based on the data collected, found that the background was not perfectly uniform but instead contained irregularities corresponding to temperature variations of 0.000 3 K. It is these small variations that provided nucleation sites for the formation of the galaxies and other objects we now see in the sky.

30.16 PROBLEMS AND PERSPECTIVES

While particle physicists have been exploring the realm of the very small, cosmologists have been exploring cosmic history back to the first microsecond of the Big Bang. Observation of the events that occur when two particles collide in an accelerator is essential in reconstructing the early moments in cosmic history. Perhaps the key to understanding the early Universe is to first understand the world of elementary particles. Cosmologists and particle physicists find that they have many common goals and are now joining efforts to attempt to study the physical world at its most fundamental level.

Our understanding of physics at short and long distances is far from complete. Particle physics is faced with many questions. Why is there so little antimatter in the Universe? Do neutrinos have a small mass, and if so, how much do they contribute to the "dark matter" holding the Universe together gravitationally? How can we understand the latest astronomical measurements, which show that the expansion of the Universe is accelerating and that there may be a kind of "antigravity force" acting between widely separated galaxies? Is it possible to unify the strong and electroweak theories in a logical and consistent manner? Why do quarks and leptons form three similar but distinct families? Are muons the same as electrons (apart from their different masses), or do they have subtle differences that have not been detected? Why are some particles charged and others neutral? Why do quarks carry a fractional charge? What determines the masses of the fundamental particles? The questions go on and on. Because of the rapid advances and new discoveries in the related fields of particle physics and cosmology, by the time you read this book some of these questions may have been resolved and others may have emerged.

An important and obvious question that remains is whether leptons and quarks have a substructure. If they do, we could envision an infinite number of deeper structure levels. However, if leptons and quarks are indeed the ultimate constituents of matter, as physicists today tend to believe, we should be able to construct a final theory of the structure of matter as Einstein dreamed of doing. In the view of many physicists, the end of the road is in sight, but how long it will take to reach that goal is anyone's guess.

SUMMARY

In **nuclear fission** and **nuclear fusion,** the total mass of the products is always less than the original mass of the reactants. Nuclear fission occurs when a heavy nucleus splits, or fissions, into two smaller nuclei. In nuclear fusion, two light nuclei combine to form a heavier nucleus.

A **nuclear reactor** is a system designed to maintain a self-sustaining chain reaction. Nuclear reactors using controlled fission events are currently being used to generate electric power.

Controlled fusion events offer the hope of plentiful supplies of energy in the future. The nuclear fusion reactor is considered by many scientists to be the ultimate energy source because its fuel is water. **Lawson's criterion** states that a fusion reactor will provide a net output power if the product of the plasma ion density, n, and the plasma confinement time, τ, satisfies the following relationships:

$$n\tau \geq 10^{14} \text{ s/cm}^3 \qquad \text{Deuterium-tritium interaction}$$

$$n\tau \geq 10^{16} \text{ s/cm}^3 \qquad \text{Deuterium-deuterium interaction}$$

[30.5]

Four fundamental forces in nature: **strong** (hadronic), **electromagnetic, weak,** and **gravitational.** The strong force is the force between nucleons that keeps the nucleus together. The weak force is responsible for beta decay. The electromagnetic and weak forces are now considered to be manifestations of a single force called the **electroweak** force.

Every fundamental interaction is said to be mediated by the exchange of field particles. The electromagnetic interaction is mediated by the photon; the weak interaction is mediated by the $W^\pm$ and Z^0 bosons; the gravitational interaction is mediated by gravitons; and the strong interaction is mediated by gluons.

An antiparticle and a particle have the same mass but opposite charge, and other properties may also have opposite values, such as lepton number and baryon number. It is possible to produce particle-antiparticle pairs in nuclear reactions if the available energy is greater than $2mc^2$, where m is the mass of the particle (or antiparticle).

Particles other than photons are classified as hadrons or leptons. **Hadrons** interact primarily through the strong force. They have size and structure and hence are not elementary particles. There are two types of hadrons, *baryons* and *mesons*. Mesons have a baryon number of 0 and have either zero or integer spin. Baryons, which generally are the most massive particles, have nonzero baryon numbers and spins of $\frac{1}{2}$ or $\frac{3}{2}$. The neutron and proton are examples of baryons.

Leptons have no structure or size and are considered truly elementary particles. Leptons interact only through the weak and electromagnetic forces. There are six leptons: the electron, e^-; the muon, μ^-; the tau, τ^-; and their associated neutrinos, ν_e, ν_μ, and ν_τ.

In all reactions and decays, quantities such as energy, linear momentum, angular momentum, electric charge, baryon number, and lepton number are strictly conserved. Certain particles have properties called **strangeness** and **charm.** These unusual properties are conserved only in those reactions and decays that occur via the strong force.

Recent theories postulate that all hadrons are composed of smaller units known as **quarks,** which have fractional electric charges and baryon numbers of $\frac{1}{3}$ and come in six "flavors": up, down, strange, charmed, top, and bottom. Each baryon contains three quarks, and each meson contains one quark and one antiquark.

According to the theory of **quantum chromodynamics,** quarks have a property called **color,** and the strong force between quarks is referred to as the **color force.**

Observation of background microwave radiation by Penzias and Wilson strongly confirmed that the Universe started with a Big Bang about 15 billion years ago. The background radiation is equivalent to that of a blackbody at a temperature of about 3 K.

CONCEPTUAL QUESTIONS

1. If high-energy electrons with the de Broglie wavelengths smaller than the size of the nucleus are scattered from nuclei, the behavior of the electrons is consistent with scattering from very dense structures much smaller in size than the nucleus—quarks. How is this similar to another classic experiment that detected small structures in atoms?

2. What factors make a fusion reaction difficult to achieve?

3. Doubly charged baryons are known to exist. Why are there no doubly charged mesons?

4. Why would a fusion reactor produce less radioactive waste than a fission reactor?

5. Atoms did not exist until hundreds of thousands of years after the Big Bang. Why?

6. Particles known as resonances have very short half-lives, of the order of 10^{-23} s. Would you guess they are hadrons or leptons?

7. Describe the quark model of hadrons, including the properties of quarks.

8. In the theory of quantum chromodynamics, quarks come in three colors. How would you justify the statement that "all baryons and mesons are colorless"?

9. Describe the properties of baryons and mesons and the important differences between them.

10. Identify the particle decays in Table 30.2 that occur by the electromagnetic interaction. Justify your answer.

11. Kaons all decay into final states that contain no protons or neutrons. What is the baryon number of kaons?

12. When an electron and a positron meet at low speeds in free space, why are *two* 0.511-MeV gamma rays produced, rather than *one* gamma ray with an energy of 1.02 MeV?

13. Two protons in a nucleus interact via the strong interaction. Are they also subject to the weak interaction?

14. Why is a neutron stable inside the nucleus? (In free space, the neutron decays in 900 s.)

15. An antibaryon interacts with a meson. Can a baryon be produced in such an interaction? Explain.

16. Why is water a better shield against neutrons than lead or steel?

17. How many quarks are there in (a) a baryon, (b) an antibaryon, (c) a meson, (d) an antimeson? How do you account for the fact that baryons have half-integral spins and mesons have spins of 0 or 1? (*Hint:* Quarks have spin $\frac{1}{2}$.)

18. Which of the four fundamental forces have effects on an electron? On a proton? On a neutron?

19. List similarities and differences between nuclear-powered electric generating stations and coal-powered ones.

PROBLEMS

1, 2, 3 = straightforward, intermediate, challenging ☐ = full solution available in Student Solutions Manual/Study Guide

web = solution posted at **http://info.brookscole.com/serway** 🔬 = biomedical application

Section 30.1 Nuclear Fission

Section 30.2 Nuclear Reactors

1. When ^{235}U absorbs a neutron, one possible fission reaction produces ^{141}Ba and ^{92}Kr as products. Write down this reaction. How many neutrons are released?

2. Find the energy released in the fission reaction

$$n + {}^{235}_{92}U \longrightarrow {}^{98}_{40}Zr + {}^{135}_{52}Te + 3\,n$$

The atomic masses of the fission products are: $^{98}_{40}$Zr, 97.912 0 u; $^{135}_{52}$Te, 134.908 7 u.

3. Find the energy released in the following fission reaction:

$$^{1}_{0}n + {}^{235}_{92}U \longrightarrow {}^{88}_{38}Sr + {}^{136}_{54}Xe + 12\,{}^{1}_{0}n$$

4. Strontium-90 is a particularly dangerous fission product of ^{235}U because it is radioactive and it substitutes for calcium in bones. What other direct fission products would accompany it in the neutron-induced fission of ^{235}U? (*Note:* This reaction may release two, three, or four free neutrons.)

5. Assume that ordinary soil contains natural uranium in amounts of 1 part per million by mass. (a) How much uranium is in the top 1.00 meter of soil on a one acre (43 560 ft^2) plot of ground, assuming the specific gravity of soil is 4.00? (b) How much of the isotope ^{235}U, appropriate for nuclear reactor fuel, is in this soil? (*Hint:* See Appendix B for the percent abundance of $^{235}_{92}$U.)

6. A typical nuclear fission power plant produces about 1.00 GW of electrical power. Assume that the plant has an

overall efficiency of 40.0% and that each fission produces 200 MeV of thermal energy. Calculate the mass of ^{235}U consumed each day.

7. Suppose that the water exerts an average frictional drag of
web 1.0×10^5 N on a nuclear-powered ship. How far can the ship travel per kilogram of fuel if the fuel consists of enriched uranium containing 1.7% of the fissionable isotope ^{235}U, and the ship's engine has a efficiency of 20%? (Assume 208 MeV released per fission event.)

8. The first atomic bomb released an energy equivalent to 20 kilotons of TNT. If 1 ton of TNT releases about 4.0×10^9 J, how much uranium was fissioned in this bomb? (Assume 208 MeV released per fission.)

9. An all-electric home uses approximately 2 000 kWh of electric energy per month. How much uranium-235 would be required to provide this house with its energy needs for 1 year? (Assume 100% conversion efficiency and 208 MeV released per fission.)

Section 30.3 Nuclear Fusion

10. Find the energy released in the fusion reaction

$$^1_1H + ^2_1H \longrightarrow ^3_2He + \gamma$$

11. When a star has exhausted its hydrogen fuel, it may fuse other nuclear fuels. At temperatures above 1.0×10^8 K, helium fusion can occur. Write the equation for the processes described below. (a) Two alpha particles fuse to produce a nucleus A and a gamma ray. What is nucleus A? (b) Nucleus A absorbs an alpha particle to produce a nucleus B and a gamma ray. What is nucleus B? (c) Find the total energy released in the sequence of reactions given in (a) and (b).

12. Another series of nuclear reactions that can produce energy in the interior of stars is the cycle described below. This process is most efficient when the central temperature in a star is above 1.6×10^7 K. Because the temperature at the center of the Sun is only 1.5×10^7 K, the cycle below produces less than 10% of the Sun's energy. (a) A high-energy proton is absorbed by ^{12}C. Another nucleus, A, is produced in the reaction, along with a gamma ray. Identify nucleus A. (b) Nucleus A decays through positron emission to form nucleus B. Identify nucleus B. (c) Nucleus B absorbs a proton to produce nucleus C and a gamma ray. Identify nucleus C. (d) Nucleus C absorbs a proton to produce nucleus D and a gamma ray. Identify nucleus D. (e) Nucleus D decays through positron emission to produce nucleus E. Identify nucleus E. (f) Nucleus E absorbs a proton to produce nucleus F plus an alpha particle. What is nucleus F? (*Note:* If nucleus F is not ^{12}C, the nucleus you started with, you have made an error and should review the sequence of events.)

13. If an all-electric home uses approximately 2 000 kWh of electric energy per month, how many fusion events described by the reaction $^2_1H + ^3_1H \rightarrow ^4_2He + ^1_0n$ would be required to keep this house running for one year?

14. To understand why plasma containment is necessary, consider the rate at which an unconfined plasma would be lost. (a) Estimate the rms speed of deuterons in a plasma at 4.00×10^8 K. (b) Estimate the order of magnitude of the time such a plasma would remain in a 10-cm cube if no steps were taken to contain it.

15. The oceans have a volume of 317 million cubic miles and contain 1.32×10^{21} kg of water. Of all the hydrogen nuclei in this water, 0.0300% of the mass is deuterium. (a) If all of these deuterium nuclei were fused to helium via the first reaction in Equation 30.4, determine the total amount of energy that could be released. (b) Present world electric power consumption is about 7.00×10^{12} W. If consumption were 100 times greater, how many years would the energy supply calculated in (a) last?

Section 30.6 Positrons and Other Antiparticles

16. Two photons are produced when a proton and an antiproton annihilate each other. What is the minimum frequency and corresponding wavelength of each photon?

17. A photon produces a proton-antiproton pair according to
web the reaction $\gamma \rightarrow p + \bar{p}$. What is the minimum possible frequency of the photon? What is its wavelength?

18. A photon with an energy of 2.09 GeV creates a proton-antiproton pair in which the proton has a kinetic energy of 95.0 MeV. What is the kinetic energy of the antiproton?

Section 30.7 Mesons and the Beginning of Particle Physics

19. When a high-energy proton or pion traveling near the speed of light collides with a nucleus, it travels an average distance of 3.0×10^{-15} m before interacting. From this information, estimate the order of magnitude of the time for the strong interaction to occur.

20. Calculate the order of magnitude of the range of the force that might be produced by the virtual exchange of a proton.

21. One of the mediators of the weak interaction is the Z^0 boson, whose mass is 93 GeV/c^2. Use this information to find the order of magnitude of the range of the weak interaction.

22. If a π^0 at rest decays into two γ's, what is the energy of each of the γ's?

Section 30.9 Conservation Laws

Section 30.10 Strange Particles and Strangeness

23. Each of the following reactions is forbidden. Determine a conservation law that is violated for each reaction.
(a) $p + \bar{p} \rightarrow \mu^+ + e^-$ (d) $p + p \rightarrow p + p + n$
(b) $\pi^- + p \rightarrow p + \pi^+$ (e) $\gamma + p \rightarrow n + \pi^0$
(c) $p + p \rightarrow p + \pi^+$

24. (a) Show that baryon number and charge are conserved in the following reactions of a pion with a proton.

$$\pi^+ + p \longrightarrow K^+ + \Sigma^+ \qquad (1)$$

$$\pi^+ + p \longrightarrow \pi^+ + \Sigma^+ \qquad (2)$$

(b) The first reaction is observed, but the second never occurs. Explain these observations.

25. Identify the unknown particle on the left side of the
web reaction

$$? + p \longrightarrow n + \mu^+$$

26. Determine the type of neutrino or antineutrino involved in each of the following processes:
(a) $\pi^+ \rightarrow \pi^0 + e^+ + ?$ (c) $\Lambda^0 \rightarrow p + \mu^- + ?$
(b) $? + p \rightarrow \mu^- + p + \pi^+$ (d) $\tau^+ \rightarrow \mu^+ + ? + ?$

27. The following reactions or decays involve one or more neutrinos. Supply the missing neutrinos.
    (a) $\pi^- \rightarrow \mu^- + ?$ (d) $? + n \rightarrow p + e^-$
    (b) $K^+ \rightarrow \mu^+ + ?$ (e) $? + n \rightarrow p + \mu^-$
    (c) $? + p \rightarrow n + e^+$ (f) $\mu^- \rightarrow e^- + ? + ?$

28. Determine which of the reactions below can occur. For those that cannot occur, determine the conservation law (or laws) that each violates.
    (a) $p \rightarrow \pi^+ + \pi^0$ (d) $\pi^+ \rightarrow \mu^+ + \nu_\mu$
    (b) $p + p \rightarrow p + p + \pi^0$ (e) $n \rightarrow p + e^- + \overline{\nu}_e$
    (c) $p + p \rightarrow p + \pi^+$ (f) $\pi^+ \rightarrow \pi^+ + n$

29. Which of the following processes are allowed by the strong interaction, the electromagnetic interaction, the weak interaction, or no interaction at all?
    (a) $\pi^- + p \rightarrow 2\eta^0$ (d) $\Omega^- \rightarrow \Xi^- + \pi^0$
    (b) $K^- + n \rightarrow \Lambda^0 + \pi^-$ (e) $\eta^0 \rightarrow 2\gamma$
    (c) $K^- \rightarrow \pi^- + \pi^0$

30. A K^0 particle at rest decays into a π^+ and a π^-. What will be the speed of each of the pions? The mass of the K^0 is 497.7 MeV/c^2 and the mass of each pion is 139.6 MeV/c^2.

31. Determine whether or not strangeness is conserved in the following decays and reactions.
    (a) $\Lambda^0 \rightarrow p + \pi^-$ (d) $\pi^- + p \rightarrow \pi^- + \Sigma^+$
    (b) $\pi^- + p \rightarrow \Lambda^0 + K^0$ (e) $\Xi^- \rightarrow \Lambda^0 + \pi^-$
    (c) $\overline{p} + p \rightarrow \overline{\Lambda}^0 + \Lambda^0$ (f) $\Xi^0 \rightarrow p + \pi^-$

32. Fill in the missing particle. Assume that (a) occurs via the strong interaction and that (b) and (c) involve the weak interaction.
    (a) $K^+ + p \rightarrow \underline{} + p$
    (b) $\Omega^- \rightarrow \underline{} + \pi^-$
    (c) $K^+ \rightarrow \underline{} + \mu^+ + \nu_\mu$

33. Identify the conserved quantities in the following processes:
    (a) $\Xi^- \rightarrow \Lambda^0 + \mu^- + \nu_\mu$ (d) $\Sigma^0 \rightarrow \Lambda^0 + \gamma$
    (b) $K^0 \rightarrow 2\pi^0$ (e) $e^+ + e^- \rightarrow \mu^+ + \mu^-$
    (c) $K^- + p \rightarrow \Sigma^0 + n$ (f) $\overline{p} + n \rightarrow \overline{\Lambda}^0 + \Sigma^-$

Section 30.12   Quarks

Section 30.13   Colored Quarks

34. The quark composition of the proton is uud, whereas that of the neutron in udd. Show that the charge, baryon number, and strangeness of these particles equal the sums of these numbers for their quark constituents.

35. Find the number of electrons, and of each species of quarks, in 1 liter of water.

36. The quark compositions of the K^0 and Λ^0 particles are d$\overline{s}$ and uds, respectively. Show that the charge, baryon number, and strangeness of these particles equal the sums of these numbers for the quark constituents.

37. Identify the particles corresponding to the following quark states: (a) suu; (b) $\overline{u}$d; (c) $\overline{s}$d; (d) ssd.

38. What is the electrical charge of the baryons with the quark compositions (a) $\overline{u}\overline{u}\overline{d}$ and (b) $\overline{u}\overline{d}\overline{d}$? What are these baryons called?

39. Analyze the first three of the following reactions at the quark level and show that each conserves the net number of each type quark. In the last reaction, identify the mystery particle.
    (a) $\pi^- + p \rightarrow K^0 + \Lambda^0$
    (b) $\pi^+ + p \rightarrow K^+ + \Sigma^+$

(c) $K^- + p \rightarrow K^+ + K^0 + \Omega^-$
(d) $p + p \rightarrow K^0 + p + \pi^+ + ?$

40. Imagine binding energies could be neglected. Find the masses of the u and d quarks from the masses of the proton and neutron.

ADDITIONAL PROBLEMS

41. A Σ^0 particle traveling through matter strikes a proton. A Σ^+ and a gamma ray emerge, as well as a third particle. Use the quark model of each to determine the identity of the third particle.

42. It was stated in the text that the reaction $\pi^- + p^+ \rightarrow K^0 + \Lambda^0$ occurs with high probability, whereas the reaction $\pi^- + p^+ \rightarrow K^0 + n$ never occurs. Analyze these reactions at the quark level and show that the first conserves the net number of each type of quark while the second reaction does not.

43. Two protons approach each other with equal and opposite velocities. What is the minimum kinetic energy of each of the protons if they are to produce a π^+ meson at rest in the reaction,

$$p + p \longrightarrow p + n + \pi^+$$

44. Name at least one conservation law that prevents each of the following reactions:
    (a) $\pi^- + p \rightarrow \Sigma^+ + \pi^0$
    (b) $\mu^- \rightarrow \pi^- + \nu_e$
    (c) $p \rightarrow \pi^+ + \pi^+ + \pi^-$

45. Find the energy released in the fusion reaction

$$\tfrac{1}{1}H + \tfrac{3}{2}He \longrightarrow \tfrac{4}{2}He + e^+ + \nu$$

46. Occasionally, high-energy muons collide with electrons and produce two neutrinos according to the reaction $\mu^+ + e^- \rightarrow 2\nu$. What kind of neutrinos are these?

47. Each of the following decays is forbidden. For each process, determine a conservation law that is violated.
    (a) $\mu^- \rightarrow e^- + \gamma$ (d) $p \rightarrow e^+ + \pi^0$
    (b) $n \rightarrow p + e^- + \nu_e$ (e) $\Xi^0 \rightarrow n + \pi^0$
    (c) $\Lambda^0 \rightarrow p + \pi^0$

48. Two protons approach each other head on, each with 70.4 MeV of kinetic energy, and engage in a reaction in which a proton and positive pion emerge at rest. What third particle, obviously uncharged and therefore difficult to detect, must have been created?

49. The atomic bomb dropped on Hiroshima on August 6, 1945, released 5×10^{13} J of energy (equivalent to that from 12 000 tons of TNT). Estimate (a) the number of $^{235}_{92}$U nuclei fissioned, and (b) the mass of this $^{235}_{92}$U.

50. A 2.0-MeV neutron is emitted in a fission reactor. If it loses one half of its kinetic energy in each collision with a moderator atom, how many collisions must it undergo in order to achieve thermal energy (0.039 eV)?

51. If baryon number were not conserved, then one possible mechanism by which a proton could decay would be

$$p \longrightarrow e^+ + \gamma$$

(a) Show that this reaction violates conservation of baryon number. (b) Assuming that this reaction occurs, and that the proton is initially at rest, determine the energy and momentum of the photon after the reaction. (*Hint:* Recall that energy and momentum must be conserved in the reaction.) (c) Determine the speed of the positron after the reaction.

52. Classical general relativity views the space-time manifold as a deterministic structure completely well defined down to arbitrarily small distances. On the other hand, quantum general relativity forbids distances smaller than the Planck length given by $L = (\hbar G/c^3)^{1/2}$. (a) Calculate the value of L. The answer suggests that after the Big Bang (when all the known Universe was reduced to a singularity), nothing could be observed until that singularity grew larger than the Planck length, L. Because the size of the singularity grew at the speed of light, we can infer that during the time it took for light to travel the Planck length, no observations were possible. (b) Determine this time (known as the Planck time, T) and compare it to the ultra-hot epoch discussed in the text. (c) Does this suggest that we may never know what happened between the time $t = 0$ and the time $t = T$?

GROUP ACTIVITIES

G.1 (a) A typical chemical reaction is one in which a water molecule is formed by combining hydrogen and oxygen. In such a reaction about 2.5 eV of energy is released. Compare this reaction to $^1_0n + ^{235}_{92}U \rightarrow ^{136}_{53}I + ^{98}_{39}Y + 2\,^1_0n$ (Take the mass of ^{136}I to be 135.914 650 u and the mass of ^{98}Y to be 97.922 300 u.) Would you expect the energy released in this reaction to be much greater, much less, or about the same as the chemical reaction? (b) Find the actual energy released in this reaction.

G.2 (a) Explain in a couple of sentences what is meant by the conservation of baryons. (b) Use conservation of baryons to show that a proton cannot decay into a positron and a neutral pion.

G.3 (a) Explain in a couple of sentences what is meant by conservation of strangeness. (b) Use conservation of strangeness to show that a reaction in which a negative pion strikes a proton producing a neutral kaon and a neutron cannot occur.

"QUARKS. NEUTRINOS. MESONS. ALL THOSE DAMN PARTICLES YOU CAN'T SEE. THAT'S WHAT DROVE ME TO DRINK. BUT NOW I CAN SEE THEM."

Mathematical Review

A.1 MATHEMATICAL NOTATION

Many mathematical symbols are used throughout this book. You are no doubt familiar with some, such as the symbol $=$ to denote the equality of two quantities.

The symbol $\propto$ denotes a proportionality. For example, $y \propto x^2$ means that y is proportional to the square of x.

The symbol $<$ means *is less than*, and $>$ means *is greater than*. For example, $x > y$ means x is greater than y.

The symbol $\ll$ means *is much less than*, and $\gg$ means *is much greater than*.

The symbol $\approx$ indicates that two quantities are *approximately equal* to each other.

The symbol $\equiv$ means *is defined as*. This is a stronger statement than a simple $=$.

It is convenient to use the notation Δx (read as "delta x") to indicate the *change in the quantity x*. (Note that Δx does not mean "the product of Δ and x.") For example, suppose that a person out for a morning stroll starts measuring her distance away from home when she is 10 m from her doorway. She then moves along a straight-line path and stops strolling 50 m from the door. Her change in position during the walk is $\Delta x = 50 \text{ m} - 10 \text{ m} = 40 \text{ m}$ or, in symbolic form,

$$\Delta x = x_f - x_i$$

In this equation x_f is the *final position* and x_i is the *initial position*.

We often have occasion to add several quantities. A useful abbreviation for representing such a sum is the Greek letter Σ (capital sigma). Suppose we wish to add a set of five numbers represented by x_1, x_2, x_3, x_4, and x_5. In the abbreviated notation, we would write the sum as

$$x_1 + x_2 + x_3 + x_4 + x_5 = \sum_{i=1}^{5} x_i$$

where the subscript i on x represents any one of the numbers in the set. For example, if there are five masses in a system, m_1, m_2, m_3, m_4, and m_5, the total mass of the system $M = m_1 + m_2 + m_3 + m_4 + m_5$ could be expressed as

$$M = \sum_{i=1}^{5} m_i$$

Finally, the magnitude of a quantity x, written $|x|$, is simply the absolute value of that quantity. The sign of $|x|$ is always positive, regardless of the sign of x. For example, if $x = -5$, $|x| = 5$; if $x = 8$, $|x| = 8$.

A.2 SCIENTIFIC NOTATION

Many quantities that scientists deal with often have very large or very small values. For example, the speed of light is about 300 000 000 m/s and the ink required to make the dot over an i in this textbook has a mass of about 0.000 000 001 kg. Obviously, it is cumbersome to read, write, and keep track of numbers such as these. We avoid this problem by using a method dealing with powers of the number 10:

$$10^0 = 1$$

$$10^1 = 10$$

$$10^2 = 10 \times 10 = 100$$

$$10^3 = 10 \times 10 \times 10 = 1\,000$$

$$10^4 = 10 \times 10 \times 10 \times 10 = 10\,000$$

$$10^5 = 10 \times 10 \times 10 \times 10 \times 10 = 100\,000$$

and so on. The number of zeros corresponds to the power to which 10 is raised, called the **exponent** of 10. For example, the speed of light, 300 000 000 m/s, can be expressed as 3×10^8 m/s.

For numbers less than one, we note the following:

$$10^{-1} = \frac{1}{10} = 0.1$$

$$10^{-2} = \frac{1}{10 \times 10} = 0.01$$

$$10^{-3} = \frac{1}{10 \times 10 \times 10} = 0.001$$

$$10^{-4} = \frac{1}{10 \times 10 \times 10 \times 10} = 0.000\,1$$

$$10^{-5} = \frac{1}{10 \times 10 \times 10 \times 10 \times 10} = 0.000\,01$$

In these cases, the number of places the decimal point is to the left of the digit 1 equals the value of the (negative) exponent. Numbers that are expressed as some power of 10 multiplied by another number between 1 and 10 are said to be in **scientific notation.** For example, the scientific notation for 5 943 000 000 is 5.943×10^9 and that for 0.000 083 2 is 8.32×10^{-5}.

When numbers expressed in scientific notation are being multiplied, the following general rule is very useful:

$$10^n \times 10^m = 10^{n+m} \qquad \text{[A.1]}$$

where n and m can be *any* numbers (not necessarily integers). For example, $10^2 \times 10^5 = 10^7$. The rule also applies if one of the exponents is negative. For example, $10^3 \times 10^{-8} = 10^{-5}$.

When dividing numbers expressed in scientific notation, note that

$$\frac{10^n}{10^m} = 10^n \times 10^{-m} = 10^{n-m} \qquad \text{[A.2]}$$

Exercises

With help from the above rules, verify the answers to the following:

1. $86\,400 = 8.64 \times 10^4$
2. $9\,816\,762.5 = 9.816\,762\,5 \times 10^6$
3. $0.000\,000\,039\,8 = 3.98 \times 10^{-8}$
4. $(4.0 \times 10^8)(9.0 \times 10^9) = 3.6 \times 10^{18}$
5. $(3.0 \times 10^7)(6.0 \times 10^{-12}) = 1.8 \times 10^{-4}$
6. $\dfrac{75 \times 10^{-11}}{5.0 \times 10^{-3}} = 1.5 \times 10^{-7}$
7. $\dfrac{(3 \times 10^6)(8 \times 10^{-2})}{(2 \times 10^{17})(6 \times 10^5)} = 2 \times 10^{-18}$

A.3 ALGEBRA

A. SOME BASIC RULES

When algebraic operations are performed, the laws of arithmetic apply. Symbols such as x, y, and z are frequently used to represent quantities that are not specified, what are called the **unknowns.**

First, consider the equation

$$8x = 32$$

If we wish to solve for x, we can divide (or multiply) each side of the equation by the same factor without destroying the equality. In this case, if we divide both sides by 8, we have

$$\frac{8x}{8} = \frac{32}{8}$$

$$x = 4$$

Next consider the equation

$$x + 2 = 8$$

In this type of expression, we can add or subtract the same quantity from each side. If we subtract 2 from each side, we get

$$x + 2 - 2 = 8 - 2$$

$$x = 6$$

In general, if $x + a = b$, then $x = b - a$.

Now consider the equation

$$\frac{x}{5} = 9$$

If we multiply each side by 5, we are left with x on the left by itself and 45 on the right:

$$\left(\frac{x}{5}\right)(5) = 9 \times 5$$

$$x = 45$$

In all cases, **whatever operation is performed on the left side of the equality must also be performed on the right side.**

The following rules for multiplying, dividing, adding, and subtracting fractions should be recalled, where a, b, and c are three numbers:

	Rule	**Example**
Multiplying	$\left(\dfrac{a}{b}\right)\left(\dfrac{c}{d}\right) = \dfrac{ac}{bd}$	$\left(\dfrac{2}{3}\right)\left(\dfrac{4}{5}\right) = \dfrac{8}{15}$
Dividing	$\dfrac{(a/b)}{(c/d)} = \dfrac{ad}{bc}$	$\dfrac{2/3}{4/5} = \dfrac{(2)(5)}{(4)(3)} = \dfrac{10}{12}$
Adding	$\dfrac{a}{b} \pm \dfrac{c}{d} = \dfrac{ad \pm bc}{bd}$	$\dfrac{2}{3} - \dfrac{4}{5} = \dfrac{(2)(5) - (4)(3)}{(3)(5)} = -\dfrac{2}{15}$

Exercises

In the following exercises, solve for x:

Answers

1. $a = \dfrac{1}{1 + x}$ $x = \dfrac{1 - a}{a}$

2. $3x - 5 = 13$ $x = 6$

3. $ax - 5 = bx + 2$ $x = \dfrac{7}{a - b}$

4. $\dfrac{5}{2x + 6} = \dfrac{3}{4x + 8}$ $x = -\dfrac{11}{7}$

B. Powers

When powers of a given quantity x are multiplied, the following rule applies:

$$x^n x^m = x^{n+m} \tag{A.3}$$

For example, $x^2 x^4 = x^{2+4} = x^6$.

When dividing the powers of a given quantity, note that

$$\frac{x^n}{x^m} = x^{n-m} \tag{A.4}$$

For example, $x^8/x^2 = x^{8-2} = x^6$.

A power that is a fraction, such as $\frac{1}{3}$, corresponds to a root as follows:

$$x^{1/n} = \sqrt[n]{x} \tag{A.5}$$

For example, $4^{1/3} = \sqrt[3]{4} = 1.587\,4$. (A scientific calculator is useful for such calculations.)

Finally, any quantity x^n that is raised to the mth power is

$$(x^n)^m = x^{nm} \tag{A.6}$$

Table A.1 summarizes the rules of exponents.

TABLE A.1

Rules of Exponents

$$x^0 = 1$$
$$x^1 = x$$
$$x^n x^m = x^{n+m}$$
$$x^n/x^m = x^{n-m}$$
$$x^{1/n} = \sqrt[n]{x}$$
$$(x^n)^m = x^{nm}$$

Exercises

Verify the following:

1. $3^2 \times 3^3 = 243$
2. $x^5 x^{-8} = x^{-3}$
3. $x^{10}/x^{-5} = x^{15}$
4. $5^{1/3} = 1.709\,975$ (Use your calculator.)
5. $60^{1/4} = 2.783\,158$ (Use your calculator.)
6. $(x^4)^3 = x^{12}$

C. Factoring

Some useful formulas for factoring an equation are

$$ax + ay + az = a(x + y + z) \qquad \text{common factor}$$
$$a^2 + 2ab + b^2 = (a + b)^2 \qquad \text{perfect square}$$
$$a^2 - b^2 = (a + b)(a - b) \qquad \text{differences of squares}$$

D. Quadratic Equations

The general form of a quadratic equation is

$$ax^2 + bx + c = 0 \tag{A.7}$$

where x is the unknown quantity and a, b, and c are numerical factors referred to as **coefficients** of the equation. This equation has two roots, given by

$$x = \frac{-b \pm \sqrt{b^2 - 4ac}}{2a} \qquad \text{[A.8]}$$

If $b^2 \geq 4ac$, the roots will be real.

Example

The equation $x^2 + 5x + 4 = 0$ has the following roots corresponding to the two signs of the square root term:

$$x = \frac{-5 \pm \sqrt{5^2 - (4)(1)(4)}}{2(1)} = \frac{-5 \pm \sqrt{9}}{2} = \frac{-5 \pm 3}{2}$$

that is,

$$x_+ = \frac{-5 + 3}{2} = \boxed{-1} \qquad x_- = \frac{-5 - 3}{2} = \boxed{-4}$$

where x_+ refers to the root corresponding to the positive sign and x_- refers to the root corresponding to the negative sign.

Exercises

Solve the following quadratic equations:

Answers

1.	$x^2 + 2x - 3 = 0$	$x_+ = 1$	$x_- = -3$
2.	$2x^2 - 5x + 2 = 0$	$x_+ = 2$	$x_- = 1/2$
3.	$2x^2 - 4x - 9 = 0$	$x_+ = 1 + \sqrt{22}/2$	$x_- = 1 - \sqrt{22}/2$

E. LINEAR EQUATIONS

A linear equation has the general form

$$y = ax + b \qquad \text{[A.9]}$$

where a and b are constants. This equation is referred to as being linear because the graph of y versus x is a straight line, as shown in Figure A.1. The constant b, called the **intercept,** represents the value of y at which the straight line intersects the y axis. The constant a is equal to the **slope** of the straight line. If any two points on the straight line are specified by the coordinates (x_1, y_1) and (x_2, y_2), as in Figure A.1, then the slope of the straight line can be expressed

$$\text{Slope} = \frac{y_2 - y_1}{x_2 - x_1} = \frac{\Delta y}{\Delta x} \qquad \text{[A.10]}$$

Note that a and b can have either positive or negative values. If $a > 0$, the straight line has a positive slope, as in Figure A.1. If $a < 0$, the straight line has a *negative* slope. In Figure A.1, both a and b are positive. Three other possible situations are shown in Figure A.2: $a > 0$, $b < 0$; $a < 0$, $b > 0$; and $a < 0$, $b < 0$.

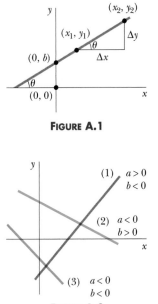

FIGURE A.1

FIGURE A.2

Exercises

1. Draw graphs of the following straight lines:
(a) $y = 5x + 3$ (b) $y = -2x + 4$ (c) $y = -3x - 6$
2. Find the slopes of the straight lines described in Exercise 1.
Answers: **(a)** 5 **(b)** -2 **(c)** -3

3. Find the slopes of the straight lines that pass through the following sets of points: **(a)** $(0, -4)$ and $(4, 2)$, **(b)** $(0, 0)$ and $(2, -5)$, and **(c)** $(-5, 2)$ and $(4, -2)$

Answers: **(a)** $\frac{3}{2}$ **(b)** $-\frac{5}{2}$ **(c)** $-\frac{4}{9}$

F. SOLVING SIMULTANEOUS LINEAR EQUATIONS

Consider an equation such as $3x + 5y = 15$, which has two unknowns, x and y. Such an equation does not have a unique solution. That is, $(x = 0, \; y = 3)$, $(x = 5, y = 0)$ and $(x = 2, y = \frac{9}{5})$ are all solutions to this equation.

If a problem has two unknowns, a unique solution is possible only if we have *two* independent equations. In general, if a problem has n unknowns, its solution requires n independent equations. In order to solve two simultaneous equations involving two unknowns, x and y, we solve one of the equations for x in terms of y and substitute this expression into the other equation.

Example

Solve the following two simultaneous equations:

$$\textbf{(1)} \quad 5x + y = -8 \qquad \textbf{(2)} \quad 2x - 2y = 4$$

Solution   From (2), we find that $x = y + 2$. Substitution of this into (1) gives

$$5(y + 2) + y = -8$$
$$6y = -18$$
$$\boxed{y = -3}$$
$$x = y + 2 = \boxed{-1}$$

Alternate Solution   Multiply each term in (1) by the factor 2 and add the result to (2):

$$10x + 2y = -16$$
$$\underline{2x - 2y = 4}$$
$$12x = -12$$
$$x = \boxed{-1}$$
$$y = x - 2 = \boxed{-3}$$

Two linear equations with two unknowns can also be solved by a graphical method. If the straight lines corresponding to the two equations are plotted in a conventional coordinate system, the intersection of the two lines represents the solution. For example, consider the two equations

$$x - y = 2$$
$$x - 2y = -1$$

These are plotted in Figure A.3. The intersection of the two lines has the coordinates $x = 5$, $y = 3$. This represents the solution to the equations. You should check this solution by the analytical technique discussed above.

Exercises

Solve the following pairs of simultaneous equations involving two unknowns:

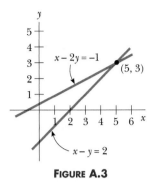

FIGURE A.3

Answers

1. $x + y = 8$ $x = 5, y = 3$
   $x - y = 2$
2. $98 - T = 10a$ $T = 65, a = 3.27$
   $T - 49 = 5a$
3. $6x + 2y = 6$ $x = 2, y = -3$
   $8x - 4y = 28$

G. LOGARITHMS

Suppose that a quantity x is expressed as a power of some quantity a:

$$x = a^y \qquad\qquad \text{[A.11]}$$

The number a is called the **base** number. The **logarithm** of x with respect to the base a is equal to the exponent to which the base must be raised in order to satisfy the expression $x = a^y$:

$$y = \log_a x \qquad\qquad \text{[A.12]}$$

Conversely, the **antilogarithm** of y is the number x:

$$x = \text{antilog}_a y \qquad\qquad \text{[A.13]}$$

In practice, the two bases most often used are base 10, called the *common* logarithm base, and base $e = 2.718 \ldots$, called the *natural* logarithm base. When common logarithms are used,

$$y = \log_{10} x \qquad (\text{or } x = 10^y) \qquad \text{[A.14]}$$

When natural logarithms are used,

$$y = \ln_e x \qquad (\text{or } x = e^y) \qquad \text{[A.15]}$$

For example, $\log_{10} 52 = 1.716$, so that $\text{antilog}_{10} 1.716 = 10^{1.716} = 52$. Likewise, $\ln_e 52 = 3.951$, so $\text{antiln}_e 3.951 = e^{3.951} = 52$.

In general, note that you can convert between base 10 and base e with the equality

$$\ln_e x = (2.302\ 585)\log_{10} x \qquad\qquad \text{[A.16]}$$

Finally, some useful properties of logarithms are

$$\log (ab) = \log a + \log b \qquad\qquad \ln e = 1$$

$$\log (a/b) = \log a - \log b \qquad\qquad \ln e^a = a$$

$$\log (a^n) = n \log a \qquad\qquad \ln \left(\frac{1}{a}\right) = -\ln a$$

A.4 GEOMETRY

Table A.2 gives the areas and volumes for several geometric shapes used throughout this text:

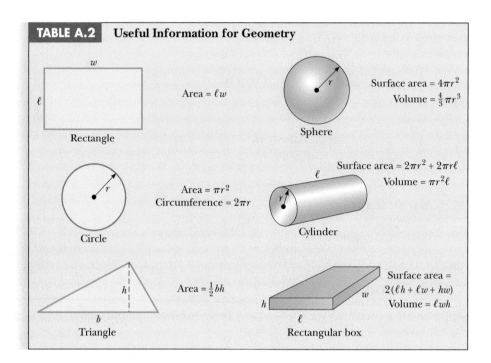

TABLE A.2	Useful Information for Geometry

Rectangle — Area $= \ell w$

Sphere — Surface area $= 4\pi r^2$, Volume $= \frac{4}{3}\pi r^3$

Circle — Area $= \pi r^2$, Circumference $= 2\pi r$

Cylinder — Surface area $= 2\pi r^2 + 2\pi r \ell$, Volume $= \pi r^2 \ell$

Triangle — Area $= \frac{1}{2} bh$

Rectangular box — Surface area $= 2(\ell h + \ell w + hw)$, Volume $= \ell w h$

A.5 TRIGONOMETRY

Some of the most basic facts concerning trigonometry are presented in Chapter 1, and we encourage you to study the material presented there if you are having trouble with this branch of mathematics. In addition to the discussion of Chapter 1, certain useful trig identities that can be of value to you follow.

$$\sin^2 \theta + \cos^2 \theta = 1$$

$$\sin \theta = \cos(90° - \theta)$$

$$\cos \theta = \sin(90° - \theta)$$

$$\sin 2\theta = 2 \sin \theta \cos \theta$$

$$\cos 2\theta = \cos^2 \theta - \sin^2 \theta$$

$$\sin(\theta \pm \phi) = \sin \theta \cos \phi \pm \cos \theta \sin \phi$$

$$\cos(\theta \pm \phi) = \cos \theta \cos \phi \mp \sin \theta \sin \phi$$

The following relationships apply to *any* triangle, as shown in Figure A.4:

$$\alpha + \beta + \gamma = 180°$$

$$a^2 = b^2 + c^2 - 2bc \cos \alpha$$

Law of cosines
$$b^2 = a^2 + c^2 - 2ac \cos \beta$$

$$c^2 = a^2 + b^2 - 2ab \cos \gamma$$

Law of sines
$$\frac{a}{\sin \alpha} = \frac{b}{\sin \beta} = \frac{c}{\sin \gamma}$$

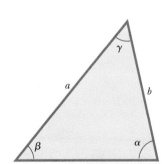

FIGURE A.4

An Abbreviated Table of Isotopes

Atomic Number, Z	Element	Symbol	Chemical Atomic Mass (u)	Mass Number (* Indicates Radioactive) A	Atomic Mass (u)[a]	Percentage Abundance	Half-Life (if Radioactive) $T_{1/2}$
0	(Neutron)	n		1*	1.008 665		10.4 min
1	Hydrogen	H	1.007 9	1	1.007 825	99.985	
	Deuterium	D		2	2.014 102	0.015	
	Tritium	T		3*	3.016 049		12.33 yr
2	Helium	He	4.002 60	3	3.016 029	0.000 14	
				4	4.002 602	99.999 86	
3	Lithium	Li	6.941	6	6.015 121	7.5	
				7	7.016 003	92.5	
4	Beryllium	Be	9.012 2	7*	7.016 928		53.3 days
				9	9.012 174	100	
5	Boron	B	10.81	10	10.012 936	19.9	
				11	11.009 305	80.1	
6	Carbon	C	12.011	11*	11.011 433		20.4 min
				12	12.000 000	98.90	
				13	13.003 355	1.10	
				14*	14.003 242		5730 yr
7	Nitrogen	N	14.006 7	13*	13.005 738		9.96 min
				14	14.003 074	99.63	
				15	15.000 108	0.37	
8	Oxygen	O	15.999 4	15*	15.003 065		122 s
				16	15.994 915	99.761	
				18	17.999 160	0.20	
9	Fluorine	F	18.998 40	19	18.998 404	100	
10	Neon	Ne	20.180	20	19.992 435	90.48	
				22	21.991 383	9.25	
11	Sodium	Na	22.989 87	22*	21.994 434		2.61 yr
				23	22.989 770	100	
				24*	23.990 961		14.96 h
12	Magnesium	Mg	24.305	24	23.985 042	78.99	
				25	24.985 838	10.00	
				26	25.982 594	11.01	
13	Aluminum	Al	26.981 54	27	26.981 538	100	
14	Silicon	Si	28.086	28	27.976 927	92.23	
15	Phosphorus	P	30.973 76	31	30.973 762	100	
				32*	31.973 908		14.26 days

(Table continues)

[a] The masses in the sixth column are atomic masses, which include the mass of Z electrons. Data are from the National Nuclear Data Center, Brookhaven National Laboratory, prepared by Jagdish K. Tuli, July 1990. The data are based on experimental results reported in *Nuclear Data Sheets* and *Nuclear Physics* and also from *Chart of the Nuclides,* 14th ed. Atomic masses are based on those by A. H. Wapstra, G. Audi, and R. Hoekstra. Isotopic abundances are based on those by N. E. Holden.

Atomic Number, Z	Element	Symbol	Chemical Atomic Mass (u)	Mass Number (* Indicates Radioactive) A	Atomic Mass (u)[a]	Percentage Abundance	Half-Life (if Radioactive) $T_{1/2}$
16	Sulfur	S	32.066	32	31.972 071	95.02	
				35*	34.969 033		87.5 days
17	Chlorine	Cl	35.453	35	34.968 853	75.77	
				37	36.965 903	24.23	
18	Argon	Ar	39.948	40	39.962 384	99.600	
19	Potassium	K	39.098 3	39	38.963 708	93.258 1	
				40*	39.964 000	0.0117	1.28×10^9 yr
20	Calcium	Ca	40.08	40	39.962 591	96.941	
21	Scandium	Sc	44.955 9	45	44.955 911	100	
22	Titanium	Ti	47.88	48	47.947 947	73.8	
23	Vanadium	V	50.941 5	51	50.943 962	99.75	
24	Chromium	Cr	51.996	52	51.940 511	83.79	
25	Manganese	Mn	54.938 05	55	54.938 048	100	
26	Iron	Fe	55.847	56	55.934 940	91.72	
27	Cobalt	Co	58.933 20	59	58.933 198	100	
				60*	59.933 820		5.27 yr
28	Nickel	Ni	58.693	58	57.935 346	68.077	
				60	59.930 789	26.223	
29	Copper	Cu	63.54	63	62.929 599	69.17	
				65	64.927 791	30.83	
30	Zinc	Zn	65.39	64	63.929 144	48.6	
				66	65.926 035	27.9	
				68	67.924 845	18.8	
31	Gallium	Ga	69.723	69	68.925 580	60.108	
				71	70.924 703	39.892	
32	Germanium	Ge	72.61	70	69.924 250	21.23	
				72	71.922 079	27.66	
				74	73.921 177	35.94	
33	Arsenic	As	74.921 6	75	74.921 594	100	
34	Selenium	Se	78.96	78	77.917 307	23.78	
				80	79.916 519	49.61	
35	Bromine	Br	79.904	79	78.918 336	50.69	
				81	80.916 287	49.31	
36	Krypton	Kr	83.80	82	81.913 481	11.6	
				83	82.914 136	11.5	
				84	83.911 508	57.0	
				86	85.910 615	17.3	
37	Rubidium	Rb	85.468	85	84.911 793	72.17	
				87*	86.909 186	27.83	4.75×10^{10} yr
38	Strontium	Sr	87.62	86	85.909 266	9.86	
				88	87.905 618	82.58	
				90*	89.907 737		29.1 yr
39	Yttrium	Y	88.905 8	89	88.905 847	100	
40	Zirconium	Zr	91.224	90	89.904 702	51.45	
				91	90.905 643	11.22	
				92	91.905 038	17.15	
				94	93.906 314	17.38	
41	Niobium	Nb	92.906 4	93	92.906 376	100	
42	Molybdenum	Mo	95.94	92	91.906 807	14.84	
				95	94.905 841	15.92	
				96	95.904 678	16.68	
				98	97.905 407	24.13	
43	Technetium	Tc		98*	97.907 215		4.2×10^6 yr
				99*	98.906 254		2.1×10^6 yr

(Table continues)

Atomic Number, Z	Element	Symbol	Chemical Atomic Mass (u)	Mass Number (* Indicates Radioactive) A	Atomic Mass (u)[a]	Percentage Abundance	Half-Life (if Radioactive) $T_{1/2}$
44	Ruthenium	Ru	101.07	99	98.905 939	12.7	
				100	99.904 219	12.6	
				101	100.905 558	17.1	
				102	101.904 348	31.6	
				104	103.905 428	18.6	
45	Rhodium	Rh	102.905 5	103	102.905 502	100	
46	Palladium	Pd	106.42	104	103.904 033	11.14	
				105	104.905 082	22.33	
				106	105.903 481	27.33	
				108	107.903 893	26.46	
				110	109.905 158	11.72	
47	Silver	Ag	107.868	107	106.905 091	51.84	
				109	108.904 754	48.16	
48	Cadmium	Cd	112.41	110	109.903 004	12.49	
				111	110.904 182	12.80	
				112	111.902 760	24.13	
				113*	112.904 401	12.22	9.3×10^{15} yr
				114	113.903 359	28.73	
49	Indium	In	114.82	115*	114.903 876	95.7	4.4×10^{14} yr
50	Tin	Sn	118.71	116	115.901 743	14.53	
				118	117.901 605	24.22	
				120	119.902 197	32.59	
51	Antimony	Sb	121.76	121	120.903 820	57.36	
				123	122.904 215	42.64	
52	Tellurium	Te	127.60	126	125.903 309	18.93	
				128*	127.904 463	31.70	$>8 \times 10^{24}$ yr
				130*	129.906 228	33.87	$\leq 1.25 \times 10^{21}$ yr
53	Iodine	I	126.904 5	127	126.904 474	100	
				129*	128.904 984		1.6×10^{7} yr
				131*	130.906 118		8.04 days
54	Xenon	Xe	131.29	129	128.904 779	26.4	
				131	130.905 069	21.2	
				132	131.904 141	26.9	
				134	133.905 394	10.4	
				136*	135.907 215	8.9	$\geq 2.36 \times 10^{21}$ yr
55	Cesium	Cs	132.905 4	133	132.905 436	100	
56	Barium	Ba	137.33	137	136.905 816	11.23	
				138	137.905 236	71.70	
				144*	143.922 673		11.9 s
57	Lanthanum	La	138.905	139	138.906 346	99.909 8	
58	Cerium	Ce	140.12	140	139.905 434	88.43	
				142*	141.909 241	11.13	$>5 \times 10^{16}$ yr
59	Praseodymium	Pr	140.907 6	141	140.907 647	100	
60	Neodymium	Nd	144.24	142	141.907 718	27.13	
				144*	143.910 082	23.80	2.3×10^{15} yr
				146	145.913 113	17.19	
61	Promethium	Pm		145*	144.912 745		17.7 yr
62	Samarium	Sm	150.36	147*	146.914 894	15.0	1.06×10^{11} yr
				149*	148.917 180	13.8	$>2 \times 10^{15}$ yr
				152	151.919 728	26.7	
				154	153.922 206	22.7	
63	Europium	Eu	151.96	151	150.919 846	47.8	
				153	152.921 226	52.2	

(Table continues)

Atomic Number, Z	Element	Symbol	Chemical Atomic Mass (u)	Mass Number (* Indicates Radioactive) A	Atomic Mass (u)[a]	Percentage Abundance	Half-Life (if Radioactive) $T_{1/2}$
64	Gadolinium	Gd	157.25	156	155.922 119	20.47	
				158	157.924 099	24.84	
				160	159.927 050	21.86	
65	Terbium	Tb	158.925 3	159	158.925 345	100	
66	Dysprosium	Dy	162.50	162	161.926 796	25.5	
				163	162.928 729	24.9	
				164	163.929 172	28.2	
67	Holmium	Ho	164.930 3	165	164.930 316	100	
68	Erbium	Er	167.26	166	165.930 292	33.6	
				167	166.932 047	22.95	
				168	167.932 369	27.8	
69	Thulium	Tm	168.934 2	169	168.934 213	100	
70	Ytterbium	Yb	173.04	172	171.936 380	21.9	
				173	172.938 209	16.12	
				174	173.938 861	31.8	
71	Lutecium	Lu	174.967	175	174.940 772	97.41	
72	Hafnium	Hf	178.49	177	176.943 218	18.606	
				178	177.943 697	27.297	
				179	178.945 813	13.629	
				180	179.946 547	35.100	
73	Tantalum	Ta	180.947 9	181	180.947 993	99.988	
74	Tungsten	W	183.85	182	181.948 202	26.3	
				183	182.950 221	14.28	
				184	183.950 929	30.7	
				186	185.954 358	28.6	
75	Rhenium	Re	186.207	185	184.952 951	37.40	
				187*	186.955 746	62.60	4.4×10^{10} yr
76	Osmium	Os	190.2	188	187.955 832	13.3	
				189	188.958 139	16.1	
				190	189.958 439	26.4	
				191*	190.960 94		15.4 days
				192	191.961 468	41.0	
77	Iridium	Ir	192.2	191	190.960 585	37.3	
				193	192.962 916	62.7	
78	Platinum	Pt	195.08	194	193.962 655	32.9	
				195	194.964 765	33.8	
				196	195.964 926	25.3	
79	Gold	Au	196.966 5	197	196.966 543	100	
80	Mercury	Hg	200.59	199	198.968 253	16.87	
				200	199.968 299	23.10	
				201	200.970 276	13.10	
				202	201.970 617	29.86	
81	Thallium	Tl	204.383	203	292.972 320	29.524	
				205	204.974 400	70.476	
		(Th C″)		208*	207.981 992		3.053 min
				210*	209.990 069		1.3 min
82	Lead	Pb	207.2	204*	203.973 020	1.4	$\geq 1.4 \times 10^{17}$ yr
				206	205.974 440	24.1	
				207	206.975 871	22.1	
				208	207.976 627	52.4	
		(Ra D)		210*	209.984 163		22.3 yr
		(Ac B)		211*	210.988 734		36.1 min
		(Th B)		212*	211.991 872		10.64 h
		(Ra B)		214*	213.999 798		26.8 min

(Table continues)

Atomic Number, Z	Element	Symbol	Chemical Atomic Mass (u)	Mass Number (* Indicates Radioactive) A	Atomic Mass (u)[a]	Percentage Abundance	Half-Life (if Radioactive) $T_{1/2}$
83	Bismuth	Bi	208.980 3	209	208.980 374	100	
		(Th C)		211*	210.987 254		2.14 min
84	Polonium	Po					
		(Ra F)		210*	209.982 848		138.38 days
		(Ra C')		214*	213.995 177		164 μs
85	Astatine	At		218*	218.008 685		1.6 s
86	Radon	Rn		222*	222.017 571		3.823 days
87	Francium	Fr					
		(Ac K)		223*	223.019 733		22 min
88	Radium	Ra		226*	226.025 402		1600 yr
		(Ms Th$_1$)		228*	228.031 064		5.75 yr
89	Actinium	Ac		227*	227.027 749		21.77 yr
90	Thorium	Th	232.038 1				
		(Rd Th)		228*	228.028 716		1.913 yr
				232*	232.038 051	100	1.40×10^{10} yr
91	Protactinium	Pa		231*	231.035 880		32.760 yr
92	Uranium	U	238.028 9	232*	232.037 131		69 yr
				233*	233.039 630		1.59×10^5 yr
		(Ac U)		235*	235.043 924	0.720	7.04×10^8 yr
				236*	236.045 562		2.34×10^7 yr
		(UI)		238*	238.050 784	99.274 5	4.47×10^9 yr
93	Neptunium	Np		237*	237.048 168		2.14×10^6 yr
94	Plutonium	Pu		239*	239.052 157		2.412×10^3 yr
				242*	242.058 737		3.73×10^5 yr
				244*	244.064 200		8.1×10^7 yr

Appendix C

Some Useful Tables

TABLE C.1	Mathematical Symbols Used in the Text and Their Meaning
Symbol	**Meaning**
$=$	is equal to
$\neq$	is not equal to
$\equiv$	is defined as
$\propto$	is proportional to
$>$	is greater than
$<$	is less than
$\gg$	is much greater than
$\ll$	is much less than
$\approx$	is approximately equal to
$\sim$	is on the order of magnitude of
Δx	change in x or uncertainty in x
$\Sigma\, x_i$	sum of all quantities x_i
$\lvert x \rvert$	absolute value of x (always a positive quantity)

TABLE C.2	Standard Symbols for Units			
Symbol	**Unit**		**Symbol**	**Unit**
A	ampere		kcal	kilocalorie
Å	angstrom		kg	kilogram
atm	atmosphere		km	kilometer
Bq	bequerel		kmol	kilomole
Btu	British thermal unit		L	liter
C	coulomb		lb	pound
°C	degree Celsius		ly	lightyear
cal	calorie		m	meter
cm	centimeter		min	minute
Ci	curie		mol	mole
d	day		N	newton
deg	degree (angle)		nm	nanometer
eV	electronvolt		Pa	pascal
°F	degree Fahrenheit		rad	radian
F	farad		rev	revolution
ft	foot		s	second
G	Gauss		T	tesla
g	gram		u	atomic mass unit
H	henry		V	volt
h	hour		W	watt
hp	horsepower		Wb	weber
Hz	hertz		yr	year
in.	inch		μm	micrometer
J	joule		Ω	ohm
K	kelvin			

TABLE C.3 The Greek Alphabet

Alpha	A	α	Nu	N	ν
Beta	B	β	Xi	Ξ	ξ
Gamma	Γ	γ	Omicron	O	o
Delta	Δ	δ	Pi	Π	π
Epsilon	E	ϵ	Rho	P	ρ
Zeta	Z	ζ	Sigma	Σ	σ
Eta	H	η	Tau	T	τ
Theta	Θ	θ	Upsilon	Y	υ
Iota	I	ι	Phi	Φ	ϕ
Kappa	K	κ	Chi	X	χ
Lambda	Λ	λ	Psi	Ψ	ψ
Mu	M	μ	Omega	Ω	ω

TABLE C.4 Physical Data Often Used[a]

Average Earth-Moon distance	3.84×10^8 m
Average Earth-Sun distance	1.496×10^{11} m
Equatorial radius of Earth	6.38×10^6 m
Density of air (20°C and 1 atm)	1.20 kg/m^3
Density of water (20°C and 1 atm)	1.00×10^3 kg/m^3
Free-fall acceleration	9.80 m/s^2
Mass of Earth	5.98×10^{24} kg
Mass of Moon	7.36×10^{22} kg
Mass of Sun	1.99×10^{30} kg
Standard atmospheric pressure	1.013×10^5 Pa

[a] These are the values of the constants as used in the text.

TABLE C.5 Some Fundamental Constants[a]

Quantity	Symbol	Value[b]
Atomic mass unit	u	$1.660\ 540\ 2(10) \times 10^{-27}$ kg
		$931.494\ 32(28)$ MeV/c^2
Avogadro's number	N_A	$6.022\ 136\ 7(36) \times 10^{23}$ (mol)$^{-1}$
Bohr radius	$a_0 = \dfrac{\hbar^2}{m_e e^2 k_e}$	$0.529\ 177\ 249(24) \times 10^{-10}$ m
Boltzmann's constant	$k_B = R/N_A$	$1.380\ 658(12) \times 10^{-23}$ J/K
Compton wavelength	$\lambda_C = \dfrac{h}{m_e c}$	$2.426\ 310\ 58(22) \times 10^{-12}$ m
Coulomb's law force constant	$k_e = \dfrac{1}{4\pi\epsilon_0}$	$8.987\ 551\ 787 \times 10^9$ N·m^2/C^2 (exact)
Electron mass	m_e	$9.109\ 389\ 7(54) \times 10^{-31}$ kg
		$5.485\ 799\ 03(13) \times 10^{-4}$ u
		$0.510\ 999\ 06(15)$ MeV/c^2
Electron volt	eV	$1.602\ 177\ 33(49) \times 10^{-19}$ J
Elementary charge	e	$1.602\ 177\ 33(49) \times 10^{-19}$ C
Gas constant	R	$8.314\ 510(70)$ J/K·mol
Gravitational constant	G	$6.672\ 59(85) \times 10^{-11}$ N·m^2/kg^2
Hydrogen ionization energy	$-E_1 = \dfrac{m_e e^4 k_e^2}{2\hbar^2} = \dfrac{e^2 k_e}{2a_0}$	$13.605\ 698(40)$ eV
Neutron mass	m_n	$1.674\ 928\ 6(10) \times 10^{-27}$ kg
		$1.008\ 664\ 904(14)$ u
		$939.565\ 63(28)$ MeV/c^2
Permeability of free space	μ_0	$4\pi \times 10^{-7}$ T·m/A (exact)
Permittivity of free space	$\epsilon_0 = 1/\mu_0 c^2$	$8.854\ 187\ 817 \times 10^{-12}$ C^2/N·m^2 (exact)
Planck's constant	h	$6.626\ 075(40) \times 10^{-34}$ J·s
	$\hbar = h/2\pi$	$1.054\ 572\ 66(63) \times 10^{-34}$ J·s
Proton mass	m_p	$1.672\ 623(10) \times 10^{-27}$ kg
		$1.007\ 276\ 470(12)$ u
		$938.272\ 3(28)$ MeV/c^2
Rydberg constant	R_H	$1.097\ 373\ 153\ 4(13) \times 10^7$ m^{-1}
Speed of light in vacuum	c	$2.997\ 924\ 58 \times 10^8$ m/s (exact)

[a] These constants are the values recommended in 1986 by CODATA, based on a least-squares adjustment of data from different measurements. For a more complete list, see Cohen, E. Richard, and Barry N. Taylor, *Rev. Mod. Phys.* **59**:1121, 1987.

[b] The numbers in parentheses for the values below represent the uncertainties in the last two digits.

Appendix D

SI Units

TABLE D.1	SI Base Units	
	SI Base Unit	
Base Quantity	**Name**	**Symbol**
Length	meter	m
Mass	kilogram	kg
Time	second	s
Electric current	ampere	A
Temperature	kelvin	K
Amount of substance	mole	mol
Luminous intensity	candela	cd

TABLE D.2	Derived SI Units			
Quantity	**Name**	**Symbol**	**Expression in Terms of Base Units**	**Expression in Terms of Other SI Units**
Plane angle	radian	rad	m/m	
Frequency	hertz	Hz	s^{-1}	
Force	newton	N	$kg \cdot m/s^2$	J/m
Pressure	pascal	Pa	$kg/m \cdot s^2$	N/m^2
Energy: work	joule	J	$kg \cdot m^2/s^2$	$N \cdot m$
Power	watt	W	$kg \cdot m^2/s^3$	J/s
Electric charge	coulomb	C	$A \cdot s$	
Electric potential (emf)	volt	V	$kg \cdot m^2/A \cdot s^3$	$W/A, J/C$
Capacitance	farad	F	$A^2 \cdot s^4/kg \cdot m^2$	C/V
Electric resistance	ohm	Ω	$kg \cdot m^2/A^2 \cdot s^3$	V/A
Magnetic flux	weber	Wb	$kg \cdot m^2/A \cdot s^2$	$V \cdot s, T \cdot m^2$
Magnetic field intensity	tesla	T	$kg/A \cdot s^2$	Wb/m^2
Inductance	henry	H	$kg \cdot m^2/A^2 \cdot s^2$	Wb/A

Answers to Quick Quizzes, Selected Questions and Problems

CHAPTER 1

Conceptual Questions

1. (a) ~ 0.1 m   (b) ~ 1 m
   (c) between 10 m and 100 m
   (d) ~ 10 m   (e) ~ 100 m
3. (a) $\sim 10^6$ beats   (b) $\sim 10^9$ beats
5. $\sim 10^9$ s
7. The length of a hand varies from person to person, so it is not a useful standard of length.
9. The ark has an approximate volume of 10^4 m^3, whereas a home has an approximate volume of 10^3 m^3.
11. $\sim \$10/\text{lb}$ (assumes a cost of \$20 000 and weight of 2 000 lb), of the same order as the cost of a good steak

Problems

1. Based on units alone, the equation might be valid.
5. $\text{m}^3/(\text{kg}\cdot\text{s}^2)$
7. (a) 3 significant figures   (b) 4 significant figures
   (c) 3 significant figures   (d) 2 significant figures
9. (a) 797   (b) 1.1   (c) 17.66
11. 115.9 m
13. 228.8 cm
15. 2×10^8 fathoms
17. 0.325 g
19. 1×10^{17} ft
21. 6.71×10^8 mi/h
23. 3.08×10^4 m^3
25. 9.82 cm
27. (a) 1.0×10^3 kg
    (b) $m_{\text{cell}} = 5.2 \times 10^{-16}$ kg, $m_{\text{kidney}} = 0.27$ kg, $m_{\text{fly}} = 1.3 \times 10^{-5}$ kg
29. 1800 balls (or $\sim 10^3$ balls) (assumes 81 games per season, 9 innings per game, an average of 10 hitters per inning, and 1 ball lost for every four hitters)
31. $\sim 10^7$ rev
33. $\sim 10^6$ balls
35. (2.1 m, 1.4 m)
37. $r = 2.2$ m, $\theta = 27°$
39. (a) 6.71 m   (b) 0.894   (c) 0.746
41. 3.46 mi (or about 18 300 ft)
43. (a) 3.00   (b) 3.00   (c) 0.800   (d) 0.800   (e) 1.33
45. 5.00/7.00, the angle itself is 35.5°
47. $\sim 10^{12}$ cells
49. The value of k, a dimensionless constant, cannot be found by dimensional analysis.
51. $\sim 10^{-9}$ m
53. $\sim 10^3$ tuners (assumes 1 tuner per 10 000 residents and a population of 7.5 million)
55. (a) 3.16×10^7 s
    (b) Between 10^{10} yr and 10^{11} yr.

CHAPTER 2

Quick Quizzes

1. (a) 200 yd   (b) 0   (c) 0
2. (a) False   (b) True   (c) True
3. The velocity-time graph (a) has a constant slope, indicating a constant acceleration, which is represented by acceleration-time graph (e).
   Graph (b) represents an object that has an always-increasing speed, but the speed is not increasing at a uniform rate. Thus, the acceleration must be increasing, and the acceleration-time graph that best indicates this is (d).
   Graph (c) depicts an object that first has a velocity that increases at a constant rate, which means its acceleration is constant. The motion then changes to one at constant speed, indicating that the acceleration of the object becomes zero. Thus, the best match to this situation is graph (f).
4. (b)
5. (a) blue graph   (b) red graph
   (c) green graph
6. (c)
7. (c)
8. (a) and (f)

Conceptual Questions

1. Yes. If the velocity of the particle is nonzero, the particle is in motion. If the acceleration is zero, the velocity of the particle is unchanging, or is a constant.
3. Yes. If this occurs, the acceleration of the car is opposite to the direction of motion and the car will be slowing down.
5. The free-fall acceleration is always directed downward, which is normally indicated by a negative sign. While the object is moving upward, its velocity is positive. After it reaches it maximum height and begins to fall downward, its velocity is negative. The acceleration is negative for the entire flight.
7. Once the objects leave the hand, both are in free fall, and both experience the same downward acceleration equal to the free-fall acceleration, $-g$.
9. Yes. Yes.
11. (a) At the maximum height, the ball is momentarily at rest (that is, has zero velocity). The acceleration remains constant, with magnitude equal to the free-fall acceleration g, and directed downward. Thus, even though the velocity is momentarily zero, it continues to change and the ball will begin to gain speed in the downward direction. (b) The acceleration of the ball remains constant in magnitude and direction throughout its free flight, from the instant it leaves the hand until the instant just before it strikes the ground. It is directed downward and has a magnitude equal to the free-fall acceleration g.
13. No. If the velocity of A is greater than that of B, car A is moving faster than B at this instant. However, car B may be picking up speed (that is, accelerating) at a greater rate than A. The driver of B may have stepped very hard on the gas pedal in the recent past, but car B has not yet reached the speed of A.
15. They are the same! You can see this for yourself by solving the kinematic equations for the two cases. The ball thrown upward will take longer to reach the ground, but when it arrives, it has the same velocity as the ball that was thrown downward.

17. Ignoring air resistance, in 16 s the pebble would fall a distance of $\frac{1}{2}gt^2 = \frac{1}{2}(9.80 \text{ m/s}^2)(16.0 \text{ s})^2 = 1250 \text{ m} = 1.25 \text{ km}$. Air resistance is an important force after the first few seconds, when the pebble has gained high speed. Also, part of the 16-s interval must be occupied by the sound returning up the well. Thus, the depth of the well is less than 1.25 km, but on the order of 10^3 m.

19. If the car is moving forward, but slowing down, its acceleration is directed in the rearward direction. If the acceleration remains constant after the car comes to rest, the car will begin to develop a velocity in the rearward direction (that is, it will back up).

Problems

1. (a) 52.9 km/h   (b) 90.0 km
3. (a) Boat A wins by 60 km   (b) 0
5. (a) 180 km   (b) 63.4 km/h
7. (a) 4.0 m/s   (b) -0.50 m/s
    (c) -1.0 m/s   (d) 0
9. (a) a, e, and f (b) c
    (c) d and f (d) f
    (e) d
11. 2.80 h, 218 km
13. 274 km/h
15. (a) 5.00 m/s   (b) -2.50 m/s
    (c) 0   (d) 5.00 m/s
17. (a) 4.0 m/s   (b) -4.0 m/s
    (c) 0   (d) 2.0 m/s
19. -4.00 m/s^2; the acceleration is in the negative x direction
21. 3.7 s
23. (a) 8.0 m/s^2 (b) about 12 m/s^2
25. 2.74×10^5 m/s^2, or 2.79×10^4 times g
27. 160 ft
29. (a) 2.32 m/s^2 (b) 14.4 s
31. (a) 8.94 s   (b) 89.4 m/s
33. -3.6 m/s^2
35. 200 m
37. (a) 1.5 m/s   (b) 32 m
39. (a) 8.2 s   (b) 1.3×10^2 m
43. (a) 31.9 m   (b) 2.55 s
    (c) 2.55 s   (d) -25.0 m/s
45. 12 m/s
47. (a) 21.1 m/s   (b) 19.6 m   (c) 18.1 m/s; 19.6 m
49. (a) 308 m   (b) 8.51 s   (c) 16.4 s
51. (a) 10.0 m/s upward   (b) 4.68 m/s downward
53. 15.0 s
55. (a) -3.50×10^5 m/s^2 (b) 2.86×10^4 s
57. 0.60 s
59. (a) 3.00 s   (b) -24.5 m/s, -24.5 m/s
    (c) 23.5 m
61. (a) 5.46 s   (b) 73.0 m
    (c) $v_{\text{Kathy}} = 26.7$ m/s, $v_{\text{Stan}} = 22.6$ m/s
63. (a) 4.47 m/s upward   (b) 6.31 s after release
    (c) 6.06 m/s downward   (d) 73.0 m
    (e) 17.1 m/s
65. (a) 5.5×10^3 ft
    (b) 3.7×10^2 ft/s
    (c) The plane would only travel 0.002 ft in the time it takes the light from the bolt to reach the eye.
67. (a) 7.82 m   (b) 0.782 s

CHAPTER 3

Quick Quizzes

1. (c)
2. (b)
3. (b)

4.

Vector	x component	y component
A	$-$	$+$
B	$+$	$-$
A + **B**	$-$	$-$

5. (a) False   (b) False
6. (c)
7. (b)
8. The answers to all three parts are the same—the acceleration is that due to gravity, 9.8 m/s^2 downward.

Conceptual Questions

1. The components of **A** will both be negative when **A** lies in the third quadrant. The components of **A** will have opposite signs when **A** lies in either the second quadrant or the fourth quadrant.
3. Assuming the ship sails in a straight line, the wrench will strike the deck at the base of the mast. The wrench has the same horizontal velocity as the ship when it is released, and will maintain that component of velocity as it falls. Thus, the wrench is always above the same spot on the deck and will strike that spot when it hits the deck. To an observer on the ship, the wrench appears to fall straight downward. To a stationary observer on shore, the wrench is seen to follow a parabolic trajectory.
5. Let v_{x0} and v_{y0} be the initial velocity components for the projectile fired on Earth. Because the projectile fired on the Moon is given the same initial velocity, its initial velocity components have the same values v_{x0} and v_{y0} as those for the projectile on Earth. From $v_y^2 = v_{y0}^2 - 2g\,\Delta y$, the maximum altitude reached by a projectile is found to be $H = v_{y0}^2/2g$. Because the free-fall acceleration g is smaller on the Moon than on Earth, the maximum altitude will be greater for the projectile fired on the Moon. If they start from the same elevation, the projectile on the Moon will have a longer time of flight and therefore a greater range. The Apollo astronauts made several long golf drives on the Moon.
7. A vector has both magnitude and direction and can be added only to other vectors representing the same physical quantity. A scalar has only a magnitude and can be added only to other scalars representing the same physical quantity.
9. (a) At the top of its flight, the velocity is horizontal and the acceleration is downward. This is the only point at which the velocity and acceleration vectors are perpendicular.
    (b) If the object is thrown straight up or down, then the velocity and acceleration will be parallel throughout the motion. For any other kind of projectile motion, the velocity and acceleration vectors are never parallel.
11. (a) The acceleration is zero, since both the magnitude and direction of the velocity remain constant.
    (b) The particle has an acceleration since the direction of **v** changes.
13. The spacecraft will follow a parabolic path. This is equivalent to a projectile thrown off a cliff with a horizontal velocity. For the projectile, gravity provides an acceleration that is always perpendicular to the initial velocity, resulting in a parabolic path. For the spacecraft, the initial velocity plays the role of the horizontal velocity of the projectile. The acceleration provided by the leaking gas provides an acceleration that plays the role of gravity for the projectile. If the orientation of the spacecraft were to change in response to the gas leak (which is by far the more likely result), then the acceleration would change direction and the motion could become quite complicated.
15. For angles $\theta < 45°$, the projectile thrown at angle θ will be in the air for a shorter time interval. For a small angle, the vertical component of the initial velocity is smaller than for the larger angle. Thus, the projectile thrown at the smaller

angle will not go as high into the air, and will spend less time in the air before landing.

17. Yes. The projectile is a freely falling body, because gravity is the only force acting on it. The vertical acceleration will be the local free-fall acceleration g; the horizontal acceleration will be zero.

19. If a projectile is to have zero speed at the top of its trajectory, it must be thrown straight upward, with zero horizontal velocity. To give a projectile a nonzero speed at the top of the trajectory, the projectile should be thrown at any angle other than 90° to the horizontal.

Problems

1. Approximately 7.9 m at 4° N of W
3. Approximately 15 m at 58° S of E
5. Approximately 310 km at 57° S of W
7. (a) Approximately 5.0 units at $-53°$
   (b) Approximately 5.0 units at $+53°$
9. 8.07 m at 42.0° S of E
11. (a) 5.00 blocks at 53.1° N of E
    (b) 13.0 blocks
13. 47.2 units at 122° from positive x axis
15. 157 km
17. 245 km at 21.4° W of N
19. 196 cm at 14.7° below the x axis
21. (a) (10.0 m, 16.0 m)
23. 6.13 m
25. 12 m/s
27. 48.6 m/s
29. 25 m
31. (a) 32.5 m from the base of the cliff
    (b) 1.78 s
33. (a) 52.0 m/s horizontally   (b) 212 m
35. 390 mi/h at 7.37° N of E
37. 153 km/h at 11.3° N of W
39. 249 ft upstream
41. 18.0 s
43. 196 cm at 14.7° below positive x axis
45. (a) 57.7 km/h at 60.0° west of vertical
    (b) 28.9 km/h downward
47. 68.6 km/h
49. (a) 1.52×10^3 m   (b) 36.1 s
    (c) 4.05×10^3 m
51. $R_{moon} = 18$ m, $R_{Mars} = 7.9$ m
53. 2.21 m/s
55. (a) 42 m/s   (b) 3.8 s
    (c) $v_x = 34$ m/s, $v_y = -13$ m/s; $v = 37$ m/s
57. 7.5 m/s in the direction the ball was thrown
61. $R/2$
63. 7.5 min
65. 10.8 m

67. (b) $y = Ax^2$ with $A = \dfrac{g}{2v_i^2}$ where v_i is the muzzle velocity

    (c) 14.5 m/s
69. 227 paces at 165° from positive x axis
71. (a) 20.0° above the horizontal
    (b) 3.05 s

CHAPTER 4

Quick Quizzes

1. (a) True   (b) False
2. (a) True   (b) True   (c) False
3. False
4. Because the value of g is smaller on the Moon than on Earth, more mass of gold would be required to represent 1 newton of weight on the Moon. Thus, your friend on the Moon is richer, by about a factor of 6!

5. (b)
6. (c); (d)
7. (c)
8. (c)
9. (b)
10. (b)
11. (b) By exerting an upward force component on the sled, you reduce the normal force of the ground and so reduce the force of kinetic friction.

Conceptual Questions

1. (a) Two external forces act on the ball. (i) One is a downward gravitational force exerted by Earth. (ii) The second force on the ball is an upward normal force exerted by the hand. The reactions to these forces are (i) an upward gravitational force exerted by the ball on Earth, and (ii) a downward force exerted by the ball on the hand.
   (b) After the ball leaves the hand, the only external force acting on the ball is the gravitational force exerted by Earth. The reaction is an upward gravitational force exerted by the ball on Earth.

3. No. If the car moves with constant acceleration, it is necessary that a net force in the direction of the acceleration be acting on it.

5. The coefficient of static friction is larger than that of kinetic friction. To start the box moving, you must counterbalance the maximum static friction force. This exceeds the kinetic friction force that you must counterbalance to maintain constant velocity motion once it is started.

7. The inertia of the suitcase would keep it moving forward as the bus stops. There would be no tendency for the suitcase to be thrown backward toward the passenger. The case should be dismissed.

9. The force causing an automobile to move is the friction exerted by the roadway on the tires as the automobile attempts to push the roadway backward. The force driving a propeller airplane forward is the reaction force exerted by the air on the propeller as the rotating propeller pushes the air backward (the action). In a rowboat, the rower pushes the water backward with the oars (the action). The water pushes forward on the oars and hence the boat (the reaction).

11. When the bus starts moving, the mass of Claudette is accelerated by the force exerted by the back of the seat on her body. Clark is standing, however, and the only force on him is the friction between his shoes and the floor of the bus. Thus, when the bus starts moving, his feet accelerate forward, but the rest of his body experiences almost no accelerating force (only that due to his being attached to his accelerating feet!). As a consequence, his body tends to stay almost at rest, according to Newton's first law, relative to the ground. Relative to Claudette, however, he is moving toward her and falls into her lap. Both performers won Academy Awards.

13. The tension in the rope is the maximum force that occurs in *both* directions. In this case, then, since both are pulling with a force of 200 N, the tension is 200 N. If the rope does not move, then the total force on each athlete must equal zero. Therefore, each athlete exerts 200 N horizontally on the ground.

15. (a) As the man takes the step, the action is the force his foot exerts on Earth; the reaction is the force exerted by Earth on his foot. (b) Here the action is the force exerted by the snowball on the girl's back; the reaction is the force exerted by the girl's back on the snowball. (c) This action is the force exerted by the glove on the ball; the reaction is the force exerted by the ball on the glove. (d) This action is the force exerted by the air molecules on the window; the reaction is the force exerted by the window on the air molecules. In each case, we could equally well interchange the terms "action" and "reaction."

17. If the brakes lock, the car will travel farther than it would travel if the wheels continued to roll because the coefficient of kinetic friction is less than that of static friction. Hence, the force of kinetic friction is less than the maximum force of static friction.

19. **(a)** The crate accelerates because of a friction force exerted by the floor of the truck on the crate. **(b)** If the driver slams on the brakes, the crate's inertia tends to keep it moving forward.

Problems

1. **(a)** 12 N   **(b)** 3.0 m/s^2
3. 2×10^4 N
5. 3.71 N, 58.7 N, 2.27 kg
7. 9.6 N
9. 1.41×10^3 N
11. **(a)** 0.200 m/s^2 **(b)** 10.0 m
    (c) 2.00 m/s
13. 4.65 m/s^2 at 35.2° above the horizontal toward the right
15. 600 N in vertical cable, 997 N in inclined cable, 796 N in horizontal cable
17. 150 N in vertical cable, 75 N in right side cable, 130 N in left side cable
19. 101 N at 13.8° to the right of vertical and downward
21. 613 N
23. 64 N
25. **(a)** 7.50×10^3 N backward   **(b)** 50.0 m
27. **(a)** 7.0 m/s^2 horizontal and to the right
    (b) 21 N
    (c) 14 N horizontal and to the right
29. 7.90 m/s
31. 1.25 kg
33. **(a)** 2.15×10^3 N forward
    (b) 645 N forward
    (c) 645 N rearward
    (d) 1.02×10^4 N at 74.1° below horizontal and rearward
35. $\mu_s = 0.38$, $\mu_k = 0.31$
37. **(a)** 0.256   **(b)** 0.509 m/s^2
39. 7.84 m/s^2, independent of the mass
41. 32.1 N
43. **(a)** 33 m/s
    (b) No. The object will speed up to 33 m/s from any lower speed and will slow down to 33 m/s from any higher speed.
45. $\mu_k = 0.287$
47. 0.465 m/s
49. 3.30 m/s^2
51. **(a)** 0.404   **(b)** 45.8 lb
53. 113 N, horizontal and at 5.20° from the centerline between the cables
55. **(a)** 84.9 N upward
    (b) 84.9 N downward
57. 50 m
59. **(a)** friction between box and truck
    (b) 2.94 m/s^2
61. **(a)** 2.22 m
    (b) 8.74 m/s down the incline
63. **(a)** 1.78 m/s^2 **(b)** 0.368
    (c) 9.37 N   **(d)** 2.67 m/s
65. **(a)** 1.7 m/s^2, 17 N   **(b)** 0.69 m/s^2, 17 N
67. 100 N, 204 N
69. **(a)** $T_1 = 78.0$ N, $T_2 = 35.9$ N   **(b)** 0.656
71. **(a)** 30.7°   **(b)** 0.843 N
73. 5.5×10^2 N
75. **(a)** 7.1×10^2 N   **(b)** 8.1×10^2 N
    (c) 7.1×10^2 N   **(d)** 6.5×10^2 N
77. 72.0 N
79. **(a)** 0.408 m/s^2 upward   **(b)** 83.3 N

CHAPTER 5

Quick Quizzes

1. (c), (a), (d), (b)
2. All three balls have the same speed the moment before they hit the ground.
3. (c)
4. (c)

Conceptual Questions

1. Since no motion is taking place, the rope experiences no displacement and no work is done on it. For the same reason, no work is being done on the pullers or the ground. Work is only being done within the bodies of the pullers. For example, the heart of each puller is applying forces on the blood to move blood through the body.

3. When the slide is frictionless, changing the length or shape of the slide will not make any difference in the final speed of the child as long as the difference in the heights of the upper and lower ends of the slide is kept constant. If friction is considered, the path length along which the friction force does negative work will be greater when the slide is made longer or given bumps. Thus, the child will arrive at the lower end with less kinetic energy (and hence less speed).

5. If we ignore the effects due to rolling friction on the tires of the car, the same amount of work would be done in driving up the switchbacks and in driving straight up the mountain, since the weight of the car is moved upward against gravity by the same vertical distance in each case. If we include friction, there is more work done in driving the switchbacks, since the distance over which the friction force acts is much longer. So why do we use switchbacks? The answer lies in the force required, not the work. The force from the engine required to follow a gentle rise is much smaller than that required to drive straight up the hill. Roadways running straight uphill would require redesigning engines to enable them to apply much larger forces. This is similar to the relative ease with which heavy objects can be moved up ramps into trucks (compared with lifting the objects vertically).

7. **(a)** The tension in the supporting cord does no work, because the motion of the pendulum is always perpendicular to the cord, and therefore to the tension force. **(b)** The air resistance does negative work at all times, because the air resistance is always acting in a direction opposite to the motion. **(c)** The force of gravity always acts downward; therefore, the work done by gravity is positive on the downswing, and negative on the upswing.

9. Because the time periods are the same for both cars, we need only to compare the work done. The sports car is moving twice as fast as the older car at the end of the time interval, so it has four times the kinetic energy. Thus, according to the work-kinetic energy theorem, four times as much work was done, and the engine must have expended four times the power.

11. During the time that the toe is in contact with the ball, the work done on the ball is given by

$$W_{\text{toe}} = \tfrac{1}{2} m_{\text{ball}} v^2 - 0 = \tfrac{1}{2} m_{\text{ball}} v^2$$

where v is the speed of the ball as it leaves the toe. After the ball loses contact with the toe, only the gravitational force and the retarding force due to air resistance continue to do work on the ball throughout its flight.

13. Yes, the total mechanical energy of the system is conserved because the only forces acting are conservative—the force of gravity and the spring force. There are two forms of potential energy in this case, gravitational potential energy and elastic potential energy stored in the spring.

15. Let's assume you lift the book slowly. In this case there are two forces on the book that are almost equal in magnitude. They are the lifting force and the force of gravity. Thus, the positive work done by you and the negative work done by gravity cancel. There is no net work performed, and no net change in the kinetic energy—the work-kinetic energy theorem is satisfied.

17. As the satellite moves in a circular orbit about Earth, its displacement along the circular path during any small time interval is always perpendicular to the gravitational force, which always acts toward the center of Earth. Therefore, the work done by the gravitational force during any displacement is zero. (Recall that the work done by a force is defined to be $F\Delta x \cos\theta$, where θ is the angle between the force and the displacement. In this case, the angle is $90°$, so the work done is zero.) Since the work-kinetic energy theorem says that the net work done on an object during any displacement is equal to the change in its kinetic energy, and the work done in this case is zero, the change in the satellite's kinetic energy is zero, hence its speed remains constant.

Problems

1. 700 J
3. 15.0 MJ
5. (a) 61.3 J   (b) -46.3 J   (c) 0
7. (a) 79.4 N   (b) 1.49 kJ   (c) -1.49 kJ
9. (a) 2.00 m/s   (b) 200 N
11. 0.265 m/s
13. (a) -5.6×10^2 J   (b) 1.2 m
15. (a) 90 J   (b) 1.8×10^2 N
17. 1.0 m/s
19. (a) 6.4 J   (b) -2.5 J
21. 1.9 J
23. 1.05 J
25. (a) 8.0×10^2 J   (b) 1.1×10^2 J
    (c) 0
27. 0.459 m
29. 6.86 m/s
31. (a) 10.9 m/s   (b) 11.6 m/s
33. (a) 544 N/m   (b) 19.7 m/s
35. 10.2 m
37. 1.9 m/s
39. 2.1 kN
41. 3.8 m/s
43. (a) 5.42 m/s   (b) 0.300   (c) 147 J
45. 289 m
47. (a) 24.5 m/s   (b) Yes   (c) 206 m
    (d) Unrealistic; the actual retarding force will vary with speed.
49. (a) 0.408 m/s   (b) 2.45 kJ
51. 5.9×10^8 W
53. (a) 2.38×10^4 W = 32.0 hp   (b) 4.77×10^4 W = 63.9 hp
55. (a) 24.0 J   (b) -3.00 J   (c) 21.0 J
57. (a) The graph is a straight line passing through the points $(0 \text{ m}, -16 \text{ N})$, $(2 \text{ m}, 0 \text{ N})$, and $(3 \text{ m}, 8 \text{ N})$.
    (b) -12.0 J
59. (a) 575 N/m   (b) 46.0 J
61. (a) 4.4 m/s   (b) 1.5×10^5 N
63. (a) 3.13 m/s   (b) 4.43 m/s   (c) 1.00 m
65. 875 W
67. (a) 423 mi/gal   (b) 776 mi/gal
69. (a) 28.0 m/s   (b) 30.0 m
    (c) 89.0 m beyond the end of the track
71. 1.68 m/s
73. (a) 6.15 m/s   (b) 9.87 m/s
75. (a) 101 J   (b) 0.410 m
    (c) 2.84 m/s   (d) -9.80 mm
    (e) 2.85 m/s
77. 914 N/m

CHAPTER 6

Quick Quizzes

1. (d)
2. (c)
3. (c) and (e)
4. (d)
5. (b)
6. (a)
7. (a) Perfectly inelastic   (b) Inelastic
   (c) Inelastic
8. (b)

Conceptual Questions

1. The passenger's body must undergo a significant momentum change as the car's velocity changes abruptly in the collision. In a front-end collision, the seat belt and shoulder harness exert the required forces on the body to give it the needed impulse (change in momentum). When a car is rammed from behind, the car and the passenger's body are given a sudden increase in forward momentum. If the head is not supported by a head rest, the head tends to continue moving at its original speed until the angle of the neck changes and the neck exerts a sufficiently large forward force on the head to give it the needed momentum change. In agreement with Newton's third law, the head exerts a rearward force on the neck. This force causes whiplash injuries.

3. If all the kinetic energy disappears, there must be no motion of either object after the collision. If nothing is moving, there is no momentum. If the final momentum of the system is zero, the initial momentum of the system must also have been zero. A situation in which this could be true would be the head-on collision of two objects, having momenta of equal magnitude but opposite direction.

5. Initially the clay has momentum directed toward the wall. When it collides and sticks to the wall, nothing appears to have any momentum. Thus, it is tempting to (wrongfully) conclude that momentum is not conserved. However, the "lost" momentum is actually imparted to the wall and Earth, causing both to move. Because of Earth's enormous mass, its recoil speed is too small to detect.

7. Before the step the momentum was zero, so afterward the net momentum must also be zero. Obviously, you have some momentum, so something must have momentum in the opposite direction. That something is Earth. Earth's enormous mass ensures that its recoil speed will be too small to detect, but if you want to make Earth move, it is as simple as taking a step.

9. As the water is forced out of the holes in the spray arm, the spray arm imparts a horizontal impulse to the water. The water exerts an equal and opposite impulse on the spray arm, causing the spray arm to rotate in the direction opposite to that of the spray.

11. By discarding empty fuel tanks and other assemblies no longer needed, the mass of the remaining system is reduced. A rocket engine is capable of providing a certain impulse—that is, the product of its thrust and the time interval that thrust can be maintained. This impulse equals the change in momentum $m(\Delta v)$ given to the rocket body. By keeping the mass m as low as possible, the increase in speed Δv provided by the engine can be maximized.

13. It will be easiest to catch the medicine ball when its speed (and kinetic energy) is lowest. The first option—throwing the medicine ball at the same velocity—will be the most difficult, because the speed will not be reduced at all. The second option, throwing the medicine ball with the same momentum, will reduce the velocity by the ratio of the masses.

Since $m_t v_t = m_m v_m$,

$$v_m = v_t \left(\frac{m_t}{m_m} \right)$$

The third option, throwing the medicine ball with the same kinetic energy, will also reduce the velocity, but only by the square root of the ratio of the masses. Since

$$\tfrac{1}{2} m_t v_t^2 = \tfrac{1}{2} m_m v_m^2, \quad \text{then } v_m = v_t \sqrt{\frac{m_t}{m_m}}$$

Thus, the slowest—and easiest—throw will be made when the momentum is held constant. If you wish to check this, try substituting in values of $v_t = 1$ m/s, $m_t = 1$ kg, and $m_m = 100$ kg. The same-momentum throw will be caught at 1 cm/s, while the same-energy throw will be caught at 10 cm/s.

15. The follow-through keeps the club in contact with the ball as long as possible, maximizing the impulse. Thus, the ball experiences a larger change in momentum than without the follow-through, and it leaves the club with a higher velocity and travels farther. With a short shot to the green, the primary factor is control, not distance. Thus, there is little or no follow-through, allowing for the golfer to have a better feel for how hard he or she is striking the ball.

17. It is the product mv that is the same for both the bullet and the gun. The bullet has a large velocity and a small mass, while the gun has a small velocity and a large mass. Furthermore, the bullet carries much more kinetic energy than the gun.

19. When you land on the ground, a certain impulse $\bar{F} \Delta t$ is required to bring your body to rest. If you land with legs relaxed and allow the knees to bend, you increase the time interval Δt over which the impulse is delivered. This reduces the average force exerted on the body. Landing with the knees locked means that the time interval will be very short and the average force very high, which is more likely to result in injury to the body.

Problems

3. **(a)** 8.35×10^{-21} kg·m/s   **(b)** 4.50 kg·m/s
   (c) 750 kg·m/s   **(d)** 1.78×10^{29} kg·m/s
5. **(a)** 31.0 m/s
   (b) the bullet, 3.38×10^3 J versus 69.7 J
7. $\sim 10^3$ N upward
9. 364 kg·m/s forward, 438 N forward
11. **(a)** 8.0 N·s   **(b)** 5.3 m/s
    (c) 3.3 m/s
13. **(a)** 12 N·s   **(b)** 8.0 N·s
    (c) 8.0 m/s, 5.3 m/s
15. **(a)** 9.60×10^{-2} s   **(b)** 3.65×10^5 N
    (c) $26.6g$
17. 800 N at 60.0° below horizontal to the right
19. 65 m/s
21. **(a)** 1.15 m/s
    (b) 0.346 m/s directed opposite to girl's motion
23. $KE_E / KE_b \sim 10^{-25}$
25. 2.66 m/s
27. **(a)** 1.80 m/s   **(b)** 2.16×10^4 J
29. 57 m
31. 15.6 m/s
33. 273 m/s
35. **(a)** -6.67 cm/s, 13.3 cm/s   **(b)** 0.889
37. 17.1 cm/s (25.0-g object), 22.1 cm/s (10.0-g object)
39. 7.94 cm
41. **(a)** 2.9 m/s at 32° N of E
    (b) 7.9×10^2 J converted into internal energy
43. 5.59 m/s north
45. **(a)** 2.50 m/s at $-60°$ **(b)** elastic collision

47. 14.8 kg·m/s in the direction of the final velocity of the ball
49. 1.78×10^3 N on truck driver, 8.89×10^3 N on car driver
51. **(a)** 8/3 m/s (incident particle), 32/3 m/s (target particle)
    (b) $-16/3$ m/s (incident particle), 8/3 m/s (target particle)
    (c) 7.1×10^{-3} J in case (a), 2.8×10^{-2} J in case (b). The incident particle loses more kinetic energy in case (a) where the target mass is 1.0 g.
53. 1.1×10^3 N (upward)
55. **(a)** 1.33 m/s   **(b)** 235 N
    (c) 0.681 s   **(d)** -160 N·s, 160 N·s
    (e) 1.82 m   **(f)** 0.454 m
    (g) -427 J   **(h)** 107 J
    (i) Equal friction forces act through different distances on man and cart to do different amounts of work on them. This is a perfectly inelastic collision in which the total work on both man and cart together is -320 J, which becomes $+320$ J of internal energy.
57. **(a)** -2.33 m/s, 4.67 m/s   **(b)** 0.277 m
    (c) 2.98 m   **(d)** 1.49 m
59. **(a)** -0.667 m/s   **(b)** 0.952 m
61. **(a)** 3.54 m/s   **(b)** 1.77 m
    (c) 3.54×10^4 N
    (d) No, the normal force exerted by the rail contributes upward momentum to the system.
63. **(a)** 0.28 or 28%
    (b) 1.1×10^{-13} J for neutron, 4.5×10^{-14} J for carbon
65. $\dfrac{2v_0^2}{9\mu_k g} - \dfrac{4d}{9}$

CHAPTER 7
Quick Quizzes

1. (c)
2. (b)
3. (b) and (d)
4. (e), (a), (b)
5. (c)
6. (b) and (c)
7. (e)
8. (a)

Conceptual Questions

1. **(a)** The head will tend to lean toward the right shoulder (that is, toward the outside of the curve).
   (b) When there is no strap, tension in the neck muscles must produce the centripetal acceleration.
   (c) With a strap, the tension in the strap performs this function, allowing the neck muscles to remain relaxed.
3. An object can move in a circle even if the total force on it is not perpendicular to its velocity, but then its speed will change. Resolve the total force into an inward radial component and a perpendicular tangential component. If the tangential force acts in the forward direction, the object will speed up, and if the tangential force acts backward, the object will slow down.
5. The speedometer will be inaccurate. The speedometer measures the number of tire revolutions per second, so its readings will be too low.
7. The car cannot round a turn at constant velocity because constant velocity means the direction of the velocity is not changing. The statement is correct if the word "velocity" is replaced by the word "speed."
9. The gravitational force exerted on the moon by the planet produces the centripetal acceleration. Mathematically,

$$m_{\text{moon}} \left(\frac{v_t^2}{r} \right) = G \frac{M m_{\text{moon}}}{r^2}$$

or the mass of the planet is $M = rv_t^2/G$. Both r and v_t can be determined by observing the motion of the moon, and the mass of the planet is then easily computed.

11. Consider an individual standing against the inside wall of the cylinder with her head pointed toward the axis of the cylinder. As the cylinder rotates, the person tends to move in a straight-line path tangent to the circular path followed by the cylinder wall. As a result, the person presses against the wall, and the normal force exerted on her provides the radial force required to keep her moving in a circular path. If the rotational speed is adjusted such that this normal force is equal in magnitude to her weight on Earth, she would not be able to distinguish between the artificial gravity of the colony and ordinary gravity.

13. The tendency of the water is to move in a straight-line path tangent to the circular path followed by the container. As a result, at the top of the circular path, the water presses against the bottom of the pail, and the normal force exerted by the pail on the water provides the radial force required to keep the water moving in its circular path.

15. Any object that moves such that the *direction* of its velocity changes has an acceleration. A car moving in a circular path will always have a centripetal acceleration.

17. When a pilot is pulling out of a dive, blood leaves the head because there is not a great enough radial force to cause it to follow the circular path of the airplane. The loss of blood from the brain can cause the pilot to black out.

Problems

1. (a) 3.2×10^8 rad   (b) 5.0×10^7 rev
3. 1.99×10^{-7} rad/s, 0.986 deg/day
5. (a) 821 rad/s²   (b) 4.21×10^3 rad
7. 36.5 rev
9. Main rotor: 179 m/s = $0.522 v_{sound}$
   Tail rotor: 221 m/s = $0.644 v_{sound}$
11. $\sim 10^{-2}$ cm
13. 3.2×10^6 m
15. (a) 3.37×10^{-2} m/s² down   (b) 0
17. (a) 0.35 m/s²   (b) 1.0 m/s
   (c) 0.35 m/s², 0.94 m/s², 1.0 m/s² at 20° forward with respect to the direction of a_c
19. (a) 1.10 kN   (b) 2.04 times her weight
21. 12 m/s
23. (a) 18.0 m/s²   (b) 900 N
   (c) 1.84; this large coefficient is unrealistic, and he will not be able to stay on the merry-go-round.
25. (a) 9.8 N   (b) 9.8 N   (c) 6.3 m/s
27. (a) 1.58 m/s²   (b) 455 N upward
   (c) 329 N upward
   (d) 397 N directed inward and 80.8° above horizontal
29. 321 N toward Earth
31. 1.1×10^{-10} N at 72° above the $+x$ axis
33. (a) 2.50×10^{-5} N toward the 500-kg object
   (b) Between the two objects and 0.245 m from the 500-kg object
35. (a) 9.58×10^6 m   (b) 5.57 h
37. 1.90×10^{27} kg
39. (a) 1.48 h   (b) 7.79×10^3 m/s
   (c) 6.43×10^9 J
41. (a) 7.27×10^{-5} rad/s   (b) 7.13 h
43. (a) 2.51 m/s   (b) 7.90 m/s²
   (c) 4.00 m/s
45. (a) 7.76×10^3 m/s   (b) 89.3 min
47. (a) $n = m\left(g - \dfrac{v^2}{r}\right)$ (b) 17.1 m/s
49. (a) $F_{g,\,true} = F_{g,\,apparent} + mR_E\omega^2$
   (b) 732 N (Equator), 735 N (either pole)
51. 0.131

55. 11.8 km/s
57. 0.75 m
59. (a) $v_0 = \sqrt{g\left(R - \dfrac{2h}{3}\right)}$ (b) $h' = \dfrac{R}{2} + \dfrac{2h}{3}$

CHAPTER 8

Quick Quizzes

1. (d)
2. (b)
3. (b)
4. (a)
5. (a) Nathan is correct.
6. (c) The box.
7. (c)
8. (a)

Conceptual Questions

1. In order for you to remain in equilibrium, your center of gravity must always be over your point of support, the feet. If your heels are against a wall, your center of gravity cannot remain above your feet when you bend forward, so you lose your balance.

3. There are two major differences between torque and work. The primary difference is that the displacement in the expression for work is directed along the force, while the important distance in the torque expression is perpendicular to the force. The second difference involves whether there is motion or not—in the case of work, work is done only if the force succeeds in causing a displacement of the point of application of the force. On the other hand, force applied at a perpendicular distance from a rotation axis results in a torque whether or not there is motion.

    As far as units are concerned, the mathematical expressions for both work and torque result in the product of newtons and meters, but this product is called a joule in the case of work and remains as a newton-meter in the case of torque.

5. No. For an object to be in equilibrium, the net external force acting on it must be zero. This is not possible when only one force acts on the object, unless that force should have zero magnitude. In that case, there is really no force acting on the object.

7. As the motorcycle leaves the ground, the friction between the tire and the ground suddenly disappears. If the motorcycle driver keeps the throttle open while leaving the ground, the rear tire will increase its angular speed and, hence, its angular momentum. The airborne motorcycle is now an isolated system, and its angular momentum must be conserved. The increase in angular momentum of the tire directed, say, clockwise must be compensated by an increase in angular momentum of the entire motorcycle counterclockwise. This rotation results in the nose of the motorcycle rising and the tail dropping.

9. In general, you want the rotational kinetic energy of the system to be as small a fraction of the total energy as possible—you want translation, not rotation. You want the wheels to have as little moment of inertia as possible, so that they represent the lowest resistance to changes in rotational motion. Disk-like wheels would have lower moments of inertia than hoop-like wheels, so disks are preferable. The lower the mass of the wheels, the less is the moment of inertia, so light wheels are preferable. The smaller the radius of the wheels, the less is the moment of inertia, so smaller wheels are preferable, within limits—you want the wheels to be large enough to be able to travel relatively smoothly over irregularities in the road.

11. The angular momentum of the gas cloud is conserved. Thus, the product $I\omega$ remains constant. As the cloud shrinks in size, its moment of inertia decreases, so its angular speed ω must increase.

13. We can assume fairly accurately that the driving motor will run at a constant angular speed and at a constant torque. Therefore, as the radius of the takeup reel increases, the tension in the tape will decrease.

$$T = \tau_{\text{const}}/R_{\text{takeup}} \tag{1}$$

$$\tau_{\text{source}} = FR_{\text{source}} = \tau_{\text{const}}R_{\text{source}}/R_{\text{takeup}} \tag{2}$$

As the radius of the source reel decreases, given a decreasing tension, the torque in the source reel will decrease even faster.

This torque will be partly absorbed by friction in the feed heads (which we assume to be small); some will be absorbed by friction in the source reel. Another small amount of the torque will be absorbed by the increasing angular speed of the source reel. However, in the case of a sudden jerk on the tape, the changing angular speed of the source reel becomes important. If the source reel is full, then the moment of inertia will be large, and the tension in the tape will be large. If it is nearly empty, then the angular acceleration will be large instead. Thus, the tape will be more likely to break when the source reel is nearly full. We see the same effect in the case of paper towels; it is easier to snap a towel free when the roll is new than when it is nearly empty.

15. The initial angular momentum of the system (mouse plus turntable) is zero. As the mouse begins to walk clockwise, its angular momentum increases, so the turntable must rotate in the counterclockwise direction with an angular momentum whose magnitude equals that of the mouse. This follows from the fact that the final angular momentum of the system must equal the initial angular momentum (zero).

17. When a ladder leans against a wall, both the wall and the floor exert forces of friction on the ladder. If the floor is perfectly smooth, it can exert no frictional force in the horizontal direction to counterbalance the wall's normal force. Therefore, a ladder on a smooth floor cannot stand in equilibrium. However, a smooth wall can still exert a normal force to hold the ladder in equilibrium against horizontal motion. The counterclockwise torque of this force prevents rotation about the foot of the ladder. So you should choose a rough floor.

Problems

1. 133 N
3. $-207 \text{ N}\cdot\text{m}$, $-145 \text{ N}\cdot\text{m}$, $-95.7 \text{ N}\cdot\text{m}$
5. $5.1 \text{ N}\cdot\text{m}$
7. $F_t = 724 \text{ N}$, $F_s = 716 \text{ N}$
9. 312 N
11. 492 N in position (a); 1.25 kN in position (b)
13. 10-year-old: 0.39 cm below midpoint (0.33% of height)
    20-year-old: 3.4 cm below midpoint (2.0% of height)
15. $T = 1.05 \text{ kN}$, $F_c = 950 \text{ N}$
17. $T = 4.2 \text{ kN}$, $\mathbf{F}_c = 4.1 \text{ kN}$ at $0.57°$ below the horizontal
19. $R = 107 \text{ N}$, $T = 157 \text{ N}$
21. $T_{\text{left wire}} = \frac{1}{3}w$, $T_{\text{right wire}} = \frac{2}{3}w$
23. $T_1 = 501 \text{ N}$, $T_2 = 672 \text{ N}$, $T_3 = 384 \text{ N}$
25. 6.15 m
27. 209 N
29. (a) $99.0 \text{ kg}\cdot\text{m}^2$ (b) $44.0 \text{ kg}\cdot\text{m}^2$
    (c) $143 \text{ kg}\cdot\text{m}^2$
31. (a) $24.0 \text{ N}\cdot\text{m}$ (b) $0.035\,6 \text{ rad/s}^2$
    (c) 1.07 m/s^2
33. (a) 34 N   (b) 33 cm
35. 177 N

37. 0.524
39. (a) 500 J   (b) 250 J   (c) 750 J
41. 276 J
43. 149 rad/s
45. (a) $7.08 \times 10^{33} \text{ J}\cdot\text{s}$ (b)$2.66 \times 10^{40} \text{ J}\cdot\text{s}$
47. 8.0 rev/s
49. 6.73 rad/s
51. $5.99 \times 10^{-2} \text{ J}$
53. (a) 0.360 rad/s counterclockwise
    (b) 99.9 J
55. (a) $\omega = \left(\dfrac{I_1}{I_1 + I_2}\right)\omega_0$ (b) $\dfrac{KE_f}{KE_i} = \dfrac{I_1}{I_1 + I_2} < 1$
59. 21.5 N
61. $36.9°$
63. $a_{\text{sphere}} = \dfrac{g\sin\theta}{1.4}$, $a_{\text{disk}} = \dfrac{g\sin\theta}{1.5}$, $a_{\text{ring}} = \dfrac{g\sin\theta}{2.0}$
    Thus, the sphere wins and the ring comes in last.
65. (a) 1.63 m/s   (b) 54.2 rad/s
67. 1.09 m
69. (a) Mvd (b) Mv^2 (c) Mvd (d) $2v$
    (e) $4Mv^2$ (f) $3Mv^2$
71. $7.5 \times 10^{-11} \text{ s}$
73. (a) $T = \dfrac{Mmg}{M + 4m}$ (b) $a_t = \dfrac{4mg}{M + 4m}$

CHAPTER 9

Quick Quizzes

1. (a)
2. (c)
3. (c)
4. (c)
5. (b)
6. (c)
7. (a)

Conceptual Questions

1. The distance of the legs below the heart increases the pumping pressure required to move blood upward from these extremities. Elastic stockings increase the pressure on the blood in the legs, helping to compensate for the needed pumping pressure.

3. She exerts enough pressure on the floor to dent or puncture the floor covering. The large pressure is caused by the fact that her weight is distributed over the very small cross-sectional area of her high heels. If you are the homeowner, you might want to suggest that she remove her high heels and put on some slippers.

5. If you think of the grain stored in the silo as a fluid, the pressure the grain exerts on the walls of the silo increases with increasing depth just as water pressure in a lake increases with increasing depth. Thus, the spacing between bands is made smaller at the lower portions to counterbalance the larger outward forces on the walls in these regions.

7. The syringe does not "draw" the blood from the vein. By moving the plunger in the syringe back, the pressure inside the syringe is lowered below the pressure inside the vein. Then, the blood pressure within the body forces blood out into the syringe, which "accepts" the blood.

9. In the ocean, the ship floats due to the buoyant force from *salt water*. Salt water is denser than fresh water. As the ship is pulled up the river, the buoyant force from the fresh water in the river is not sufficient to support the weight of the ship, and it sinks.

11. The balance will not be in equilibrium—the lead side will be lower. Despite the fact that the weights on both sides of the balance are the same, the Styrofoam, due to its larger volume, experiences a larger buoyant force from the sur-

rounding air. Thus, the net force of the weight and the buoyant force is larger, in the downward direction, for the lead than for the Styrofoam.

13. The two cans displace the same volume of water and hence experience buoyant forces of equal magnitude. The total weight of the diet cola can must be less than this buoyant force, whereas the total weight of the regular cola can is greater than this force. This is possible even though the two containers are identical and contain the same volume of liquid. Because of the difference in the quantities and densities of the sweeteners used, volume V of the diet mixture has less mass than an equal volume of the regular mixture.

15. As the truck passes, the air between your car and the truck is squeezed into the channel between you and the truck and moves at a higher speed than when your car is in the open. According to Bernoulli's principle, this high-speed air has a lower pressure than the air on the outer side of your car. The difference in pressure provides a net force on your car toward the truck.

17. Opening the windows results in a smaller pressure difference between the exterior and interior of the house, and therefore less tendency for severe damage to the structure due to the Bernoulli effect.

Problems

1. ~ 1 cm
3. 1.8×10^6 Pa
5. 3.5×10^8 Pa
7. 4.4 mm
9. 0.024 mm
11. 36 nm
13. (a) 1.88×10^5 Pa   (b) 2.65×10^5 Pa
15. 1.33×10^3 kg/m^3
17. 3.4×10^2 m
19. 1.33 m
21. 271 kN horizontally toward the cellar
23. 1.05×10^5 Pa
25. 2.1 N·m
27. 9.41 kN
29. 1.4
31. 2.67×10^3 kg
33. 5.57 N
35. (a) 1.46×10^{-2} m^3 (b) 2.10×10^3 kg/m^3
37. 1.07 m/s^2
39. 17.3 N (upper scale), 31.7 N (lower scale)
41. (a) 80 g/s   (b) 0.27 mm/s
43. 12.6 m/s
45. 9.00 cm
47. 1.47 cm
49. (a) 28.0 m/s   (b) 28.0 m/s
    (c) 2.11 MPa
51. (b) For any y less than $y_{max} = P_0/\rho g$
53. 8.3×10^{-2} N/m
55. 5.6×10^{-2} N/m
57. 8.6 N
59. 2.1 MPa
61. 2.8 μm
63. 90 cm
65. $RN = 4.3 \times 10^3$; turbulent flow
67. 1.8×10^{-3} kg/m^3
69. 1.4×10^{-5} N·s/m^2
71. 9.28 MPa
73. 2.5×10^7 capillaries
75. (a) 1.57 kPa, 1.55×10^{-2} atm, 11.8 mm of Hg
    (b) The fluid level in the tap should rise.
    (c) Blockage of flow of the cerebrospinal fluid
77. 4.14×10^3 m^3
79. 15 m
81. 2.25 m above the level of point B

83. 6.4 m
85. 1.3 cm
87. 17.0 cm above the floor
89. 532 cm^3

CHAPTER 10

Quick Quizzes

1. **(b)**
2. **(c)**
3. **(c)**
4. **(c)**
5. **(a)**
6. **(b)**

Conceptual Questions

1. An ordinary glass dish will usually break because of stresses that build up as the glass expands when heated. The expansion coefficient for Pyrex glass is much lower than that of ordinary glass. Thus, a Pyrex dish expands much less than the dish of ordinary glass, and does not normally develop sufficient stress to cause breakage.

3. The accurate answer is that it doesn't matter! Temperatures on the Kelvin and Celsius scales differ by only 273 degrees. This difference is insignificant for temperatures on the order of 10^7 degrees. If we imagine that the temperature is given in kelvins, and ignore any problems with significant figures, then the Celsius temperature is $1.499\,972\,7 \times 10^7$ °C.

5. Mercury must have the larger coefficient of expansion. As the temperature of a thermometer rises, both the mercury and the glass expand. If they both had the same coefficient of linear expansion, the mercury and the cavity in the glass would both expand by the same amount, and there would be no apparent movement of the end of the mercury column relative to the calibration scale on the glass. If the glass expanded more than the mercury, the reading would go down as the temperature went up! Now that we have argued this conceptually, we can look in a table and find that the coefficient for mercury is about 20 times as large as that for glass, so that the expansion of the glass can sometimes be ignored.

7. We can think of each bacterium as being a small bag of liquid containing bubbles of gas at a very high pressure. If the bacterium is raised rapidly to the surface, the ideal gas law tells us that the volume must increase dramatically. In fact, this increase in volume is sufficient to rupture the bacterium.

9. Velocity is a vector quantity, so direction must be considered. If the same number of particles move to the right along the x direction as move to the left along the $-x$ direction, the x component of velocity will be zero.

11. The chip bags contain a sealed sample of air. When the bags are taken up the mountain, the external atmospheric pressure on the bags is reduced. As a result, the difference between the pressure of the air inside the bags and the reduced pressure outside results in a net force pushing the plastic of the bag outward.

13. Additional water vaporizes into the bubble, so that n increases.

Problems

1. (a) -460°C   (b) 37.0°C   (c) -280°F
3. (a) -423°F, 20.28 K   (b) 68°F, 293 K
9. (a) 1 337 K, 2 933 K   (b) 1 596°C, 1 596 K
11. 31 cm
13. 55.0°C
15. (a) -179°C (attainable)
    (b) -376°C (below 0 K, unattainable)
17. 11.507 cm (lid), 11.506 cm (jar)

19. 2.7×10^2 N
21. 1.1 L (0.29 gal)
25. (a) increases   (b) 1.603 cm
27. 8.208×10^{-2} L·atm/mol·K
29. (a) 2.5×10^{19} molecules   (b) 4.1×10^{-21} mol
31. 5.87 atm
33. 7.1 m
35. 16.0 cm^3
37. 6.21×10^{-21} J
39. 6.64×10^{-27} kg
41. (a) 8.76×10^{-21} J
    (b) $v_{rms, He} = 1.62$ km/s, $v_{rms, Ar} = 514$ m/s
43. 16 N
45. 0.663 mm at 78.2° below the horizontal
47. 3.55 L
49. 6.57 MPa
51. The expansion of the mercury is almost 20 times that of the flask (assuming Pyrex glass).
53. shorter, by 0.061 mm
55. 2.74 m
57. 1.61 MPa
59. 0.417 L
61. (a) $\theta = \dfrac{(\alpha_2 - \alpha_1) L_0 (\Delta T)}{\Delta r}$
    (c) The bar bends in the opposite direction.

CHAPTER 11

Quick Quizzes

1. (a) Water, glass, iron.   (b) Iron, glass, water.
2. (d)
3. (c)
4. (c)
5. (e)

Conceptual Questions

1. Having sweat glands only in the pads of their feet, dogs rely heavily on the evaporation of fluid on the tongue for cooling. Panting involves rapid shallow breathing that accelerates evaporation of fluid from the tongue. Dogs have a rich blood supply in their tongues, and energy produced in the dog's muscles and tissues is transported to the tongue where is it eliminated through panting.
3. When you rub the surface, you increase the temperature of the rubbed region. With the metal surface, some of this energy is transferred away from the rubbing site by conduction. Thus, the temperature in the rubbed area is not as high for the metal as it is for the wood, and it feels relatively cooler than the wood.
5. The fruit loses energy into the air by radiation and convection from its surface. Before ice crystals can form inside the fruit to rupture cell walls, all of the liquid water on the skin will have to freeze. The resulting time delay may prevent damage within the fruit throughout a frosty night. Furthermore, a surface film of ice provides some insulation to slow subsequent energy loss by conduction from within the fruit.
7. The operation of an immersion coil depends on the convection of water to maintain a safe temperature. As the water near a coil warms up, the warmed water floats to the top due to Archimedes's principle. The temperature of the coil cannot go higher than the boiling temperature of water, 100°C. If the coil is operated in air, the convection process is reduced, and the upper limit of 100°C is removed. As a result, the coil can become hot enough to be damaged. If the coil is used in an attempt to warm a thick liquid like stew, the convection process cannot occur fast enough to carry energy away from the coil, so that it again may become hot enough to be damaged.

9. One of the ways that objects transfer energy is by radiation. If we consider a mailbox, the top of the box is oriented toward the clear sky. Radiation emitted by the top of the mailbox goes upward and into space. There is little radiation coming down from space to the top of the mailbox. Radiation leaving the sides of the mailbox is absorbed by the environment. Radiation from the environment (tree, houses, cars, etc.), however, can enter the sides of the mailbox, keeping them warmer than the top. As a result, the top is the coldest portion and frost forms there first.
11. The tile is a better conductor of energy than carpet. Thus, the tile conducts energy away from your feet more rapidly than does the carpeted floor.
13. The large amount of energy stored in the concrete during the day as the sun falls on it is released at night resulting in an overall higher average temperature than the countryside. The heated air in a city rises as it is displaced by cooler air moving in from the countryside. Thus, evening breezes tend to blow from country to city.
15. The fingers are wetted to create a layer of steam between the fingers and the molten lead. The steam acts as an insulator and prevents serious burns. That said, the molten lead demonstration is still dangerous, and we do not recommend it.
17. The increase in the temperature of the ethyl alcohol will be about twice that of the water.
19. When water is heated at atmospheric pressure, the temperature of the water does not exceed its boiling temperature until all the water has evaporated. If done carefully, the rate of energy transfer to the water through the paper is sufficient to keep the temperature of the paper near that of the water, and hence at or below the boiling temperature of water. This temperature is below that required to ignite the paper.

Problems

1. 1.03 kJ
3. (a) 1.67×10^{18} J   (b) 53.1 yr
5. 0.258°C
7. 2.85 km
9. 176°C
11. 88 W
13. 23°C
15. 80 g
17. 1.8×10^3 J/kg·°C
19. 0.26 kg
21. 65°C
23. 21 g
25. 2.3 km
27. 16°C
29. Liquid lead at 805°C if the specific heat is constant
31. (a) all ice melts, $T_f = 40$°C   (b) 8.0 g melts, $T_f = 0$°C
33. (a) 0.22 kW   (b) 13 mW   (c) 56 mW
35. 402 MW
37. 14 ft^2·°F·h/Btu
39. 9.0 cm
41. 0.11 kW
43. $\sim 10^3$ W
45. 16
47. 12 kW
49. 6.0×10^3 kg
51. 2.3 kg
53. 29°C
55. 30.3 kcal/hr
57. 109°C
59. 51.2°C
61. (a) 74 stops
    (b) Assumes no energy loss to surroundings and that all internal energy generated stays with the brakes.
63. (b) 2.7×10^3 J/kg·°C

CHAPTER 12

Quick Quizzes

1. (a) near 0, +, +
   (b) +, near 0, +
   (c) near 0, −, −
2. A is isovolumetric, B is adiabatic, C is isothermal, *D* is isobaric.
3. C, B, A
4. (b)
5. The number 7 is the most probable outcome. The numbers 2 and 12 are the least probable outcomes.

Conceptual Questions

1. The carbon dioxide undergoes a large increase in volume as it emerges from the container and spends considerable energy doing work on the outside air. Because the spent energy is not immediately replaced by heat, there is a large decrease in the internal energy, and hence the temperature, of the gas.
3. If there is no change in internal energy, then, according to the first law, the heat is equal to the negative of the work done on the gas (and thus equal to the work done *by* the gas). $Q = -W = W_{by gas}$.
5. The energy that is leaving the body by work and heat is replaced by means of biological processes that transform chemical energy in the food that the individual ate into internal energy. Thus, the temperature of the body can be maintained.
7. The statement shows a misunderstanding of the concept of heat. Heat is energy in the process of being transferred, not a form of energy that is held or contained. If you wish to speak of energy that is "contained," you speak of **internal energy,** not heat. The following are correct statements: (1) "Given any two objects in thermal contact, the one with the higher temperature will transfer energy by heat to the other." (2) "Given any two objects of equal mass, the one with the higher product of absolute temperature and specific heat contains more internal energy."
9. Although no energy is transferred into or out of the system by heat, work is done on the system as the result of the agitation. Consequently, both the temperature and the internal energy of the coffee increase.
11. A decrease in temperature indicates a decrease in internal energy. This can occur, at least for a short duration, if the system does a quantity of work that exceeds the heat input during some time interval.
13. Practically speaking, it is not possible to create a heat engine on Earth that creates no thermal pollution, because there must be both a hot heat source (energy reservoir) and a cold heat sink (low-temperature energy reservoir). The heat engine will warm the cold heat sink, and will cool down the heat source. If either of those two events is undesirable, then there will be thermal pollution.

   There are some circumstances where the thermal pollution would be negligible. For example, suppose a satellite in space were to run a heat pump between its sunny side and its dark side. The satellite would intercept some of the energy that gathered on one side and would dump it to the dark side. Since neither of those effects would be particularly undesirable, it could be said that such a heat pump produced no thermal pollution.
15. The rest of the Universe must have an entropy change of +8.0 J/K or more.
17. The first law is a statement of conservation of energy that says that we cannot devise a cyclic process that produces more energy than we put into it. If the cyclic process takes in energy by heat and puts out work, we call the device a heat engine. In addition to the first law's limitation, the second law says that during the operation of a heat engine, some energy must be rejected as heat to the environment. As a result, it is theoretically impossible to construct a engine that will work with 100% efficiency.

Problems

3. 1.1×10^4 J
5. (a) -810 J (b) -507 J (c) -203 J
7. (a) -6.1×10^5 J (b) 4.6×10^5 J
9. (a) 1.09×10^3 K (b) -6.81 kJ
11. (a) $Q < 0, W = 0, \Delta U < 0$ (b) $Q > 0, W = 0, \Delta U > 0$
13. (a) 567 J (b) 167 J
15. (a) -88.5 J (b) 722 J
17. 720 J from the system
19. (a) -9.12×10^{-3} J (b) -333 J
21. (a) 0.95 J (b) 3.2×10^5 J
    (c) 3.2×10^5 J
23. 0.489 (or 48.9%)
25. (a) 0.333 (or 33.3%) (b) $2/3$
27. (a) 0.672 (or 67.2%) (b) 58.8 kW
29. (a) 0.294 (or 29.4%) (b) 500 J
    (c) 1.67 kW
31. $\frac{1}{8}$
33. $0.49°$C
35. (a) -1.2 kJ/K (b) 1.2 kJ/K
37. 57 J/K
39. 3.27 J/K
41. (a)

End Result	Possible Tosses	Total Number of Same Result
All H	HHHH	1
1T, 3H	HHHT, HHTH, HTHH, THHH	4
2T, 2H	HHTT, HTHT, THHT, HTTH, THTH, TTHH	6
3T, 1H	TTTH, TTHT, THTT, HTTT	4
All T	TTTT	1

Most probable result = 2H and 2T

   (b) all H or all T (c) 2H and 2T
43. The maximum efficiency possible with these reservoirs = 50%; claim is invalid.
45. (a) $-|Q_h|/T_h$ (b) $|Q_c|/T_c$
    (c) $|Q_h|/T_h - |Q_c|/T_c$ (d) 0
47. $2.8°$C
51. (a) 12.2 kJ (b) 4.05 kJ (c) 8.15 kJ
53. (a) 26 J (b) 9.0×10^5 J
    (c) 9.0×10^5 J
55. (a) 2.49 kJ (b) 1.50 kJ (c) -990 J
57. (a) 0.554 (or 55.4%)
    (b) The Carnot efficiency is 0.749 (or 74.9%)
59. (a) 2.6×10^3 metric tons/day
    (b) $\$7.7 \times 10^6$/yr (c) 4.1×10^4 kg/s

CHAPTER 13

Quick Quizzes

1. (d)
2. (c)
3. (b)
4. (d)
5. (b) decreases when elevator accelerates upward
   (c) remains the same when elevator moves with constant velocity

6. (a)
7. (b)

Conceptual Questions

1. It will increase. The speed of the wave varies inversely with the mass per unit length of the rope, so it is higher in the lighter rope. The frequency is unchanged as the wave passes from one rope to the other one. The wavelength is $\lambda = v/f$, so with the frequency constant and the speed increasing, the wavelength must increase.

3. No. Because the total energy is $E = \frac{1}{2}kA^2$, changing the mass of the object while keeping A constant has no effect on the total energy. When the object is at displacement x from equilibrium, the potential energy is $\frac{1}{2}kx^2$, independent of mass, and the kinetic energy is $KE = E - \frac{1}{2}kx^2$, also independent of mass.

5. When the spring with two objects on opposite ends is set into oscillation in space, the coil in the exact center of the spring does not move. Thus, we can imagine clamping the center coil in place without affecting the motion. If we do this, we have two separate oscillating systems, one on each side of the clamp. The half-spring on each side of the clamp has twice the spring constant of the full spring, as shown by the following argument. The force exerted by a spring is proportional to the separation of the coils as the spring is extended. Imagine that we extend a spring by a given distance and measure the distance between coils. We then cut the spring in half. If one of the half-springs is now extended by the same distance, the coils will be twice as far apart as they were for the complete spring. Thus, it takes twice as much force to stretch the half-spring, from which we conclude that the half-spring has a spring constant that is twice that of the complete spring. So our clamped system of objects on two half-springs will vibrate with a frequency that is higher than f by a factor of the square root of 2.

7. The bouncing ball is not an example of simple harmonic motion. The ball does not follow a sinusoidal function for its position as a function of time. The daily movement of a student is also not simple harmonic motion, because the student stays at a fixed location (school) for a long period of time. If this motion were sinusoidal, the student would move more and more slowly as she approached her desk, and as soon as she sat down at the desk, she would start to move back toward home again.

9. We assume that the buoyant force acting on the sphere is negligible in comparison to its weight, even when the sphere is empty. We also assume that the bob is small compared to the pendulum length. Then, the frequency of the pendulum is

$$f = \frac{1}{T} = \frac{1}{2\pi}\sqrt{\frac{g}{L}}$$

independent of mass. Thus, the frequency does not change as the water leaks out.

11. As the temperature increases, the length of the pendulum increases, due to thermal expansion, and with a greater length, the period of the pendulum increases. Thus, it takes longer to execute each swing, so that each second according to the clock will take longer than an actual second. Thus, the clock will run slow.

13. A pulse in a long line of people is longitudinal, because the movement of people is parallel to the direction of propagation of the pulse. The speed is determined by the reaction time of the people and the speed with which they can move, once a space opens up. There is also a psychological factor, in that people may not want to fill a space that opens up in front of them too quickly, so as not to intimidate the person in front of them. The "wave" at a stadium is transverse, because the fans stand up vertically as the wave sweeps past

them horizontally. The speed of this pulse depends on the limits of the fans' abilities to rise and sit rapidly, and on psychological factors associated with the anticipation of seeing the pulse approach the observer's location.

15. A wave on a massless string would have an infinite speed of propagation because the linear mass density of the string is zero.

17. The kinetic energy is proportional to the square of the speed, whereas the potential energy is proportional to the square of the displacement. Therefore, both must be positive quantities.

19. From $v = \sqrt{F/\mu}$, we see that increasing the tension by a factor of 4 doubles the wave speed.

Problems

1. **(a)** 24 N toward the equilibrium position
   (b) 60 m/s^2 toward the equilibrium position
3. **(b)** 1.81 s
   (c) No, the force is not of Hooke's law form.
5. 1.23 cm
7. **(a)** 60 J   **(b)** 49 m/s
9. 2.94×10^3 N/m
11. **(a)** $PE = E/4$ $KE = 3E/4$
    (b) $x = A/\sqrt{2}$
13. 0.478 m
15. **(a)** 0.28 m/s   **(b)** 0.26 m/s
    (c) 0.26 m/s   **(d)** 3.5 cm
17. 39.2 N
19. **(a)** You observe uniform circular motion projected on a plane perpendicular to the motion.
    (b) 0.628 s
21. The horizontal displacement is described by $x(t) = A \cos \omega t$, where A is the distance from the center of the wheel to the crank pin.
23. 0.63 s
25. 3.14×10^6 Pa
27. **(a)** 11.0 N toward the left   **(b)** 0.881 oscillations
29. $v = \pm \omega A \sin \omega t, a = -\omega^2 A \cos \omega t$
31. 1.001 5
33. **(a)** slow   **(b)** 9:47 A.M.
35. **(a)** $L_{Earth} = 25$ cm, $L_{Mars} = 9.4$ cm,
    (b) $m_{Earth} = m_{Mars} = 0.25$ kg
37. **(a)** 9.00 cm   **(b)** 20.0 cm
    (c) 40.0 ms   **(d)** 5.00 m/s
39. **(a)** 11.4 ns   **(b)** 3.41 m
41. 31.9 cm
43. 80.0 N
45. **(a)** 30.0 N   **(b)** 25.8 m/s
47. 28.5 m/s
49. **(a)** 0.051 kg/m   **(b)** 20 m/s
51. 13.5 N
53. **(a)** Constructive interference gives $A = 0.50$ m.
    (b) Destructive interference gives $A = 0.10$ m.
55. 6.62 cm
57. **(a)** 588 N/m   **(b)** 0.700 m/s
59. 1.1 m/s
61. **(a)** 15.8 rad/s   **(b)** 5.23 cm
63. $F_{tangential} = -[(\rho_{air} - \rho_{He})Vg/L]s, T = 1.40$ s
67. **(a)** 0.50 m/s   **(b)** 8.6 cm

CHAPTER 14

Quick Quizzes

1. **(c)**
2. **(a)** 10 dB   **(b)** 13 dB
3. **(c)**
4. **(b)**

Conceptual Questions

1. Most speech contains frequencies below about 4 000 Hz. Thus, the spoken word is not greatly affected. However, music contains frequencies spanning the gamut of 20 Hz to 20 000 Hz, and listeners may find their enjoyment of music less than optimal.

3. The camera is designed such that it assumes the speed of sound is 345 m/s, the speed of sound at a room temperature of 23°C. If the temperature should decrease to, say, 0°C, the speed will also decrease, and the camera will note that it takes longer for the sound to make its round trip. Thus, it will think the object is farther away than it really is.

    It is of interest to note that bats use echo-sounding like this to locate insects or to avoid obstacles in front of them, something that they must do because of poor eyesight and high speed. Likewise, blue whales use this technique to avoid objects in their path. The need here is obvious because a typical whale has a mass of 10^5 kg and travels at a relatively fast speed of 20 mi/h. Thus, it takes a long time for it to stop its motion or to change direction.

5. Sophisticated electronic devices break the frequency range of about 60 to 4 000 Hz used in telephone conversations into several frequency bands and then mix them in a predetermined pattern so that they become unintelligible. The descrambler, of course, moves the bands back into their proper order.

7. A rise in temperature will increase the dimensions of the wind instrument much less than it increases the speed of sound in the enclosed air. This effect will raise the resonant frequencies that are produced by the instrument. The instrument goes sharp as the temperature increases and goes flat as the temperature decreases.

9. The echo is Doppler shifted, and the shift is like both a moving source and a moving observer. The sound that leaves your horn in the forward direction is Doppler shifted to a higher frequency, because it is coming from a moving source. As the sound reflects back and comes toward you, you are a moving observer, so there is a second Doppler shift to an even higher frequency. If the sound reflects from the spacecraft coming toward you, there is a different moving source shift to an even higher frequency. The reflecting surface of the spacecraft acts as a moving source.

11. The center of the string is a node for the second harmonic, as well as for every even-numbered harmonic. By placing the finger at the center and plucking, the guitarist is eliminating any harmonic that does not have a node at that point, which is all the odd harmonics. The even harmonics can vibrate relatively freely with the finger at the center because they exhibit no displacement at that point. The result is a sound with a mixture of frequencies that are integer multiples of the second harmonic, which is one octave higher than the fundamental.

13. The bow string is pulled away from equilibrium and released, similar to the way that a guitar string is pulled and released when it is plucked. Thus, standing waves will be excited in the bow string. If the arrow leaves from the exact center of the string, then a series of odd harmonics will be excited. Even harmonics will not be excited because they have a node at the point where the string exhibits its maximum displacement.

15. At the instant at which there is no displacement of the string, the string is still moving. Thus, the energy is present at that instant entirely as kinetic energy of the string.

17. As the child is approaching you, the sound is Doppler shifted to a frequency higher than the natural frequency of the whistle. You will momentarily hear the natural frequency just as the child comes parallel with you and is in the act of passing. As soon as the child is past, you will hear sound that is Doppler shifted to a frequency lower than the natural frequency of the whistle. As the frequency of the sound steps down from one constant value to a second constant value, the intensity of the sound varies continuously. The loudness smoothly increases to a maximum and then decreases.

19. The two engines are running at slightly different frequencies, thus producing a beat frequency between the two.

Problems

1. 5.56 km
3. 32°C
5. 1.43 km/s
7. 1.99 km
9. (a) 1.00×10^{-2} W/m^2 (b) 105 dB
11. 20 km
13. 9 additional machines
15. 1.0×10^{-7} W/m^2
17. (a) 1.3×10^2 W (b) 96 dB
21. (a) 75.2-Hz drop (b) 0.953 m
23. 595 Hz
25. 0.391 m/s
27. 19.3 m
29. 48°
31. 800 m
33. (a) 0.240 m (b) 0.855 m
35. (a) Nodes at 0, 2.67 m, 5.33 m, and 8.00 m
    Antinodes at 1.33 m, 4.00 m, and 6.67 m
    (b) 18.6 Hz
37. At 0.089 1 m, 0.303 m, 0.518 m, 0.732 m, 0.947 m, and 1.16 m from one speaker
39. (a) 79 N (b) 2.1×10^2 Hz
41. 19.976 kHz
43. 58 Hz
45. 3.0 kHz
47. (a) 0.552 m (b) 317 Hz
49. 5.26 beats/s
51. 3.79 m/s toward the station,
    3.88 m/s away from the station
53. (a) 1.98 beats/s (b) 3.40 m/s
55. 1.76 cm
57. 262 kHz
59. 64 dB
61. 439 Hz and 441 Hz
63. 32.9 m/s
65. 3.97 beats/s
67. 1.34×10^4 N
69. 1 204 Hz

CHAPTER 15

Quick Quizzes

1. d
2. b
3. c
4. b
5. a
6. c and d
7. *A*, *B*, and *C*
8. (b) and (d)

Conceptual Questions

1. Because of lower moisture content during winter, the air acts as a better insulator and allows larger static charges to build up on objects during winter than during summer. As a result, shocks from static electricity are more severe in winter months than in summer months.

3. The configuration shown is inherently unstable. The negative charges repel each other. If there is any slight rotation of one of the rods, the repulsion can result in further rota-

tion away from this configuration. There are three conceivable final configurations shown below. Configuration (a) is stable—if the positive upper ends are pushed toward each other, they will repel and move the system back to the original configuration. Configuration (b) is an equilibrium configuration, but it is unstable—if the lower ends are moved toward each other, the attraction of the lower ends will be larger than that of the upper ends and the configuration will shift to (c). Configuration (c) is another possible stable configuration.

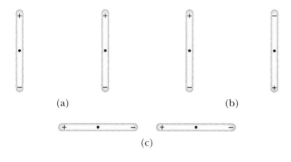

5. Move an object A with a net positive charge so it is near, but not touching, a neutral metallic object B that is insulated from the ground. The presence of A will polarize B, causing an excess negative charge to exist on the side nearest A and an equal magnitude excess positive charge to exist on the side farthest from A. While A is still near B, touch B with your hand. This allows additional electrons to flow from ground through your body and onto B. With A continuing to be near but not touching B, remove your hand from B, thus trapping the excess electrons on B. When A is now removed, B is left with excess electrons causing a net negative charge. This negative charge, by means of mutual repulsion, will now spread uniformly over the entire surface of B.

7. An object's mass decreases very slightly (immeasurably) when it is given a positive charge, because it loses electrons. When the object is given a negative charge, its mass increases slightly because it gains electrons.

9. Electric field lines start on positive charges and end on negative charges. Thus, if the fair weather field is directed into the ground, the ground must have a negative charge.

11. The two charged plates create a region of uniform electric field between them, directed from the positive toward the negative plate. Once the ball is disturbed so as to touch one plate, say the negative one, some negative charge will be transferred to the ball and it will experience an electric force that will accelerate it to the positive plate. Once the charge touches the positive plate, it will release its negative charge, acquire a positive charge, and accelerate back to the negative plate. The ball will continue to move back and forth between the plates until it has transferred all their net charge, thereby making both plates neutral.

13. The electric shielding effect of conductors depends on the fact that there are two kinds of charge—positive and negative. As a result, charges can move within the conductor so that the combination of positive and negative charges establishes an electric field that exactly cancels the external field within the conductor and any cavities inside the conductor. There is only one type of gravitational charge, however—there is no "negative mass." As a result, gravitational shielding is not possible.

15. The electric field patterns of each of these three configurations do not have sufficient symmetry to make the calculations practical. Gauss's law is only useful for calculating the electric field of highly symmetric charge distributions, such as uniformly charged spheres, cylinders, and sheets.

17. No. The balloon induces charge of opposite sign in the wall, causing the balloon and the wall to be attracted to each other. The balloon eventually falls because its charge slowly diminishes after leaking to ground. Some of the balloon's charge could also be lost due to positive ions in the surrounding atmosphere, which would tend to neutralize the negative charges on the balloon.

19. When the comb is nearby, charges separate on the paper, resulting in the paper being attracted. After contact, charges from the comb are transferred to the paper so that it has the same type charge as the comb. It is thus repelled.

Problems

1. 1.1×10^{-8} N (attractive)
3. 91 N (repulsion)
5. (a) 36.8 N   (b) 5.54×10^{27} m/s^2
7. 5.12×10^5 N
9. (a) 2.2×10^{-5} N (attraction)
    (b) 9.0×10^{-7} N (repulsion)
11. 1.38×10^{-5} N at 77.5° below $-x$ axis
13. 0.872 N at 30.0° below $+x$ axis
15. 7.2 nC
17. 5.5×10^{11} N/C (away from the proton)
19. 7.20×10^5 N/C (downward)
21. 1.63×10^5 N/C
23. (a) 6.12×10^{10} m/s^2 (b) 19.6 μs
    (c) 11.8 m   (d) 1.20×10^{-15} J
25. zero
27. 1.8 m to the left of the -2.5 μC charge
33. (a) 0
    (b) $+5$ μC inside, -5 μC outside
    (c) 0 inside, -5 μC outside
    (d) 0 inside, -5 μC outside
35. 1.3×10^{-3} C
37. (a) 4.8×10^{-15} N   (b) 2.9×10^{12} m/s^2
39. (a) 858 N·m^2/C
    (b) 0   (c) 657 N·m^2/C
41. 4.1×10^6 N/C
43. (a) 0   (b) $k_e q / r^2$ outward
47. 57.5 N
49. 24 N/C in the $+x$ direction
51. (a) $E = 2k_e qb(a^2 + b^2)^{-3/2}$ in $+x$ direction
    (b) $E = 2k_e Qb(a^2 + b^2)^{-3/2}$ in $+x$ direction
53. (a) 0
    (b) 7.99×10^7 N/C (outward)
    (c) 0
    (d) 7.34×10^6 N/C (outward)
55. 3.55×10^5 N·m^2/C
57. 4.4×10^5 N/C
59. (a) 10.9 nC   (b) 5.43×10^{-3} N

CHAPTER 16

Quick Quizzes

1. (b)
2. Either (a) or (b) might be true
3. (a) and (b)
4. (c)
5. (a) C decreases.   (b) Q stays the same.
    (c) E stays the same.   (d) ΔV increases.
    (e) The energy stored increases.
6. (a) C increases.   (b) Q increases.
    (c) E stays the same.   (d) ΔV remains the same.
    (e) The energy stored increases.

Conceptual Questions

1. The statement is false. A uniform field simply means that the field is constant in magnitude and direction. Any electric

field does work on a charge as that charge undergoes a displacement in the direction of the field. Thus, a change in potential occurs when one undergoes any displacement along the direction of the field, even if the field is uniform. If we define the potential to be zero at one point, it will be different from zero at neighboring points that are downhill and uphill in the field.

3. The work done in pulling the capacitor plates farther apart is transferred into additional electric energy stored in the capacitor. The charge is constant and the capacitance decreases, but the potential difference between the plates increases, which results in an increase in the stored electric energy.

5. The power line, if it makes electrical contact with the metal of the car, will raise the potential of the car to 20 kV. It will also raise the potential of your body to 20 kV, because you are in contact with the car. In itself, this is not a problem. If you step out of the car, your body at 20 kV will make contact with the ground, which is at zero volts. As a result, a current will pass through your body, and you will likely be injured. Thus, it is best to stay in the car until help arrives.

7. If two points on a conducting object were at different potentials, then free charges in the object would move, and we would not have static conditions, in contradiction to the initial assumption. (Free positive charges would migrate from higher to lower potential locations; free electrons would move rapidly from lower to higher potential locations.) The charges would continue to move until the potential became equal everywhere in the conductor.

9. The capacitor often remains charged long after the voltage source is disconnected. This residual charge can be lethal. The capacitor can be safely handled after discharging the plates by short-circuiting the device with a conductor, such as a screwdriver with an insulating handle.

11. Field lines represent the direction of the electric force on a positive test charge. If electric field lines were to cross, then at the point of crossing, there would be an ambiguity regarding the direction of the force on the test charge, because there would be two possible forces. Thus, electric field lines cannot cross. It is possible for equipotential surfaces to cross. (However, equipotential surfaces at different potentials cannot intersect.) For example, suppose two identical positive charges are at diagonally opposite corners of a square, and two negative charges of equal magnitude are at the other two corners. Then the planes perpendicular to the sides of the square at their midpoints are equipotential surfaces. These two planes cross each other, at the line perpendicular to the square at its center.

13. (a) From $Q = C(\Delta V)$, if you double the potential difference across a capacitor having a fixed capacitance, you will double the charge stored in the capacitor. (b) The energy stored in a capacitor may be expressed as $W = Q^2/2C$. Assuming a parallel-plate capacitor, the capacitance is $C = \kappa \epsilon_0 A/d$, and the stored energy is $W = Q^2 d/2\kappa \epsilon_0 A$. Thus, doubling the plate separation while holding the charge constant will double the energy stored.

15. You should use a dielectric-filled capacitor whose dielectric constant is very large. Furthermore, you should make the dielectric as thin as possible, keeping in mind that dielectric breakdown must also be considered.

17. Inserting a dielectric into the capacitor increases the capacitance of that device. (a) From $W = C(\Delta V)^2/2$, it is seen that increasing the capacitance while keeping the potential difference constant will increase the energy stored in the capacitor.
(b) From $W = Q^2/2C$, it is seen that increasing the capacitance while holding the charge on the capacitor constant decreases the energy stored in the capacitor.

Problems

1. (a) 6.40×10^{-19} J   (b) -6.40×10^{-19} J
   (c) -4.00 V
3. 1.4×10^{-20} J
5. 1.7×10^6 N/C
7. (a) 1.13×10^5 N/C   (b) 1.80×10^{-14} N
   (c) 4.38×10^{-17} J
9. (a) 0.500 m   (b) 0.250 m
11. (a) 1.44×10^{-7} V   (b) -7.19×10^{-8} V
13. (a) 2.67×10^6 V   (b) 2.13×10^6 V
15. (a) 103 V
    (b) -3.85×10^{-7} J; positive work must be done to separate the charges.
17. -11.0 kV
19. 2.74×10^{-14} m
21. 0.719 m, 1.44 m, 2.88 m. No. The equipotentials are not uniformly spaced. Instead, the radius of an equipotential is inversely proportional to the potential.
23. (a) 1.1×10^{-8} F   (b) 27 C
25. (a) 11.1 kV/m toward the negative plate   (b) 3.74 pF
    (c) 74.7 pC and -74.7 pC
27. 2.26×10^{-5} m^2
29. (a) 13.3 μC on each   (b) 20.0 μC, 40.0 μC
31. (a) 2.00 μF
    (b) $Q_3 = 24.0$ μC, $Q_4 = 16.0$ μC, $Q_2 = 8.00$ μC, $(\Delta V)_2 = (\Delta V)_4 = 4.00$ V, $(\Delta V)_3 = 8.00$ V
33. (a) 5.96 μF
    (b) $Q_{20} = 89.5$ μC, $Q_6 = 63.2$ μC, $Q_3 = Q_{15} = 26.3$ μC
35. $Q_1 = 16.0$ μC, $Q_5 = 80.0$ μC, $Q_8 = 64.0$ μC, $Q_4 = 32.0$ μC
37. (a) $Q_{25} = 1.25$ mC, $Q_{40} = 2.00$ mC
    (b) $Q'_{25} = 288$ μC, $Q'_{40} = 462$ μC, $\Delta V = 11.5$ V
39. $Q'_1 = 3.33$ μC, $Q'_2 = 6.67$ μC
41. 83.6 μC
43. 2.55×10^{-11} J
45. 3.2×10^{10} J
47. $\kappa = 4.0$
49. (a) 8.13 nF   (b) 2.40 kV
51. (a) volume 9.09×10^{-16} m^3, area 4.54×10^{-10} m^2
    (b) 2.01×10^{-13} F
    (c) 2.01×10^{-14} C; 1.26×10^5 electronic charges
55. $v = 1.67$ m/s upward, $y = 6.24$ cm above bottom plate
57. 6.25 μF
59. 4.47 kV
61. 0.75 mC on C_1, 0.25 mC on C_2
65. 50 N

CHAPTER 17

Quick Quizzes

1. d, b = c, a
2. c, d
3. b
4. b, d
5. a
6. $I_a = I_b > I_c = I_d > I_e = I_f$
7. B, B

Conceptual Questions

1. It is made possible by the very high density of charge carriers ($\sim 10^{28}$/m^3) in metallic conductors.
3. The gravitational force pulling the electron to the bottom of a piece of metal is much smaller than the electrical repulsion pushing the electrons apart. Thus, free electrons stay distributed throughout the metal. The concept of charges residing on the surface of a metal is true for a metal with an excess charge. The number of free electrons in an electrically neutral piece of metal is the same as the number of positive ions — the metal has zero net charge.

5. A voltage is not something that "surges through" a completed circuit. A voltage is a potential difference that is applied across a device or a circuit. What goes through the circuit is current. Thus, it would be more correct to say, "1 ampere of electricity surged through the victim's body." Although this current would have disastrous results on the human body, a value of 1 (ampere) doesn't sound as exciting for a newspaper article as 10 000 (volts). Another possibility is to write, "10 000 volts of electricity were applied across the victim's body," which still doesn't sound quite as exciting.

7. We would conclude that the conductor is nonohmic.

9. The shape, dimensions, and the resistivity affect the resistance of a conductor. Because temperature and impurities affect the conductor's resistivity, these factors also affect resistance.

11. The radius of wire B is the square root of three times the radius of wire A, to make its cross-sectional area three times larger.

13. Drift velocity might increase steadily as time goes on, because collisions between electrons and atoms in the wire would be essentially nonexistent and the conduction electrons would move with constant acceleration. The current would rise steadily without bound also, because I is proportional to the drift velocity.

Problems

1. 3.00×10^{20} electrons move past in the direction opposite to the current.

3. 3.0 mA

5. 1.05 mA

7. 27 yr

9. **(a)** n is unaffected **(b)** v_d is doubled

11. 32 V, is 200 times larger than 0.16 V

13. 0.17 mm

15. **(a)** 30 Ω **(b)** 4.7×10^{-4} Ω·m

17. silver ($\rho = 1.59 \times 10^{-8}$ Ω·m)

19. 2.50 mA decrease

21. 1.98 A

23. 26 mA

25. **(a)** 5.89×10^{-2} Ω **(b)** 5.45×10^{-2} Ω

27. **(a)** 3.0 A **(b)** 2.9 A

29. **(a)** 1.2 Ω
    (b) 8.0×10^{-4} (a 0.080% increase)

31. 5.00 A, 24.0 Ω

33. 18 bulbs

35. 11.2 min

37. 34.4 Ω

39. 1.6 cm

41. 295 metric tons/h

43. 26 cents

45. 23 cents

47. $1.2

49. 1.1 km

51. 1.47×10^{-6} Ω·m, differs by 2.0% from value in Table 17.1

53. 0.400 μA

55. **(a)** 667 A **(b)** 50.0 km

57. 3.77×10^{28}/m³

59. **(a)** 144 Ω **(b)** 26 m
    (c) To fit the required length into a small space.
    (d) 25 m

61. 37 MΩ

63. 0.48 kg/s

65. **(a)** 2.6×10^{-5} Ω **(b)** 76 kg

CHAPTER 18

Quick Quizzes

1. Bulb R_1 becomes brighter. Bulb R_2 goes out.

2. **(b)**

3. **(a)**

4. **(a)** decrease; **(b)** decrease; **(c)** increase; **(d)** decrease; **(e)** decrease.

Conceptual Questions

1. No. When a battery serves as a source and supplies current to a circuit, the conventional current flows through the battery from the negative terminal to the positive one. However, when a source having a larger emf than the battery is used to charge the battery, the conventional current is forced to flow through the battery from the positive terminal to the negative one.

3. The total amount of energy delivered by the battery will be less than W. Recall that a battery can be considered to be an ideal, resistanceless battery in series with the internal resistance. When charging, the energy delivered to the battery includes the energy necessary to charge the ideal battery plus the energy that goes into raising the battery's temperature due to I^2r heating in the internal resistance. This latter energy is not available during the discharge of the battery. During discharge, part of the reduced available energy again transforms into internal energy in the internal resistance, further reducing the available energy below W.

5. The starter in the automobile draws a relatively large current from the battery. This large current causes a significant voltage drop across the internal resistance of the battery. As a result, the terminal voltage of the battery is reduced, and the headlights dim accordingly.

7. An electrical appliance has a given resistance. Thus, when it is attached to a power source with a known potential difference, a definite current will be drawn. The device can be labeled with both the voltage and the current. Batteries, however, can be applied to a number of devices. Each device will have a different resistance, so the current from the battery will vary with the device. As a result, only the voltage of the battery can be specified.

9. Connecting batteries in parallel does not increase the emf. A high-current device connected to batteries in parallel can draw currents from both batteries. Thus, connecting the batteries in parallel does increase the possible current output and, therefore, the possible power output.

11. The lightbulb will glow for a very short while as the capacitor is being charged. Once the capacitor is almost totally charged, the current in the circuit will be nearly zero and the bulb does not glow.

13. The bird is at rest on a wire whose electric potential is nearly constant along its length. In order to be electrocuted, a large potential difference is required between the bird's feet. The potential difference between the bird's feet is too small to harm the bird.

15. The junction rule is a statement of conservation of charge. It says that the amount of charge that enters the junction in some time interval must equal the charge that leaves the junction in that time interval. The loop rule is a statement of conservation of energy. It says that the potential increases and decreases around a closed loop in a circuit must add to zero.

17. A few of the factors involved are as follows: the conductivity of the string (is it wet or dry?); how well you are insulated from ground (are you wearing thick rubber- or leather-soled shoes?); the magnitude of the potential difference between you and the kite; the type and condition of the soil under your feet.

19. She will not be electrocuted if she holds onto only one high-voltage wire, because she is not completing a circuit. There is no potential difference across her body, as long as she clings to only one wire. However, she should immediately release the wire once it breaks, because she will become part of a closed circuit when she reaches the ground or comes into contact with another object.

21. **(a)** The intensity of both lamps increases because lamp C is short-circuited and there is current (which increases) only

in lamps A and B. **(b)** The intensity of lamp C goes to zero because the current in this branch goes to zero. **(c)** The current in the circuit increases because the total resistance decreases from $3R$ (with the switch open) to $2R$ (after the switch is closed). **(d)** The voltage drop across lamps A and B increases, while the voltage drop across lamp C becomes zero. **(e)** The power dissipated increases from $\mathcal{E}^2/3R$ (with the switch open) to $\mathcal{E}^2/2R$ (after the switch is closed).

Problems

1. 4.92 Ω
3. 73.8 W; your circuit diagram will consist of two 0.800-Ω resistors in series with the 192-Ω resistance of the bulb.
5. **(a)** 17.1 Ω
   (b) 1.99 A for 4.00 Ω and 9.00 Ω, 1.17 A for 7.00 Ω, 0.818 A for 10.0 Ω
7. 9.8 Ω
9. **(a)** 0.227 A   **(b)** 5.68 V
11. 55 Ω
13. 0.43 A
15. **(a)** Connect two 50-Ω resistors in parallel, and then connect this combination in series with a 20-Ω resistor. **(b)** Connect two 50-Ω resistors in parallel, connect two 20-Ω resistors in parallel, and then connect these two combinations in series with each other.
17. 0.846 A down in 8.00-Ω resistor; 0.462 A down in the middle branch; 1.31 A up in the right-hand branch
19. **(a)** 3.00 mA   **(b)** -19.0 V   **(c)** 4.50 V
21. 10.7 V
23. **(a)** 0.385 mA, 3.08 mA, 2.69 mA
    (b) 69.2 V with c at the higher potential
25. $I_1 = 3.5$ A, $I_2 = 2.5$ A, $I_3 = 1.0$ A
27. $I_{30} = 0.353$ A, $I_5 = 0.118$ A, $I_{20} = 0.471$ A
29. $\Delta V_2 = 3.05$ V, $\Delta V_3 = 4.57$ V, $\Delta V_4 = 7.38$ V, $\Delta V_5 = 1.62$ V
31. **(a)** 12 s   **(b)** 1.2×10^{-4} C
33. 1.3×10^{-4} C
35. 0.982 s
37. **(a)** heater, 10.8 A; toaster, 8.33 A; grill, 12.5 A
    (b) $I_{total} = 31.6$ A, so a 30-A breaker is insufficient.
39. **(a)** 6.25 A   **(b)** 750 W
41. **(a)** 1.2×10^{-9} C, 7.3×10^9 K$^+$ions. Not large, only $1e/320$ Å^2
    (b) 1.7×10^{-9} C, 1.0×10^{10} Na$^+$ ions
    (c) 0.83 μA   **(d)** 7.5×10^{-12} J
43. 11 nW
45. 7.5 Ω
47. **(a)** 15 Ω
    (b) $I_1 = 1.0$ A, $I_2 = I_3 = 0.50$ A, $I_4 = 0.30$ A, and $I_5 = 0.20$ A
    (c) $(\Delta V)_{ac} = 6.0$ V, $(\Delta V)_{ce} = 1.2$ V,
    $(\Delta V)_{ed} = (\Delta V)_{fd} = 1.8$ V, $(\Delta V)_{cd} = 3.0$ V,
    $(\Delta V)_{db} = 6.0$ V
    (d) $\mathcal{P}_{ac} = 6.0$ W, $\mathcal{P}_{ce} = 0.60$ W, $\mathcal{P}_{ed} = 0.54$ W, $\mathcal{P}_{fd} = 0.36$ W, $\mathcal{P}_{cd} = 1.5$ W, $\mathcal{P}_{db} = 6.0$ W
49. **(a)** 12.4 V   **(b)** 9.65 V
51. $I_1 = 0$, $I_2 = I_3 = 0.50$ A
53. 112 V, 0.200 Ω
55. **(a)** $R_x = R_2 - (\frac{1}{4})R_1$
    (b) $R_x = 2.8$ Ω (inadequate grounding)
59. $\mathcal{P} = \dfrac{(144 \text{ V}^2)R}{(R + 10.0 \text{ Ω})^2}$

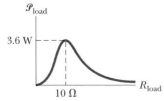

CHAPTER 19

Quick Quizzes

1. a or c might be true
2. c
3. a
4. c
5. b

Conceptual Questions

1. No. In a uniform magnetic field, the poles of the bar magnetic experience forces that are equal in magnitude but oppositely directed. Thus, the net force exerted on the magnet by the magnetic field is zero.

3. The proton moves in a circular path upward on the page. After completing half the circle, it exits the field and moves in a straight-line path back in the direction from whence it came. An electron behaves similarly, but the direction of traversal of the circle is downward, and the radius of the circular path is smaller.

5. The magnetic force on a moving charged particle is always perpendicular to the direction of motion. There is no magnetic force on the charge when it moves parallel to the direction of the magnetic field. However, the force on a charged particle moving in an electric field is never zero and is always parallel to the direction of the electric field. Therefore, by projecting the charged particle in different directions, it is possible to determine the nature of the field.

7. The magnetic field produces a magnetic force on the electrons moving toward the screen that produce the image. This magnetic force deflects the electrons to regions on the screen other than the ones to which they are supposed to go. The result is a distorted image.

9. Such levitation would never occur. At the North Pole where Earth's magnetic field is downward, toward the equivalent of a buried south pole, a coffin would be repelled if its south magnetic pole were directed downward. However, equilibrium would be only transitory as any slight disturbance would upset the balance between the magnetic force and the gravitational force.

11. If you are moving along with the electrons, you will measure zero current for the electrons, so the electrons would not produce a magnetic field according to your observations. However, the fixed positive charges in the metal are now moving backward relative to you and creating a current equivalent to the forward motion of the electrons when you were stationary. Thus, you will measure the same magnetic field as when you were stationary, but it will be due to the positive charges presumed to be moving from your point of view.

13. A compass does not detect currents in wires near light switches for two reasons. Because the cable to the light switch contains two wires, with one carrying current to the switch and the other away from the switch, the net magnetic field is very small and falls off rapidly. The second reason is that the current is alternating at 60 Hz with increasing distance. As a result, the magnetic field is oscillating at 60 Hz, also. This frequency is too fast for the compass to follow, so the effect on the compass reading averages to zero.

15. The levitating wire is stable with respect to vertical motion—if it is displaced upward, the repulsive force weakens, and the wire drops back down. Conversely, if it drops lower, the repulsive force increases, and it moves back up. The wire is not stable, however, with respect to lateral movement. If it moves away from the vertical position directly over the lower wire, the repulsive force has a sideways component, which pushes the wire away.

   In the case of the attracting wires, the hanging wire is not stable for vertical movement. If it rises, the attractive force increases, and the wire moves even closer to the upper wire.

If the hanging wire falls, the attractive force weakens, and the wire falls farther. If the wire moves to the right, it moves farther from the upper wire and the attractive force decreases. Although there is a restoring force component pulling it back to the left, the vertical force component is not strong enough to hold it up and it falls.

17. Each coil of the Slinky becomes a magnet, because a coil acts as a current loop. Because the sense of rotation of the current is the same in all coils, each coil becomes a magnet with the same orientation of poles. Thus, all of the coils attract, and the Slinky compresses.

19. There is no net force on the wires, but there is a torque. To understand this, imagine a fixed vertical wire, and a free horizontal wire (Fig. ANS Q19.19). The vertical wire carries an upward current, and creates a magnetic field that circles the wire. Each segment of the horizontal wire (of length ℓ) also carries a current that interacts with the magnetic field according to the equation $F = BI\ell \sin \theta$. Applying the right-hand rule, we see that the horizontal wire experiences an upward force on one side, and an equal downward force on the other. The forces cancel, creating a torque around the point at which the wires cross.

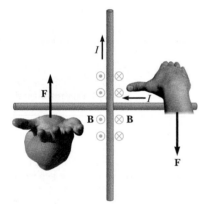

Problems

1. **(a)** horizontal and due east
   (b) horizontal and 30°N of E
   (c) horizontal and due east
   (d) zero force
3. **(a)** into the page   **(b)** toward the right
   (c) toward the bottom of the page
5. $F_g = 8.93 \times 10^{-30}$ N (downward), $F_e = 1.60 \times 10^{-17}$ N (upward), $F_m = 4.80 \times 10^{-17}$ N (downward)
7. 2.83×10^7 m/s west
9. 0.021 T in the $-y$ direction
11. 8.0×10^{-3} T in the $+z$ direction
13. **(a)** to the left   **(b)** into the page
    (c) out of the page   **(d)** toward top of page
    (e) into the page   **(f)** out of the page
15. 7.50 N
17. 0.131 T (downward)
19. 0.20 T directed out of the page
21. ab: 0, bc: 0.040 0 N in the $-x$ direction, cd: 0.040 0 N in the $-z$ direction, da: 0.040 0 $\sqrt{2}$ N parallel to the xz plane and at 45° with both $+x$ and $+z$ directions
23. 9.05×10^{-4} N·m tending to make the left-hand side of the loop move toward you and the right-hand side move away.
25. **(a)** 3.97°   **(b)** 3.39×10^{-3} N·m
27. 2.0×10^{-12} kg
31. 1.77 cm
33. **(a)** 5.00 cm   **(b)** 8.78×10^6 m/s
35. 20.0 μT
37. 2.0 cm
39. 20.0 μT toward the bottom of the page

41. 0.167 μT out of the page
43. **(a)** 4.00 m   **(b)** 7.50 nT   **(c)** 1.26 m   **(d)** zero
45. 4.5 mm
47. 31.8 mA
49. 2.26×10^{-4} N away from the center, zero torque
51. 1.7 N·m
53. **(a)** 0.500 μT out of the page
    (b) 3.89 μT parallel to the xy plane and at 59.0° clockwise from the $+x$ direction
55. 2.13 cm
57. **(a)** 1.33 m/s
    (b) the sign of the emf is independent of charge
59. 1.41×10^{-6} N
61. 13.0 μT toward the bottom of the page
63. **(a)** 53 μT toward the bottom of the page
    (b) 20 μT toward the bottom of the page
    (c) 0
65. **(a)** -8.00×10^{-21} kg·m/s   **(b)** 8.90°

CHAPTER 20

Quick Quizzes

1. *b, c, a*
2. The left wing tip on the west side of the airplane.
3. b
4. c
5. b

Conceptual Questions

1. According to Faraday's law, an emf is induced in a wire loop if the magnetic flux through the loop changes with time. In this situation, an emf can be induced by either rotating the loop around an arbitrary axis or by changing the shape of the loop.
3. As the spacecraft moves through space, it is apparently moving from a region of one magnetic field strength to a region of a different magnetic field strength. The changing magnetic field through the coil induces an emf and corresponding current in the coil.
5. If the bar were moving to the left, the magnetic force on the negative charges in the bar would be upward, causing an accumulation of negative charge on the top, and positive charges at the bottom. Hence, the electric field in the bar would be upward.
7. If, for any reason the magnetic field should change rapidly, a large emf could be induced in the bracelet. If the bracelet were not a continuous band, this emf would cause high-voltage arcs to occur at any gap in the band. If the bracelet were a continuous band, the induced emf would produce a large induced current and result in resistance heating of the bracelet.
11. As the aluminum plate moves into the field, eddy currents are induced in the metal by the changing magnetic field at the plate. The magnetic field of the electromagnet interacts with this current producing a retarding force on the plate, slowing it down. In a similar fashion, as the plate leaves the magnetic field, a current is induced, and once again there is an upward force to slow the plate.
13. The energy stored in an inductor carrying a current I is equal to $PE_L = \frac{1}{2}LI^2$. Therefore, doubling the current quadruples the energy stored in the inductor.
15. If an external battery is acting to increase the current in the inductor, an emf is induced in a direction to oppose the increase of current. Likewise, if we attempt to reduce the current in the inductor, the emf set up tries to support the current. Thus, the induced emf always acts to oppose the change occurring in the circuit, or it acts in the "back" direction to the change.

Problems

1. $5.9 \times 10^{-2}\,\text{T}\cdot\text{m}^2$
3. $1.72 \times 10^{-6}\,\text{T}\cdot\text{m}^2$, 0, $9.15 \times 10^{-7}\,\text{T}\cdot\text{m}^2$
5. $\Phi_{B,\,\text{net}} = 0$ **(b)** 0
7. **(a)** $3.1 \times 10^{-3}\,\text{T}\cdot\text{m}^2$ **(b)** $\Phi_{B,\,\text{net}} = 0$
9. 9.82 mV
11. 160 A
13. 2.7 T/s
15. **(a)** $4.0 \times 10^{-6}\,\text{T}\cdot\text{m}^2$ **(b)** 18 μV
17. 11.6 μV
19. 0.763 V
21. **(a)** toward the east   **(b)** 4.58×10^{-4} V
23. **(a)** left to right   **(b)** right to left
25. left to right
27. into the page
29. **(a)** right to left   **(b)** right to left
     (c) left to right   **(d)** left to right
31. 28.8 V
33. **(a)** 11 Ω **(b)** 76 V
35. **(a)** 60 V   **(b)** 57 V   **(c)** 0.13 s
37. 100 V
39. **(a)** 2.0 mH   **(b)** 38 A/s
43. 12 mH
45. 1.92 Ω
47. 0.140 J
49. **(a)** 18 J   **(b)** 7.2 J
51. negative ($V_a < V_b$)
53. **(a)** 20.0 ms   **(b)** 37.9 V   **(c)** 1.52 mV   **(d)** 51.8 mA
55. 1.20 μC
57. **(a)** 0.500 A   **(b)** 2.00 W   **(c)** 2.00 W
59. 115 kV
61. **(a)** 0.157 mV (end B is positive)
     (b) 5.89 mV (end A is positive)
63. **(a)** 9.00 A   **(b)** 10.8 N
     (c) b is at the higher potential   **(d)** No

CHAPTER 21

Quick Quizzes

1. c
2. b
3. b
4. b, c

Conceptual Questions

1. For best reception, the length of the antenna should be parallel to the orientation of the oscillating electric field. Because of atmospheric variations and reflections of the wave before it arrives at your location, the orientation of this field may be in different directions for different stations.

3. The primary coil of the transformer is an inductor. When an AC voltage is applied, the back emf due to the inductance will limit the current flow through the coil. If DC voltage is applied, there is no back emf and the current can rise to a higher value. It is possible that this increased current will deliver so much energy to the resistance in the coil that its temperature rises to the point at which insulation on the wire can burn.

5. An antenna that is a conducting line responds to the electric field of the electromagnetic wave—the oscillating electric field causes an electric force on electrons in the wire along its length. The movement of electrons along the wire is detected as a current by the radio and amplified. Thus, a line antenna must have the same orientation as the broadcast antenna. A loop antenna responds to the magnetic field in the radio wave. The varying magnetic field induces a varying current in the loop (Faraday's law), and this signal is amplified. The loop should be in the vertical plane containing the sight line to the broadcast antenna, so the magnetic field lines go through the area of the loop.

7. The flashing of the light according to Morse code is a drastic amplitude modulation—the amplitude is changing from a maximum to zero. In this sense, it is similar to the on-and-off binary code used in computers and compact discs. The carrier frequency is that of the light, on the order of 10^{14} Hz. The signal frequency depends on the skill of the signal operator, but it is on the order of a single hertz, as the light is flashed on and off. The broadcasting antenna for this modulated signal is the filament of the lightbulb in the signal source. The receiving antenna is the eye.

9. The sail should be as reflective as possible so that the maximum momentum is transferred to the sail from the reflection of sunlight.

11. Suppose the extraterrestrial looks around your kitchen. Lightbulbs and the toaster glow brightly in the infrared. Somewhat fainter are the back of the refrigerator and the back of the television set, while the television screen is dark. The pipes under the sink show the same weak glow as the walls until you turn on the faucets. Then the pipe on the right gets darker and that on the left develops a gleam that quickly runs up along its length. The food on the plates shines, as does human skin, the same color for all races. Clothing is dark as a rule, but your seat and the chair seat glow alike after you stand up. Your face appears lit from within, like a jack-o-lantern; your nostrils and the openings of your ear canals are bright; brighter still are the pupils of your eyes.

13. Radio waves move at the speed of light. They can travel around the curved surface of the Earth, bouncing between the ground and the ionosphere, which has an altitude that is small compared to the radius of the Earth. The distance across the lower 48 states is approximately 5 000 km, requiring a travel time of $(5 \times 10^6\,\text{m})/(3 \times 10^8\,\text{m/s}) \sim 10^{-2}$ s. Likewise, radio waves take only 0.07 s to travel halfway around the Earth. In other words, a speech can be heard on the other side of the world (in the form of radio waves) before it is heard at the back of the room (in the form of sound waves).

15. No. The wire will emit electromagnetic waves only if the current varies in time. The radiation is the result of accelerating charges, which can only occur when the current is not constant.

17. The resonance frequency is determined by the inductance and the capacitance in the circuit. It is not affected by the resistance in the circuit, so doubling the resistance does not alter the resonance frequency.

Problems

1. **(a)** 141 V   **(b)** 20.0 A
     (c) 28.3 A   **(d)** 2.00 kW
3. 70.7 V, 2.95 A
5. 6.76 W
9. 4.0×10^2 Hz
11. 17 μF
15. 3.14 A
17. 0.450 T·m^2
19. **(a)** 205 mA   **(b)** 154 V   **(c)** 54.3 V   **(d)** 90.0°
     (e)

$\Delta V_L = 154$ V

$\Delta V = \Delta V_L - \Delta V_C = 100$ V

90°

$\Delta V_C = 54.3$ V

21. (a) $1.4 \text{ k}\Omega$ (b) 0.10 A (c) $51°$
   (d) voltage leads current
23. (a) 89.6 V (b) 109 V
25. 1.88 V
27. (a) 103 V (b) 150 V (c) 127 V (d) 23.6 V
29. (a) $208 \,\Omega$ (b) $40.0 \,\Omega$ (c) 0.541 H
31. (a) $1.8 \times 10^2 \,\Omega$ (b) 0.71 H
33. $2.29 \,\mu\text{H}$
35. $C_{\text{min}} = 4.9 \text{ nF}$, $C_{\text{max}} = 51 \text{ nF}$
37. 0.242 J
39. (a) $1\,600$ (b) 30 A
41. (a) $1.1 \times 10^3 \text{ kW}$ (b) $3.1 \times 10^2 \text{ A}$
   (c) $8.3 \times 10^3 \text{ A}$
43. $1\,000 \text{ km}$, there will always be better use for tax money.
45. 733 nT
47. $2.94 \times 10^8 \text{ m/s}$
49. $E_{\text{max}} = 1.01 \times 10^3 \text{ V/m}$, $B_{\text{max}} = 3.35 \times 10^{-6} \text{ T}$
51. (a) 188 m to 556 m (b) 2.78 m to 3.4 m
53. 60.0 km
55. (a) 18 turns   (b) 3.6 W
57. 99.6 mH
59. 1.7 cents
61. (a) resistor and inductor   (b) $R = 10 \,\Omega$, $L = 30 \text{ mH}$
63. (a) $6.7 \times 10^{-16} \text{ T}$ (b) $5.3 \times 10^{-17} \text{ W/m}^2$
   (c) $1.7 \times 10^{-14} \text{ W}$
65. (a) 0.536 N (b) $8.93 \times 10^{-5} \text{ m/s}^2$
   (c) 33.9 days
67. $4.47 \times 10^{-9} \text{ J}$

CHAPTER 22

Quick Quizzes

1. a
2. Beams 2 and 4 are reflected; beams 3 and 5 are refracted.
3. b
4. c

Conceptual Questions

1. Sound radiated upward at an acute angle with the horizontal is bent back toward Earth by refraction. This means that the sound can reach the listener by this path as well as by a direct path. Thus, the sound is louder.
3. The color will not change, for two reasons. First, despite the popular statement that color depends on wavelength, it actually depends on the *frequency* of the light, which does not change under water. Second, when the light enters the eye, it travels through the fluid within the eye. Thus, even if color did depend on wavelength, the important wavelength is that of the light in the ocular fluid, which does not depend on the medium through which the light traveled to reach the eye.
5. A prism creates a multicolored spectrum when white light passes through it because the white light contains all visible wavelengths. Because the index of refraction varies with wavelength, the different wavelengths emerge from the prism traveling in slightly different directions. This produces the observed rainbow of colors. If only one wavelength of light is allowed to enter the prism, as in the case of the second prism, a single wavelength (and hence color) will emerge from it.
7. No, the catalog information is incorrect. The index of refraction is given by $n = c/v$, where c is the speed of light in vacuum and v is the speed of light in the material. Because light travels faster in a vacuum than any other material, it is impossible for the index of refraction of any material to have a value less than 1.
9. There is no dependence of the angle of reflection on wavelength, because the light does not enter deeply into the material during reflection—it reflects from the surface.

11. A ball covered with mirrors sparkles by reflecting light from its surface. On the other hand, a faceted diamond lets in light at the top, reflects it by total internal reflection in the bottom half, and sends the light out through the top again. Because of its high index of refraction, the critical angle for diamond in air for total internal reflection, $\theta_c = \sin^{-1}(n_{\text{air}}/n_{\text{diamond}})$, is small. Thus, light rays enter through a large area and exit through a very small area with a much higher intensity. When a diamond is immersed in carbon disulfide, the critical angle is increased to $\theta_c = \sin^{-1}(n_{\text{carbon disulfide}}/n_{\text{diamond}})$. As a result, the light is emitted from the diamond over a larger area and appears less intense.
13. The index of refraction of water is 1.333, quite different from that of air, which has an index of refraction of about 1. On the other hand, the index of refraction of liquid helium happens to be much closer to that of air. Consequently, light undergoes less refraction in helium than it does in water.
15. The diamond acts like a prism in dispersing the light into its spectral components. Different colors are observed as a consequence of the manner in which the index of refraction varies with the wavelength.
17. Light travels through vacuum at a speed of 3×10^8 m/s. Thus, an image we see from a distant star or galaxy must have been generated some time ago. For example, the star Altair is 16 lightyears away; if we look at an image of Altair today, we know only what Altair looked like 16 years ago. This may not initially seem significant; however, astronomers who look at other galaxies can get an idea of what galaxies looked like when they were much younger. Thus, it does make sense to speak of "looking backward in time."

Problems

1. $3.00 \times 10^8 \text{ m/s}$
3. 114 rad/s for a maximum intensity of returning light
5. (b) $3.000 \times 10^8 \text{ m/s}$
7. $19.5°$ above the horizontal
9. (a) 1.52 (b) 417 nm
   (c) $4.74 \times 10^{14} \text{ Hz}$ (d) $1.98 \times 10^8 \text{ m/s}$
11. (a) 584 nm (b) 1.12
13. $111°$
15. $16.5°$
17. five times from the right-hand mirror and six times from the left
19. 0.388 cm
21. $\theta = 30.4°$, $\theta' = 22.3°$
23. 6.39 ns
25. $\theta = \tan^{-1}(n_g)$
27. 3.39 m
29. $\theta_{\text{red}} = 48.22°$, $\theta_{\text{blue}} = 47.79°$
31. (a) $\theta_{1i} = 30°$, $\theta_{1r} = 19°$, $\theta_{2i} = 41°$, $\theta_{2r} = 77°$
   (b) first surface: $\theta_{\text{reflection}} = 30°$, second surface:
      $\theta_{\text{reflection}} = 41°$
33. (a) $31.3°$ (b) $44.2°$ (c) $49.8°$
35. (a) $33.4°$ (b) $53.4°$
37. 1.38
39. $1.000\,08$
41. (a) $10.7°$ (b) air
   (c) Sound falling on the wall from most directions is 100% reflected.
43. $27.5°$
45. $22.0°$
47. (a) $53.1°$ (b) $\geq 38.7°$
49. (a) $38.5°$ (b) ≥ 1.44
53. $24.7°$
55. 1.93
59. $\theta = \sin^{-1}(\sqrt{n^2 - 1}\sin\phi - \cos\phi)$

CHAPTER 23

Quick Quizzes

1. At C
2. c
3. (a) False   (b) False   (c) True
4. b
5. An infinite number
6. (a) False   (b) True   (c) False

Conceptual Questions

1. You will not be able to focus your eyes on both the picture and your image at the same time. To focus on the picture, you must adjust your eyes so that an object several centimeters away (the picture) is in focus. Thus, you are focusing on the mirror surface. But, your image in the mirror is as far behind the mirror as you are in front of it. Thus, you must focus your eyes beyond the mirror, twice as far away as the picture to bring the image into focus.

3. A single flat mirror forms a virtual image of an object due to two factors. First, the light rays from the object are necessarily diverging from the object, and second, the lack of curvature of the flat mirror cannot convert diverging rays to converging rays. If another optical element is first used to cause light rays to converge, then the flat mirror can be placed in the region in which the converging rays are present, and it will simply change the direction of the rays so that the real image is formed in a different location. For example, if a real image is formed by a convex lens, and the flat mirror is placed between the lens and the image position, the image formed by the mirror will be real.

5. The ultrasonic range finder sends out a sound wave and measures the time for the echo to return. Using this information, the camera calculates the distance to the subject and sets the camera lens. When the camera is facing a mirror, the ultrasonic signal reflects from the mirror surface and the camera adjusts its focus so that the mirror surface is at the correct focusing distance from the camera. But your image in the mirror is twice this distance from the camera, so it is blurry.

7. Light rays diverge from the position of a virtual image just as they do from an actual object. Thus, a virtual image can be as easily photographed as any object can. Of course, the camera would have to be placed near the axis of the lens or mirror in order to intercept the light rays.

9. We consider the two trees to be two separate objects. The far tree is an object that is farther from the lens than the near tree. Thus, the image of the far tree will be closer to the lens than the image of the near tree. The screen must be moved closer to the lens to put the far tree in focus.

11. If a converging lens is placed in a liquid having an index of refraction larger than that of the lens material, the direction of refractions at the lens surfaces will be reversed, and the lens will diverge light. A mirror depends only on reflection, which is independent of the surrounding material, so a converging mirror will be converging in any liquid.

13. This is a possible scenario. When light crosses a boundary between air and ice, it will refract in the same manner as it does when crossing a boundary of the same shape between air and glass. Thus, a converging lens may be made from ice as well as glass. However, ice is such a strong absorber of infrared radiation that it is unlikely you will be able to start a fire with a small ice lens.

15. The focal length for a mirror is determined by the law of reflection from the mirror surface. The law of reflection is independent of the material of which the mirror is made and of the surrounding medium. Thus, the focal length de-

pends only on the radius of curvature and not on the material. The focal length of a lens depends on the indices of refraction of the lens material and surrounding medium. Thus, the focal length of a lens depends on the lens material.

17. If parallel rays strike a flat mirror, they are reflected as parallel rays. Therefore, the focal length of a flat mirror is infinite, $f \rightarrow \infty$. This is consistent with the mirror equation

$$\frac{1}{p} + \frac{1}{q} = \frac{1}{f}$$

Because the image is always as far behind a flat mirror as the object is in front of it, then $q = -p$ and the mirror equation yields $f \rightarrow \infty$.

Problems

1. on the order of 10^{-9} s younger
3. 10.0 ft, 30.0 ft, 40.0 ft
5. 0.268 m behind the mirror; virtual, upright, and diminished; $M = 0.026\ 8$
7. (a) 13.3 cm in front of mirror, real, inverted, $M = -0.333$
   (b) 20.0 cm in front of mirror, real, inverted, $M = -1.00$
   (c) No image is formed. Parallel rays leave the mirror.
9. behind the worshipper, 3.33 m from the deepest point in the niche
11. 5.00 cm
13. 1.0 m
15. 8.05 cm
17. -20.0 cm
19. (a) concave with focal length $f = 0.83$ m
    (b) Object must be 1.0 m in front of the mirror.
21. 38.2 cm below the upper surface of the ice
23. 3.8 mm
25. (a) 120 cm   (b) -24.0 cm
    (c) -8.00 cm   (d) -3.43 cm
27. 20.0 cm
29. (a) 40.0 cm beyond the lens, real, inverted, $M = -1.00$
    (b) No image is formed. Parallel rays leave the lens.
    (c) 20.0 cm in front of the lens, virtual, upright, $M = +2.00$
31. (a) 13.3 cm in front of the lens, virtual, upright, $M = +\frac{1}{3}$
    (b) 10.0 cm in front of the lens, virtual, upright, $M = +\frac{1}{2}$
    (c) 6.67 cm in front of the lens, virtual, upright, $M = +\frac{2}{3}$
33. (a) either 9.63 cm or 3.27 cm
    (b) 2.10 cm
35. (a) 39.0 mm   (b) 39.5 mm
37. at distance $2|f|$ in front of lens
39. 40.0 cm
41. 30.0 cm in front of the second lens, $M = -3.00$
43. 7.47 cm in front of the second lens; 1.07 cm; virtual, upright
45. from 0.224 m to 18.2 m
47. real image, 5.71 cm in front of the mirror
49. -40.0 cm
51. 160 cm to the left of the lens, inverted, $M = -0.800$
53. $q = 10.7$ cm
55. 32.0 cm to the right of the second surface (real image)
57. (a) 20.0 cm to the right of the second lens; $M = -6.00$
    (b) inverted
    (c) 6.67 cm to the right of the second lens; $M = -2.00$; inverted
59. (a) 1.99
    (b) 10.0 cm to the left of the lens
    (c) inverted
61. (a) 5.45 m to the left of the lens
    (b) 8.24 m to the left of the lens
    (c) 17.1 m to the left of the lens
    (d) by surrounding the lens with a medium having a refractive index greater than that of the lens material

CHAPTER 24

Quick Quizzes

1. b
2. The compact disc.

Conceptual Questions

1. You will not see an interference pattern from the automobile headlights for two reasons. The first is that the headlights are not coherent sources and are therefore incapable of producing sustained interference. Also, the headlights are so far apart in comparison to the wavelengths emitted that, even if they were made into coherent sources, the interference maxima and minima would be too closely spaced to be observable.

3. The result of the double slit is to redistribute the energy arriving at the screen. Although there is no energy at the location of a dark fringe, there is four times as much energy at the location of a bright fringe as there would be with only a single narrow slit. The total amount of energy arriving at the screen is twice as much as with a single slit, as it must be according to the conservation of energy.

5. One of the materials has a higher index of refraction than water, the other lower. The material with the higher index of refraction than water appears black as it approaches zero thickness. There is a 180° phase shift for the light reflected from the upper surface, but no such phase change from the lower surface, because the index of refraction for water on the other side is lower than that of the film. Thus, the two reflections are out of phase and interfere destructively. The material with index of refraction lower than water experiences a phase change for the light reflected from both upper and lower surfaces, so that the reflections from the zero-thickness film will be back in phase, and the film will appear bright.

7. For normal incidence, the extra path length followed by the reflected ray is twice the thickness of the film. For destructive interference, this must be a distance of half a wavelength of the light in the material of the film. For a film in air no 180° phase change occurs in these reflections, so the thickness of the film must be one-quarter wavelength, which is the same as the condition for constructive interference of reflected light. This means that the transmitted light is a minimum when the reflected light is a maximum, and vice-versa.

9. Since the light reflecting at the lower surface of the air film experience a 180° phase change whereas light reflecting from the upper surface of the film does not undergo a phase change, the central spot (where the film has near zero thickness) is dark. If the observed rings are not circular, the curved surface of the lens does not have a true spherical shape.

11. For regional communication at Earth's surface, radio waves are typically broadcast from currents oscillating in tall, vertical towers. These waves have vertical planes of polarization. Light originates from the vibrations of atoms, or electronic transitions within atoms, which represent oscillations in all possible directions. Thus, light is not generally polarized.

13. Yes. In order to do this, first measure the radar-reflectivity of the metal of your airplane. Then, choose a light, durable material that has approximately half the radar-reflectivity of the metal in your plane. Measure its index of refraction, and place onto the metal a coating equal in thickness to one quarter of 3 cm, divided by that index. Sell it quick and then you can sell to the supposed enemy new radars operating at 1.5 cm, which the coated metal will reflect with extra-high efficiency.

15. If you wish to perform an interference experiment, you need monochromatic coherent light. To obtain it, you must first pass light from an ordinary source through a prism or diffraction grating to disperse different colors into different directions. Using a single narrow slit, select a single color and make that light diffract to cover both slits for Young's experiment. The procedure is much simpler with a laser because its output is already monochromatic and coherent.

17. Audible sound has wavelengths on the order of meters or centimeters, whereas visible light has wavelengths on the order of half a micrometer. Sound, therefore, diffracts around walls and doorways (roughly meter-sized apertures). Visible light diffracts only through very small angles as it passes ordinary sized objects or apertures, because $\sin\theta = m\lambda/a$ by Equation 24.11, and λ/a is extremely small.

19. Strictly speaking, the ribs do act as a diffraction grating, but the separation distance of the ribs is so much larger than the wavelength of the x-rays that there are no observable effects.

Problems

1. 1.58 cm
3. (a) 2.62 mm (b) 2.62 mm
5. (a) 36.2° (b) 5.08 cm (c) 5.08×10^{14} Hz
7. (a) 55.7 m (b) 124 m
9. 75.0 m
11. 11.3 m
13. 148 m
15. 91.9 nm
17. Any odd-integer multiple of 85.4 nm
19. 0.500 cm
21. (a) 238 nm (b) λ will increase (c) 328 nm
23. 4.35 μm
25. 4.75 μm
27. No, the wavelengths intensified are 276 nm, 138 nm, 92.0 nm, . . .
29. 4.22 mm
31. (a) 1.1 m (b) 1.7 mm
33. 1.20 mm, 1.20 mm
35. (a) 479 nm, 647 nm, 698 nm (b) 20.5°, 28.3°, 30.7°
37. 5.49° in first order; 12.3° in second order; and 24.9° in third order
39. 44.5 cm
41. 9.13 cm
43. (a) 25.6° (b) 19.0°
45. (a) 1.11 (b) 42.0°
47. (a) 56.7° (b) 48.8°
49. 31.2°
53. 6.89 units
55. (a) 413.7 nm, 409.7 nm (b) 8.6°
57. 0.156 mm
59. 2.50 mm
61. 58.1°
63. (a) 16.6 m (b) 8.28 m
65. 1.31
67. 0.350 mm
69. 115 nm

CHAPTER 25

Answers to Quick Quizzes

1. You should choose a blue filter.

Conceptual Questions

1. The observer is not using the lens as a simple magnifier. To use a lens as a simple magnifier, the object distance must be less than the focal length of the lens. Also, a simple magnifier produces a virtual image at the normal near point of the eye, or at a point about $q = -25$ cm. With a large object distance and a relatively short image distance, the magnitude of the magnification by the lens would be considerably less than one. Most likely, the lens is part of a lens combination used as a telescope.

3. The image formed on the retina by the lens and cornea is already inverted.

5. There will be an effect on the interference pattern—it will be distorted. The high temperature of the flame changes the index of refraction of air for the arm of the interferometer in which the match is held. As the index of refraction varies turbulently, the wavelength of the light in that region will also vary turbulently. As a result, the effective difference in length between the two arms varies, resulting in a wildly varying interference pattern.

7. Large lenses are difficult to manufacture and machine with accuracy. Their large weight leads to sagging, which produces a distorted image. In reflecting telescopes, light does not pass through glass; hence, problems associated with chromatic aberrations are eliminated. Large-diameter reflecting telescopes are also technically easier to construct. Some designs use a rotating pool of mercury as the reflecting surface.

9. In order to "see" an object, the wavelength of the light in the microscope must be smaller than the size of the object. An atom is much smaller than the wavelength of light in the visible spectrum, so an atom can never be seen using visible light.

Problems

1. 30.0 cm beyond the lens, $M = -1/5$
3. Yes, the largest image dimension is only 19 mm.
5. $f/1.4$
7. $f/8.0$
9. 40.0 cm
11. 23.2 cm
13. (a) -2.00 diopters   (b) 17.6 cm
15. $+17.0$ diopters
17. $m = +3.50$
19. (a) 4.07 cm   (b) $m = +7.14$
21. (a) $|M| = 1.22$ (b) $\theta/\theta_0 = 6.08$
23. 2.1 cm
25. $m = -115$
27. $f_o = 90$ cm, $f_e = 2.0$ cm
29. (b) $-fh/p$ (c) -1.07 mm
31. (a) $m = 1.50$ (b) $m = 1.90$
33. 492 km
35. 0.40 μrad
37. 3.09 m
39. 9.8 km
41. 50.4 μm
43. 449 nm, blue
45. 98 fringe shifts
47. No. A resolving power of 2.0×10^5 is needed and that available is only 1.8×10^5.
49. (a) $+2.67$ diopters   (b) 0.16 diopters too low
51. (a) $+44.6$ diopters   (b) 3.03 diopters
53. (a) 1.0×10^3 lines   (b) 3.3×10^2 lines
55. $m = 10.7$

CHAPTER 26

Quick Quizzes

1. b
2. The answer to both questions is no.
3. a, e
4. (a) False   (b) False   (c) True   (d) False
5. a

Conceptual Questions

1. One must be careful using such phrases because events that occur at different locations and appear to occur simultaneously to one observer will not appear simultaneous to observers in motion relative to the first. This lack of simultaneity becomes more pronounced at very high speeds.

3. This scenario is not possible with light. Light waves are described by the principles of special relativity. As you detect the light wave ahead of you and moving away from you (which would be a pretty good trick—think about it!), its velocity relative to you is c. Thus, you will not be able to catch up to the light wave.

5. Present knowledge of the Universe shows that even the "fixed" stars are in motion relative to us and in motion relative to each other. Thus, the stars are not really fixed at all and do not provide us with an absolute frame of reference.

7. The light from the quasar moves at 3.00×10^8 m/s. The speed of light is independent of the motion of the source or observer.

9. For a wonderful fictional exploration of this question, get a "Mr. Tompkins" book by George Gamow. All of the relativity effects would be obvious in our lives. Time dilation and length contraction would both occur. Driving home in a hurry, you would push on the gas pedal not to increase your speed very much, but to make the blocks shorter. Big Doppler shifts in wave frequencies would make red lights look green as you approached and make car horns and radios useless. High-speed transportation would be both very expensive, requiring huge fuel purchases, as well as dangerous, because a speeding car could knock down a building. When you got home, hungry for lunch, you would find that you had missed dinner; there would be a five-day delay in transit when you watch a live TV program originating in Australia. Finally, we would not be able to see the Milky Way, because the fireball of the Big Bang would surround us at the distance of Rigel, or Deneb.

11. This is not a violation of the concepts of relativity. The theory of relativity states that the result of any measurement of the vacuum speed of light will be independent of the motion of the observer or source. It makes no claim that light must have the same speed in all media.

13. Your assignment: Measure the length of a rod as it slides past you. Mark the position of its front end on the floor and have an assistant mark the position of the back end. Then measure the distance between the two marks. This distance will represent the length of the rod only if the two marks were made simultaneously in your frame of reference.

Problems

1. (a) $t_{OB} = 1.67 \times 10^3$ s, $t_{OA} = 2.04 \times 10^3$ s
     (b) $t_{BO} = 2.50 \times 10^3$ s, $t_{AO} = 2.04 \times 10^3$ s
     (c) $\Delta t = 90$ s
3. 5.0 s
5. $c(\sqrt{3}/2)$
7. (a) 1.3×10^{-7} s   (b) 38 m
     (c) 7.6 m
9. (a) 2.2 μs   (b) 0.65 km
11. $0.950c$
13. yes, with 19 m to spare
15. (a) 39.2 μs   (b) accurate to one digit
17. 3.3×10^5 m/s
19. $0.285c$
21. $0.54c$ to the right
23. $0.357c$
25. $0.998c$ toward the right
27. (a) 54 min   (b) 52 min
29. $c(\sqrt{3}/2)$
31. $0.272c$
33. 18.4 g/cm^3
35. 1.98 MeV
37. 2.27×10^{23} Hz, 1.32 fm for each photon
39. (a) 3.10×10^5 m/s   (b) $0.758c$
41. 1.42 MeV/c
43. (a) $0.80c$ (b) 7.5×10^3 s
     (c) 1.4×10^{12} m, $0.38c$

45. $0.37c$ in $+x$ direction
47. (a) $v/c = 1 - 1.12 \times 10^{-10}$ (b) 6.00×10^{27} J
    (c) $\$2.17 \times 10^{20}$
49. $0.80c$
51. (a) $0.946c$ (b) 0.160 ly
    (c) 0.114 yr   (d) 7.50×10^{22} J
53. (a) 7.0 μs   (c) 1.1×10^4 muons

CHAPTER 27

Quick Quizzes

1. b
2. c
3. c
4. c
5. b

Conceptual Questions

1. The shape of an object is determined by observing the light reflecting from its surface. In a kiln, the objects will be very hot and will be glowing red. The emitted radiation is far stronger than the reflected radiation, and the thermal radiation emitted is only slightly dependent on the material from which the objects are made. Thus, we have a collection of objects, including the kiln walls, glowing equally with emitted radiation and only weak reflected light compared to the emitted light. This will result in indistinct outlines of the objects when viewed.

3. The "blackness" of a blackbody refers to its ideal property of absorbing all radiation incident on it. If an observed room-temperature object in everyday life absorbs all radiation, we describe it as (visibly) black. The black appearance, however, is due to the fact that our eyes are sensitive only to visible light. If we could detect infrared light with our eyes, we would see the object emitting radiation. If the temperature of the blackbody is raised, Wien's law tells us that the emitted radiation will move into the visible. Thus, the blackbody could appear possibly as black, red, white, or blue, depending on its temperature.

5. All objects do radiate energy, but at room temperature, this energy is primarily in the infrared region of the electromagnetic spectrum, which our eyes cannot detect. (Pit vipers have sensory organs that are sensitive to infrared radiation; thus they can seek out their warm-blooded prey in what we would consider absolute darkness.)

7. Most metals have cutoff frequencies corresponding to photons in or near the visible range of the electromagnetic spectrum. AM radio wave photons will have far too little energy to eject electrons from the metal.

9. We can picture higher-frequency light as a stream of photons of higher energy. In a collision, one photon can give all of its energy to a single electron. The kinetic energy of such an electron is measured by the stopping potential. The reverse voltage (stopping voltage) required to stop the current is proportional to the frequency of the incoming light. More intense light consists of more photons striking a unit area each second, but atoms are so small that one emitted electron never gets a "kick" from more than one photon. Increasing the light intensity will generally increase the size of the current but will not change the energy of the individual ejected electrons. Thus, the stopping potential remains constant.

11. Wave theory predicts that the photoelectric effect should occur at any frequency, provided that the light intensity is high enough. However, as seen in photoelectric experiments, the light must have sufficiently high frequency for the effect to occur.

13. The photons will transfer the same energy to all photoelectrons in the absorption process. However, different electrons may start from different potential energy levels and emerge from the surface with different kinetic energies. The energy of the incident photons determines only the maximum kinetic energy of the liberated electrons.

15. No. Suppose that the incident light frequency at which you first observed the photoelectric effect is above the cutoff frequency of the first metal but less than the cutoff frequency of the second metal. In that case, the photoelectric effect would not be observed at all in the second metal.

17. In a collision with an electron or other scattering particle, the photon will impart some energy to that particle. Therefore, the energy of the photon will be decreased and its wavelength increased.

Problems

1. (a) ≈ 3000 K   (b) $\approx 20\,000$ K
3. (a) 999 nm
    (b) Peak radiation is in the infrared region.
5. (a) 2.49×10^{-5} eV   (b) 2.49 eV
    (c) 249 eV
7. 2.27×10^{30} photons/s
9. (a) 2.3×10^{31} (b) $\Delta E/E = 4.3 \times 10^{-32}$
11. (a) 2.24 eV   (b) 555 nm
    (c) 5.41×10^{14} Hz
13. (a) 296 nm   (b) 1.01×10^{15} Hz
    (c) 2.71 V
15. 148 days, incompatible with observation
17. 4.8×10^{14} Hz, 2.0 eV
19. 1.2×10^2 V and 1.2×10^7 V, respectively
21. 41.4 kV
23. 0.078 nm
25. 0.281 nm
27. 1.78 eV, 9.47×10^{-28} kg·m/s
29. $70°$
31. 1.18×10^{-23} kg·m/s, 478 eV
33. (a) 1.2 eV   (b) 6.5×10^5 m/s
35. (a) 1.46 km/s   (b) 7.28×10^{-11} m
37. (a) $\sim 10^2$ MeV
    (b) No. With kinetic energy much larger than the magnitude of the negative potential energy, the electron would immediately escape.
39. 3.58×10^{-13} m
41. (a) 15 keV   (b) 1.2×10^2 keV
43. 2.1×10^{-32} m/s
45. 116 m/s
47. $\approx 5\,200$ K; clearly, a firefly is not at this temperature, so this cannot be blackbody radiation.
49. $18.2°$
51. 1.36 eV
53. 2.00 eV
55. (a) $0.0220c$ (b) $0.9992c$
57. (b) 3.72 km/s
59. (b) 2.60×10^{-16} m

CHAPTER 28

Quick Quizzes

1. d
2. (a) 9   (b) 16
3. (a) 5 values of ℓ
    (b) 9 values of m_ℓ
4. b

Conceptual Questions

1. If the energy of the hydrogen atom were proportional to n (or any power of n), then the energy would become infinite

n	3	3	3	3	3	3	3	3	3	3	3	3	3	3	3
ℓ	2	2	2	2	2	2	2	2	2	2	2	2	2	2	2
m_ℓ	+2	+2	+2	+1	+1	+1	0	0	0	−1	−1	−1	−2	−2	−2
m_s	+1	0	−1	+1	0	−1	+1	0	−1	+1	0	−1	+1	0	−1

as n grew to infinity. But the energy of the atom is inversely proportional to n^2. Thus, as n grows to infinity, the energy of the atom approaches a value that is above the ground state by a finite amount, namely the ionization energy 13.6 eV. As the electron falls from one bound state to another, its energy loss is always less than the ionization energy. The energy and frequency of any emitted photon are finite.

3. The characteristic x-rays originate from transitions within the atoms of the target, such as an L shell electron making a transition to a vacancy in the K shell. This vacancy in the K shell is caused when an accelerated electron in the x-ray tube supplies energy to the K shell electron to eject it from the atom. If the energy of the bombarding electrons were to be increased, the K shell electron will be ejected from the atom with more remaining kinetic energy. But the energy difference between the K and L shell has not changed, so the emitted x-ray has exactly the same wavelength.

5. A continuous spectrum without characteristic x-rays is possible. At a low-accelerating potential difference for the electron, the electron may not have enough energy to eject an electron from a target atom. As a result, there will be no characteristic x-rays. The change in speed of the electron as it enters the target will result in the continuous spectrum.

7. The hologram is an interference pattern between light scattered from the object and the reference beam. If anything moves by a distance comparable to the wavelength of the light (or more), the pattern will wash out. The effect is just like making the slits vibrate in Young's experiment, to make the interference fringes vibrate wildly so that a photograph of the screen displays only the average intensity everywhere.

9. If the Pauli exclusion principle were not valid, the elements and their chemical behavior would be grossly different because every electron would end up in the lowest energy level of the atom. All matter would therefore be nearly alike in its chemistry and composition, since the shell structures of each element would be identical. Most materials would have a much higher density, and the spectra of atoms and molecules would be very simple, resulting in the existence of less color in the world.

11. The three elements have similar electronic configurations, with filled inner shells, plus a single electron in an s orbital. Since atoms typically interact through their unfilled outer shells, and the outer shell of each of these atoms is similar, the chemical interactions of the three atoms is also similar.

13. Each of the eight electrons must have at least one quantum number different from each of the others. They can differ (in m_s) by being spin-up or spin-down. They can differ (in ℓ) in angular momentum and in the general shape of the wave function. Those electrons with $\ell = 1$ can differ (in m_ℓ) in orientation of angular momentum.

15. The way the total energy of an atom, and in fact other energies, can be negative is through an arbitrary choice of the zero point for potential energies. The electrostatic potential energy between charged particles is arbitrarily chosen to be zero when the particles are separated by an infinite distance. When two particles having opposite signs, such as the electron and proton in a hydrogen atom, are a finite distance apart, the potential energy is less than zero, or negative. If they are sufficiently close, the magnitude of the negative potential energy may exceed the positive kinetic energy, making the total energy negative.

17. Stimulated emission is the reason laser light is coherent and tends to travel in a well-defined parallel beam. When a photon passing by an excited atom stimulates that atom to emit a photon, the emitted photon is in phase with the original photon and travels in the same direction. As this process is repeated many times, an intense, parallel beam of coherent light is produced. Without stimulated emission, the excited atoms would return to the ground state by emitting photons at random times and in random directions. The resulting light would not have the useful properties of laser light.

Problems

1. 656 nm, 486 nm, and 434 nm
3. (a) 2.3×10^{-8} N
   (b) -14 eV
5. (a) 1.6×10^6 m/s
   (b) No. $v/c = 5.3 \times 10^{-3} \ll 1$
   (c) 0.46 nm
   (d) Yes. The wavelength is roughly the same size as the atom.
7. (a) 0.212 nm   (b) 9.95×10^{-25} kg·m/s
   (c) 2.11×10^{-34} J·s
   (d) 3.40 eV   (e) -6.80 eV   (f) -3.40 eV
11. 8.22×10^{-8} N
13. (a) 0.967 eV   (b) 0.266 eV
15. (a) 122 nm, 91.1 nm   (b) 1.87×10^3 nm, 820 nm
17. 97.2 nm
19. (a) 488 nm   (b) 0.814 m/s
21. (d) $n = 2.53 \times 10^{74}$
   (e) No. At such large quantum numbers the allowed energies are essentially continuous.
23. (a) 2.47×10^{15} Hz, $f_{\text{orb}} = 8.23 \times 10^{14}$ Hz
   (b) 6.59×10^3 Hz, $f_{\text{orb}} = 6.59 \times 10^3$ Hz. For large n, classical theory and quantum theory approach each other in their results.
25. 4.42×10^4 m/s
27. (a) -122 eV   (b) 1.76×10^{-11} m
29. (a) 0.026 5 nm
   (b) 0.017 6 nm
   (c) 0.013 2 nm
31. 1.33 nm
33. $n = 3, \ell = 1, m_\ell = +1, m_s = \pm 1/2; n = 3, \ell = 1, m_\ell = 0, m_s = \pm 1/2; n = 3, \ell = 1, m_\ell = -1, m_s = \pm 1/2$
35. The table at the top of the page summarizes 15 possible states:
37. (a) 30 possible states   (b) 36
39. (a) $n = 4$ and $\ell = 2$
   (b) $m_\ell = (0, \pm 1, \pm 2)$, $m_s = \pm \frac{1}{2}$
   (c) $1s^2 2s^2 2p^6 3s^2 3p^6 3d^{10} 4s^2 4p^6 4d^2 5s^2 = $ [Kr] $4d^2 5s^2$
41. 0.160 nm
43. L shell: 11.7 keV; M shell: 10.0 keV; N shell: 2.30 keV
45. (a) 10.2 eV   (b) 7.88×10^4 K
47. (a) -8.18 eV, -2.04 eV, -0.904 eV, -0.510 eV, -0.325 eV
   (b) 1.09×10^3 nm and 609 nm
49. (a) The four lowest energies are -10.39 eV, -5.502 eV, -3.687 eV, and -2.567 eV.
   (b) The wavelengths of the emission lines are 158.5 nm, 185.0 nm, 253.7 nm, 422.5 nm, 683.2 nm, and 1 107 nm.
   (c) 1.31×10^6 m/s
51. (a) 4.24×10^{15} W/m^2 (b) 1.20×10^{-12} J
55. (a) $E_n = (-1.49 \times 10^4 \text{ eV})/n^2$
   (b) $n = 4 \rightarrow n = 1$

57. (a) 9.03×10^{22} m/s^2 (b) -4.63×10^{-8} W
    (c) $\sim 10^{-11}$ s

CHAPTER 29

Answers to Quick Quizzes

1. c
2. b
3. a
4. a and b
5. b

Conceptual Questions

1. Isotopes of a given element correspond to nuclei with different numbers of neutrons. This results in a variety of different physical properties for the nuclei, including the obvious one of mass. The chemical behavior, however, is governed by the electrons. All isotopes of a given element have the same number of electrons and, therefore, the same chemical behavior.

3. An alpha particle contains two protons and two neutrons. Because a hydrogen nucleus only contains one proton, it cannot emit an alpha particle.

5. In alpha decay, there are only two final particles—the alpha particle and the daughter nucleus. There are also two conservation principles—energy and momentum. As a result, the alpha particle must be ejected with a discrete energy to satisfy both conservation principles. However, beta decay is a three-particle decay—the beta particle, the neutrino (or antineutron), and the daughter nucleus. As a result, the energy and momentum can be shared in a variety of ways among the three particles while still satisfying the two conservation principles. This allows a continuous range of energies for the beta particle.

7. The larger rest energy of the neutron means that a free proton in space will not spontaneously decay into a neutron and a positron. When the proton is in the nucleus, however, the important question is that of the total rest energy of the nucleus. If it is energetically lower for the nucleus to have one less proton and one more neutron, then the decay process will occur to achieve this lower energy.

9. Carbon dating cannot generally be used to estimate the age of a stone, because the stone was not alive to take up carbon from the environment. Only the ages of artifacts that were once alive can be estimated with carbon dating.

11. The protons, although held together by the nuclear force, are repulsed by the electrical force. If enough protons were placed together in a nucleus, the electrical force would overcome the nuclear force, which is based on the number of particles, and cause the nucleus to fission.

    The addition of neutrons prevents such fission. The neutron does not increase the electrical force, being electrically neutral, but does contribute to the nuclear force.

13. The statement is false. Both patterns show monotonic decrease over time, but with very different shapes. For radioactive decay, maximum activity occurs at time zero. Cohorts of people now living will be dying most rapidly perhaps forty years from now. Everyone now living will be dead within less than two centuries, while the mathematical model of radioactive decay tails off exponentially forever. A radioactive nucleus never gets old. It has constant probability of decay however long it has existed.

15. Since the two samples are of the same radioactive nuclide, they have the same half-life; the 2:1 difference in activity is due to a 2:1 difference in the mass of each sample. After five half-lives, each will have decreased in mass by a power of $2^5 = 32$. However, since this simply means that the mass of each is 32 times smaller, the ratio of the masses will still be $(\frac{2}{32}):(\frac{1}{32})$, or 2:1. Therefore, the ratio of their activities will *always* be 2:1.

17. The photon and the neutrino are similar in that both particles have zero mass and zero charge. Both must travel at the speed of light and are capable of transferring both energy and momentum. They differ in that the photon has spin (intrinsic angular momentum) of $\hbar$ and is involved in electromagnetic interactions, whereas the neutrino has spin $\hbar/2$ and is closely related to beta decays.

Problems

1. $A = 2$, $r = 1.5$ fm; $A = 60$, $r = 4.7$ fm; $A = 197$, $r = 7.0$ fm; $A = 239$, $r = 7.4$ fm

3. 1.8×10^2 m

5. (a) 27.6 N away from the carbon nucleus
   (b) 4.16×10^{27} m/s^2 away from the nucleus
   (c) 1.73 MeV

7. (a) 1.9×10^7 m/s   (b) 7.1 MeV

9. 8.66 MeV/nucleon for $^{93}_{41}$Nb, 7.92 MeV/nucleon for $^{197}_{79}$Au

11. 3.54 MeV

13. 0.210 MeV/nucleon greater for $^{23}_{11}$Na, attributable to less proton repulsion

15. 0.46 Ci

17. (a) 9.98×10^{-7} s^{-1} (b) 1.9×10^{10} nuclei

19. (a) 3.18×10^{-7} mol   (b) 1.92×10^{17} nuclei
    (c) 1.08×10^{14} Bq   (d) 8.96×10^6 Bq

21. 4.31×10^3 yr

23. (a) 5.58×10^{-2} h^{-1}, 12.4 h   (b) 2.39×10^{13} nuclei
    (c) 1.9 mCi

25. $^{208}_{81}$Tl, $^{95}_{37}$Rb, $^{144}_{60}$Nd

27. $^{40}_{20}$Ca, $^{94}_{42}$Mo, $^4_2$He

29. e^+ decay, $^{56}_{27}$Co $\rightarrow$ $^{56}_{26}$Fe $+ e^+ + \nu$

31. 71.4 keV

33. 18.6 keV

35. 4.22×10^3 yr

37. (a) $^{30}_{15}$P   (b) -2.64 MeV

39. (a) $^{21}_{10}$Ne   (b) $^{144}_{54}$Xe   (c) X = e^+, X' = ν

41. (a) $^{13}_6$C   (b) $^{10}_5$B

43. 1.00 MeV

45. (a) $^1_0$n   (b) Fluorine mass = 18.000 953 u

47. 18.8 J

49. 24 d

51. (a) 8.97×10^{11} electrons   (b) 0.100 J   (c) 100 rad

53. 46.5 d

55. $Q = 3.27$ MeV > 0, no threshold energy required

57. (a) 2.52×10^{24} (b) 2.29×10^{12} Bq
    (c) 1.07×10^6 yr

59. (a) 4.0×10^9 yr
    (b) It could be no older. The rock could be younger if some ^{87}Sr were initially present.

61. 54 μCi

63. 2.3×10^2 yr

65. 4.4×10^{-8} kg/h

CHAPTER 30

Answers to Quick Quizzes

1. c
2. a
3. b
4. c, e

Conceptual Questions

1. The experiment described is a nice analogy to the Rutherford scattering experiment. In the Rutherford experiment, alpha particles were scattered from atoms, and the scattering was consistent with a small structure in the atom containing the positive charge.

3. The largest charge of a quark is $2e/3$, so a combination of only two particles, a quark and an antiquark forming a meson, could not possibly have electric charge up to $+2e$. Only

particles containing three quarks, each with a charge of $2e/3$, can combine to produce a total charge of 2e.

5. Until about 700 000 years after the Big Bang, the temperature of the Universe was high enough for any atoms that formed to be ionized by ambient radiation. Once the average radiation energy dropped below the hydrogen ionization energy of 13.6 eV, hydrogen atoms could form and remain as neutral atoms for a relatively long period of time.

7. In the quark model, all hadrons are composed of smaller units called quarks. Quarks have a fractional electric charge and a baryon number of $\frac{1}{3}$. There are six flavors of quarks; up (u) down (d), strange (s), charmed (c), top (t), and bottom (b). All baryons contain three quarks, and all mesons contain one quark and one antiquark. Section 30.12 has a more detailed discussion of the quark model.

9. Baryons and mesons are hadrons, interacting primarily through the strong force. They are not elementary particles, being composed of either three quarks (baryons), or a quark and an antiquark (mesons). Baryons have a nonzero baryon number with a spin of either $\frac{1}{2}$ or $\frac{3}{2}$. Mesons have a baryon number of zero and a spin of either 0 or 1.

11. All stable particles other than protons and neutrons have baryon number zero. Since the baryon number must be conserved, and the final states of the kaon decay contain no protons or neutrons, the baryon number of all kaons must be *zero*.

13. Yes, but the strong interaction predominates.

15. Unless the particles have enough kinetic energy to produce a baryon-antibaryon pair, the answer is *no*. Antibaryons have a baryon number of -1; baryons have a baryon number of $+1$; mesons have a baryon number of 0. If such an interaction were to occur, and produce a baryon, the baryon number would not be conserved.

17. Baryons and antibaryons contain three quarks, whereas mesons and antimesons contain two quarks. Quarks have a spin of $\frac{1}{2}$; thus three quarks in a baryon can only combine to form a net spin that is half-integral. Likewise, two quarks in a meson can only combine to form a net spin of 0 or 1.

19. Nuclear and coal-fired power plants are similar in that both produce high-pressure steam to drive turbines and generate electricity. They differ in the type and amount of fuel consumed, the ways they affect the environment, and in some safety issues. A moderate size nuclear-powered plant consumes a few kilograms of fissionable materials per day, whereas a comparable size coal-powered plant requires a thousand tons or more of coal each day. One byproduct of nuclear plants is thermal pollution in the waters used for cooling. In addition to thermal pollution, coal-fired plants produce emissions that cause acid rain and contribute to the greenhouse effect. Both types of plants have risks of steam explosions. In addition, nuclear-powered plants raise questions on how to dispose of radioactive waste materials.

Problems

1. $^{1}_{0}n + ^{235}_{92}U \rightarrow ^{141}_{56}Ba + ^{92}_{36}Kr + 3 \, ^{1}_{0}n$, 3 neutrons are released
3. 126 MeV
5. (a) 16.2 kg   (b) 117 g
7. 2.9×10^{3} km ($\approx$ 1 800 miles)
9. 1.01 g
11. (a) $^{8}_{4}Be$ (b) $^{12}_{6}C$ (c) 7.27 MeV

13. 3.07×10^{22} events/yr
15. (a) 3.44×10^{30} J   (b) 1.56×10^{8} yr
17. (a) 4.53×10^{23} Hz   (b) 0.622 fm
19. $\sim 10^{-23}$ s
21. $\sim 10^{-18}$ m
23. (a) conservation of electron-lepton number and conservation of muon-lepton number
    (b) conservation of charge
    (c) conservation of baryon number
    (d) conservation of baryon number
    (e) conservation of charge
25. $\overline{\nu}_\mu$
27. (a) $\overline{\nu}_\mu$ (b) ν_μ (c) $\overline{\nu}_e$ (d) ν_e
    (e) ν_μ (f) ν_μ and $\overline{\nu}_e$
29. (a) not allowed; violates conservation of baryon number
    (b) strong interaction   (c) weak interaction
    (d) weak interaction
    (e) electromagnetic interaction
31. (a) not conserved   (b) conserved
    (c) conserved   (d) not conserved
    (e) not conserved   (f) not conserved
33. (a) charge, baryon number, L_e, L_τ
    (b) charge, baryon number, L_e, L_μ, L_τ
    (c) charge, L_e, L_μ, L_τ, strangeness number
    (d) charge, baryon number, L_e, L_μ, L_τ, strangeness number
    (e) charge, baryon number, L_e, L_μ, L_τ, strangeness number
    (f) charge, baryon number, L_e, L_μ, L_τ, strangeness number
35. 3.34×10^{26} electrons, 9.36×10^{26} up quarks, 8.70×10^{26} down quarks
37. (a) Σ^+ (b) π^- (c) K^0 (d) Ξ^-
39.

Reaction	At Quark Level	Net Quarks (Before and After)
$\pi^- + p \rightarrow K^0 + \Lambda^0$	$\overline{u}d + uud \rightarrow d\overline{s} + uds$	1 up, 2 down, 0 strange
$\pi^+ + p \rightarrow K^+ + \Sigma^+$	$u\overline{d} + uud \rightarrow u\overline{s} + uus$	3 up, 0 down, 0 strange
$K^- + p \rightarrow$ $K^+ + K^0 + \Omega^-$	$\overline{u}s + uud \rightarrow$ $u\overline{s} + d\overline{s} + sss$	1 up, 1 down, 1 strange

(d) The mystery particle is a Λ^0 or a Σ^0.
41. a neutron, udd
43. 70.45 MeV
45. 18.8 MeV
47. (a) electron-lepton and muon-lepton numbers not conserved
    (b) electron-lepton number not conserved
    (c) charge not conserved
    (d) baryon and electron-lepton numbers not conserved
    (e) strangeness violated by 2 units
49. (a) 1.5×10^{24} nuclei   (b) ≈ 0.6 kg
51. (a) 1 baryon before and zero baryons after decay. Baryon number is not conserved.
    (b) 469 MeV, 469 MeV/c
    (c) 0.999 999 4c

Index

Page numbers in italics indicate illustrations; page numbers followed by "t" indicate tables.